Hydraulic Systems Volume 2

Electro-Hydraulic Components and Systems

Dr. Medhat Kamel Bahr Khalil, Ph.D, CFPHS, CFPAI.
Director of Professional Education and Research Development,
Applied Technology Center, Milwaukee School of Engineering,
Milwaukee, WI, USA.

CompuDraulic LLC
www.CompuDraulic.com

CompuDraulic LLC

Hydraulic System Volume 2

Electro-Hydraulic Components and Systems

ISBN: 978-0-9977634-2-3

Printed in the United States of America
First Published by 2017
Revised by October 2019

Disclaimer

It is always advisable to review the relevant standards and the recommendations from the system manufacturer. However, the content of this book provides guidelines based on the author's experience.

Any portion of information presented in this book might not be suitable for some applications due to various reasons. Since errors can occur in circuits, tables, and text, the author/publisher assumes no liability for the safe and/or satisfactory operation of any system designed based on the information in this book.

The author/publisher does not endorse or recommend any brand name product by including such brand name products in this book. Conversely the author/publisher does not disapprove any brand name product not included in this book. The publisher obtained data from catalogs, literatures, and material from hydraulic components and systems manufacturers based on their permissions. The author/publisher welcomes additional data from other sources for future editions. This disclaimer is applicable for the workbook (if available) for this textbook.

Hydraulic Systems Volume 2
Electro-Hydraulic Components and Systems

Chapter 2: Hydro-Mechanical versus Electro-Hydraulic Solutions, 43

PREFACE

Questions being asked that are: "should a machine built based on electrical or hydraulic power transmission and control system?" and "what are the benefits of converting a legacy hydro-mechanical system into an electro-hydraulic system?"

This book is intended for industry professionals who would like to achieve a deeper understanding of the principles of electro-hydraulic control, and who would like to improve their practical skills in building an EH-driven machine. Unlike similar books, the contents of this book are presented in a unique way. It fills the gap between the very academic style of fluid power books and the very commercial style of books that are produced by fluid power manufacturers basically to promote their products.

The book contains *Animated Circuits Files* to download. The files are produced by Automation Studio to presents examples of electrical circuits that drive switching valves. Other circuits are to present examples of open-loop and closed-loop electro-hydraulic control systems.

Hydraulic system builders and users will find this book beneficial in understanding the construction and the operating principles of the electro-hydraulic systems. The book is written to cover the knowledge of electro-hydraulic components including solenoid operated valves, proportional valves, servo valves and amplifiers. The book also covers the technicalities of in-field tuning of open-loop and closed-loop electro-hydraulic systems. The book also presents guideline to select a valve for an application and how to read data sheet of a valve. The book covers the basic functions contained in the electronic control units that drive EH valves such as gain adjustor, maximum current limiter, ramp generator, pulse width modulation, dead band eliminator, null adjustment, and much more.

With more than twenty-five years of experience in teaching fluid power for industry professionals, the author had effectively applied his solid understanding to the subject and his post-doctoral level of academic education in the preparation of this book. The book features in brief are easy language, brand non-biased, practical oriented, associated with a workbook, colored, and demonstrative.

The author wants to continue on his goal of supporting fluid power and motion control professional education by developing the following series of volumes:

- Hydraulic Systems Volume 1: Introduction to Hydraulics for Industry Professionals.
- Hydraulic Systems Volume 2: Electro-Hydraulic Components and Systems.
- Hydraulic Systems Volume 3: Hydraulic Fluids and Contamination Control.
- Hydraulic Systems Volume 4: Hydraulic Fluids Conditioning.
- Hydraulic Systems Volume 5: Best Practices for Safety and Maintenance.
- Hydraulic Systems Volume 6: Troubleshooting and Failure Analysis.
- Hydraulic Systems Volume 7: Hydraulic Systems Modeling and Simulation for Application Engineers.
- Hydraulic Systems Volume 8: Design Strategies of Hydraulic Systems.
- Hydraulic Systems Volume 9: Design Strategies of Electro-Hydraulic Systems.
- Hydraulic Systems Volume 10: Hydraulic Components Modeling and Simulation.

ACKNOWLEDGEMENT

All praise is to Allah who granted me the knowledge, resources and health to finish this work.

To the soul of my parents who taught me the values of ISLAM

To my family: wife, sons, daughters in law, and grandson "Adam"

To my best teachers and supervisors

To friends who were always supportive

The author wishes to thank these companies: (Bosch Rexroth, Assofluid, Atos, Wandfluh, Hydraforce, Moog, Parker, and others) who permit us to use some of their copyrighted material to make this book more demonstrative.

Lastly, the author particularly wishes to thank the following gentlemen for their effective reviews of the book language:

- ***Tom Wanke, Director of Fluid Power Institute at Milwaukee School of Engineering.***
- ***Kamara Sheku, Dean of Applied Researches at Milwaukee School of Engineering.***

ABOUT THE BOOK

Book Description:

The book is a learning package for students and professionals who are looking to advance their fluid power careers. The package includes colored ***textbook***, electronic files ***for the animated hydraulic circuits***, and a ***colored workbook*** (separate price) that contains printed power point slides, chapter reviews and assignments. The book is the second in a series that the author plans to publish to offer complete and comprehensive teaching and design tools for the fluid power industry. This book is an attempt to fill the gap between the very academic style of fluid power books and the very commercial style of books that are produced by fluid power manufacturers basically to promote their products.

The book presents constructional and operational qualitative analogies between electrical, hydraulic and electro-hydraulic systems. The book considers presenting real-life examples for various types of electro-hydraulic valves. The book also covers the technicalities of in-field tuning of open-loop and closed-loop electro-hydraulic systems. The book also presents guideline to select a valve for an application and how to read data sheet of a valve. The book covers the basic functions contained in the electronic control units that drive EH valves such as gain adjustor, maximum current limiter, ramp generator, pulse width modulation, dead band eliminator, null adjustment current, and much more.

The textbook is produced in letter size (8.5 x 11) inches, and weighs 3.5 lb. The textbook contains a total of ten chapters distributed on 498 pages. The workbook is produced in the same size and weighs 2 lb. The associated software is online downloadable.

Book Objectives:

Chapter 1: Hydraulic versus Electrical Systems

This chapter presents structural and operational comparisons between hydraulic and electrical power transmission and control systems. Similarities and differences between the inductive, capacitive, and resistive elements are discussed.

Chapter 2: Hydro-Mechanical versus Electro-Hydraulic Solutions

This chapter explores the features and challenges of electro-hydraulic systems. Additionally, the chapter introduces the benefits of converting the classical hydro-mechanical systems into electro-hydraulic ones. Pressure control, flow control, power control, and sequence control are among the systems discussed.

Chapter 3: Switching Valves Construction and Operation

This chapter covers the principle of operation and the construction of various types of switching (ON/OFF) solenoids that are used to actuate hydraulic valves. The chapter discusses, qualitatively, the pros and cons of wet type versus dry type switching solenoids and DC type versus AC type switching solenoids. The chapter also discusses the undesirable effects of using AC switching solenoids such as AC hum, eddy current, and inrush current, and constructional considerations to minimize their effects. The chapter concludes by presenting examples of using switching solenoids in directional, pressure, and control functions.

Chapter 4: Electrical Circuits for Switching Valves

This chapter covers the basic safety precautions that must be considered when building an electrohydraulic circuit that drives a switching valve. The chapter also presents the electrical devices and their symbols that are most commonly used with switching valves including: switches, buttons, relays, and PLCs. The chapter also covers the rules to read and write various forms of electrical circuit diagrams including: Joint Industrial Council (JIC) schematic diagrams, wiring diagram, and sequence diagram. The chapter concludes with various electrohydraulic circuits that simulate typical applications.

Chapter 5: Proportional Valves

This chapter introduces the technology of proportional valves and discusses the construction differences as compared to conventional switching valves. The chapter also introduces the conceptual construction of force-controlled versus stroke-controlled types of proportional valves. Additionally, the chapter presents the control schemes when a proportional valve is used in open-loop and closed-loop control system. More important, the chapter concludes by exploring examples of actual proportional directional, pressure, and flow control valves from various suppliers.

Chapter 6: Servo Valves

This chapter covers the construction and wiring methods of the main electric components in servo valves including torque motors. The chapter also covers the conceptual construction and the operation of flapper-nozzle and jet-pipe servo valves. The chapter introduces examples of typical valves with mechanical or electrical feedback.

Chapter 7: Electro-Hydraulic Valve Selection Criteria

This chapter presents the various criteria to select a specific valve for an application. These criteria include, valve type, spool design, operating conditions, static and dynamic characteristics. The chapter also provides examples of the current valves that are produced by existing manufacturers.

Chapter 8: Open-Loop versus Closed-Loop EH Applications

This chapter explores the differences between electro-hydraulic open-loop and closed-loop systems. Several examples have been presented to discuss various ways to control load, speed and position of a hydraulic actuator. The chapter concludes by introducing examples from industrial and mobile applications.

Chapter 9: Control Electronics for Electro-Hydraulic Valves

This chapter explores the differences between electro-hydraulic open-loop and closed-loop systems. Several examples have been presented to discuss various ways to control load, speed and position of a hydraulic actuator. The chapter concludes by introducing examples from industrial and mobile applications.

Chapter 10: Electro-Hydraulic Valves Commissioning and Maintenance

This chapter introduces guidelines for commissioning and maintenance of EH valves.

Book Statistics:

The table shown below contains interesting statistical date about the textbook:

Chapter #	*Pages*	*Figures*	*Animated Circuits*	*Equations*	*Tables*	*Lines*	*Words*	*Characters*
Chapter 1	27	26	1	11	0	522	3086	21373
Chapter2	40	31	16	0	0	1310	6046	38834
Chapter 3	50	75	0	6	3	1086	7477	46084
Chapter 4	57	48	17	0	1	1361	6746	39681
Chapter 5	69	76	0	0	0	1106	7022	38398
Chapter 6	49	41	0	5	1	917	5484	34027
Chapter 7	83	83	0	5	3	1459	7349	46195
Chapter 8	45	45	0	1	0	908	6104	38171
Chapter 9	77	58	0	1	1	891	4249	22092
Chapter 10	9	7	0	1	0	209	1157	6316
Other	45	-	-	-	-	-	-	-
Total	**550**	**490**	**34**	**30**	**9**	**9769**	**54720**	**331171**

ABOUT THE AUTHOR

Medhat Khalil, Ph.D., Director of Professional Education & Research Development at the Applied Technology Center, Milwaukee School of Engineering, Milwaukee, WI, USA. Medhat got his bachelor's degree in mechanical engineering from Military Technical College (MTC), Cairo, Egypt. He got his master's degree in Mechanical Engineering from Cairo University, Cairo, Egypt. Medhat has been granted his Ph.D. in Mechanical Engineering and Post-Doctoral Industrial Research Fellowship from Concordia University in Montreal, Quebec, Canada. Medhat, published several fluid power textbooks. He participated in many technical conferences, published several reviewed technical papers, and is in the process of registering a number of patents. Medhat has been certified by the International Fluid Power Society (IFPS) as: Certified Fluid Power Hydraulic Specialist (CFPHS) and Certified Fluid Power Accredited Instructor (CFPAI). Medhat is a member of many grand institutions such as Center for Compact and Efficient Fluid Power Engineering Research Center (CCEFP), listed Fluid Power Consultant by the National Fluid Power Association (NFPA) and listed professional instructor by the American Society of Mechanical Engineers (ASME) and National American Die Casting Association (NADCA). Medhat has been assigned as the chair of the education committee for the International Fluid Power Exposition (IFPE 2017 and 2020). Medhat developed and taught various courses for industry professionals. He has a balanced academic and industrial experience. Medhat has a deep working experience in the field of Mechanical Engineering; more specifically in fluid power and motion control. Medhat had worked for several world-wide recognized industrial organizations such as Rexroth in Germany and CAE in Canada. Medhat has designed several hydraulic systems and developed analytical and educational software. Medhat also has vast experience in modeling and simulation of dynamic systems using Matlab-Simulink. Medhat was the designer and founder of the Universal Fluid Power Trainers. Medhat was the recipient of the "Otto Maha Pioneers in Fluid Power" award in 2012.

Chapter 1

Hydraulic versus Electrical Systems

Objectives:

This chapter presents structural and operational comparisons between hydraulic and electrical power transmission and control systems. Similarities and differences between the inductive, capacitive, and resistive elements are discussed.

Brief Contents:

Chapter 1 – Hydraulic versus Electrical Systems

Hydraulic and electrical power transmission and control systems, among other systems, are the two most comparable systems because both:

- Have various controlling and actuating elements.
- Can transmit the same magnitude of power.
- Can be used in *Industrial* and *Mobile* applications.

This chapter presents some analogies between the two systems.

1.1- Structural Analogy between Hydraulic and Electrical Systems

Figure 1.1 shows the structural similarity between hydraulic and electrical power transmission and control systems. The figure shows that each system is composed of the same sections.

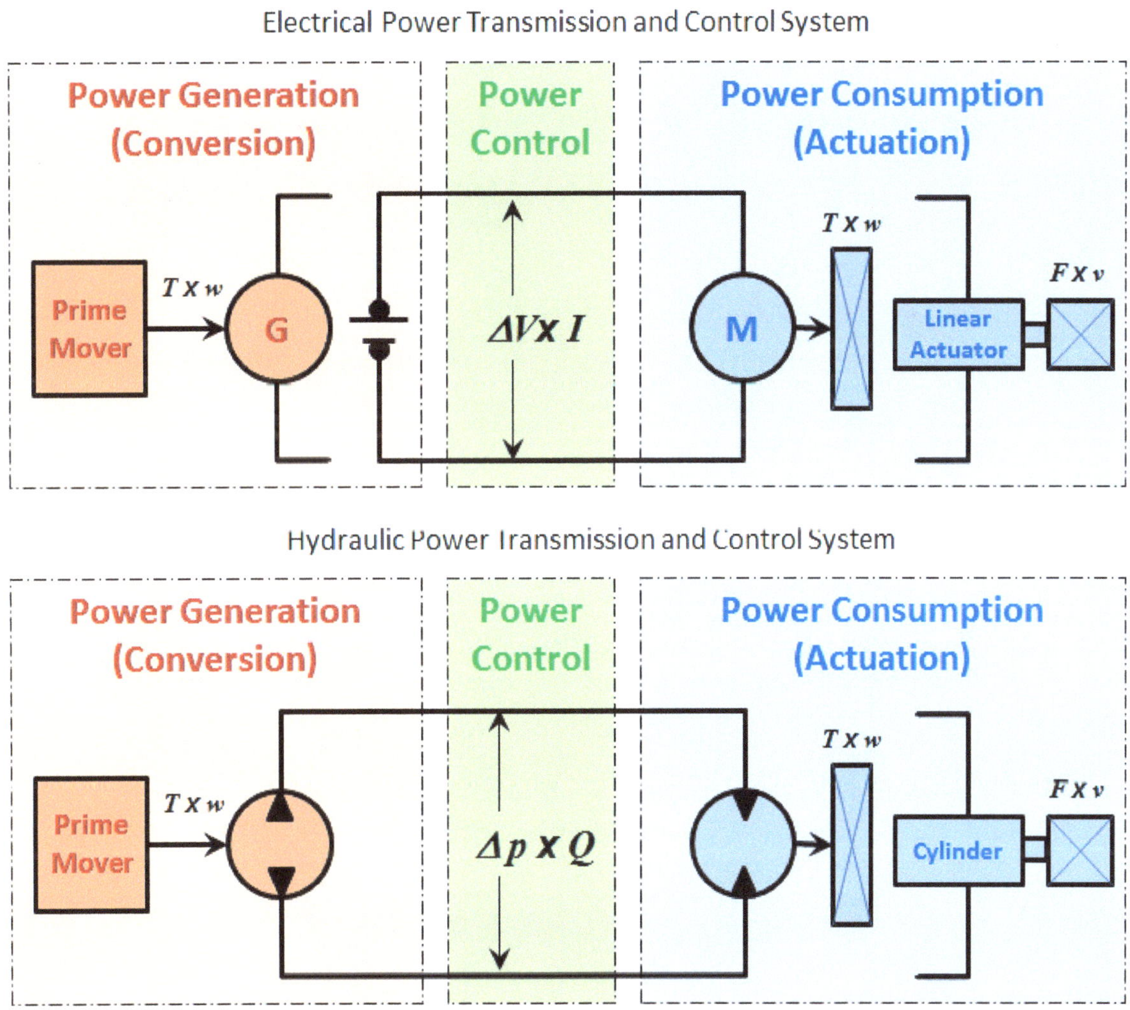

Fig. 1.1 - Structure Similarity between Hydraulic and Electrical Systems

1.1.1- Power Generation (Conversion)

Power Generators in both systems are also referred as *Power Convertors*. These are used to convert the power from the prime movers into a form of power that can be transmitted through power lines to the power consumers.

1.1.1.1- Electrical Power Generators

Figure 1.2 shows different sources that supply electrical power. Commonly electrical power is supplied through power wall outlets (1) connected to the electric grid. Alternatively, electrical power can be generated locally by electrical generators that convert the input mechanical power (Torque $\boldsymbol{T}$ x Angular Speed $\boldsymbol{\omega}$) into output electrical power (*Voltage Difference* ΔV x *Electrical Current* $\boldsymbol{I}$). Small electrical generators (2) can be driven by a belt. In medium to large sized generators (3 & 4) the generator is integrated as one unit with a prime-mover that is typically a thermal engine.

Fig. 1.2 - Electrical Power Generators

1.1.1.2- Hydraulic Power Generators

A hydraulic power generator (pump) converts the input mechanical power (Torque $\boldsymbol{T}$ x Angular Speed $\boldsymbol{\omega}$) into output hydraulic power (Differential Pressure $\boldsymbol{\Delta p}$ x flow $\boldsymbol{Q}$). A pump must be driven by a prime-mover. A prime-mover is typically an electric motor if the hydraulic system is used in an industrial application. DC type electric motors are commonly used for self-contained hydraulic power units (1). AC type electric motors are used for stationary large hydraulic power units (1). Thermal engines are typically used for hydraulic systems in mobile applications.

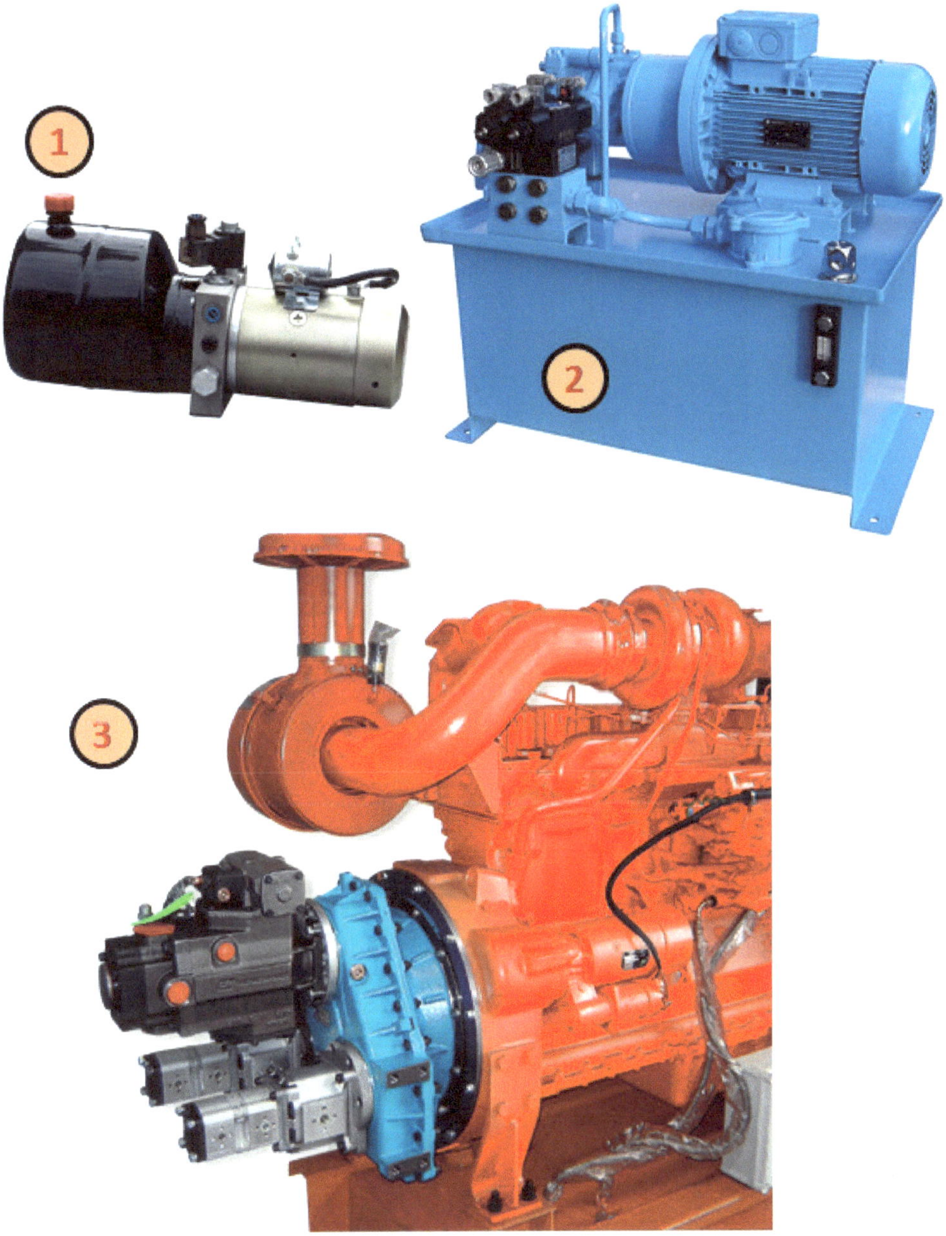

Fig. 1.3 - Hydraulic Power Generators (Pumps)

1.1.2- Power Control

Figure 1.4 shows more detailed structural similarity between hydraulic and electric systems. Each of the systems contains sections to control the magnitude and direction of the transmitted power as follows: Effort variable controller (1), flow variable controller (2), directional controller (3), energy storage element (4), and one-way element (5).

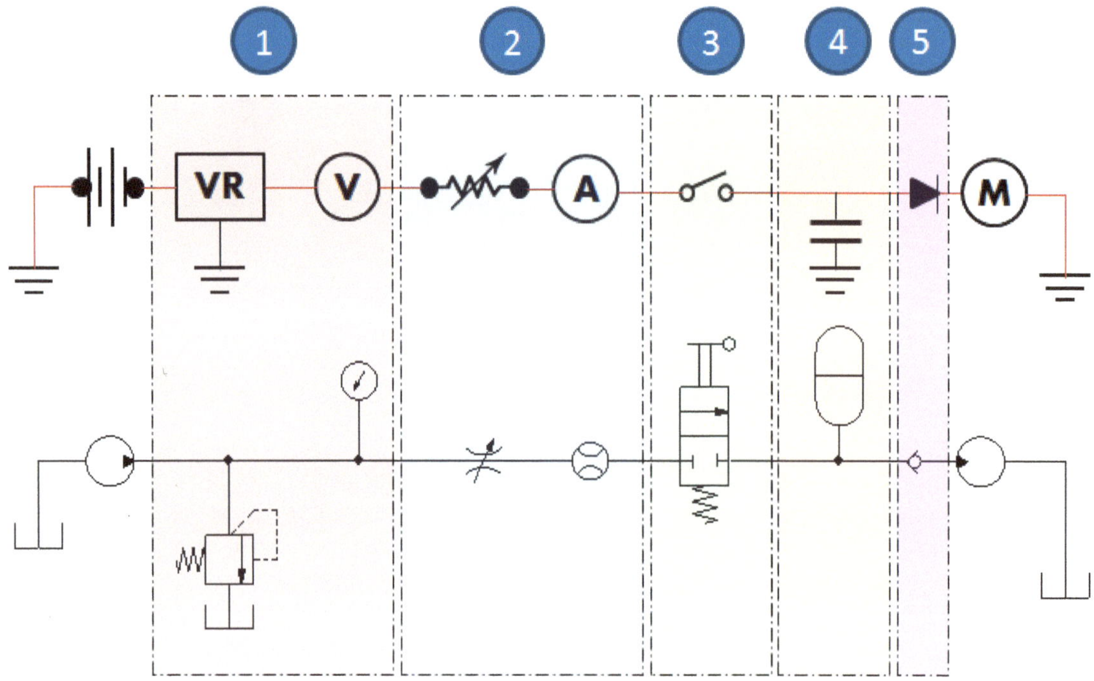

Fig. 1.4 - Common Control Elements

1.1.2.1- Power Quantification

Equations 1.1 and 1.2 show that, in both systems, the transmitted power is quantified similarly by the product of the *Effort Variable* and the *Flow Variable*.

$$\textbf{Electrical Power: } P_E\,(kW) = \frac{\Delta V\,(V) \times I(A)}{1000} \qquad \textbf{1.1}$$

$$\textbf{Hydraulic Power: } P_H\,(kW) = \frac{\Delta p\,(bar) \times Q(\frac{lit}{min})}{600} \qquad \textbf{1.2}$$

Where:

- Electrical effort variable = ΔV = voltage difference measured in Volts (V).
- Electrical flow variable = I = electrical current measured in Amperes (A).
- Hydraulic effort variable = Δp = differential pressure measured in bar.
- Hydraulic flow variable = Q = Volume flow rate measured in lit/min.

1.1.2.2- Effort Variable Controllers (Overload Protection)

Each of the systems has elements to control the effort variable and to protect the system against overloading conditions. As shown in Fig. 1.5, electrical systems use *Voltage Regulators (1)* to regulate the voltage. *Circuit Breakers* and *Fuses* (2) are used to protect the system against overload. Voltage is measured by Voltmeters (3). Hydraulic systems use various *Pressure Control Valves* (4) to regulate the pressure. Pressure Relief Valves protect hydraulic systems against overpressure. Pressure is measured by pressure gauges (5)

Fig. 1.5 - Effort Variable Controllers

1.1.2.3- Flow Variable Controllers (Resistive Elements)

Each of the systems has *Resistive Elements* to control the flow variable. As shown in Fig. 1.6, electrical systems use fixed resistors (1) or variable resistors (2) to control the electrical current that is measured by Ammeters (3). Hydraulic systems use *Fixed Orifices* (4) or *Flow Control Valves* (5) to regulate the flow. Fluid flow is measured by flow meters (6).

Fig. 1.6 - Flow Variable Controllers

Fig. 1.7 shows the behavioral similarity of an electric resistor and the resistivity of a hydraulic line. The figure shows the linear proportionality between the electrical current ***I*** and the voltage difference ***ΔV*** across an electrical resistor. For the same voltage, small resistance passes electrical current larger than largr resistance. Similarly, the figure shows the linear proportionality between the fluid flow ***Q*** and the differential pressure ***Δp*** across a hydraulic line. A hydraulic line with large diameter is less resistive than a hydraulic line with small diameter. So, for same differential pressure, large diameter line passes flow rate larger than small diameter line.

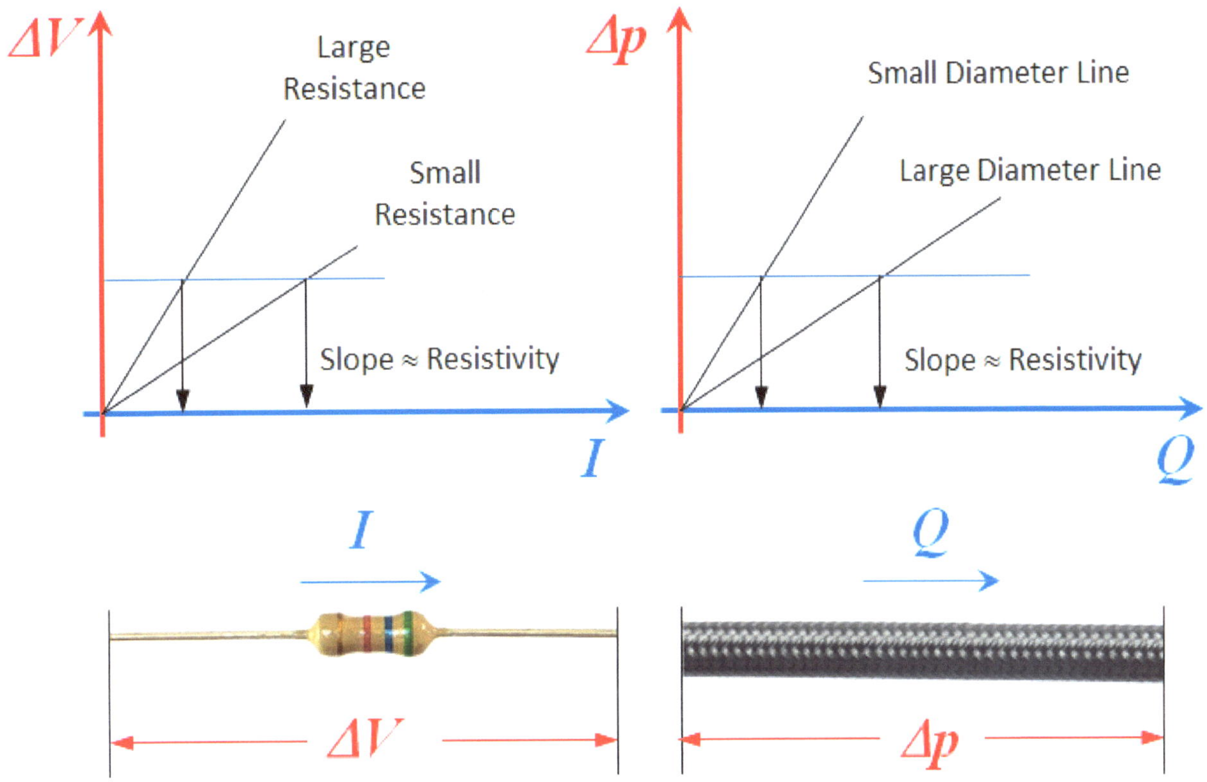

Fig. 1.7 - Similarity Characteristics of Electrical Resistance and Hydraulic Lines

Equations 1.3 and 1.4 confirm the linear proportionality between the effort variable and the *flow* Variable in both elements.

$$\Delta V = R\, I \qquad 1.3$$
$$\Delta p = R\, Q \qquad 1.4$$

Where:

- ***R*** is the resistance that depends on the dimensional parameter of the resistive element.
- Electrical resistance has the units of Ohm (Ω).
- Hydraulic line resistance has the units of [bar / (lit/min)].

The following section shows the similarity in finding the equivalent resistance of a number of resistive elements.

Linear Resistances in Series:

Figure 1.8 shows three electrical resistors in series and three hydraulic lines in series. Equations 1.5 and 1.6 show how to calculate the equivalent resistivity in both cases.

The calculation in both cases is based on the following two facts:

- The flow variable through the resistive elements in series is the same
- The total effort variable equals the sum of the effort variables across each element in series.

$$\Delta V_{Total} = \Delta V_1 + \Delta V_2 + \Delta V_3 = IR_1 + IR_2 + IR_3 = IR_{total} \Rightarrow R_{total} = \sum_{i=1}^{n} R_n \quad 1.5$$

$$\Delta p_{Total} = \Delta p_1 + \Delta p_2 + \Delta p_3 = QR_1 + QR_2 + QR_3 = QR_{total} \Rightarrow R_{total} = \sum_{i=1}^{n} R_n \quad 1.6$$

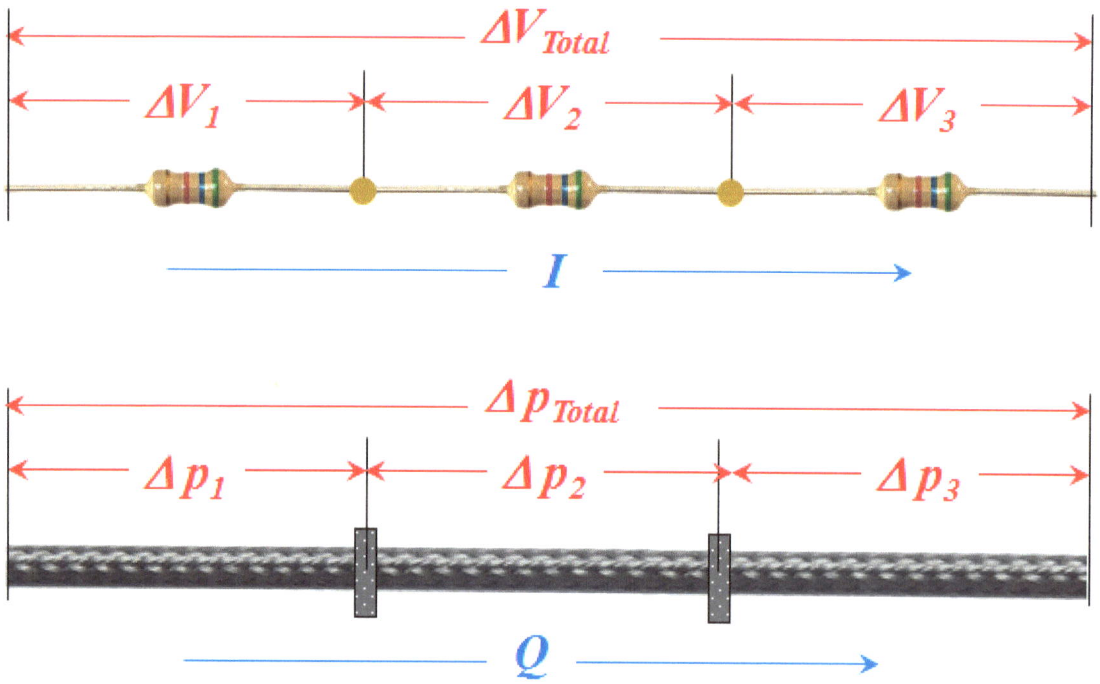

Fig. 1.8 - Linear Resistances in Series

Linear Resistances in Parallel:

Figure 1.9 shows three electrical resistors in parallel and three hydraulic lines in parallel. Equations 1.7 and 1.8 show how to calculate the equivalent resistance in both cases.

The calculation in both cases is based on the following two facts:

- The effort variable across the resistive elements in parallel is the same.
- The total flow variable equals the sum of the flow variables through each element in parallel.

$$I_{Total} = I_1 + I_2 + I_3 = \frac{\Delta V}{R_1} + \frac{\Delta V}{R_2} + \frac{\Delta V}{R_3} = \frac{\Delta V}{R_{Total}} \Rightarrow \frac{1}{R_{Total}} = \sum_{i=1}^{n} \frac{1}{R_n} \quad 1.7$$

$$Q_{Total} = Q_1 + Q_2 + Q_3 = \frac{\Delta p}{R_1} + \frac{\Delta p}{R_2} + \frac{\Delta p}{R_3} = \frac{\Delta p}{R_{Total}} \Rightarrow \frac{1}{R_{Total}} = \sum_{i=1}^{n} \frac{1}{R_n} \quad 1.8$$

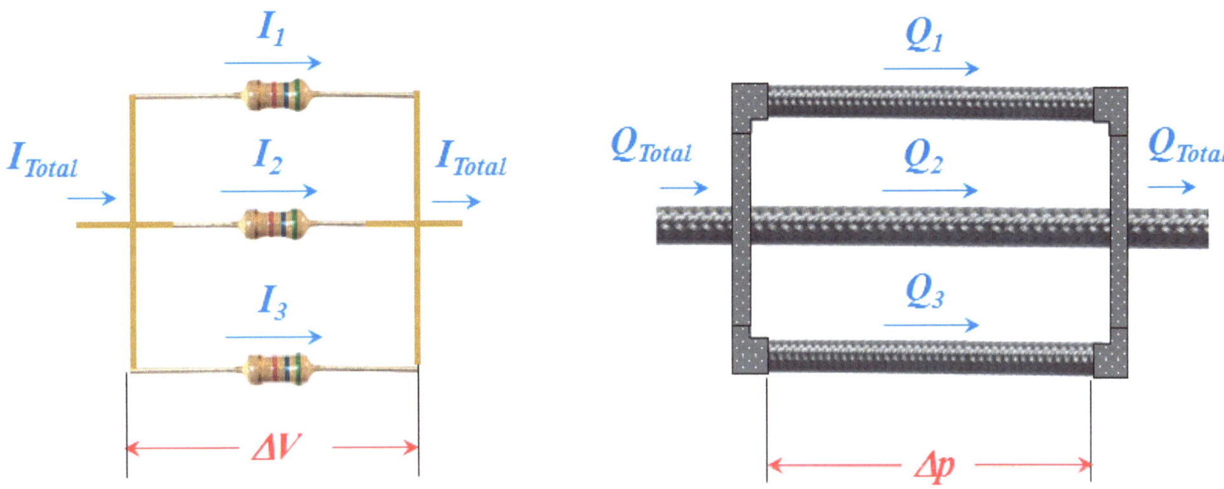

Fig. 1.9 - Linear Resistances in Parallel

Nonlinear Hydraulic Resistances:

When a hydraulic valve performs its control function, fluid flow through the valve is accompanied by a pressure drop. Valve manufacturers express the resistivity of a hydraulic valve by a term called *Valve Coefficient*. Equation 1.9 shows the nonlinearity between the differential pressure Δp across the valve and the flow rate Q of hydraulic fluid that has a specific gravity SG. A large valve coefficient means the valve passes a large flow for small pressure drop.

$$Q = C_v \sqrt{\frac{\Delta p}{SG}} \qquad 1.9$$

Units of the valve coefficient is [(units of flow rate)/(units of pressure)$^{-0.5}$]

Hydraulic Nonlinear Resistances in Series:

Figure 1.10 shows three hydraulic valves in series. Equations 1.10A, 1.10B and 1.10C show how to calculate the *Equivalent Valve Coefficient* for valves in series C_{ves}.

The calculation is based on the following two facts:

- The flow through the valves in series is the same.
- The total differential pressure equals the sum of the differential pressures across each valve in series.

$$\Delta p_{Total} = \Delta p_1 + \Delta p_2 + \Delta p_3 = Q^2.SG\left(\frac{1}{C_{v1}^2} + \frac{1}{C_{v2}^2} + \frac{1}{C_{v3}^2}\right) \qquad 1.10A$$

$$Q = \sqrt{\frac{\Delta p_{Total}}{\left(\frac{1}{C_{v1}^2} + \frac{1}{C_{v2}^2} + \frac{1}{C_{v3}^2}\right) \times SG}} = C_{ves}\sqrt{\frac{\Delta p_{Total}}{SG}} \qquad 1.10B$$

$$C_{ves} = \sqrt{\frac{1}{\sum_1^n \frac{1}{C_{vn}^2}}} \qquad 1.10C$$

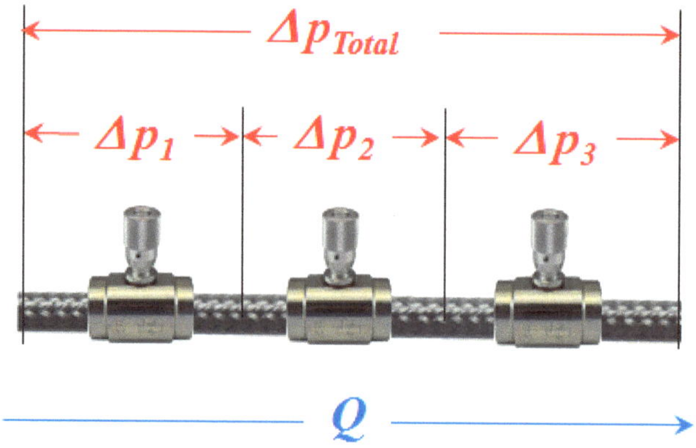

Fig. 1.10 - Nonlinear Hydraulic Resistances in Series

Hydraulic Nonlinear Resistances in Parallel:

Figure 1.10 shows three hydraulic valves in parallel. Equations 1.11A, 1.11B and 1.11C show how to calculate the equivalent valve coefficient for valves in parallel $\boldsymbol{C_{vep}}$. The calculation is based on the following two facts:

- The differential pressure across the valves in parallel is the same.
- The total flow through the valves equals the sum of the flow through each valve in parallel.

$$Q_{Total} = Q_1 + Q_2 + Q_3 = (C_{v1+} C_{v2+} C_{v3}) \sqrt{\frac{\Delta p}{SG}} = C_{vep} \sqrt{\frac{\Delta p}{SG}} \qquad 1.11A$$

$$C_{vep} = \sum_{1}^{n} C_{vn} \qquad 1.11B$$

$$\Delta p = Q_{Total}^2 \times SG \times \frac{1}{C_{vep}^2} \qquad 1.11C$$

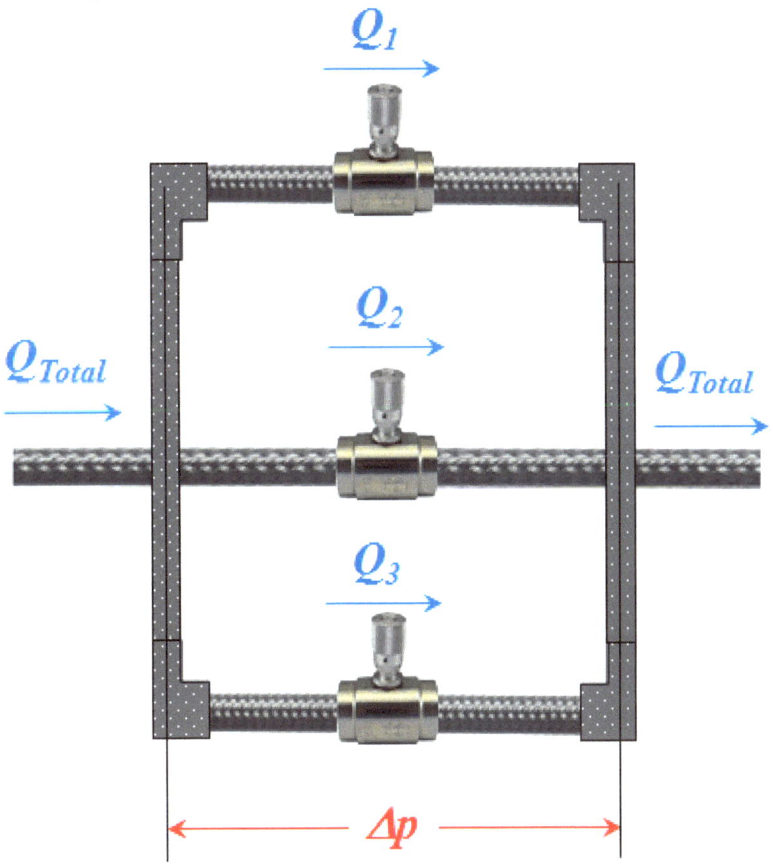

Fig. 1.11 - Nonlinear Hydraulic Resistances in Parallel

1.1.2.4- Directional Controllers

As shown in Fig. 1.12, switches and buttons are used in electrical systems. Directional control valves are used in hydraulic systems.

Fig. 1.12 - Directional Controllers

As shown in Fig. 1.13, the term Normally-Closed means an electrical switch closes the circuit and electrical current passes to drive the electrical motor. On the other hand the same term means a hydraulic valve terminates the flow to stop the hydraulic motor.

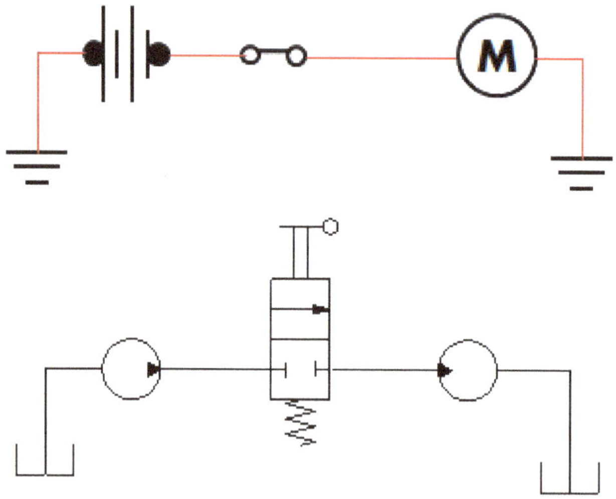

Fig. 1.13 - Electrical Switch versus Directional Valve

1.1.2.5- Energy Storage Elements (Capacitive Elements)

Each of the two systems has *Capacitive Elements* of different sizes to store energy. As shown in Fig. 1.14, *Capacitors* (1) or *Batteries* (2) are used to store electrical energy. Accumulators (3) are used to store hydraulic energy. The amount of stored energy and the charge time is proportional to the size of the capacitive element in both systems. It is known for fact that electrical capacitive elements have larger *Energy Density* than hydraulic accumulators.

Fig. 1.14 - Energy Storage Elements

1.1.2.6- Non-Return Elements

Each of the two systems has a *Non-Return Element* of different sizes to prevent flow in the reverse direction. As shown in Fig. 1.15, *Diodes* (1) are used in electrical systems and *Check Valves* (2) are used in hydraulic systems. Both diodes and check valves require a minimum value of the effort variable across the element in order for flow to pass through the element in the forward direction. Minimum *Cracking Pressure* is required to open a check valve. *Barrier Voltage* is required to open the diode.

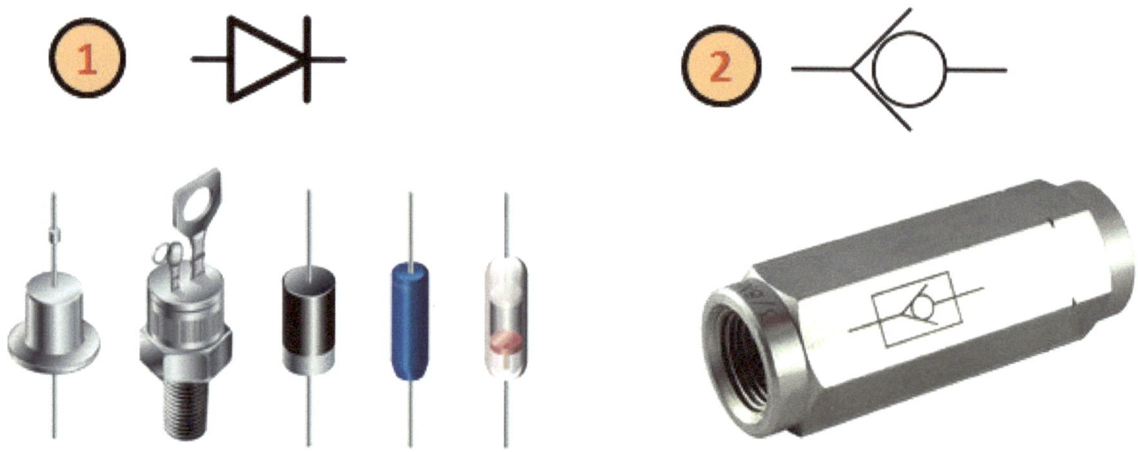

Fig. 1.15 - Non-Return Elements

1.1.2.7- Electrical versus Hydraulic Power Control Elements

From what has been previously discussed, electrical control elements have the following advantages over the hydro-mechanical valves:

- **Controllability:** because of the digital control capabilities, electrical control elements perform expanded, reprogrammable and smarter control functions than hydro-mechanical valves.
- **Closed-Loop Control:** hydro-mechanical components can't be used to build closed loop control system with electrical feedback. They can perform limited closed control with mechanical feedback such as in pressure compensated, power limited and load sense pumps.
- **Size:** electrical control elements are smaller in size for the same power.
- **Material:** electrical control elements are produced from less amount of material.
- **Production Process:** electrical control elements require less physical production process. Hydraulic valves require casting, machining, forging, heat treatment, etc.

1.1.3- Power Consumption (Actuation)

Each of the systems has linear and rotational actuators. A *Rotational Actuator* reconverts the corresponding input power (hydraulic or electrical) back to the mechanical form to drive a rotational torque $\boldsymbol{T}$ with an angular speed $\boldsymbol{\omega}$. A *Linear Actuator* reconverts the corresponding power to a mechanical form to drive a translational force $\boldsymbol{F}$ with a speed $\boldsymbol{v}$.

1.1.3.1- Rotational Actuators

Figure 1.16, shows constant speed electrical motor (1), fixed displacement hydraulic motor (2), variable speed electrical motor (3), and variable displacement hydraulic motor (4). The following section explores the main features of *Rotational Actuators* in both systems:

Similarities:

- Both systems have rotational actuators referred to "*Motors*" of different sizes.
- Both systems have constant speed motors (1 & 2) and variable speed motors (3 & 4).

Differences:

- **Types:** Electrical motors could be of AC or DC type.
- **Nmax:** Maximum speed of electrical motors is higher than hydraulic motors. Fluid friction limits the maximum speed of a hydraulic motor.
- **Nmin:** Minimum speed of electrical motors is less than hydraulic motors because the hydraulic motors are inefficient at low speed.
- **Power Density:** Hydraulic motors have larger *Power Density* than electrical motors.
- **Saturation:** Hydraulic motors are not saturated as electric motors. A Hydraulic motor generates torque proportional to differential pressure across the motor and its size. The only saturation limits of a hydraulic motor are maximum working pressure and maximum torque of the motor's shaft.
- **Starting Torque:** Hydraulic motors have better starting torque and better dynamic response as compared to the electrical motors.
- **Cost:** Electric motors are commonly less expensive for the same power.

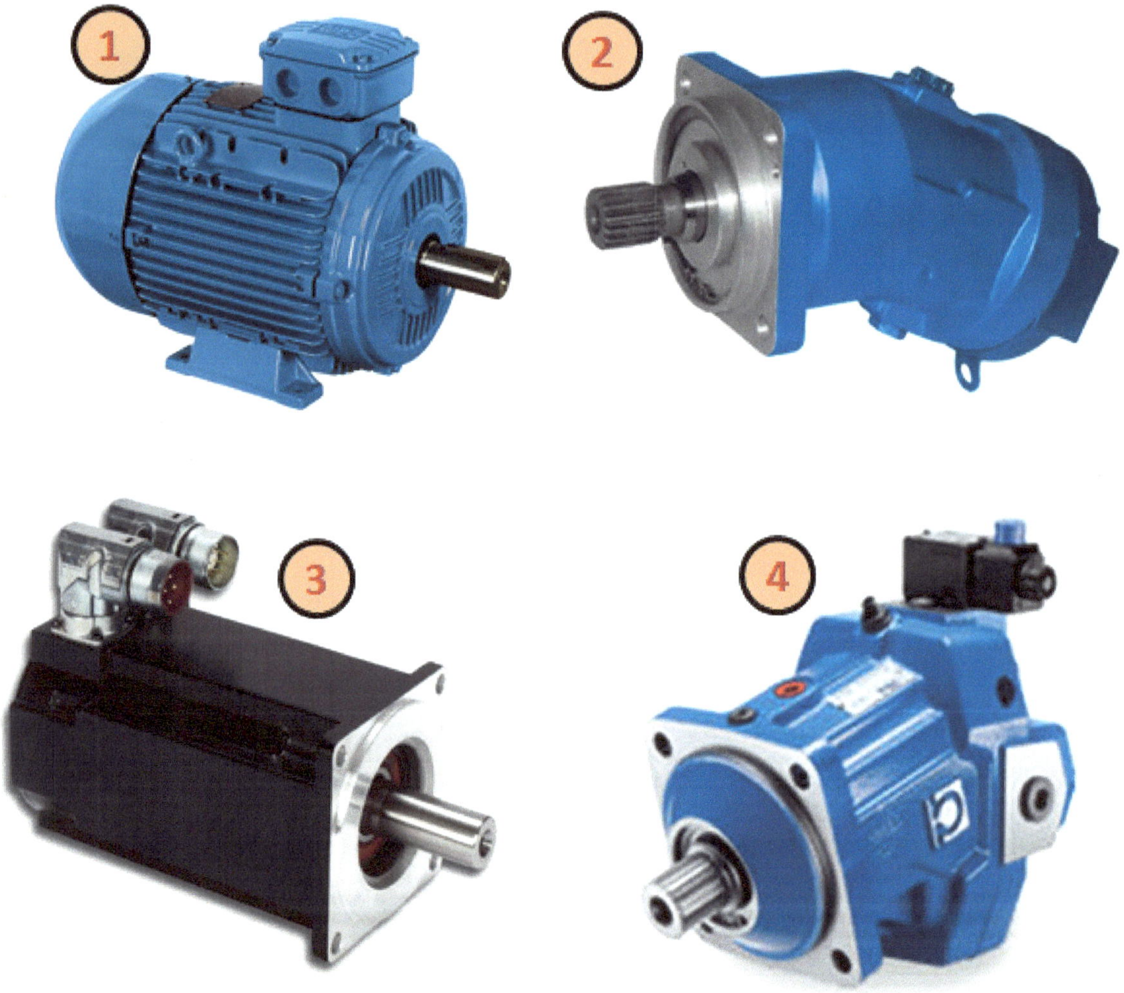

Fig. 1.16 - Rotational Actuators

1.1.3.2- Linear Actuators

Figure 1.17, shows a linear electrical actuator (1), hydraulic cylinder (2), servo-controlled linear electrical actuator (3), and servo-controlled hydraulic cylinder (4). The following section explores the main features of *Linear Actuators* in both systems:

Similarities:

- Both systems have linear actuators of different sizes. They are called "*Cylinders*" in hydraulic systems.
- Both systems have servo-controlled linear actuators.

Differences:

- **Mechanisms:** A hydraulic cylinder produces the linear motion without the use of any intermediate mechanism as in electrical linear actuators.
- **Strokes:** Hydraulic cylinders offer longer strokes than electrical linear actuators.
- **Power Density:** For the same physical size actuator, hydraulic cylinders carry higher loads than electrical linear actuators.
- **Cost:** Hydraulic cylinders are commonly less expensive than the equivalent electrical linear actuators that have the same size.

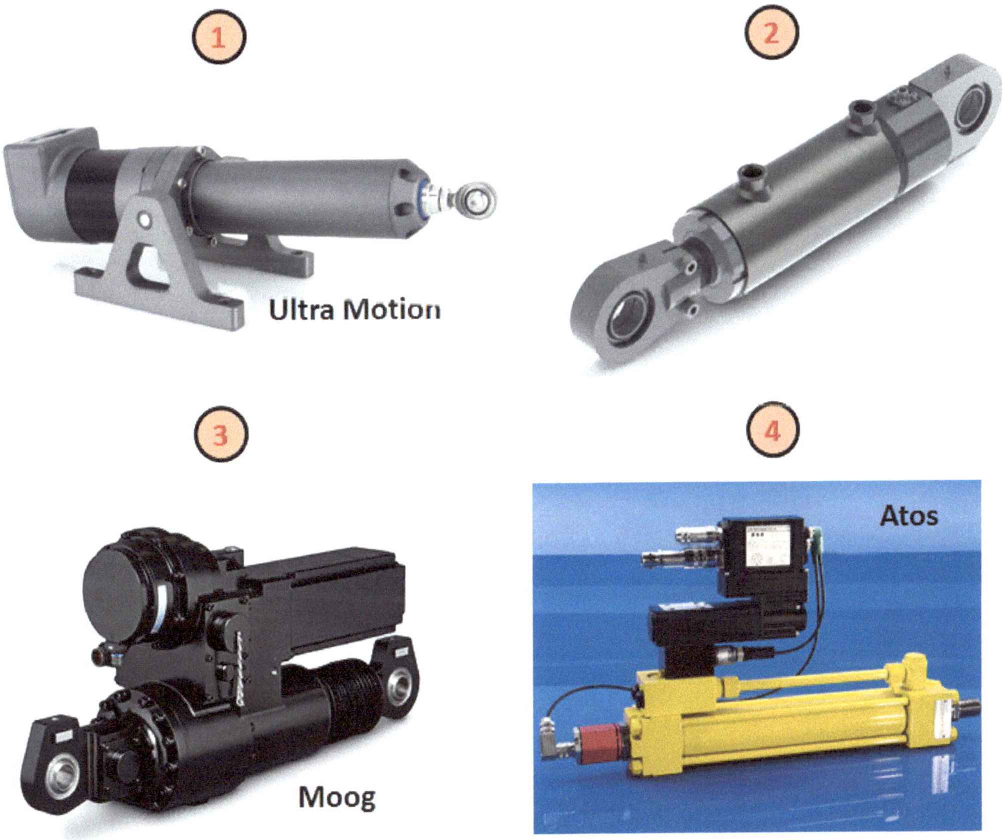

Fig. 1.17 - Linear Actuators

1.2- Operational Analogy between Hydraulic and Electrical Systems

1.2.1- Energy Transmission in a Circuit

As shown in Fig. 1.18, electrical and hydraulic systems are structured using the concept of a circuit that has the following features:

- **Flow Variable:** Each circuit has a flow variable that starts at a ground point, circulates through the elements of the circuit sequentially, and ends at the same ground point. A hydraulic circuit has the oil reservoir as the ground point versus the earth point in an electrical circuit.
- **Effort Variable:** As the flow variable goes through each of the elements of the circuit, it will be accompanied by the effort variable.
- **Power of Each Element:** The power P_E (gained, consumed, or wasted) by any of the elements is quantified by the product of the flow variable through the element and the effort variable across the element.
- **Circuit Power:** The power gained at the power generator equals the sum of the power consumed or wasted by the rest of elements in the circuit plus the line losses.

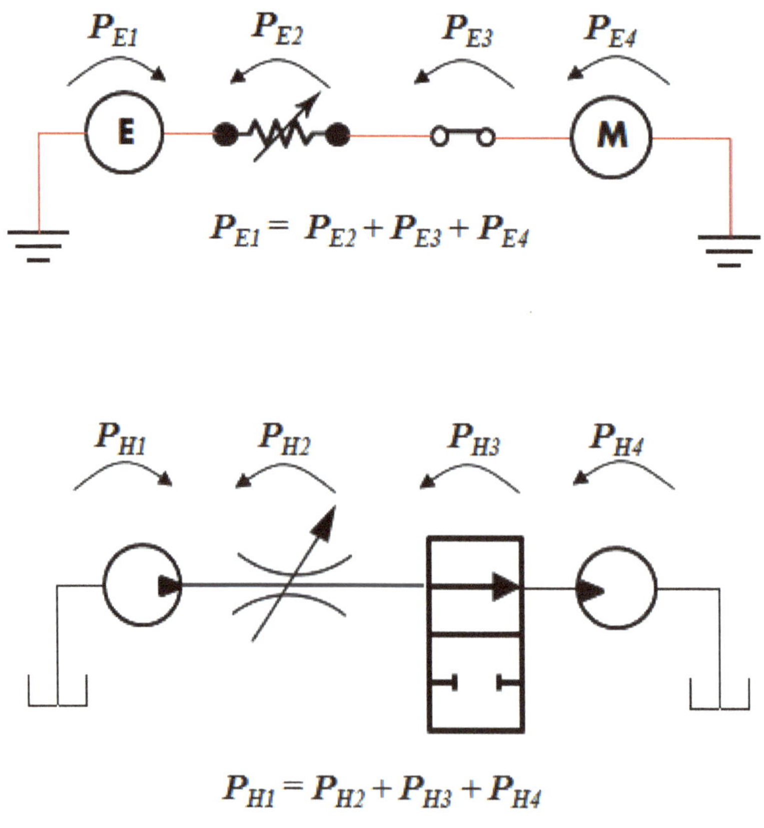

Fig. 1.18 - Concept of a Circuit

1.2.2- Energy Transmission Efficiency

As shown in Fig. 1.19, both systems experience losses while energy is transmitted along the forward path of the system. An electrical system generally wastes less energy for the same transmission distance and same amount of transmitted energy because:

Energy Carrier:
A hydraulic fluid, as the energy carrier in a hydraulic system, has internal friction due to its viscosity. Electrons, as energy carrier in an electrical system, waste less energy.

Transmission Lines:
Pipes, tubes and hoses, as the hydraulic transmission lines have large resistance because of their friction and the fluid dynamics inside the lines. Wires, as the transmission line in electrical systems, have less resistance than hydraulic lines.

Power Generators and Power Consumers:
Hydraulic pumps and motors are less efficient than equivalent elements in an electrical system. Additionally, efficiencies of pumps and motors are highly affected by the operating conditions of the system, particularly the operating pressure, temperature and speed.

Control Elements:
Hydraulic systems commonly use valves to perform control and regulation functions. Use of valves is always accompanied by pressure drop.

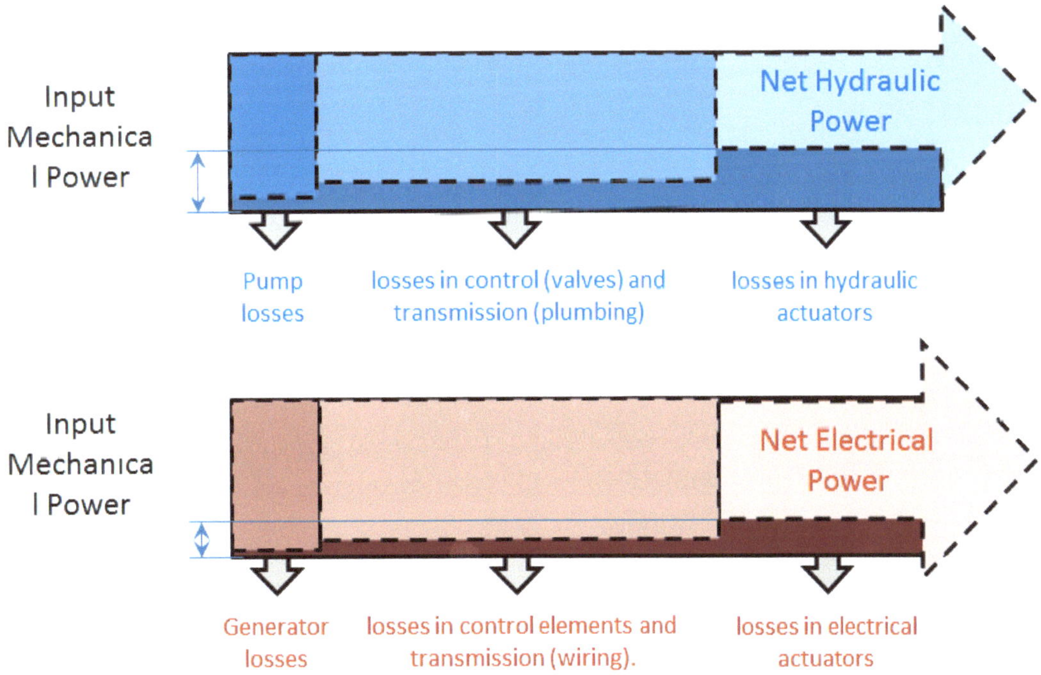

Fig. 1.19 - Energy Transmission Efficiency

1.2.3- Energy Transmission Cost Factor

Figure 1.20 shows the result of an experiment that was conducted at the University of Kassel in Germany to show the energy requirement for different linear actuators that perform the same duty cycle. The results show clearly that the actuation cost using linear electrical actuators is the least.

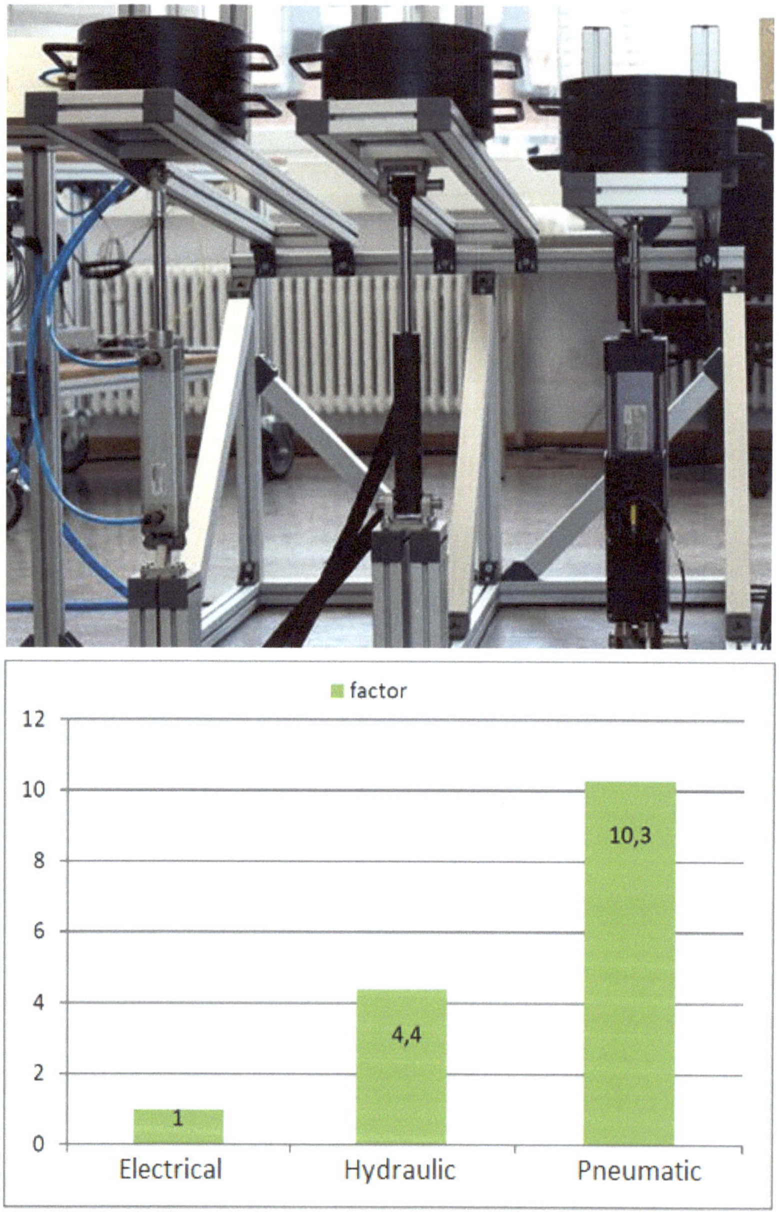

Fig. 1.20 - Energy Transmission Cost Factor (Courtesy of Exlar)

1.2.4- Energy Transmission Distance

Obviously an electrical control system can transmit the energy, between the power generator and the power consumer, to a distance that is much longer than a hydraulic system can. As shown in Fig. 1.21, electrical systems use wire cables instead of pipes, tubes, and hoses that are used in hydraulic systems.

Electrical Power Lines

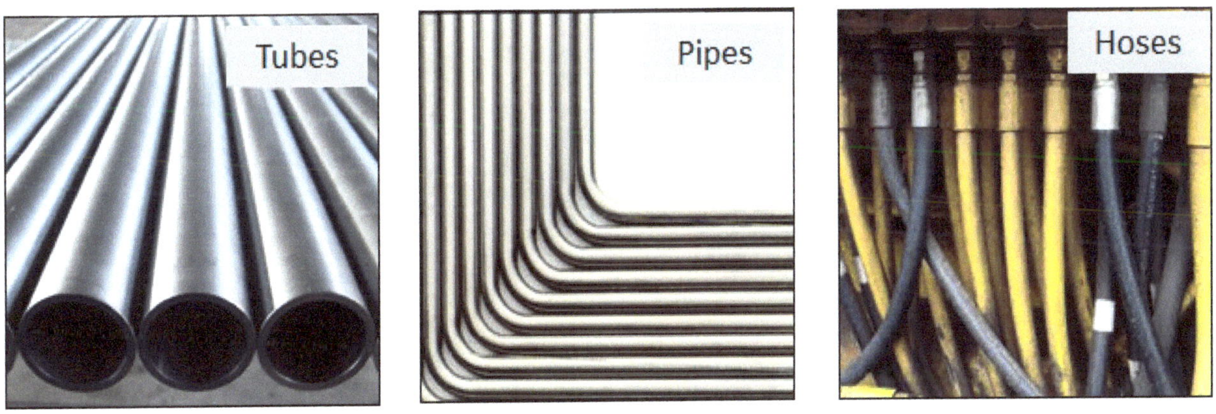

Hydraulic Power Lines

Fig. 1.21 - Electrical versus Hydraulic Power Lines

1.2.5- Readiness of Operation

Because the electrical power can commonly be introduced on spot, installing and commissioning an electrically driven machine is much easier than a hydraulically driven machine. As shown in Fig. 1.22, quick connect cables install in a small fraction of the time it takes to plumb a hydraulic system. A hydraulic system requires a lot of preparation, e.g. a reservoir, plumbing, cooling water, filter, etc.

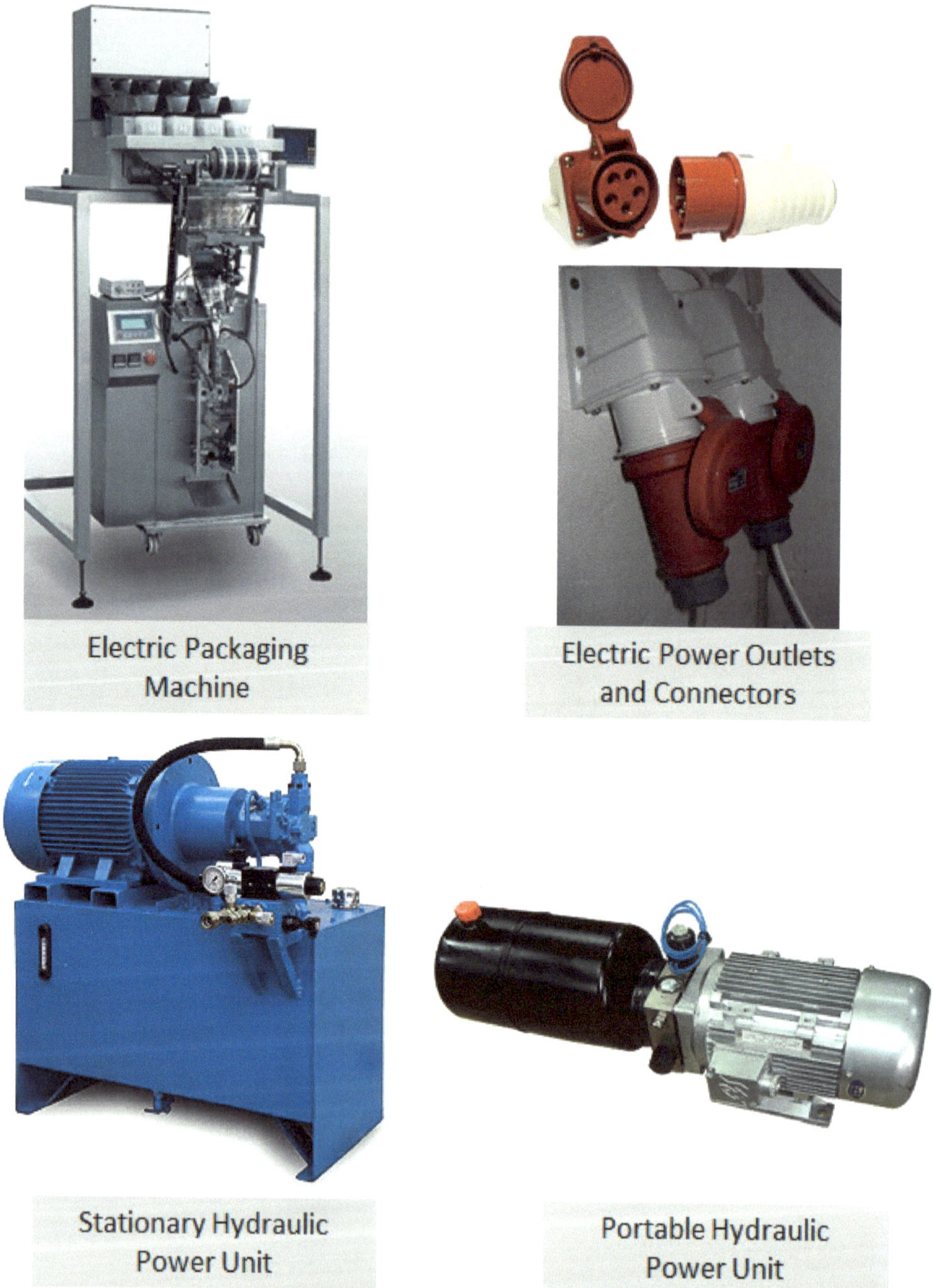

Fig. 1.22 - Commissioning an Electrical Machine versus Hydraulic Driven Machine

1.2.6- Cleanliness of Operation

The work environment of an electrically-driven machine is mostly cleaner than the work environment of a hydraulic driven machine. A hydraulically driven machine, as shown in Fig. 1.23, may experience oil contamination (1), oil leakage (2), and dirt on the surfaces of the hydraulic components (3). Agricultural machines should use biodegradable hydraulic fluids (4). Hydraulic fluids, other than the biodegradable ones, are not friendly to the environment and must be recycled (5). Possible toxicity of hydraulic fluid restricts such systems from some applications, e.g. food and pharmaceutical industries.

Fig. 1.23 - Cleanliness of Hydraulic-Driven Machine

1.2.7- Maintenance of Operation

Maintenance requirements of an electrically driven machine are much less than a hydraulically driven machine. Many of the electrical components are maintenance-free. Examples are shown in Fig. 1.24, maintenance free motors (1) and maintenance free hoists (2).

Since 80 % of hydraulic system faults are due to the oil contamination, it is important to keep such systems well maintained. Additionally, hydraulic fluids are expensive and require replacement at a specific number of working hours. The figure shows a dirty oil filter (3) that must be replaced.

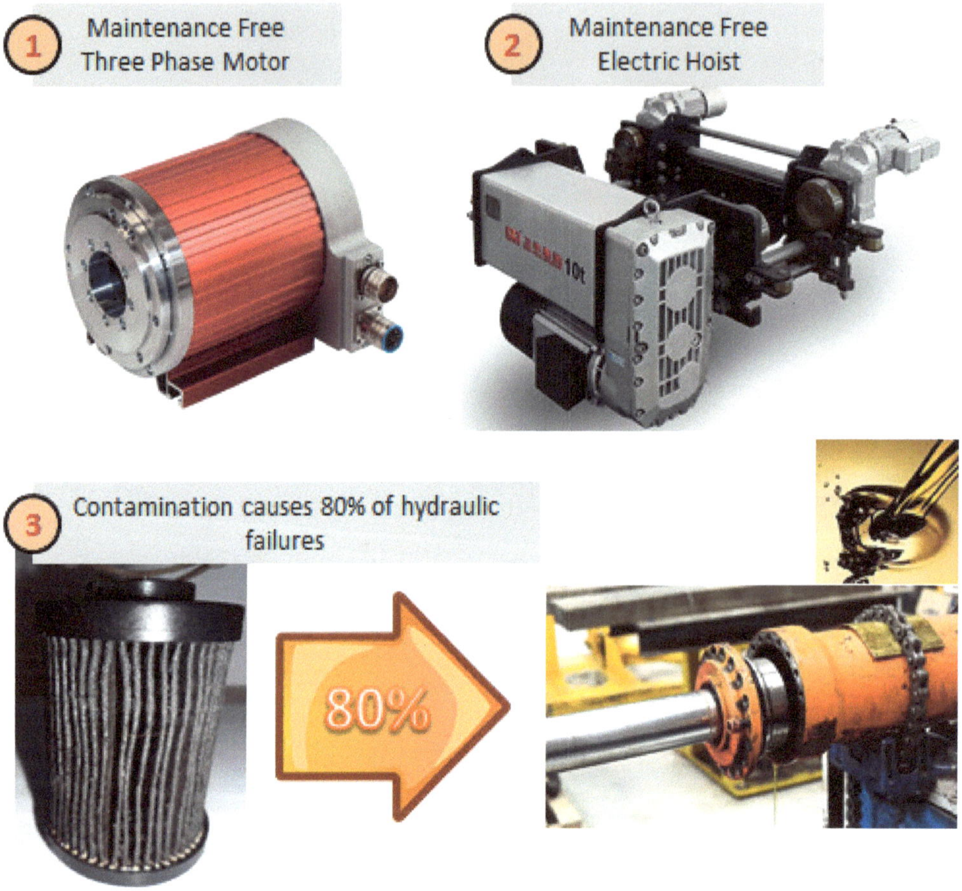

Fig. 1.24 – Maintenance Requirements for Electric and Hydraulic Machines

1.2.8- Safety of Operation under High Temperature

If an electrically driven machine can work under high temperature, it will offer a safer operating condition than a hydraulically driven machine since there is no fire risk from flammable oils. Hydraulic systems that work in a fire hazard environment require special hydraulic fluids. Examples of that are shown in Fig. 1.25.

Fig. 1.25 - Hydraulic Systems under High Temperature

1.2.9- Noise of Operation

The noise level of an electrical system is typically 30% lower than that of a hydraulic system. Reducing noise of a machine is a major requirement for health and safety standards.

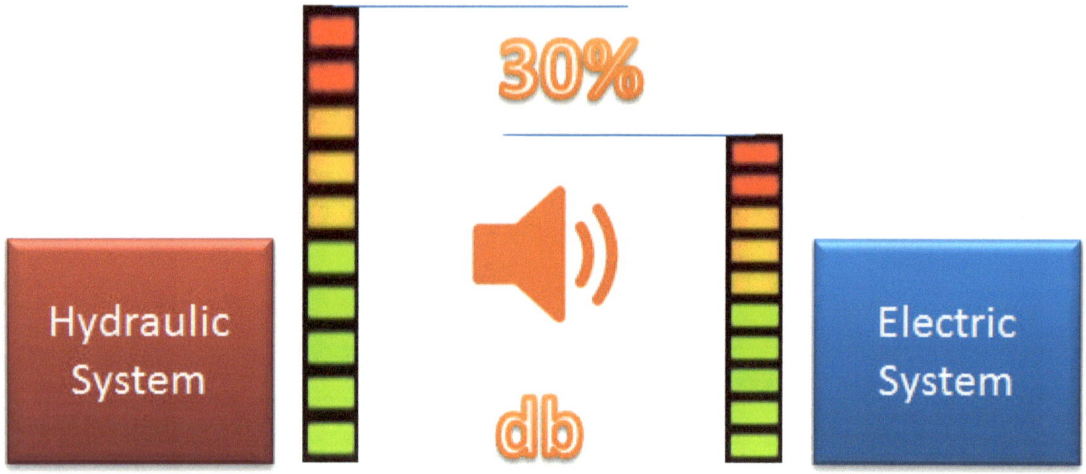

Fig. 1.26 - Noise Level in Hydraulic Systems versus Electric Systems

Chapter 2

Hydro-Mechanical versus Electro-Hydraulic Solutions

Objectives:

This chapter explores the features and challenges of electro-hydraulic systems. Additionally, the chapter introduces the benefits of converting the classical hydro-mechanical systems into electro-hydraulic ones. Pressure control, flow control, power control, and sequence control are among the systems discussed.

Brief Contents:

Chapter 2 – Hydro-Mechanical versus Electro-Hydraulic Solutions

2.1- Features and Challenges of Electro-Hydraulic Systems

After briefly discussing the structural and operational analogies between the electric and hydraulic power transmission and control systems, what happens if both systems are merged into one system? The aforementioned discussion in Chapter 1 justifies why in a power transmission and control system, it is recommended to assign the control function to electrical components and the actuating functions to hydraulic components. As shown in Fig. 2.1, the system that results by merging a hydraulic and an electrical system is called an *Electro-Hydraulic System*. Such a system will be referred to as an EH system in the text whenever necessary. Such a system combines the advantages of both systems and has the following main advantageous features:

- **Step-less Analog Settings:** because of the use of proportional and servo valves.
- **Remote Setting:** because of the ability of providing and transmitting electrical signals.
- **Remote Monitoring**: because of the use of sensors and transducers.
- **Compactness:**
  - Because of the use of a few electro-hydraulic components.
  - Electro-hydraulic components are lighter than hydro-mechanical components.
  - Replacing plumbing by cable wires.
- **Energy Efficiency:** because of possible use of electro-hydraulic variable displacement pumps to control the actuator directly.
- **Smartness:** because of the expanded control capabilities of the electrical control device and the use of digital controllers.
- **Automatic Closed-Loop Control:** because of the use of proportional and servo valves.

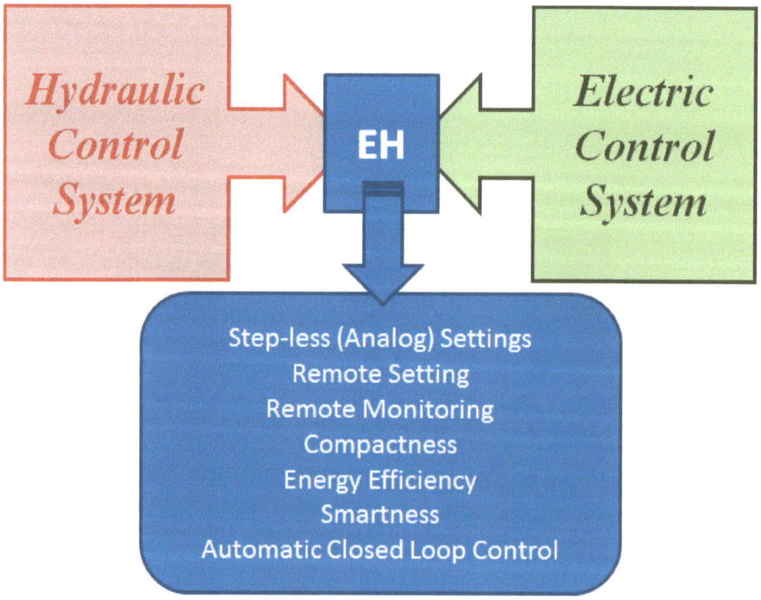

Fig. 2.1 - Advantageous Features of Electro-Hydraulic Systems

Although an electro-hydraulic system has a number of advantageous features, it also has the following challenges:

Reliability: electro-hydraulic devices are highly sensitive to contamination.

Know-How: programming, diagnostics, troubleshooting, and know how it works.

System Tuning: adjustment of the closed-loop gain, and other performance enhancement functions, requires certain minimum level of experience and education.

Single Point Failure: an electro-hydraulic system commonly consists of a small number of components. As a result, if any of these devices fail, the rest of the system fails.

Cost: cost is another barrier to convert a hydro-mechanical system to an electro-hydraulic one.

The following sections have been written to explore the advantages that can be achieved by converting a classical hydro-mechanical system to an electro-hydraulic one.

2.2- Pressure Control Solutions

2.2.1- Pressure Control using Hydro-Mechanical Pressure Relief Valve

Fig. 2.2 shows the simplest solution to control the maximum pressure in a hydraulic system by using a hydro-mechanical pressure relief valve (PRV). Such a system has the following features:

Step-less (Analog) Settings: The PRV can be used to set the maximum system pressure through a continuous range of values within the maximum range.

Remote Setting: The valve is set manually in the field.

Remote Monitoring: Not possible since the PRV does not offer any electrical communication.

Compactness: The system is compact since just one valve is used.

Energy Efficiency: Use of a PRV causes energy losses.

Smartness: There is no possibility because the PRV can't be set electrically.

Automatic Closed-Loop Control: A PRV can't be used to build a closed-loop pressure control.

Features	Status
Step-less (Analog) Settings	✓
Remote Setting	×
Remote Monitoring	×
Compactness	✓
Energy Efficiency	×
Smartness	×
Automatic Closed-Loop Control	×

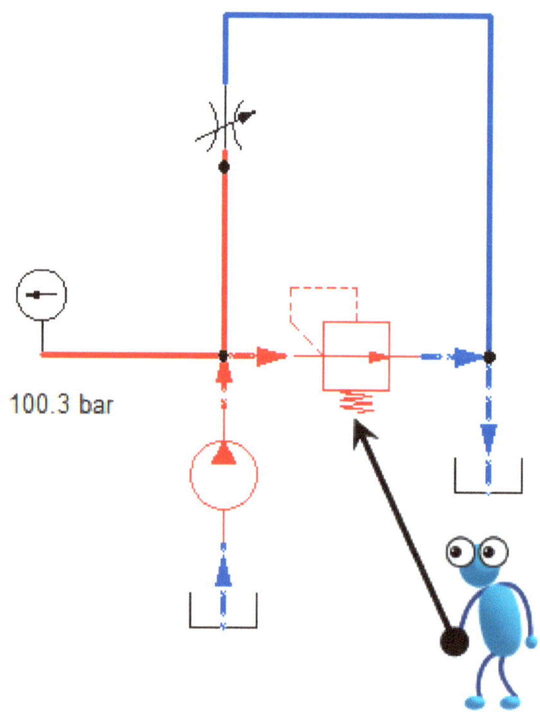

Fig. 2.2 - Pressure Control using a Hydro-Mechanical Pressure Relief Valve

2.2.2- Pressure Control using Hydro-Mechanical Pressure Compensated Pump

In this solution, as shown in Fig. 2.3, a hydro-mechanical *pressure compensated pump* replaces the fixed pump and the PRV. Such a system has the following features:

Step-less (Analog) Settings: The pressure compensator on the pump can be used to set the maximum system pressure seamlessly through a continuous range of values within the maximum range.

Remote Setting: The pressure compensator is set manually in the field.

Remote Monitoring: Not possible since the pump does not offer any electrical communication.

Compactness: The system is compact since the pressure compensated pump replaces two elements.

Energy Efficiency: As compared with the previous solution, the use of the hydro-mechanical pressure compensated pump improves the energy efficiency of the system. As shown in the figure, the pump de-strokes itself at the maximum pressure and minimizes the energy losses.

Smartness: There is no possibility because such a pump can't be set electrically.

Automatic Closed-Loop Control: Such a pump can't be used to build an automatic closed-loop pressure control.

Features	Status
Step-less (Analog) Settings	✓
Remote Setting	×
Remote Monitoring	×
Compactness	✓
Energy Efficiency	✓
Smartness	×
Automatic Closed-Loop Control	×

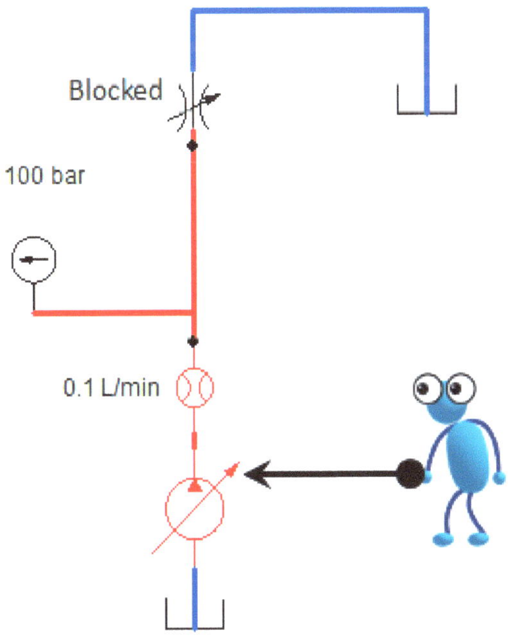

Fig. 2.3 - Pressure Control using Hydro-Mechanical Pressure Compensated Pump

2.2.3- Multiple Pressure Settings using Electro-Hydraulic Switching Valves

Figure 2.4 shows an example of multiple pressure selection. The figure shows that two additional pilot stages are connected to the pilot operated pressure relief valve via normally-closed, 2/2, and solenoid-operated (switching) directional valves. In addition, the figure shows that adding an extra switching valve connected directly to the tank is used to completely unload the pump. Such a system has the following features:

Step-less (Analog) Settings: Changing the maximum pressure setting in this system is in a discrete way. Maximum pressure jumps from one value to another depending on the combination of the status of the switching valves. Since the pressure is set so that $\mathbf{p_1 > p_2 > p_3}$, the pressure setting can be changed based on the truth table shown in the figure. As an example, if Y3 is set as True, then the pump is unloaded no matter what the status of the other valves are. Hypothetically, an infinite number of pilot stages are needed in such a system to achieve step-less pressure control.

Remote Setting: A set of pushbuttons or a programmable logic controller (PLC) can be used to control, remotely, the status of the switching valves (True or False) so that maximum system pressure is set accordingly.

Remote Monitoring: Since a PLC is used, the system pressure can be now monitored remotely.

Compactness: Obviously the system implements a large number of components including hydraulic lines, manifolds, PLC, etc. that will make the system large and heavy.

Energy Efficiency: Use of pressure relief valve results in energy losses.

Smartness: A microprocessor controller can be programmed to control the system pressure interactively and repeatedly during a duty cycle of the machine or based on certain operating conditions.

Automatic Closed-Loop Control: Such a pressure relief valve can't be used to build a closed-loop automatic pressure control.

Pmax	Y1	Y2	Y3
p1	F	F	F
p2	T	F	F
p3		T	F
p4			T

Features	Status
Step-less (Analog) Settings	×
Remote Setting	✓
Remote Monitoring	✓
Compactness	×
Energy Efficiency	×
Smartness	✓
Automatic Closed-Loop Control	×

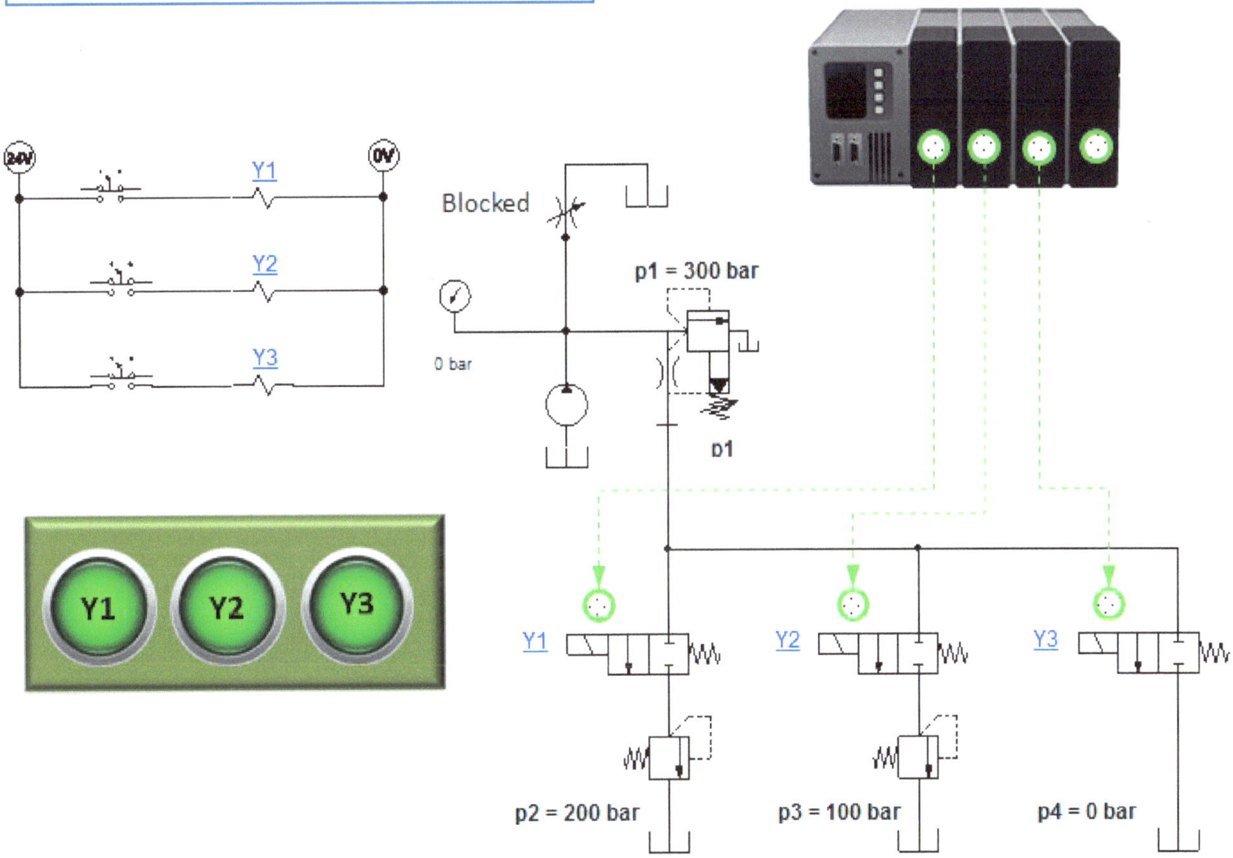

Fig. 2.4 - Multiple Pressure Settings using Solenoid Operated Valves

2.2.4- Pressure Control using an Electro-Hydraulic Continuous Valve

Figure 2.5 shows an example of controlling the system pressure using a *proportional pressure relief valve*. Doing so, as shown in the figure, increases the number of features in the table. Such system has the following features:

Step-less (Analog) Settings: The proportional pressure relief valve can be used to set the maximum pressure, in an analog fashion, through a continuous range of values within the maximum range.

Remote Setting: A manual potentiometer or a PLC can be used to control remotely the adjustment of the proportional pressure relief valve so that maximum system pressure is set accordingly.

Remote Monitoring: Since a PLC is used, the system pressure can be now monitored remotely.

Compactness: The system is compact, since just one valve is used.

Energy Efficiency: Use of pressure relief valves, even proportional ones, causes energy losses.

Smartness: A microprocessor controller can be programmed to control the system pressure interactively and repeatedly during a duty cycle of the machine or based on certain operating conditions.

Automatic Closed-Loop Control: Continuous valves can be used to build a closed-loop automatic pressure control.

Features	Status
Step-less (Analog) Settings	✓
Remote Setting	✓
Remote Monitoring	✓
Compactness	✓
Energy Efficiency	×
Smartness	✓
Automatic Closed-Loop Control	✓

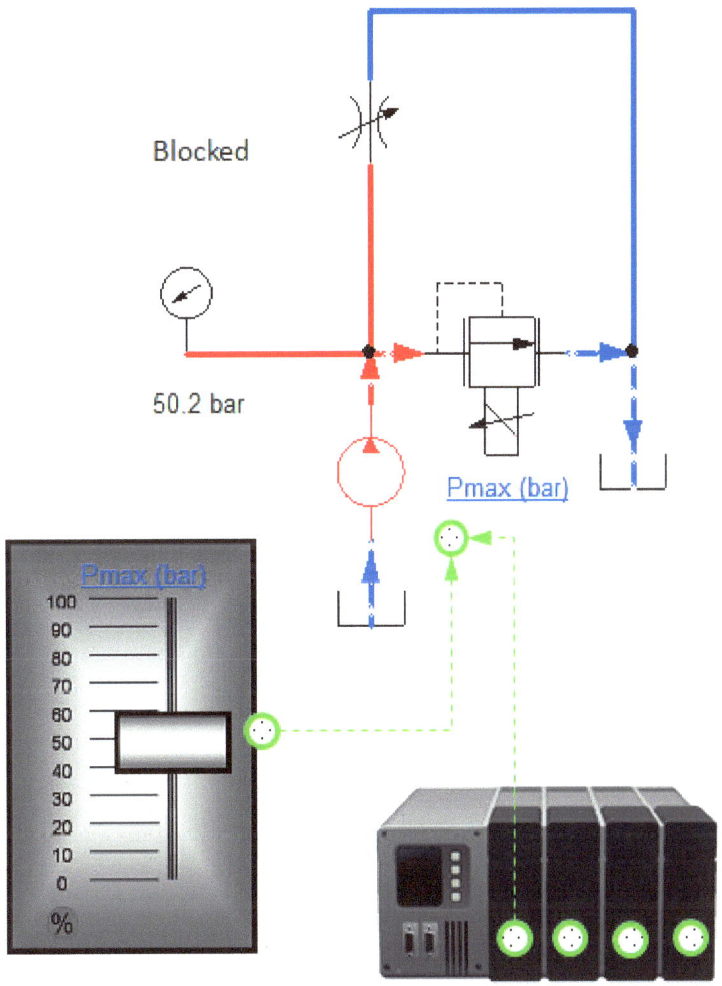

Fig. 2.5 - Pressure Control using Proportional Pressure Relief Valve

2.2.5- Pressure Control using an Electro-Hydraulic Pressure Compensated Pump

Figure 2.6 shows an example of controlling the system pressure using an electro-hydraulic pressure compensated pump. Doing so, as shown in the figure, removes all the shortcomings of the previous pressure control solutions. The only limitation to use such a solution is the cost and the know-how. Such a system has the following features:

Step-less (Analog) Settings: The electro-hydraulic pressure compensated pump can be used to set the maximum pressure, in an analog fashion, through a continuous range of values within the maximum range.

Remote Setting: A manual potentiometer or a PLC can be used to control, remotely, the adjustment of the pump pressure compensator so that maximum system pressure is set accordingly.

Remote Monitoring: Since a PLC is used, the system pressure can be now monitored remotely as well.

Compactness: The system is compact since just the pressure control is localized at the pump.

Energy Efficiency: The use of the electro-hydraulic pressure compensated pump improves the energy efficiency of the system. The electro-hydraulic pressure compensated pump performs exactly like the hydro-mechanical one except that it is controlled electrically rather than manually.

Smartness: A microprocessor controller can be programmed to control the system pressure interactively and repeatedly during a duty cycle of the machine or based on certain operating conditions.

Automatic Closed-Loop Control: electro-hydraulic pressure compensated pump can be used to build a closed-loop automatic pressure control.

Features	Status
Step-less (Analog) Settings	✓
Remote Setting	✓
Remote Monitoring	✓
Compactness	✓
Energy Efficiency	✓
Smartness	✓
Automatic Closed-Loop Control	✓

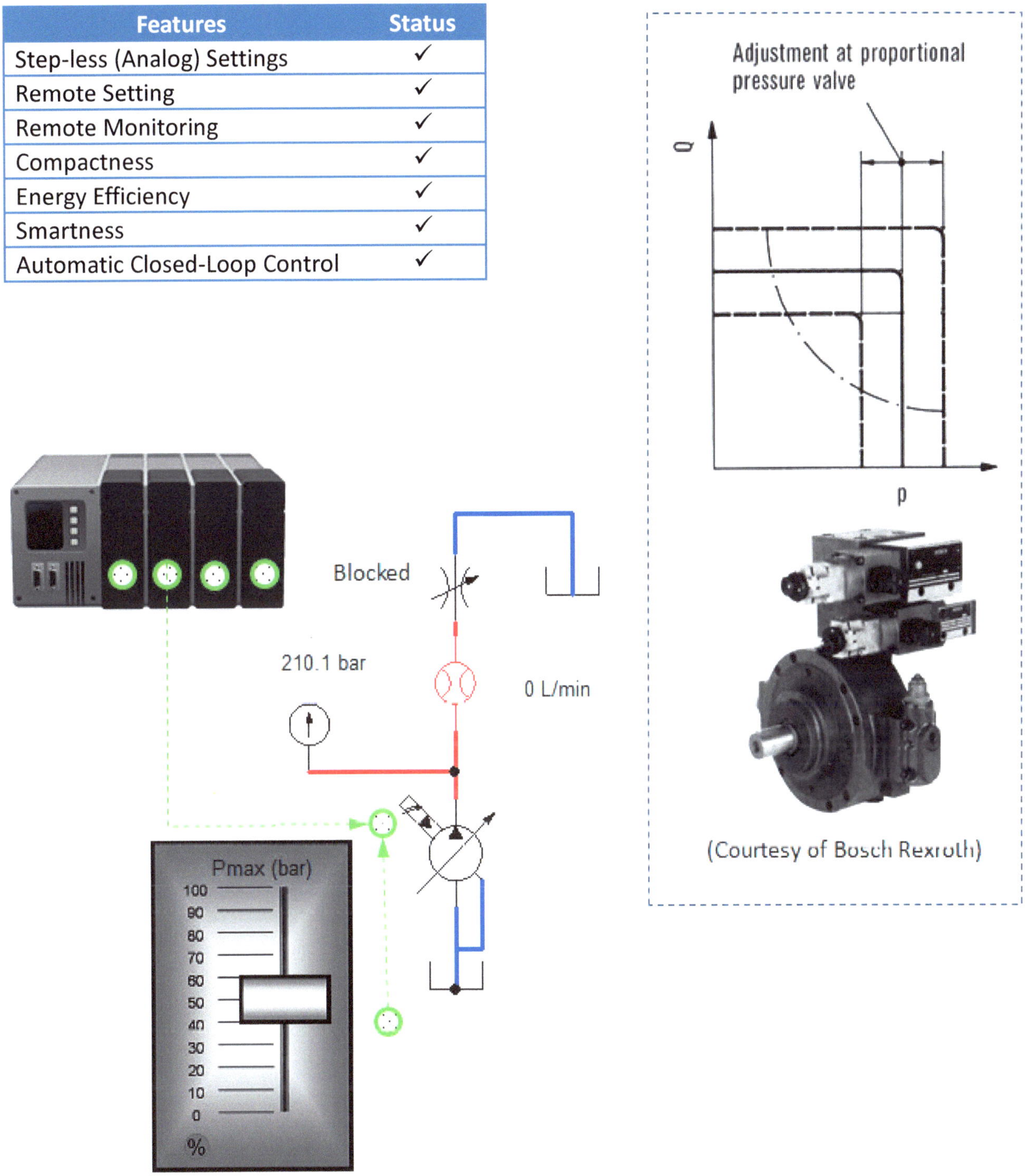

Fig. 2.6 - Pressure Control using Proportional Pressure Compensated Pump

2.3- Flow Control Solutions

2.3.1- Flow Control using Hydro-Mechanical Flow Control Valve

Fig. 2.7 shows the simplest solution to control the flow in a hydraulic system using a throttle valve. Such a system has the following features:

Step-less (Analog) Settings: The throttle valve can be used to set the flow, in an analog fashion, through a continuous range of values within the maximum range.

Remote Setting: The valve is set manually in the field.

Remote Monitoring: Not possible since the throttle valve does not offer any electrical communication.

Compactness: The system is compact since just one valve is used.

Energy Efficiency: Use of a throttle valve results in with energy losses.

Smartness: There is no possibility because the throttle valve can't be set electrically.

Automatic Closed-Loop Control: A throttle valve can't be used to build a closed-loop pressure control.

Features	Status
Step-less (Analog) Settings	✓
Remote Setting	×
Remote Monitoring	×
Compactness	✓
Energy Efficiency	×
Smartness	×
Automatic Closed-Loop Control	×

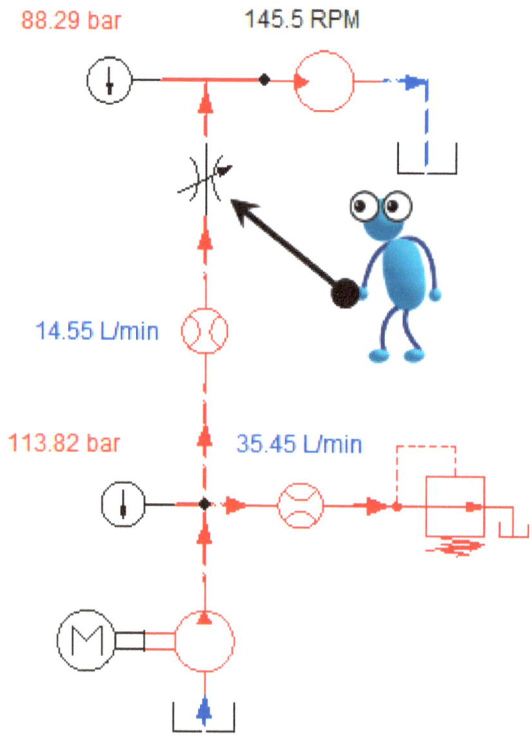

Fig. 2.7 - Flow Control using a Throttle Valve

2.3.2- Flow Control using Hydro-Mechanical Displacement Controlled Pump

In this solution, as shown in Fig. 2.8, a hydro-mechanical displacement-controlled pump replaces the fixed pump and the flow control valve. Such a system has the following features:

Step-less (Analog) Settings: Such a pump can be used to set the flow, in an analog fashion, through a continuous range of values within the maximum range.

Remote Setting: The flow controller is set manually in the field.

Remote Monitoring: Not possible since the pump does not offer any electrical communication.

Compactness: The system is compact since the displacement-controlled pump replaces two elements (fixed pump and a flow control valve).

Energy Efficiency: As compared with the previous solution, the use of the hydro-mechanical displacement-controlled pump improves the energy efficiency of the system. The pump controller is set so that only the required flow is generated from the pump.

Smartness: There is no possibility because such a pump can't be set electrically.

Automatic Closed-Loop Control: Such a pump can't be used to build an automatic closed-loop flow control.

Features	Status
Step-less (Analog) Settings	✓
Remote Setting	×
Remote Monitoring	×
Compactness	✓
Energy Efficiency	✓
Smartness	×
Automatic Closed-Loop Control	×

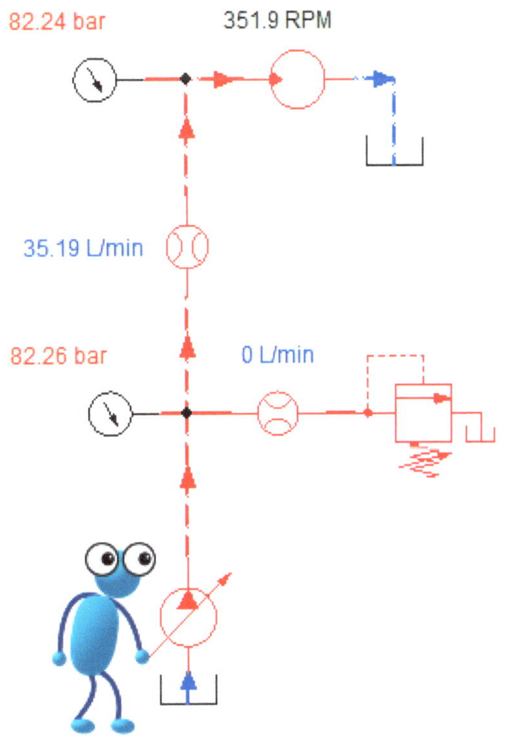

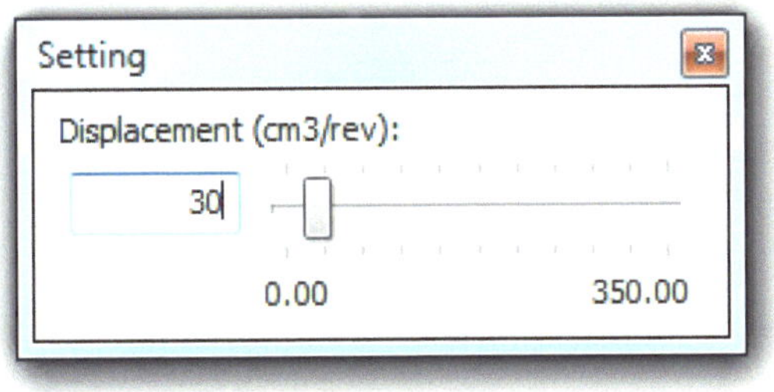

Fig. 2.8 - Flow Control using Hydro-Mechanical Displacement Controlled Pump

2.3.3- Multiple Flow Settings using Electro-Hydraulic Switching Valves

Figure 2.9 shows an example of multiple flows selection. The figure shows that three flow control valves are connected in parallel to the inlet side of the actuator via normally-closed, 2/2, and solenoid-operated (switching) directional valves. As has been discussed in Chapter 1, Eq. 1.11B shows that multiple throttle valves in parallel can be replaced by one throttle valve with an equivalent valve coefficient. Such a system has the following features:

Step-less (Analog) Settings: Changing the flow setting in this system is in a discrete way. Flow is jumped from one value to another depending on the combination of the switching valves statuses. Equation 2.1 shows that the number of combinations z out of a number $\boldsymbol{n}$ of parallel throttles. In the example shown in the figure, since three parallel throttles are used, then seven different equivalent valve coefficients can be achieved as per the truth table shown in the figure.

$$z = 2^n - 1 \qquad 2.1$$

Hypothetically, an infinite number of throttle valves are needed in such a system to achieve step-less pressure control.

Remote Setting: A set of pushbuttons or a PLC can be used to control, remotely, the status of the switching valves (True or False) so that the flow supplied to the actuator is set accordingly.

Remote Monitoring: Since a PLC is used, the flow can now be monitored remotely.

Compactness: Obviously the system implements a large number of components including plumping, manifolds, PLC, etc. that will make the system large and heavy.

Energy Efficiency: Use of throttle valve results in energy losses.

Smartness: A microprocessor controller can be programmed to control the flow interactively and repeatedly during a duty cycle of the machine or based on certain operating conditions.

Automatic Closed-Loop Control: A throttle valve can't be used to build a closed-loop automatic pressure control.

Features	Status
Step-less (Analog) Settings	×
Remote Setting	✓
Remote Monitoring	✓
Compactness	×
Energy Efficiency	×
Smartness	✓
Automatic Closed-Loop Control	×

C_{veq}	Y1	Y2	Y3
C_{v1}	T	F	F
C_{v2}	F	T	F
C_{v3}	F	F	T
C_{v12}	T	T	F
C_{v13}	T	F	T
C_{v23}	F	T	T
C_{v123}	T	T	T

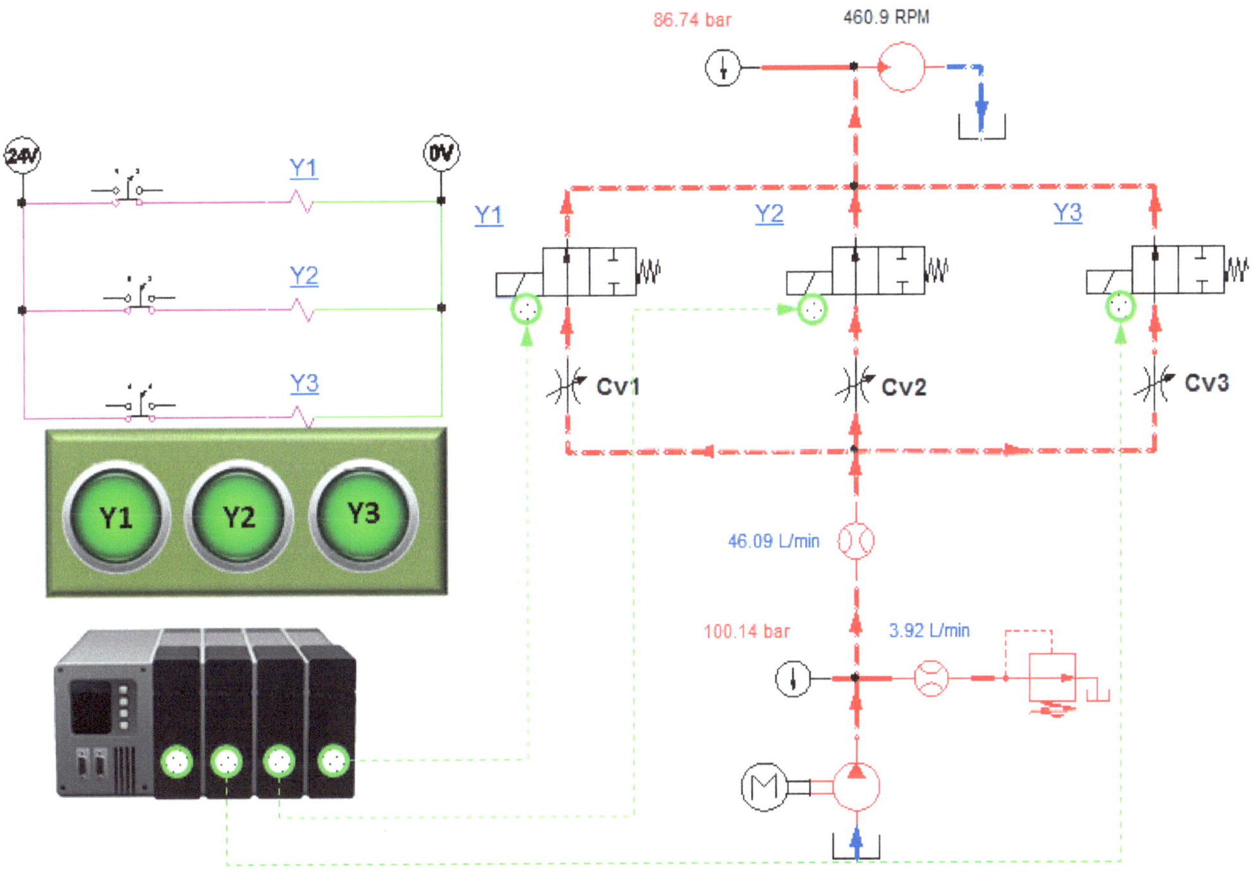

Fig. 2.9 - Multiple Flow Settings using Solenoid Operated Valves

2.3.4- Flow Control using an Electro-Hydraulic Continuous Valve

Figure 2.10 shows an example of controlling the system flow using a *proportional flow control valve*. Doing so, as shown in the figure, increases the number of features in the table. Such a system has the following features:

Step-less (Analog) Settings: The proportional flow control valve can be used to set the flow, in an analog fashion, through a continuous range of values within the maximum range.

Remote Setting: A manual potentiometer or a PLC can be used to control, remotely, the adjustment of the proportional flow control valve.

Remote Monitoring: Since a PLC is used, the system flow can now be monitored remotely.

Compactness: The system is compact since just one valve is used.

Energy Efficiency: Use of flow control valves, even proportional ones, causes energy losses.

Smartness: A microprocessor controller can be programmed to control the system flow interactively and repeatedly during a duty cycle of the machine or based on certain operating conditions.

Automatic Closed-Loop Control: Continuous valves can be used to build a closed-loop automatic flow control.

Features	Status
Step-less (Analog) Settings	✓
Remote Setting	✓
Remote Monitoring	✓
Compactness	✓
Energy Efficiency	×
Smartness	✓
Automatic Closed-Loop Control	✓

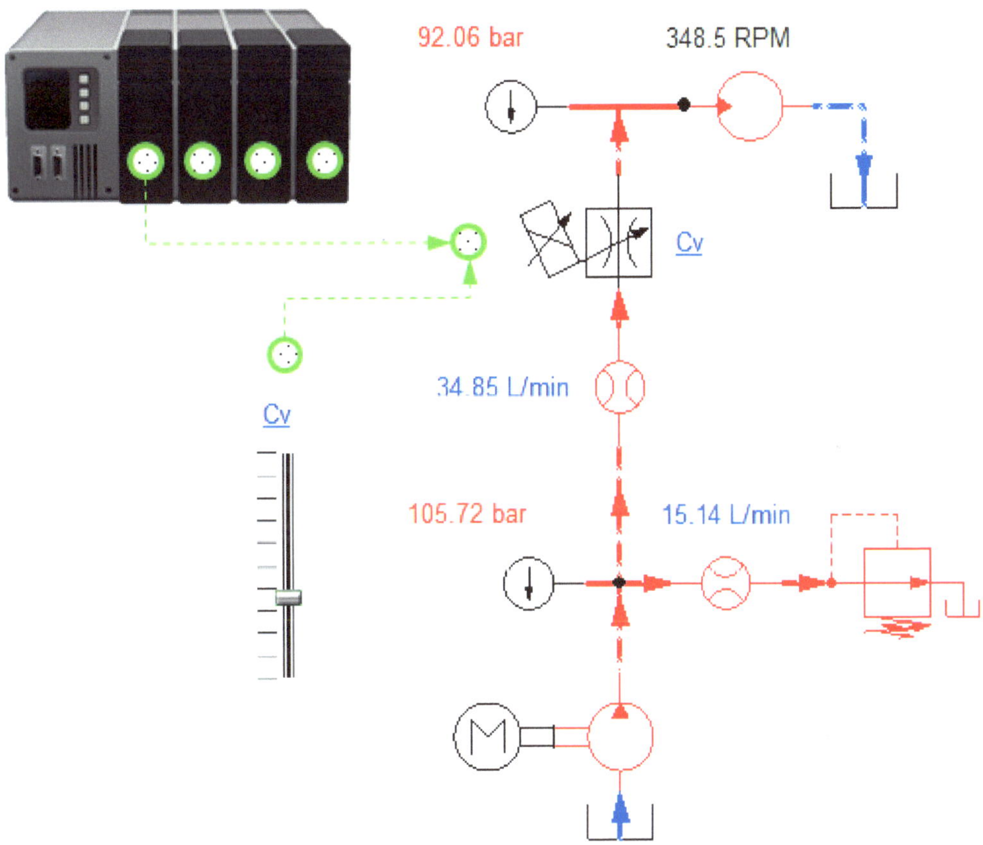

Fig. 2.10 - Flow Control using Proportional Flow Control Valve

2.3.5- Flow Control using an Electro-Hydraulic Displacement Controlled Pump

Figure 2.11 shows an example of controlling the system flow using an electro-hydraulic displacement-controlled pump. Doing so, as shown in the figure, removes all the shortcomings of the previous flow control solutions. The only limitation to use such a solution is the cost and the know-how. Such a system has the following features:

Step-less (Analog) Settings: The electro-hydraulic *displacement-controlled pump* can be used to set the system flow, in an analog fashion, through a continuous range of values within the maximum range.

Remote Setting: A manual potentiometer or a PLC can be used to control, remotely, the adjustment of the pump controller so that pump flow is set accordingly.

Remote Monitoring: Since a PLC is used, the flow can now be monitored remotely.

Compactness: The system is compact since the flow control is built in with the pump.

Energy Efficiency: The use of the electro-hydraulic displacement-controlled pump improves the energy efficiency of the operation. The electro-hydraulic displacement-controlled pump performs exactly like the hydro-mechanical one, except that it is controlled electrically rather than manually.

Smartness: A microprocessor controller can be programmed to control the pump flow interactively and repeatedly during a duty cycle of the machine or based on certain operating conditions.

Automatic Closed-Loop Control: electro-hydraulic displacement-controlled pump can be used to build a closed-loop automatic flow control.

Features	Status
Step-less (Analog) Settings	✓
Remote Setting	✓
Remote Monitoring	✓
Compactness	✓
Energy Efficiency	✓
Smartness	✓
Automatic Closed-Loop Control	✓

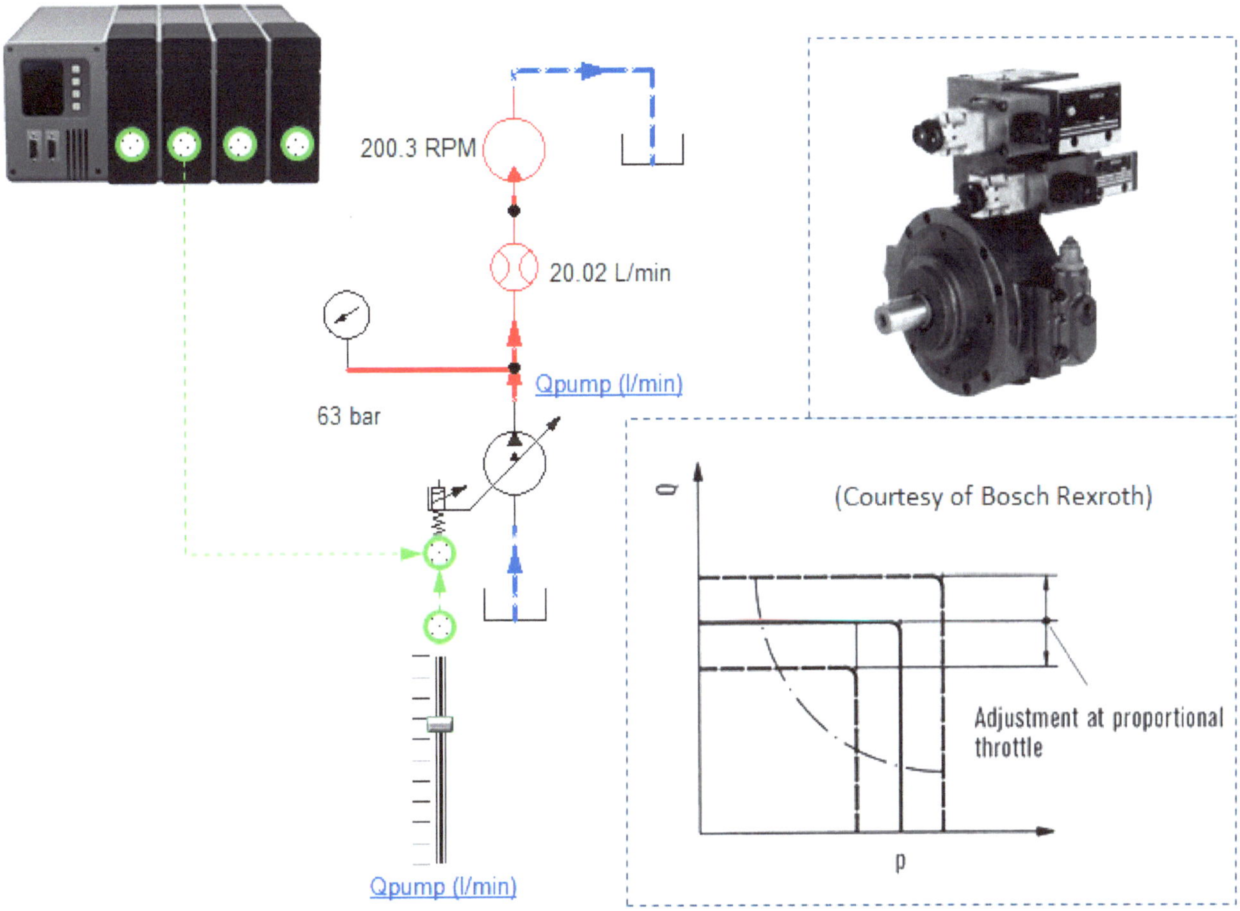

Fig. 2.11 - Flow Control using an Electro-Hydraulic Displacement Controlled Pump

2.3.6- Flow Control using a Fixed Pump Driven by Variable Speed Motor

Figure 2.12 shows an example of controlling the system flow using a gear pump driven by a *Variable Speed Motor.* Such a system has the following features:

Step-less (Analog) Settings: Such a system can be used to set the system flow, in an analog fashion, through a continuous range of values within the maximum range.

Remote Setting: The motor controller can be used to control, remotely, the flow of the pump.

Remote Monitoring: Since a PLC is used, the flow can be monitored remotely as well.

Compactness: The system is even more compact than an electro-hydraulic displacement-controlled pump since a fixed pump is used.

Energy Efficiency: The use of such a system improves the energy efficiency of the operation. The only loss is due to the pump inefficiency, which might be bad at low speed.

Smartness: The servo motor controller can be programmed to control the pump flow interactively and repeatedly during a duty cycle of the machine or based on certain operating conditions.

Automatic Closed-Loop Control: Such a system can be used to build a closed-loop automatic flow control.

In addition to the previous features, as compared to the electro-hydraulic displacement-controlled pump, this system has the following features:

- **Advantages:**
- The system can be used to control pressure in a closed-loop via controlling the amount of flow through an orifice.
- Has low mass moment of inertia of the internal gear pump.
- Requires no valves to control the pump flow.
- **Limitations:**
- Has slower dynamics than a electro-hydraulic displacement controlled pump. That is because the flow control is occurred via controlling the motor speed, which has slower dynamics than the variable displacement pump controller.
- Results in inefficient pump at low speed. There is a speed at which the pump outlet flow is equal to its internal leakage. If this happened at high pressure, the pump will overheat in few minutes. Therefore, this system should not drive the pump below certain specified minim speed.

Features	Status
Step-less (Analog) Settings	✓
Remote Setting	✓
Remote Monitoring	✓
Compactness	✓
Energy Efficiency	✓
Smartness	✓
Automatic Closed-Loop Control	✓

Technical Data:

- Maximum power per drive = 250 kW
- Maximum flow per pump = 625 l/min
- Maximum accuracy of pressure control = ± 1 bar
- Maximum operating pressure = 345 bar

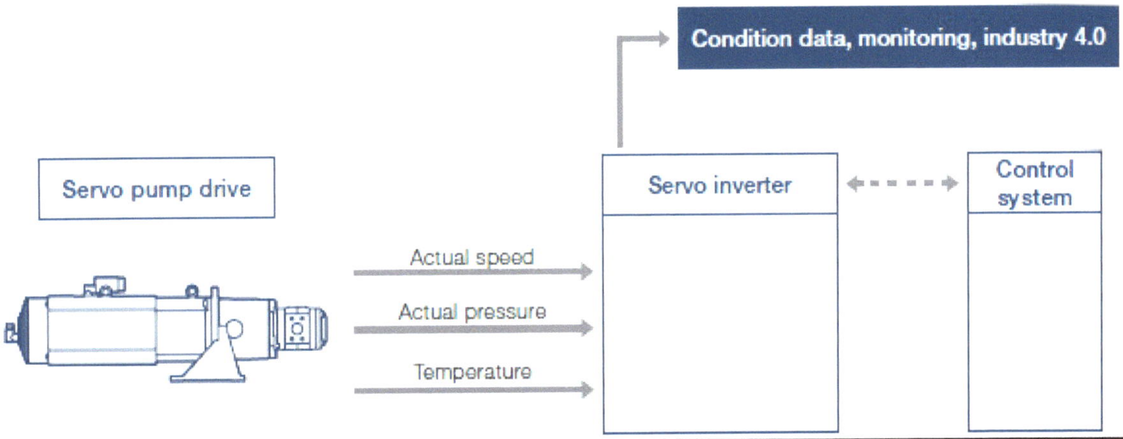

Fig. 2.12 - Flow Control using a Fixed Pump Driven by Variable Speed Motor (www.VOITH.com)

2.4- Power Control Solutions

The controller of such a pump serves to limit the maximum power that the pump can withdraw from a prime mover. The magnitude of the hydraulic power $\boldsymbol{P}$ out of a pump is the product of the pump outlet pressure $\boldsymbol{p}$ and the flow $\boldsymbol{Q}$ discharged by the pump. Since the working pressure $\boldsymbol{p}$ is a load dependent variable, then the hydraulic power $\boldsymbol{P}$ is controlled by controlling the pump flow. The controller senses the pressure based on which it adjusts the pump flow $\boldsymbol{Q}$. Figure 2.13 shows a set of three $\boldsymbol{p}$ -$\boldsymbol{Q}$ hyperbolic curve that represents the ideal constant power operation of the pump. Curves A, B, and C go from a minimum power to intermediate to maximum power. At any point on any of these curves, e.g. 1, 2, and 3, the product of the working pressure $\boldsymbol{p}$ and the corresponding pump flow $\boldsymbol{Q}$ has the same value of the power. On point 4 and 5 the controller sets maximum flow and maximum pressure; respectively regardless the available power.

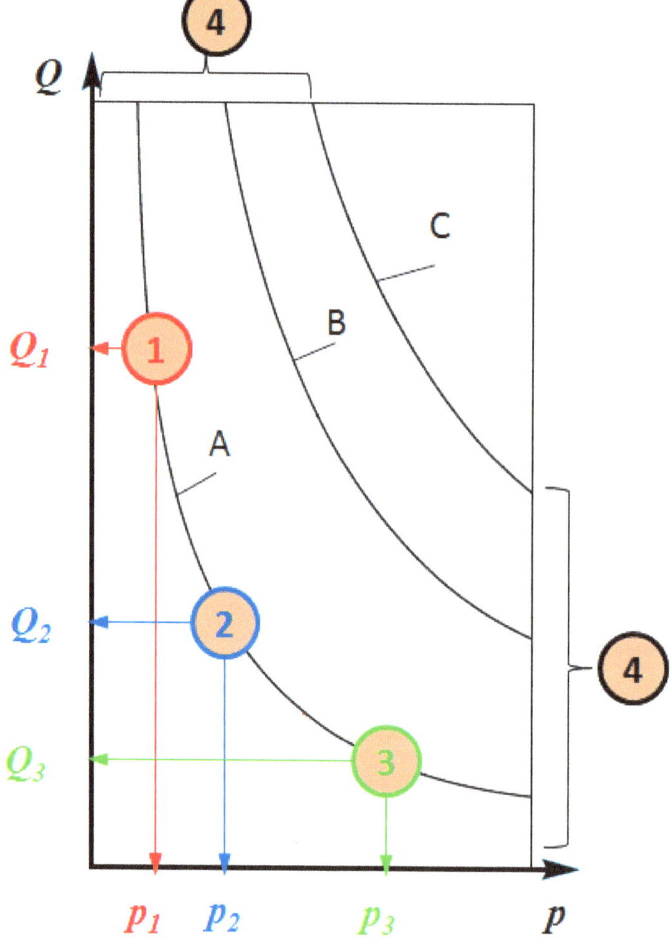

$$Q_1 \times p_1 = Q_2 \times p_2 = Q_3 \times p_3 = \textit{Constant}$$

Fig. 2.13 - Constant Power Operation of a Variable Displacement Pump

2.4.1- Hydro-Mechanical Power Controlled Pumps

As shown in Fig. 2.14, the hydro-mechanical power controller varies the pump displacements against two coaxially coincided coil springs. The springs have different lengthens and each, as known, follows Hooke's law. At low pressure, when the longer spring 1 is compressed, p-Q relation follows the part 1 of the curve. When both springs 1 and 2 are compressed simultaneously, they act like a new tougher spring. In such a case, p-Q relation follows the part 2 of the curve.

The resulting overall curve, a combination of two broken lines, approximates the ideal hyperbolic curve. The red shaded areas under the ideal curve are available power that can't be achieved because of the spring characteristics.

Figure 2.15 shows three springs are used rather than two. The resulting curve is closer to the ideal curve. Hypothetically an infinite number of springs are used to get the ideal constant power operation. As shown in the figure, such a hydro-mechanical power controller has the following features:

Step-less (Analog) Settings: Such a controller does not offer power adjustment but by replacing the spring package. Other hydro-mechanical power controllers, not presented in this textbook, allows step-less power adjustment by setting a spring force externally.

Remote Setting: Power must be set manually at the pump.

Remote Monitoring: Not possible since the controller does not offer electrical communication.

Compactness: The system is compact since the controller is built on the pump

Energy Efficiency: Using such a controller improves the system efficiency since the generated power is fully utilized to do the mechanical work.

Smartness: Such a controller does not offer any smart control options.

Automatic Closed-Loop Control: Such a controller can't be used to build a closed-loop power control.

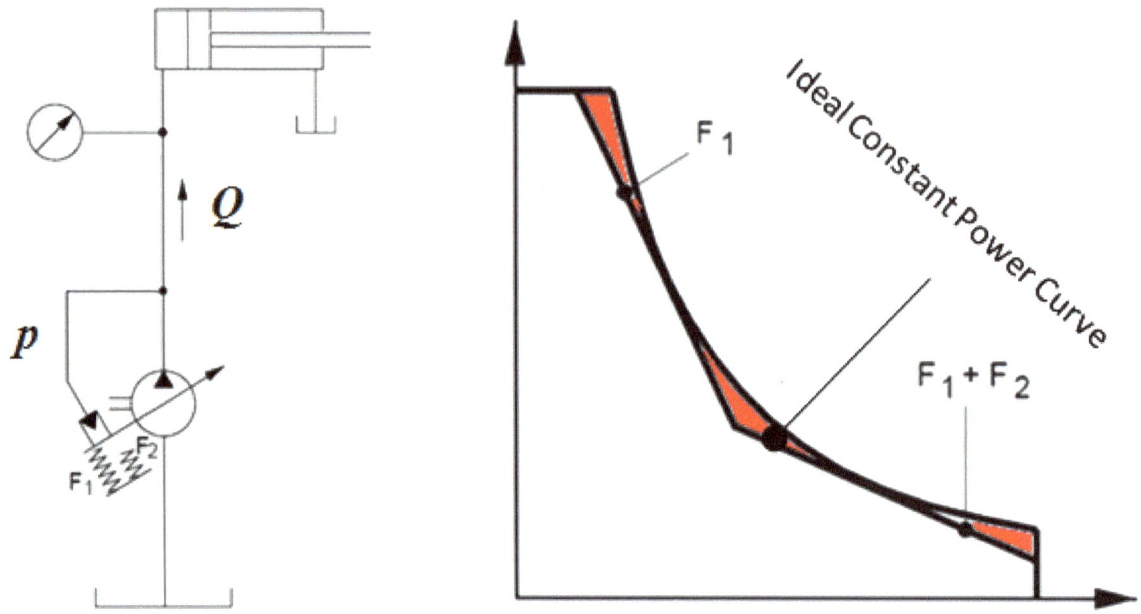

Fig. 2.14 - Hydro-Mechanical Constant Power Controller

Features	Status
Step-less (Analog) Settings	✓
Remote Setting	×
Remote Monitoring	×
Compactness	✓
Energy Efficiency	✓
Smartness	×
Automatic Closed-Loop Control	×

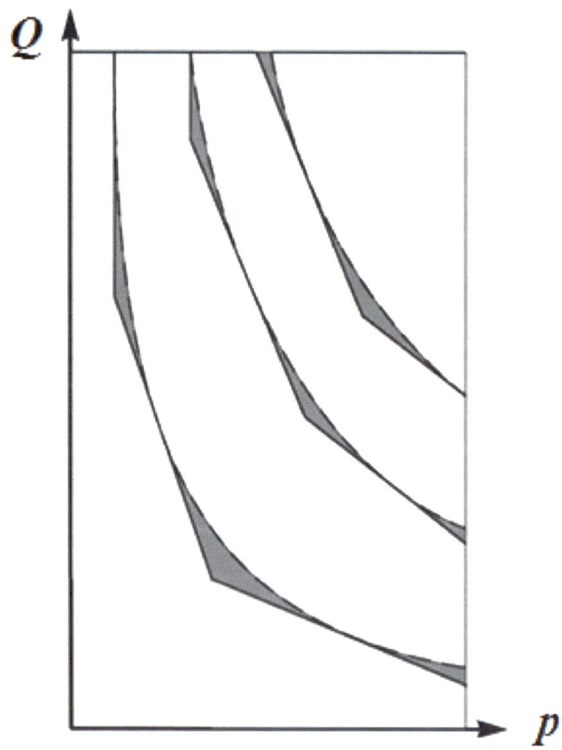

Fig. 2.15 - Hydro-Mechanical Constant Power Controller Characteristics

2.4.2- Electro-Hydraulic Power Controlled Pumps

As shown in Fig. 2.16, pump displacement is controlled automatically in a closed-loop based on the pump pressure variation to keep the power constant. Variables shown on the figure are defined as follows:

- P_{max} = Maximum Power.
- Q_{max} = Maximum Flow.
- p_{max} = Maximum Pressure.
- p_p = Actual Pump Pressure.
- Q_p = Actual Pump Flow.
- Q_R = Reference Flow.
- Q_E = Error Flow.
- Q_C = Control Signal.

As shown in the figure, an Electric Control Unit (ECU) is fed by three characterizing operational variables; P_{max}, Q_{max}, ***and*** p_{max}. The ECU calculates first Q_R based P_{max} and p_p. The ECU then checks Q_R based on Q_{max} and p_{max}. Q_R is used as a reference value for the closed-loop. The negative feedback closed-loop checks the difference between Q_R and Q_p. The resulted flow error Q_E is manipulated by the controller that generates the control signal Q_C. The latter is used to drive the servo valve that adjusts the pump displacement to make Q_p equal to Q_R.

As shown in Fig. 2.17, by using such a controller, the ideal hyperbolic constant power operation can be achieved and the system should have the following features:

Step-less (Analog) Settings: Using such a controller, the maximum power, maximum flow and maximum pressure can be seamlessly controlled through a PLC or by a potentiometer.

Remote Setting: The controller offers electrical communication so that it can be set remotely.

Remote Monitoring: As long as both the pump pressure and flow are measured electronically for the sake of power control, power can be monitored remotely. Alternatively, a power meter can be used to measure the prime mover power and communicated remotely with the controller.

Compactness: The system is compact since the controller is built on the pump.

Energy Efficiency: Using such a controller improves the system efficiency since the generated power is fully utilized to do the required work. However, there are certain power losses in the control valve.

Smartness: Such a controller can control the pump power within entire machine duty cycle.

Automatic Closed-Loop Control: Such a controller can be used to build a closed-loop power control.

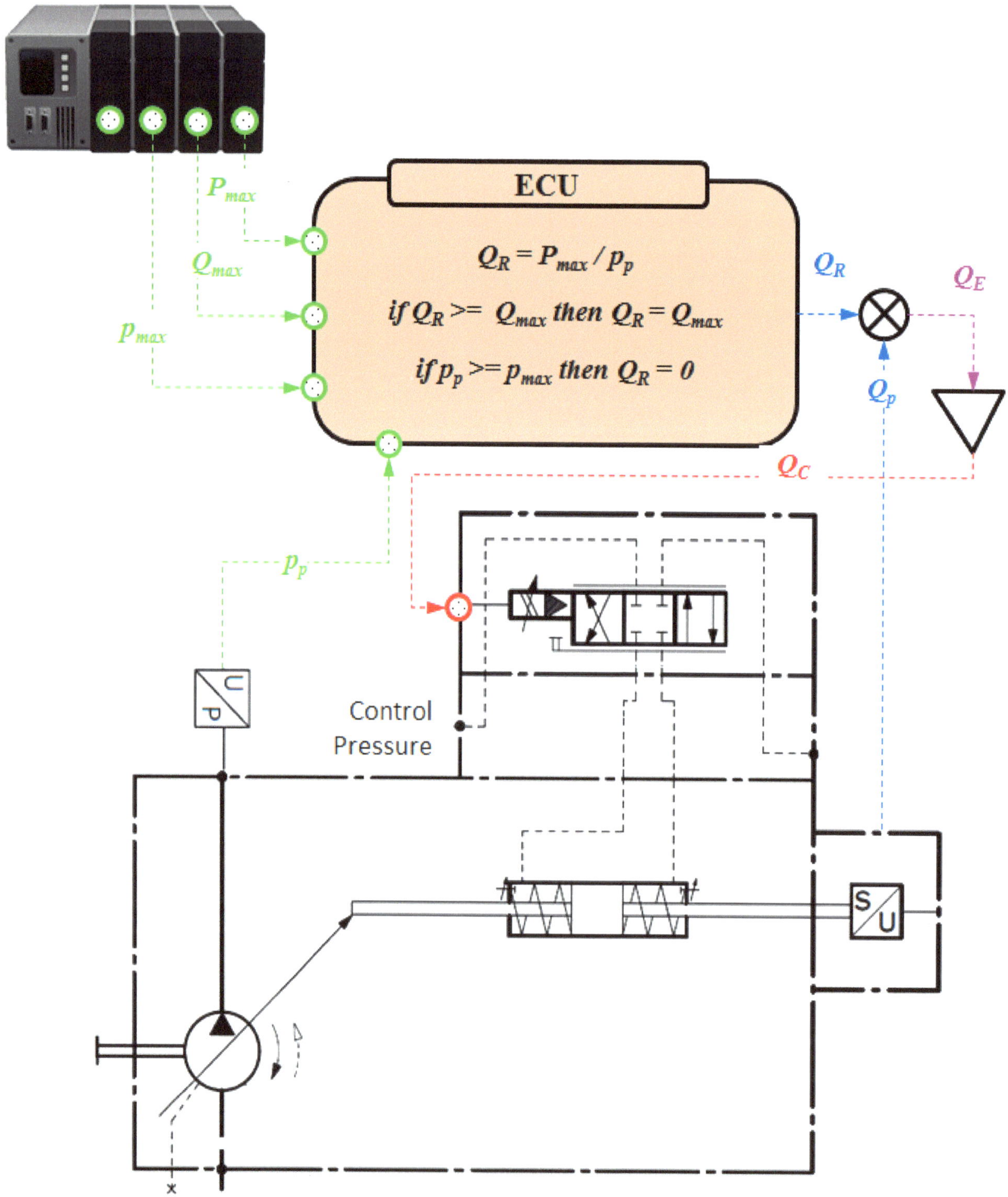

Fig. 2.16 - Electro-Hydraulic Constant Power Controller

Features	Status
Step-less (Analog) Settings	✓
Remote Setting	✓
Remote Monitoring	✓
Compactness	✓
Energy Efficiency	✓
Smartness	✓
Automatic Closed-Loop Control	✓

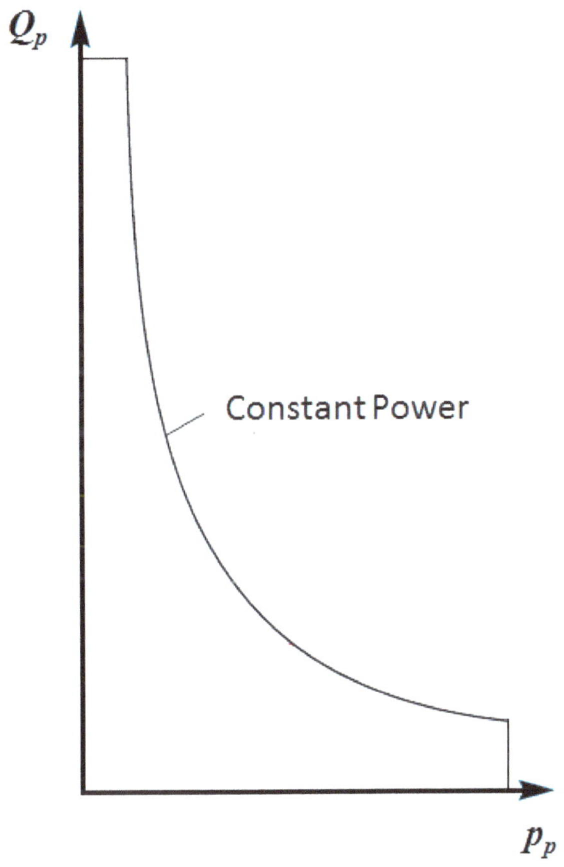

Fig. 2.17 - Electro-Hydraulic Constant Power Controller Characteristics

2.5- Sequence Control Solutions

2.5.1- Hydro-Mechanical Sequence Control

Figure 2.18 shows a classical sequence control circuit, in operational mode, using hydro-mechanical sequence valves. As shown in the figure, such a solution does not offer any electrical communication so that the system can't be set or monitored remotely. The sequence of the operation depends on the manual setting of the sequence valves.

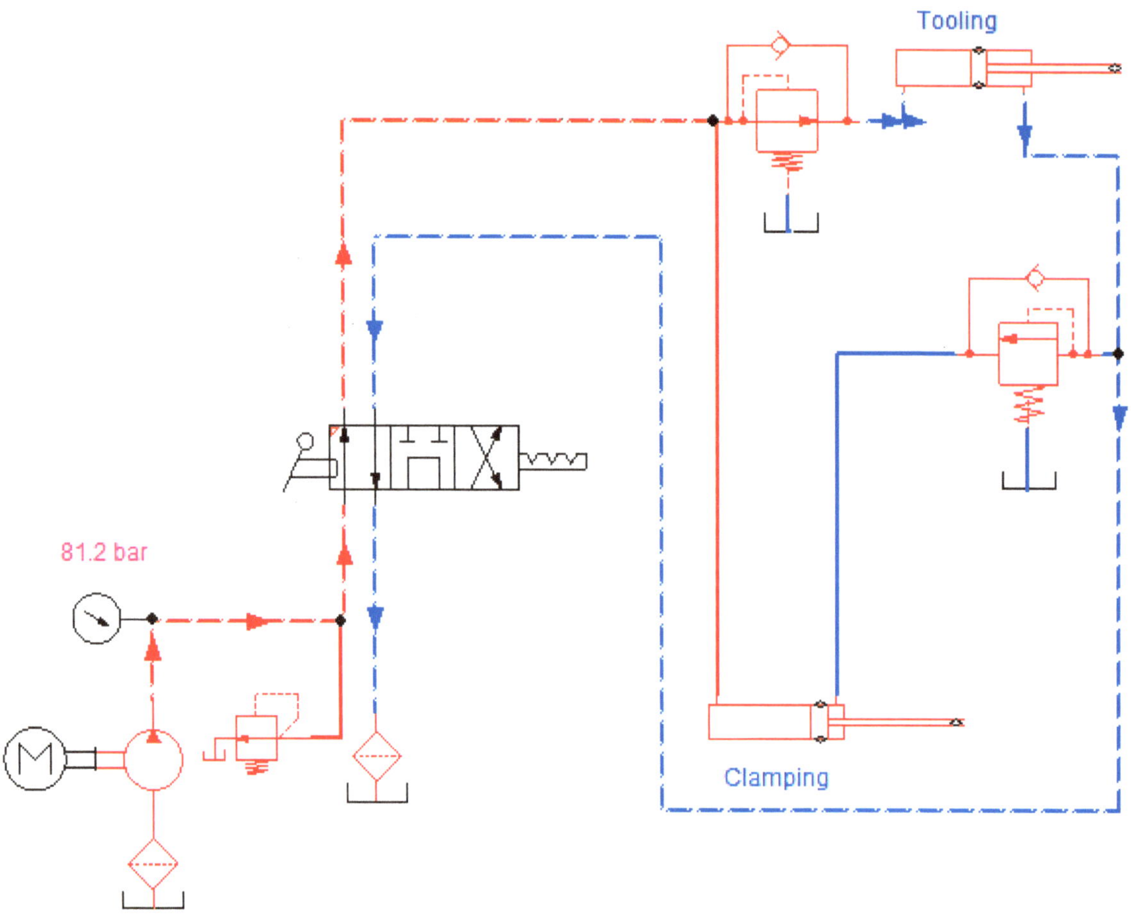

Fig. 2.18 - Hydro-Mechanical Sequence Control

2.5.2- Electro-Hydraulic Sequence Control

In the system shown in Fig. 2.19, the two hydro-mechanical sequence valves have been replaced by two EH solenoid-operated directional valves. In this first solution, the cycle starts by energizing the valve 1 that drives the clamping cylinder. When the clamping cylinder reaches a certain position, at which the clamping function is satisfied, it hits a *position switch* that energizes the second directional valve so that the tooling cylinder is extended. It is to be noted that the electrical circuit is only used on the extension stroke of the cylinder.

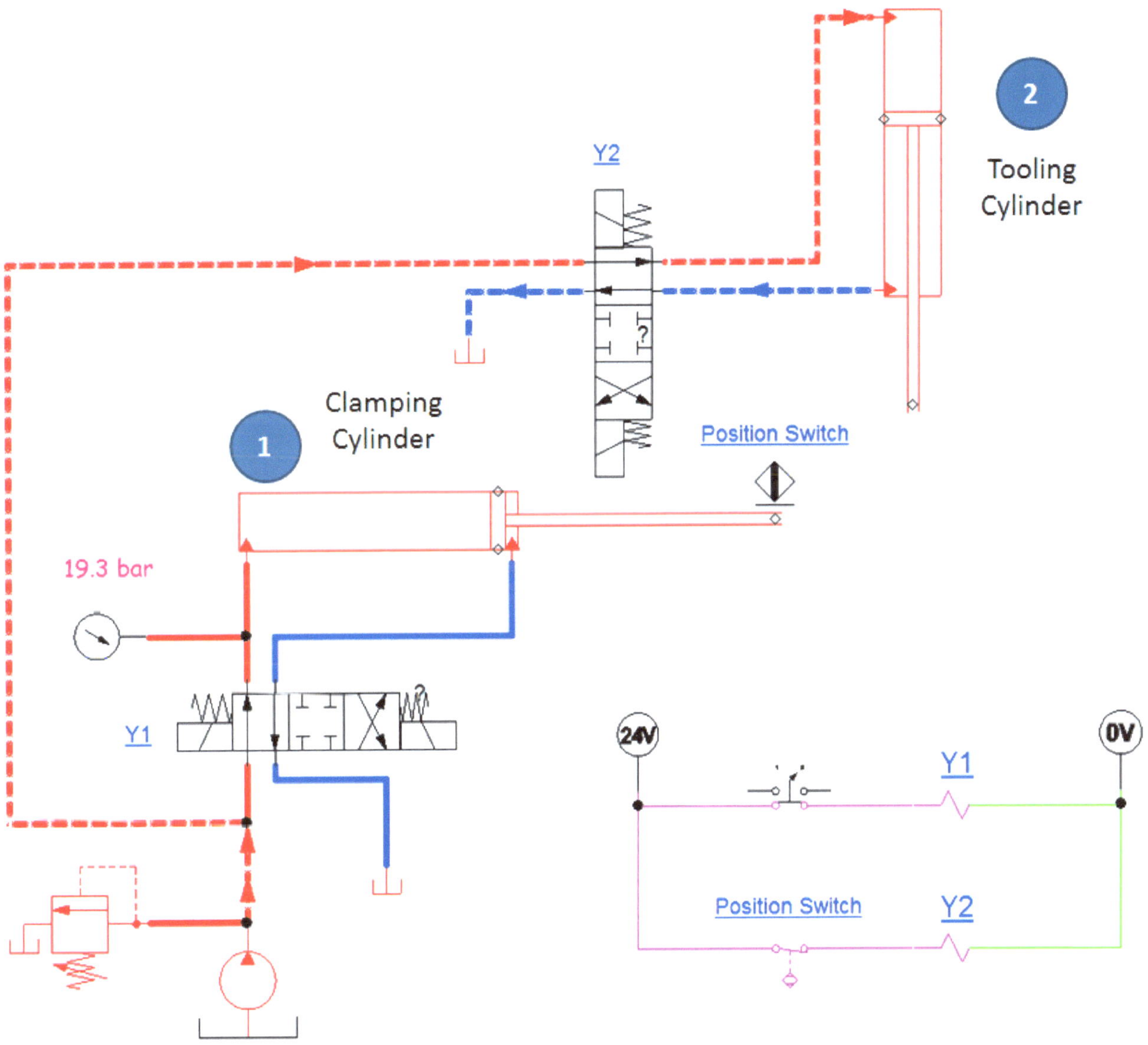

Fig. 2.19 - Electro-Hydraulic Sequence Control, Solution 1

The second solution shown in Fig. 2.20 has smarter operation. The tooling cylinder is extended based on two signals. The first signal is generated from the position switch to assure that the clamping cylinder rod touched the work piece. The second signal is generated from a *pressure switch* to assure that the clamping force has been applied.

It is to be noted that the electrical circuit is to serve only used on the extension stroke of the cylinder where the second rung forms an AND function from both the position and the pressure switches.

Since both solutions 1 and 2 are driven by an EH valve, they can be remotely and interactively operated within an entire automated machine. The pressure drop across the EH valves is less than in the hydro-mechanical sequence valves. The pressure switch can be set to vary the clamping force in a step-less way.

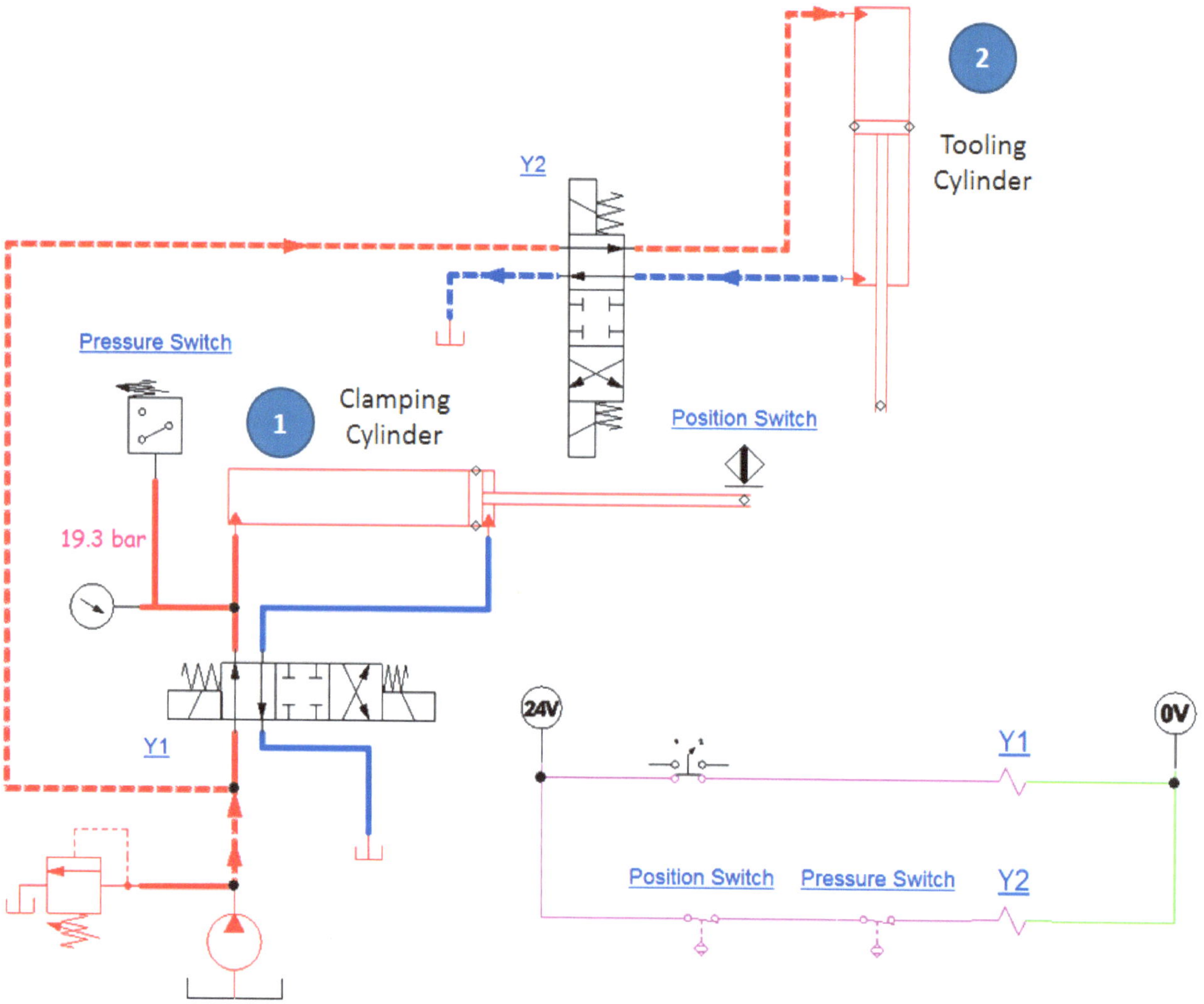

Fig. 2.20 - Electro-Hydraulic Sequence Control, Solution 2

2.6- Hydraulic Deceleration System

2.61- Hydro-Mechanical Deceleration System

If the cylinder speed exceeds 20 cm/sec, even the cushioning device won't be able to absorb the kinetic energy of the equivalent moving mass. In such a case, an external deceleration system must be used. Figure 2.21 shows a double-acting cylinder (1) that has the cylinder rod connected to a cam surface (2). A 4-way 3-position DCV (3) is used to drive the cylinder. In the early stage of the cylinder extension stroke, the flow regulator (4) is bypassed by a normally-open, roller-operated 2/2 DCV (5). When the cylinder approaches the end of the stroke, the cam surface gradually closes the DCV (5) so that the pump flow is forced to pass through the flow regulator (4). The cylinder is retracted by maximum speed because the flow regulator is bypassed by the check valve.

The system does not offer any electrical communication. The start of the cylinder deceleration and the rate of deceleration are due to the design of the cam surface. The only available in field setting is the minimum speed of the cylinder by manual setting of the flow regulator (4).

As it has been stated before, using valves to perform any control function will be always associated with energy losses.

The system is obviously not a compact one. Attaching a cam surface to the cylinder rod may add side loads on the cylinder rod, affecting the rod seals, and requires greasing and mechanical adjustment every once while.

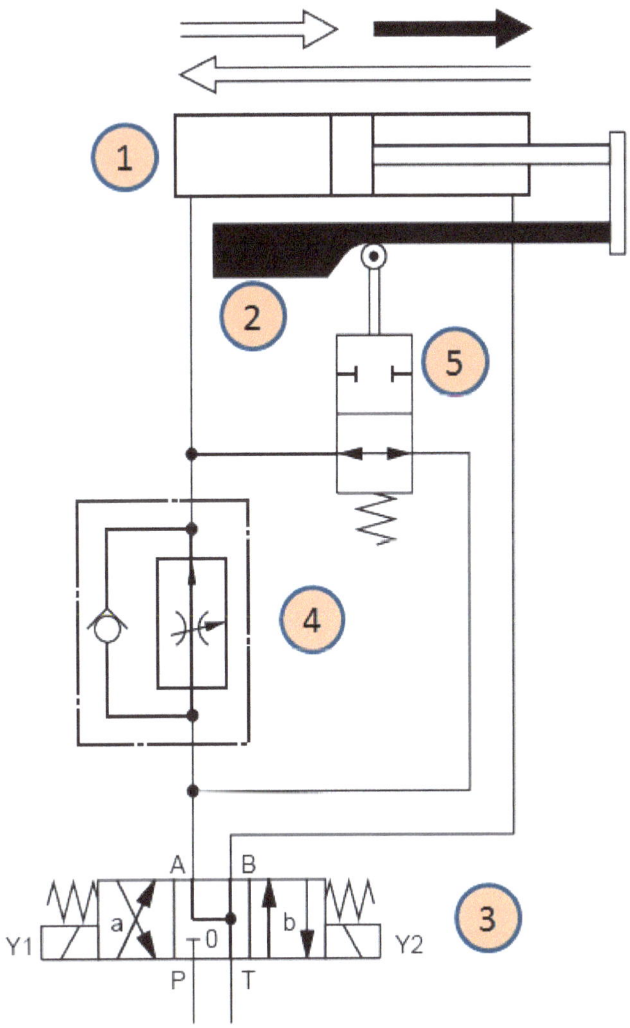

Fig. 2.21- Hydro-Mechanical Cylinder Deceleration Circuit (Courtesy of Bosch Rexroth)

2.6.2- Electro-Hydraulic Deceleration System

The hydro-mechanical system shown in the previous figure is replaced by the electro-hydraulic system shown in Fig. 2.22. The figure shows a double-acting cylinder (1) that has the cylinder rod and a set of three limit switches (2). A 4-way 3-position DCV (3) is used to drive the cylinder. In the early stage of the cylinder extension stroke, between the switches S1 and S2, the normally-closed 2/2 DCV (5) is energized so that the flow regulator (4) is bypassed. When the cylinder approaches the end of the stroke, between S2 and S3, the DCV (5) is de-energized so that the cylinder speed is stepped down based on the setting of the flow regulator (4). If a gradual deceleration is required, then a proportional valve is to be used instead of the ON/OFF DCV (5). The cylinder is retracted by maximum speed because the flow regulator is bypassed by the check valve. The operation of such a system can be remotely and interactively operated and monitored within an entire automated machine. Repositioning the limit switches offer flexibility for where the deceleration is to start.

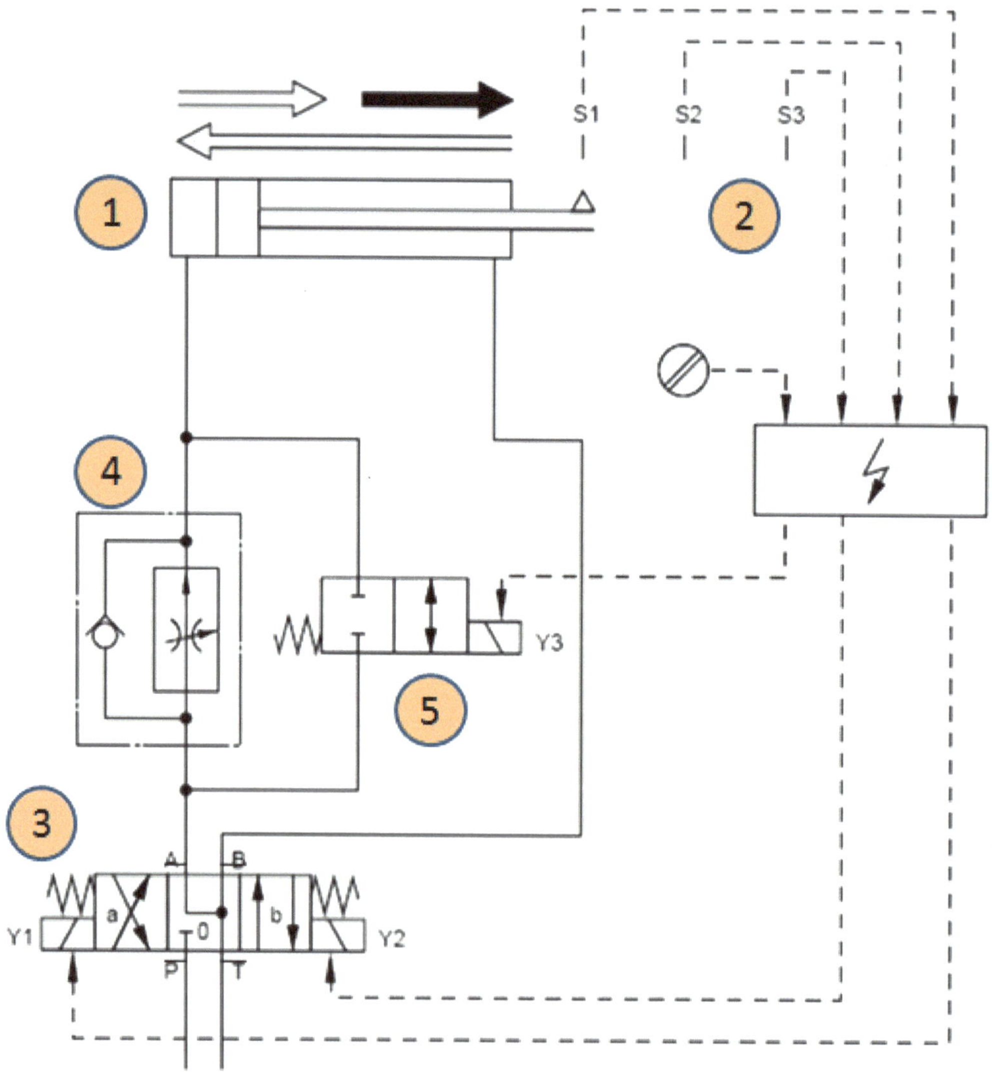

Fig. 2.22- Electro-Hydraulic Cylinder Deceleration Circuit (Courtesy of Bosch Rexroth)

2.7- Cylinder Speed Synchronization

2.7.1- Hydro-Mechanical Cylinder Speed Synchronization

As shown in Fig. 2.23, a rotary type flow divider is used to synchronize the speed of two identical cylinders regardless of the load distribution on the cylinders. Such a system is good as long as 1% margin of error is acceptable. If one of the cylinders deadheaded before the other, the 2/2 valve is used to drive the other cylinder manually to the end of its stroke. Such a system does not offer any electrical communication so that the stem can't be set or monitored remotely.

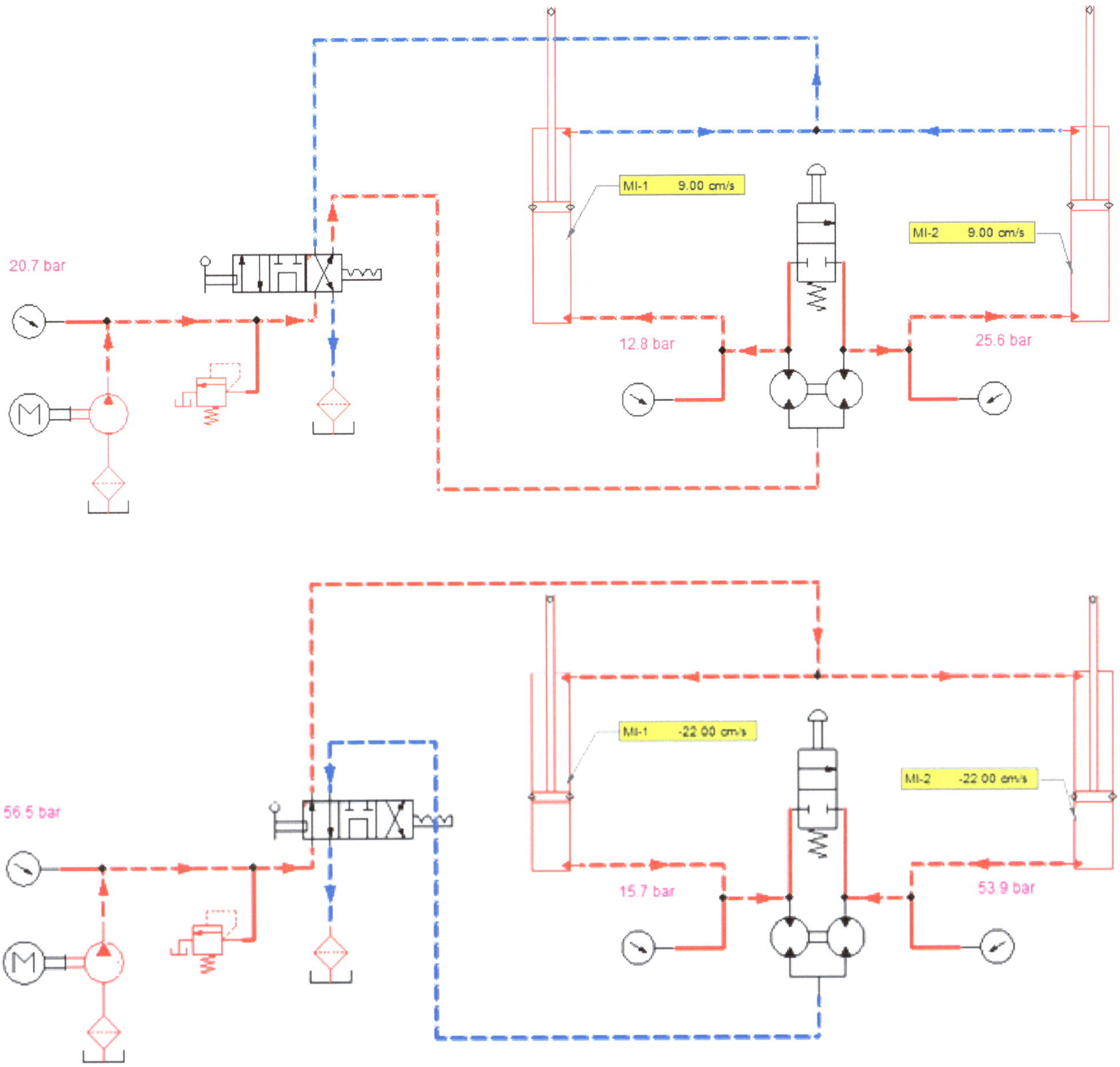

Fig. 2.23- Hydro-Mechanical Cylinder Speed Synchronization using a Flow Divider

2.7.2- Electro-Hydraulic Cylinder Speed Synchronization

Figure 2.24 shows how an electro-hydraulic system can be used to synchronize the speed of two identical cylinders with error margin less than 1%. The system consists of a leading cylinder (1) and a follower cylinder (2). The leading cylinder (1) is driven, in open-loop architecture, by a solenoid operated valve (3). The extension and the retraction speed of the leading cylinder (1) are controlled by the same flow regulator (4). In order to use one flow regulator (4) to control the speed of the leading cylinder (1) in both directions, the flow regulator (4) is sub-plated on a rectifier block (5).

The speed of the leading cylinder is used as an input (reference) value for the closed-loop control system that controls the follower cylinder (2). A continuous valve (6) is used to drive the follower cylinder (2).

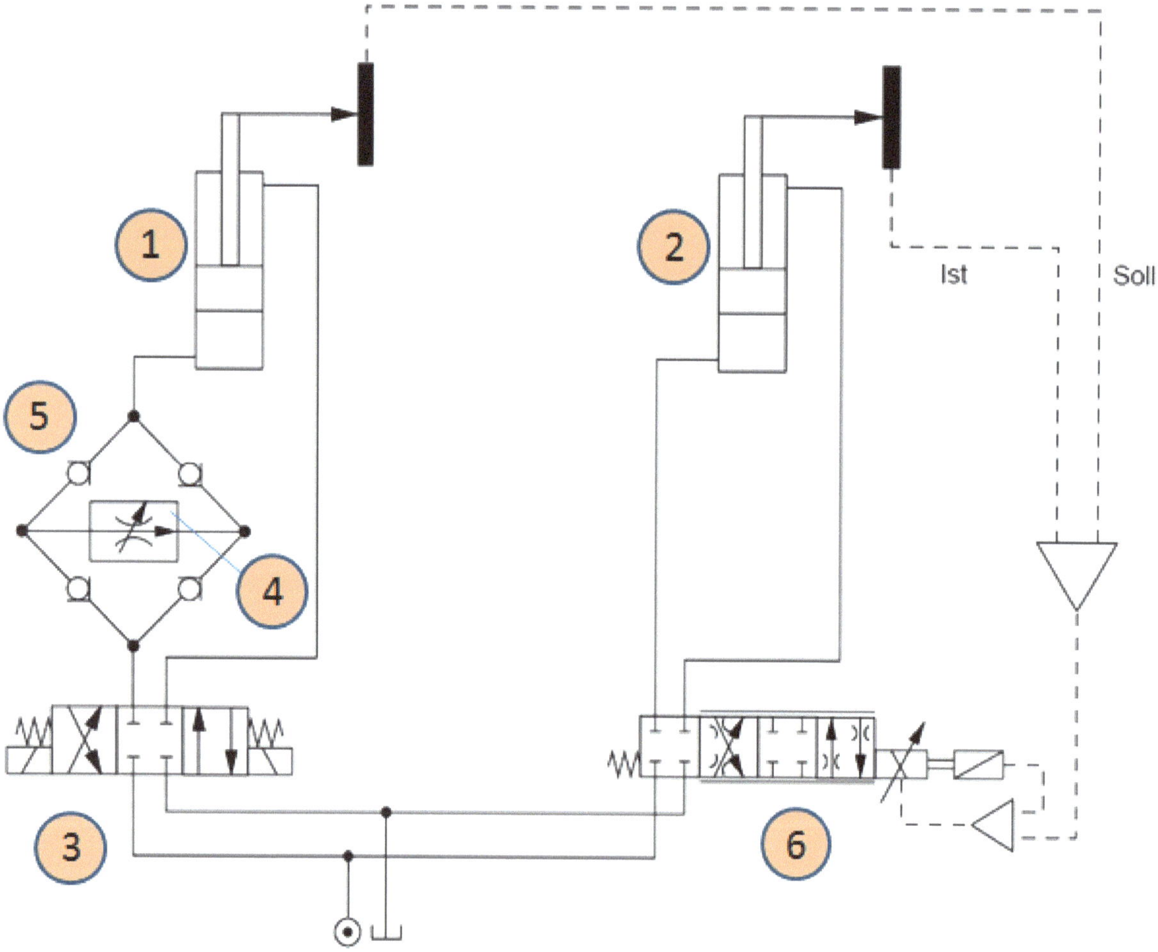

Fig. 2.24- Electro-Hydraulic Cylinder Speed Synchronization using Closed-Loop Control (Courtesy of Bosch Rexroth)

2.8- Accumulator Charging

2.8.1- Hydro-Mechanical Accumulator Charging

Figure 2.25 shows an accumulator continuously charged and discharged between maximum and minimum working pressures that are set by a hydro-mechanical accumulator charging valve.

When the system power is turned ON for first time, the valve blocks the tank line, the pump is loaded to charge the accumulator, and pressure rises. When the pressure hits the maximum pressure, the check valve is closed, and the charging valve unloads the pump to the tank.

When the stored energy is withdrawn from the accumulator towards the actuator, the pressure falls down. When the pressure hits the minimum value, the valve loads the pump to charge the accumulator.

Such a system does not offer any electrical communication so that the system can't be set or monitored remotely. The two pressure limits of the valve are set manually by adjusting the two knobs of the valve.

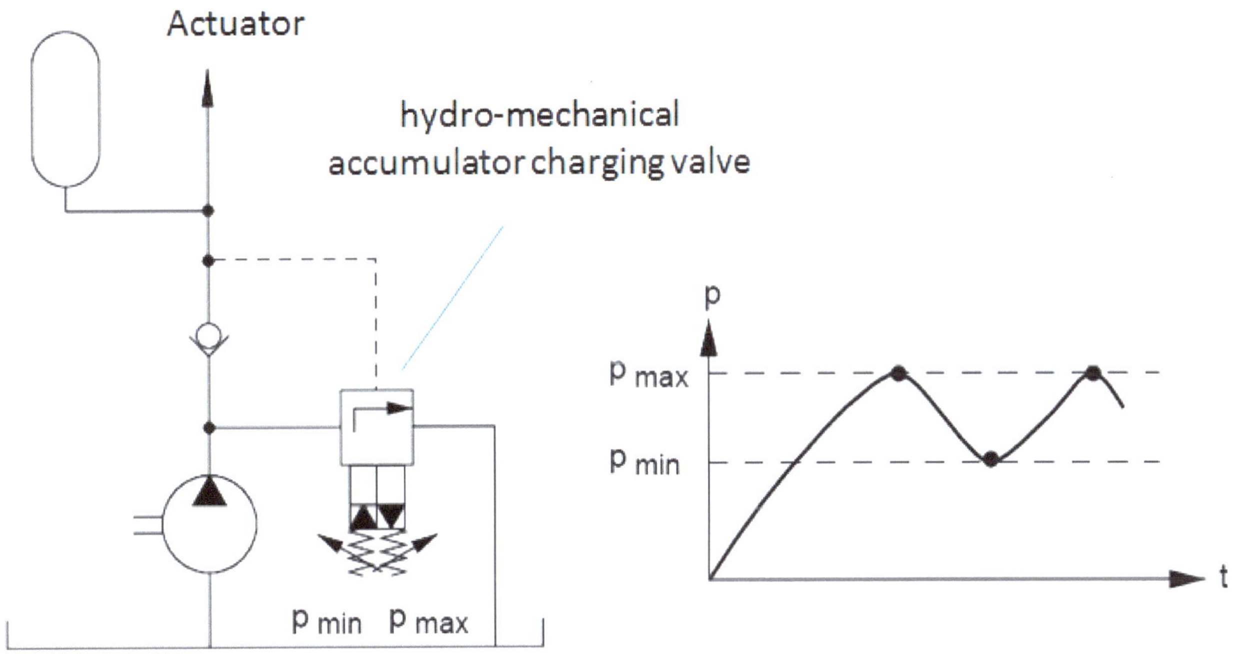

**Fig. 2.25- Hydro-Mechanical Accumulator Charging System
(Courtesy of Bosch Rexroth)**

2.8.2- Electro-Hydraulic Accumulator Charging

The hydro-mechanical accumulator charging valve in the previous system has been replaced by an electro-hydraulic system shown in Fig. 2.26. The system employs a pressure switch that has a pressure range Pmax-Pmin. The pressure switch is activated (changes its normal state) when the pressure rises above Pmax or fallen below Pmin. Such a system can be part of an automatic control process in which the accumulator can be controlled and monitored remotely. The following section explains how the circuit works.

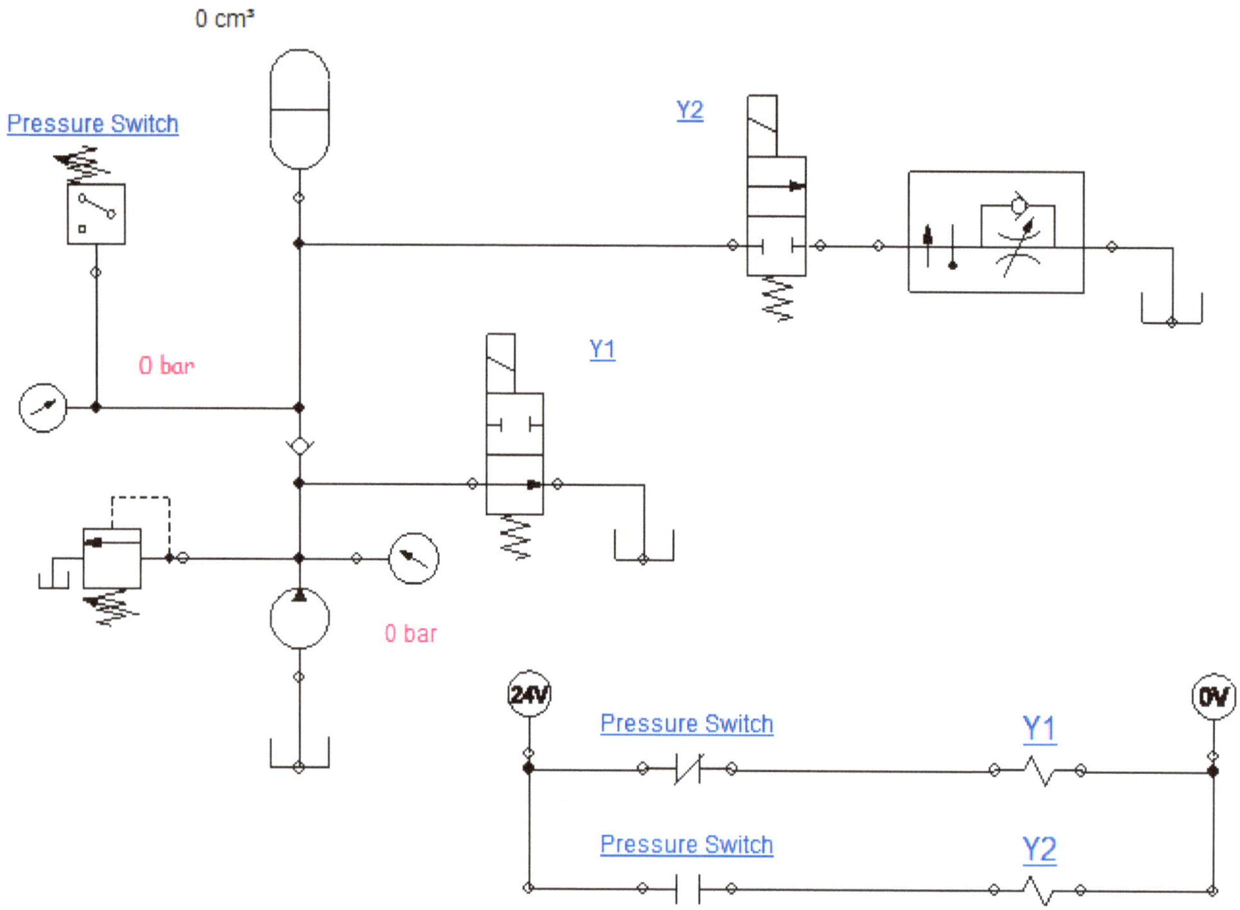

Fig. 2.26A- Electro-Hydraulic Accumulator Charging System

<u>Accumulator Charging (Fig. 2.26B):</u>
When the system power is turned ON, system pressure is below Pmin so that the pressure switch is in its initial (deactivated) state. Because the pressure switch is normally-closed, the DCV valve solenoid Y1 is energized right away and the valve closes the tank line. As a result, the pump is loaded to charge the accumulator. The accumulator is kept charging and the pressure rises. When the system pressure exceeds Pmax, the pressure switch goes to its alternate (active - open) state, the solenoid valve is de-energized, and the pump is unloaded.

Accumulator Discharging (Fig. 2.26C):

Shifting the valve (Y2) directs the energy stored in the accumulator into the system. As the energy in the accumulator is consumed, system pressure is reduced. When the system pressure falls below Pmin, pressure switch goes back to its initial (closed) state, solenoid Y1 is energized, and the pump get back to charge the accumulator.

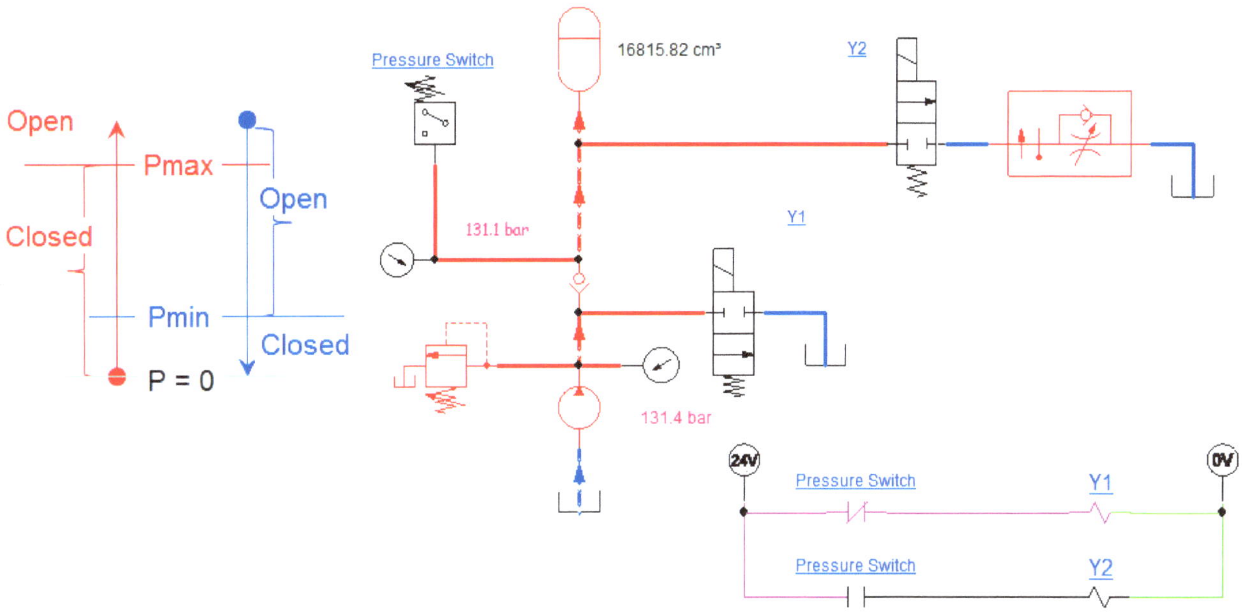

Fig. 2.26B- Accumulator Charging

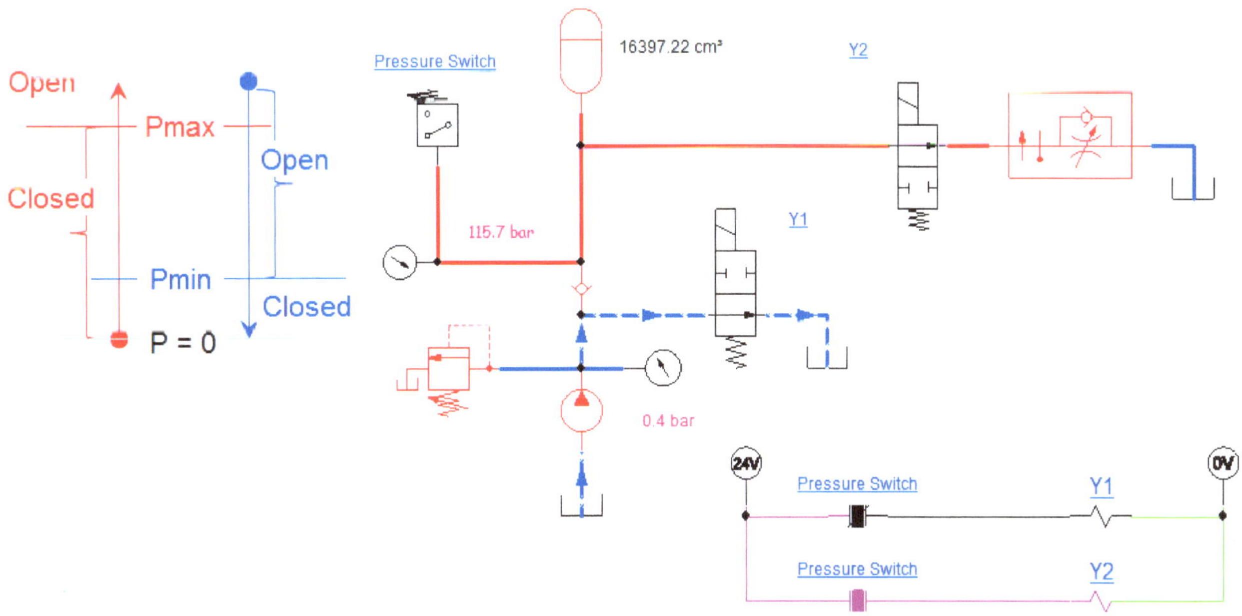

Fig. 2.26C- Accumulator Discharging

2.9- Control of Overrunning Loads

2.9.1- Hydro-Mechanical Overrunning Load Control

Figure 2.27 shows one example of an actuator working against a combined load. The actuator is a double acting cylinder that is oriented vertically to lift and lower a mass. While lifting the mass, the pump power is supplied to the rod side of the cylinder against the resistance of the load. While lowering the load, the power is supplied to the piston side while the load becomes negative. In this stroke, one of three valves can be used as shown in the figure. Depending on the valve used, the following operative features are achieved:

- Prevents overrunning the load.
- Gives a chance for the pump to make up the oil in the piston side of the cylinder.
- Avoids creating cavitation in the piston side of the cylinder.
- Controls the lowering speed of the load.
- If the pump fails, the load will be suspended with creep-free conditions.

Control valves in such systems are set manually and do not offer electrical communication.

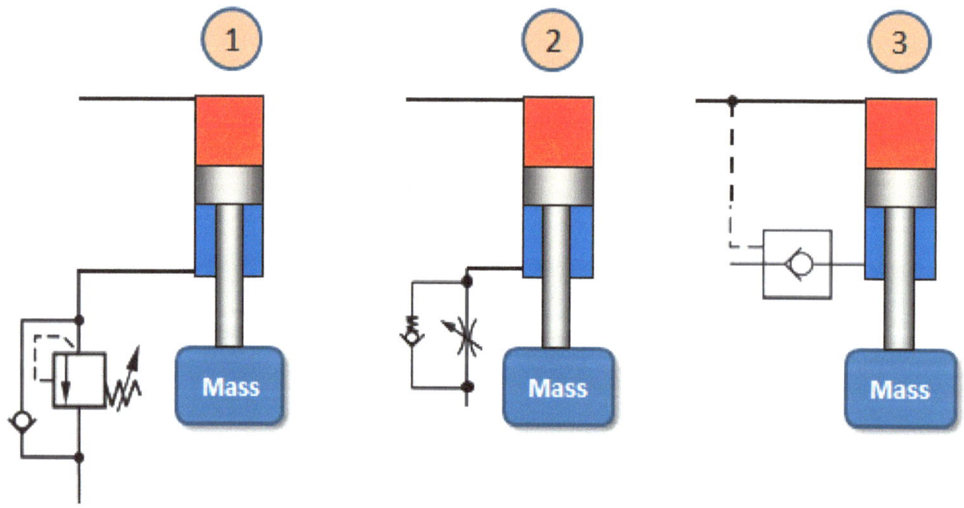

Fig. 2.27 - Hydro-Mechanical Overrunning Load Control

2.9.2- Electro-Hydraulic Overrunning Load Control

There are different ways to control an overrunning load electro-hydraulically. Figure 2.28 shows the operational mode of an electro-hydraulic system to control an overrunning load. The figure shows the use of a continuous valve to control the vertical cylinder. By moving the joystick, the continuous valve moves proportionally to drive the cylinder. Such a system can be used to create open-loop control or closed-loop control architecture. The system can be controlled and monitored remotely. The system can work as a part of an entire automated machine to perform a predefined duty cycle.

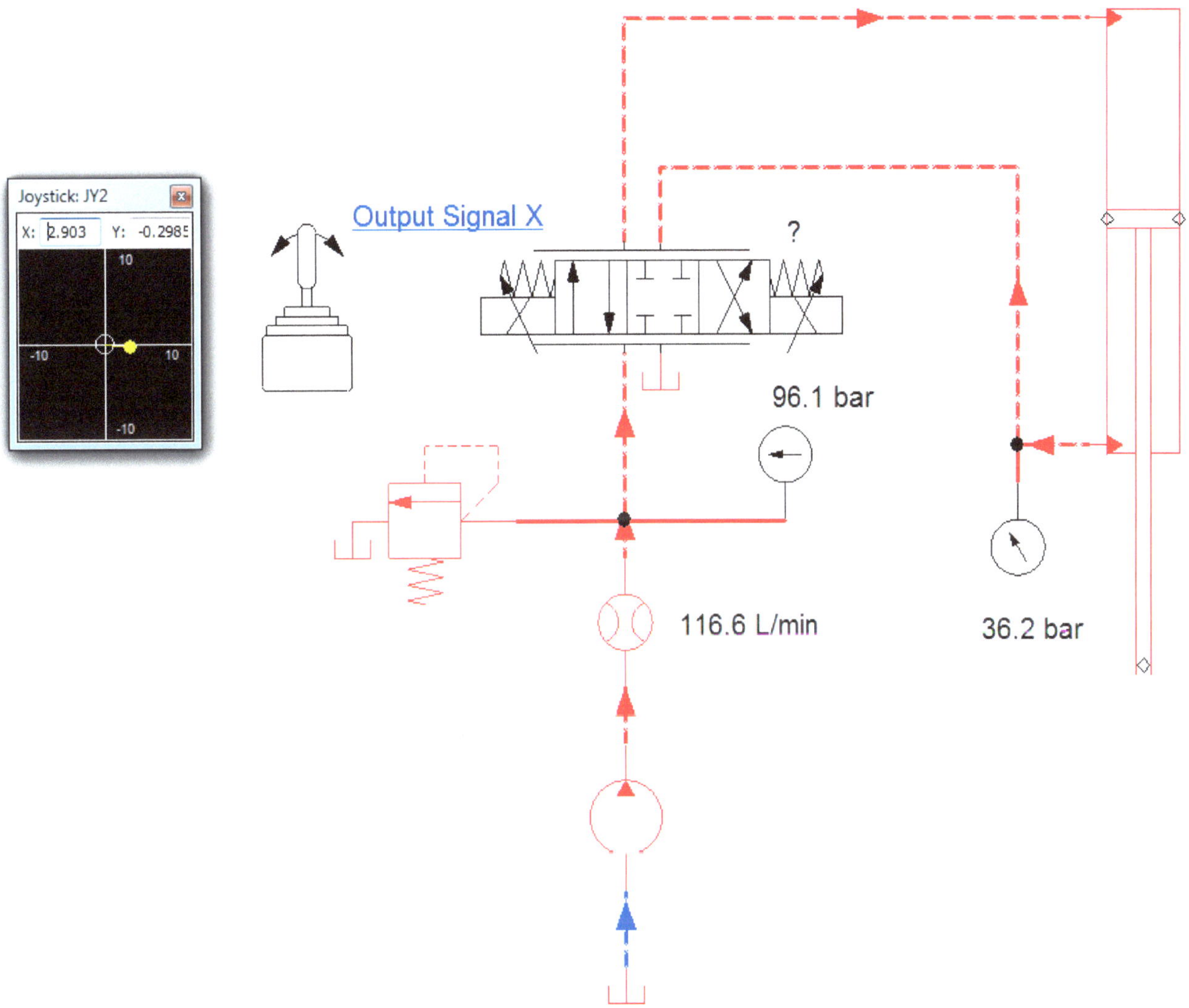

Fig. 2.28 - Electro-Hydraulic Overrunning Load Control

2.10- Control Block Actuation in Mobile Applications

2.10.1- Hydro-Mechanical Control Block Actuation

Figure 2.29 shows a hydraulic circuit of a two-axis hydro-mechanical *Pilot-Controlled* valve block. Fig. 2.30 shows a four-port pilot valve and its typical hydraulic circuit. It is obvious in such a system many hydraulic lines and fittings are used.

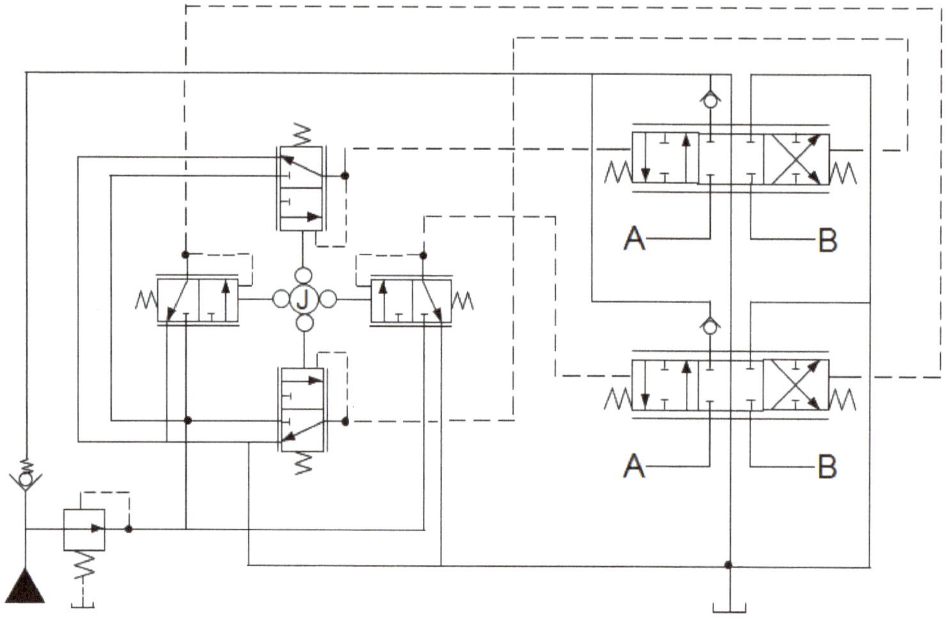

Figure 2.29- Hydraulic-Actuated (Pilot-Controlled) Control Block

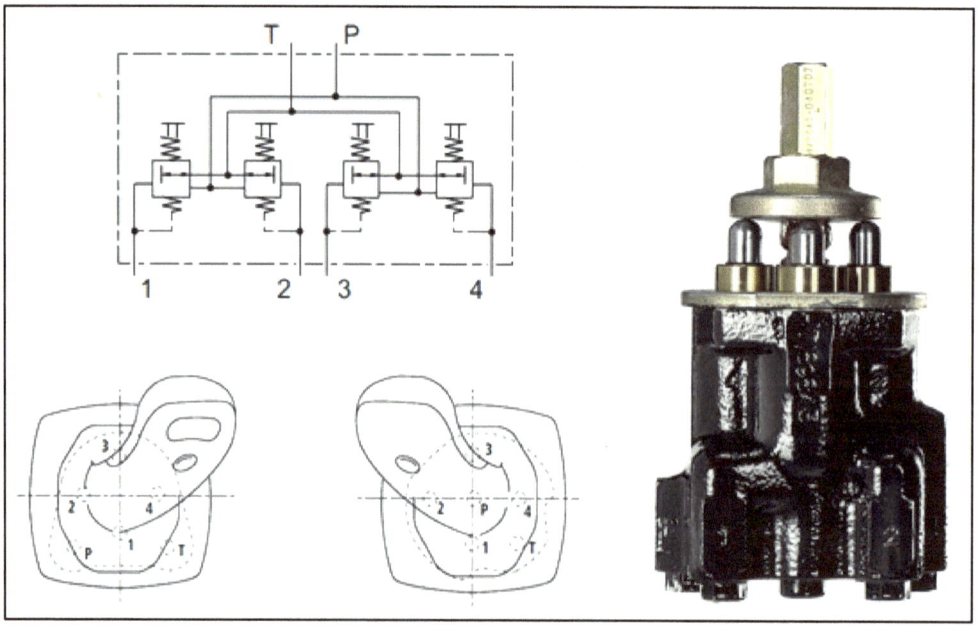

Fig. 2.30- Four-Port Pilot Valve

2.10.2- Electro-Hydraulic Control Block Actuation

Electrically-Actuated Valve Blocks are also referred to as (*Drive-by-Wire) Control Blocks*. Figure 8.31 shows an example of an electric-actuated control block and the actuation valves. Signals are transmitted electronically. Electrically-actuated control blocks have the following features:

- The most space saving and flexibility in laying out the system.
- Less energy consumption as compared to hydraulic-operated systems.
- Requires less labor in assembling the system.
- Upgradable to features of artificial intelligence.
- High maintenance cost due to the lack of know-how.

Fig. 2.31- Electrically-Actuated Control Block (Courtesy of Bosch Rexroth)

Chapter 3

Switching Valves Construction and Operation

Objectives:

This chapter covers the principle of operation and the construction of various types of switching (ON/OFF) solenoids that are used to actuate hydraulic valves. The chapter discusses, qualitatively, the pros and cons of wet type versus dry type switching solenoids and DC type versus AC type switching solenoids. The chapter also discusses the undesirable effects of using AC switching solenoids such as AC hum, eddy current, and inrush current, and constructional considerations to minimize their effects. The chapter concludes by presenting examples of using switching solenoids in directional, pressure, and control functions.

Brief Contents:

Chapter 3 - Switching Valves Construction and Operation

3.1- Basic Electro-Magnetic Concepts

3.1.1- Magnetic Field around a Conducting Wire

As shown in Fig.3.1, when electric current flows in a piece of conducting wire, it creates a magnetic field around the wire. As shown in the figure, the direction of the magnetic flux lines follows the *Right Hand Rule*. Such phenomena occur also oppositely. If the magnetic field intensity varies around a piece of conducting wire, it produces a flow of electrical current in the wire.

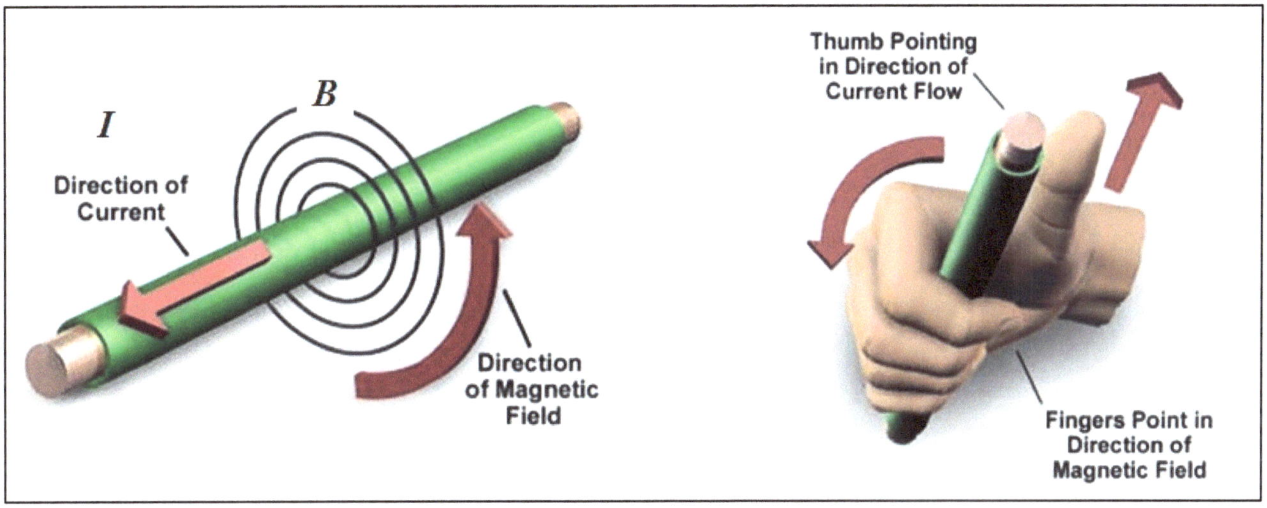

Fig. 3.1 - Magnetic Field around a Conducting Wire (www.micro.magnet.fsu.edu)

3.1.2- Magnetic Field Measuring Units

The units of measurement for the magnetic field is a *Tesla* (T) or *Gauss* (G) where 1T = 10^4 G. These units are a measure of the force created by a moving charge flowing through a given length of conducting wire. Equation 3.1 (Courtesy of Hydraforce) illustrates this principle.

$$T\left(\frac{N}{Am}\right) = \frac{Force\ (Newton)}{Current\ (Amplere) \times Lengthof\ Wire\ (Meter)} \qquad 3.1$$

Magnetic field is typically measured in Gauss because the unit of a Tesla is very large. To illustrate how large the Tesla is; the magnetic field of the Earth is 0.6 G or 0.00006T.

3.1.3- Magnetic Field Intensity

Figure 3.2 and Eq. 3.2 describe how to calculate the intensity of a magnetic field. The magnetic field intensity in a cylindrical space of radius ***r*** around a conducting wire is a function of the current ***I***, the distance between the point of measuring and the wire, and the permeability μ_0 of the medium that surrounds the wire.

$$B\,(T) = \mu_0 \frac{I\,(A)}{2\pi\, r\,(m)} \qquad 3.2$$

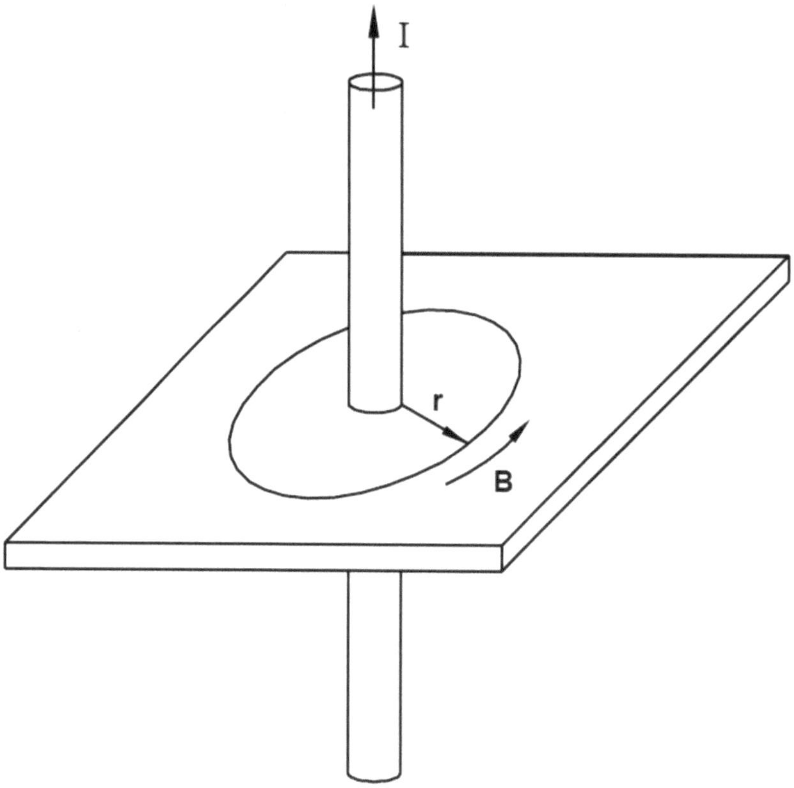

Fig. 3.2 - Magnetic Field Intensity (Courtesy of Hydraforce)

The following example (Courtesy of Hydraforce) shows how to apply the previous equation.

If μ_0 for air = 4π x 10^{-7} N/A^2, ***I*** = 1A, and ***r*** = 2.54 cm (1 inch), then:

B = (4π x 10^{-7}) (1) / (2π) (0.0254m) = 0.00000394 ***T*** = 0.0394 ***G***

3.1.4- Magnetic Field around a Coil

If a conducting wire is wound into a circle, the magnetic field becomes stronger. As shown in Fig. 3.3, by circling the wire several times to form a coil, the resulting magnetic field will be substantial.

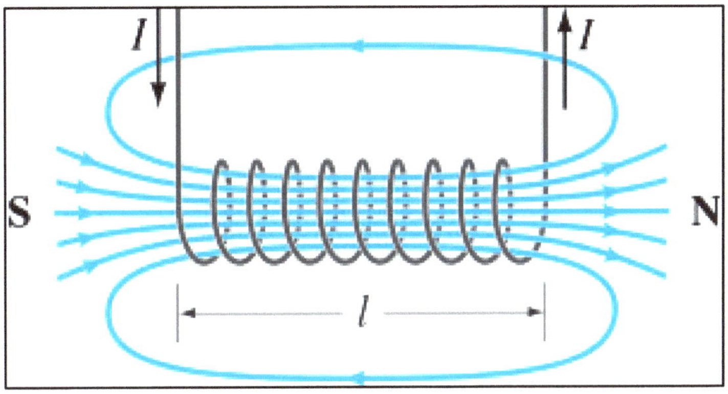

Fig. 3.3 - Magnetic Field around a Coil (www.miniphysics.com)

Equation 3.3 and the following example (Courtesy of Hydraforce) show how to calculate the magnetic field inside a coil.

$$\boldsymbol{B}\ (\boldsymbol{T}) = \frac{\mu_0(\boldsymbol{nI})}{\boldsymbol{l}} \qquad \mathbf{3.3}$$

where $\boldsymbol{n}$ is the number of turns in the coil and $\boldsymbol{l}$ is the length as shown in the figure.

Example: If $\boldsymbol{n}$ − 1000, $\boldsymbol{I}$ − 2 A and $\boldsymbol{l}$ = 0.04 m, then

$\boldsymbol{B = (4\pi \times 10^{-7})\ (1000)\ (2)\ /\ 0.04 = 0.063\ T = 630\ G}$

3.1.5- Electro-Magnetic Force

As shown in Fig. 3.4, if a coil is wound around an iron *Core*, the line of electromagnetic forces will be concentrated and lined up with the central line of the core. The core is also referred to as an "*Armature*" or a "*Plunger*" and must be made of a ferromagnetic material such as iron. Once the current passes through the coil, the armature will become magnetized.

If the assembly of the coil and the armature is left unframed, it will be self-oriented towards the geographical directions. The direction of magnetization depends on the polarity of the current.

The right-hand rule is also applicable for coils. While the right-hand fingers point out to the direction of the current, the thumb points out to the magnetic North Pole that is also the direction of the resulted electromagnetic forces.

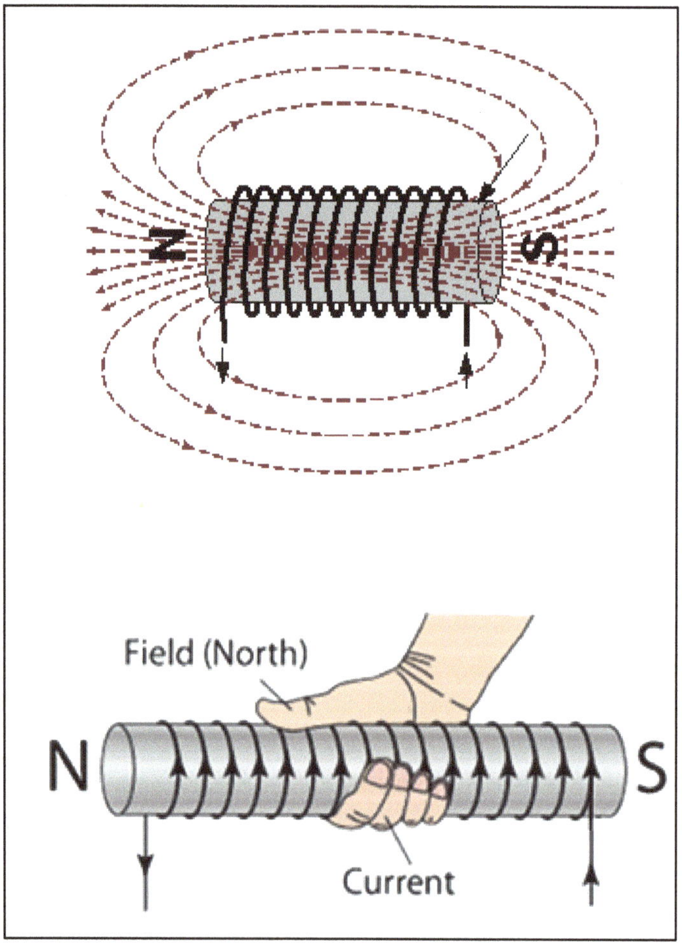

Fig. 3.4 - Magnetic Field around a Coil

3.2- Switching Solenoids

3.2.1- Switching Solenoid Basic Structure, Operation, and Functions

The *Switching solenoid* is simply a framed coil-armature assembly. It is also referred to as an *ON/OFF Solenoid.* As shown in Fig. 3.5, a switching solenoid consists of a coil wound around a bobbin. The armature is placed inside the sleeve of the bobbin. The bobbin-armature assembly is enclosed by a frame. A supporting spring is used to maintain the gap when the switching solenoid is de-energized.

When the valve is de-energized, the armature is not seated completely underneath the coil because the supporting spring is uncompressed and pushes the armature out of the frame. As a result, a certain axial gap is formed between the armature and the coil. When the switching solenoid is energized, the magnetic field starts to develop with its weakest condition because of the maximum axial gap. As the armature gets more under the coil, the axial gap is reduced, the magnetic field intensity and the electromagnetic force grows rapidly. As a result, the armature is pulled rapidly to be completely seated under the coil against the supporting spring.

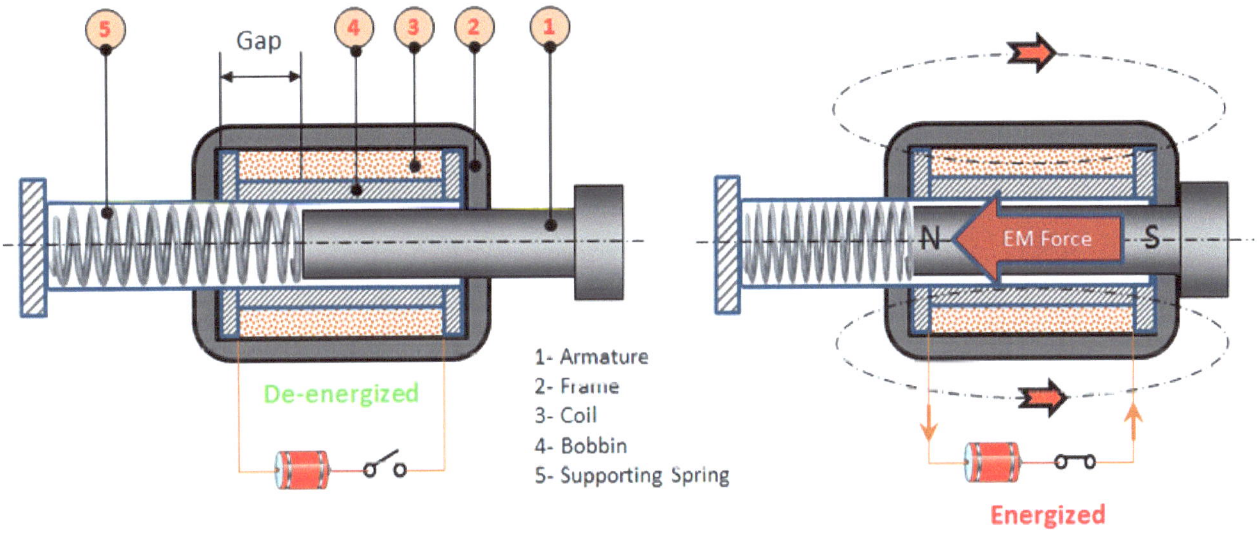

Fig. 3.5 - Basic Structure of a Switching Solenoid

Figure 3.6 shows various typical switching solenoids that can be used for multiple purposes such as door locks, vacuum valves, starter motor engagement mechanisms, gasoline injector switching, automatic transmission clutch controls, and electrohydraulic valves.

Fig. 3.6 - Various Typical Switching Solenoids

The following section, as a Courtesy of Hydraforce, presents in more details the interior structure of a standard switching solenoid. As shown in Fig. 3.7, the switching solenoid frame is referred to as *Yoke*. The material which encases the interior elements is referred to as the *Encapsulant*. The tube around which the coil is turned is referred to as the *Bobbin*, Fig. 3.8 shows the construction of the bobbin and its main dimensions.

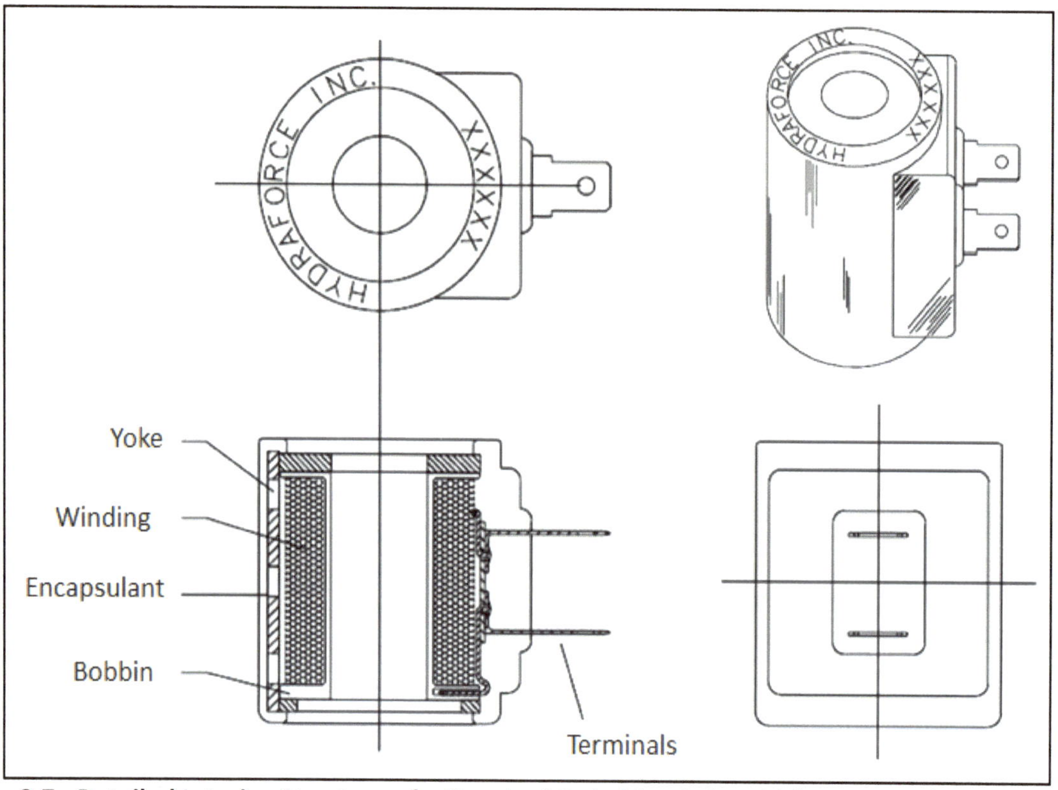

Fig. 3.7 - Detailed Interior Structure of a Standard Switching Solenoid (Courtesy of Hydraforce)

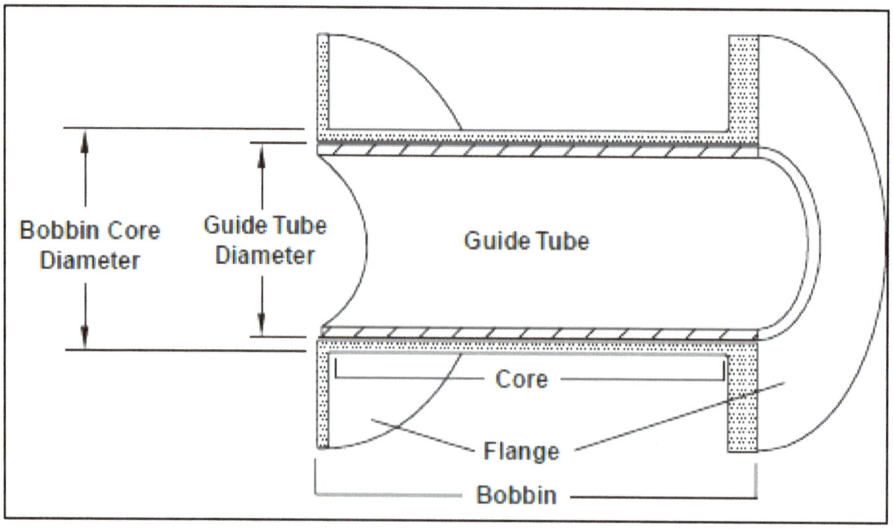

Fig. 3.8 - The Bobbin (Courtesy of Hydraforce)

The coil is referred to as the *Winding*. The winding consists of copper wire known as magnet wire. It is called magnet wire because it is processed specifically for use in electromagnetic actuators such as switching solenoids. In order to wind the copper in this manner, each turn must be separated from the next by insulating material as shown in Fig. 3.9. This prevents the current from jumping from one turn to the next, rather than flowing through the winding. Each turn is separated from the next by insulation on the wire. Insulation is a material which does not allow electricity to pass through.

There are several classes of insulation available based on their thermal rating as listed in the Table 3.1. Each subsequent level is either a little thicker than the previous or uses a higher-grade material. As the level of insulation increases, the level of protection increases and unfortunately, the cost increases as well. The current industry standard is class H. Since the winding temperature of the majority of the applications does not exceed 180° C, the cost of using a higher-class wire is not justified.

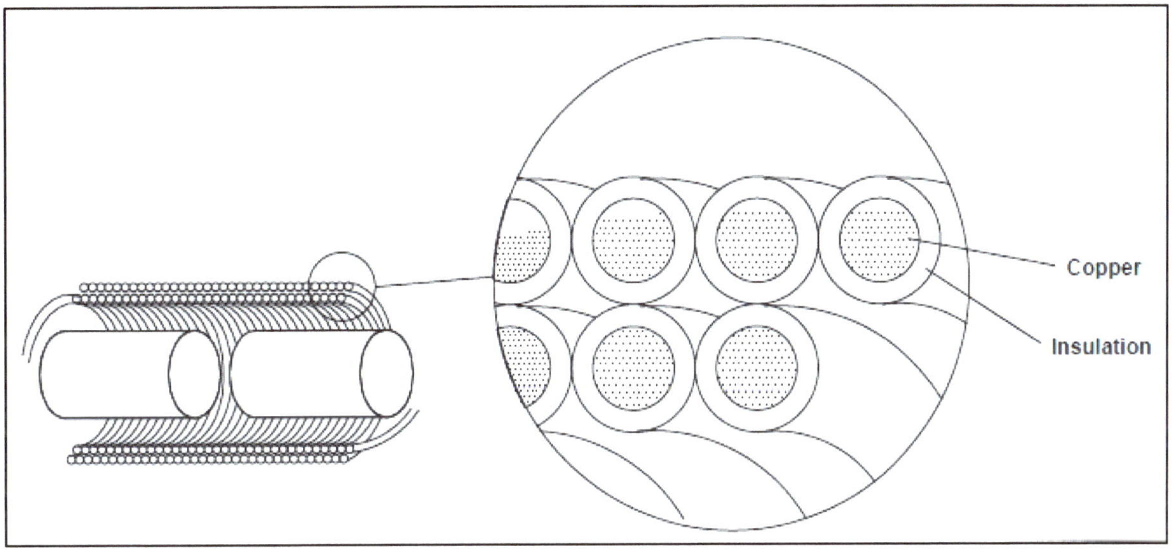

Fig. 3.9 - Winding Insulation (Courtesy of Hydraforce)

Thermal Rating (^{O} C)	Insulation Class
105	A
130	B
155	F
180	H
200	N
220	C

Table 3.1 - Classes of Winding Insulation (Courtesy of Hydraforce)

3.2.2- Wiring Methods of a Coil

The number of turns of wire which fit onto a bobbin is determined by the core diameter and the flange diameter. The diameter of the core defines the innermost diameter of the winding. The length of the core determines how many turns of copper can be laid next to each other. Figure 3.10 illustrates this concept.

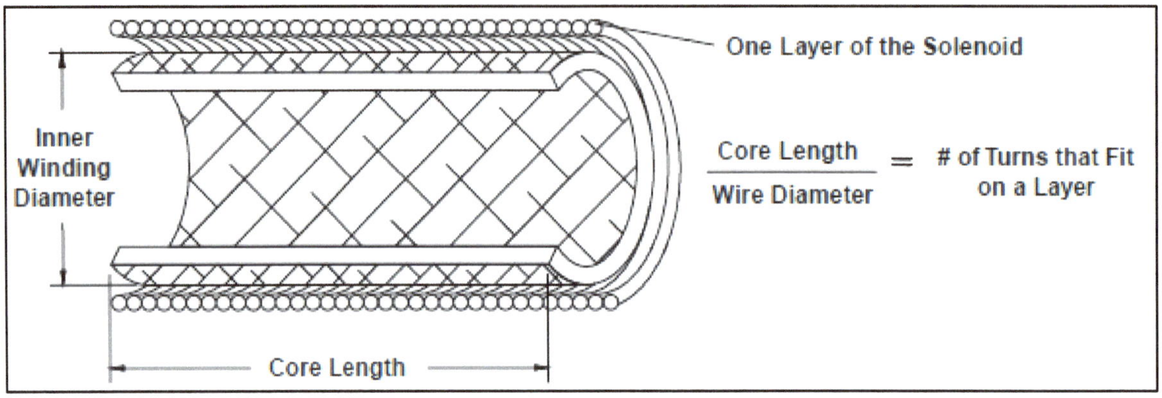

Fig. 3.10 - Number of Turns in One Layer in a Coil (Courtesy of Hydraforce)

As shown in Fig. 3.11, the area between the core diameter and the outside diameter of the flange, determines the number of layers that can fit onto the bobbin.

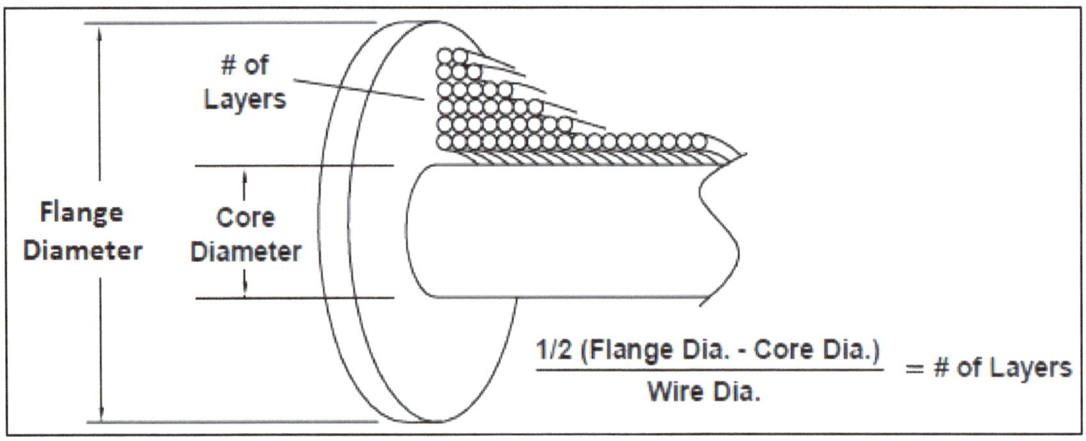

Fig. 3.11 - Number of Layers in a Coil (Courtesy of Hydraforce)

The total number of turns in a coil then is the product of the number of turns in one layer and the number of layers.

There are two methods for building a winding, *Perfect Wound* and *Random Wound method.*

Perfect Wound Method: The perfect wound method shown in Fig. 3.12 is also referred to as *Layer Wound* method. Notice that the bobbin is different from the ones shown in previous diagrams. There are small grooves named *Ridges*, to ensure that the first layer on the bobbin fills the space completely. The subsequent layers of wire are wound between the previous ones in a tightly packed form as shown in the figure. Modern machines are designed to wind perfect wound coils. The machines are programmed with the wire diameter so that they will move smoothly across the length of the core, minimizing the space between each turn of wire.

Advantage: Perfect wound method ensures the wires were tightly packed and develop more turns in the same space.

Disadvantage: A special bobbin is needed for each voltage. The reason for this is that a different thickness of wire is required for each different voltage or winding and therefore the size of the grooves varies. The perfect layer wound coil has the potential to become random after 15-20 layers.

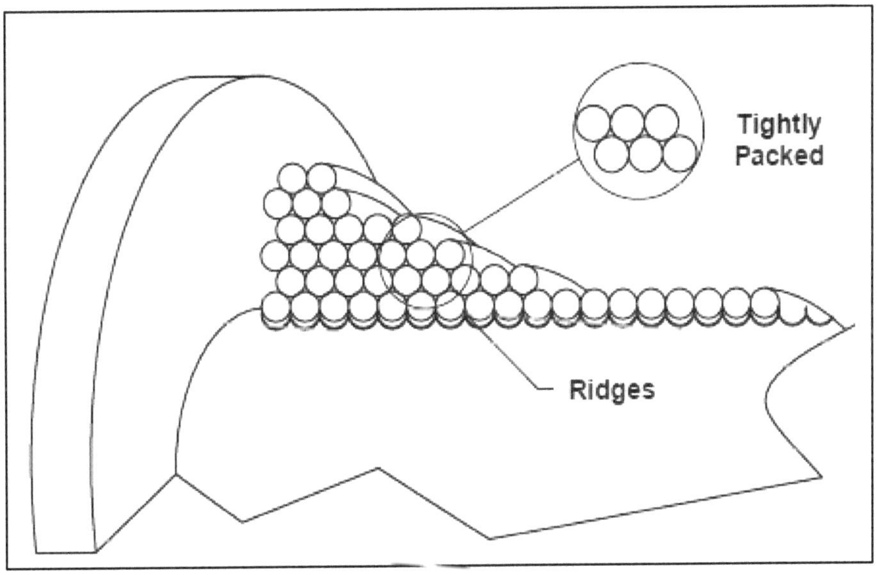

Fig. 3.12 - Perfect Wound Method (Courtesy of Hydraforce)

Random Wound Method: The turns in a random wound coil are just that, random. They do not necessarily lay tightly inside the previous row, as seen in Fig. 3.13. The winding process does not become random until the first seven to ten layers.

Advantage:

The cost of the bobbin is lower with the random wound process.

- One bobbin is used for any size wire.
- The random wound process provides perfect layer wound coils for at least the first 7-10 layers.

Disadvantage:

Loosely packed wires could vibrate and rub against one another in the application. This could potentially cause the insulation to be rubbed off, causing the coil to short.

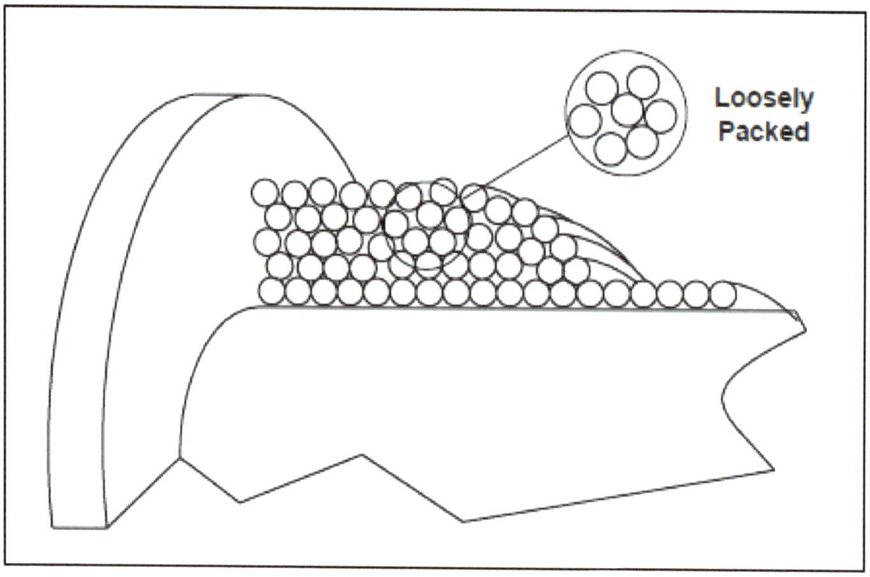

Fig. 3.13 - Random Wound Method (Courtesy of Hydraforce)

3.2.3- Switching Solenoids Force-Stroke Relationship

As it has been discussed before, once the switching solenoid is switched ON, an initial electromagnetic force is developed and continues to grow rapidly when the gap is reduced. Equation 3.4 is used to calculate the electromagnetic force.

$$F = \frac{1}{2}(nI)^2 \times \frac{\mu_0 A}{X^2} \qquad 3.4$$

where n is the number of turns in the coil and I is the length as shown in the Fig. 3.14.

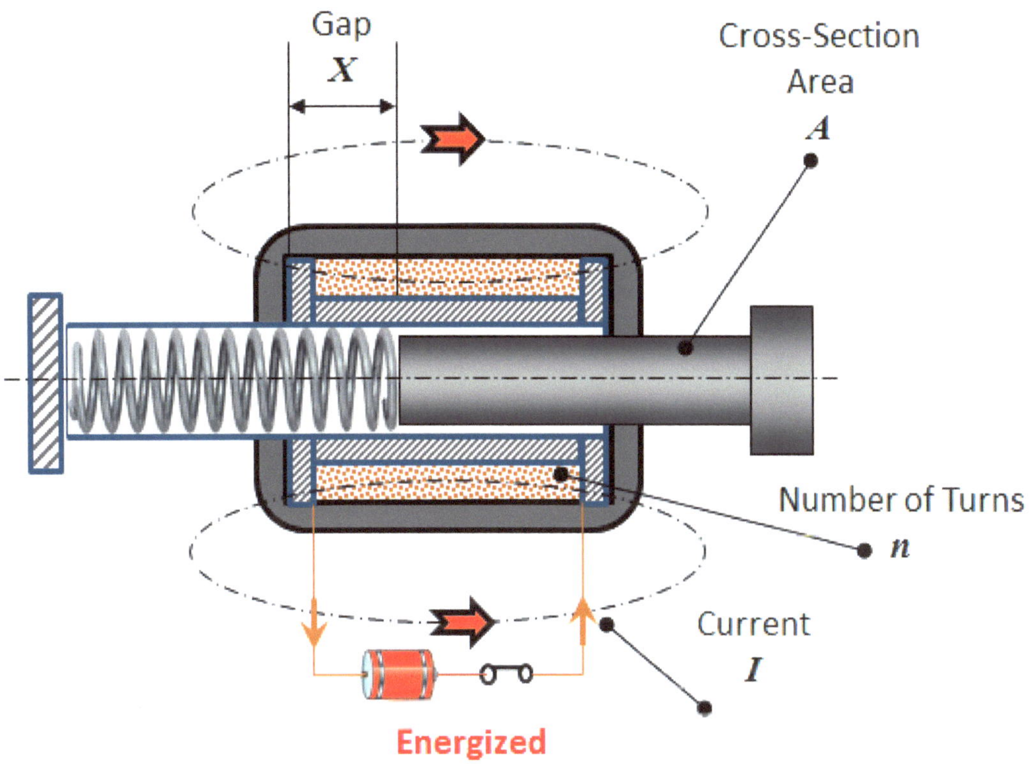

Fig. 3.14 - Electromagnetic Force-Stroke Relationship

Example:

For a coil of 1000 turns, initial air gap of 1 cm, initial current of 2 Amp, and a plunger of 4 cm^2 cross-section area, the *Initial Electromagnetic Force* is:

$$F = \frac{1}{2} \times \frac{(1000\,x\,2)^2 \times (4\pi\,x\,10^{-7})\,4}{1} = 10.048\,N = \frac{10.048}{4.45} = 2.26\,lb$$

As indicated in Equation 3.4 and shown in Fig. 3.15, the following matter of facts can be concluded about the electromagnetic forces of a switching solenoid:

- It is inversely and nonlinearly proportional to the gap.
- It starts with an initial value and increases rapidly with the armature moving under the coil.
- It reaches its maximum value at the rated current and the smallest gap.
- Maximum value is called the *Holding Force*.
- Available commercial switching solenoids produce forces within 60-70 N (13-16) lb.
- The current must be rated so that the electromagnetic force can overcome all forces through the full stroke of the armature.

The example shown in Fig. 3.15, shows that the correct rated current is 2 Amps. If the current falls down to 1.5 Amps, the armature will move a little bit before it balances with the spring force. If the current falls down to 1 Amp, the armature won't move because the initial electromagnetic force is less than the initial spring force.

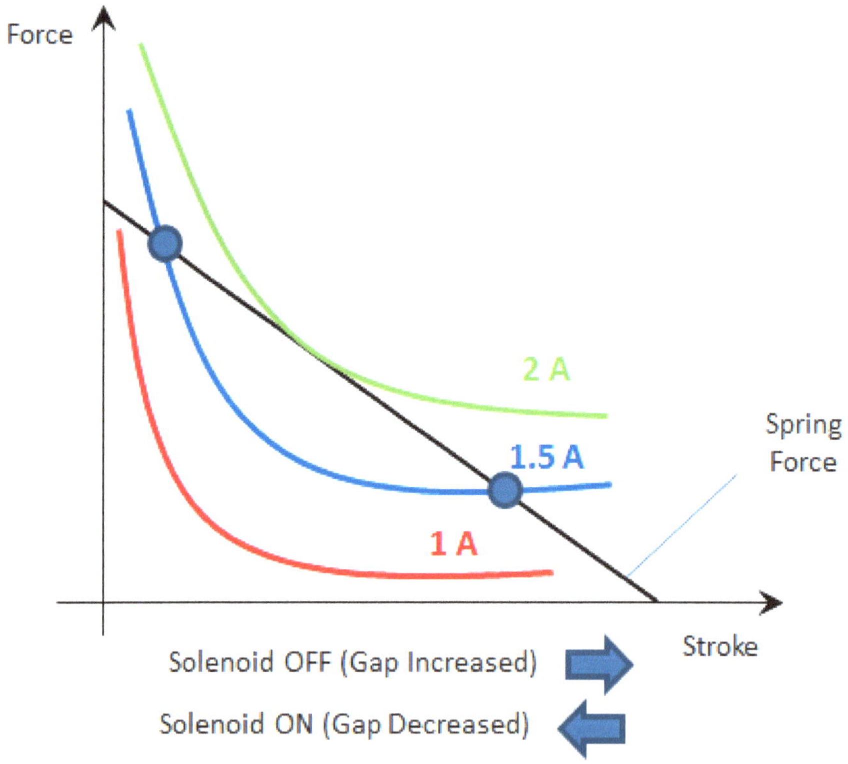

Fig. 3.15 - Switching Solenoid Force-Stroke Relationship

3.2.4- Switching Solenoid Performance Terminologies

This section presents the common terminologies for switching solenoids. To help identify the correct switching solenoid for an application, the following specifications are needed:

Voltage Type: it means DC or AC type of voltage.

Nominal (Rated) Voltage (U_N): is the voltage for which the switching solenoid has been designed.

Voltage Tolerance (%): is the % deviation from the rated voltage.

Nominal (Rated) Current (I_N): is the current based on the nominal voltage and coil resistance, at 20 ^{O}C coil temperature and nominal frequency.

Nominal (Rated) Power (P_N): is the power consumption at rated voltage and a coil temperature of 20 ^{O}C.

Nominal (Rated) Frequency (f_N): is the frequency (60 Hz in North America or 50 Hz in Europe) for which an AC switching solenoid has been designed when supplied by the nominal voltage.

Holding Force: is the switching solenoid force that is effective at the end of its stroke.

Residual Force: is the force that still applies due to any remaining magnetic field after turning the switching solenoid OFF.

Response (Delay) Time: is the time the coil takes to form a magnetic field and an electromagnetic force high enough to move the armature from its initial position. So, it is the time between the current application and the initial movement of the armature.

Stroke Time: is the time the armature takes to move from its initial position to the end of its stroke.

Switching Time: is the sum of the response time and the stroke time. It is also referred as *Pull-in Time* or *Rise Time*.

Switching (Cycling) Rate: in reference to Fig. 3.16, cycling rate means how many cycles the switching solenoid can complete per unit time.

Duty Cycle: it is the ratio between the ON period and the complete cycle time in %.

Continuous Duty: it means the coil can be turned ON continuously without fear of burning out the coil under normal climate conditions. It is to be noted that the coils do heat up during the continuous duty operation that results in loss of power after hours of operation. This internal heat and the heat from external sources increase the resistance of the coil and thus, decreases the current ($V = IR$) because the voltage is constant.

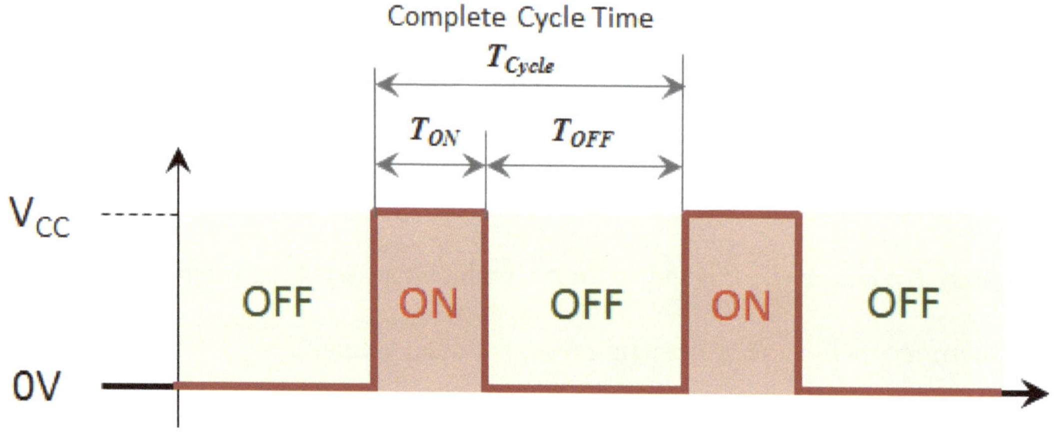

Fig. 3.16 - Cycling of a Switching Solenoid

Figure 3.17 shows an example of published data for a switching solenoid used for a size 6 directional valve. The figure shows the use of many of the aforementioned specifications. The figure shows the switching frequency equal 7200 cycle per hour (2 Hz). The figure shows the insulation class is F that means maximum temperature is 155 ^{O}C.

Rexroth
Bosch Group

Directional spool valves, direct operated, smoothly switching, with solenoid actuation

electric		
Available voltages [2)]	V	12, 24, 96, 205
Voltage tolerance (nominal voltage)	%	±10
Power consumption	W	30
Duty cycle	%	100
Switching time according to ISO 6403	ms	Approx. 3 to 4 times longer than standard valve
Maximum switching frequency	1/h	7200
Maximum surface temperature of the coil [3)]	°C *[°F]*	150 *[302]*
Protection class according to DIN EN 60529		IP 65 (with mating connector mounted and locked)
Insulation class VDE 0580		F
Electrical protection		Every solenoid must be protected individually, using a suitable fuse with tripping characteristic K (inductive loads).
Behavior in case of errors		The specified solenoid surface temperature may be exceeded.

Fig. 3.17 – Specifications for a Typical Switching Solenoid (Courtesy of Rexroth)

3.3- Switching Solenoids for Electro-Hydraulic Valves

Figure 3.18 shows that switching solenoids for EH valves are classified, based on the medium that surrounding the coil, as *Dry (Air) Switching solenoids* and *Wet (Oil) Switching solenoids*. They are also classified, based on the current, as *DC Switching solenoids* and AC Switching solenoids. The following sections present the structures of the different types and a brief qualitative comparison between each two type.

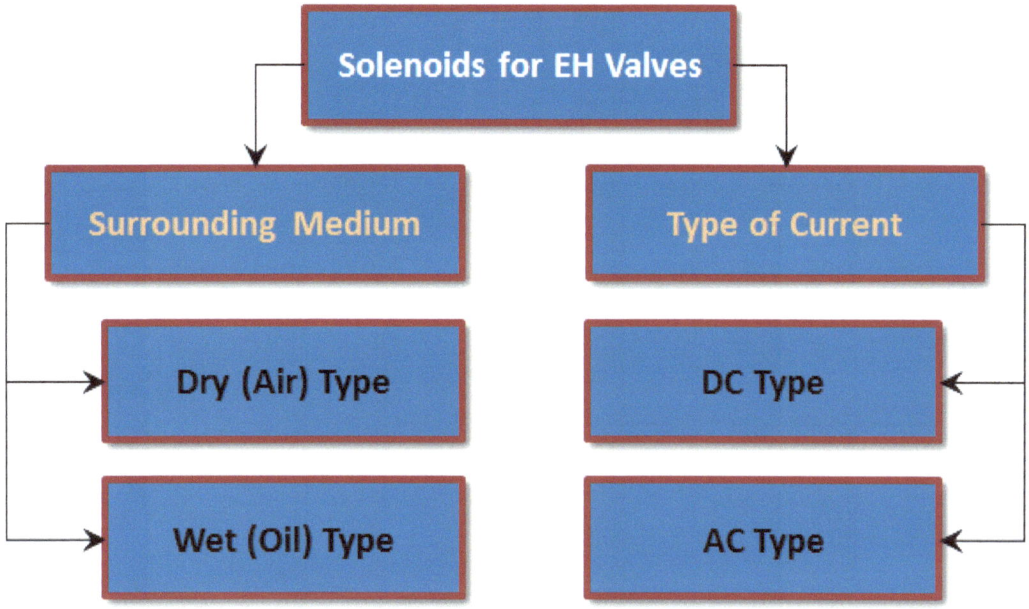

Fig. 3.18 - Classification of Switching Solenoids for EH Valves

3.3.1 Dry (Air) Type Switching Solenoids for EH Valves

As shown in Fig. 3.19, the dry (air) type switching solenoid consists almost of the same elements as the switching solenoid that has been generically presented in Fig. 3.5. The differences here are that the armature is T-shaped, and the frame is C-shaped.

A standard electrical terminal is used to connect the switching solenoid to the electrical circuit. The armature is connected mechanically to a push pin that passes through a dynamic seal to push the spool of the valve. The purpose of the dynamic seal is to seal the hydraulic side of the valve from the switching solenoid so that the gap is kept dry.

As shown in Fig. 3.20, a dry type switching solenoid has a possible chance to leak. In normal operation the oil accumulation in the switching solenoid cavity is one drop for every 3000-5000 cycles. Switching solenoid leakage increased due to contamination, high working temperature and pressure, seal wear, and pin misalignment.

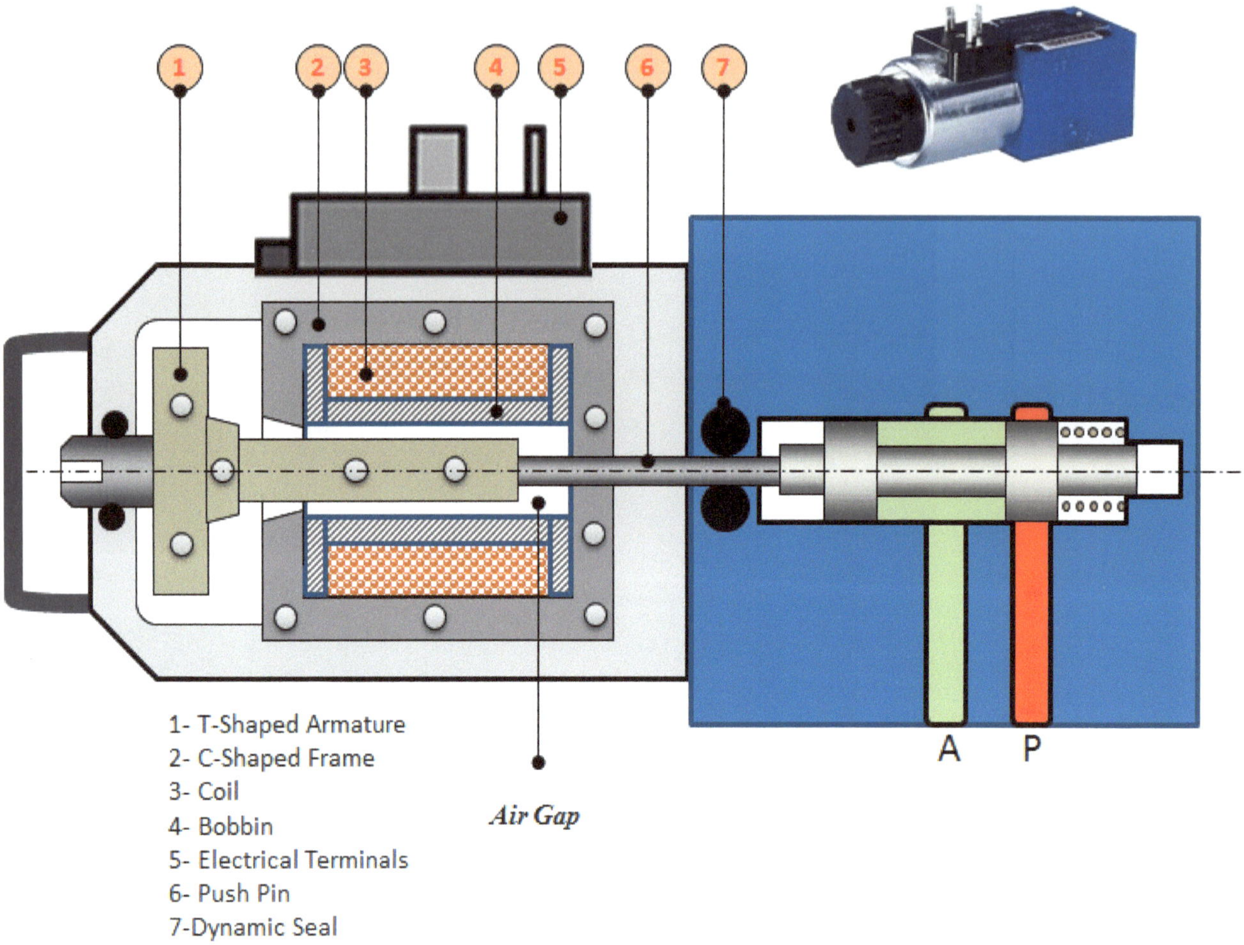

Fig. 3.19 - Dry Type Switching Solenoid for EH Valves

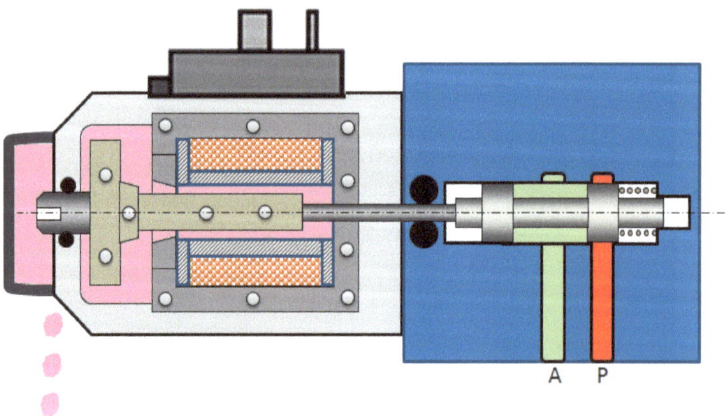

Fig. 3.20 - Leakage from a Dry Type Switching Solenoid for EH Valves

3.3.2- Wet (Oil) Type Switching Solenoids for EH Valves

As shown in Fig. 3.21, a wet type switching solenoid has the armature placed inside a tube that is open to the valve so that the armature moves in oil bath. A static seal has been used to prevent external leakage.

Some of the armatures may have a damping hole drilled as shown in the figure. Such a hole helps to damp the motion of the armature when the switching solenoid is turned ON or OFF so that the noise generated by cycling the valve is reduced.

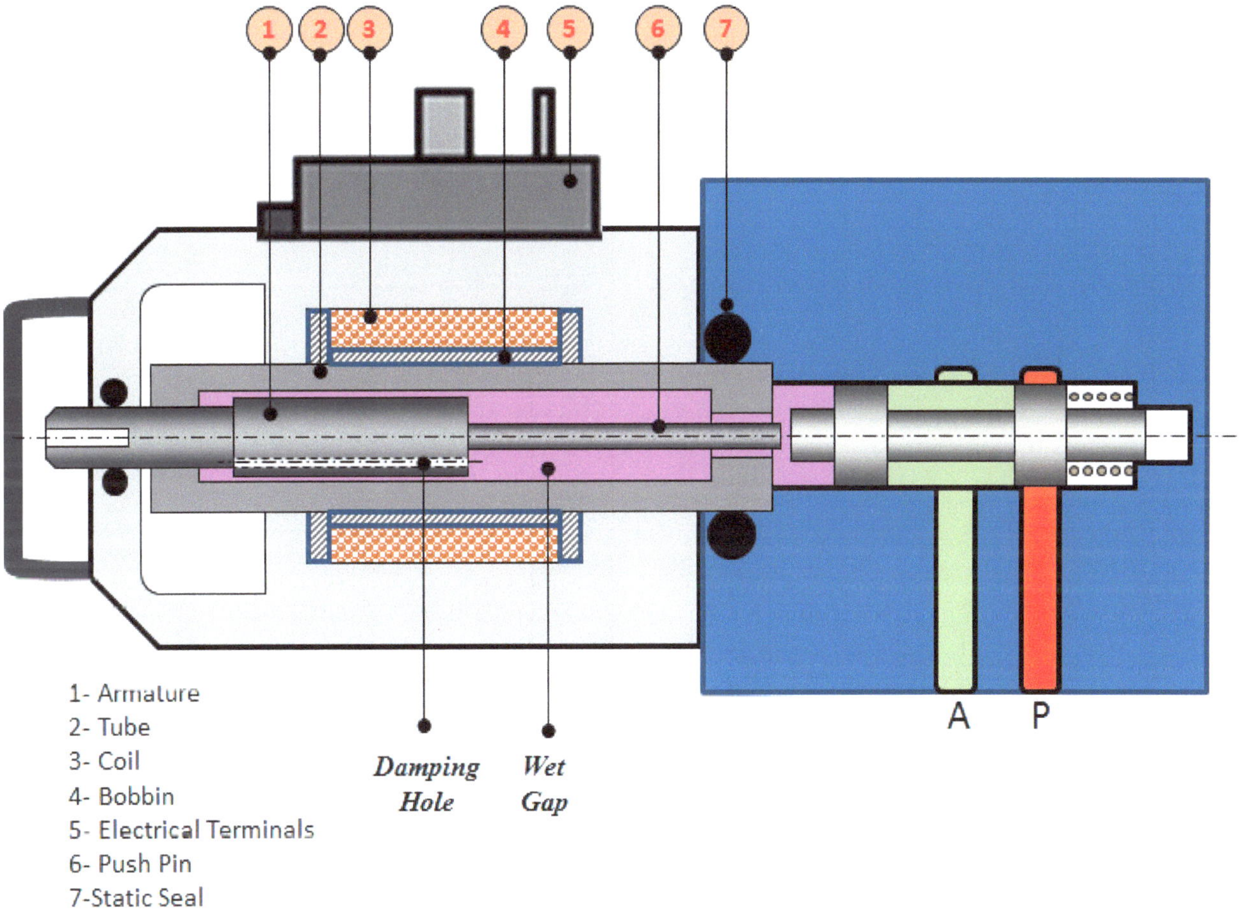

Fig. 3.21 - Wet Type Switching Solenoid for EH Valves

3.3.3- Dry versus Wet Type Switching Solenoids for EH Valves

Table 3.2 presents a qualitative comparison between dry and wet types of switching solenoids. The table shows that the wet type switching solenoid offers better cooling and heat dissipation due to the oil bath around the armature. The oil around the armature also works like a first line of defense preventing the air and the dust from reaching into inside the hydraulic valve. Statistically a wet type switching solenoid is more reliable and can work for a number of cycles that is higher than the dry type switching solenoid. The static seal in a wet type switching solenoid lasts longer than the dynamic seal in a dry type switching solenoid. A wet type switching solenoid produces less noise because the oil bath around the armature absorbs the sound better.

The oil bath in a wet type switching solenoid offers higher resistance against the magnetic field. Therefore, it consumes more power to produce the same holding force. That is why the only advantage seen for a dry type switching solenoid is less power consumption. It is a considerable advantage only if the application uses hundreds of these valves. One example is a ceramic factory that has ten production cells.

From what has been presented, it can be concluded that a wet type switching solenoid is more popular to use.

Item of Comparison	*Dry Type Switching solenoid*	*Wet Type Switching solenoid*
Better coil cooling and heat dissipation.	×	✓
Better valve sealing against air and dust.	×	✓
Better reliability and longer service life.	×	✓
Sealing element last longer.	×	✓
Less noise generation by the switching solenoid.	×	✓
Lower power consumption.	✓	×

✓ Advantage × Disadvantage

Table 3.2 - Dry versus Wet Type Switching Solenoid

3.3.4- DC Switching Solenoids for EH Valves

3.3.4.1- Construction and Operation of DC Switching Solenoids for EH Valves

DC Switching solenoids are simple in construction. Operating a DC switching solenoid is not associated with any undesirable phenomenon, so it requires no special constructional arrangement.

Figure 3.22 shows the characteristics of DC switching solenoid operation. When the switching solenoid is OFF, no current will pass through the coil. Once the coil is turned ON, a *Response (Delay) Time* T_R will elapse before the armature starts to move. During this time the current rises rapidly to a level I_{PEAK} that is high enough to overcome the opposing forces. At point 1, the armature starts to move. The armature movement reduces the gap, strengthens the intensity of the magnetic field, and induces back resistance in the switching coil. This causes a brief reduction in the switching solenoid current. The *Stroke Time* T_S is the time the armature takes to complete its movement. At point 2, the armature has moved completely, and the current gets reduced to I_{VALLEY}. The time T is *Pull-in Time* that is the total switching time for the solenoid to respond to a signal and move to the end of its stroke. After complete movement of the armature, the switching solenoid current continues to rise until it reaches its steady state nominal value at point 3 that is limited by the coil DC resistance.

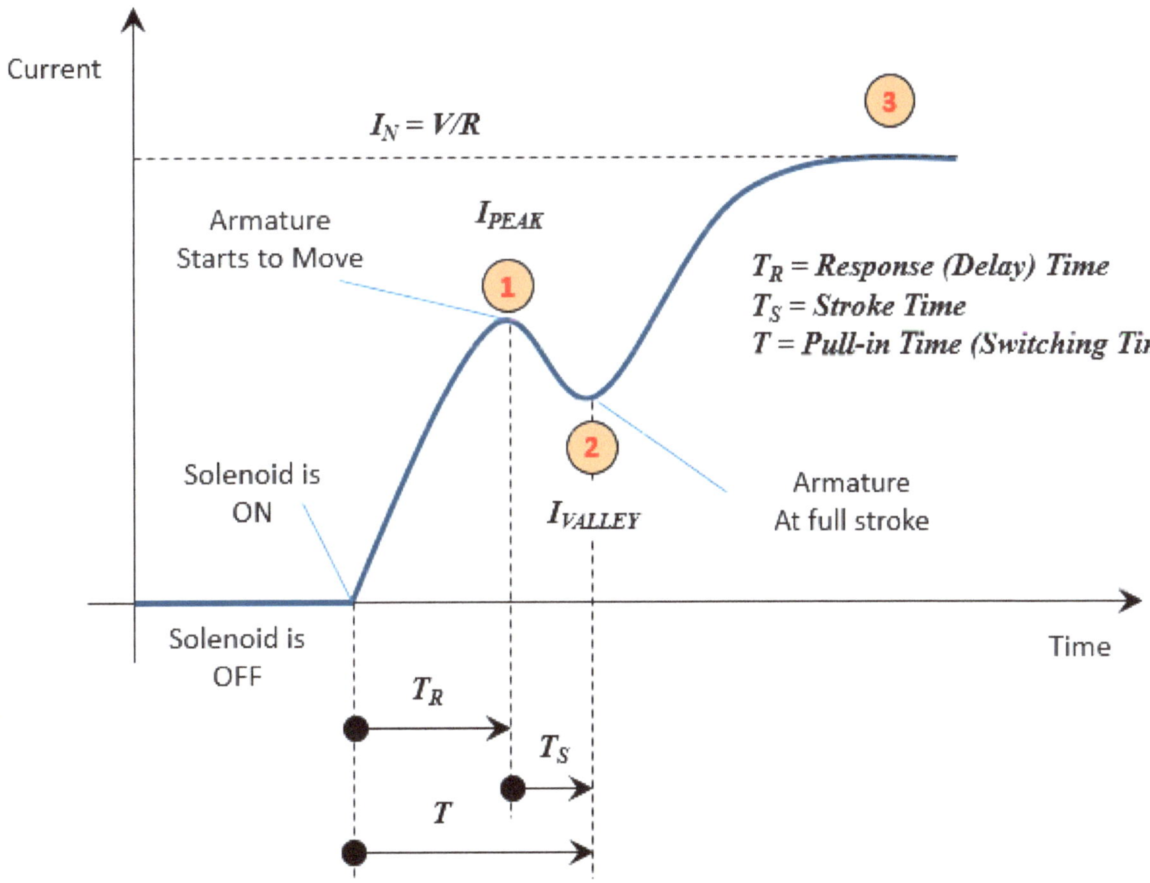

Fig. 3.22 - Current Characteristics of a DC Switching Solenoid for EH Valves

Figure 3.23 shows the detailed symbol used to represent the DC switching solenoid.

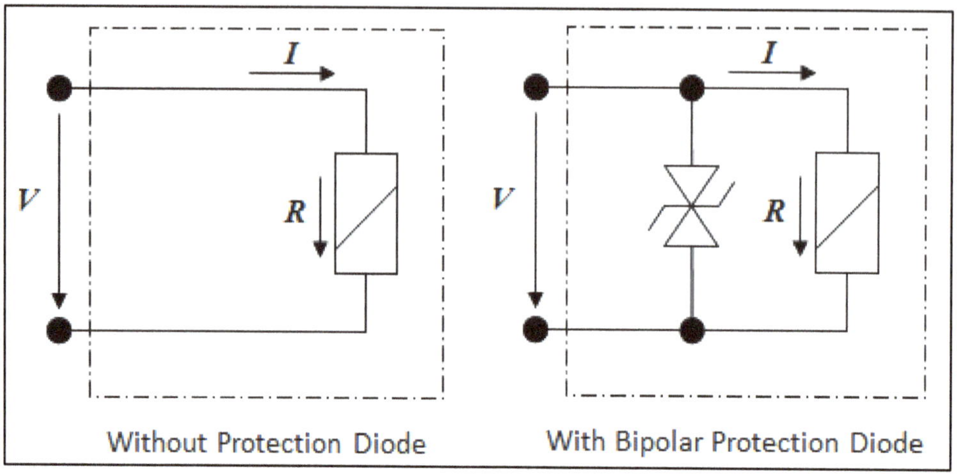

Fig. 3.23 - Symbols of a DC Switching Solenoid for EH Valves

3.3.4.2- Performance of DC Switching Solenoids for EH Valves

Power Consumption: With its relatively slow buildup of current, a DC switching solenoid consumes relatively more power than an AC switching solenoid. As shown in Equation 3.5, the magnitude of the power of a DC switching solenoid is determined by the product of voltage times current:

$$\boldsymbol{Power}\ (\boldsymbol{Watt}) = \boldsymbol{V}\ (\boldsymbol{Volt}) \times \boldsymbol{I}\ (\boldsymbol{Amp}) \qquad \mathbf{3.5}$$

Heat Generation: A DC switching solenoid generates heat due to the resistivity of the coil that has a constant resistivity. Switching solenoids are designed to dissipate the small amount of heat resulting from the low holding current under continuous duty operation so that the coil temperature does not rise above the maximum rated working temperature. If, for any reason, the coil temperature rises above the rated temperature, the electromagnetic force will be reduced as shown in Fig. 3.24. Figure 3.25 shows the force-stroke characteristics of a typical switching solenoid. The figure shows the reduction of the force produced by the switching solenoid due to the increase of the coil temperature.

Switching Time: As it has been explained, applying DC voltage to a switching solenoid coil increases the current asymptotically until it reaches the nominal value that is equal to voltage divided by the coil resistance. The switching time of a DC switching solenoid is relatively longer than in AC switching solenoids.

Switching Rate: Despite the longer switching time of a DC switching solenoid, it can usually be cycled at a higher rate than an AC switching solenoid. The reason is that a DC switching solenoid does not experience the inrush current phenomenon so that they have better chance to cool down faster during the Off time in the cycle.

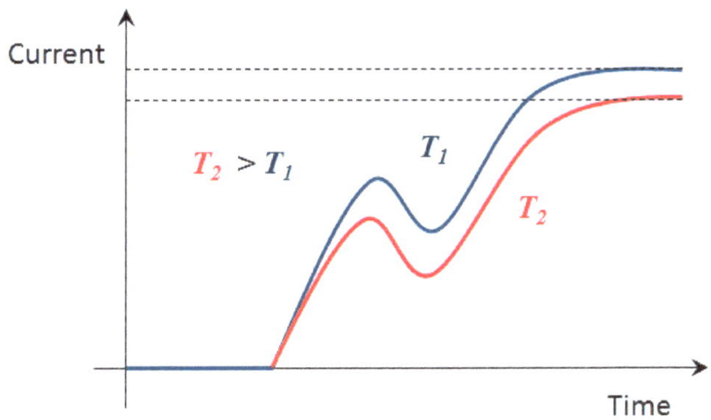

Fig. 3.24 - Effect of Temperature on a DC Switching solenoid for EH Valves

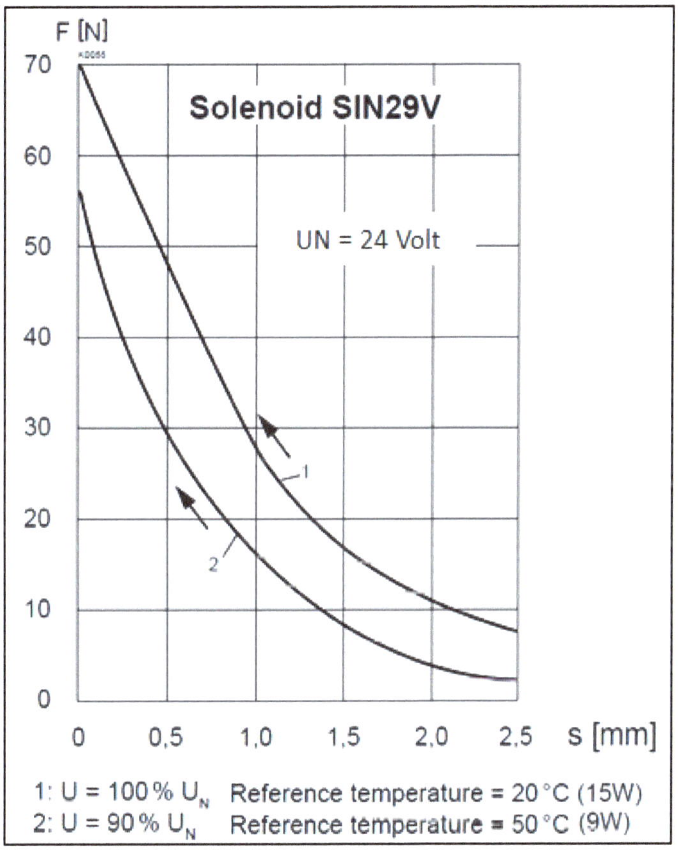

Fig. 3.25 - Typical Force-Stroke Characteristics of a DC Switching Solenoid (Courtesy of Wandfluh)

3.3.4.3- Applications of DC Switching solenoids for EH Valves

DC switching solenoid operation requires DC power to be available. It is more common in mobile applications since DC power availability is more common. Even in industrial applications nowadays, extending DC power lines, hot water, compressed air, and internet communication have become culture in building a factory.

3.3.5- AC Switching Solenoids for EH Valves

3.3.5.1- Construction and Operation of AC Switching Solenoids for EH Valves

Applying the same principles of DC to an AC type the electro valve would be always opening and closing. Then the question here is, how is that a steady electromagnetic force can be generated from AC current? That is simply happening by having the switching solenoid fabricated with built-in *Full Wave Rectifier* as shown in Fig 3.26. The figure shows that the AC voltage is received on the inlet side of the rectifier and the coil is hooked to the outlet side where the wave is rectified. As a result, the electromagnetic force acts on one side only and pulsates with double the frequency of the input AC. An equivalent steady average electromagnetic force pushes the armature in one side only.

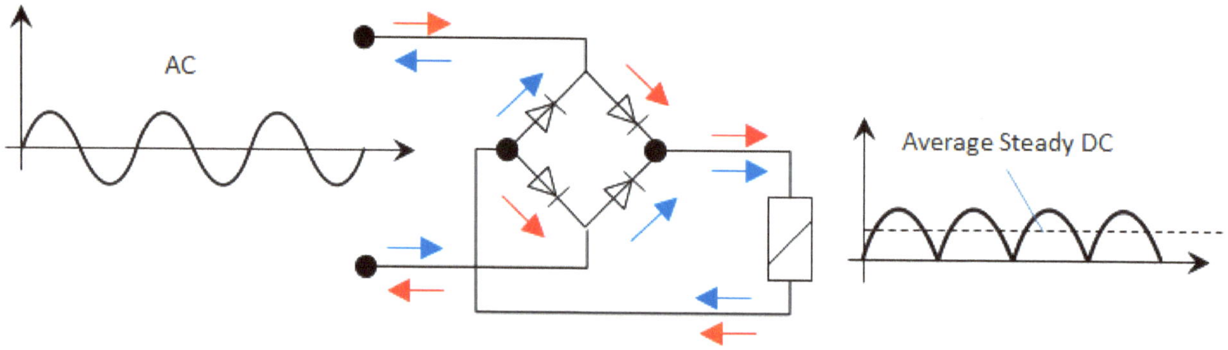

Fig. 3.26 - Current Rectification in AC Switching Solenoids for EH Valves

Figure 3.27 shows the symbol of an AC switching solenoid. When an AC switching solenoid is switched ON or OFF, an extremely high induced voltage spike can result. Therefore, these high-voltage transients can cause the push button switch contacts to burn out quickly. They also represent an extreme hazard in circuits where solid-state components are used. *Surge Suppression* is the method used to protect sensitive components from such spikes.

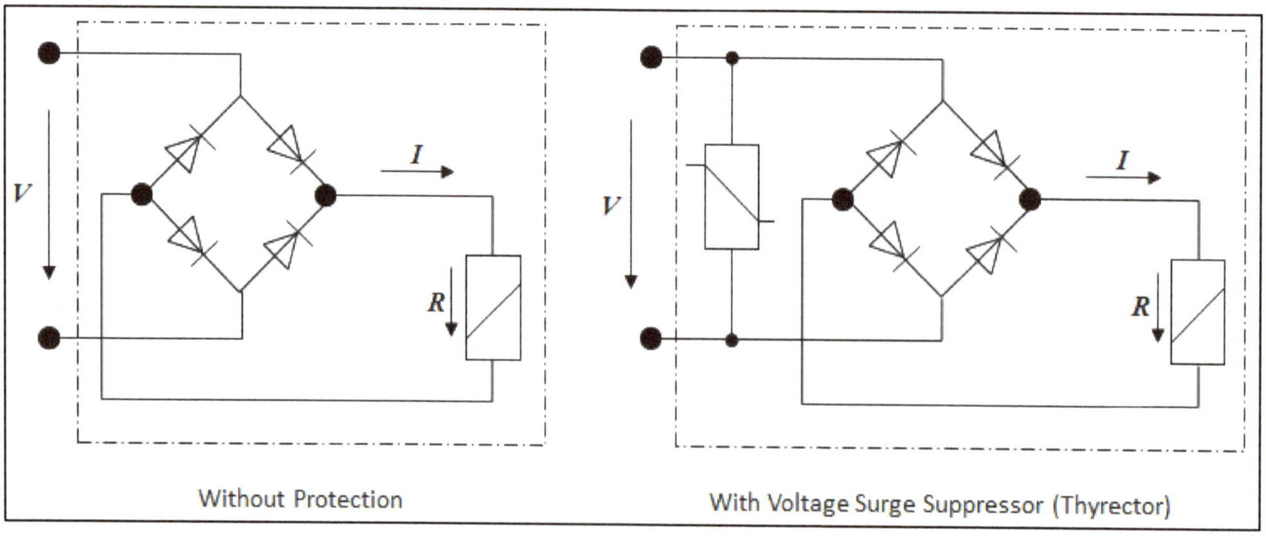

Fig. 3.27 - Symbols of an AC Switching Solenoid for EH Valves

Use of AC current impacts the construction of AC switching solenoids because of the undesirable operational effects of AC current. The following sections present such effects and how they have been resolved by retrofitting the conventional switching solenoid construction.

3.3.5.2- AC Hum

With the use of a full wave built in rectifier, the electromagnetic force pulsates at double the frequency of the applied voltage. As a result, the switching solenoid produces a humming or buzzing sound that is known as *AC Hum*. As shown in Fig. 3.28, by adding a *Shading Coil* to the dry type switching solenoid or *Shading Ring* to the wet type switching solenoid, the AC hum is greatly reduced.

With a shading element, the built-in rectifier isn't fundamentally necessary since these shading elements hold the movable iron core in place when the AC current goes to zero. As shown in the figure, placing the shading element (coil or ring) induces a counter magnetic field with phase shift of 90°. The resulting electromagnetic force of the coil and the shading element is always above zero. The pulsation of the resulted electromagnetic force is then reduced and consequently the humming sound of the coil is reduced.

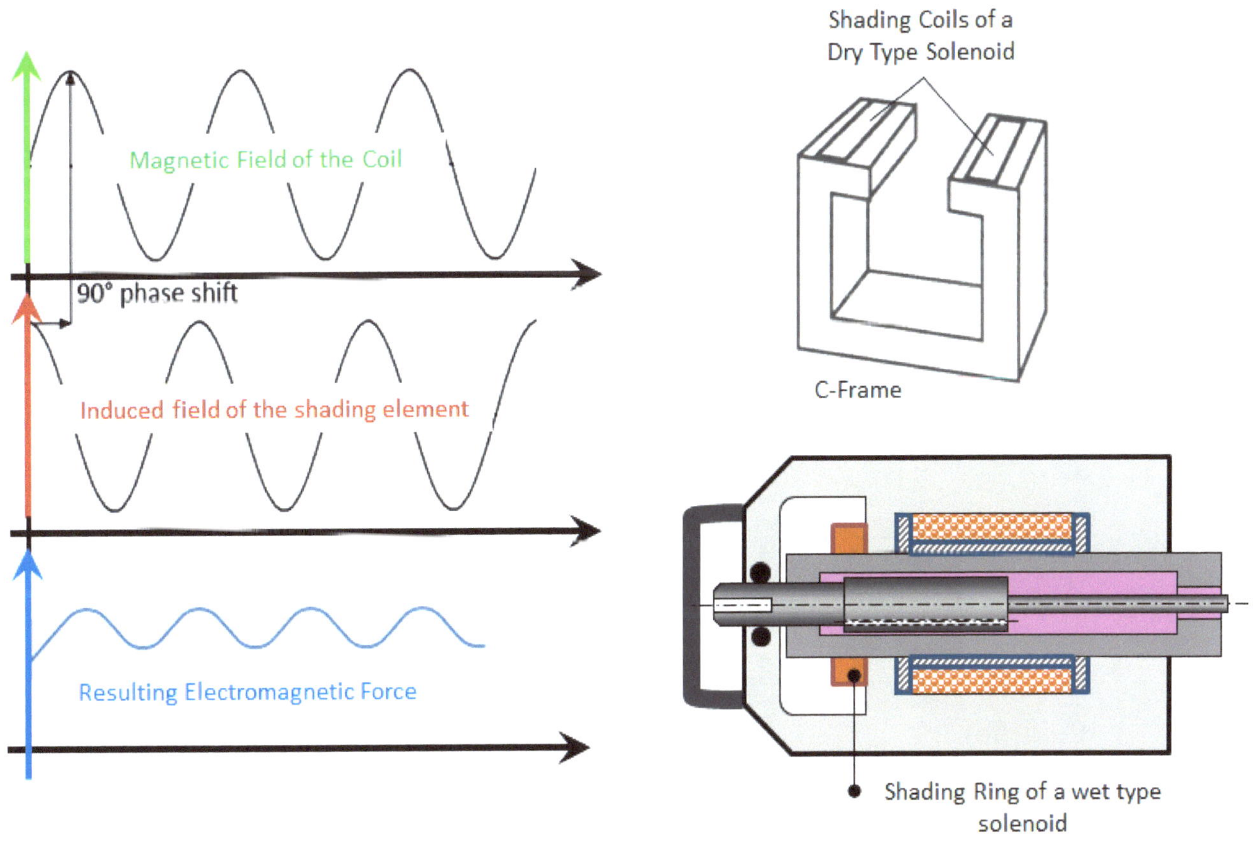

Fig. 3.28 - Resolving the AC Hum in a Switching Solenoid for EH Valves

Figure 3.29 shows the shading ring (3) among the other elements in a wet type AC switching solenoid. In the design of an AC switching solenoid valve, the positions of the shading element and the plunger must be accurately determined. If contaminants found their way around the armature, the alignment is no longer optimal and as a result the armature vibrates accordingly.

The figure shows also a tube-shaped component called *'Pole Piece'* (17) on which the core rests during the excitation and whose task is to improve the gap closing.

The perforated disc (12), held in position by the opposing spring (11), allows a suitable amount of fluid to flow from the outlet port into the sleeve in order to lubricate and cool down the switching solenoid.

The rest of the numbered elements in the figure are similar to the ones explained previously in the standard conceptual switching solenoid designs.

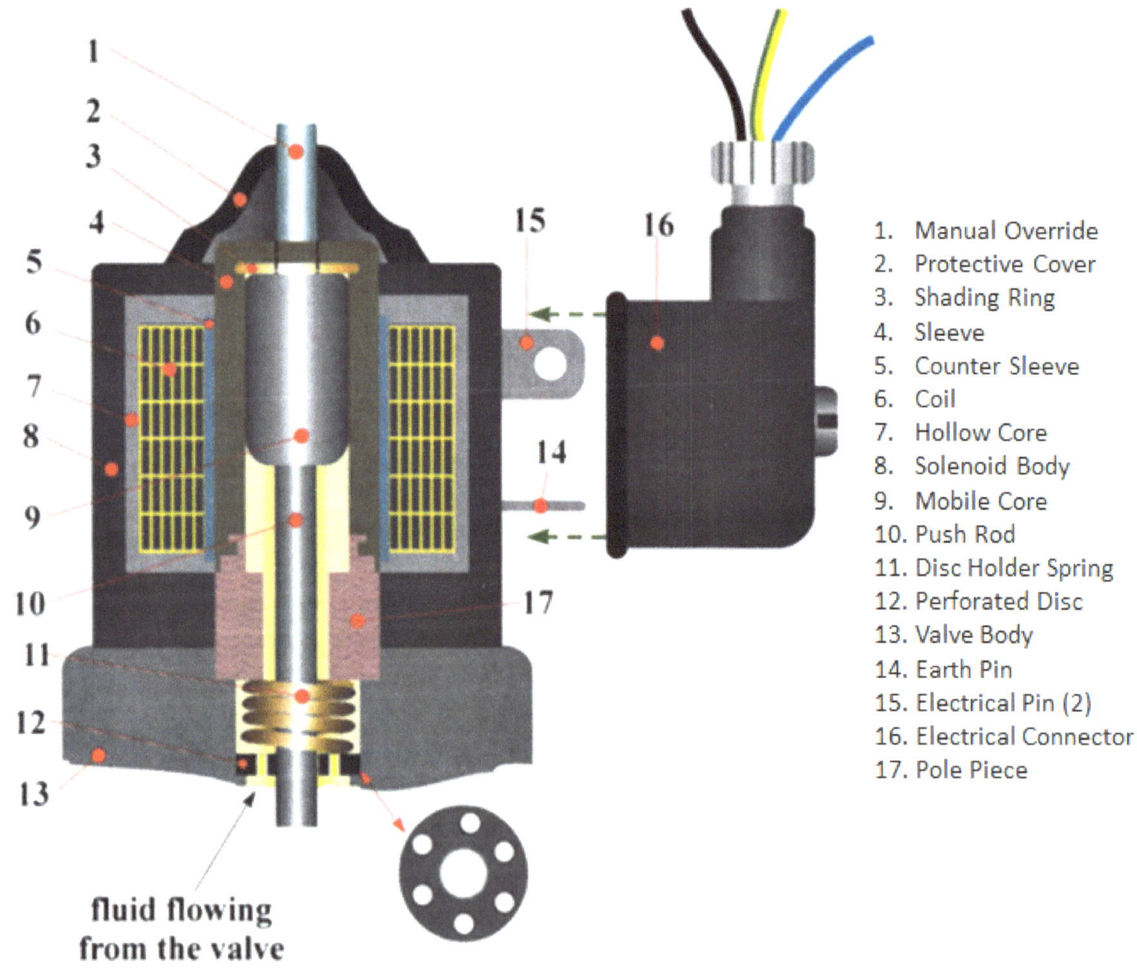

Fig. 3.29 - Schematic Drawing of a Wet Type AC Switching Solenoid for EH Valves (Courtesy of ASSOFLUID)

3.3.5.3- Eddy Current

Again because of the frequency of the applied voltage, the magnetic field goes up and down accordingly. As a result, a small amount of current is developed and strays through the metallic elements that form the switching solenoid. Such current is named *Eddy Current* and causes power consumption and heat generation. As shown in the Fig. 3.30, the C-Frame and the T-plunger in a dry type switching solenoid are made of a stack of laminated thin sheets and each lamination is coated with insulation. The purpose of this construction is to prevent the eddy currents from leaking between the laminations. The figure shows also, in a wet type switching solenoid, the tube can be made of a stack of laminated layers as well but the armature of a wet type switching solenoid can't be laminated. Eddy currents are not a problem in DC circuits, so DC switching solenoids are often made with solid iron parts.

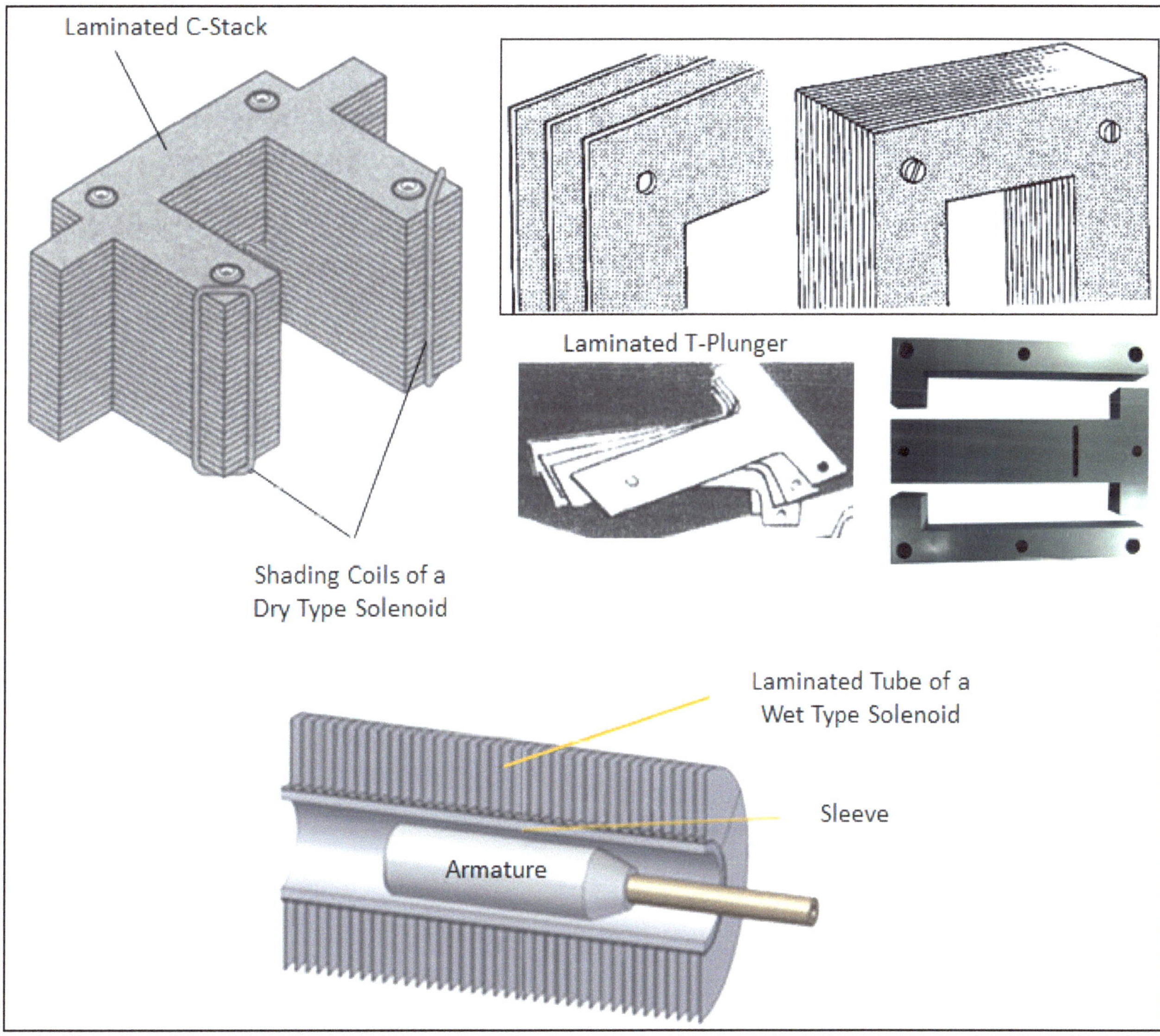

Fig. 3.30 - AC Switching Solenoid Construction to Minimize the Eddy Current

3.3.5.4- Inrush Current

Inrush Current Development: As it has been explained, in a DC switching solenoid, the current builds up slowly until it reaches the nominal value that depends on the coil resistance. AC switching solenoid experience *Inrush Current* phenomenon. The inrush current is also known as *Surge Current.*

As shown in Fig. 3.31, resistivity to electric current flow in the coil is due to the coil wire itself and the magnetic field. The resistivity of the coil is constant because it depends on the design parameters of the coil; namely length, wire diameter, and material of the coil.

The resistivity of the magnetic field is directly proportional to its intensity. A strong magnetic field offers more resistivity to the flow of the current. The intensity of the magnetic field depends on the radial gap that is constant. It also depends on the axial gap that varies with the armature movement.

As shown in Fig. 3.32, right after energizing the switching solenoid, the magnetic field starts with its weakest condition because of the axial gap has maximum value. So, in this early stage, the magnetic field offers its least resistivity against the flow of the electrical current. Consequently, the current rushes into the coil and the armature start to move forward. As the armature gets more under the coil, the axial gap is reduced; the magnetic field intensity and the electromagnetic force grow proportionally. As a result, the armature is pulled rapidly like a shuttle until it is completely seated under the coil. In this case, the magnetic field offers its maximum resistivity, and the current reaches its safe nominal value.

Figure 3.33 shows that the inrush current can reach 4-10 ten times the nominal current.

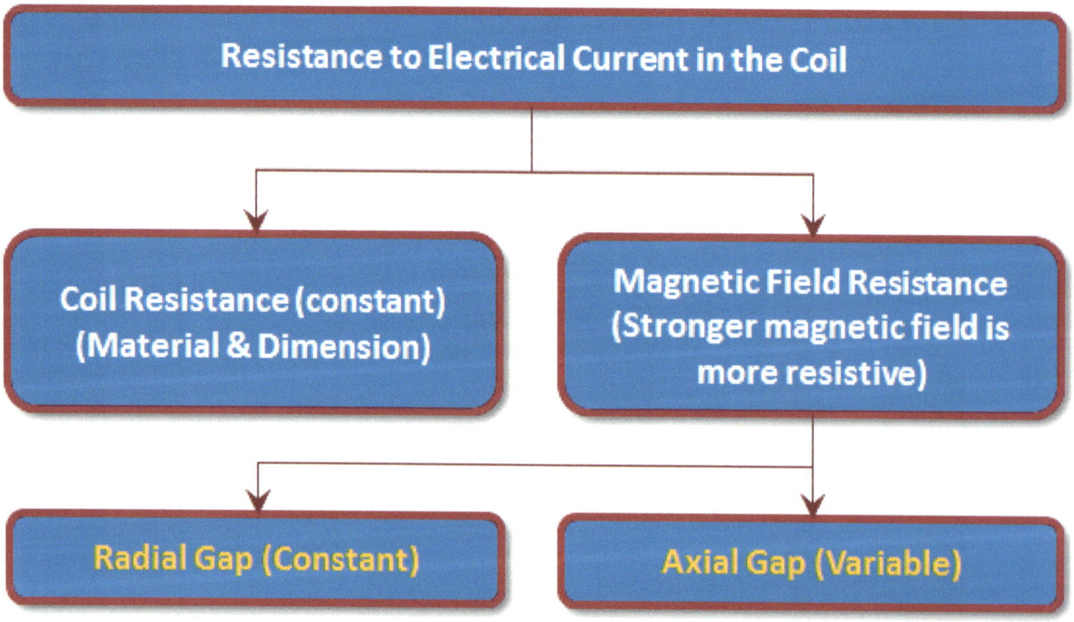

Fig. 3.31 - Resistance to Electrical Current in the Coil

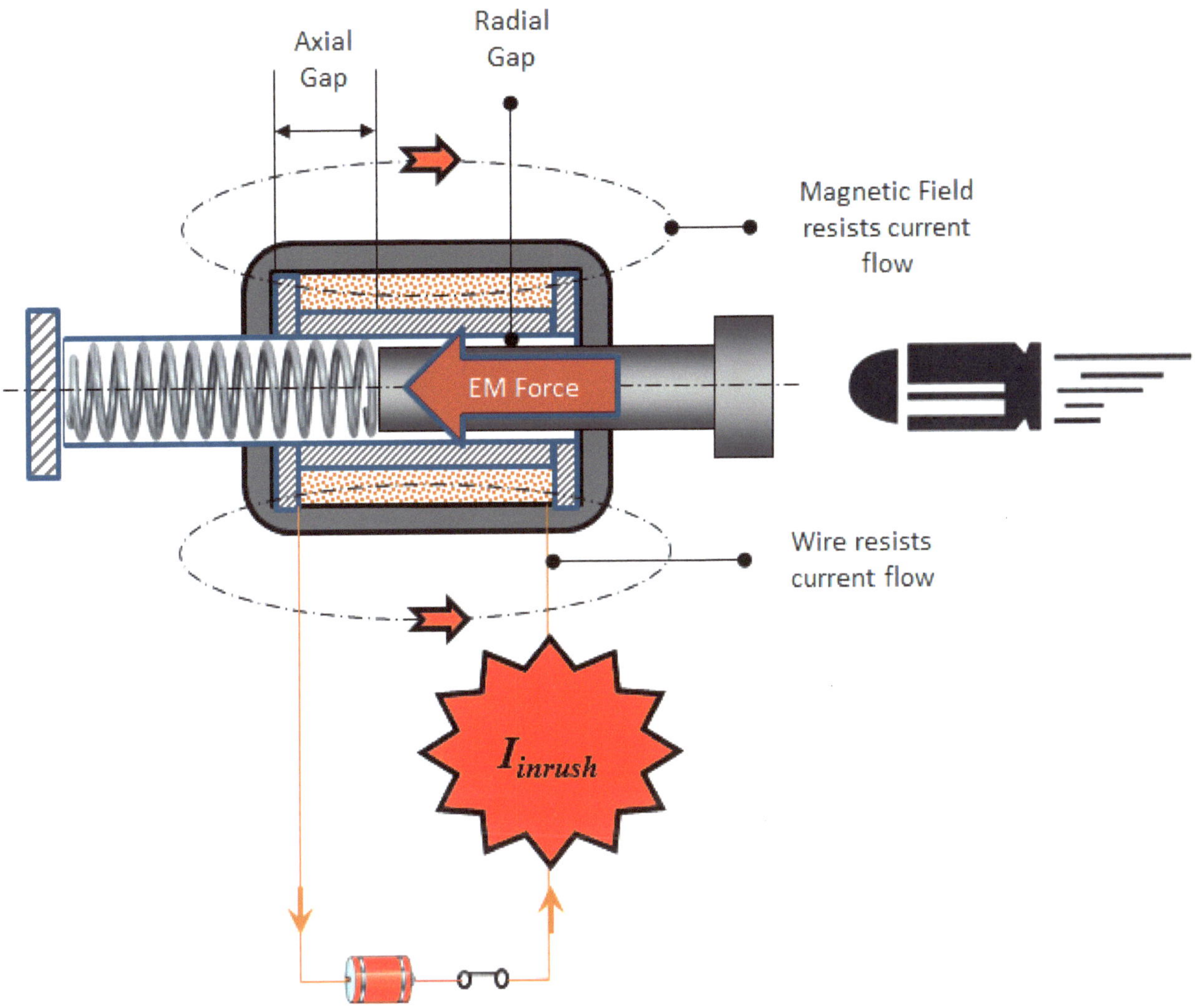

Fig. 3.32 - Inrush Current Development in an AC Switching Solenoid

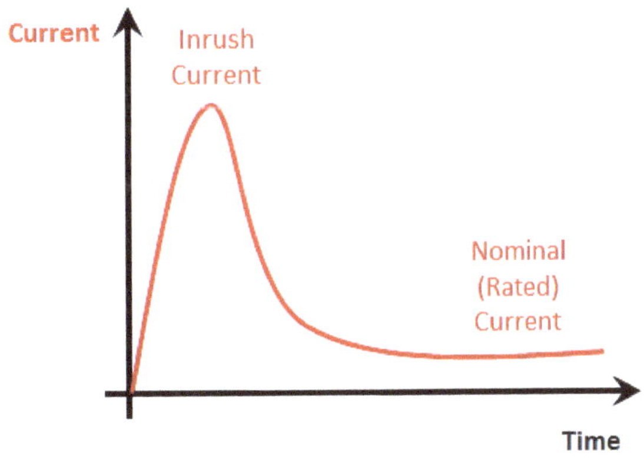

Fig. 3.33 - Inrush Current Characteristics

Consequences of the Inrush Current: In some cases, the armature may not be able to move fully into the coil and close the gap due to some external reasons such as mechanical failure, contamination seizure, and working pressure exceeds the rated value. In such a case, during the early stage of the armature movement where the magnetic field is still weak, the current will continue to rush into the coil. The temperature of the coil will then rise rapidly until it exceeds the rated temperature of the winding insulation. This will eventually create a short circuit across the turns of the coil winding leading to a catastrophic failure of the coil. As shown in Fig 3.34, the resulted heat could melt the insulation and burn the coil in minutes. It also boils the oil around the core in a wet type switching solenoid.

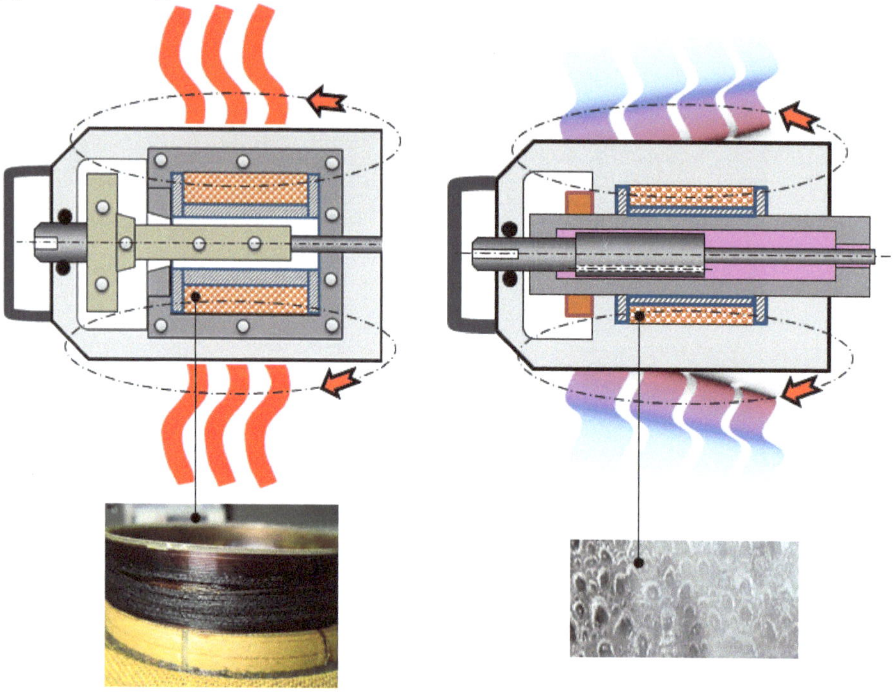

Fig. 3.34 - Coil Overheating Due To Inrush Current

How to minimize the effect of the Inrush Current:

During Valve Design, switching solenoid is designed to:

- Achieve sufficient shifting force to overcome all the opposing forces including the return spring force, the flow force and the viscous friction force.
- Shift the armature in shortest possible time.
- Produce holding force with as little current as possible by designing the coil with sufficient AC resistance which is known as impedance.

During Valve Operation:

- The valve must be energized using the right type of voltage.
- The valve must be maintained periodically to avoid seizing the spool.
- The outside surfaces of the valve as well as the oil must be kept clean as per recommendations.
- In three position valves, avoid energizing opposing solenoids simultaneously.

3.3.5.5- Performance of AC Switching Solenoids for EH Valves

Power Consumption: With its relatively fast buildup of current, an AC switching solenoid is more efficient than a DC switching solenoid. As shown in Equation 3.6, the magnitude of the power of an AC switching solenoid is determined by the product of voltage times the current times the power factor ($\cos\varphi$), where φ is the complex power phase angle.

$$Power\ (Watt) = V\ (Volt) \times I\ (Amp)\ \times \cos\varphi \qquad 3.6$$

Heat Generation: The electrical resistance of an AC coil is lower than the resistance of a DC coil at the same voltage. With AC coils, the current is not only determined by the resistance of the windings, but also by the inductive resistance (reactance). Therefore, an AC switching solenoid generates heat during the inrush current period more than a DC switching solenoid. During the continuous duty operation, similar to the DC switching solenoid, the AC switching solenoid to dissipates the small amount of heat resulting from the low holding current so that the coil temperature does not rise above the maximum rated working temperature.

Switching Time: As it has been explained previously, due to the inrush current phenomenon, the movement of the armature of an AC switching solenoid is faster than a DC switching solenoid. Consequently, the shifting time is shorter than a DC switching solenoid.

Switching Rate: The high inrush current problem is also a limiting factor in the cycling rate of an AC switching solenoid. Despite the shorter switching time of an AC switching solenoid, it can usually be cycled at a lower rate than a DC switching solenoid. The reason is an AC switching solenoid generates more heat by the inrush current so that it has no chance to cool down between cycles. Consider, for instance, a switching solenoid with a 25% duty cycle as shown in Figure 3.35, the area under each energizing pulse represents heat generation in the switching solenoid. So if the heat can be dissipated, there is no problem. Otherwise, if the heat generated by cycling the switching solenoid exceeds its ability to dissipate the heat, the switching solenoid will overheat and fail. For continuous duty operation, the switching solenoid is capable of dissipating the small amount of heat resulting from the low holding current. The inrush current is the real problem.

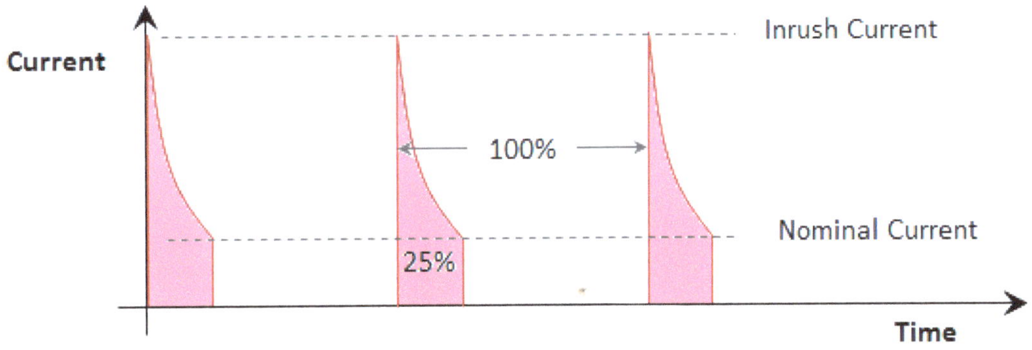

Fig. 3.35 - Switching Rate of an AC Switching solenoid

3.3.5.6- Applications of AC switching Solenoids for EH Valves

AC voltage is very common and readily available to industrial applications. So, AC switching solenoid valves are commonly used in industrial applications.

3.3.6- DC versus AC Switching Solenoid

Table 3.3 presents a qualitative comparison between DC and AC types of switching solenoids.

Item of Comparison	*DC Switching solenoid*	*AC Switching solenoid*
Low Power Consumption and higher Holding Force.	×	✓
Faster Switching Time.	×	✓
Higher Switching Rate.	✓	×
Better Reliability and less possible coil overheating.	✓	×
Less noise generation by the switching solenoid.	✓	×
Tolerance to contamination.	✓	×

✓ Advantage × **Disadvantage**

Table 3.3 - DC versus AC Switching Solenoid

From Table 3.2 and 3.3, it can be concluded that a wet type DC switching solenoid is more popular and recommended to use whenever possible

3.3.7- Electrical Ratings for Switching Solenoids

Nominal Voltages: Switching solenoids are designed for voltages of DC (12V, 24V, and 48V) and AC (110V and 220V). Within limits, DC switching solenoids can also be operated on AC power. Again, this is not recommended by switching solenoid manufacturers. The primary problem here is in the difference in the construction of the two types of switching solenoids. A DC solenoid is not constructed to limit the eddy current and the AC Hum. So, if a DC solenoid operated by AC power, it could be easily burned out.

Nominal Currents: The current absorbed depends on nominal voltage and the size of switching solenoid valves. For EH valves, the current ranges from 1-3 Amps.

Nominal Frequencies for AC Switching solenoid: In the United States, Canada, and some South American countries, the AC power supplies operate at 60 Hz. In most other countries, 50 Hz is used.

Dual-Frequency Switching Solenoids: Ideally, a switching solenoid should be designed and fabricated for a specific frequency. If a coil is rated for use on a 60 Hz supply is used on 50 Hz, it will draw excess current. If a 50 Hz coil is used on a 60 Hz supply, it will draw less than its rated current, producing a lower electromagnetic force, inrush current period is extended, and the coil becomes less reliable.

Most manufacturers develop dual-frequency switching solenoids that are rated for both 50 and 60 Hz systems. Data sheets for such dual-frequency switching solenoids must reflect that.

3.3.8- Manual Override

Optionally, EH valves can be equipped with *Manual Override* feature. This means that, as shown in Fig. 3.36, a little push pin is integrated to the armature so that the armature can be manually forced to move in cases like the following:

- During emergency situations (e.g. power supply failure): the valve spool can be manually shifted to complete the cycle or to release pressure in the valve.
- During system troubleshooting: the valve spool can be shifted manually to check if the problem is from the electrical signal or from a mechanical failure.

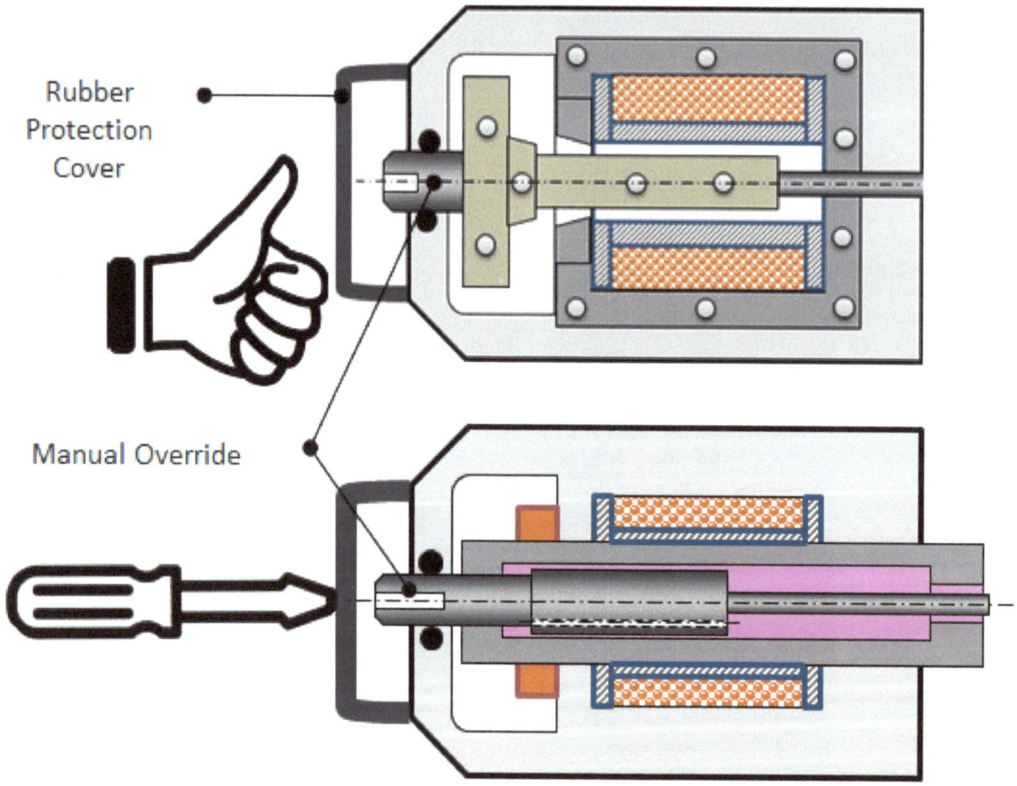

Fig. 3.36 - Manual Override

3.3.9- Switching Solenoids for Cartridge Valves

In addition to spool-type valves, there are *Cartridge-Type* valves. They are also referred to as *Screw-In* valves because they are assembled by screwing them into a manifold. Figure 3.37 shows conceptually the construction of a switching solenoid for *Cartridge-Type Valves*. The figure shows the same elements found in the standard switching solenoid. The figure shows also that the armature in this design consists of two parts. First part is a fixed part and named the *Pole Piece*. The second part is a movable part and named the *Plunger*. The pole piece acts as a magnet when the current is applied to the winding. It must be made of a ferromagnetic material.

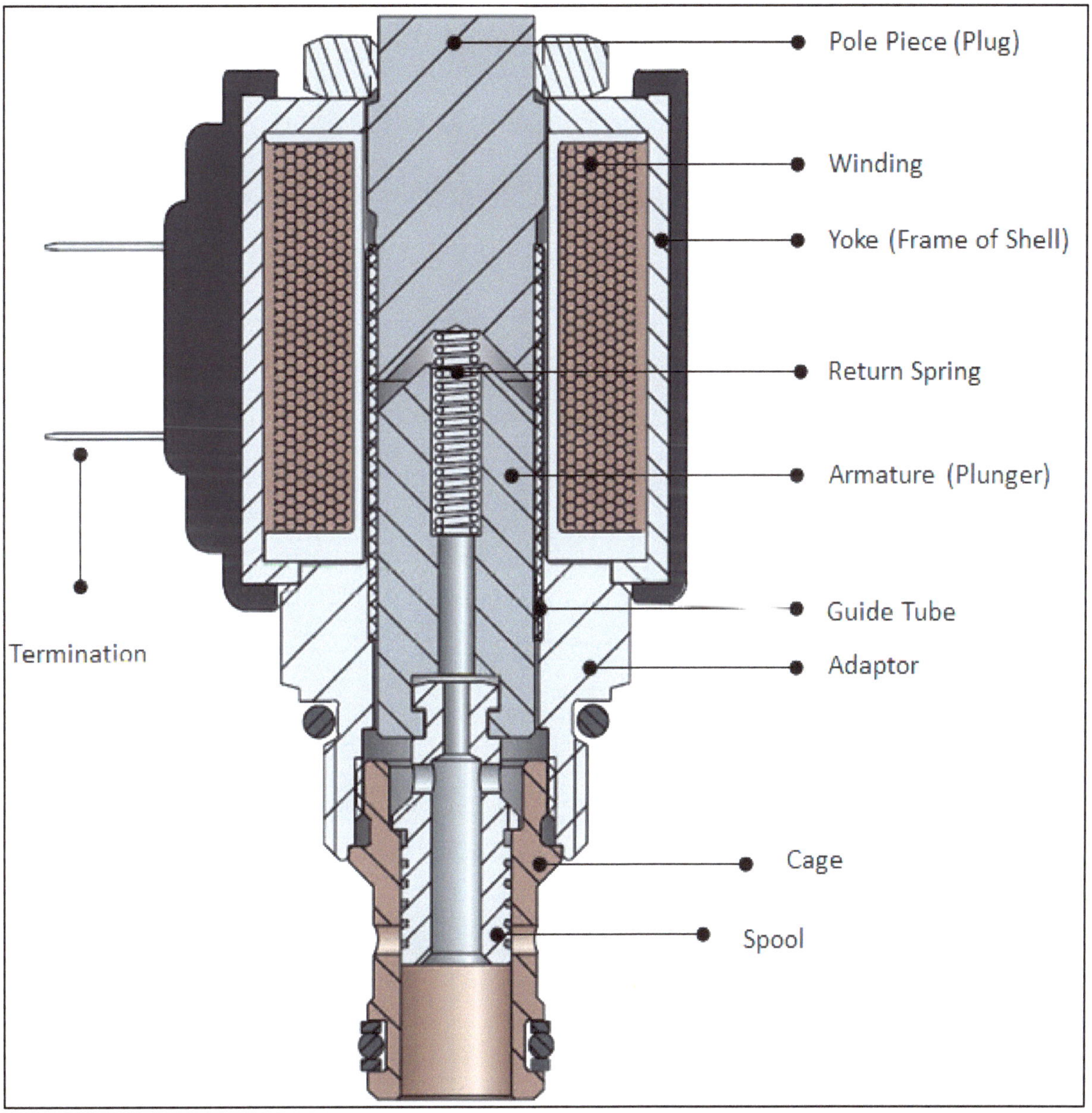

Fig. 3.37 - Switching Solenoid for Cartridge Valves in a De-energized Status (Courtesy of Hydraforce)

3.3.10- Standard Electrical Terminations for Switching Solenoids

The most popular method of electrical termination, particularly for spool type EH valves, is shown in Fig. 3.38. It is the standard DIN 43650 type of connector. The figure shows different styles of the same connector, e.g. with a transparent or opaque body, with led indicator for better system troubleshooting, and with cable holder.

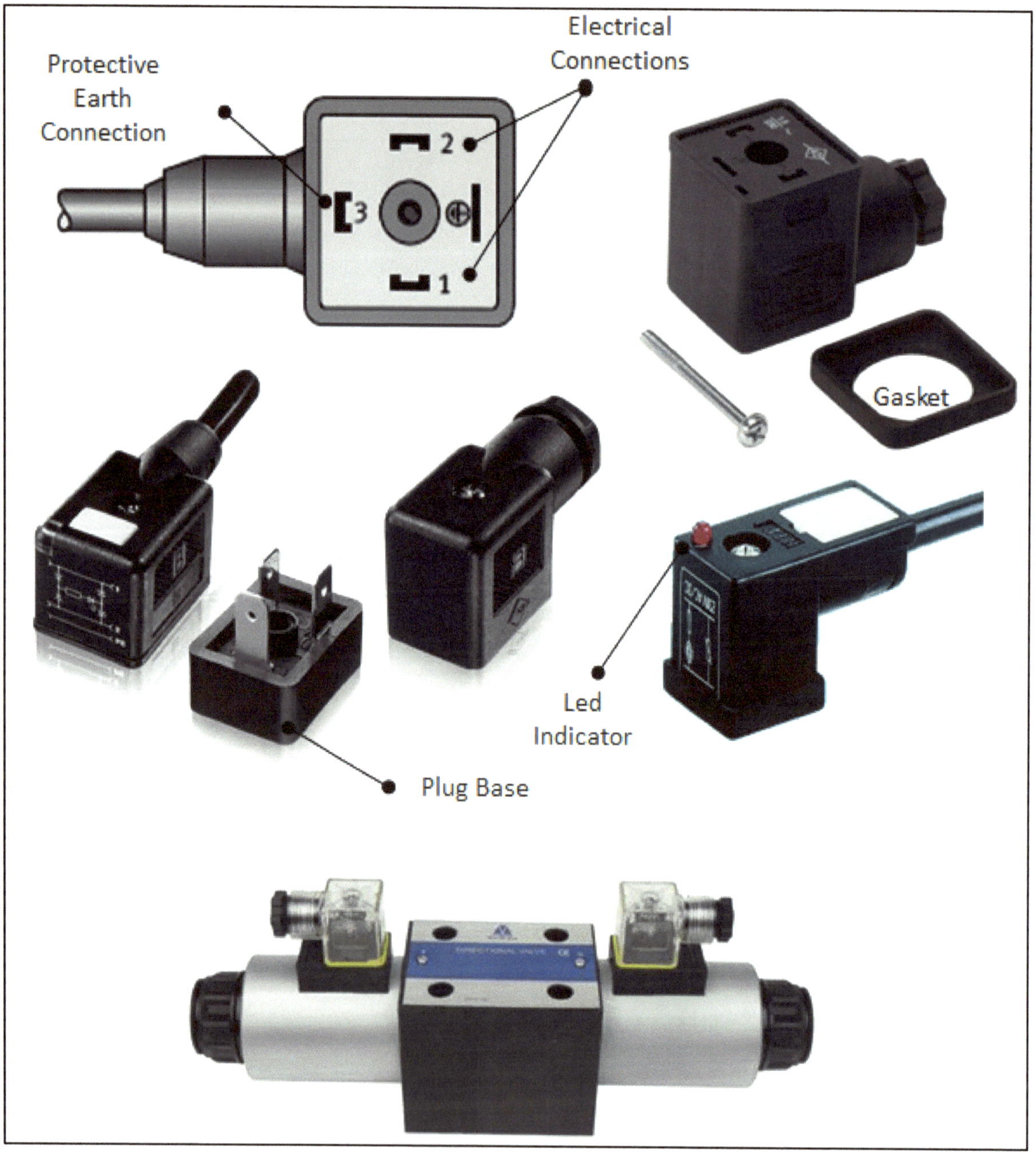

Fig. 3.38 - DIN 43650 Electrical Connector for EH Valves

However, the electrical connectors for EH switching solenoid operated valves can be standardized to one of the connectors shown in Figs. 3.39.

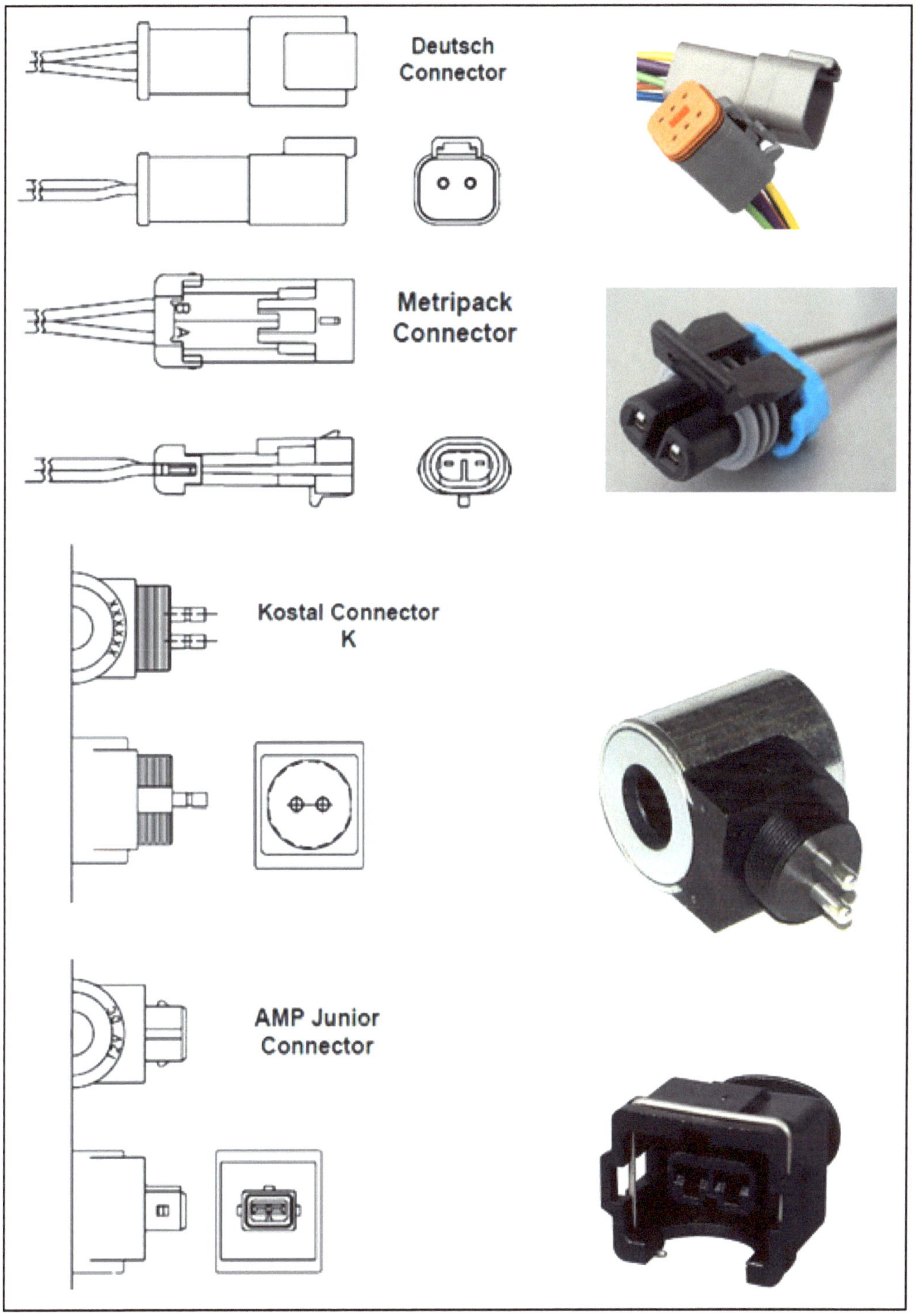

Fig. 3.39 - Different Mating Connectors for Switching Solenoids

3.4- Electro-Hydraulic Switching Directional Control Valves

The set of symbols shown in Fig. 3.40 represent various types of switching solenoid operated spool-type direct-operated directional control valves (DCV) as follows:

1- 2/2, normally-closed, switching solenoid-actuated, spring-returned, direct-operated DCV.
2- 3/2, normally-open, switching solenoid-actuated, spring-returned, direct-operated DCV.
3- 4/2, closed-center, switching solenoid-actuated, spring-centered, direct-operated DCV. An example of conceptual construction of such valve is shown in Fig. 3.41.
4- 4/3, closed-center, switching solenoid-actuated, spring-centered, direct-operated DCV. An example of conceptual construction of such valve is shown in Fig. 3.42.
5- Switching solenoid actuation with manual override.

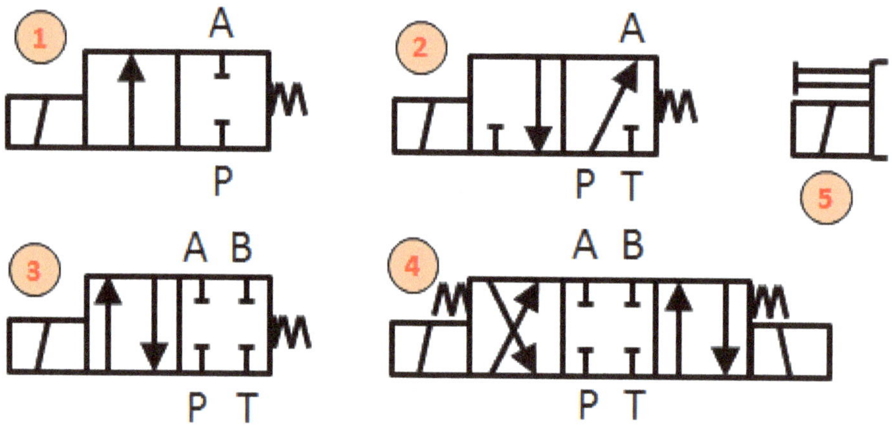

Fig. 3.40 - Symbols of Direct-Operated Spool-Type Solenoid-Actuated Directional Control Valves

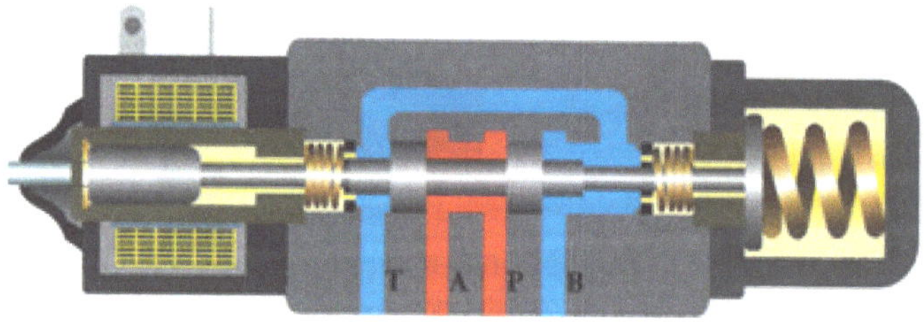

Fig. 3.41 - Direct-Operated 4/2 Spool-Type Solenoid-Actuated DCV (Courtesy of ASSOFLUID)

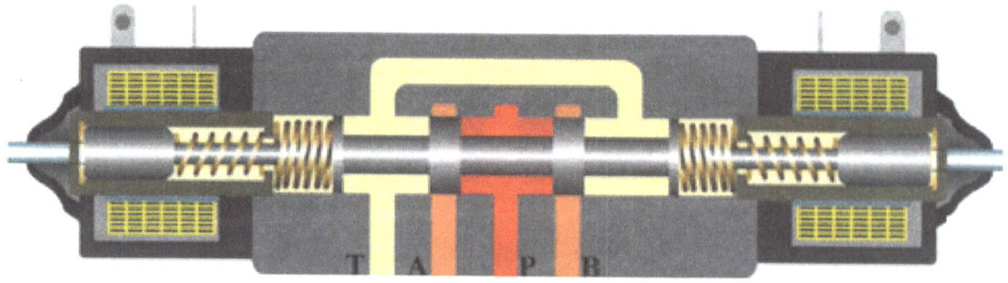

Fig. 3.42 - Direct-Operated 4/3 Spool-Type Solenoid-Actuated DCV (Courtesy of ASSOFLUID)

Figure 3.43 shows a portion of published data for a typical direct-operated 4/3 spool-type solenoid-actuated DCV. The published data confirms the fact that an AC switching solenoid can operated faster (10-20 ms) but cycles at a frequency (7200 cycle per hour = 2 Hz) lower than a DC switching solenoid. 100% duty cycle means solenoid can receive signal on continuous basis.

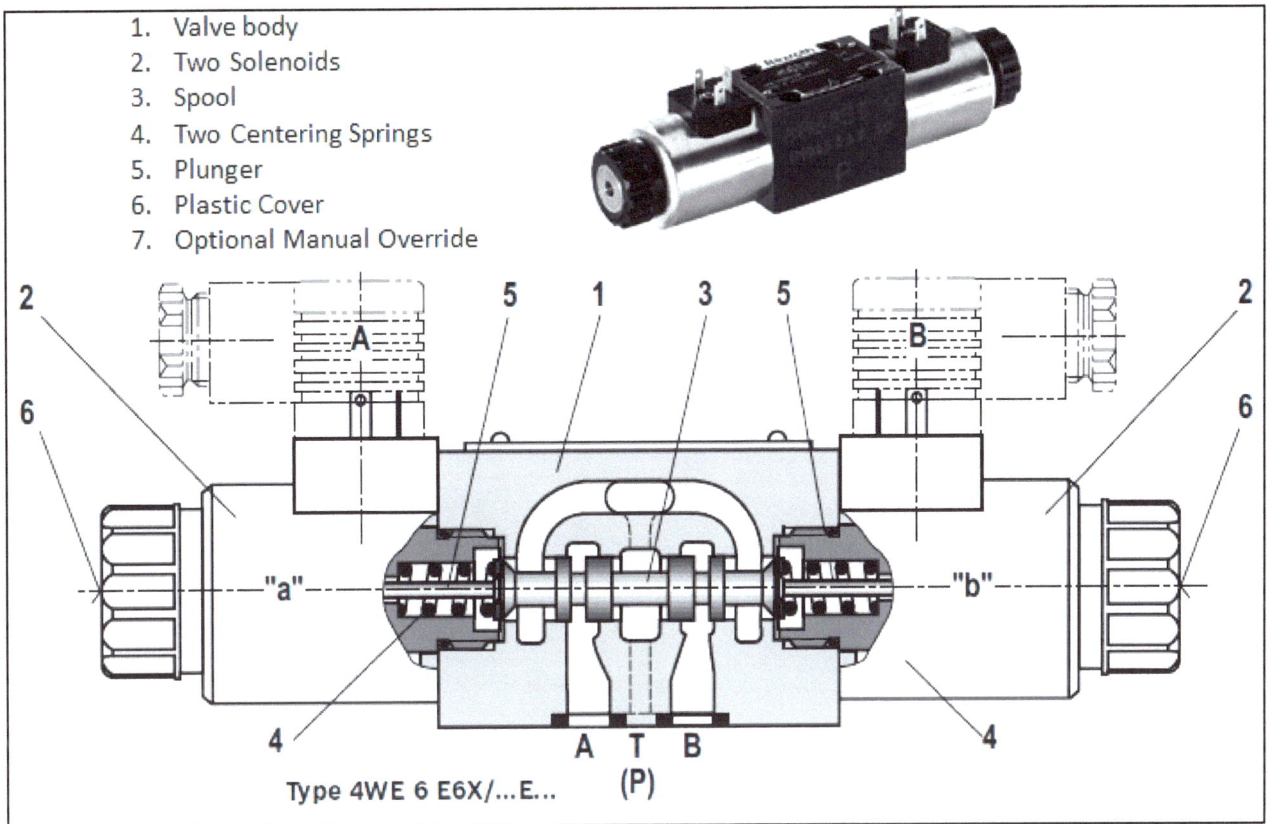

Technical data

electric

Voltage type		Direct voltage	Alternating voltage 50/60 Hz
Available voltages	V	12, 24, 96, 205	110, 230
Voltage tolerance (nominal voltage)	%	±10	
Power consumption	W	30	–
Holding power	VA	–	50
Switch-on power	VA	–	220
Duty cycle	%	100	
Switching time according to ISO 6403 – ON	ms	25 ... 45	10 ... 20
Switching time according to ISO 6403 – OFF	ms	10 ... 25	15 ... 40
Maximum switching frequency	1/h	15000	7200
Maximum surface temperature of the coil [4]	°C [°F]	120 *[248]*	180 *[356]*

Fig. 3.43 - Published Data for a Typical Direct-Operated 4/3 Spool-Type Solenoid-Actuated DCV (Courtesy of Bosch Rexroth)

Figure 3.44 shows conceptually the construction and the detailed symbol of a 4/3, closed center, solenoid-actuated, spring-centered, pilot-operated, direct-controlled DCV. The figure presents the valve when the left switching solenoid is energized. The figure shows the pilot (control) stage is a 4/3, float-center, switching solenoid-actuated, spring-centered DCV. The operation of a pilot-operated DCV has been explained in Volume 1 of this series of Textbooks. The figure shows also that the valve is equipped by a *Switching Time Adjustor*. A pilot-operated DCV drives a large flow and drives a large load. When the pilot stage is energized, the main spool is switched very fast and the main pressure port is connected suddenly to the load port so that the load moves in a jerky way. To ease the connection with the load, two meter out throttle-check valves are used to restrict the vented port of the main spool so that the time it takes to switch between different positions is tuned in the field. Figure 3.45 shows the simplified symbol of a closed-center pilot-operated, direct-controlled DCV. It is to be reminded that the simplified symbol presents the central position of the main stage because the pilot stage is a float-center type.

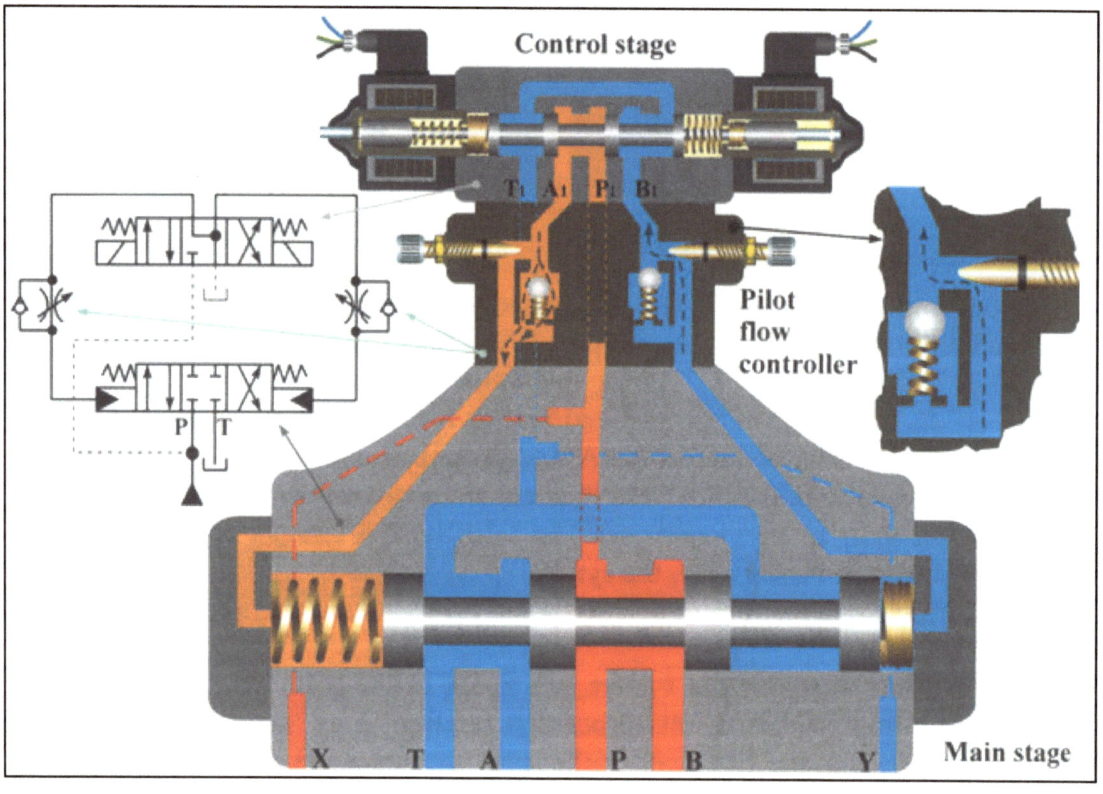

Fig. 3.44 - Pilot-Operated 4/3 Spool-Type Solenoid-Actuated DCV (Courtesy of ASSOFLUID)

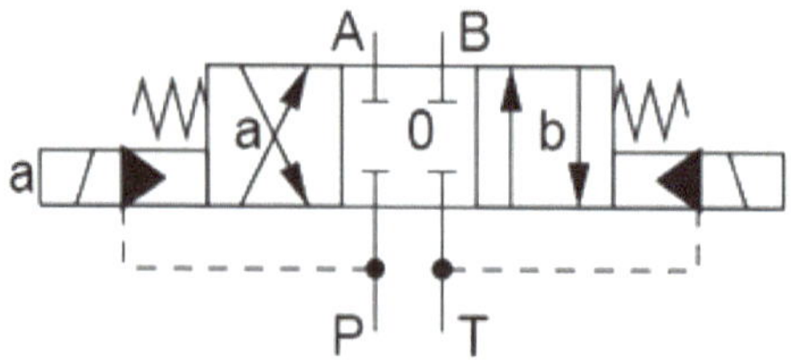

Fig. 3.45 - Simplified Symbol of a Pilot-Operated 4/3 DCV

Figure 3.46 shows a portion of published data for a typical pilot-operated 4/3 spool-type solenoid-actuated DCV. Switching time of such a valve depends on the specifications of the pilot valve, the size of the main valve, and the working pressure. The data sheet reports that the switching time is reduced by 20 ms if AC switching solenoids are used.

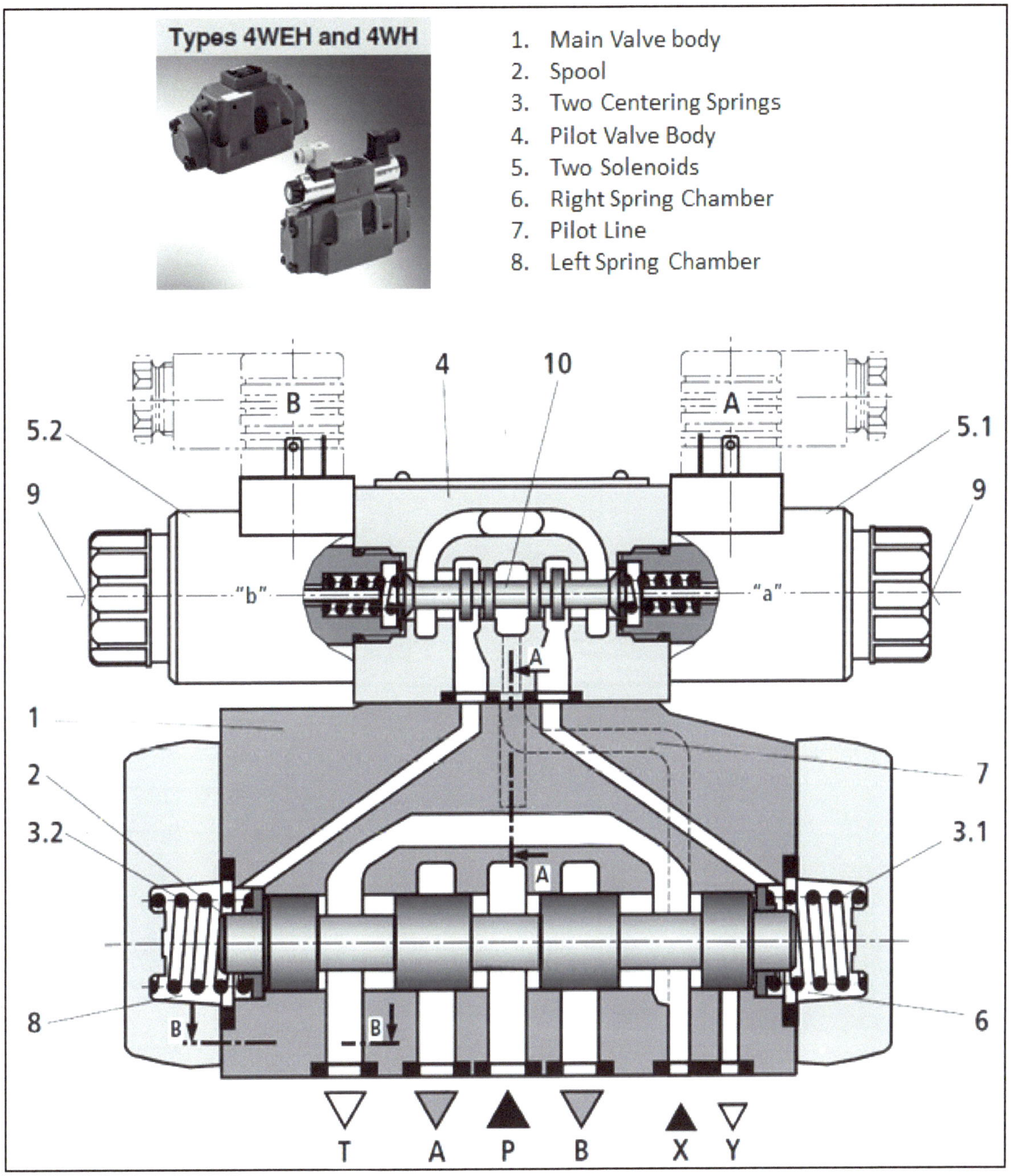

Fig. 3.46 - Cross Sectional View of a Typical Pilot-Operated 4/3 Spool-Type Solenoid-Actuated DCV (Courtesy of Bosch Rexroth)

Figure 3.47 show typical characteristics of a DCV as an example of what is published by the manufacturer to identify the valve performance. The figure shows a family of curves for various spool types. It is to be noted that the manufacturer should declare the conditions under which these characteristics have been developed such as the type of oil and oil temperature.

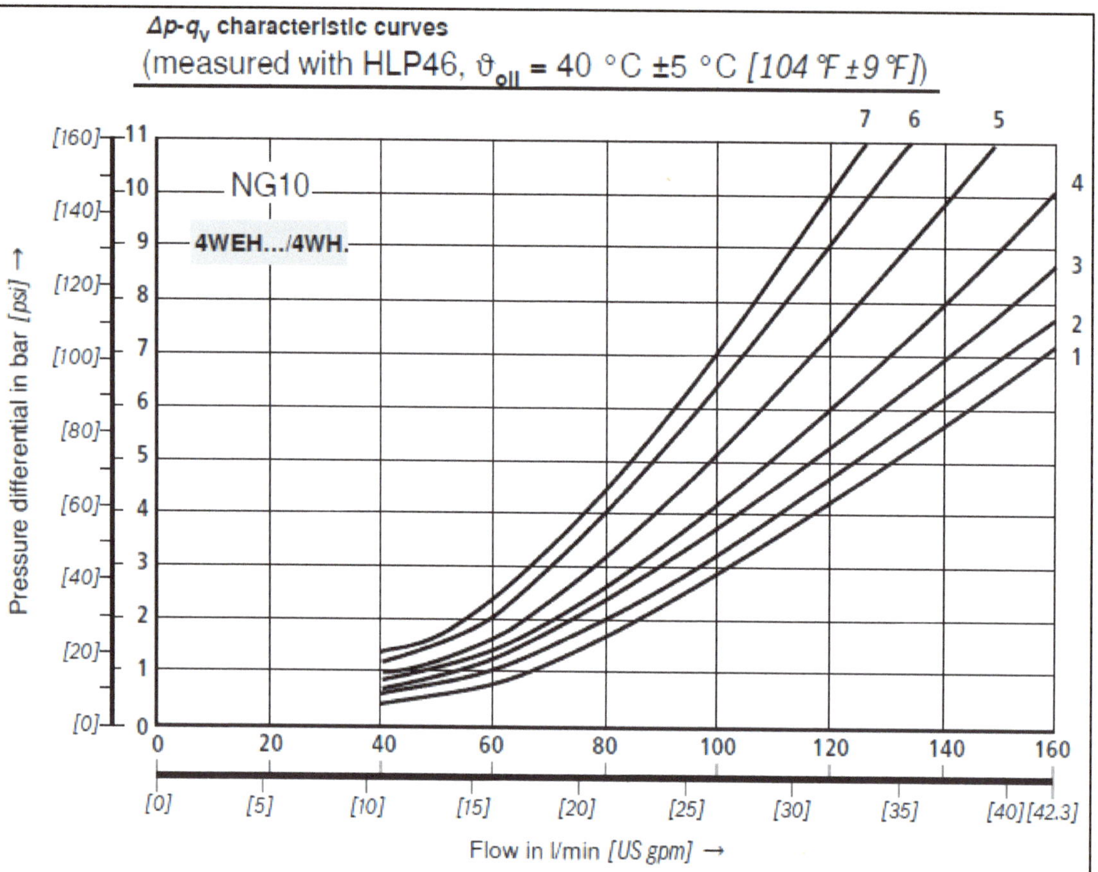

Spool	Spool position			
	P – A	P – B	A – T	B – T
E, Y, D	2	2	4	5
F	1	4	1	4
G, T	4	2	2	6
H, C	4	4	1	4
J, K	1	2	1	3
L	2	3	1	4
M	4	4	3	4
P	4	1	3	4
Q, V, W, Z	2	2	3	5
R	2	2	3	–
U	3	3	3	4
A, B	2	2	–	–

Figure 3.47 - Example of DCV Performance Curves (Courtesy of Bosch Rexroth)

Figure 3.48 shows an example of using switching solenoids to actuate a spool-type control block of a mobile machine. Figure 3.49 shows an example of using switching solenoids to actuate a cartridge-type control block of a mobile machine.

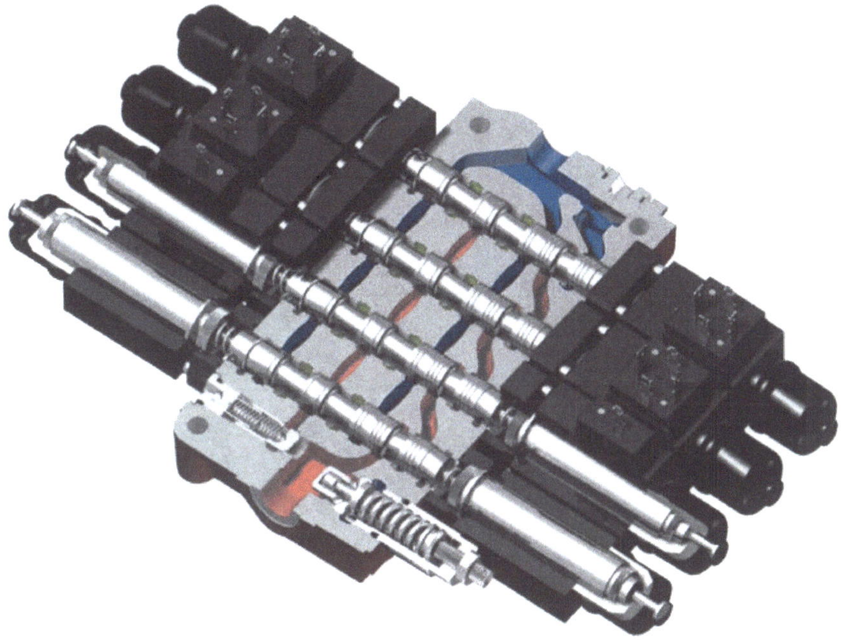

Fig. 3.48 - Switching Solenoid-Actuated Spool-Type Mobile Control Blocks (Courtesy of ASSOFLUID)

Fig. 3.49 - Switching Solenoid-Actuated Cartridge-Type Mobile Control Blocks (Courtesy of Hydraforce)

Figure 3.50 shows a conceptual construction of 2/2 solenoid-actuated cartridge-type DCV. The coil is usually direct current (DC), with a standard voltage of 12, 24, 48 Volts; It demands a particular connector with a circuit rectifier with alternating current (AC) 24, 48, 110,220 Volts. Dissipated power varies from 18 to 33 Watt. According to the size, they can sustain standard pressures up to 210 or 350 bar and flows ranging from 1 to 200 liters per minute. They are available in NO or NC. When the solenoid is energized, it pulls up the core so that the ports communicate accordingly.

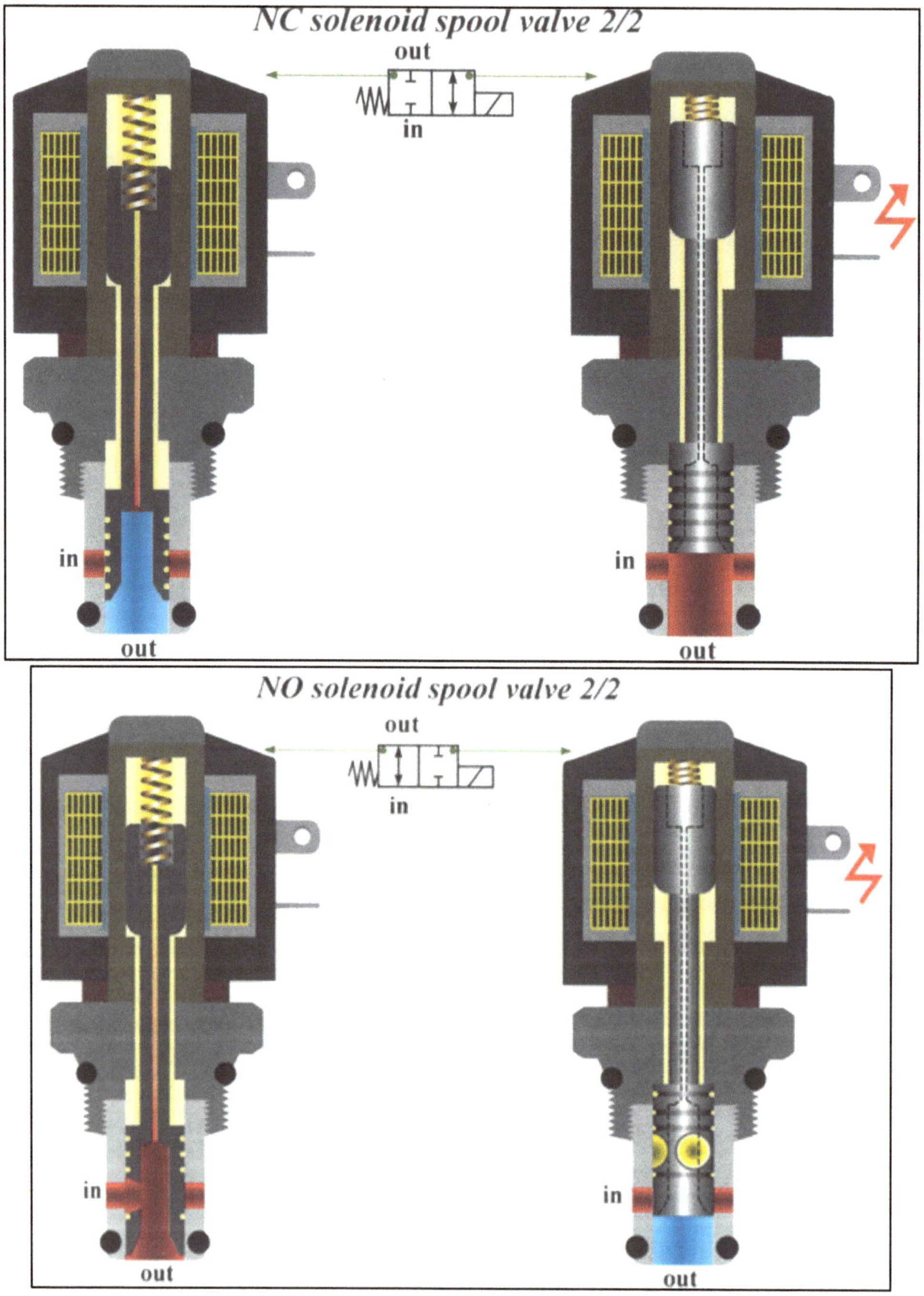

Fig. 3.50 - Direct-Operated 2/2 Cartridge-Type Solenoid-Actuated DCV (Courtesy of ASSOFLUID)

For actuating single-acting cylinders or unidirectional hydraulic motors, a 3/2 valve is used. Figure 3.51 shows a conceptual construction of 3/2 solenoid-actuated cartridge-type DCV.

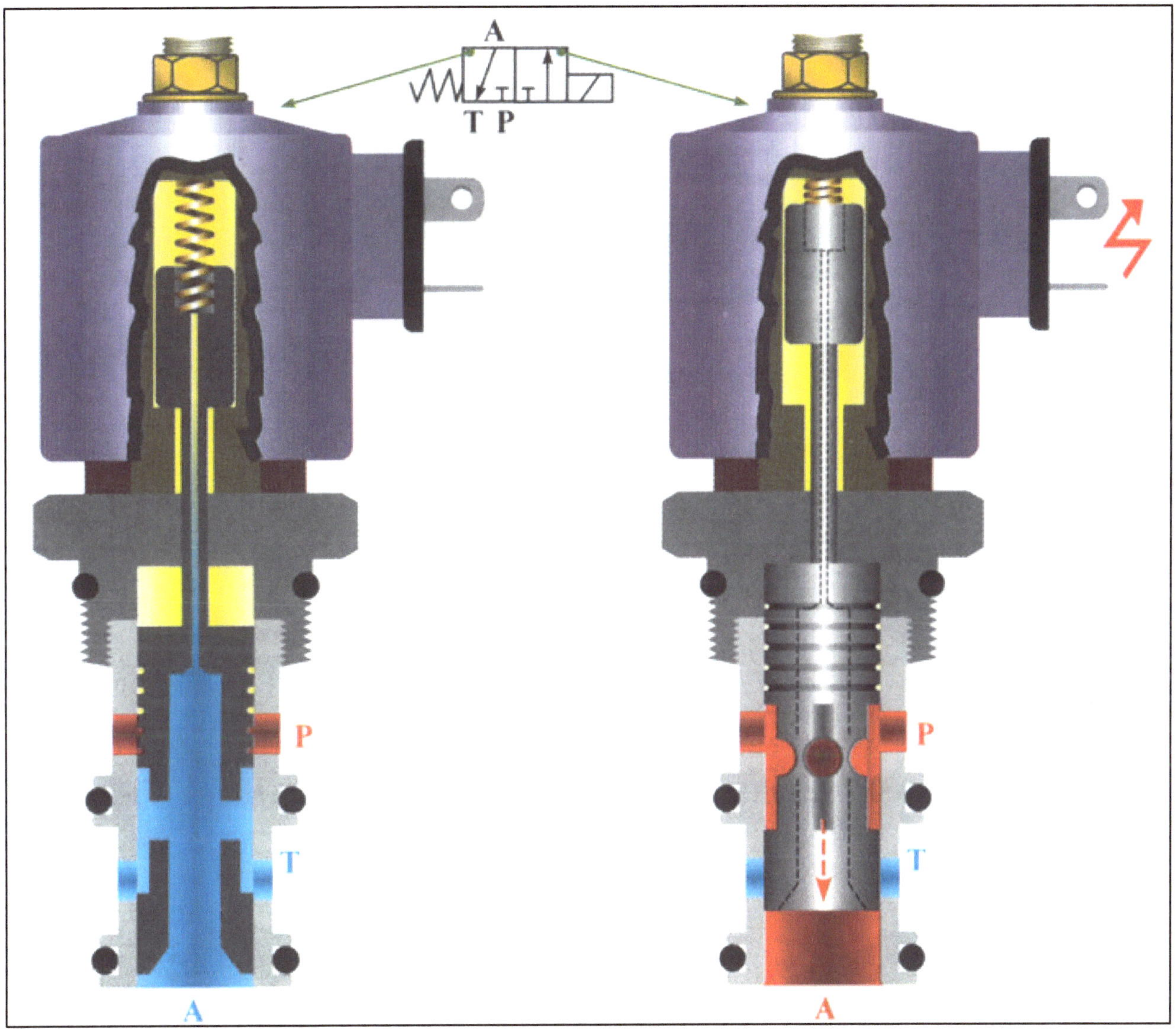

Fig. 3.51 - Direct-Operated Normally-Closed 3/2 Cartridge-Type Solenoid-Actuated DCV (Courtesy of ASSOFLUID)

For actuating double-acting cylinders or bidirectional hydraulic motors, a four-way valve is used. Figure 3.52 shows a conceptual construction of 4/2 solenoid-actuated cartridge-type DCV.

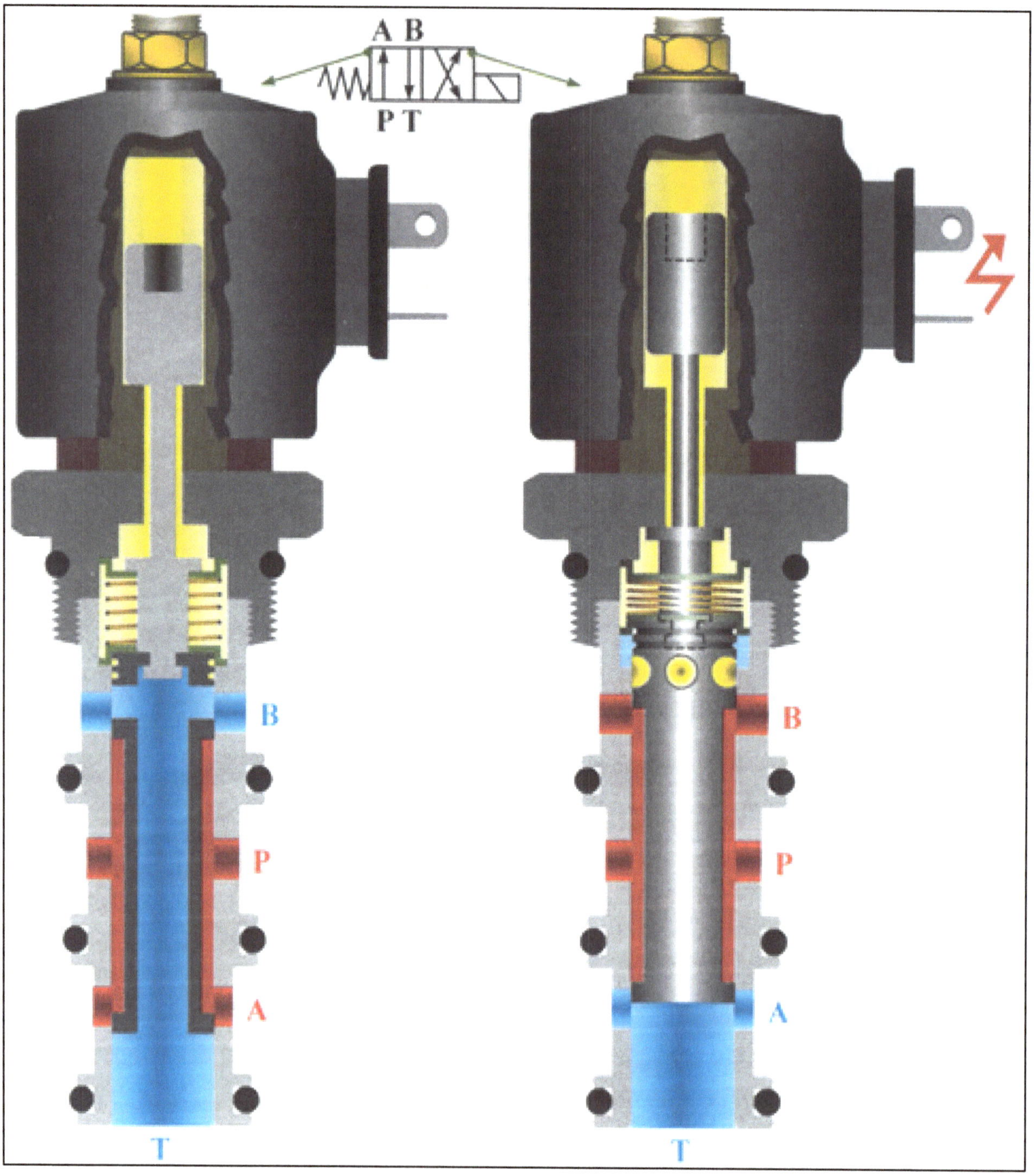

Fig. 3.52 - Direct-Operated 4/2 Cartridge-Type Solenoid-Actuated DCV (Courtesy of ASSOFLUID)

Figure 3.53 shows a conceptual construction of 4/3 solenoid-actuated cartridge-type DCV. Two solenoids are placed end-to-end as shown in the figure.

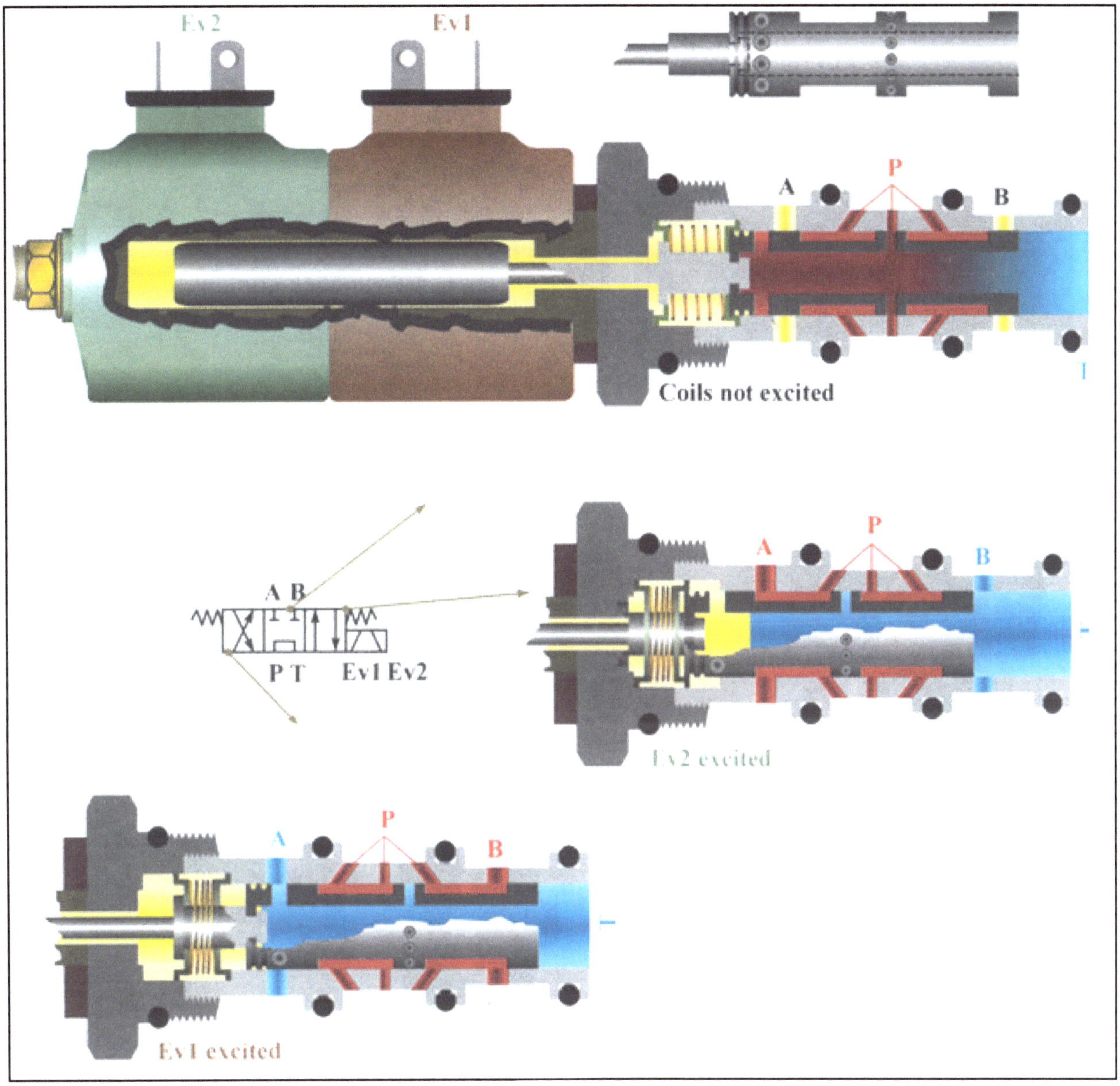

Fig. 3.53 - Direct-Operated 4/3 Cartridge-Type Solenoid-Actuated DCV (Courtesy of ASSOFLUID)

Figure 3.54 shows a portion of published data for a typical direct-operated 4/3 cartridge-type solenoid-actuated DCV. The electrical ratings show that the valve has a continuous duty operation with ±15% voltage tolerance. Such a valve also works under a wide range of temperatures.

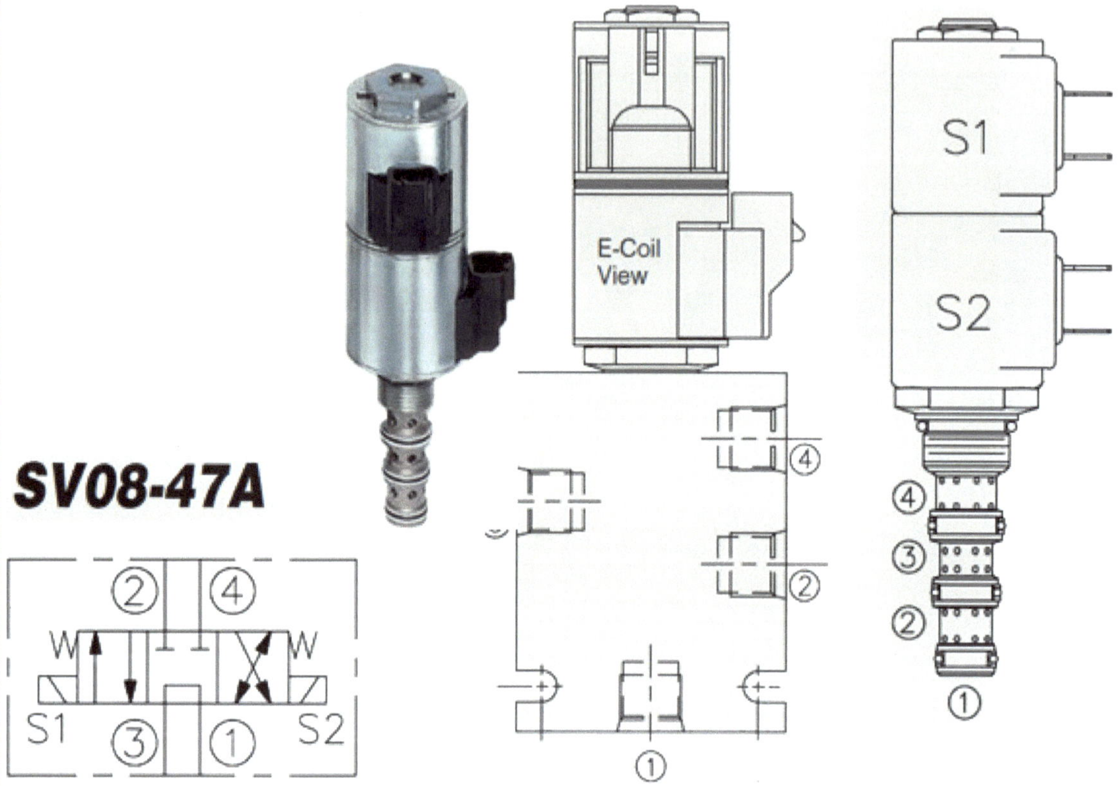

RATINGS

Operating Pressure: 210 bar (3000 psi)

Flow: 11.4 lpm (3.0 gpm) max.

Internal Leakage: 278 cc/minute (17 cu. in./minute) max. at 207 bar (3000 psi)

Temperature: -40 to 120 °C with standard Buna seals

Coil Duty Rating: Continuous from 85% to 115% of nominal voltage

Initial Coil Current Draw at 20 °C: Standard Coil: 1.2 A at 12 Vdc; 0.13 A at 115 Vac (full wave rectified). E-Coil: 1.4 A at 12 Vdc; 0.7 A at 24 Vdc

Minimum Pull-in Voltage: 85% of nominal at 207 bar (3000 psi)

Fig. 3.54 - Solenoid-Actuated Cartridge-Type DCV (Courtesy of Hydraforce)

3.5- Electro-Hydraulic Switching Pressure Control Valves

Figure 3.55 shows an example of using a switching solenoid with a pressure relief valve. The figure shows the use of a 2/2, normally closed, switching solenoid actuated, spring returned, direct operated directional valve (1). If the switching solenoid directional valve (1) is energized, it connects the pressure line (2) to the tank and bypasses the pilot stage (3). The pilot flow (4) is drained to tank at low pressure, and the main poppet (5) opens to unload the pump.

The initial position of the directional valve can to be selected as normally-open (unloading feature is activated) or normally-closed (unloading feature is deactivated) for two reasons. First is to minimize the period of time for which the switching solenoid is energized during a duty cycle of a machine operation. Second is to select specifically normally-open valve to unload the pump in case of power outage as part of fail-safe design strategy.

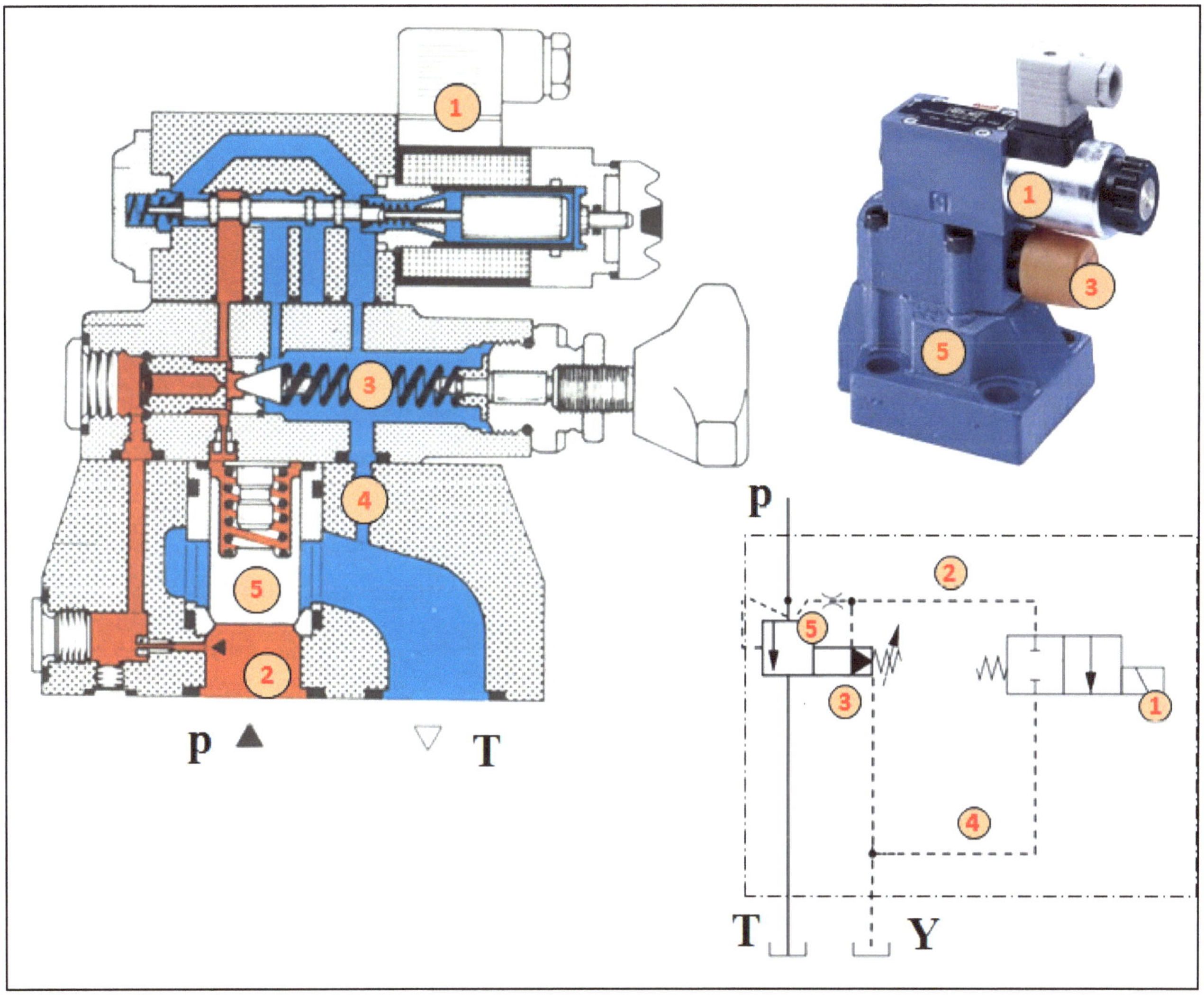

Fig. 3.55 - Pilot Operated Pressure Relief Valve with Unloading Feature (Courtesy of Bosch Rexroth)

Figure 3.56 shows typical characteristics of a switching pressure relief valve. The figure shows a family of curves for various spool types.

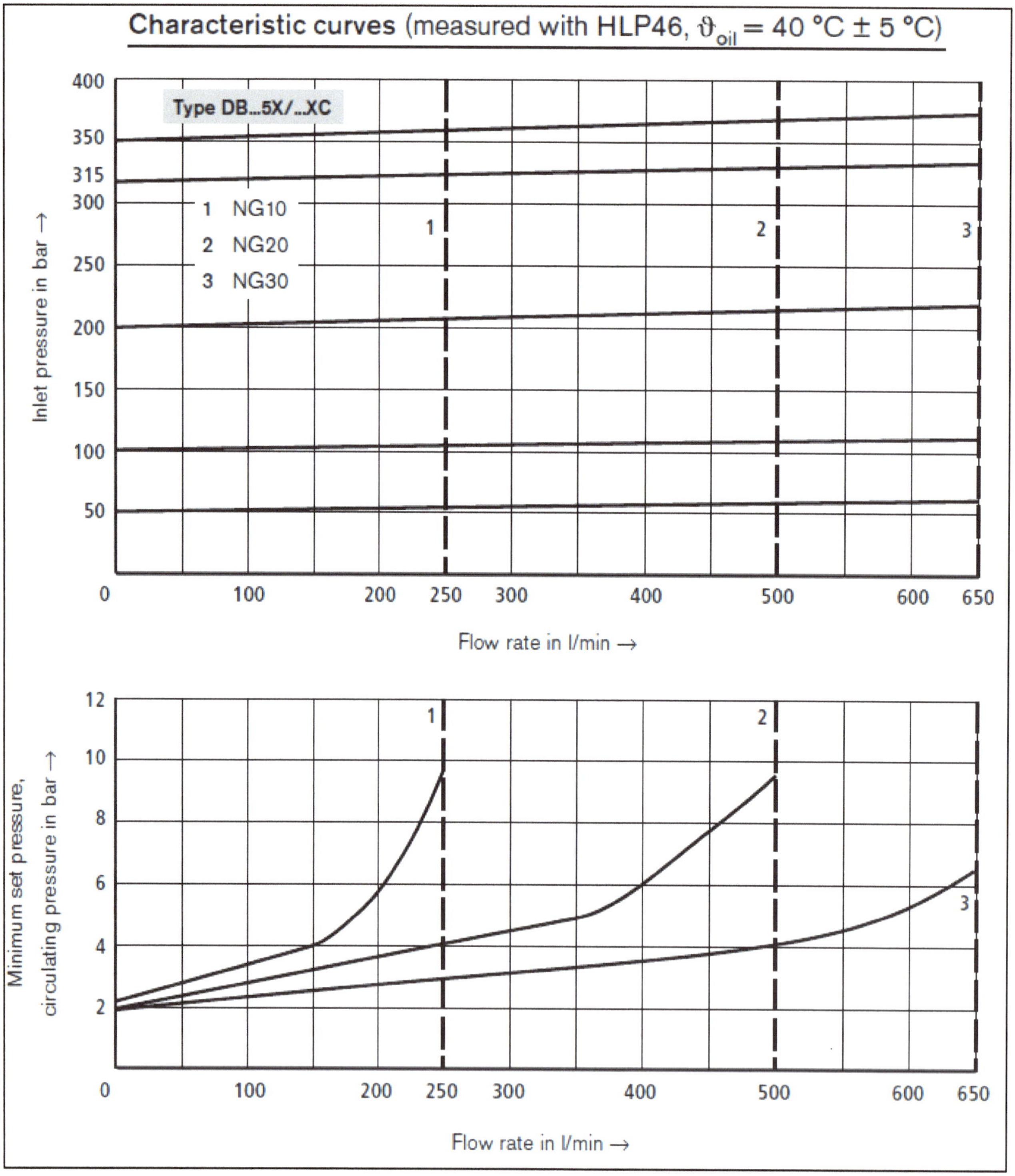

Fig. 3.56 - Example of PRV Performance Curves (Courtesy of Bosch Rexroth)

3.6- Electro-Hydraulic Switching Flow Control Valves

A switching solenoid can also be seen as integrated with a flow control function as shown in Fig. 3.57. It can be integrated with spool type flow control valve to choose between fully closed orifice and metered orifice (1). It can be used to convert the manually-operated *Shut-Off* valve into a switching solenoid-actuated one (2).

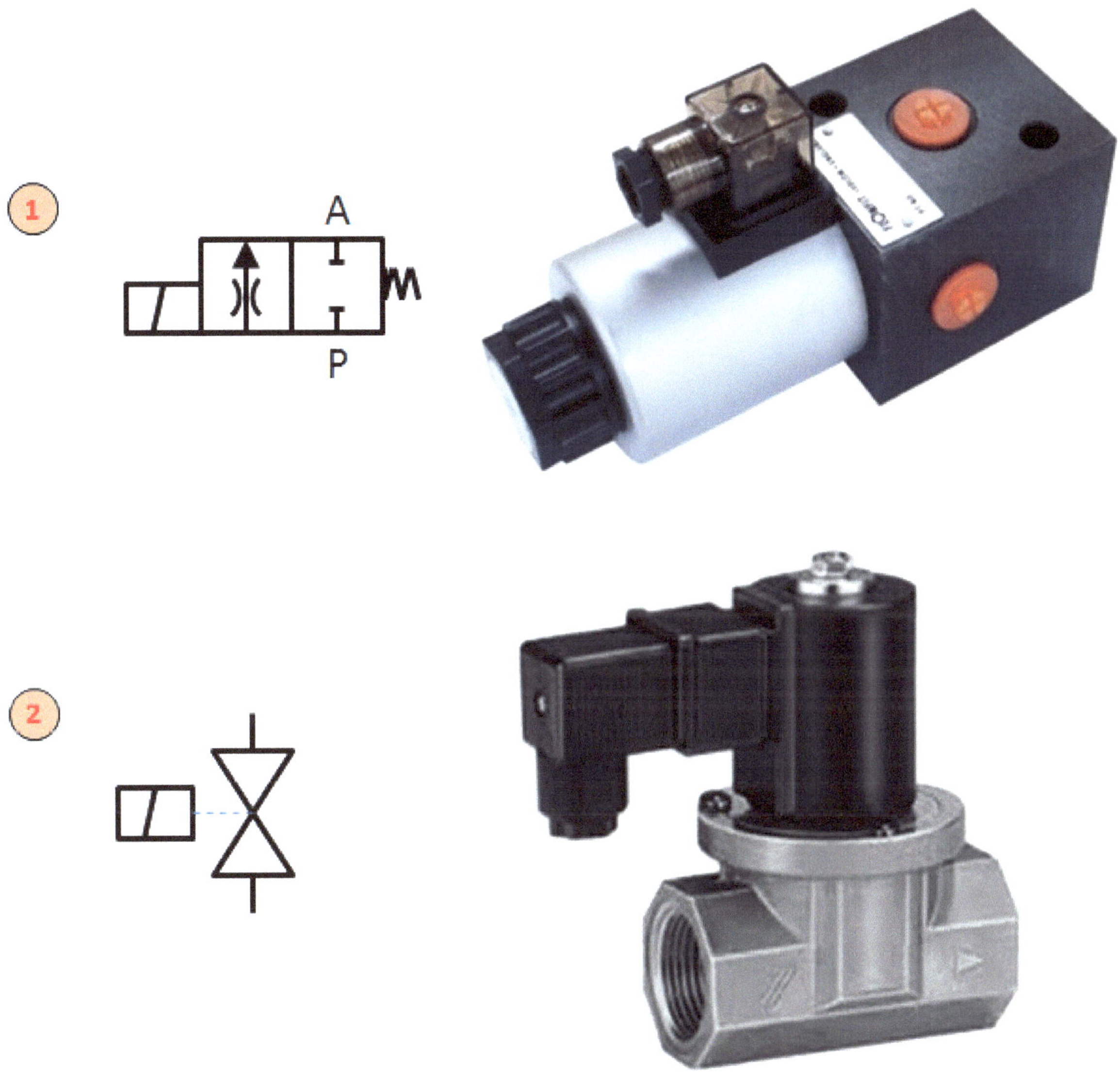

Fig. 3.57 - Use of a Switching Solenoid for a Flow Control Function

Chapter 4

Electrical Circuits for Switching Valves

Objectives:

This chapter covers the basic safety precautions that must be considered when building an electrohydraulic circuit that drives a switching valve. The chapter also presents the electrical devices and their symbols that are most commonly used with switching valves including: switches, buttons, relays, and PLCs. The chapter also covers the rules to read and write various forms of electrical circuit diagrams including: Joint Industrial Council (JIC) schematic diagrams, wiring diagram, and sequence diagram. The chapter concludes with various electrohydraulic circuits that simulate typical applications.

Brief Contents:

Chapter 4 - Electrical Circuits for Switching Valves

4.1- Best Practices for Safe Operation of Electro-Hydraulic Systems

In order to drive the electro-hydraulic valves safely, this section presents the basic and the most common safety design tips that must be considered in building the associated electrical circuits.

Emergency Switch: It is very important to consider having an emergency switch located in an accessible and visible place in the machine as shown in Fig. 4.1. The head of the emergency switch should be colored in red as predetermined by ISO standard.

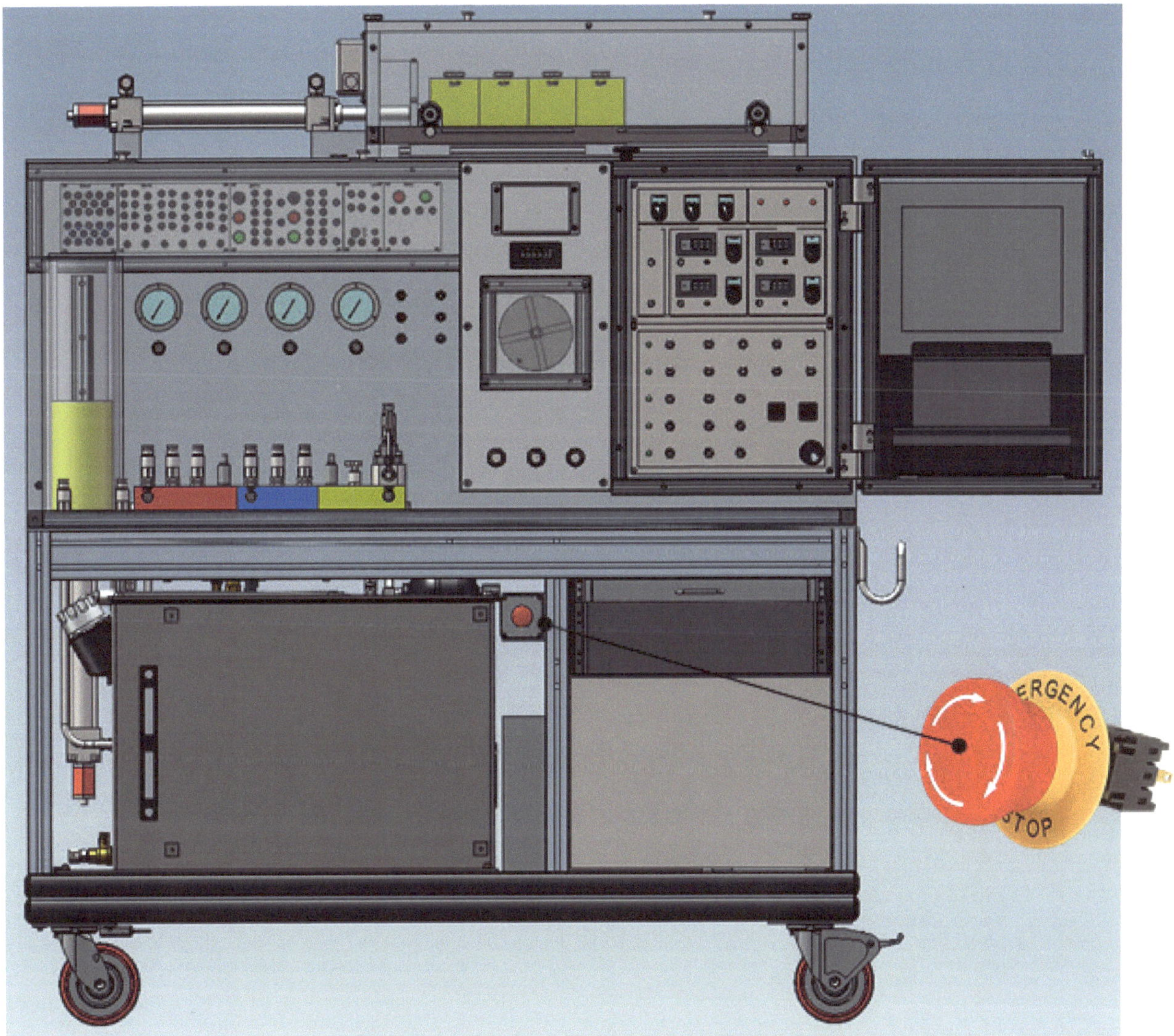

Fig. 4.1 - Use of Emergency Switches for Electro-Hydraulic Systems

Electrical Cable Shielding: As shown in Fig. 4.2, electrical cables that transmit signals to EH valves must be shielded in order to avoid possible sparking and to minimize the effect of surrounding noise on the transmitted circuits.

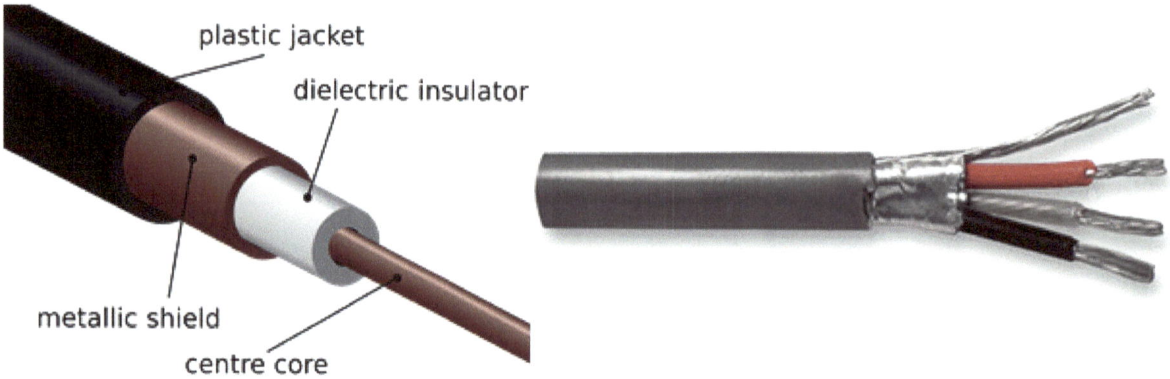

Fig. 4.2 - Cable Shielding

Electrical Cable Routing: Electrical cables must be routed properly to avoid intersection with hydraulic lines. Figure 4.3 shows various types of cable clamps, fixtures, clips and ducts used for proper electrical cables routing.

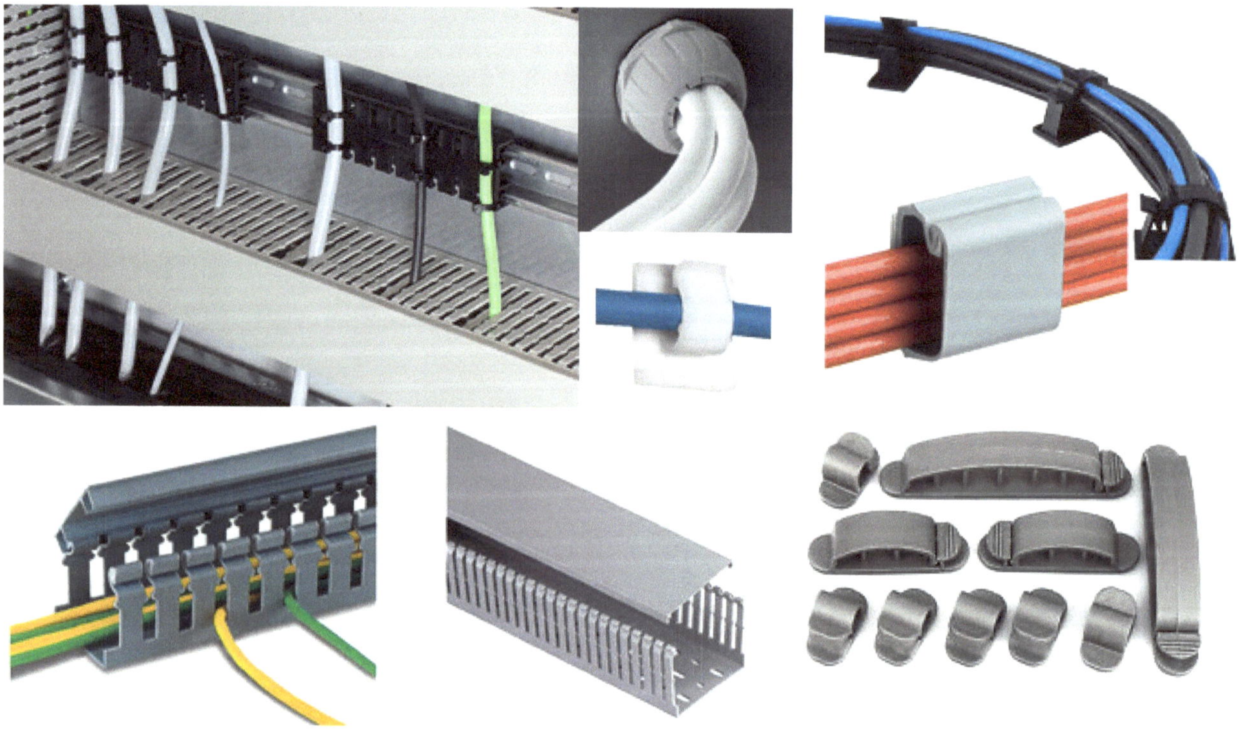

Fig. 4.3 - Accessories for Cable Routing

Standard Electrical Diagrams: Work based on standard schematic diagrams. Figure 4.4 shows an example of *Joint Industrial Council* (JIC) functional schematic diagrams as it will be explained later.

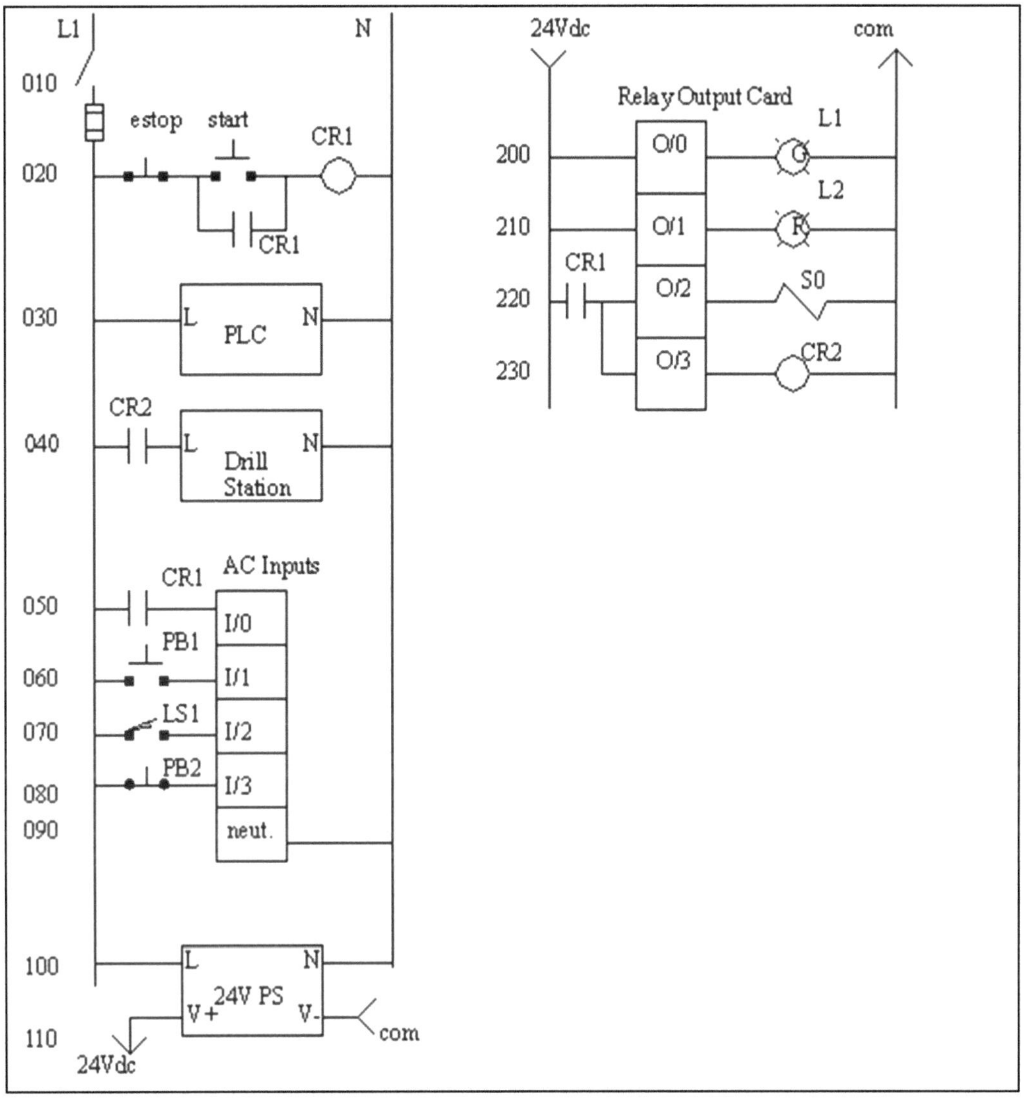

Fig. 4.4 - Standard JIC Electrical Diagrams (http://engineeronadisk.com)

Avoid Short Circuiting: As shown in Fig 4.5, short circuit means failure in a circuit caused by accidental contact between two points. Avoid short circuiting by making sure the wires are properly covered and secure, not loosely connected to a terminal box.

Fig. 4.5 - Hazard of Short Circuit

Proper Grounding: As shown in Fig. 4.6, electrical circuits must be properly grounded so that no power will be transmitted through the body of the machine operator.

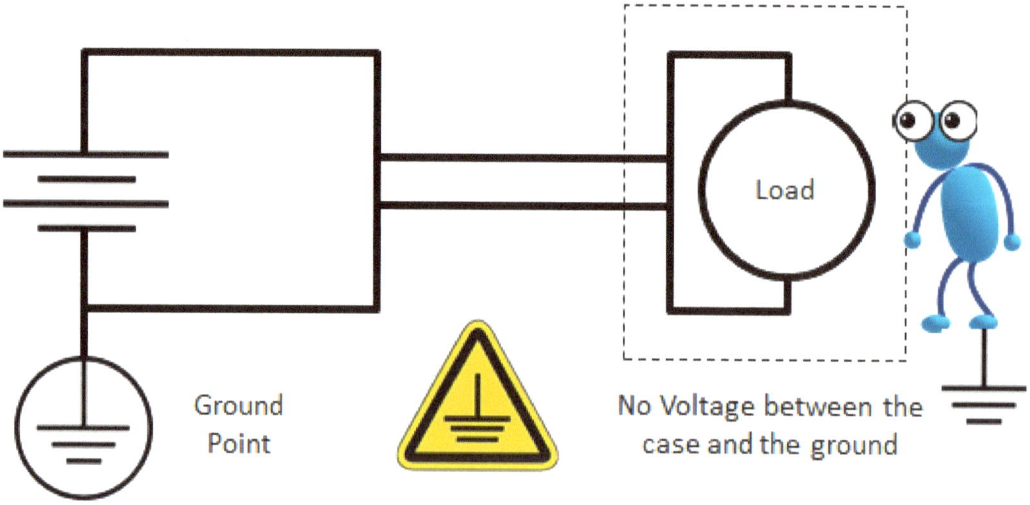

Fig. 4.6 - Proper Grounding of an Electrical Circuit

Proper Powering of Cable Ends: As shown in Fig. 4.7, cables have male and female ends. Never power the male ends to avoid electrical shocks. Female receptacles are powered for operator safety.

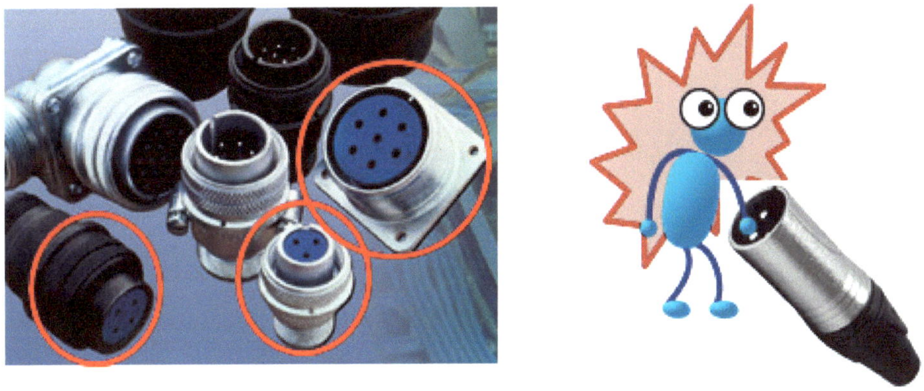

Fig. 4.7 - Powering of Cable Ends

Overload Protection: an electrical circuit that drives an EH valve must be protected based on the maximum current of the valve. There are many methods for open load protection. The most popular ones are fuses or circuit breakers as shown in Fig. 4.8.

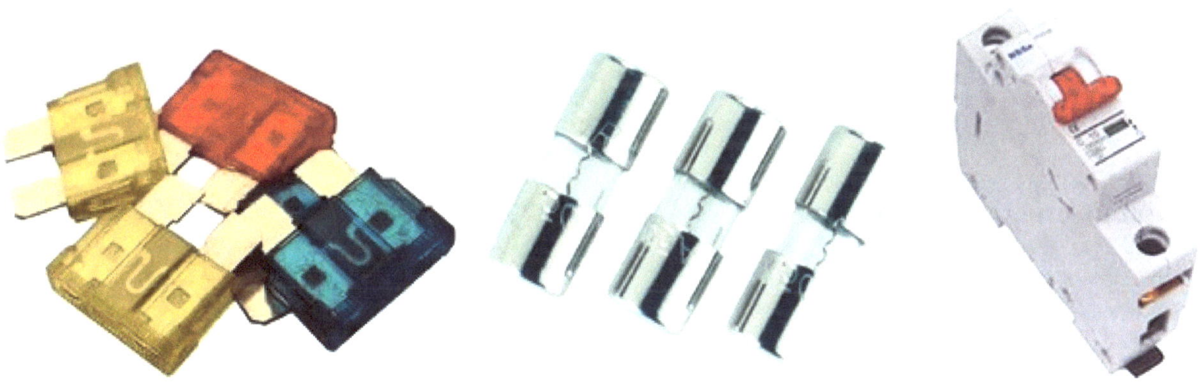

Blade Type Fuse | Cylindrical Glass Type Fuse | Circuit Breaker

Fig. 4.8 - Methods for Overload Protection

4.2- Basic Electrical Symbols

Figure 4.9 shows some symbols for electrical devices, selected from JIC standard, that are commonly used in electro-hydraulic systems.

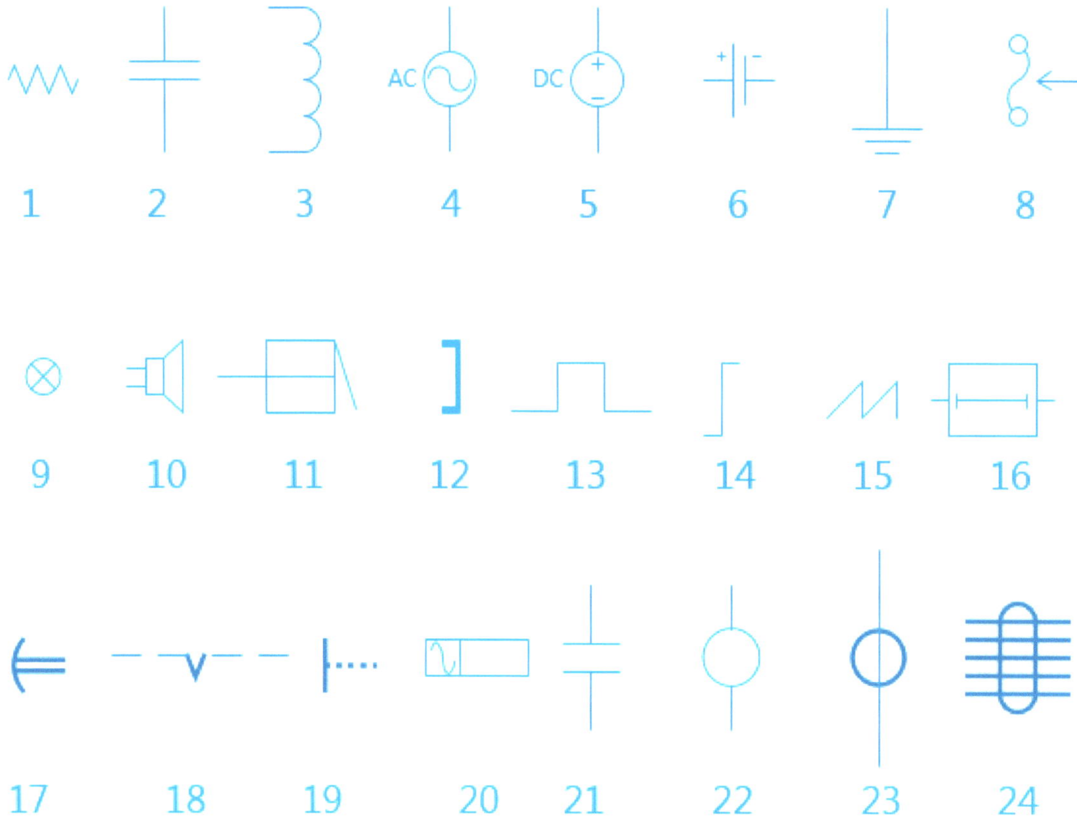

1	Resistor	9	Light Indicator	17	Delay Action
2	Capacitor	10	Speaker	18	Detent (Mechanical Latching)
3	Inductor	11	Buzzing	19	Manual Control
4	AC Source	12	Permanent Magnet	20	Relay
5	DC Source	13	Pulse	21	NO Relay Contactor
6	Battery	14	Step Function	22	Relay Coil
7	Ground Connection	15	Saw Tooth	23	Line (Cable)
8	Fuse	16	Delay Element	24	Cable Group

Fig. 4.9 - Symbols for Electrical Devices Commonly Used in EH Systems

The term *Switch* means a device that produces a digital (binary) signal like 0 or 1, TRUE or False, etc. The term *Sensor* or *Transducer* is a device that produces an analog (continuous) signal. While switches are used to drive switching (ON/OFF) valves, sensors are used with continuous (proportional and servo) valves.

Figure 4.10 shows some JIC "Joint Industrial Council" selected symbols for switches and relays that are commonly used in electro-hydraulic control systems. Other standards are found such as NEMA "National Electrical Manufacturer Association".

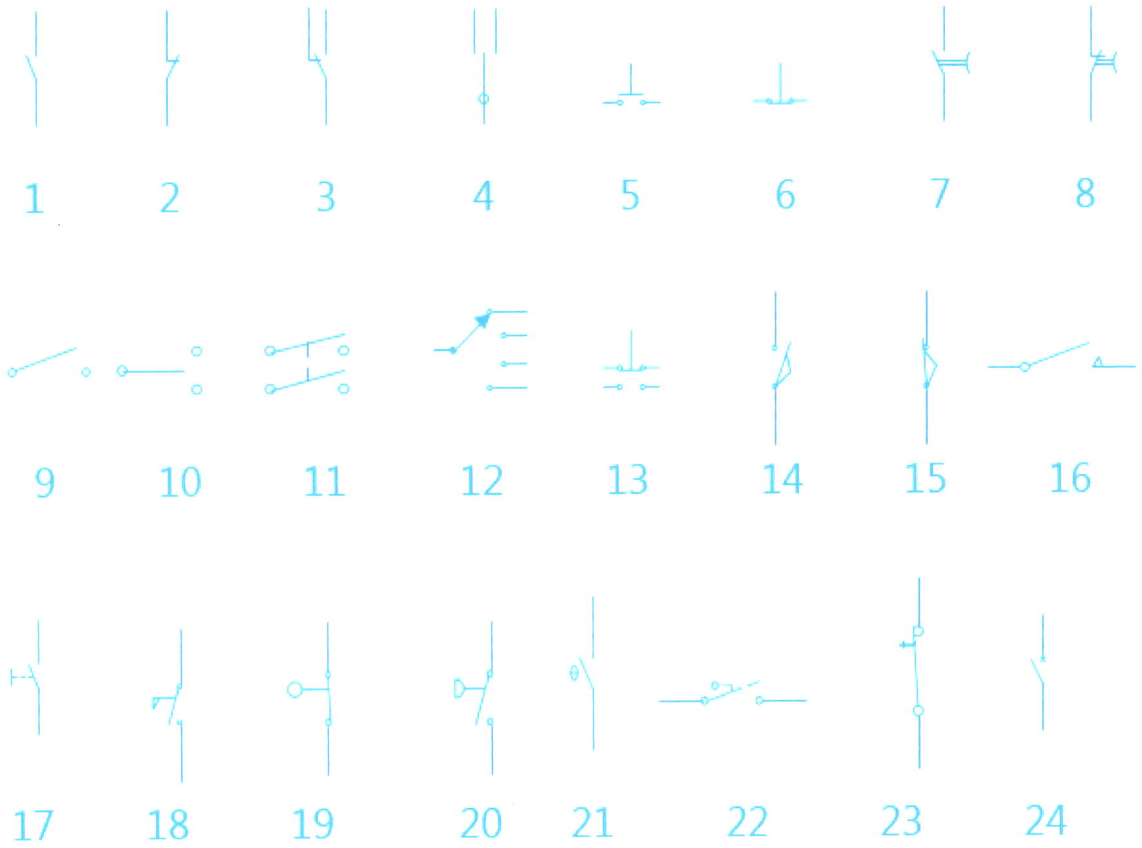

1	Make Contact	9	SPST (Single Pole Single Throw)	17	Manual Switch
2	Break Contact	10	SPDT (Single Pole Double Throw)	18	Flow Switch
3	2-Way Contact	11	DPST (Double Pole Single Throw)	19	Liquid Level Switch
4	Change-Over Contact	12	Selector Switch	20	Pressure Switch
5	Make Pushbutton	13	2-Circuits Pushbutton	21	Temperature Switch
6	Break Pushbutton	14	Limit Switch NO	22	Inertia Switch
7	Make Time Delay	15	Limit Switch NC	23	Thermostat
8	Break time Delay	16	Spring Return	24	Circuit Breaker

Fig. 4.10 - Symbols for Switches and Relays Commonly Used for EH Systems

4.3- Basic Electrical Devices

4.3.1- Measuring Instruments

Figure 4.11 shows some selected panel mounted electrical measurement devices. A Galvanometer is used to measure very small currents in the order of milliamps. All measuring gauges are also found in digital forms.

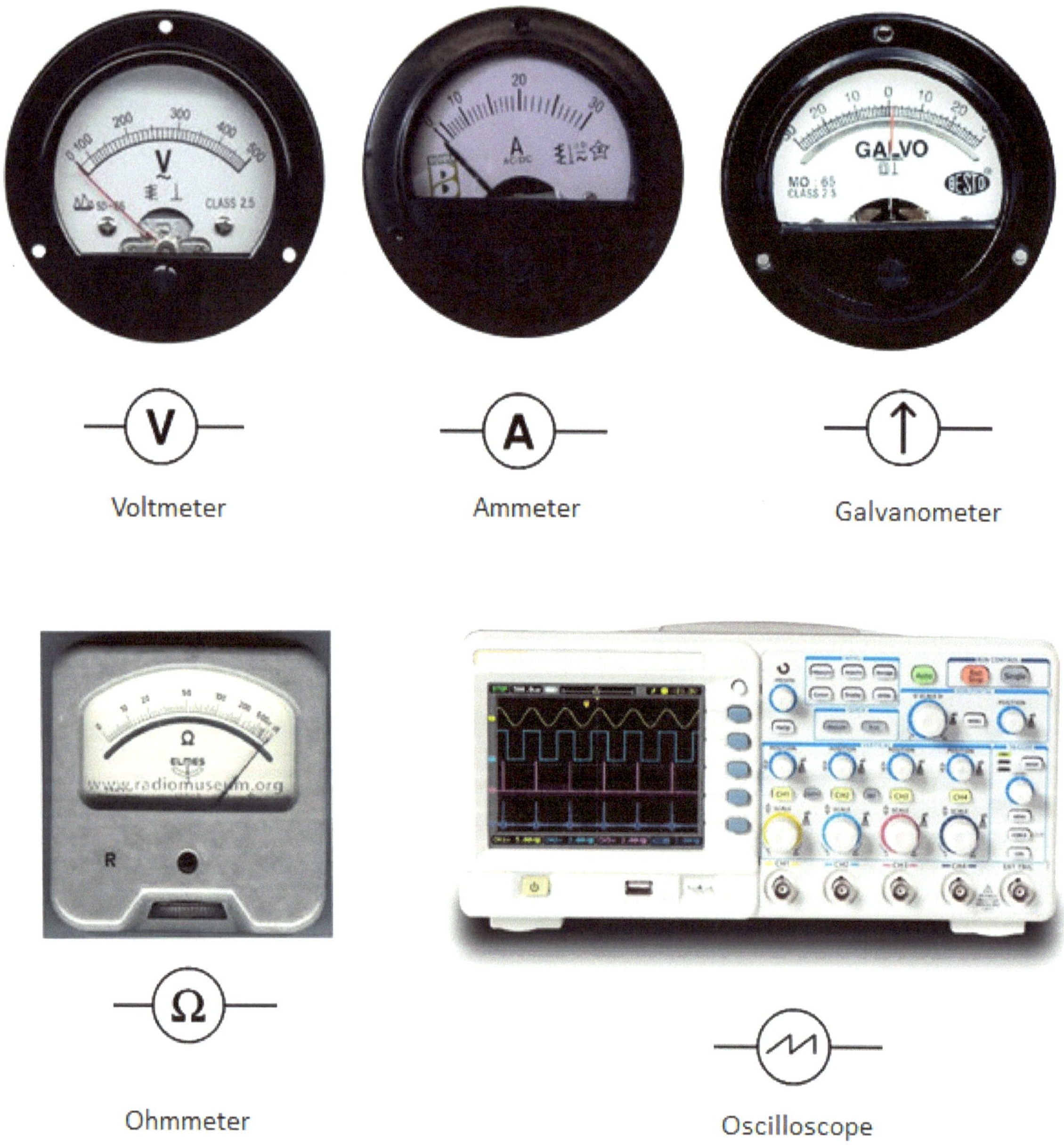

Fig. 4.11 - Electrical Measurement Devices

Figure 4.12 shows various brands of electrical *Multi-meters*. The figure shows the load positioning when current and voltage are measured. Load must be in series when measuring current, loads must be in parallel when measuring voltage.

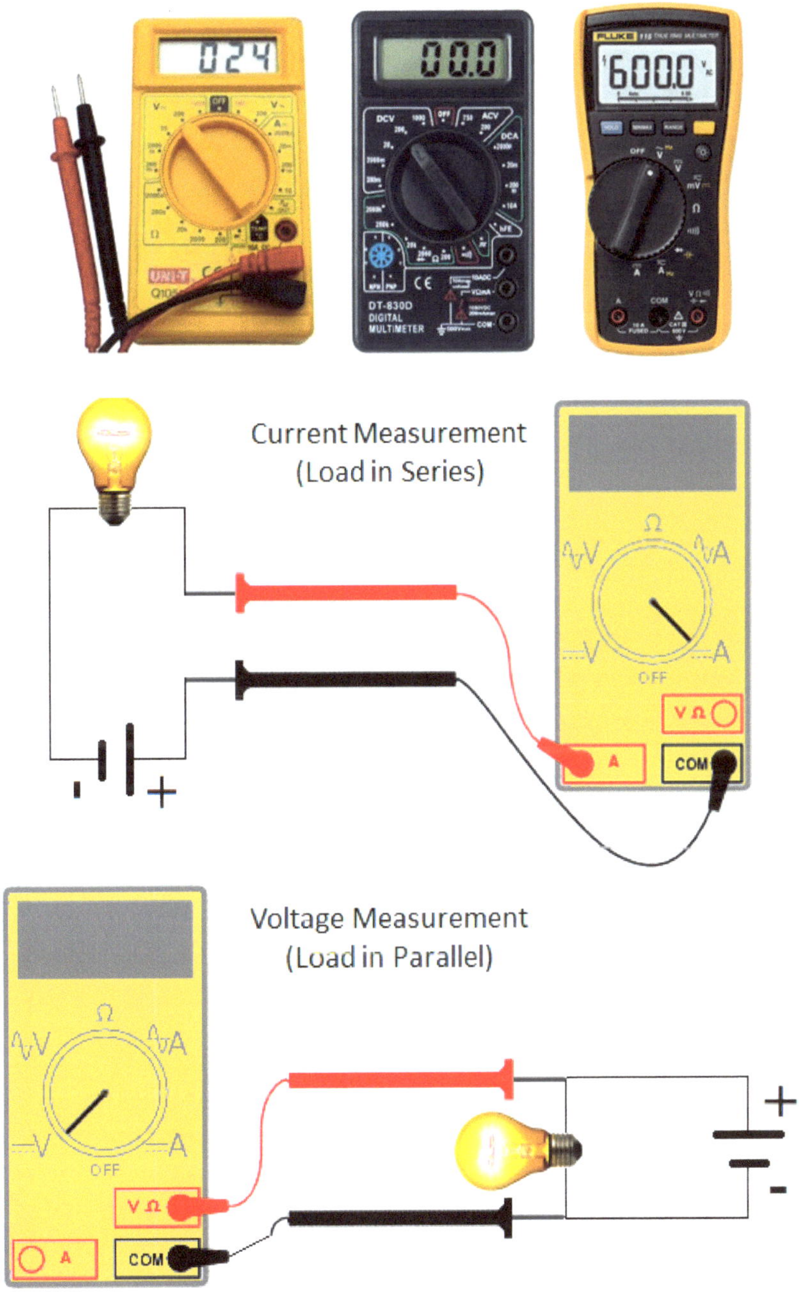

Fig. 4.12 - Electrical Multi-meter

4.3.2- Electrical Pushbuttons and Switches

Figure 4.13 illustrates different styles of industrial type momentary contact *Electrical Pushbuttons and Switches*. Single or double pushbuttons, each one with spring return, are designed for mounting in a control panel or standalone. Pushbuttons are available also with a key lock. If two positions are not sufficient, rotary selector switches are used. For functions like "Forward - Reverse", "Up - Down", "Start Stop", combined function pushbuttons are used. The figure also shows *Toggle Switches* are used with fluid power systems to operate hydraulic actuators.

Single Pushbutton

Double Pushbutton

Selector Switch

Combined Function Pushbutton

Key-Locked Pushbutton

Finger Toggle Switch

Push-Pull Toggle Switch

Fig. 4.13 - Industrial Electrical Pushbuttons and Switches

4.3.3- Limit Switches

Figure 4.14 shows various styles and brands of electro-mechanical industrial *Limit Switches* that can be used for electro-hydraulic control systems. Such switches are mounted either on fixed or movable surfaces so that they can energize or de-energize switching valves.

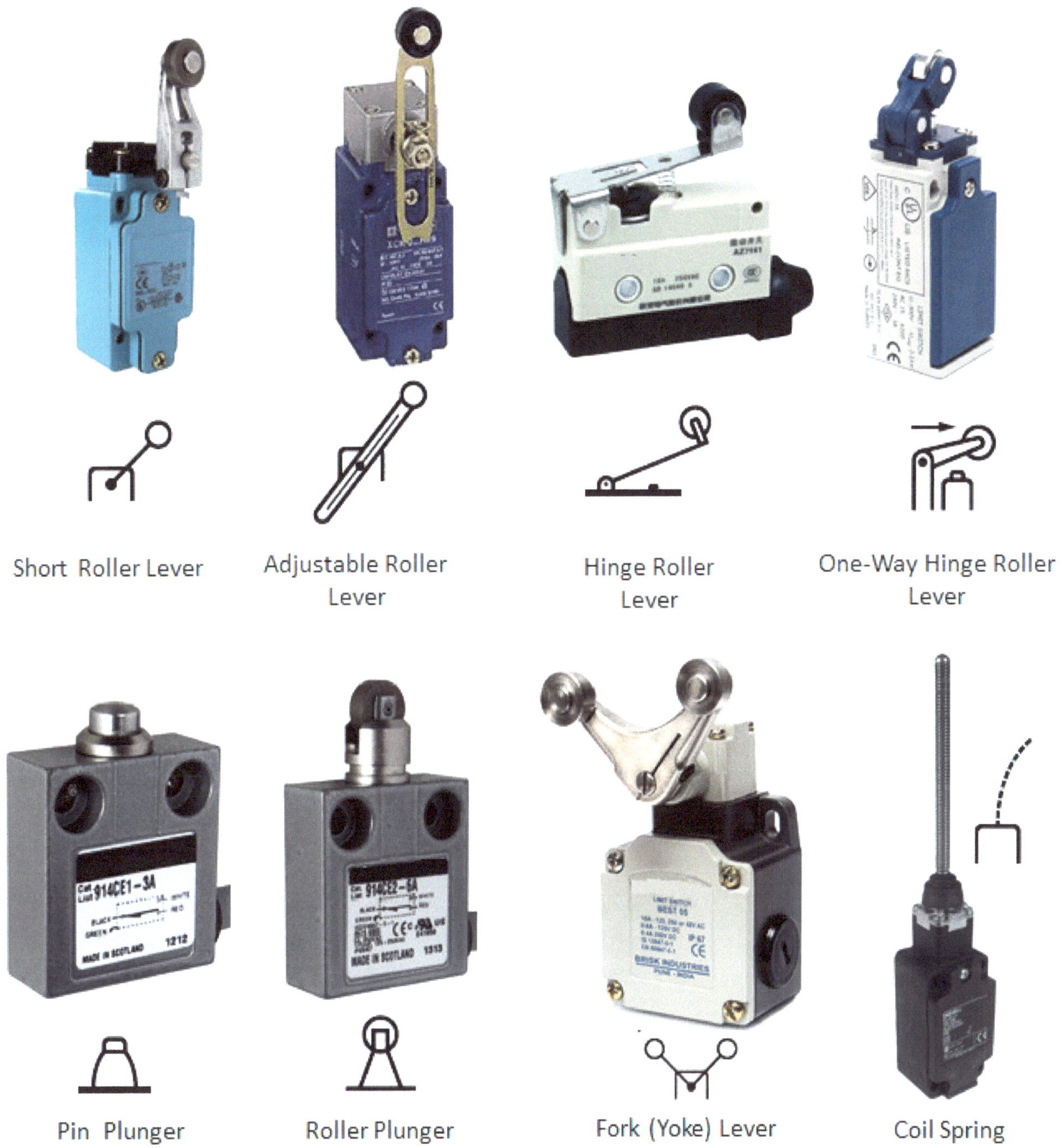

Fig. 4.14 - Electro-Mechanical Industrial Limit Switches

Figure 4.15 shows various styles and brands of *Foot-Operated Switches* that can be actuated by an operator's foot. They are found in single or double styles.

Single Foot-Operated Switch | Double Foot-Operated Switch

Fig. 4.15 - Foot Operated Industrial Electro-Mechanical Switches

4.3.4- Proximity Switches

Figure 4.16 shows some selected symbols for proximity switches that are commonly used in electro-hydraulic systems. A *Proximity Switch* is a non-contact switch actuated when an object approaches it. *Inductive Proximity Switches* detect ferrous targets. *Capacitive Proximity Switches* can detect both metallic and non-metallic targets. An *Optical Proximity Switch* is actuated by an object that obstructs a beam of light. *Magnetic Proximity Switches* detect permanent magnet targets.

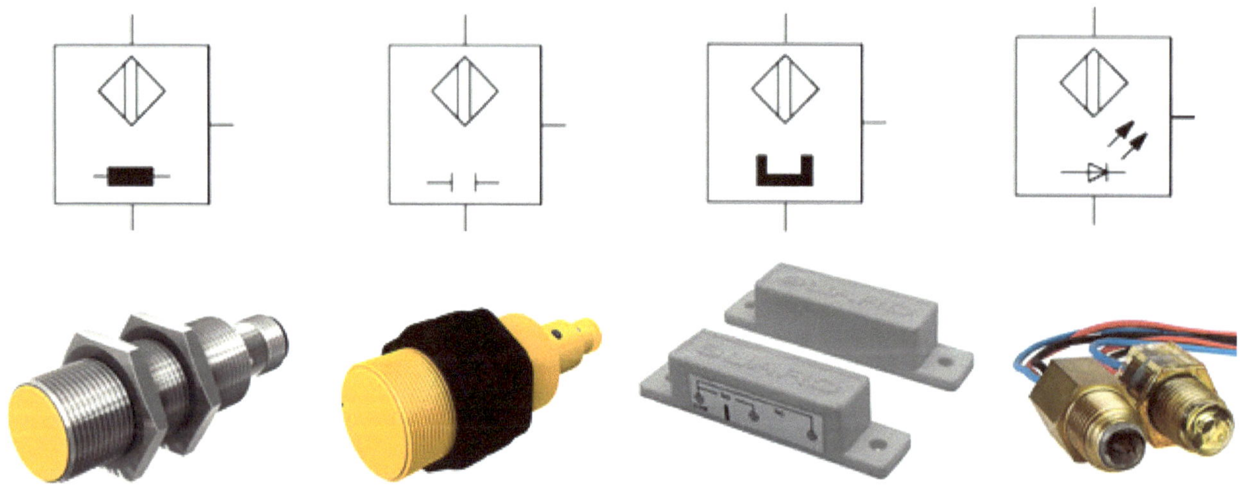

Inductive | Capacitive | Magnetic | Optical

Fig. 4.16 - Symbols for Proximity Switches Commonly Used for EH Systems

4.3.5- Pressure Switches

Pressure Switches have many different operating principles, the simplest of which is the *Piston Type Pressure Switch* that is shown in Fig. 4.17. Pressure switches could be with or without a case drain. They can also be designed as normally-closed or normally-open. A *Differential Pressure Switch* generates a signal based on the pressure difference between two points. *Vacuum Switch* is used as a proactive method to protect a pump against cavitation.

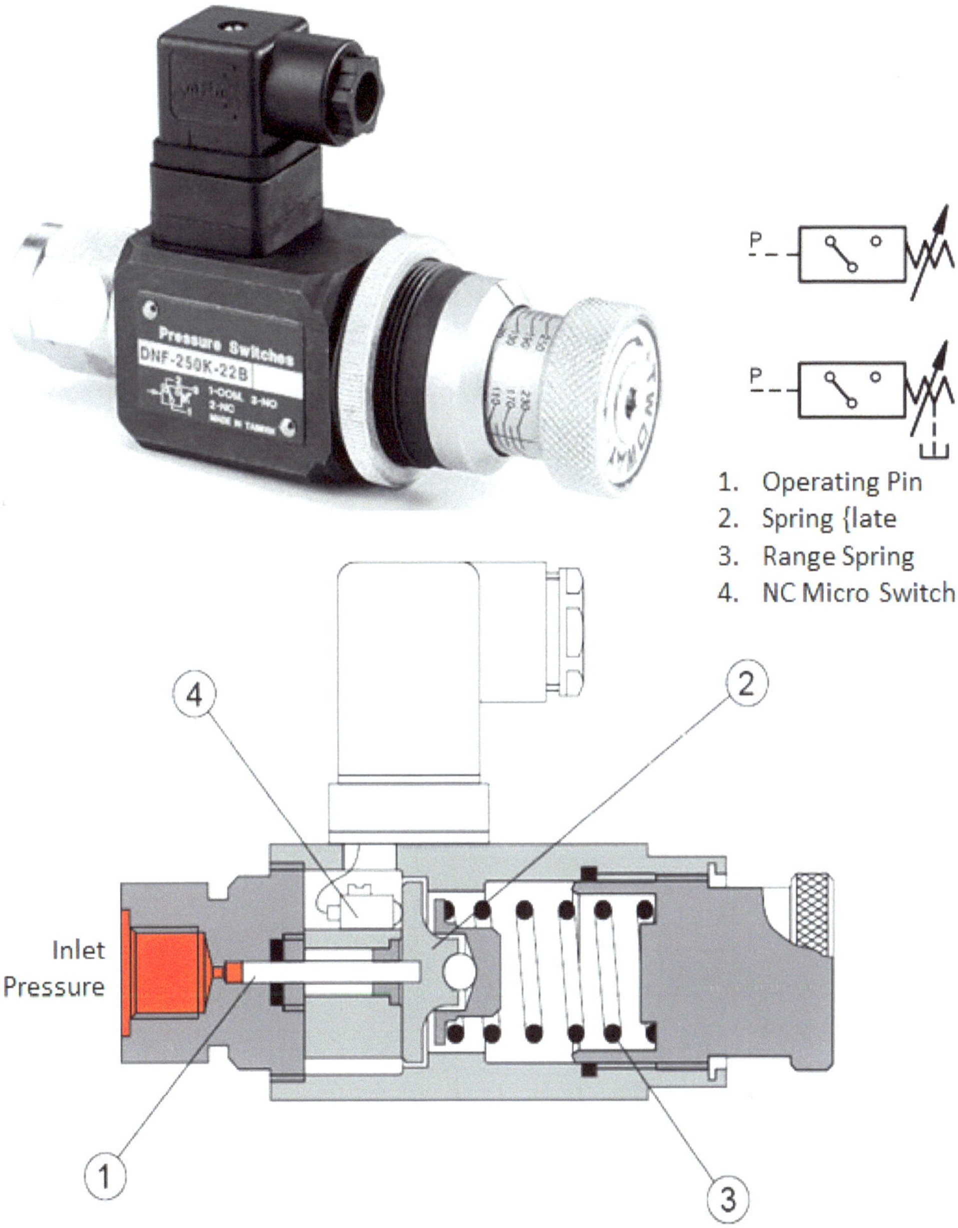

Fig. 4.17 - Piston Type Pressure Switch (Courtesy of Bosch Rexroth)

4.3.6- Fluid Flow Switches

Flow switches have many different operating principles, the simplest of which is the *Differential Pressure Flow Switch* that is shown in Fig. 4.18. Differential pressure increases with the increase of the fluid flow. Differential pressure is used to generate an electrical signal. The switch is adjusted to change its state when the differential pressure reaches certain value. It can be designed as Normally-opened or normally closed.

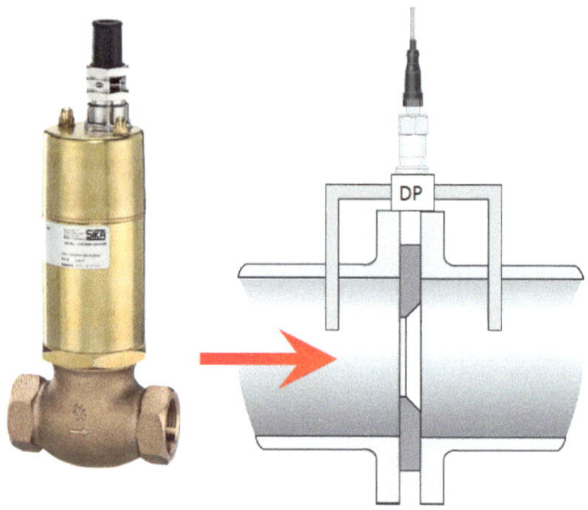

Fig. 4.18 - Differential Pressure Flow Switch

4.3.7- Fluid Level Switches

Fluid level switches have many different operating principles, the simplest of which is the *Float Type Level Switch.* Figure 4.19 shows a piece of floating element which moves with the fluid surface. The moving floating element is used to closes an electric circuit or to approach a proximity switch so that an electrical signal is generated when the oil level reaches a certain minimum level.

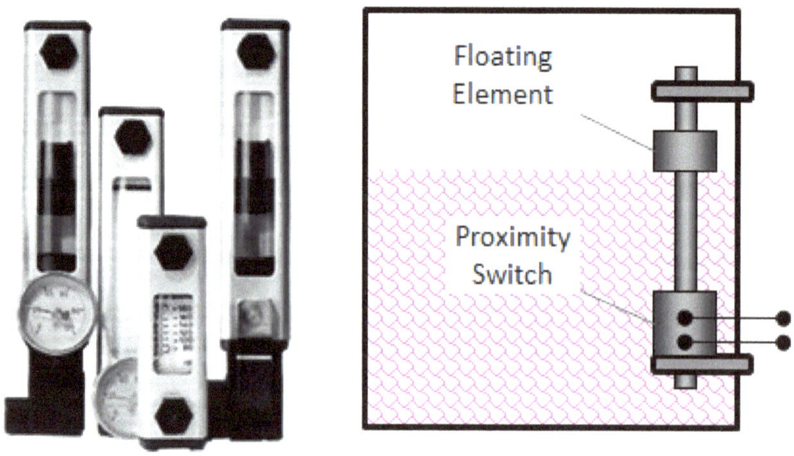

Fig. 4.19 - Float Type Level Switch

4.3.8- Temperature Switches

Controlling the temperature of a hydraulic system can be done by an open or closed-loop analog control circuits or by an ON/Off control circuit where a *Temperature Switch* is used. It is also referred as *Thermo-Switch.* Figure 4.20 shows the simplest operating principle for a normally-open thermos-switch. A bi-metal strip takes a convex shape when it is subjected to heat so that it lifts the arm. When the two contact points get in contact, an electrical circuit is closed and a signal is generated.

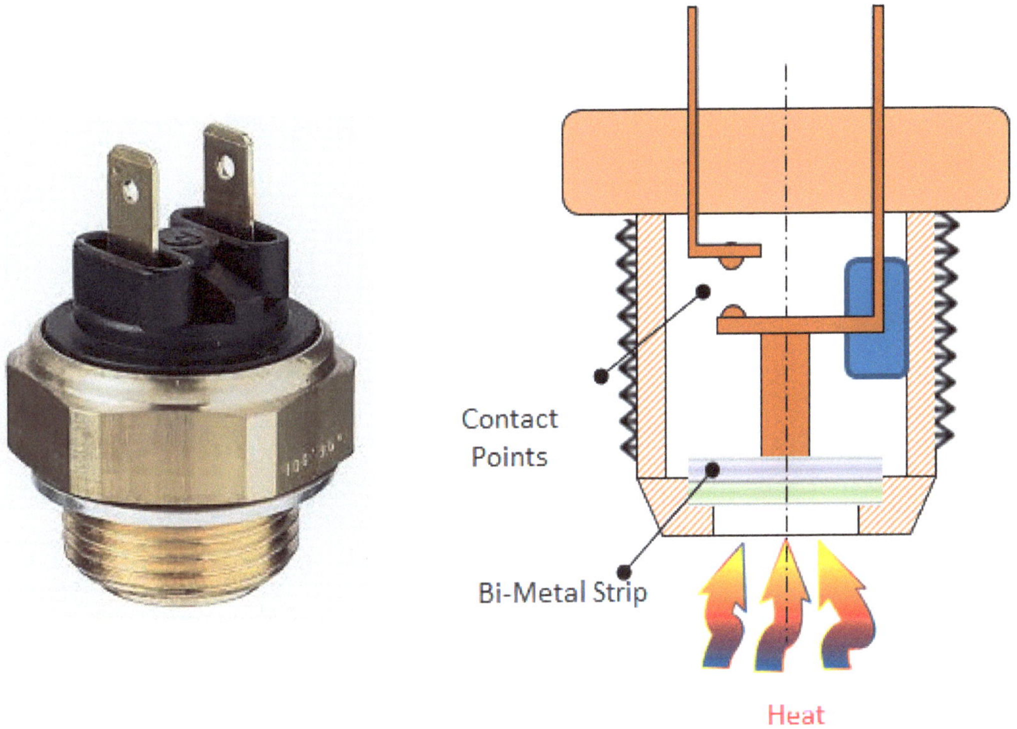

Fig. 4.20 - Temperature Switch

4.3.9- Control Relays

Figure 4.21 shows the operating principles of *Normally-Open (NO)* and *Normally-Closed (NC) Control Relays*. The standard JIC schematic symbols are shown beside each one. For a normally-open control relay, when the coil is energized, the relay contact closes and the solenoid is turned ON. The normally-closed works oppositely.

A designer should use good judgement in selecting relays for a control circuit. The criterion of selecting NO or NC is to minimize the tie of energizing the coil and the solenoid so that the heat generated is reduced. For example, if in a duty cycle of a machine a cylinder must be kept extended most of the time, then the cylinder should be connected hydraulically to the valve to be extended while the solenoid is in OFF (de-energized) state. Then the relay and the solenoid will be energized only when the cylinder is required to extend. A designer must specify coil voltage and frequency (Hz) if the relay is used for AC current.

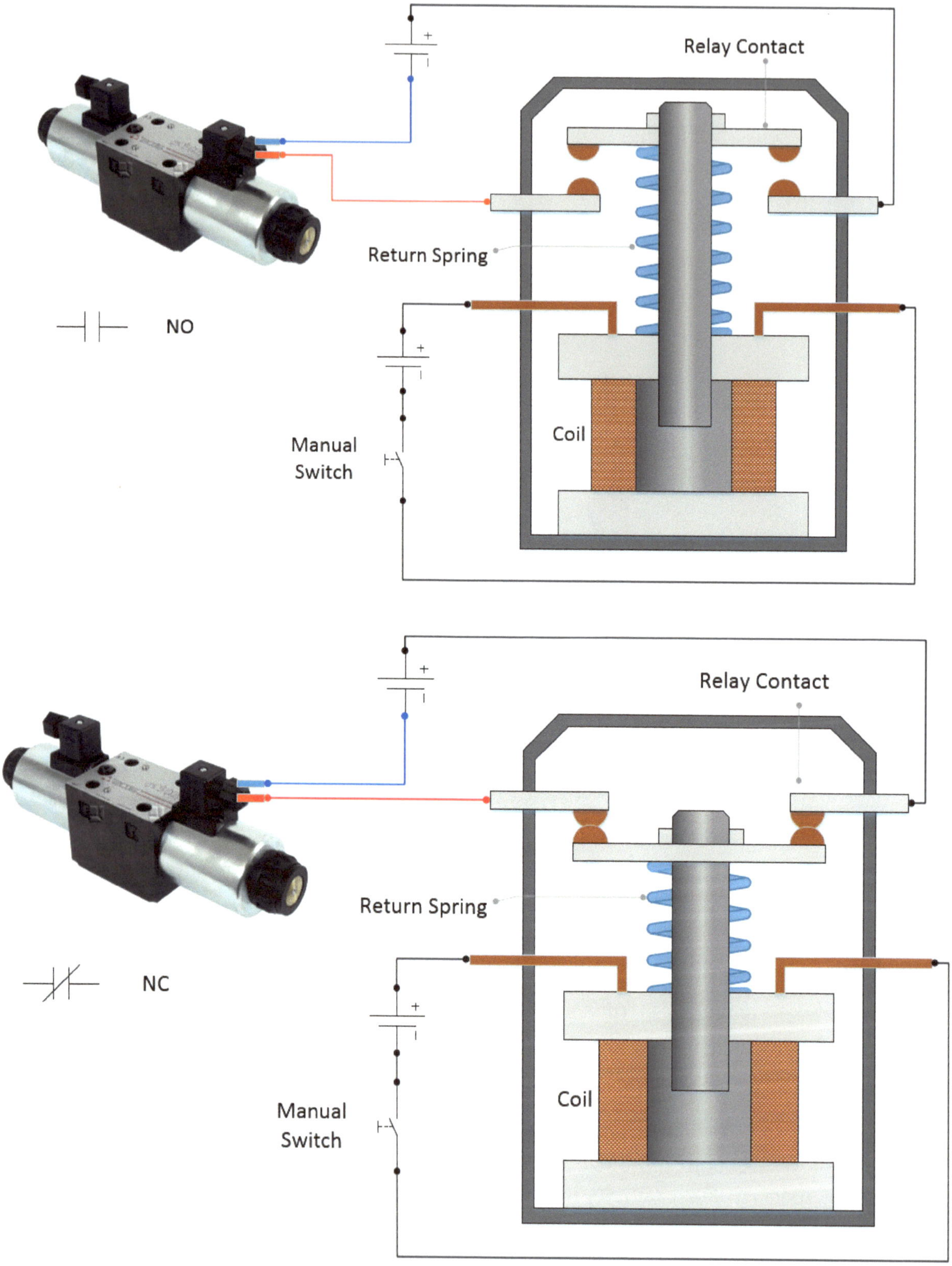

Fig. 4.21 - Normally-Open and Normally-Closed Control Relay

For digital control of a solenoid actuated hydraulic valve, a solid state relay is used. Figure 4.22 shows a solid state relay in which the coil is replaced by an integrated circuit. It is triggered by a low power signal to change the state of a high power circuit. It could be provided as NO or NC as well. The low power triggering signal can be generated by a PLC or a PC using the adequate interface.

Solid state relays have the following advantages over the previously shown electro-mechanical relays:

- Lasts for millions of cycles.
- No mechanical contacts of less reliability.
- Has higher switching frequency.

Fig. 4.22 - Connecting a Solid State Relay to a Solenoid-Actuated Directional Control Valve

4.3.10 - Programmable Logic Controllers (PLCs)

A *Programmable Logic Controller (PLC)* is a solid-state electrical device that replaces the "hard wired" electric logic components, and can be programmed to perform an infinite variety of logic functions. It is also considered as one of various ways to build an interface between an operator and a machine or for machines to communicate with each other.

A detailed description of the structure, operation and programming of PLCs is out of scope of this textbook. This section presents only a brief definition of that device as one of the most common controllers used to control hydraulic switching valves.

Use of a PLC for machine automation is very common in industrial applications. Mobile applications usually implement custom controllers.

PLCs range from basic to advanced configurations. As shown in Fig. 4.23, in the simplest form, a PLC consists of an input interface unit through which signals are received and an output interface unit through which signals are sent out. The output interface could send out signals that are powerful enough to drive the output devices directly or through relays.

A *Central Processing Unit (CPU)* contains a digital processor to run predefined programs and a memory unit for data storage.

Various peripheral units are used to communicate with the CPU, such as a Human Machine Interface (HMI) or computer.

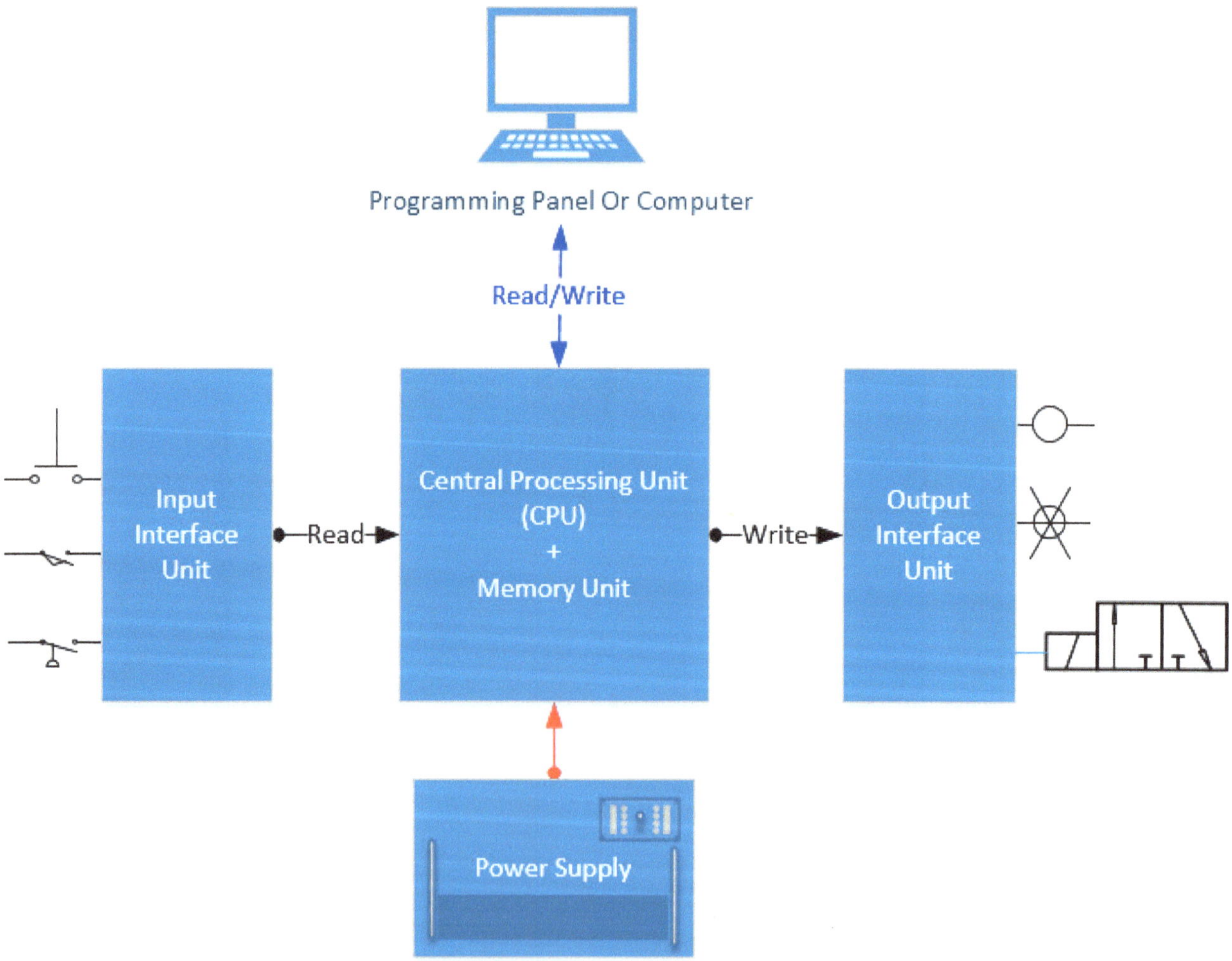

Fig. 4.23 - Basic Structure of Programmable Logic Controllers

4.4- Electrical Schematic Diagrams

4.4.1- JIC Schematic Diagrams

JIC Stands for "*Joint Industrial Council*". It is a hardwired system that means the components are connected with wires. To ease wiring the components, wiring boards or terminals are used as shown in Fig. 4.24.

The JIC Schematic diagram shows only the lay out of the circuit, the components used, and how they are connected to each other. A JIC schematic diagram does give information about the size of the components unless an associated table is provided with such information.

Fig. 4.24 - Electrical Terminal Blocks

Figure 4.25 shows an example of JIC diagram to operate a two-position solenoid-actuated hydraulic valve. The following bullets explain how to read the diagram:

- A vertical ladder format is used.
- The two vertical lines are called "*Buses*" and represent the power lines.
- They are generally labeled "+" and "-" for DC circuits.
- They are generally labeled "L1" and "L2" or H (high) and N (neutral) for AC circuits.
- The left vertical line represents the "hot" or ungrounded side of the power line.
- All switches must be connected to this bus.
- The right vertical line represents the grounded side of the power line.
- The horizontal lines are called "*Rungs*."
- As many rungs as necessary are hung between these uprights
- Also reference numbers to the left of each rung are given, such as R1, R2, etc.
- The diagram is read from left to right and from the top bottom.
- The devices are arranged and numbered in the order in which they are actuated
- The output device is always the last item on the rung and is always drawn close to the right (low voltage) bus.

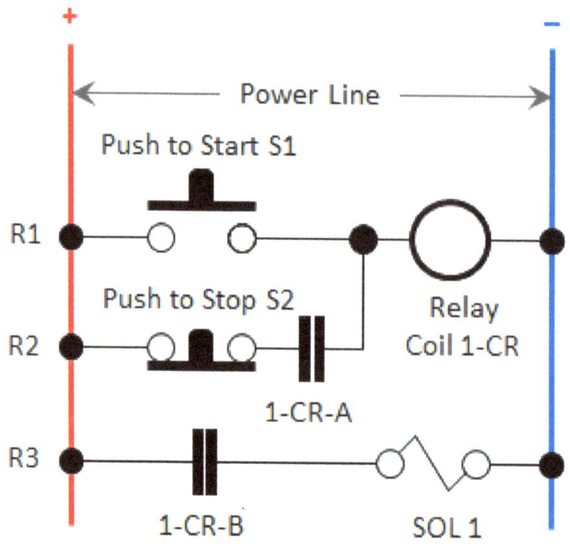

Fig. 4.25 - Example of a JIC Schematic Diagram

A *Truth Table* can be provided with the JIC schematic diagram to ease understanding of the logic functions in the circuit. The following concepts are used to build the truth table:

- For Switches and Buttons: 0 represents "Initial State" and 1 means "Alternate State".
- For Solenoids and Relay Coils: 0 means "De-energized state" and 1 means "Energized state".
- For Relay Contacts: 0 means "Open Contacts" state and 1 means "Closed Contacts".

Table 4.1 shows an example of applying the previous concepts on the previous JIC diagram

S1	S2	1-CR	1-CR-A	1-CR-B	SOL 1
1/0	0	1	1	1	1
0	1/0	0	0	0	0

Table 4.1 - Example of Truth Table

The previous truth table is read as follows:

- When S1 is pressed and released, the relay coil 1-CR is energized, the relay contact 1-CR-A is closed, the relay contact 1-CR-B is closed, and the solenoid is energized.
- When S2 is pressed and released, the relay coil 1-CR is de-energized, the relay contact 1-CR-A is opened, the relay contact 1-CR-B is opened, and the solenoid is de-energized.

Figure 4.26 explains common mistakes in drawing JIC schematic diagrams as follows:

- DO NOT place an input device across a rung without ending with an output device. An example of that is, in R1, pressing S1 results in short circuit.

- DO NOT place output devices in series on the same rung even if they have the same power rating. An example of that is, in R2, pressing S2 results in splitting the 24 Volt evenly between coil 1 and 2 that have the same power ratings. As a result, each of the coils receives 12 Volt instead of the original 24 Volts. The situation will be worse if the coils have different power rating.

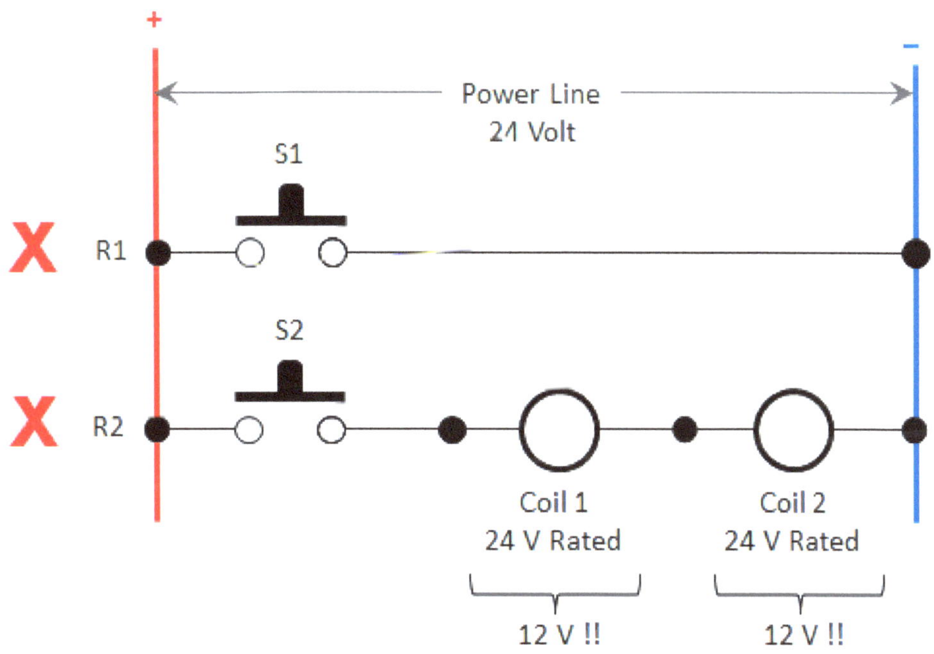

Fig. 4.26 - Common Mistakes in Drawing a JIC Schematic Diagram

4.4.2- Wiring Diagrams

As shown in Fig. 4.27, based on a JIC schematic diagram, a wiring diagram can be developed to illustrate the distribution of the components on a chassis or inside an enclosure.

Components should be drawn in the form of blocks or pictures with enough detail or markings to make them recognizable. Each component should be drawn approximately to relative size and should be placed in its approximate position relative to other components. Terminal locations in a wiring diagram are also used to show the routing of connecting wires.

The wiring diagram shown in this figure is developed for the JIC schematic diagram shown in Fig. 5.25.

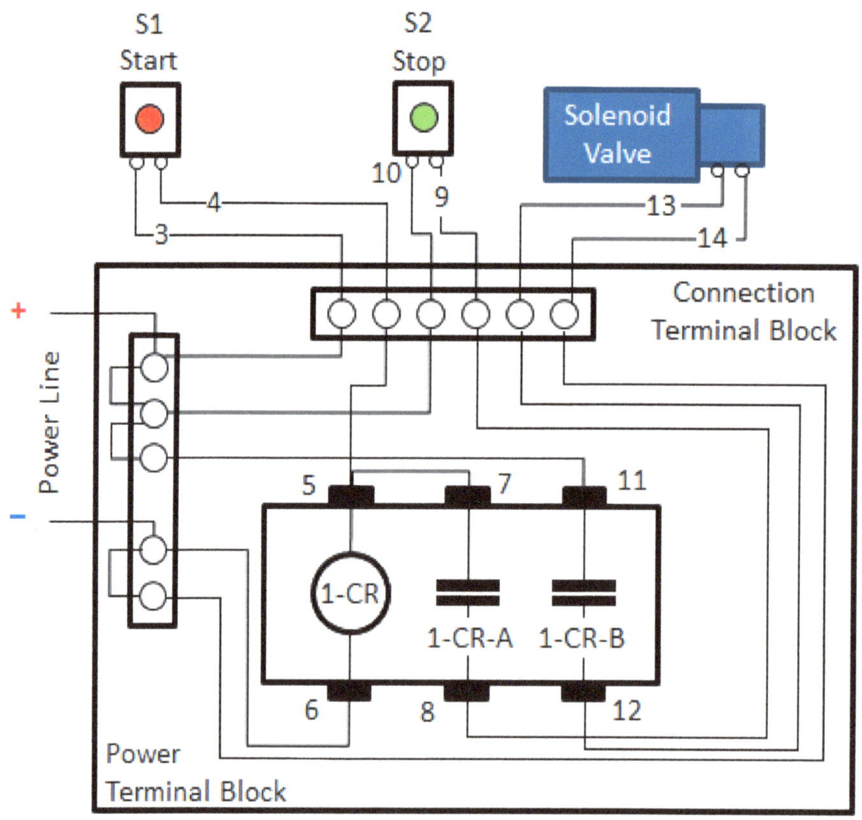

Fig. 4.27 - Wiring Diagram

4.4.3- Sequence Diagrams

Similar to a truth table, a *Sequence Diagram* is an auxiliary diagram that can be developed to ease understanding the logic function of a JIC schematic diagram. Such auxiliary diagram and the truth table are helpful tools for system troubleshooters. The sequence diagram shown in Fig. 4.28 is developed for the JIC schematic diagram shown in Fig. 4.25. For a period of one complete cycle, a sequence diagram demonstrates the electrical state of each component. There are no standard rules for drawing a sequence diagram. A sequence diagram is typically composed of horizontal lines, each represent the status of a component over the period of the cycle time.

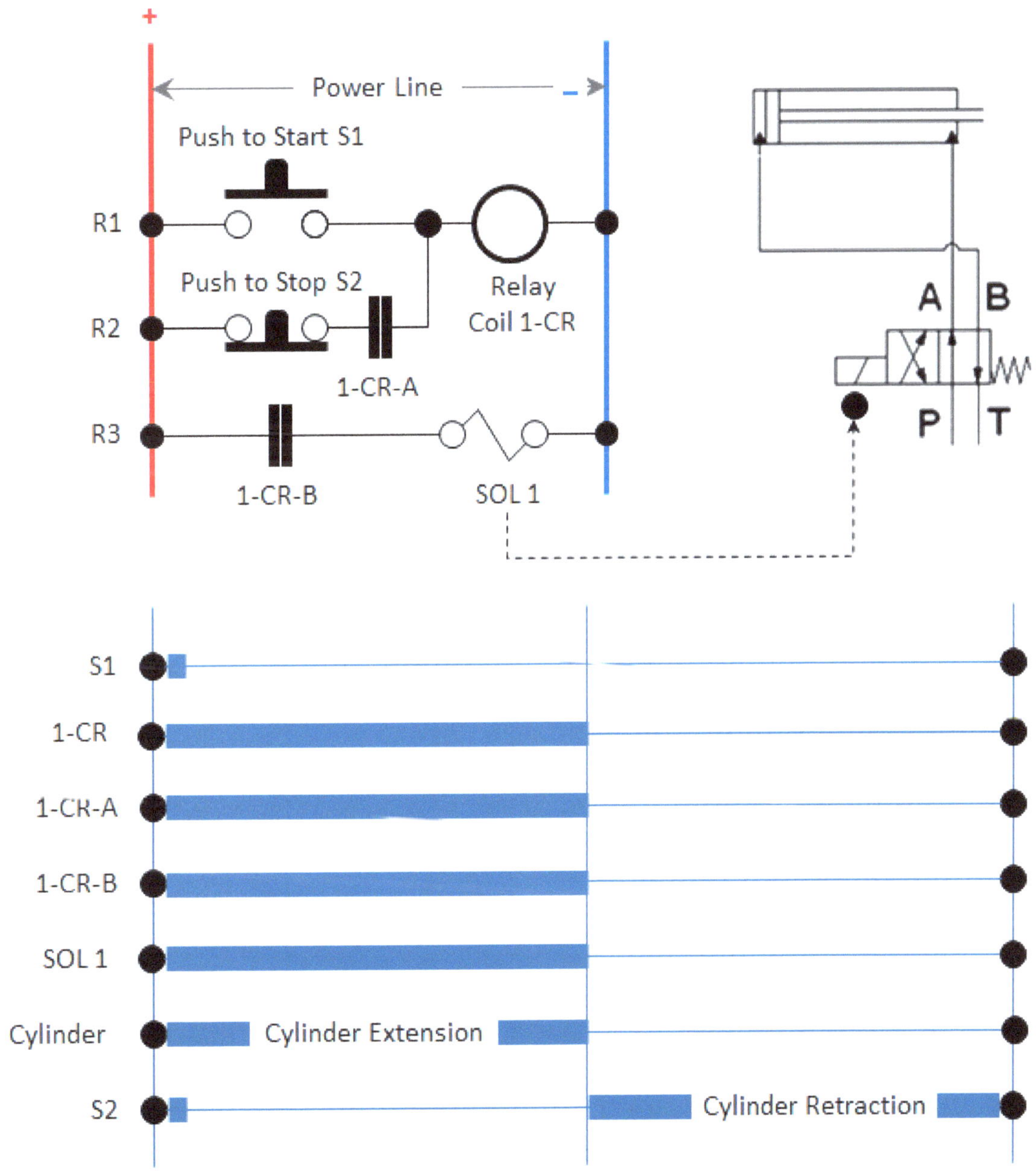

Fig. 4.28 - Sequence Diagram

4.5- Electrical Circuits for Applications of Switching Valves

In this section, the following abbreviations are used:

- PS: Pressure Switches.
- LS: Limit or Position Switches (Mechanical).
- TS: Toggle Switches.
- PX: Proximity Switches.
- S: Pushbuttons.
- K1, K2, etc.: Relay Coils.
- K1A, K1B, etc.: Relay Contacts.

4.5.1- Identity (YES Function) by Direct Activation

Figure 4.29 shows an example of *Identity (YES) Function*. In such a function the output device follows the same state as the input device. A practical example of this function is by pressing the normally-open pushbutton S1, the solenoid Y1 is energized and the pump is unloaded.

Direct Activation means that both the input and the output devices are on the same rung. So, whatever the power required to actuate the output device will also be received by the input device. In the example shown below, if the solenoid is rated for 2 Amps, the pushbutton must be rated the same.

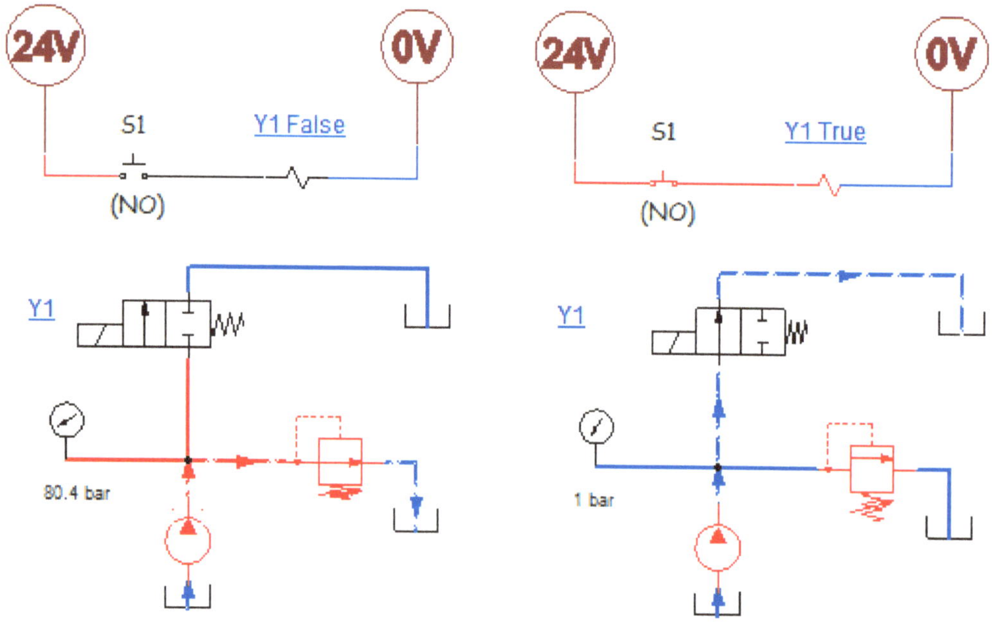

S1	Y1	Pump
0	0	Loaded
1	1	Unloaded

Fig. 4.29 - Identity (YES Function) by Direct Activation

4.5.2- Negation (NOT Function) by Direct Activation

Figure 4.30 shows an example of *Negation (NOT) Function*. In such a function, the output device has the opposite state with respect to the input device. A practical example of this function is by pressing the normally-closed pushbutton S1, the solenoid Y1 is de-energized, and the pump is loaded.

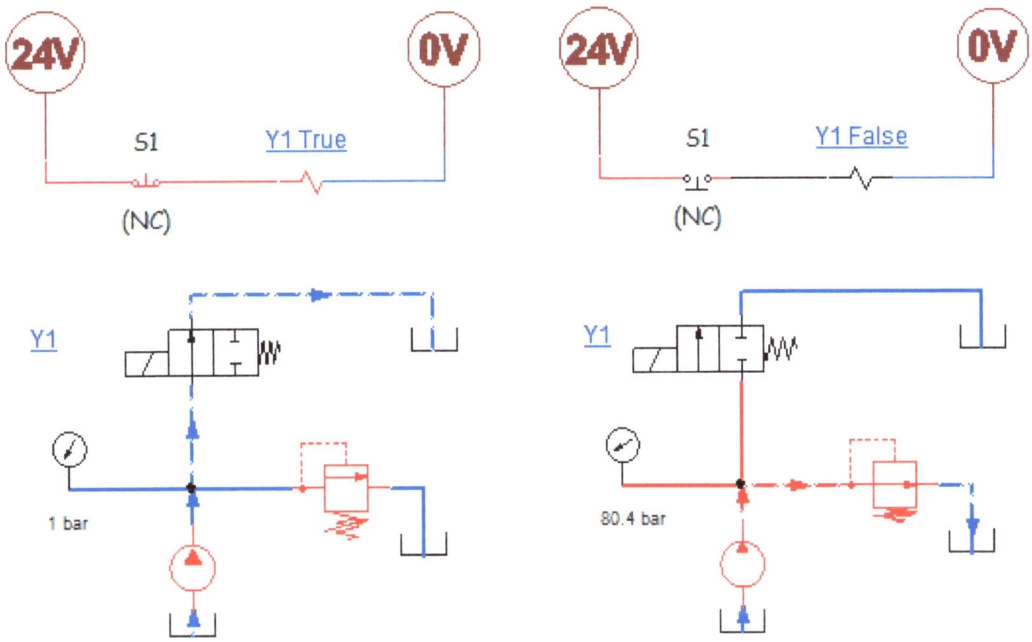

S1	Y1	Pump
0	1	Unloaded
1	0	Loaded

Fig. 4.30 - Negation (NOT Function) by Direct Activation

4.5.3- Identity (YES Function) by Indirect Activation

Figure 4.31 shows an example of Identity (YES) Function by *Indirect Activation*. It is the same YES function that has been discussed previously except that the control circuit that contains the input device has been separated from the power circuit that contains the output device. The figure shows a practical example in which by pressing the normally-open pushbutton S1, the relay coil K1 is energized, the normally-open relay contact K1-A is closed, the solenoid Y1 is energized, and the cylinder extends. The advantages of indirect activation are:

- Separation of the control circuits from the power circuits eases troubleshooting.
- Input devices rated for lower power.
- Remote control.

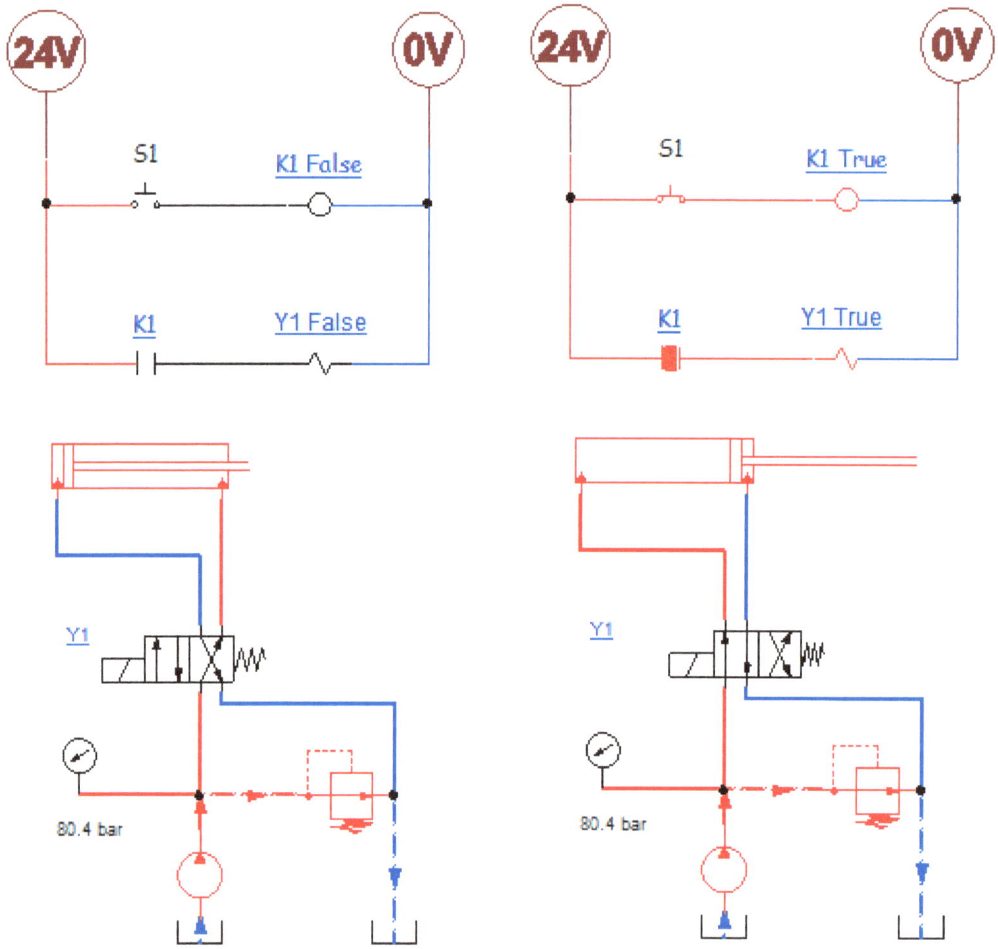

S1	K1	K1A	Y1	Cylinder
0	0	0	0	Retracts
1	1	1	1	Extends

Fig. 4.31 - Identity (YES Function) by Indirect Activation

4.5.4- Negation (NOT Function) by Indirect Activation

Figure 4.32 shows an example of Negation (NOT) Function by Indirect Activation. It is the same NOT function that has been previously discussed except that the solenoid is indirectly actuated through a relay.

The figure shows a practical example in which by pressing the normally-closed pushbutton S1, the relay coil K1 is de-energized, the normally-open relay contact K1A opens, the solenoid Y1 is de-energized, and the cylinder retracts.

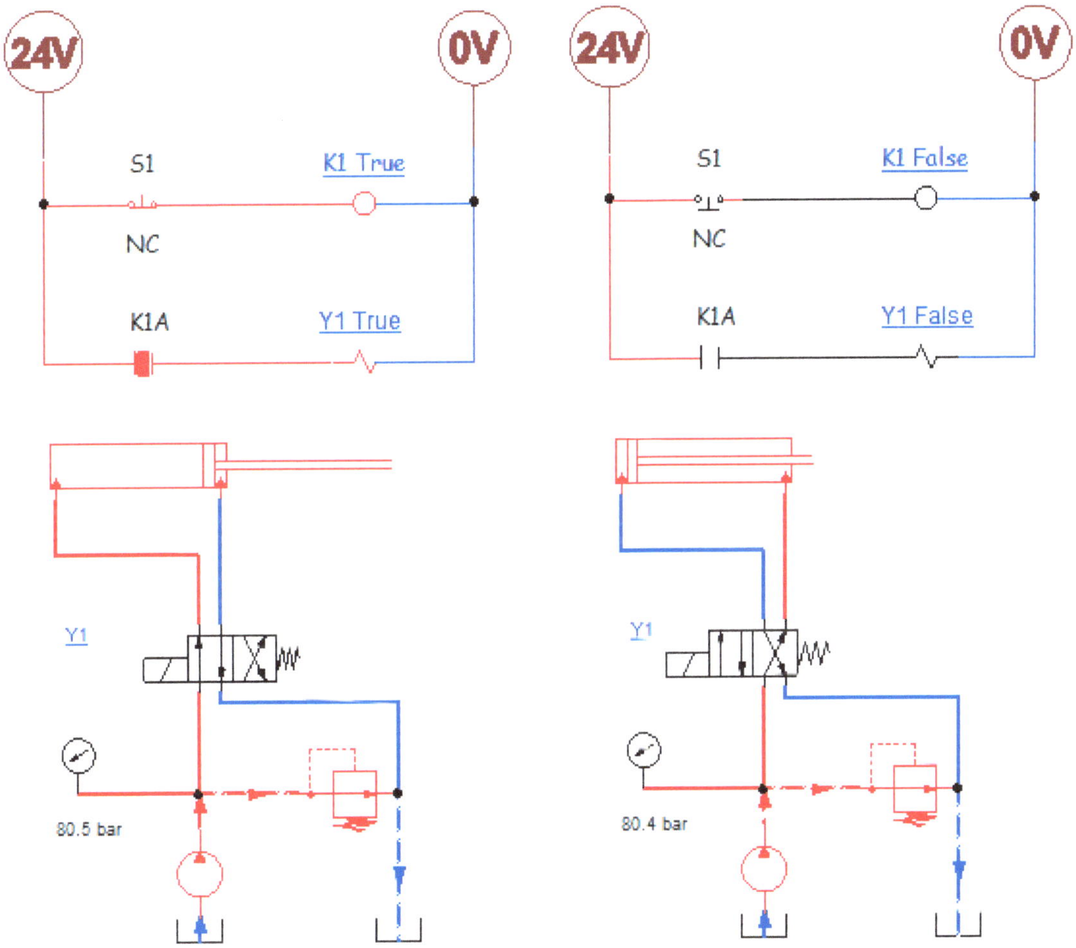

S1	K1	K1A	Y1	Cylinder
0	1	1	1	Extends
1	0	0	0	Retracts

Fig. 4.32 - Negation (NOT Function) by Indirect Activation

4.5.5- Signal Storage by Electrical Latching

In the previous two systems, the pushbutton S1 must be kept pressed in order for the operator to change the state of the cylinder. An input signal can be stored electrically by a *Latching Circuit* as shown in Fig. 4.33. The right part of the figure shows that, by pressing the pushbutton S1, the relay coil K1 is energized, the normally-open relay contacts K1A are closed, the solenoid Y1 is energized, and the cylinder extends. Even after releasing S1, K1B continues to pass the power to K1 so that the signal is stored. At any time, during the cylinder extension or even after the cylinder is deadheaded, by pressing/releasing the normally-closed pushbutton S2, the latching line is terminated, K1 is de-energized, Y1 is de-energized and the cylinder retracts. Functionally, a toggle switch can replace the latching circuit, but the latching circuit is helpful in automatic retraction of the cylinder if S2 is replaced by a limit switch as will be explained later.

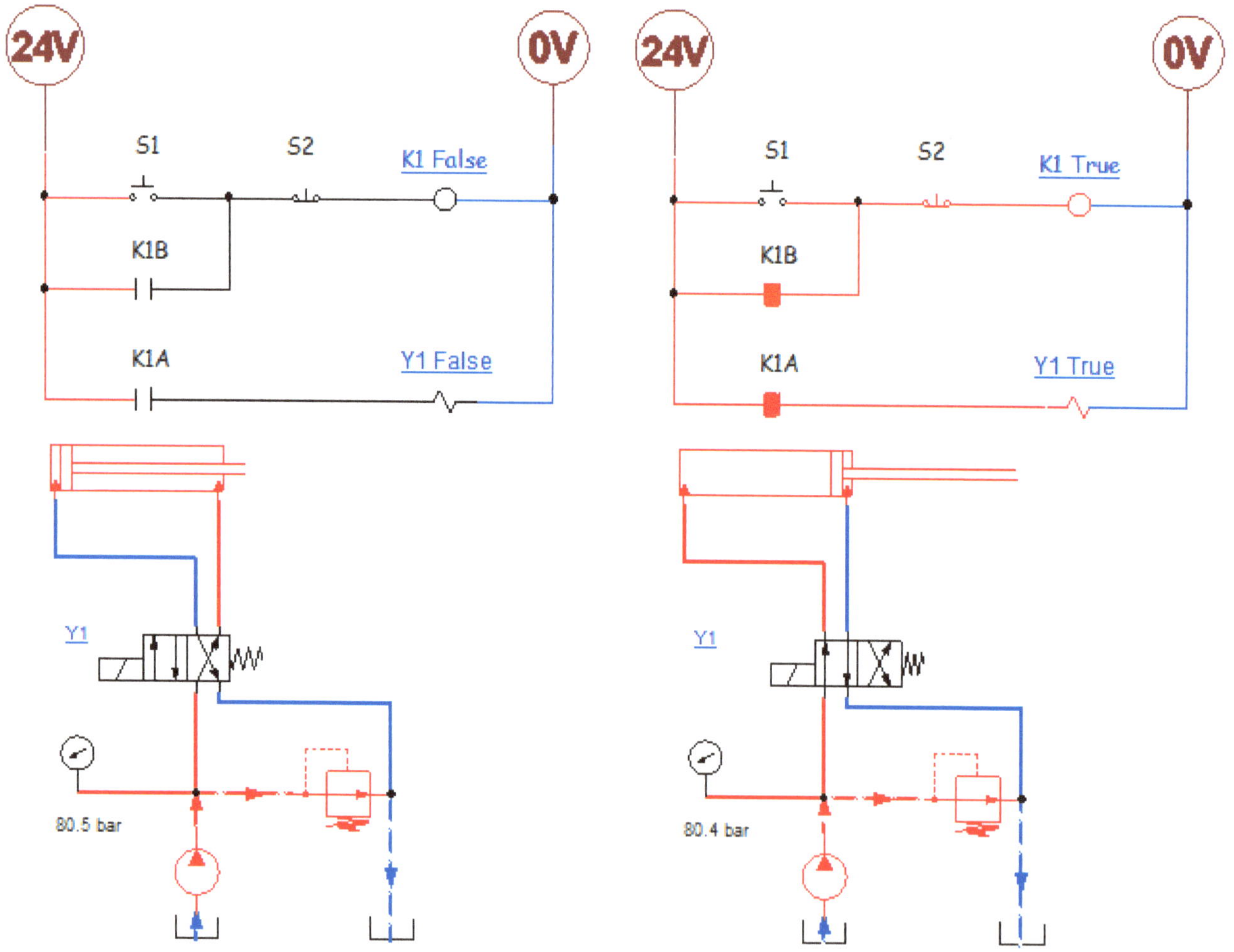

S1	S2	K1	K1A	K1B	Y1	Cylinder
1/0	0	1	1	1	1	Extends
0	1/0	0	0	0	0	Retracts

Fig. 4.33 - Signal Storage by Electrical Latching

4.5.6- Electromechanical Protection of a Valve with Two Solenoids

With a 3-position directional valve, the two solenoids must not be energized simultaneously. Otherwise, the solenoids may burn each other out as has been explained in the previous chapter. The circuit shown in Fig. 4.34 presents an electromechanical way to offer such protection. As shown in the figure, if both toggle switches TS1 and TS2 are pressed or depressed simultaneously, the two solenoids Y1 and Y2 are de-energized; the pump is unloaded through the tandem-center directional valve, and the motor stops. If one of the pushbuttons is pressed while the other is left un-pressed, only one solenoid is energized, and the motor rotates accordingly as shown in the associated truth table.

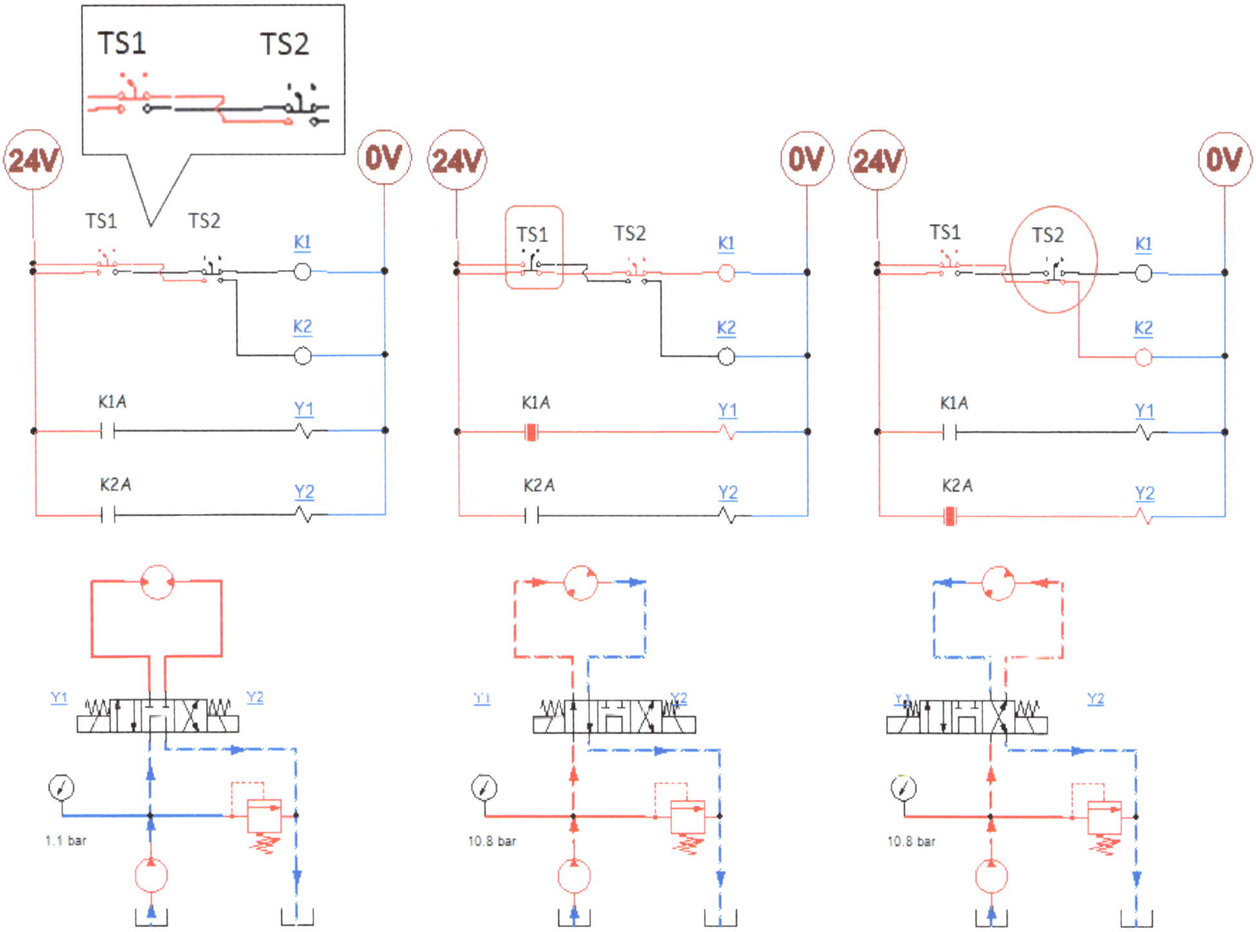

S1	S2	K1	K2	K1A	K2A	Y1	Y2	Motor
0	0	0	0	0	0	0	0	Stop
1	1	0	0	0	0	0	0	Stop
1	0	1	0	1	0	1	0	CW
0	1	0	1	0	1	0	1	CCW

Fig. 4.34 - Electromechanical Protection of a Valve with Two Solenods

4.5.7- Electrical Protection of a Valve with Two Solenoids

An alternative way to protect two solenoids in a directional valve is the electrical circuit diagram shown in Fig. 4.35A. The circuit consists of:

- Two latching circuits; each one takes care of one of the solenoids.
- 1st latching circuit contains: a pushbutton S1, a relay coil K1, a normally-open relay contact K1B, and a solenoid Y1.
- 2nd latching circuit contains: a pushbutton S2, a relay coil K2, a normally-open relay contact K2B, and a solenoid Y2.
- In order to protect the two solenoids, from being energized simultaneously, the following elements have been added to the previous latching circuits: a normally-closed relay contact K2C has been placed upstream of K1 and a normally-closed relay contact K1C has been placed upstream of K2.
- A reset pushbutton S3.

The figure and the associated truth table show the reset mode of the circuit where K1 and K2 are de-energized.

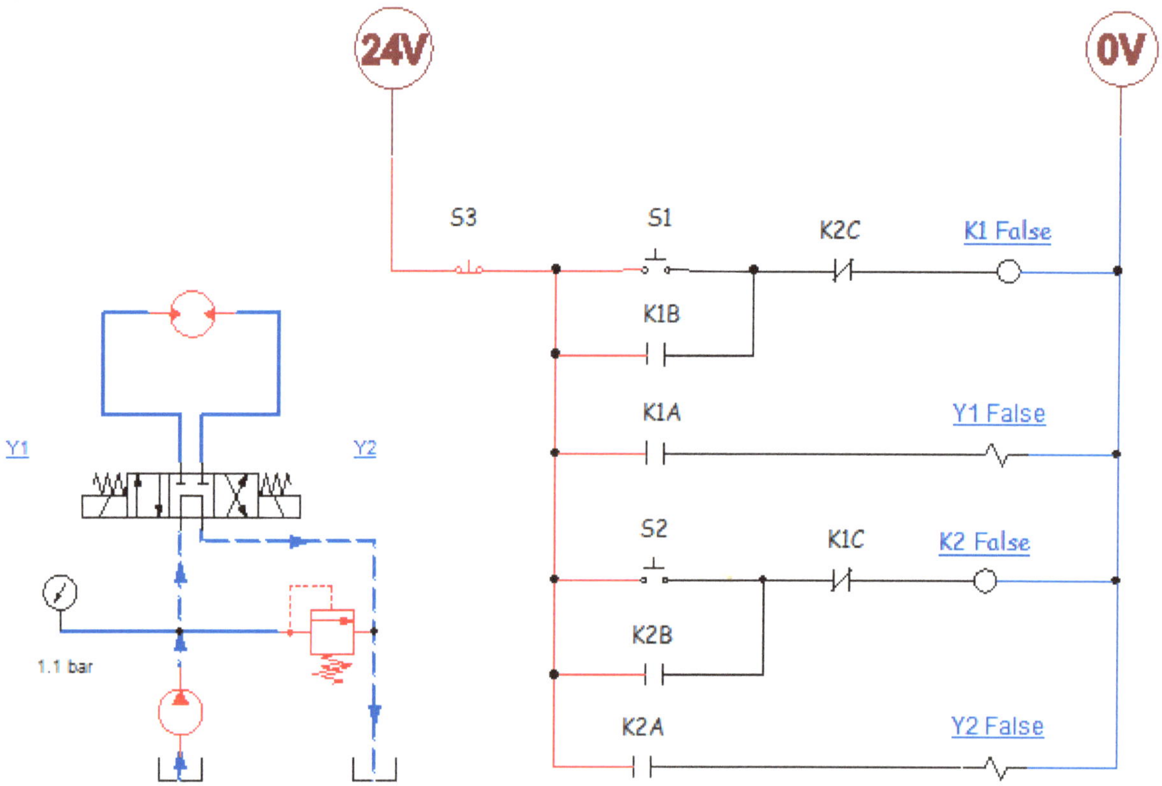

S1	S2	S3	K1	K1A	K1B	K1C	K2	K2A	K2B	K2C	Y1	Y2	Motor
0	0	1	0	0	0	0	0	0	0	0	0	0	Stop

Fig. 4.35A - Electrical Protection of a Valve with Two Solenoids

Figure 4.35B and the associated truth table show the operating mode of the circuit where Y1 is energized and the motor rotates clockwise. The sequence of operation is as follows:

1. S1 is pressed & released.
2. K1 is energized.
3. K1B latches the circuit and continues to pass the power to K1.
4. Y1 is energized, the directional valve is shifted accordingly, and the motor rotates clockwise.
5. Simultaneously, K1C is open so that pressing S2 will not energize K2.

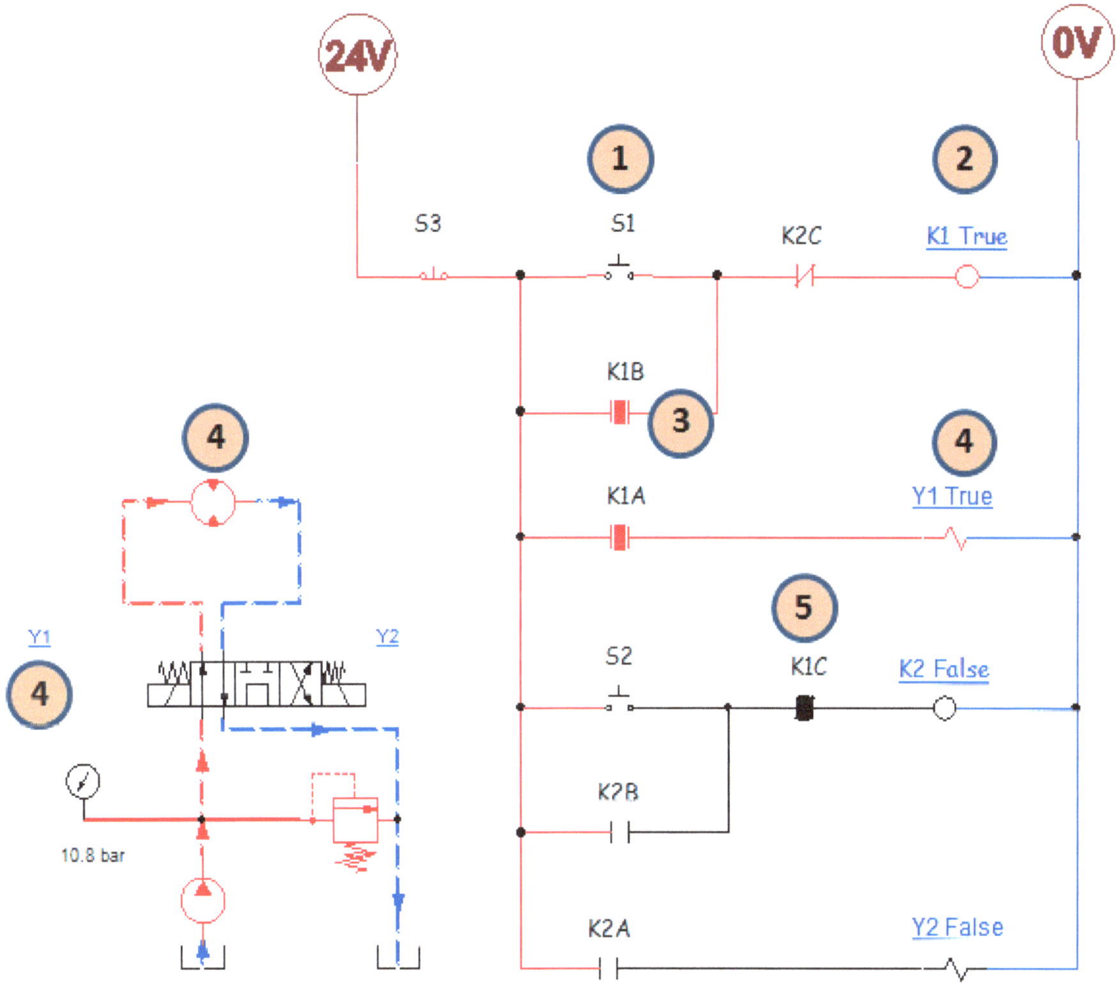

S1	S2	S3	K1	K1A	K1B	K1C	K2	K2A	K2B	K2C	Y1	Y2	Motor
1/0	0	0	1	1	1	0	0	0	0	1	1	0	CW

Fig. 4.35B - Operating Conditions of the Circuit, Motor CW Rotation

Figure 4.35C and the associated truth table show the operating mode of the circuit where Y2 is energized and the motor rotates counter clock clockwise. The sequence of operation is as follows:

1. S3 must be pressed and released first to reset the system.
2. S2 is pressed & released.
3. K2 is energized.
4. K2B latches the circuit and continues to pass the power to K2.
5. Y2 is energized, the directional valve is shifted accordingly, and the motor rotates clockwise.
6. Simultaneously, K2C is open so that pressing S1 will not change the situation.

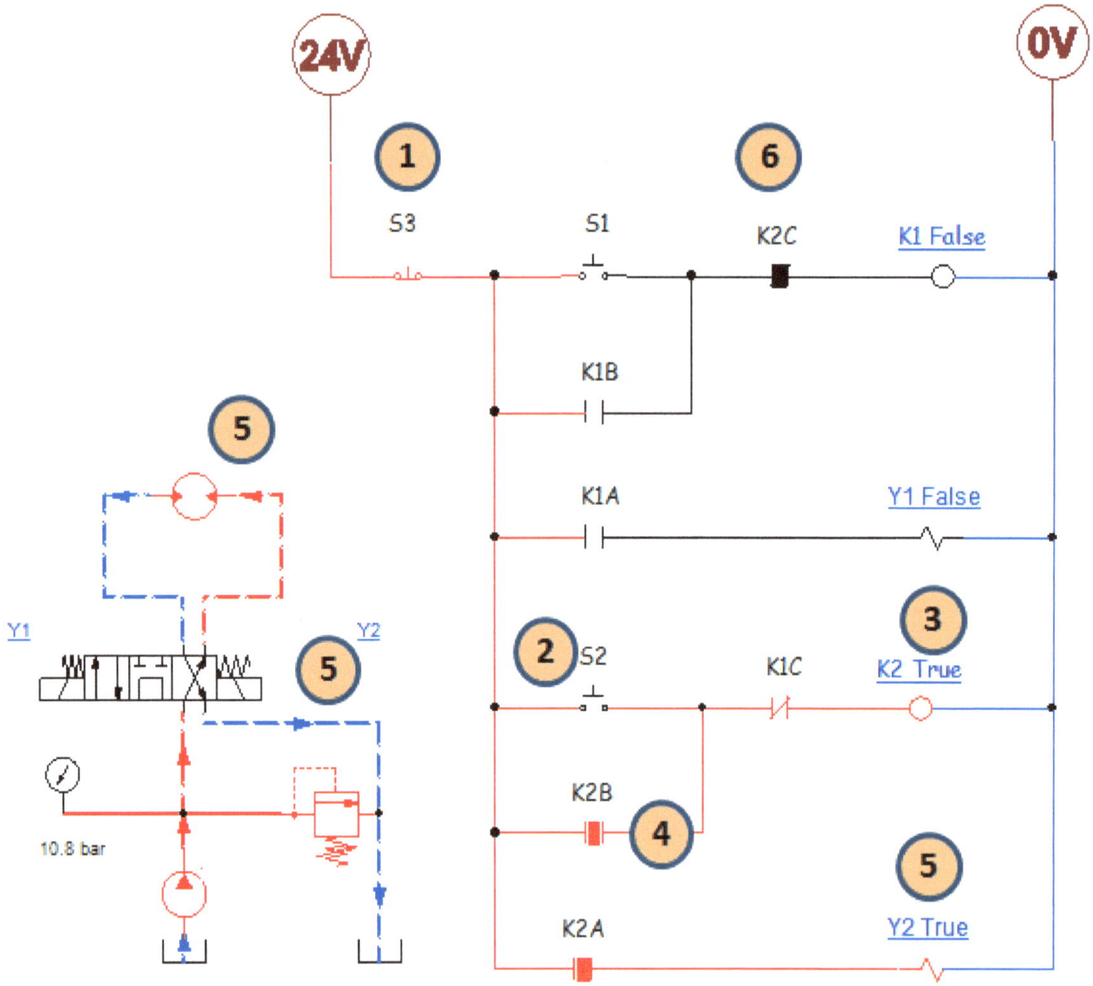

S1	S2	S3	K1	K1A	K1B	K1C	K2	K2A	K2B	K2C	Y1	Y2	Motor
0	1/0	0	0	0	0	1	1	1	1	1	0	1	CCW

Fig. 4.35C - Operating Conditions of the Circuit, Motor CCW Rotation

4.5.8- Position-Dependent Cylinder Deceleration

By adding some more elements to the previous diagram, a new function can be developed. As it has been stated in Volume 1 of this series of books, if cylinder speed is between 10-20 cm/s, the cylinder should be equipped with cushioning devices. If the cylinder speed exceeds 20 cm/s, the cylinder must be decelerated towards the end of its stroke in order to reduce the impact on the end caps. One way is to decelerate the cylinder mechanically. Alternatively, a hydraulic deceleration system can be used. The hydraulic deceleration system, shown in the Fig. 4.36A is used to decelerate the cylinder after it passes 50% of its extension stroke. The associated electric schematic diagram contains the same circuit shown in Fig 4.35A that protects and actuates the three-position directional valve with two solenoids.

To add the functionality of cylinder deceleration, the following components have been added to the lower part of the circuit:

- A normally-open proximity switch PX. It is placed at 50% of the cylinder stroke.
- A relay coil K3 and a corresponding normally-open relay contact K3A.

The figure and the associated truth table shows the reset mode of the circuit where K1, K2 and K3 are de-energized.

Figure 4.36B and the associated truth table shows the operating mode of the circuit where Y1 and Y3 are energized and the cylinder extends. The sequence of operation is as follows:

1. S1 is pressed & released.
2. K1 is energized.
3. K1B latches the circuit and continues to pass power to K1.
4. Y1 is energized, the directional valve is shifted accordingly, and the cylinder extends with maximum speed.
5. Simultaneously, K1C is opened so that pressing S2 will not energize K2.
6. Once the cylinder reaches 50% of its extension, PX energizes K3, K3A closes, and the solenoid Y3 is energized. Since the flow has nowhere to go but through the flow control valve, the cylinder decelerates.

Figure 4.36C and the associated truth table shows the operating mode of the circuit where Y2 is energized and the cylinder retracts. The sequence of operation is as follows:

1. S3 must be pressed and released first to reset the system.
2. S2 is pressed & released.
3. K2 is energized.
4. K2B latches the circuit and continues to supply power to K2.
5. Y2 is energized, the directional valve is shifted accordingly, and the cylinder retracts with maximum speed.
6. Simultaneously, K2C is opened so that pressing S1 will not energize K1.
7. PX will get back to its 0 state only when the cylinder retracts to less than 50% of its stroke.

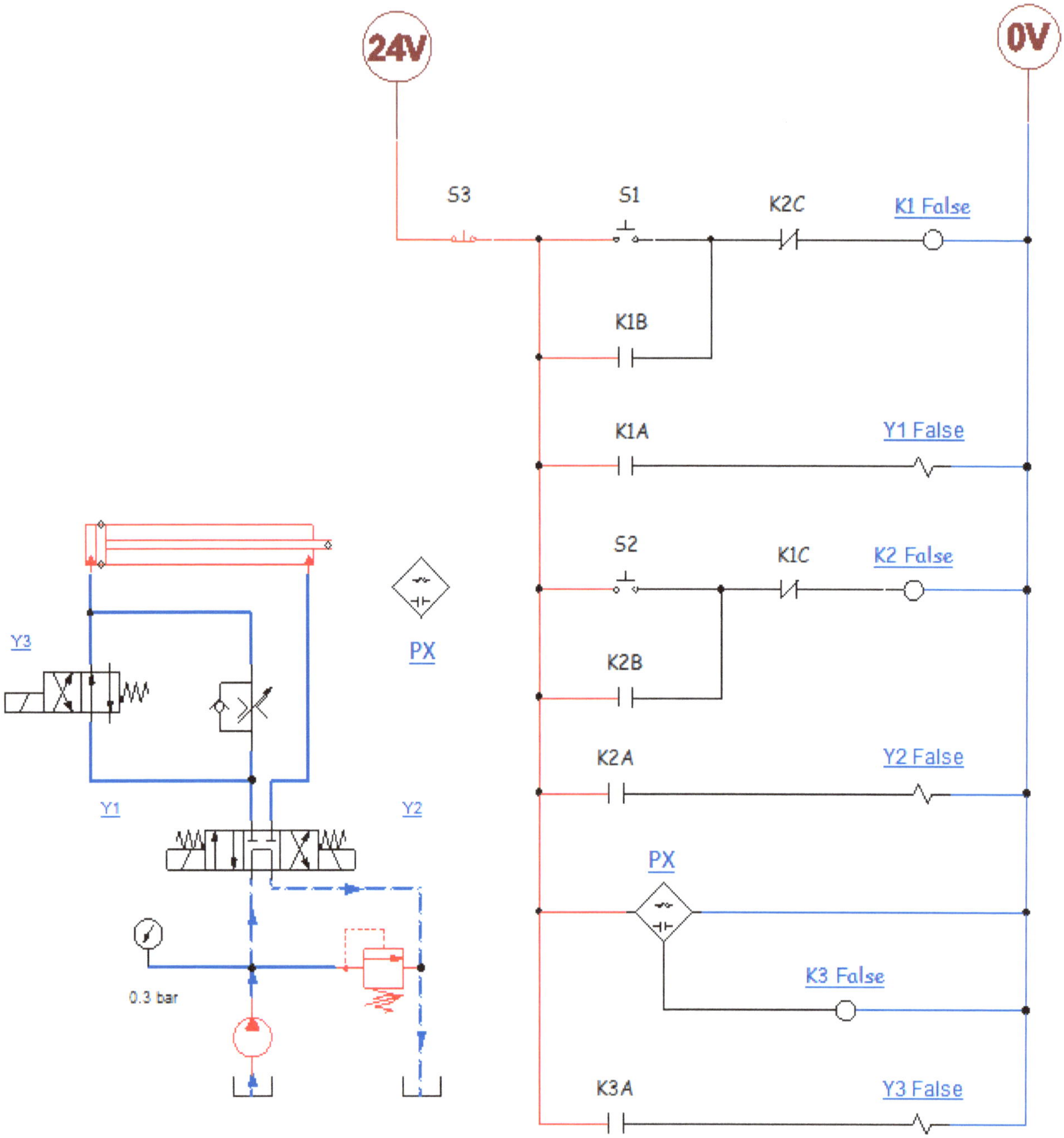

S1	S2	S3	K1	K1A	K1B	K1C	K2	K2A	K2B
0	0	1	0	0	0	1	0	0	0

K2C	Y1	Y2	PX	K3	K3A	Y3	Cylinder
1	0	0	0	0	0	0	Stop

Fig. 4.36A - Reset State of a Cylinder Deceleration System

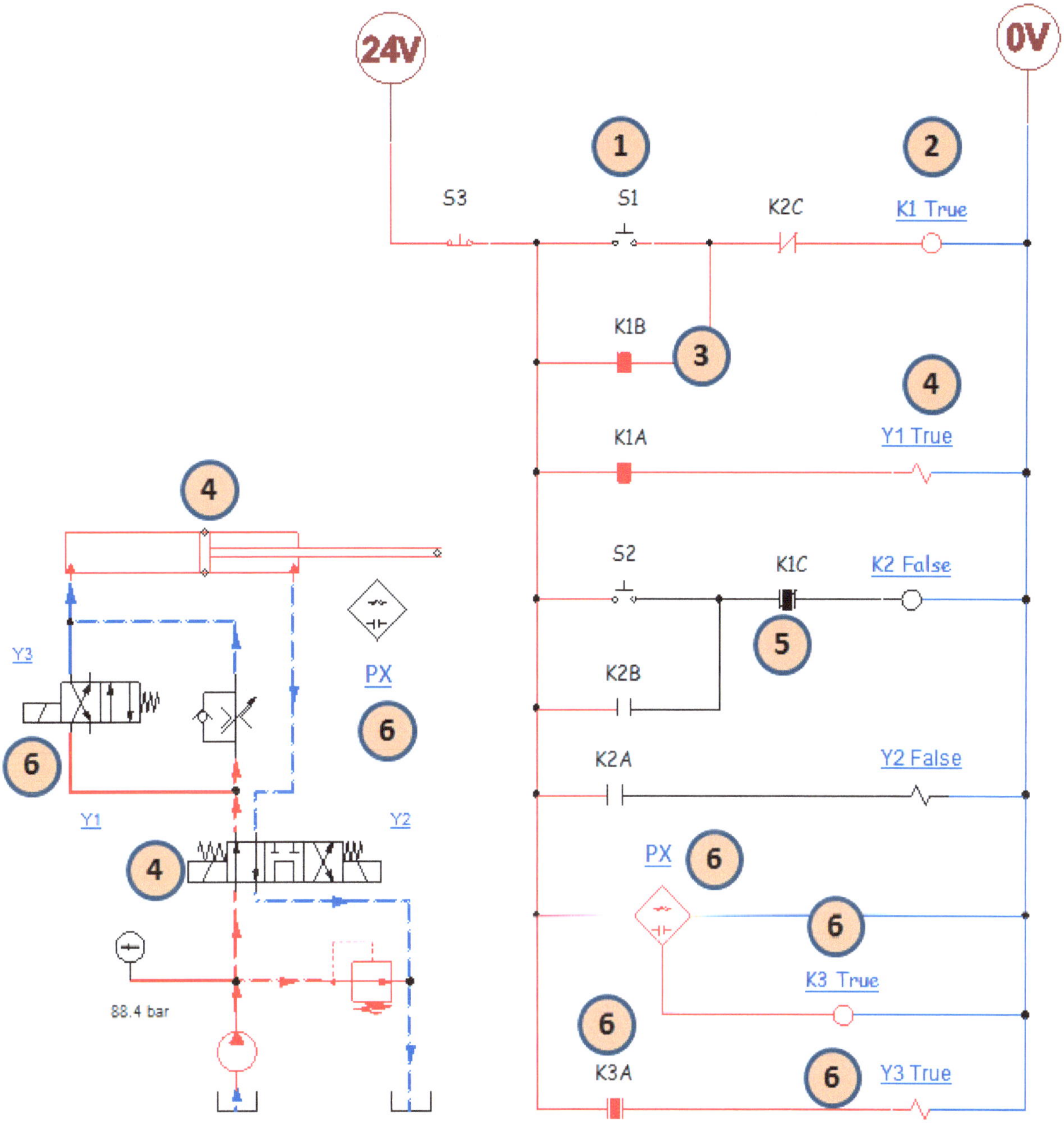

S1	S2	S3	K1	K1A	K1B	K1C	K2	K2A	K2B
1/0	0	0	1	1	1	0	0	0	0

K2C	Y1	Y2	PX	K3	K3A	Y3	Cylinder
1	1	0	1	1	1	1	Extends

Fig. 4.36B - Cylinder Deceleration System during Cylinder Extension

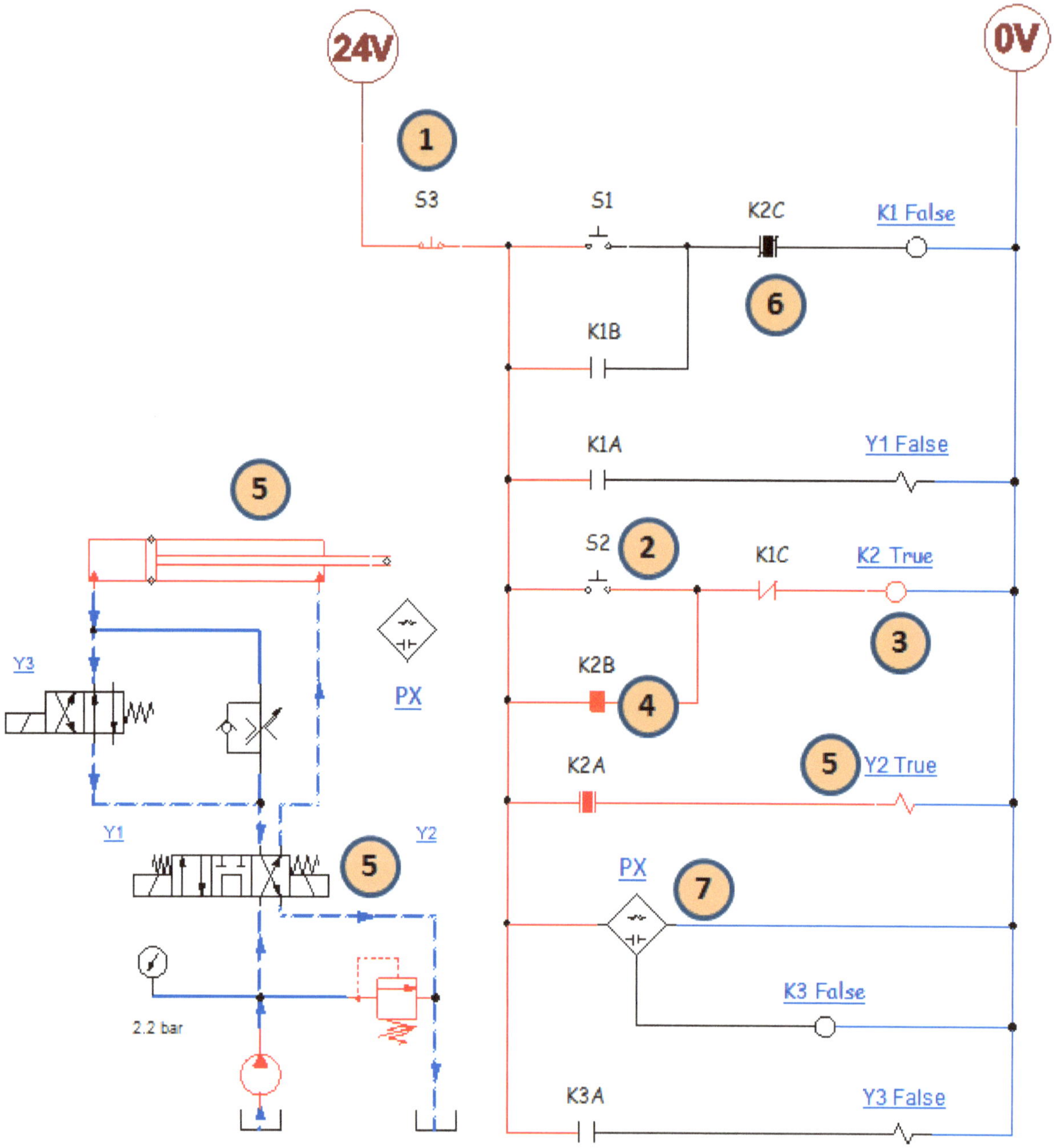

S1	S2	S3	K1	K1A	K1B	K1C	K2	K2A	K2B
0	0/1	0	0	0	0	1	1	1	1

K2C	Y1	Y2	PX	K3	K3A	Y3	Cylinder
0	0	1	0	0	0	0	Extends

Fig. 4.36C - Cylinder Deceleration System during Cylinder Retraction

4.5.9- One-Cycle Hydraulic Cylinder Reciprocation

One-cycle cylinder reciprocation means extending the cylinder by pressing/releasing a pushbutton then the cylinder retracts automatically. The simple latching circuit shown in Fig. 4.37 is used to extend the cylinder by pressing/releasing S1. A normally-closed X component is used to retract the cylinder when it is triggered. The figure shows various components that can do such a function as follows:

- A pressure switch (1) retracts the cylinder based on the cylinder pressure.
- A proximity or limit switch (2) retracts the cylinder based on the cylinder position.
- A break time delay switch (3) retracts the cylinder based on elapsed time after triggering the switch.

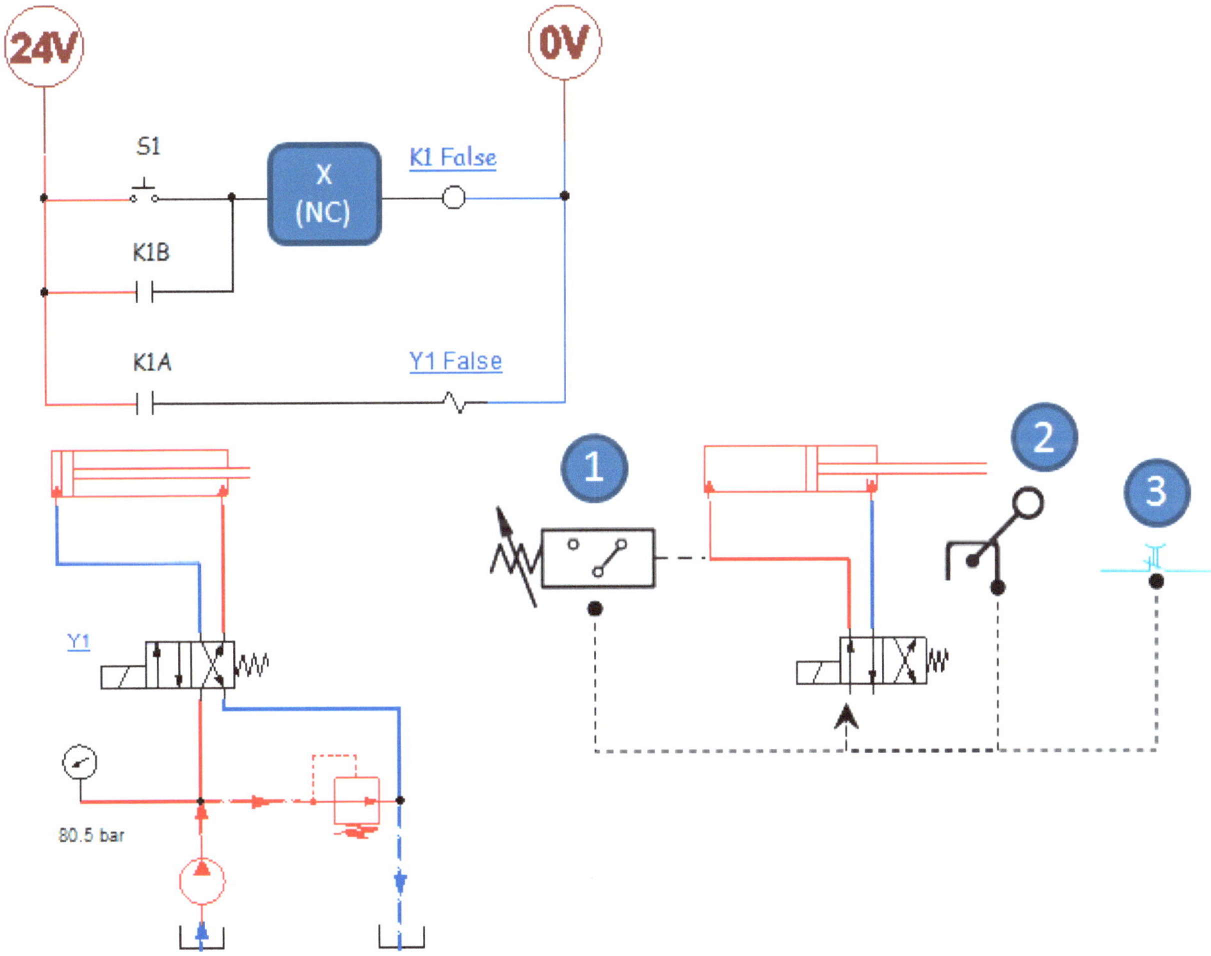

Fig. 4.37 - One Cycle Cylinder Reciprocation

Figure 4.38 shows a typical example of automatic cylinder reversal using a pressure switch PS. When the machine operator presses the momentary pushbutton S1, the cylinder makes a forward stroke. The signal is stored by the latching relay contact K1B. When the cylinder is deadheaded, piston side pressure increases, PS is triggered, the normally-closed contact K2A opens terminating the latching line, and the cylinder is retracted. The cycle can be started over by pressing S1. At any time or at any position of the cylinder, cycle can be terminated and the cylinder intentionally reversed by pressing S2.

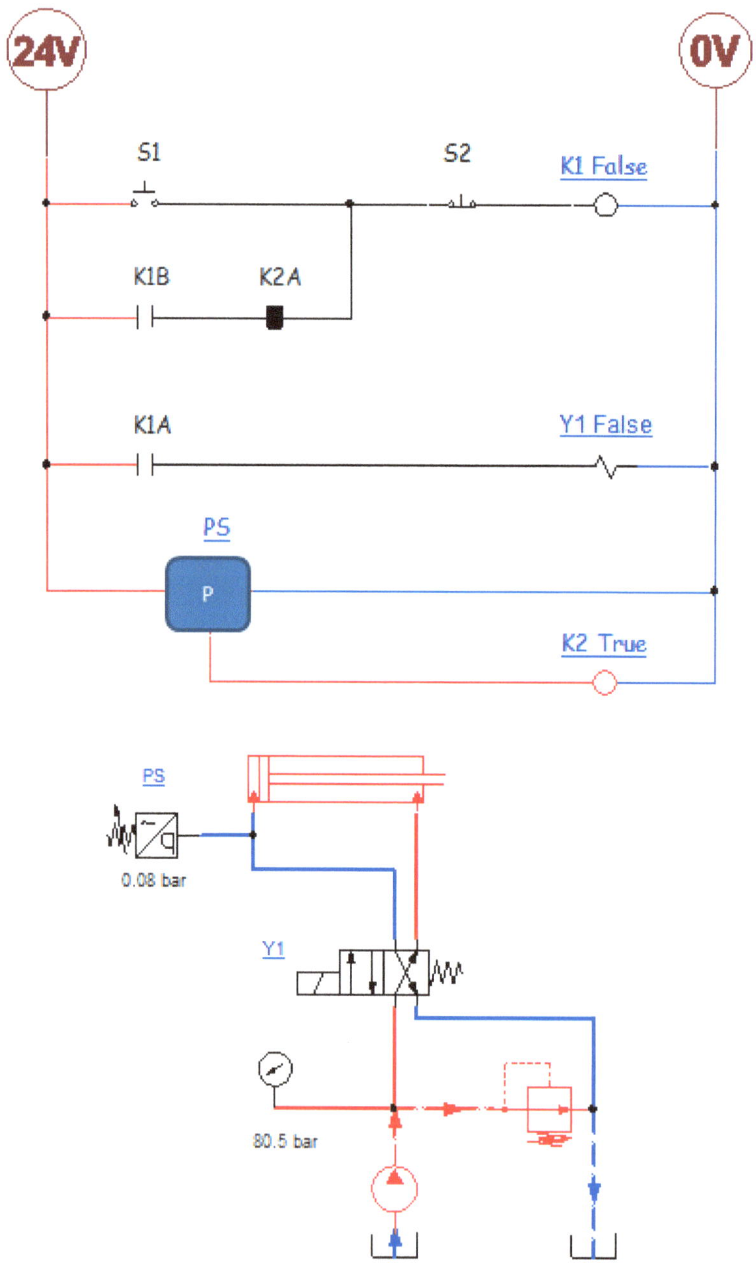

Fig. 4.38 - One Cycle Cylinder Reciprocation using a Pressure Switch

4.5.10- Continuous Cylinder Reciprocation

Figure 4.39A shows a typical example of continuous cylinder reciprocation using two limit switches. Positioning of the limit switches defines the stroke along which the cylinder reciprocates. Similar circuits can be built using proximity or pressure switches. The figure shows the reset mode of the circuit where K1 is de-energized. The circuit consists of:

- Two normally-closed limit switches, LS1 and LS2.
- One toggle switch TS1 and one reset pushbutton S2.
- A relay coil K1, with two normally-open contacts K1A and K1C, and a normally closed contact K1B.

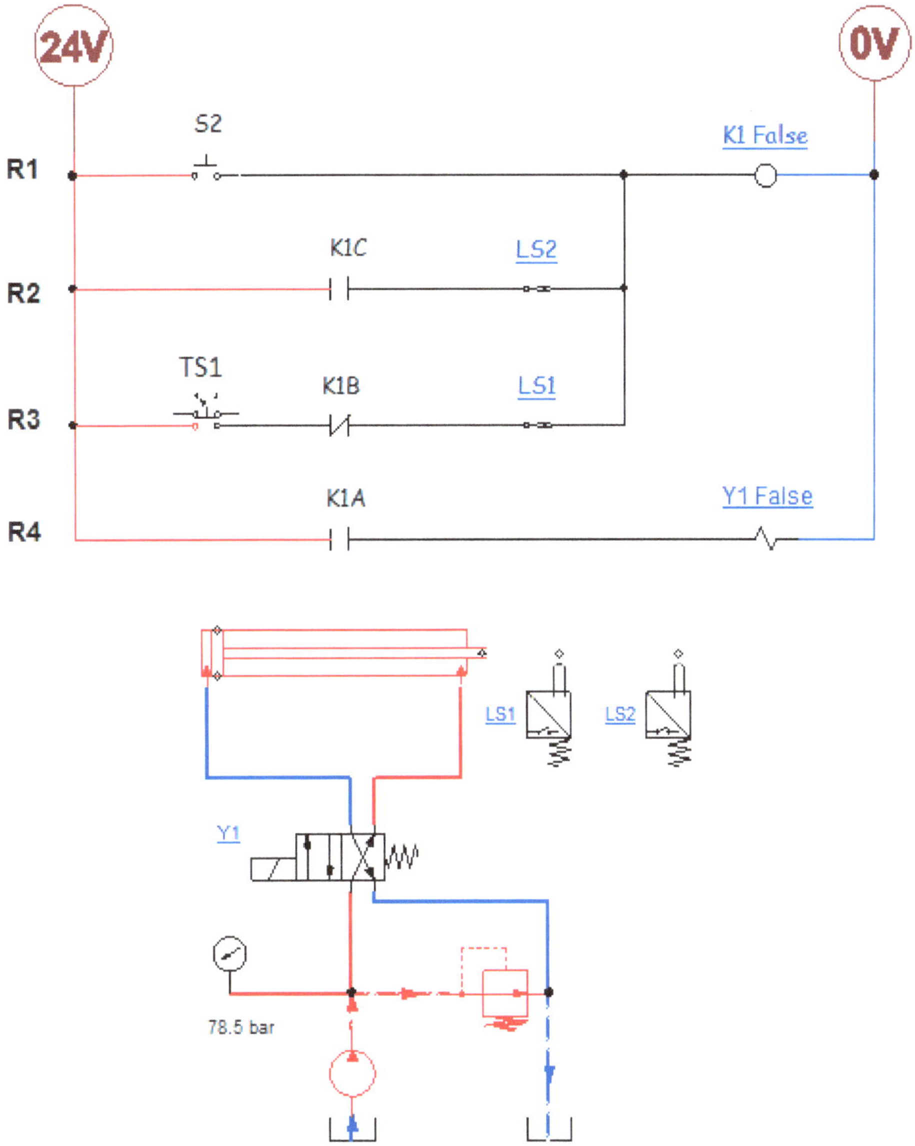

Fig. 4.39A - Continuous Reciprocation using a Limit Switches

Figure 4.39B shows the operating mode of the circuit where Y1 is energized and the cylinder extends. The sequence of operation is as follows:

1. TS1 is in the closed position.
2. Power passes through the rung R2, K1 is energized.
3. K1C latches the circuit and continues to pass the power to K1.
4. Simultaneously K1B opens.
5. Simultaneously K1A closes, Y1 is energized, the directional valve is shifts accordingly, and the cylinder extends.
6. When the cylinder contacts the first limit switch LS1, the corresponding contact opens.

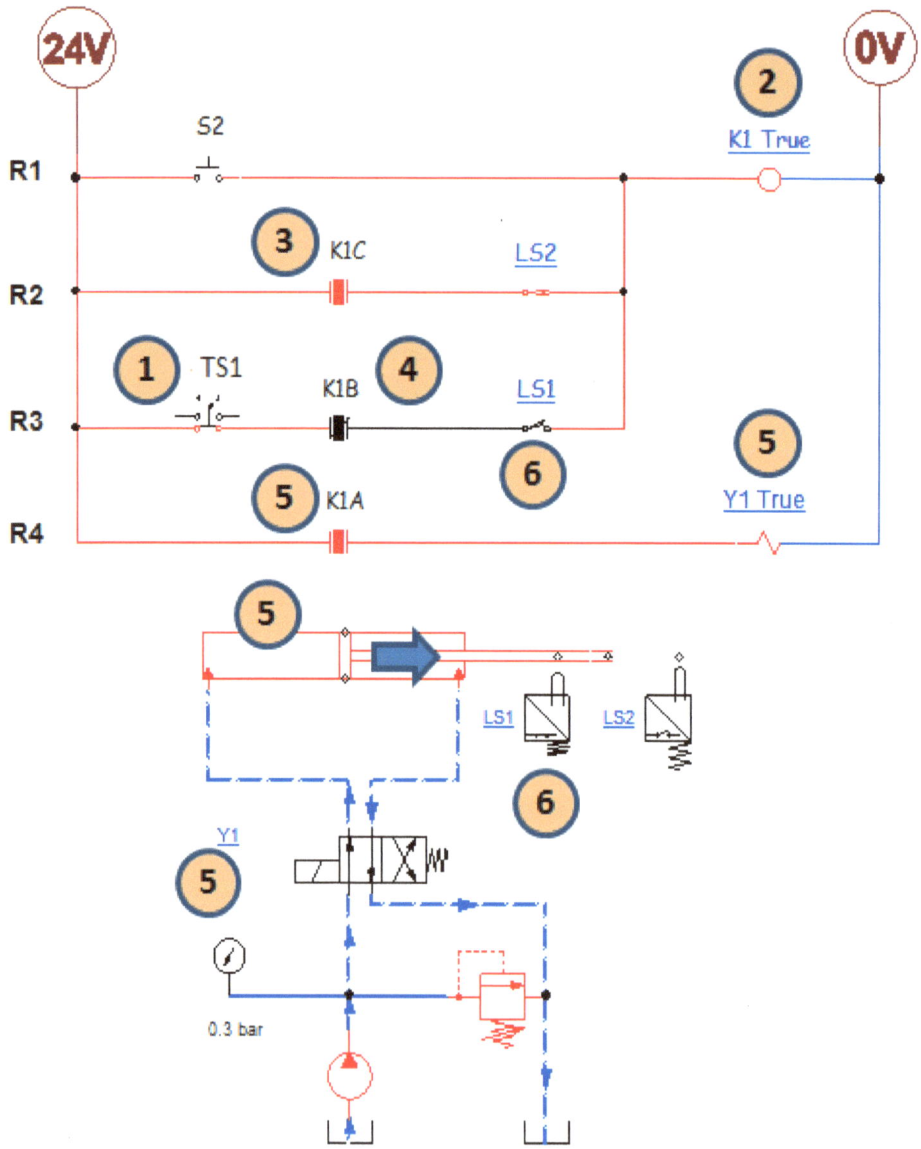

Fig. 4.39B - Cylinder Extension during Continuous Reciprocation

Figure 4.39C shows the operating mode of the circuit where Y1 is de-energized and the cylinder retracts. The sequence of operation is as follows:

1. Once the cylinder contacts the second limit switch LS2, the corresponding contact on R2 terminates the latching circuit.
2. K1 is de-energized and K1A opens.
3. Y1 is de-energized, directional valve shifts back to its initial position, and the cylinder retracts.
4. When the cylinder clears its contact with LS2 and LS1, the corresponding contacts on R2 and R3 returns to their initial closed states.
5. The circuit is now reset and the cylinder continues to reciprocate as long as TS1 is closed.
6. If TS1 is open, pressing and releasing S2 will cause the cylinder to perform one cycle only.

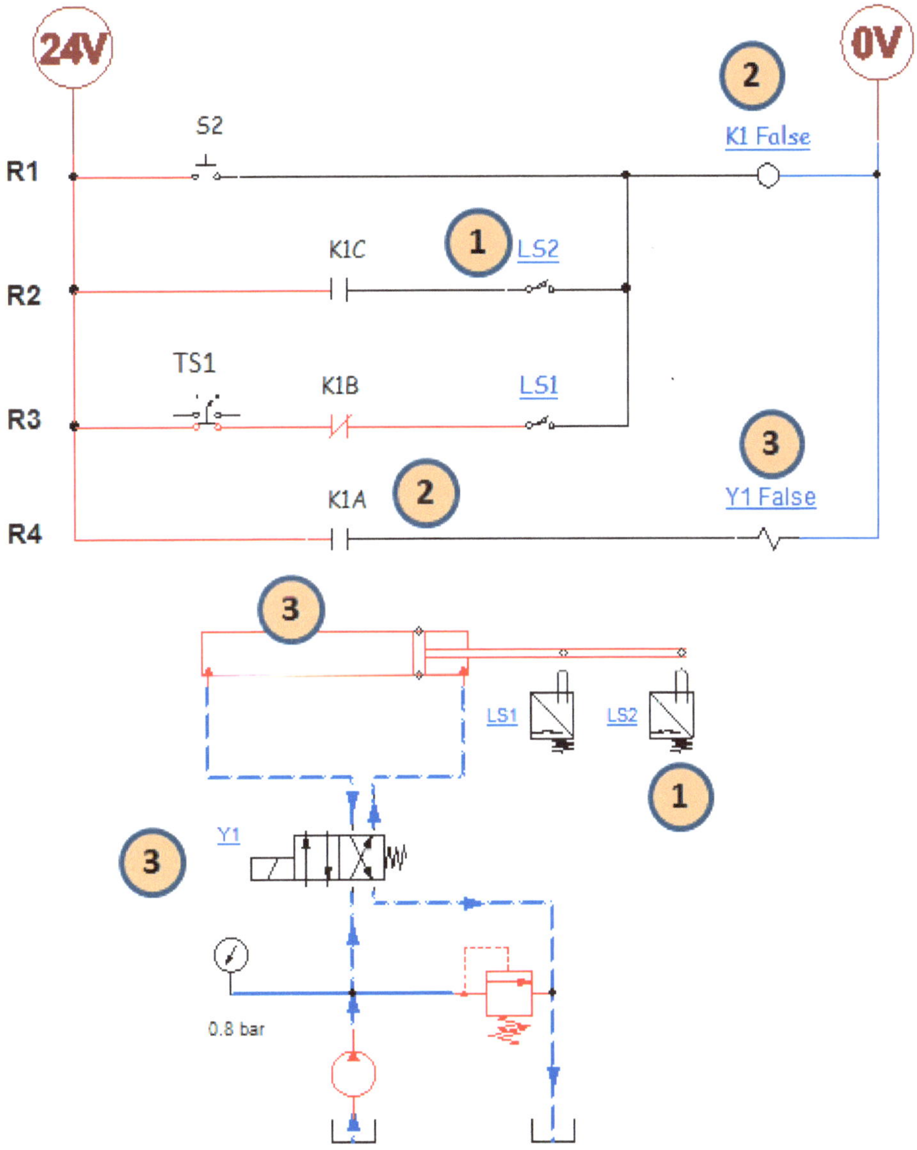

Fig. 4.39C - Cylinder Retraction during Continuous Reciprocation

4.5.11- Panic Circuit

Panic devices, such as warning lights and buzzing sounds, can be added to the basic electrical circuit to improve the safety of machine operation. Figure 4.40 shows a typical example of using panic devices. The figure shows a hydraulic elevator that is actuated by a toggle switch TS1.

By switching TS1, the cylinder extends lifting the load. Simultaneously, both the red warning light and the buzzing sound are turned ON. Putting TS1 back to its original state, retracts the cylinder lowering the load, red light and buzzing sound are turned OFF, and green light is turned ON.

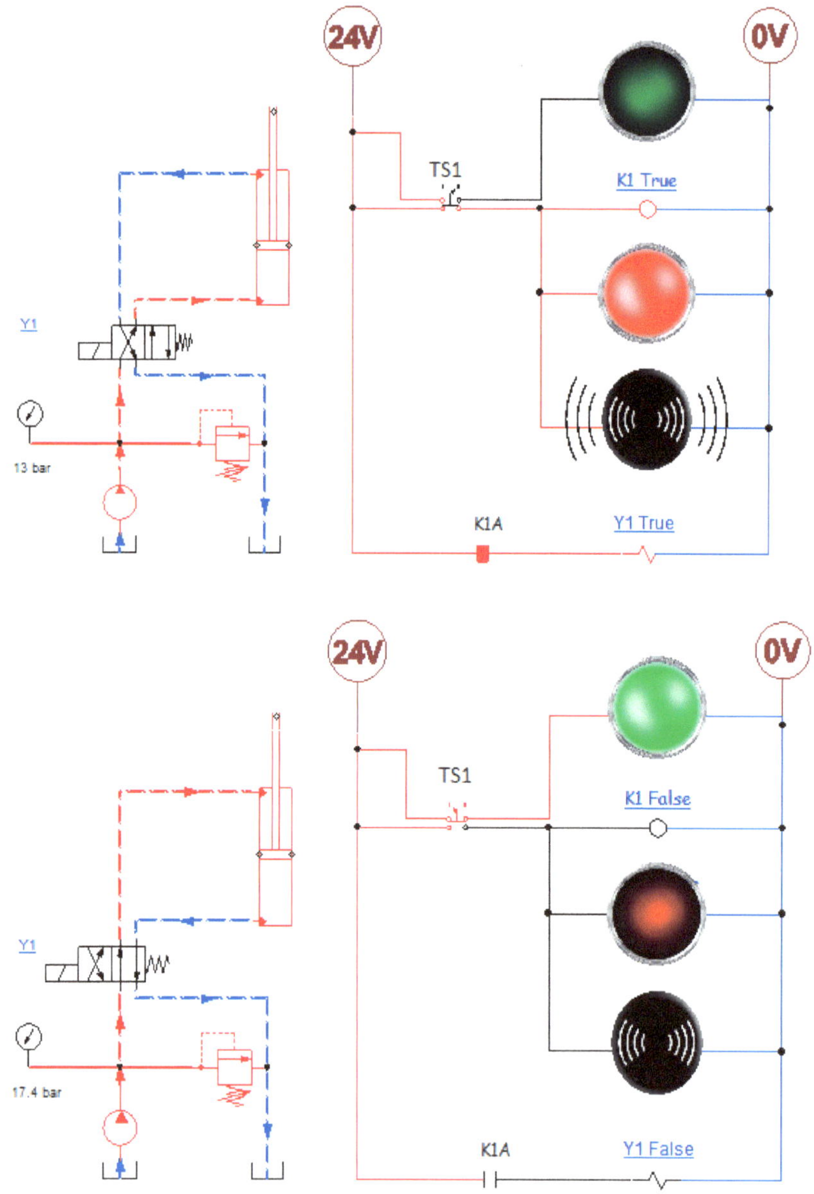

Fig. 4.40 - Panic Circuit

4.5.12- Conjunction Functions

In order to make sure the output device is actuated based on triggering more than one input device; a logic function must be used. Such a logic function is named *Conjunction* function. The conjunction function is primarily used for safety purposes to ensure appropriate inputs are triggered before driving a load.

Figure 4.41 shows a typical example of changing state of a load based on signal generated by two sources. As shown in the upper row of the figure, two normally-open pushbuttons are placed in series upstream of the load on the same rung. Both pushbuttons must be pressed in order to energize the rung in the logic diagram. This Conjunction function (*AND for YES*) is commonly known as an "*AND*" function.

As shown in the lower row of the figure, two normally-closed pushbuttons are placed in parallel upstream of the load on the same rung. Both pushbuttons must be pressed in order to de-energize the rung in the logic diagram. This Conjunction function (*AND for NOT*) is commonly known as an "*NAND*" function.

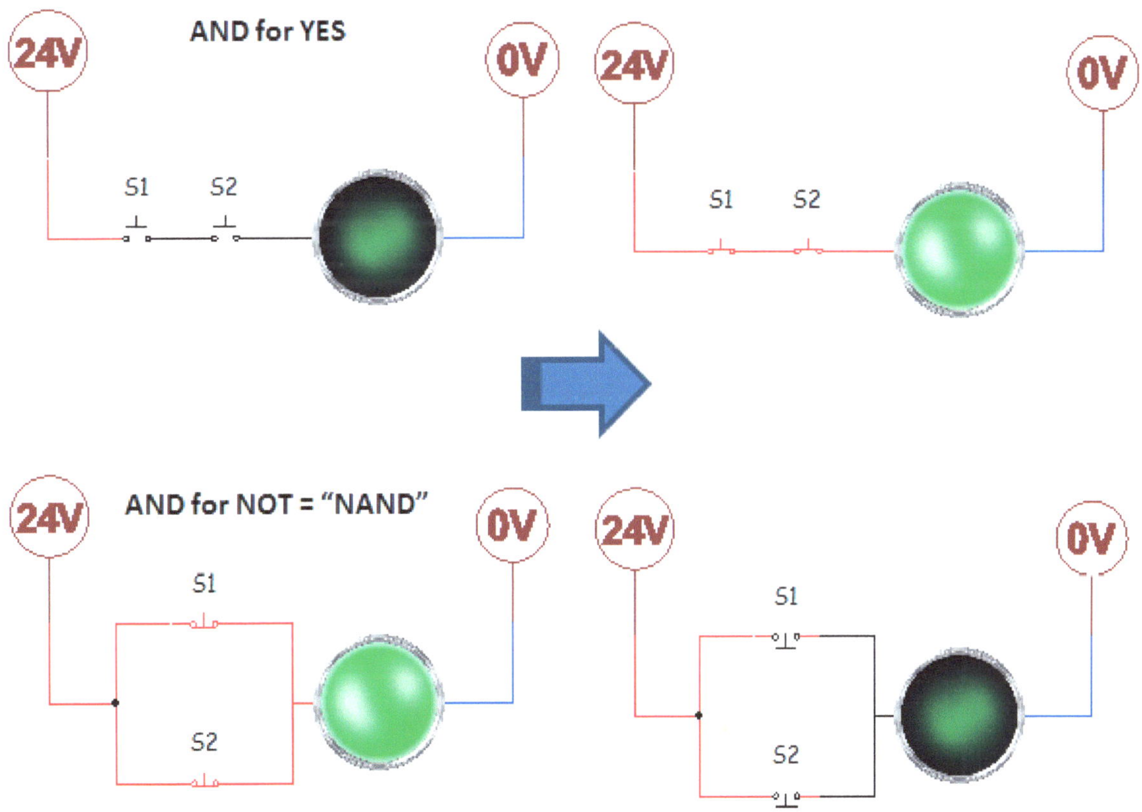

Fig. 4.41 - Conjunction Functions

Figure 4.42A represents a practical application of a NAND function in which a pressure switch PS and a proximity switch PX must be actuated in order to retract the cylinder. The figure shows also a way of numbering the terminals of the devices for the sake of building a wiring diagram.

The figure shows the circuit in the reset mode where K1 is de-energized. The circuit consists of:

- A pushbutton S1 and a reset pushbutton S2.
- A relay coil K1, two normally-open relay contacts K1A and K1B, and a solenoid Y1.
- A pressure switch PS, corresponding relay coil K2, and a normally-closed contact K2A.
- A proximity switch PX, corresponding relay coil K3, and a normally-closed contact K3A.

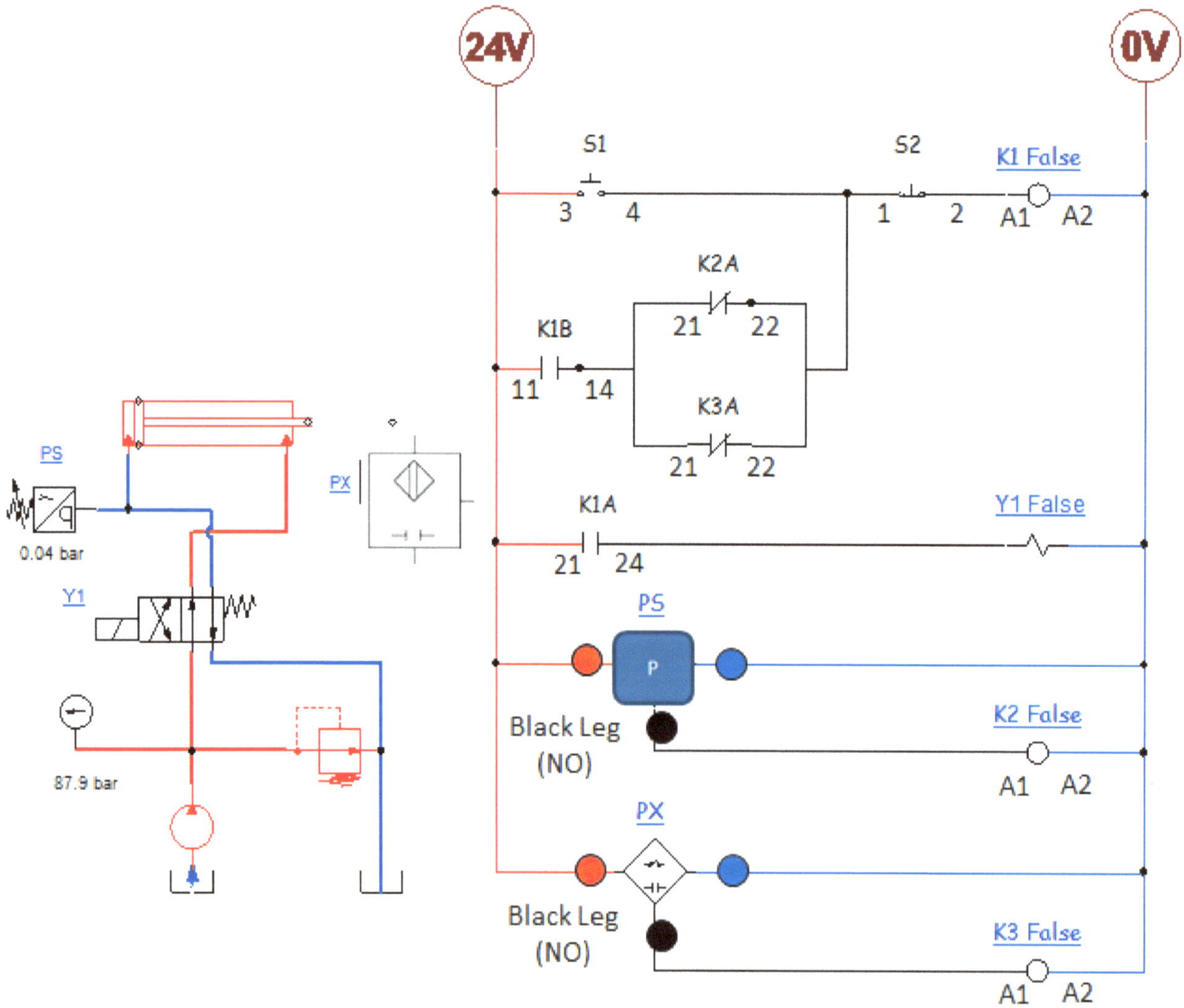

Fig. 4.42A - Conjunction (NAND) Function

Figure 4.42B shows the operating mode of the circuit where Y1 is energized and the cylinder extends. The sequence of operation is as follows:

1. S1 is pressed & released and K1 is energized.
2. K1B latches the circuit and continues to pass power to K1 through the normally-closed K2A and K3A.
3. Y1 is energized, the directional valve is shifted accordingly, and the cylinder extends.

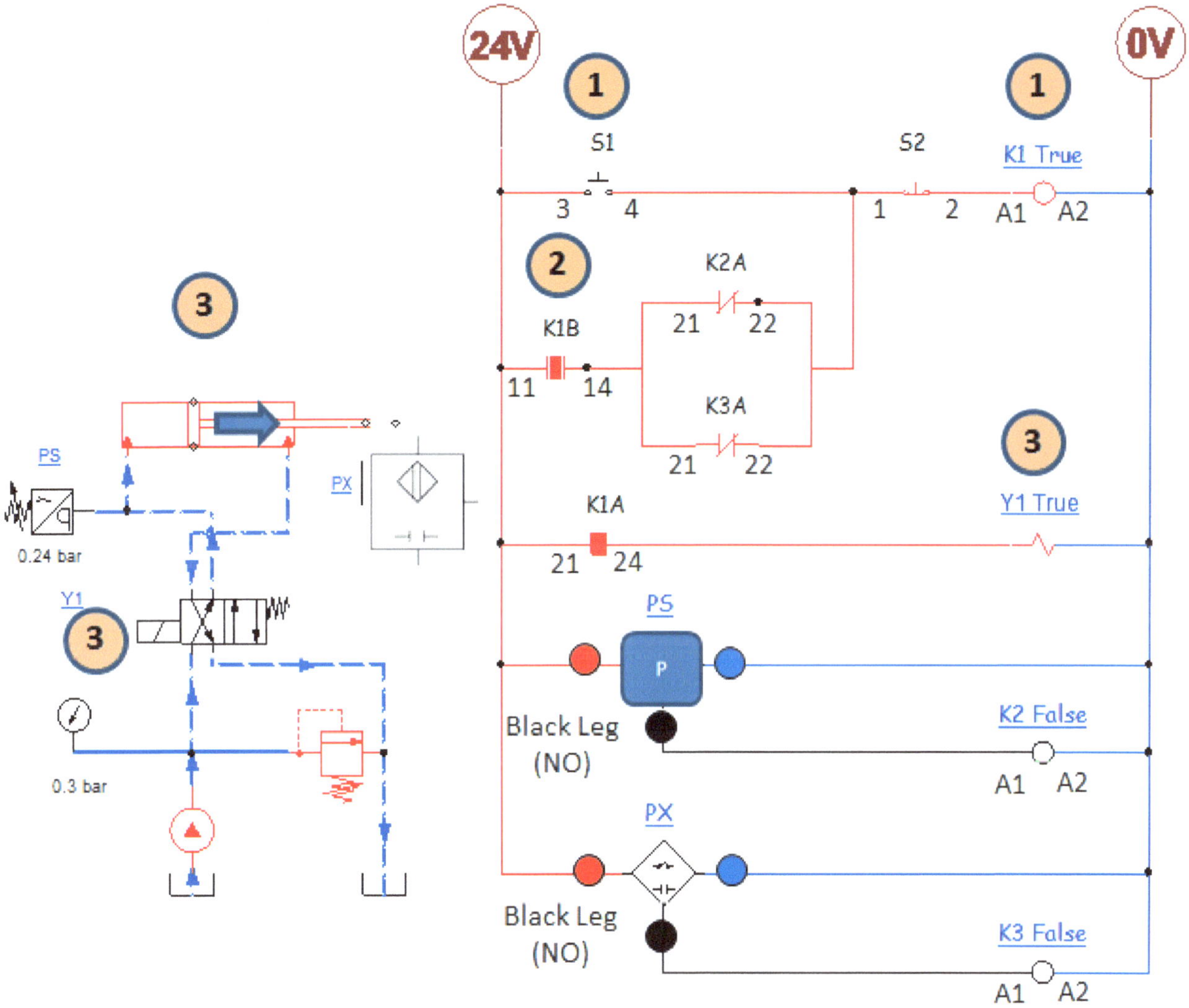

Fig. 4.42B - Cylinder Extension in a Conjunction (NAND) Function

Figure 4.42C shows the operating mode of the circuit where Y1 is de-energized and the cylinder retracts. The sequence of operation is as follows:

1. During the cylinder extension, it actuates first PX so that K3 is energized and K3A opens.
2. The cylinder continues to extend, if it faces a heavy load or deadheaded, pressure at the piston side increases so that the K2 is energized and K2A opens.
3. When K2A AND K3A are open, the latching circuit is terminated, K1 is de-energized.
4. Y1 is de-energized and the directional valve shifts back to its initial position.
5. The cylinder retracts, and the circuit goes back to the reset mode.
6. At any time, pressing S2 will turn K1 Off and the cylinder retracts.

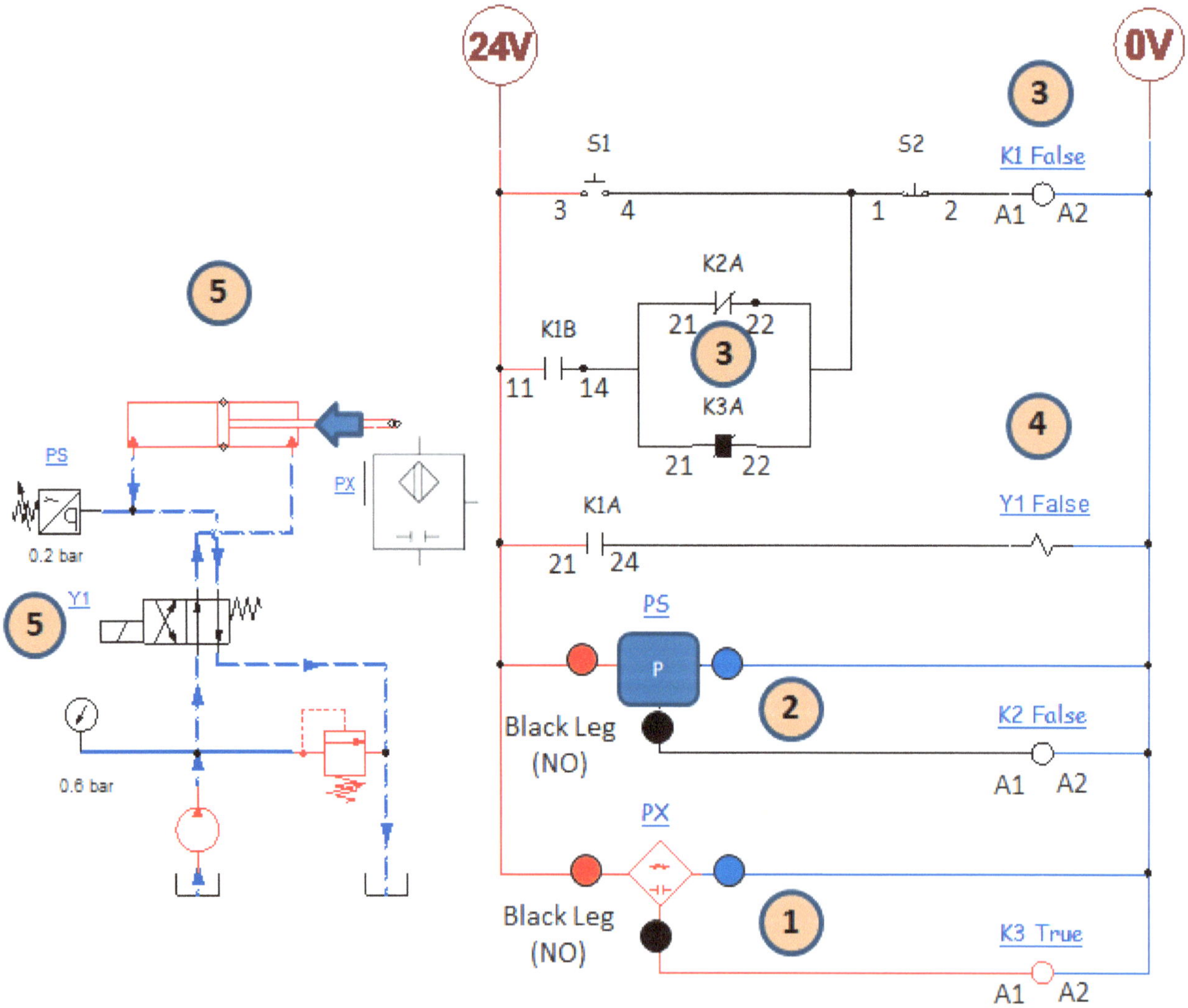

Fig. 4.42C - Cylinder Retraction in a Conjunction (NAND) Function

4.5.13- Disjunction Functions

In order to actuate the output device from different locations using more than one input device; a logic function must be used. Such a logic function is named *Disjunction* function. The Disjunction function is primarily used for remote control purposes.

Figure 4.43 shows a typical example of changing state of a load based on signal from either one of two sources. As shown in the upper row of the figure, two normally-open pushbuttons are placed in parallel upstream of the load on the same rung. Pressing either S1 or S2 will energize the circuit. This Disjunction function (*OR for YES*) is commonly known as an "*OR*" function.

As shown in the lower row of the figure, two normally-closed pushbuttons are placed in series upstream of the load on the same rung. Pressing either S1 or S2 will de-energize the circuit. This Disjunction function (*OR for NOT*) is commonly known as an "*NOR*" function.

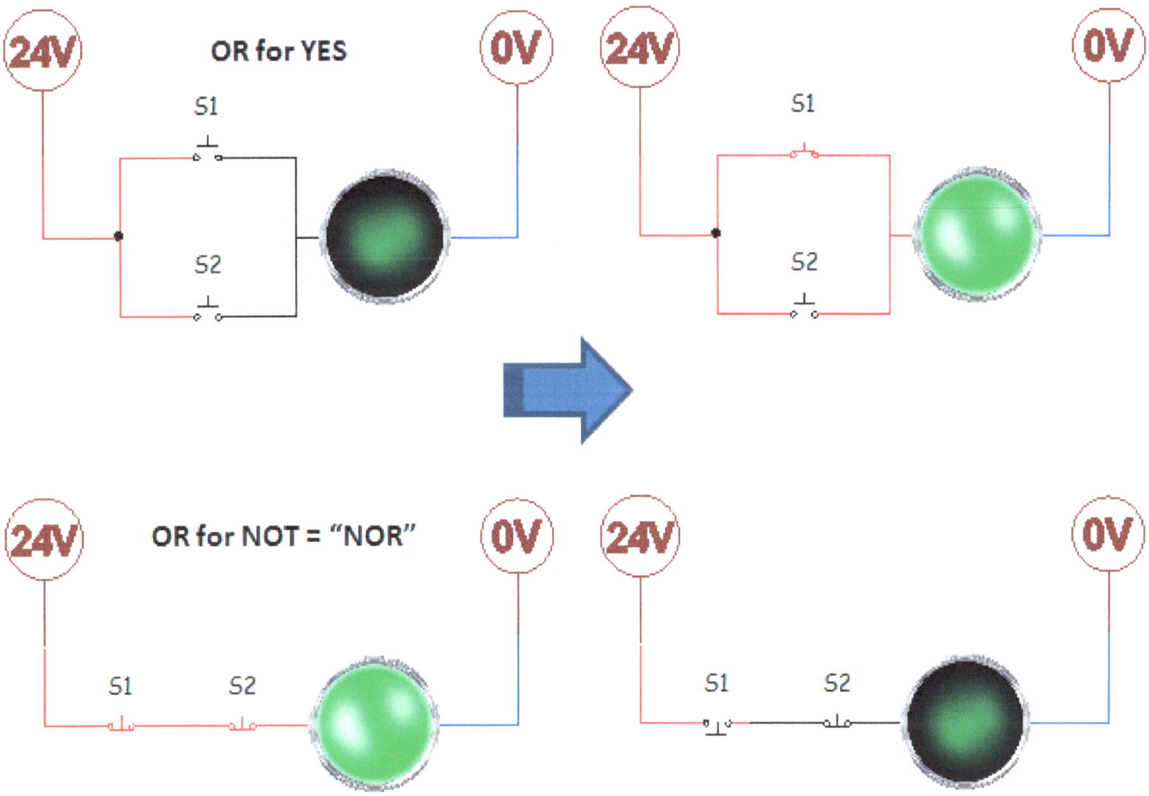

Fig. 4.43 - Disjunction Functions

Figure 4.44 represents a practical application of a NOR function in which a pressure switch PS or proximity switch PX must be actuated in order to retract the cylinder. The figure shows also a way of numbering the terminals of the devices for the sake of building a wiring diagram.

The circuit shown here contains the same elements that are contained in Fig. 4.42A. In the circuit shown here, K2A and K3A are placed in series on the latching line. The cylinder will continue to extend until either PS OR PX is actuated, whichever is actuated first. If one of the switches actuates, the latching line is terminated and the cylinder reverses.

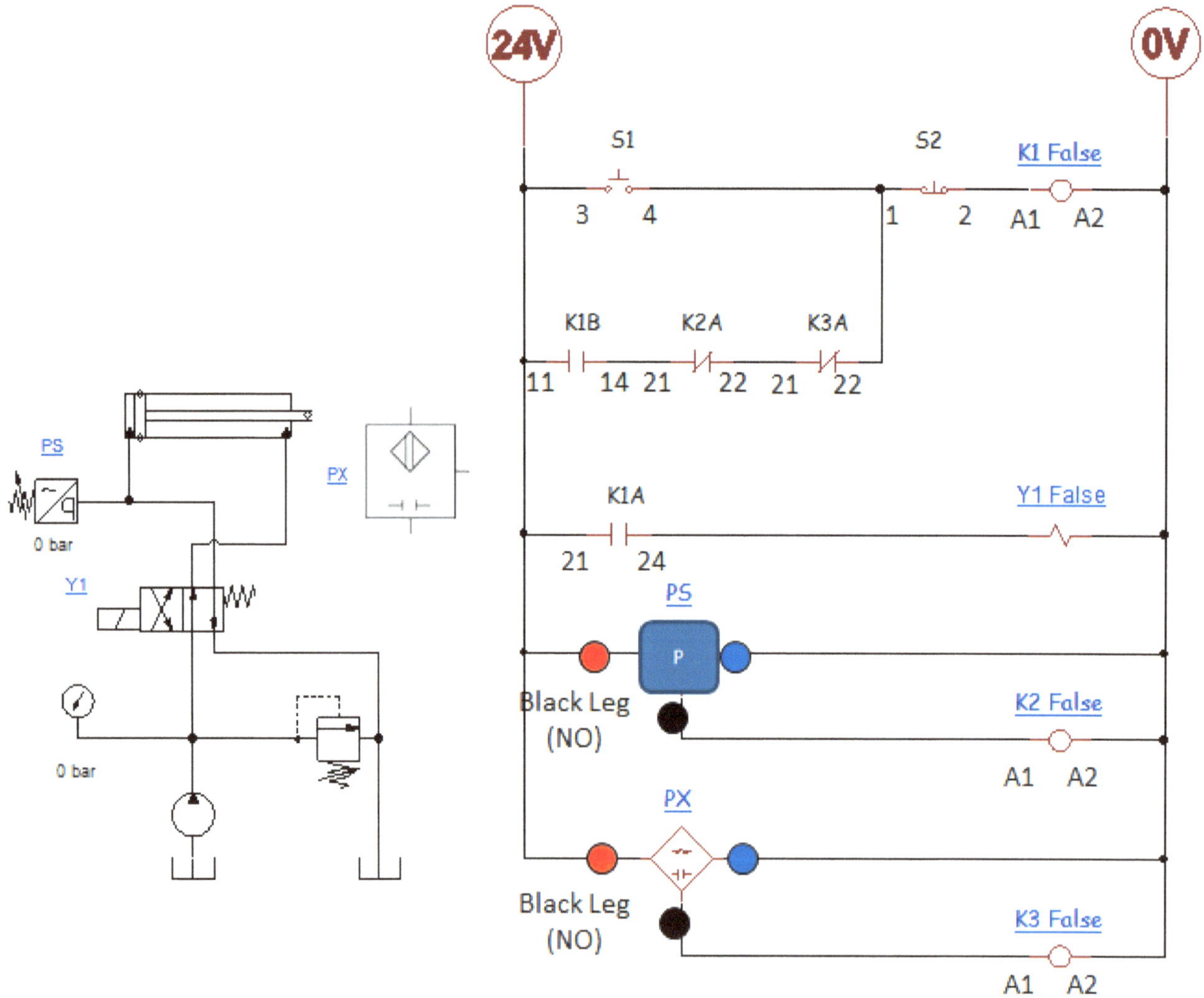

Fig. 4.44 - Disjunction (NOR) Function

4.5.14- Timer Circuits

Time-based sequence control is a common application in industrial applications. This can be accomplished in different ways.

Figure 4.45 shows an application in which the cylinder is reversed automatically based on elapsed time set by a normally-closed contact with ON-delay K2A. When pushbutton S1 is pressed/released the relay coil K1 is energized. Therefore, the latching line is initiated passing power through the relay contact K1B and K2A. As a result, solenoid Y1 is energized, and the cylinder extends. Once the cylinder contacts the proximity switch PX, the relay coil K2 is energized, K2A is triggered to count down the delay time. After the delay time has elapsed, K2A opens terminating the latching line and the cylinder is retracted.

At any time, pressing the reset pushbutton S2 will reverse the cylinder from any position it is in.

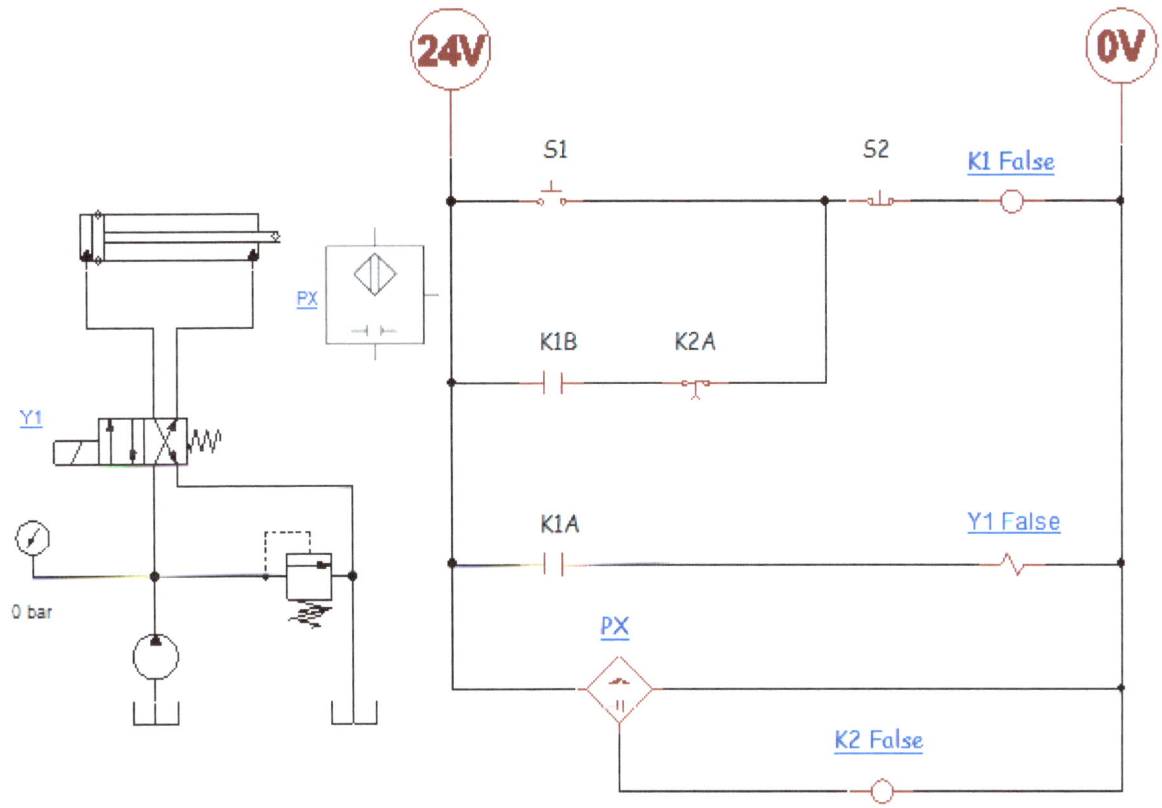

Fig. 4.45 - Time Delay Circuit I

The same function can be achieved by using an analog timer and a normally-closed timer contact as shown in Fig. 4.46. The same sequence of operation applies as the previous circuit. The only difference is the elements that are used.

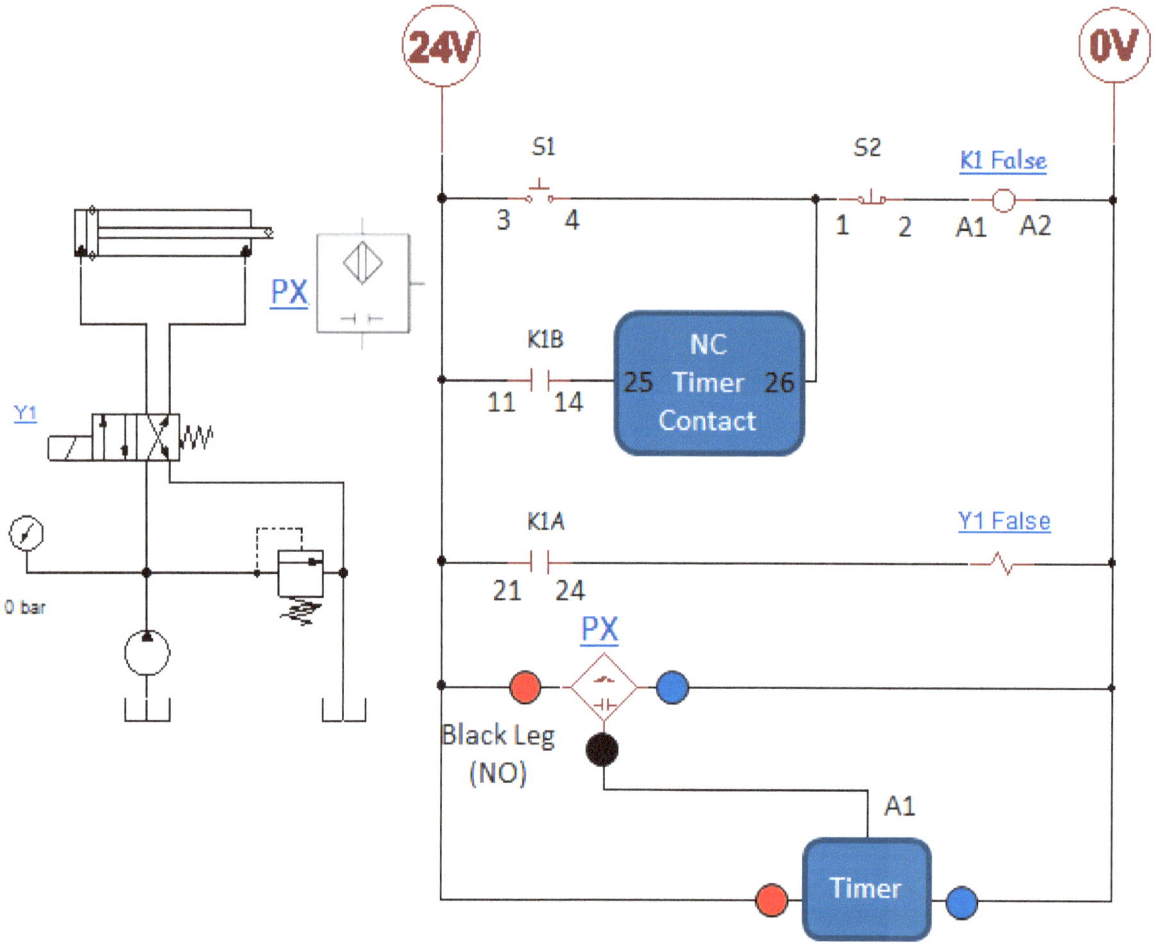

Fig. 4.46 - Time Delay Circuit II

4.5.15- Sequence Control

There are various solutions to accomplish a sequence control. Figure 5.47 shows a schematic diagram of a hydraulic-driven drilling machine tool.

A horizontal cylinder must be extended first to clamp the work piece. Then the vertical cylinder extends to perform the drilling function. When the drilling function is completed, the drilling cylinder must be retracted first before the work piece is unclamped.

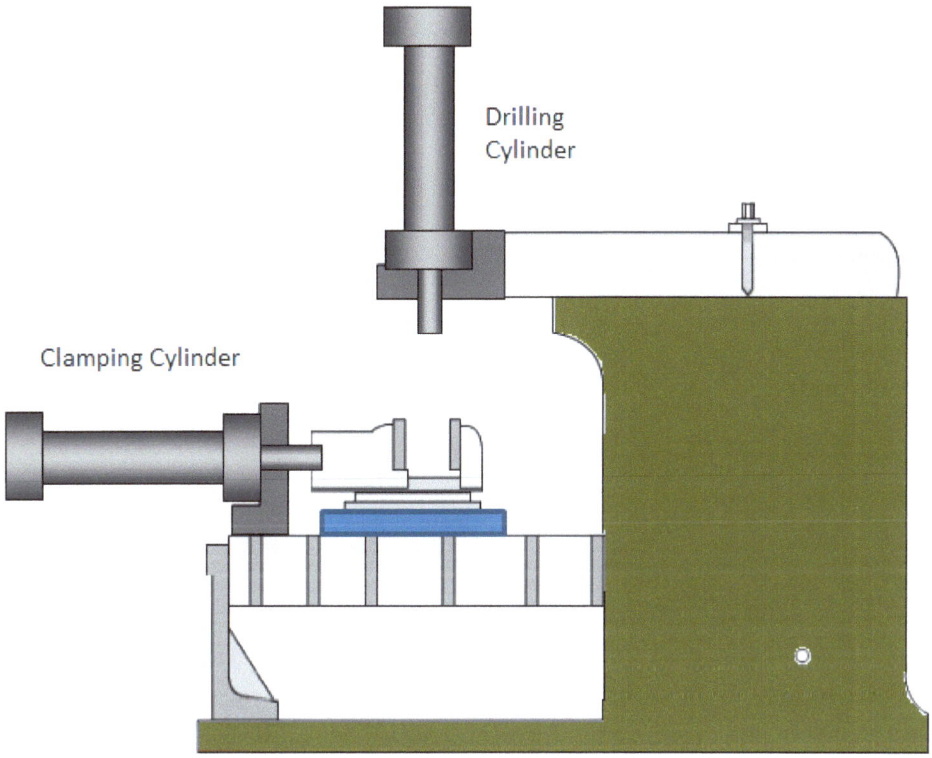

Fig. 4.47 - Hydraulic Driven Drilling Machine

Figure 4.48A shows one of the solutions to control this operation. The circuit consists of:

- A start pushbutton S1 and a reset pushbutton S2.
- A pressure switch PS and a corresponding normally-closed relay contact.
- A limit switch LS and a corresponding normally-closed relay contact.
- A relay coil K1 and two normally-open relay contacts K1A and K1B.

The figure shows the circuit in a reset mode where K1 is de-energized.

The hydraulic circuit contains one two-position solenoid-operated directional valve Y1 and two hydro-mechanical sequence valves SV1 and SV2.

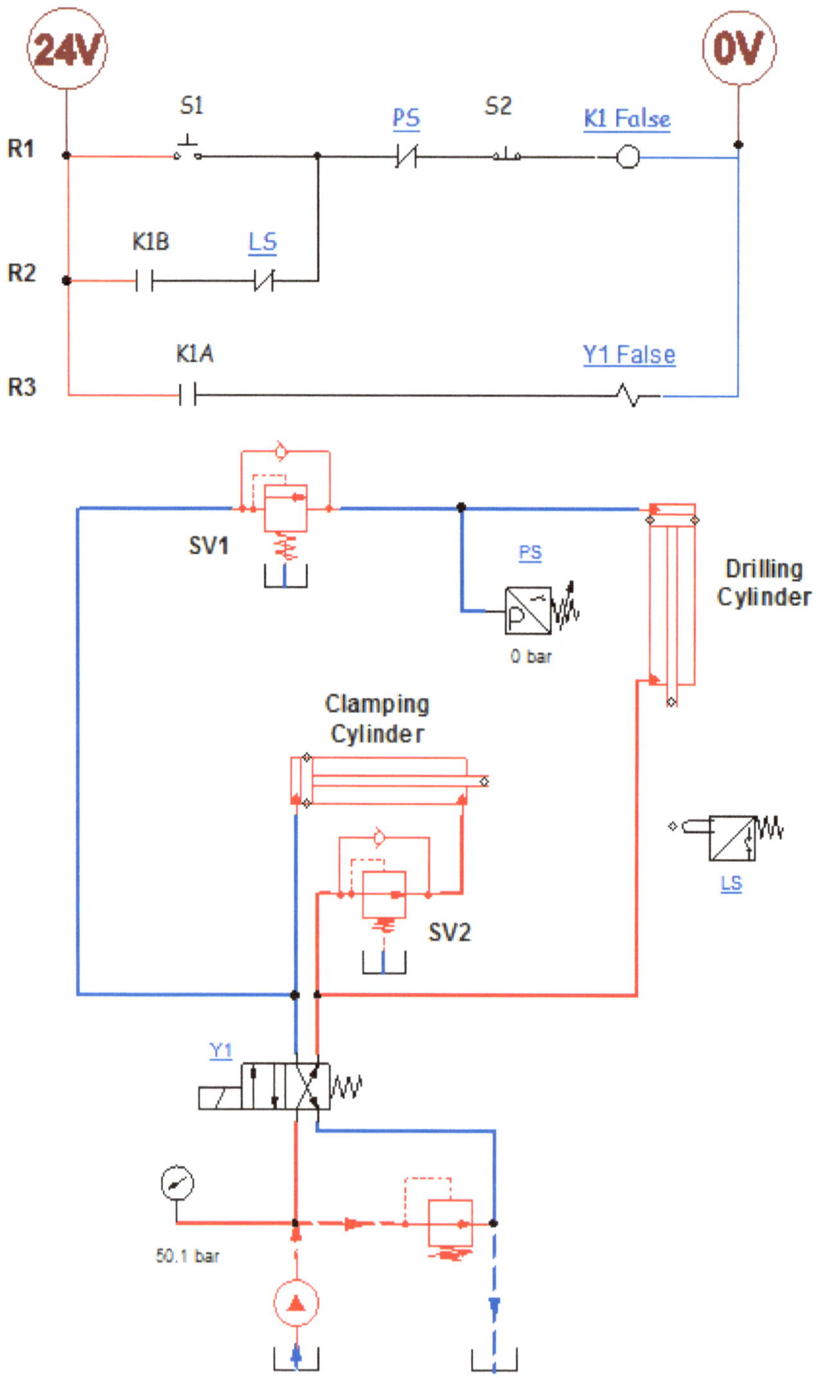

Fig. 4.48A - Reset Conditions for the Sequence Circuit

As shown in Fig. 4.48B, by pressing S1, the latching line R2 is activated. As a result, Solenoid Y1 on R3 is energized, and the clamping cylinder extends.

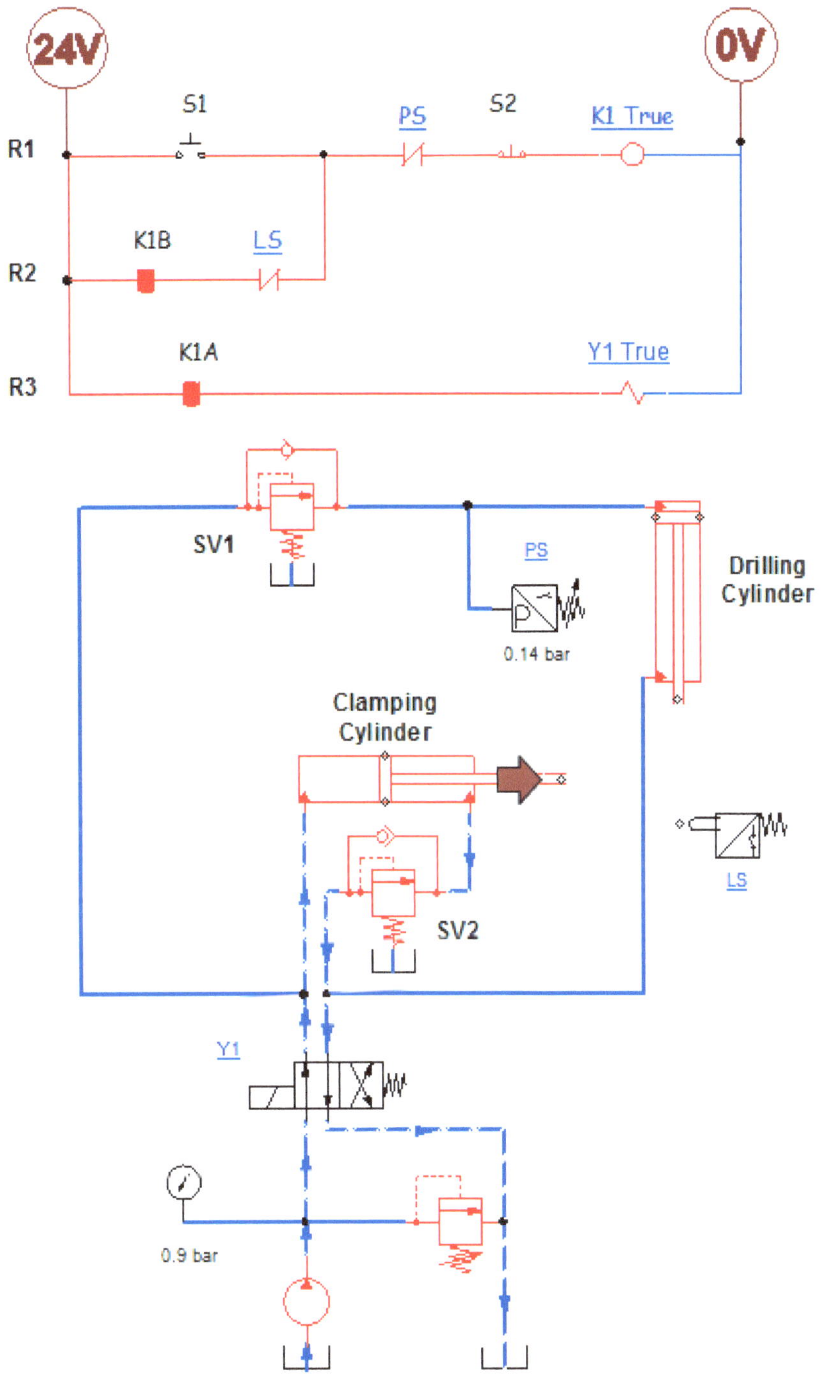

Fig. 4.48B - Clamping Cylinder Extends

SV1 takes care of clamping the work piece before drilling with enough force based on the clamping cylinder size and setting of SV1. As shown in Fig. 4.48C, when the work piece is well clamped, SV1 passes the power to the drilling cylinder to perform the tooling function. At any portion of the drilling function, if the drilling force exceeds the value set by the pressure switch PS, the normally-closed contact PS on R1 will open. As a result, the whole circuit will reset, and the two cylinders will be reversed.

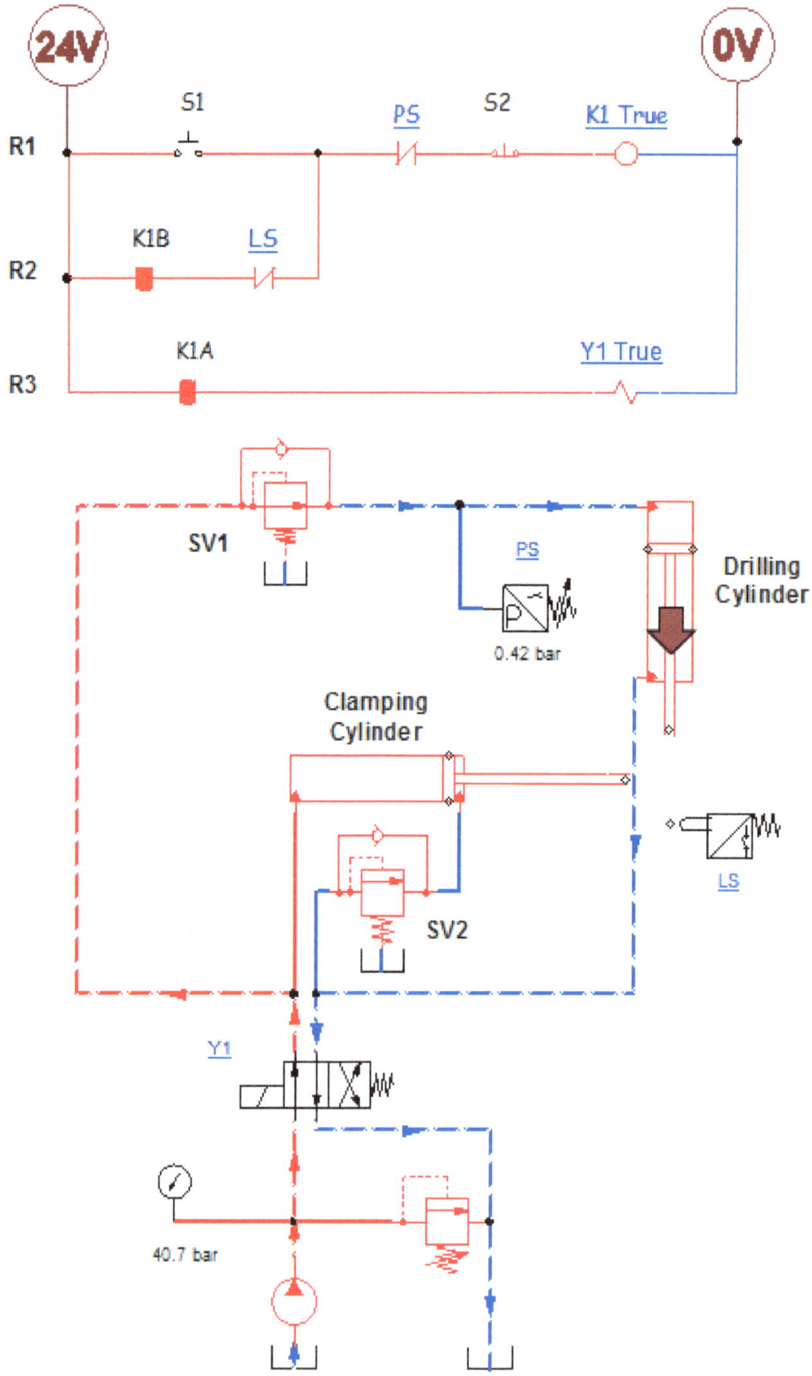

Fig. 4.48C - Drilling Cylinder Extends

As shown in Fig. 4.48D, once the drilling cylinder finishes the drilling function, it contacts the limit switch LS so that the latching line R2 is deactivated. As a result, the drilling cylinder retracts.

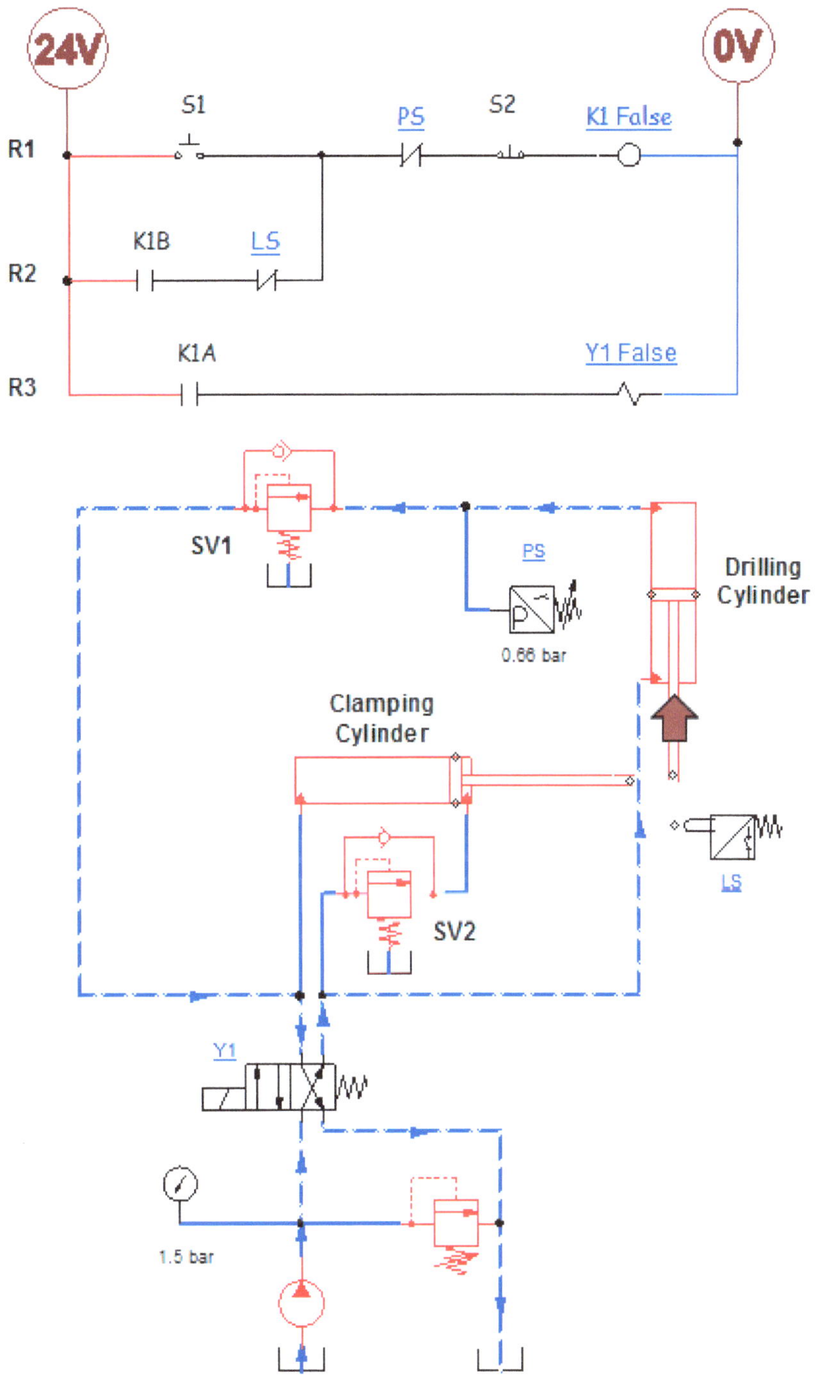

Fig. 4.48D - Drilling Cylinder Retracts

As shown in Fig. 4.48E, when the drilling cylinder is fully retracted, SV2 passes power to unclamp the work piece. SV2 takes care of making sure that the drilling cylinder is fully retracted before unclamping the work piece.

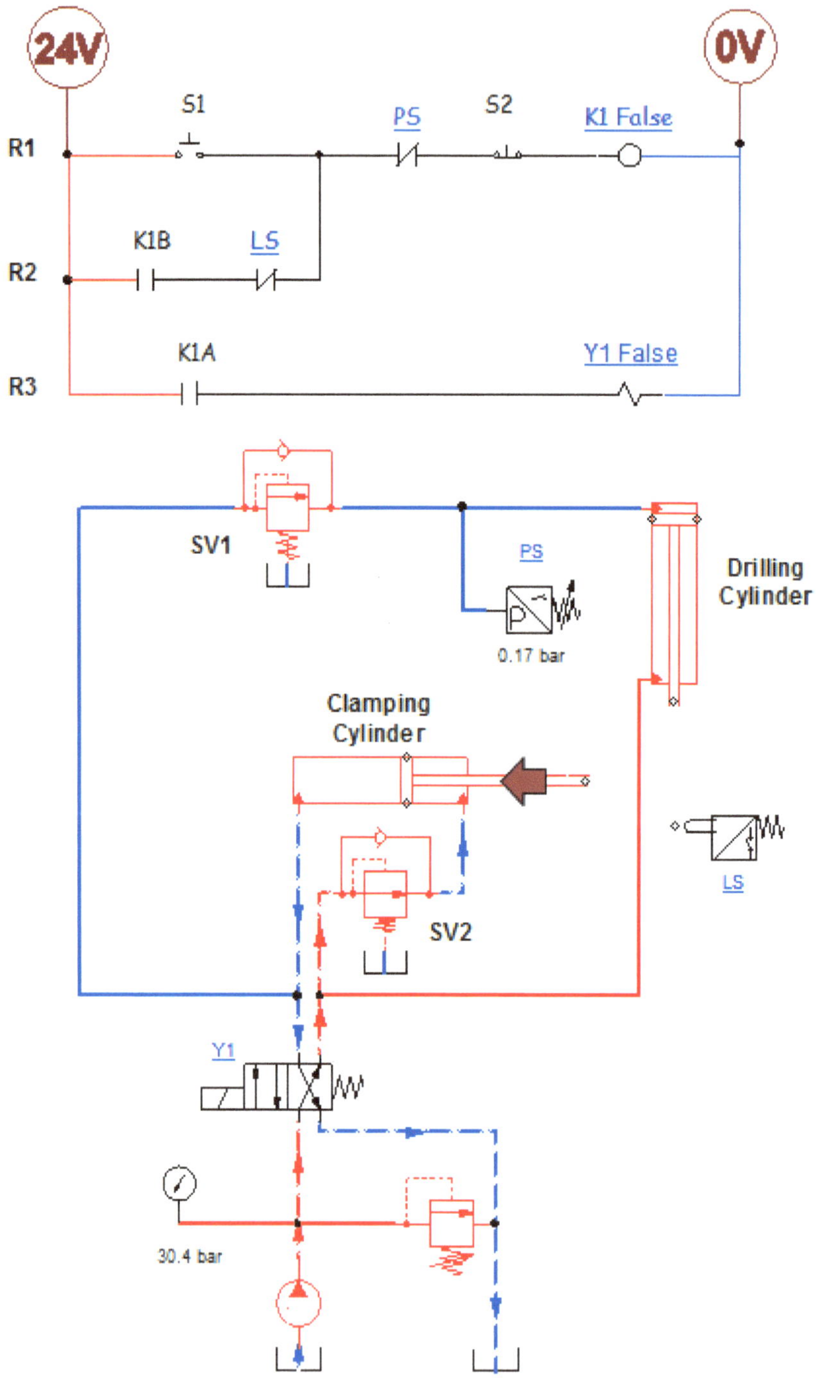

Fig. 4.48E - Clamping Cylinder Retracts

Chapter 5

Proportional Valves

Objectives:

This chapter introduces the technology of proportional valves and discusses the construction differences as compared to conventional switching valves. The chapter also introduces the conceptual construction of force-controlled versus stroke-controlled types of proportional valves. Additionally, the chapter presents the control schemes when a proportional valve is used in open-loop and closed-loop control system. More important, the chapter concludes by exploring examples of actual proportional directional, pressure, and flow control valves from various suppliers.

Brief Contents:

Chapter 5 - Proportional Valves

5.1- Introduction to Proportional Valves

Technology of electro-hydraulic *Proportional Valves* has a history that goes back to the time of World War II. As shown in Fig. 5.1, control surfaces of small planes were conventionally controlled using mechanical linkages and ropes.

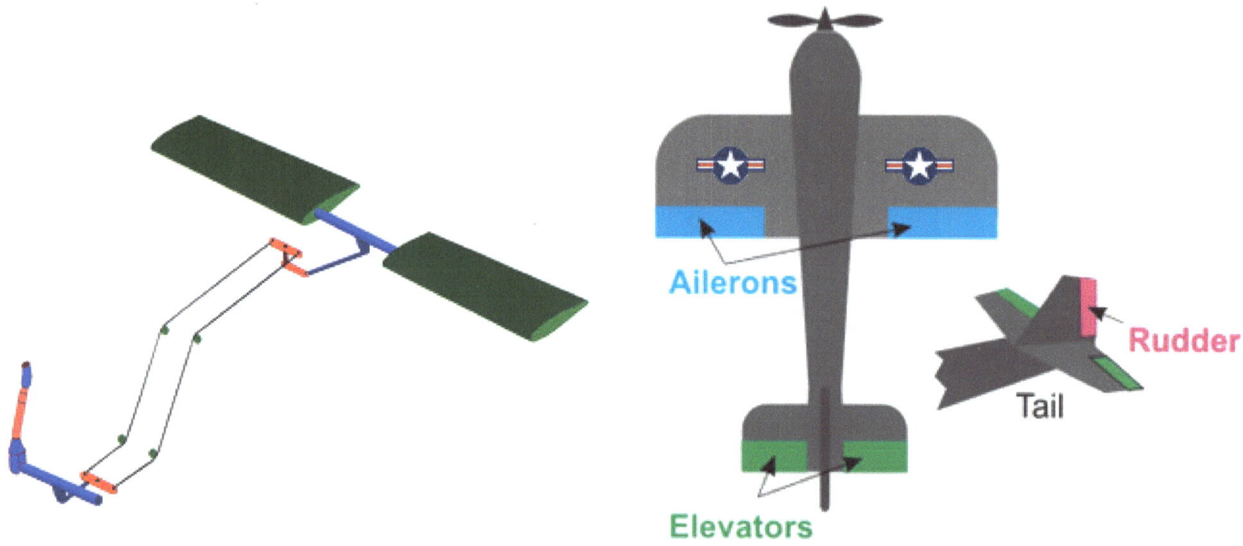

Fig. 5.1 - Mechanical Flight Control System

When large jets and powerful propeller planes came into service, the aerodynamic forces that act on the moving surfaces of such powerful planes were considerably increased. Application engineers at the US Air Force were in a position of having to upgrade the conventional flight control system by a new one that is:

- Power assisted.
- Reliable.
- Offers input-output linear relationship.
- Fast.
- Precise
- Repeatable.
- Automatically compensate the effect of the external disturbances.

The first transition, as shown in Fig. 5.2, was replacing the mechanical linkages and ropes by a hydraulic system that has a manually or foot actuated hydraulic valve and a hydraulic actuator. Such a system fulfills the first three bullets of the aforementioned requirements of flight control systems. The hydraulic power assists the pilot to overcome the large aerodynamic forces. The hydraulic system is reliable as long as it is kept clean. The figure shows open-loop architecture, in which the pilot himself controls the flying modes.

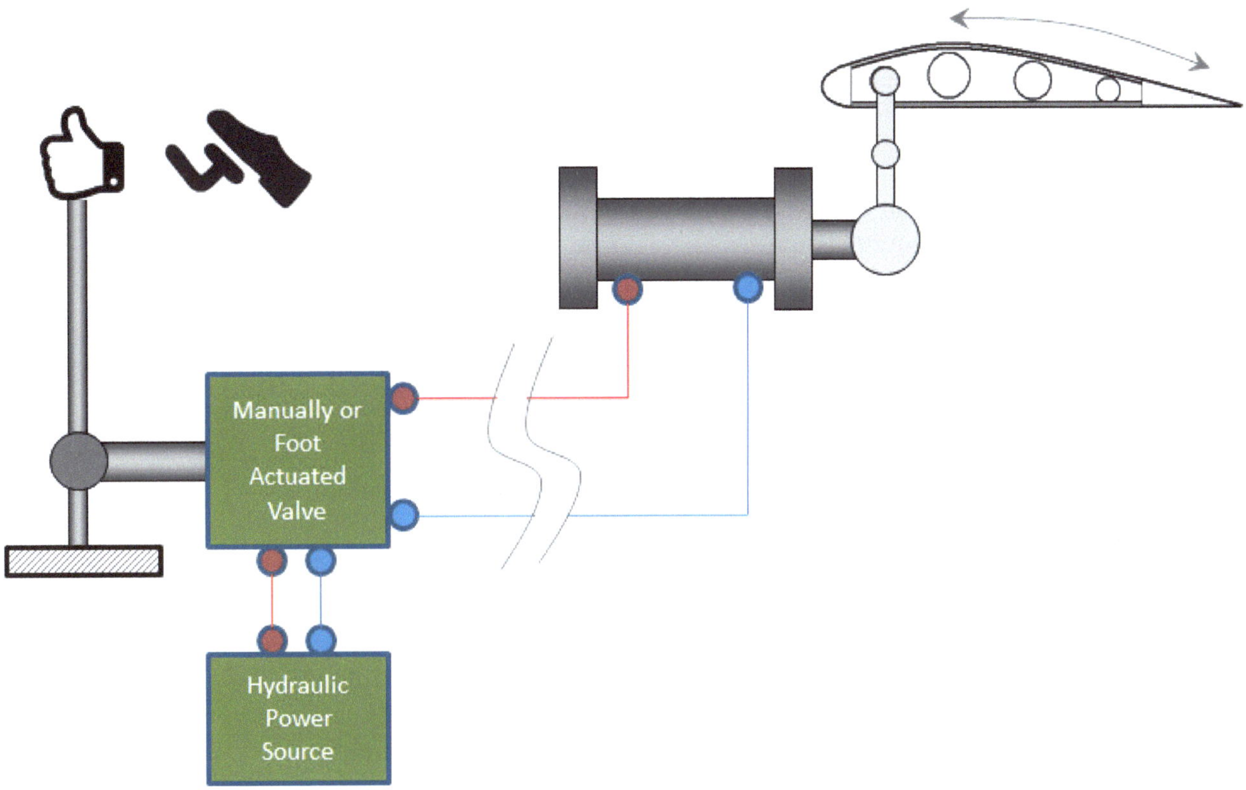

Fig. 5.2 - Open-Loop Hydro-Mechanical and Manually Operated Flight Control System

In order to fulfill the whole requirement, positioning the control surfaces should be performed through closed-loop control architecture. Electro-hydraulic closed-loop control requires replacing the manually operated hydraulic valve by a continuous, electro-hydraulic valve.

Figure 5.3 shows an advanced *Electro-Hydraulic Actuator* (EHA). Technology used in flight controls for the next generation of military and commercial aircraft.

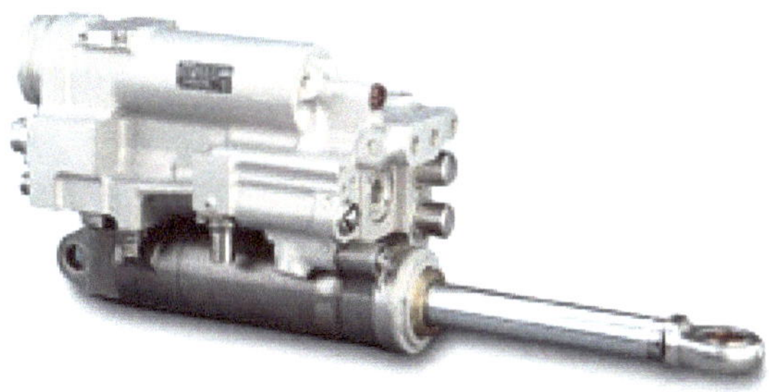

Fig..5.3 - Advanced Flight Control Systems (Courtesy of Moog)

Figure 5.4 (left side) shows a skilled operator controlling a hydraulic motor using a manually-actuated hydraulic directional valve. The operator should stand at the place of the valve, observe the motor speed in a virtual closed-loop, and manually correct it to meet the desired value.

The right side of the figure shows that, by using a proportional valve, the operation has been improved. To achieve such improvement, a proportional valve should be designed to be:

- Similar to switching solenoid valves, simple in construction and operation.
- Similar to manually-actuated valves, the spool can take infinite positions.
- Similar to servo valves:
    - Controlled remotely manually by potentiometers or joysticks.
    - Controlled automatically by a PLC or a controller embedded in the machine.
    - Be able to correct the controlled parameter if being used in a closed-loop.

The lower side of the figure reveals that a proportional valve fills the gaps between several valves.

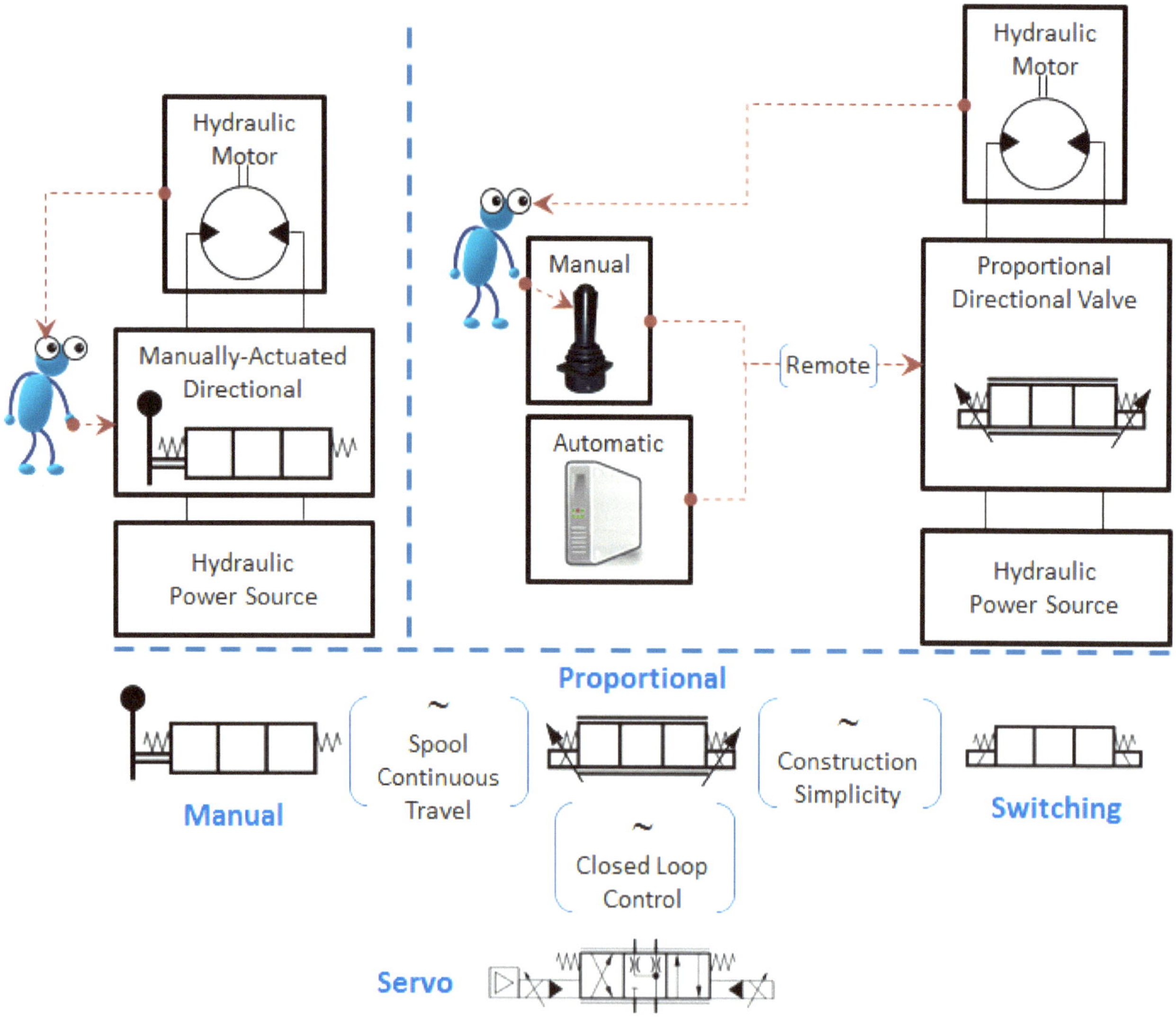

Fig. 5.4 - Features of Proportional Valves

Figure 5.5 reveals that, a proportional valve simply consists of a proportional solenoid and the hydraulic part of the valve. The figure also indicates that, unlike a servo valve, there are dedicated proportional valves for each control function such as direction, pressure, and flow control. A proportional valve may have one proportional solenoid, such as flow control and pressure control valves, or it may have two proportional solenoids such as three position directional valves.

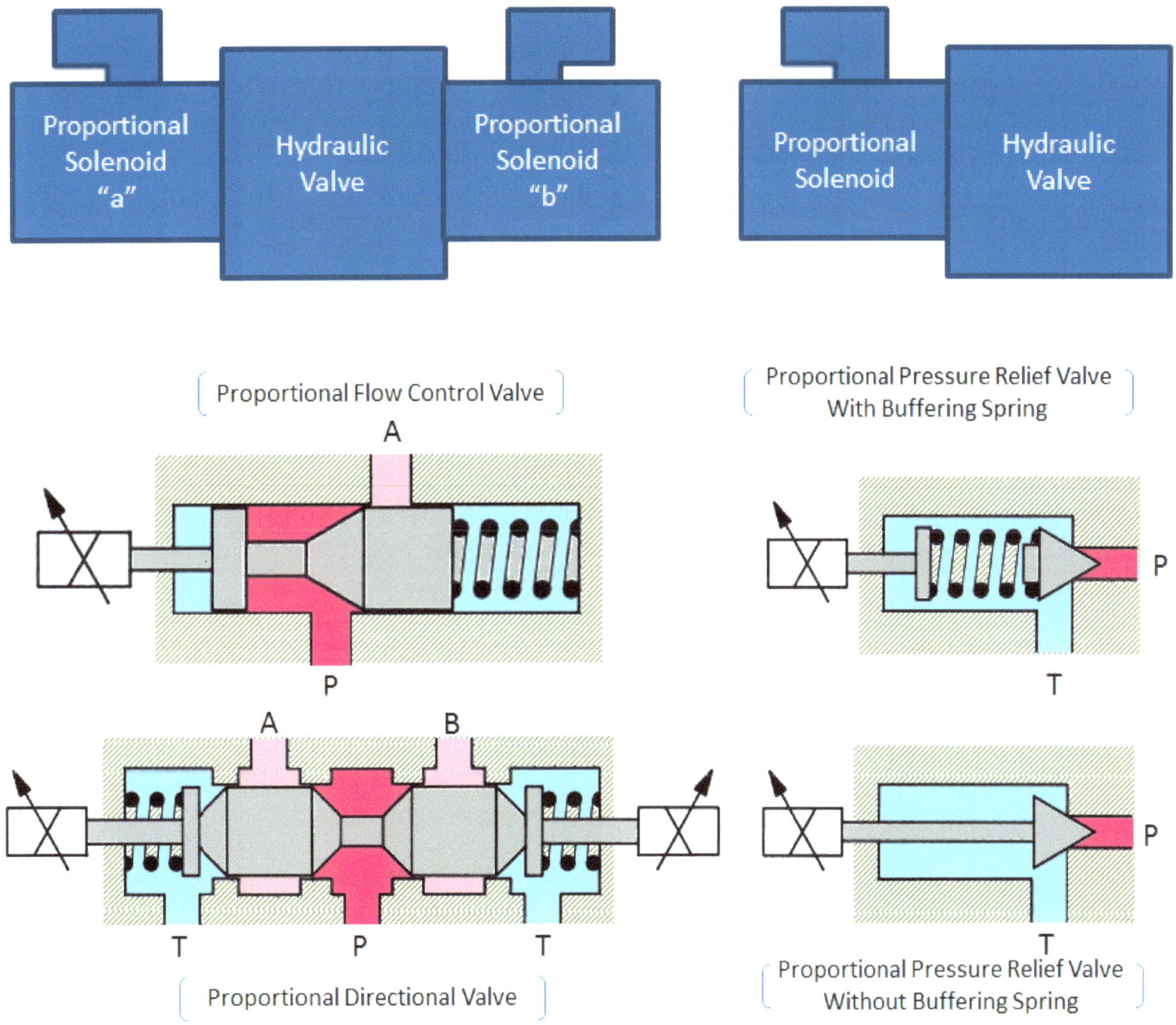

Fig. 5.5 - Basic Construction of a of Proportional Valves

5.2- Proportional Solenoids

A proportional solenoid forms an interface between the control section and the hydraulic power section in a hydraulic system. It is to be noted that a proportional solenoid is only DC-powered and wet-type.

5.2.1- Force-Controlled Proportional Solenoid

Basic versions of *Proportional Solenoids* are named *Force-Controlled*. This means that, as shown in Fig. 5.6, the plunger is positioned based on equilibrium between the input signal and the balancing spring. The plunger of the solenoid moves in an open-loop with no feedback on the spool position.

As it has been previously discussed, an open-loop does not compensate from the effect of external disturbances. For example, valve spool will not be positioned accurately if any of the following conditions are changed

- Balancing spring stiffness.
- Proportional solenoid current-force characteristics.
- Spool-Sleeve viscos friction due to contamination.

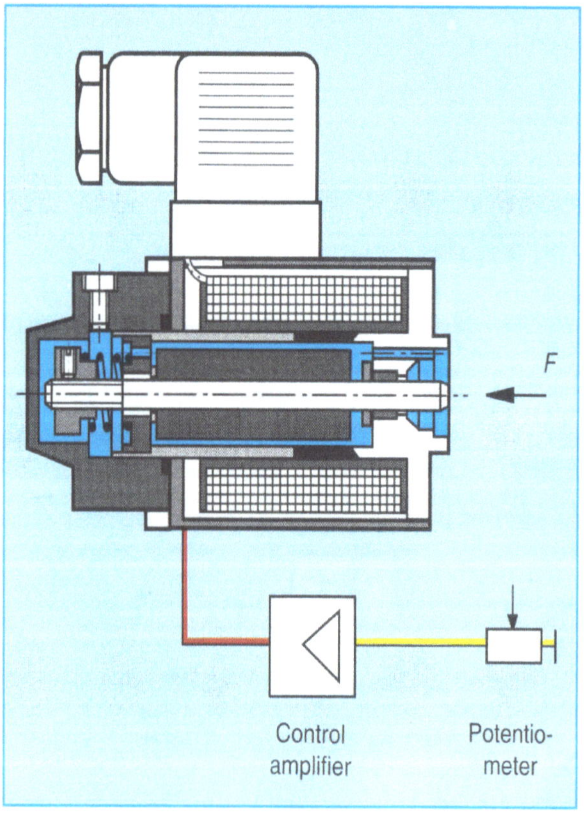

Fig. 5.6 - Operation of Force-Controlled Proportional Solenoid (Courtesy of Bosch Rexroth)

Force-Controlled proportional solenoids have the following features:

- Less expensive.
- Simple in construction and looks similar to the switching solenoids.
- Compact due to no feedback elements.
- Short stroke, about 1.5 mm.

Figure 5.7 shows, in block diagram form, an electro-hydraulic open-loop control system using a force-controlled proportional directional valve. Figure 6.8 shows, in block diagram form, an electro-hydraulic closed-loop control system using force-controlled proportional directional valve.

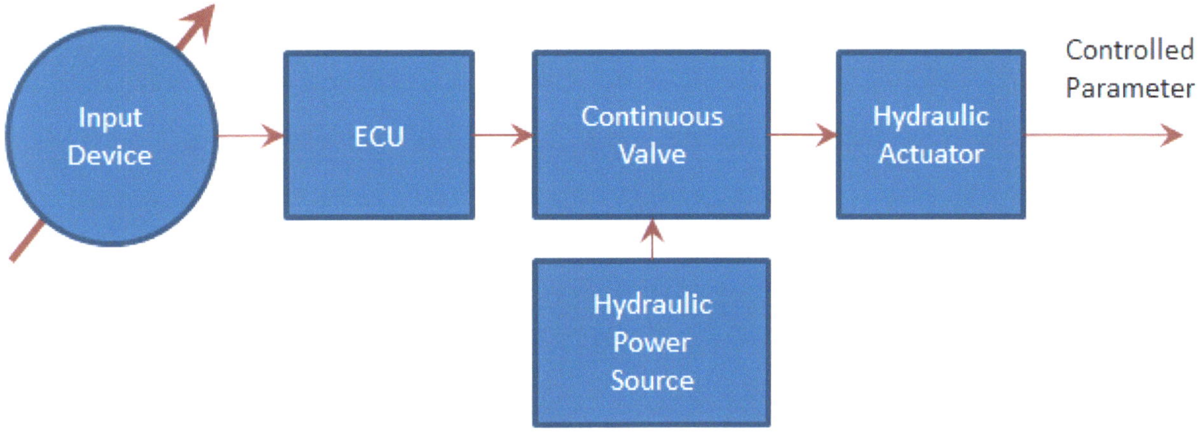

Fig. 5.7 - EH Open-Loop Control System using Force-Controlled Proportional Valve

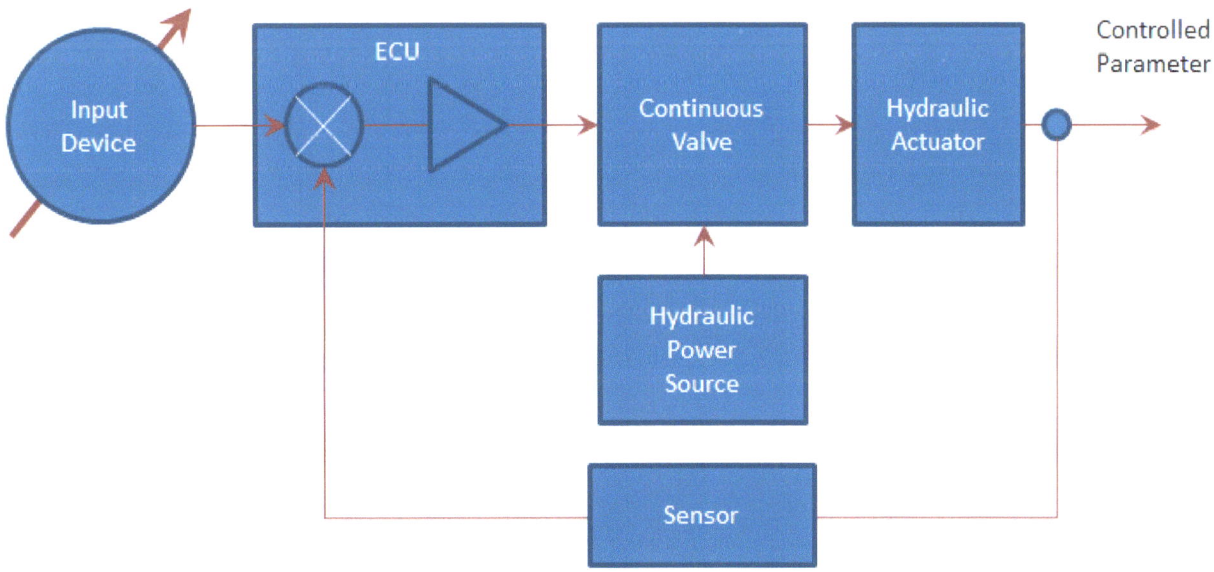

Fig. 5.8 - EH Closed-Loop Control System using Force-Controlled Proportional Valve

Figure 5.9 shows the construction of a wet-type force-controlled proportional solenoid. The *Non-Magnetic Disk* # 12 is the basic part that converts the conventional switching solenoid into a proportional solenoid. Figure 6.10 shows a typical force-controlled proportional solenoid and its design characteristics.

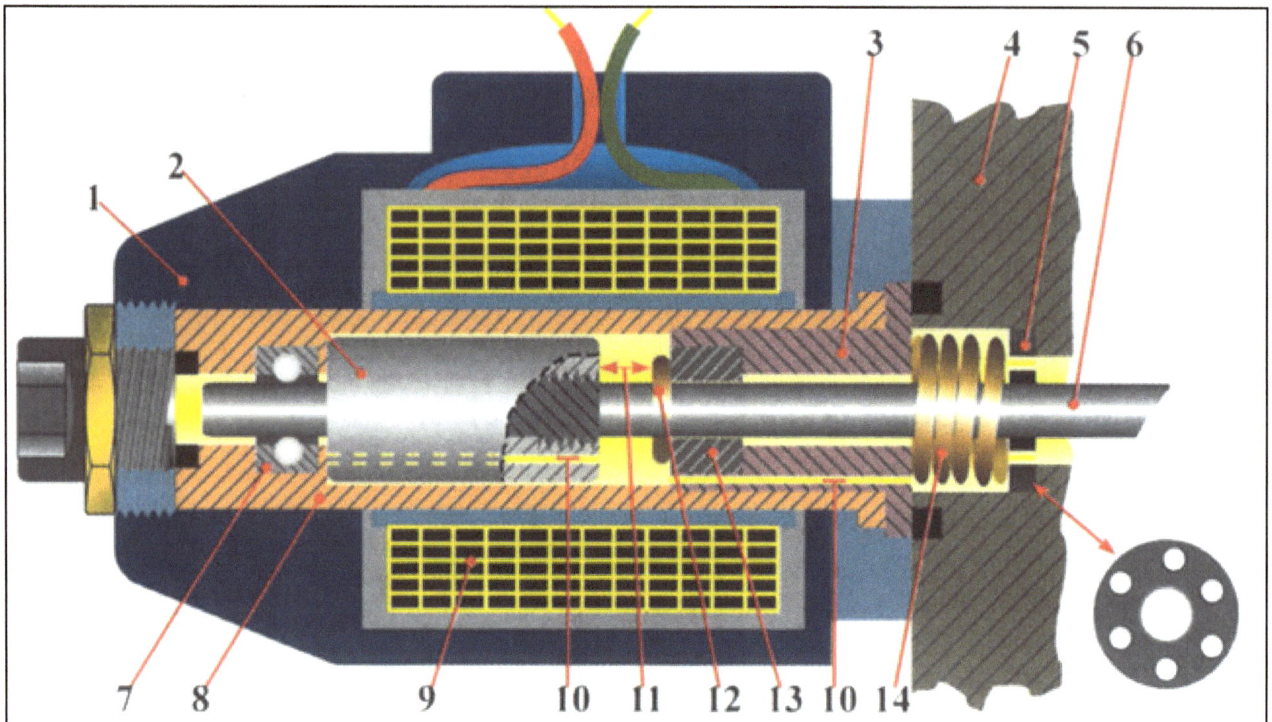

1. Solenoid Body
2. Mobile Core
3. Pole Expansion
4. Valve Body
5. Perforated Disc
6. Push Rod
7. Bearing
8. Sleeve
9. Coil
10. Fluid Passage
11. Core Stroke
12. Non-Magnetic Disc
13. Bushing
14. Spring

Fig. 6.9 - Construction of a Force-Controlled Proportional Solenoid (Courtesy of ASSOFLUID)

Proportional Solenoids for Hydraulic Application Type G RC Y
3 sizes: Ø (mm) 37, 45, 63
pressure-proof armature space up to max. 350 bar (= 5075 psi) static
stroke (mm) 2 up to 4
magnetic force (N) 47 up to 112 (at working stroke)
fastening with central thread
pulling design on request

Fig. 5.10 - Typical Force-Controlled Proportional Solenoid (www.magnet-schultz.com)

As shown in Fig.5.11, for a given input signal, curve # 1 represents the electromagnetic force that is generated by the conventional design of the switching solenoid. Curve # 2 represents the electromagnetic force that is due to adding the non-magnetic disc. Curve # 3 is the sum of curves 1 and 2. Curve # 3 shows that the resultant force is kept constant along the working (effective) core stroke. The figure also shows that the magnitude of the force is proportional to the input current.

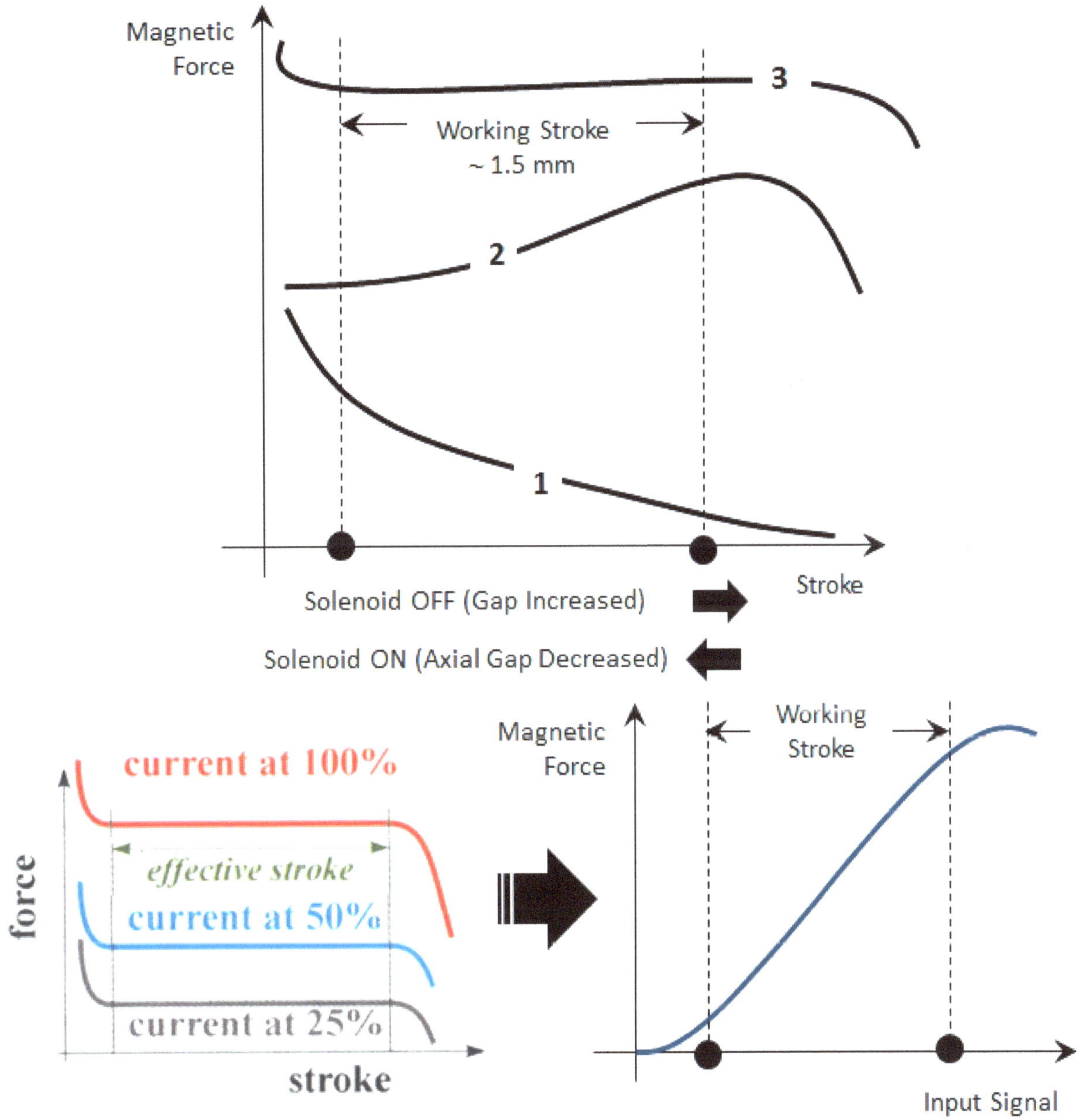

Fig. 5.11 - Characteristic Curves of a Stroke-Controlled Proportional Solenoid

5.2.2- Stroke-Controlled Proportional Solenoid

Advanced versions of proportional solenoids are named *Stroke-Controlled.* This means that, as shown in Fig. 5.12, positioning of the solenoid plunger is controlled by a closed-loop. The solenoid construction contains a built-in position feedback device for the purpose of closed-loop control. The closed-loop compensates the aging conditions of the valve and minimizes the spool position error.

Stroke-Controlled proportional solenoids have the following features:

- More expensive and complex in construction.
- Larger in size due to adding feedback elements.
- Longer stroke, about 3-5 mm.
- Less hysteresis.
- Better accuracy in spool positioning.

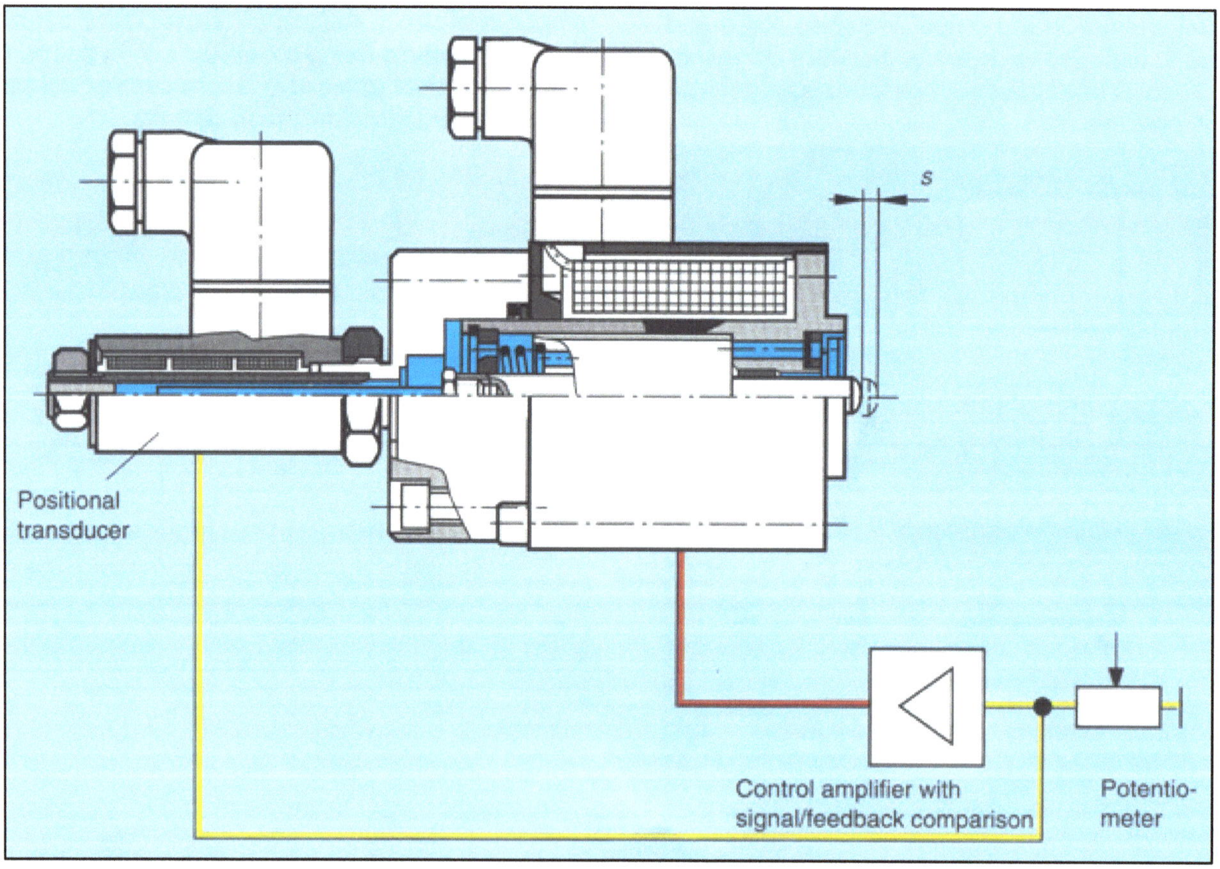

Fig. 5.12 - Operation of Stroke-Controlled Proportional Solenoid (Courtesy of Bosch Rexroth)

Figure 5.13 shows, in block diagram form, an electro-hydraulic open-loop control system using stroke-controlled proportional directional valve. Figure 5.14 shows, in block diagram form, an electro-hydraulic closed-loop control system using stroke-controlled proportional directional valve.

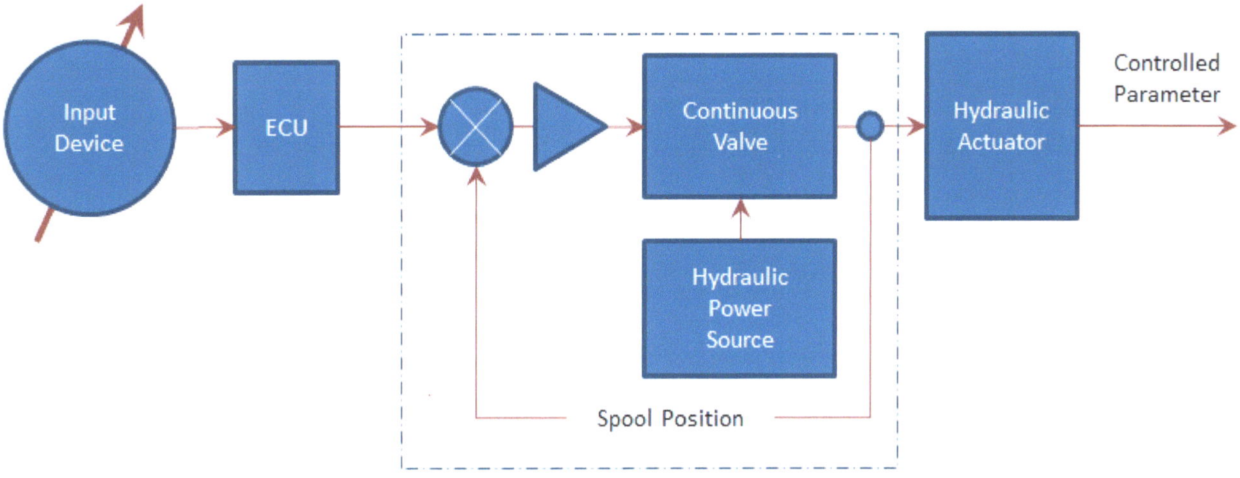

Fig. 5.13 - EH Open-Loop Control System using Stroke-Controlled Proportional Valve

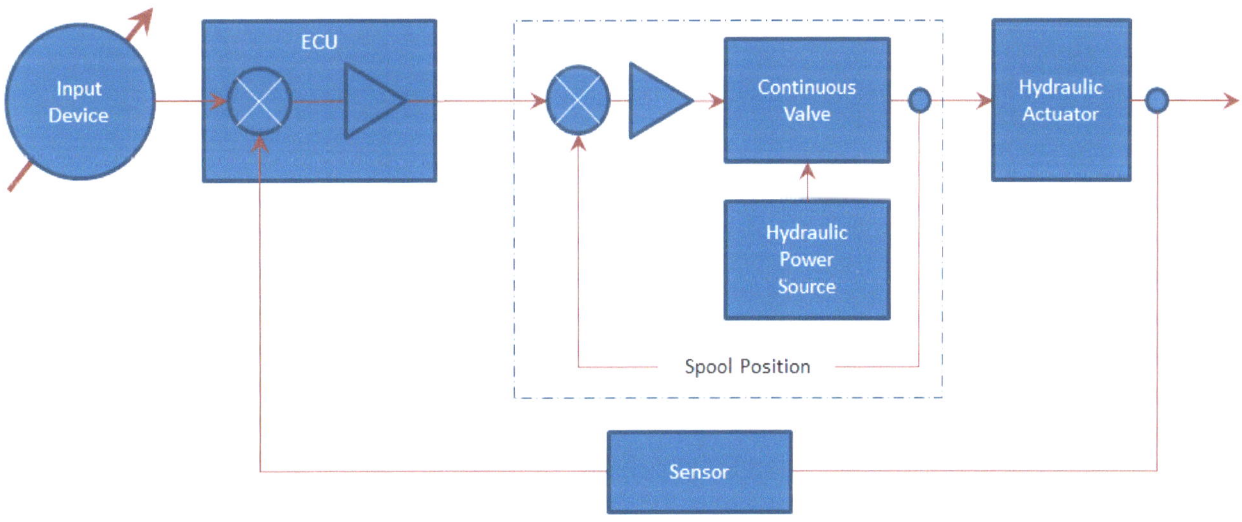

Fig. 5.14 - EH Closed-Loop Control System using Stroke-Controlled Proportional Valve

Figure 5.15 shows the construction of a wet-type stroke-controlled proportional solenoid. The figure shows a built-in spool position sensor. The solenoid plug and the sensor plug are connected to a separate electric control unit (ECU). In some cases, particularly with On-Board Electronics (OBE), the position sensor may be contained inside the valve and connected to the ECU internally.

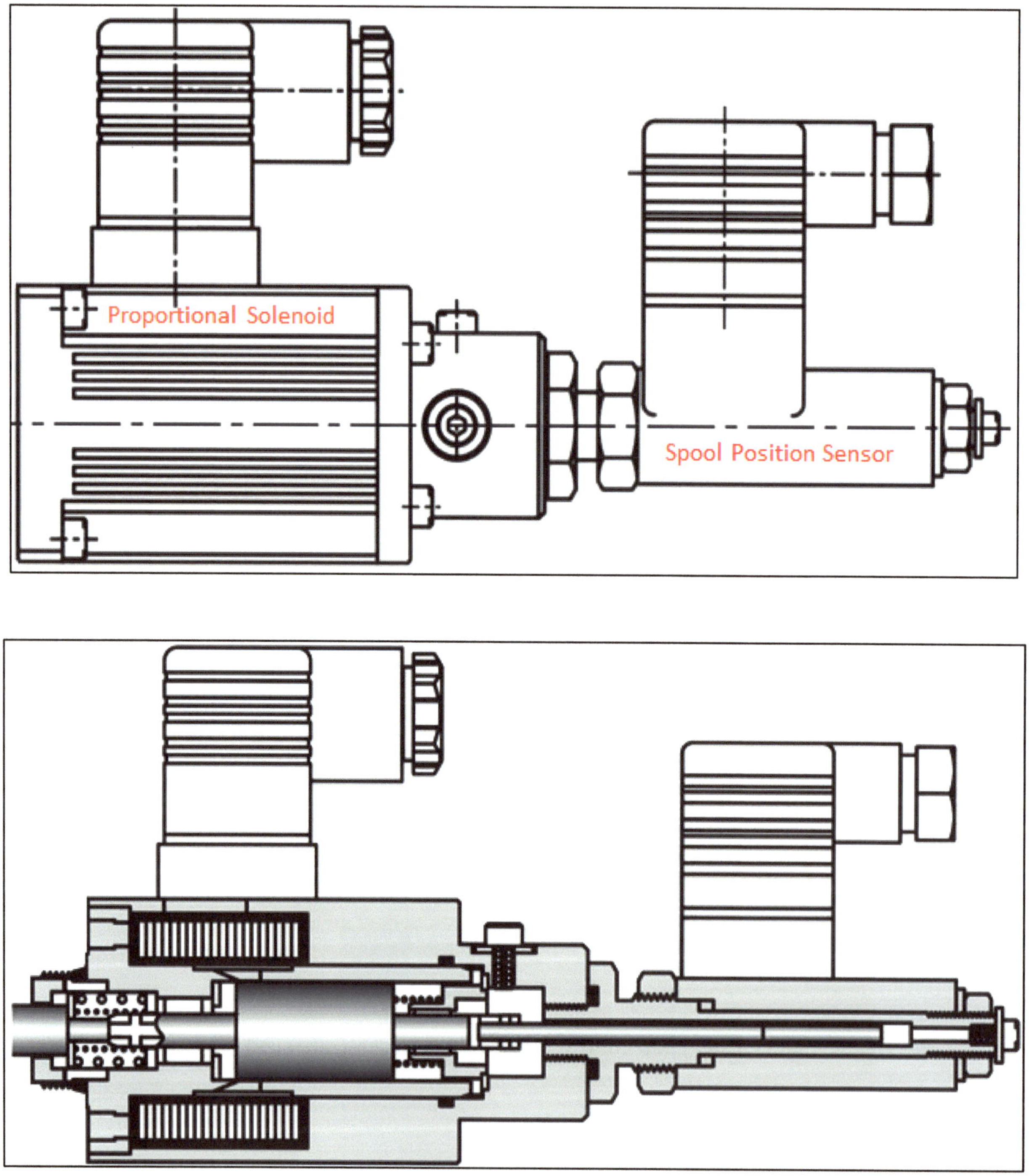

Fig. 5.15 - Construction of a Stroke-Controlled Proportional Solenoid (www.yuken.co.uk)

Figure 5.16 shows a typical stroke-controlled proportional solenoid and its design characteristics.

Rated Current[A] : 2.5
Normal Resistance[Ω]: 3.7
Rated Attraction Force [N] : 160
Rated Stroke[mm] : 4
Total Stroke [mm]: ≥9
Force Hysteresis [%]: ≤5
Current Hysteresis [%]: ≤3
Repeat Accuracy [%]: ≤1
Operating Oil Temperature[℃]: -90

Fig. 5.16 - Typical Stroke-Controlled Proportional Solenoid (http://yfsolenoid.com)

Figure 5.17 shows another typical configuration of a stroke-controlled proportional solenoid and its design characteristics.

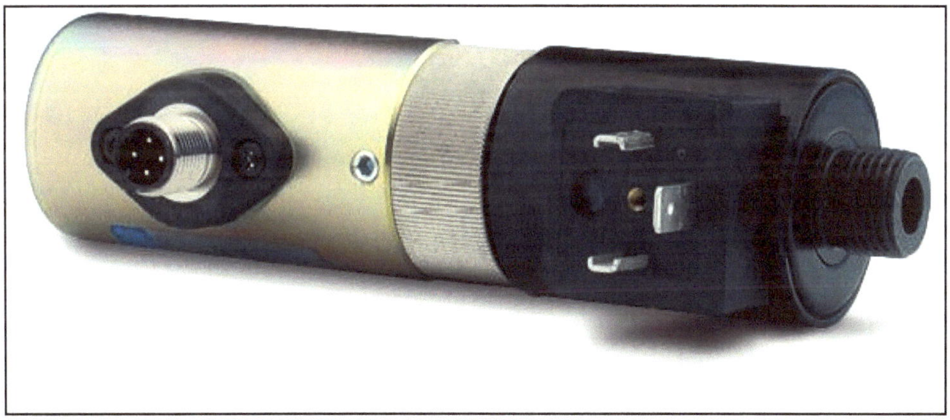

Proportional Solenoids for Hydraulic Application with Transducer type G RC...A62
3 sizes: Ø (mm) 37, 45, 63
pressure-proof armature space
up to max. 320 bar (= 4640 psi-static)
stroke (mm) 5.5 up to 9
magnetic force (N) 43 up to 120
fastening with central thread
extendable to a complete closed
loop control circuit

Fig. 5.17 - Typical Stroke-Controlled Proportional Solenoid (www.magnet-schultz.com)

5.3- Proportional Directional Control Valves

Proportional Directional Valves (*PDV*) can be configured differently depending on the number of positions, number of stages, controlling the spool position, and the associated ECU.

5.3.1- Interpretation of Symbols for Proportional Directional Valves

Figure 5.18 explores symbols for various proportional directional valves. The symbols show the following features:

.

- The two parallel lines on top and bottom of the symbol reveal the infinite positioning of the PDV spool.
- The arrow on the proportional solenoid indicates that the solenoid receives variable signal.
- A four-way PDV could have any form of central position such as tandem, closed, open, float, regenerative, or even custom.
- A two-way or three-way PDV could be configured, similar to the switching valves, as normally-open or normally-closed.
- A Two-position PDV versus a three-position PDV; (1 versus 2).
- A direct-operated PDV versus a pilot-operated PDV; (2 versus 3).
- A PDV with a separate ECU versus a PDV with an OBE; (3 versus 4).
- A Force-controlled PDV versus a stroke-controlled PDV; (4 versus 5).
- A PDV without fail-safe position and with two solenoids versus a PDV with fail-safe position and a single solenoid; (5 versus 6).

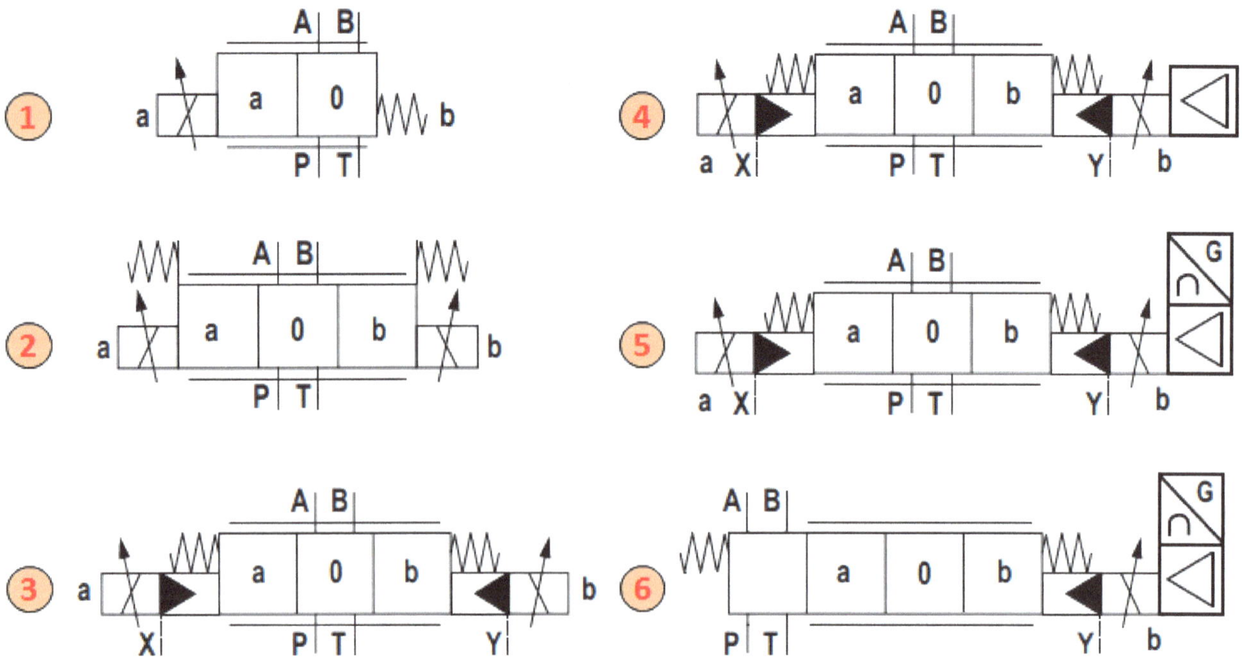

Fig. 5.18 - Symbols for Proportional Directional Valves

5.3.2- Hydraulic Static Characteristics of Proportional Directional Valves

Because of the ability to control how much the valve opens, a four-way PDV is not only used to control the direction of a hydraulic actuator, but also to control the flow/speed and the pressure/load of the actuator. Similar to hydro-mechanical or solenoid actuated switching valves; the hydraulic part of a PDV has steady state characteristics that should be published by the manufacturer. It is to be noted that the manufacturer should declare the conditions under which these characteristics have been developed such as the type of oil and oil temperature.

Figures 5.19, 5.20, and 5.21 show typical characteristics of a PDV as examples of what is published by the manufacturer to identify the valve performance.

If a PDV is used to control flow/speed of an actuator, then the *Flow Gain Characteristics* are needed. As shown in Fig. 6.19, flow through a PDV depends on the valve opening and the differential pressure across the valve.

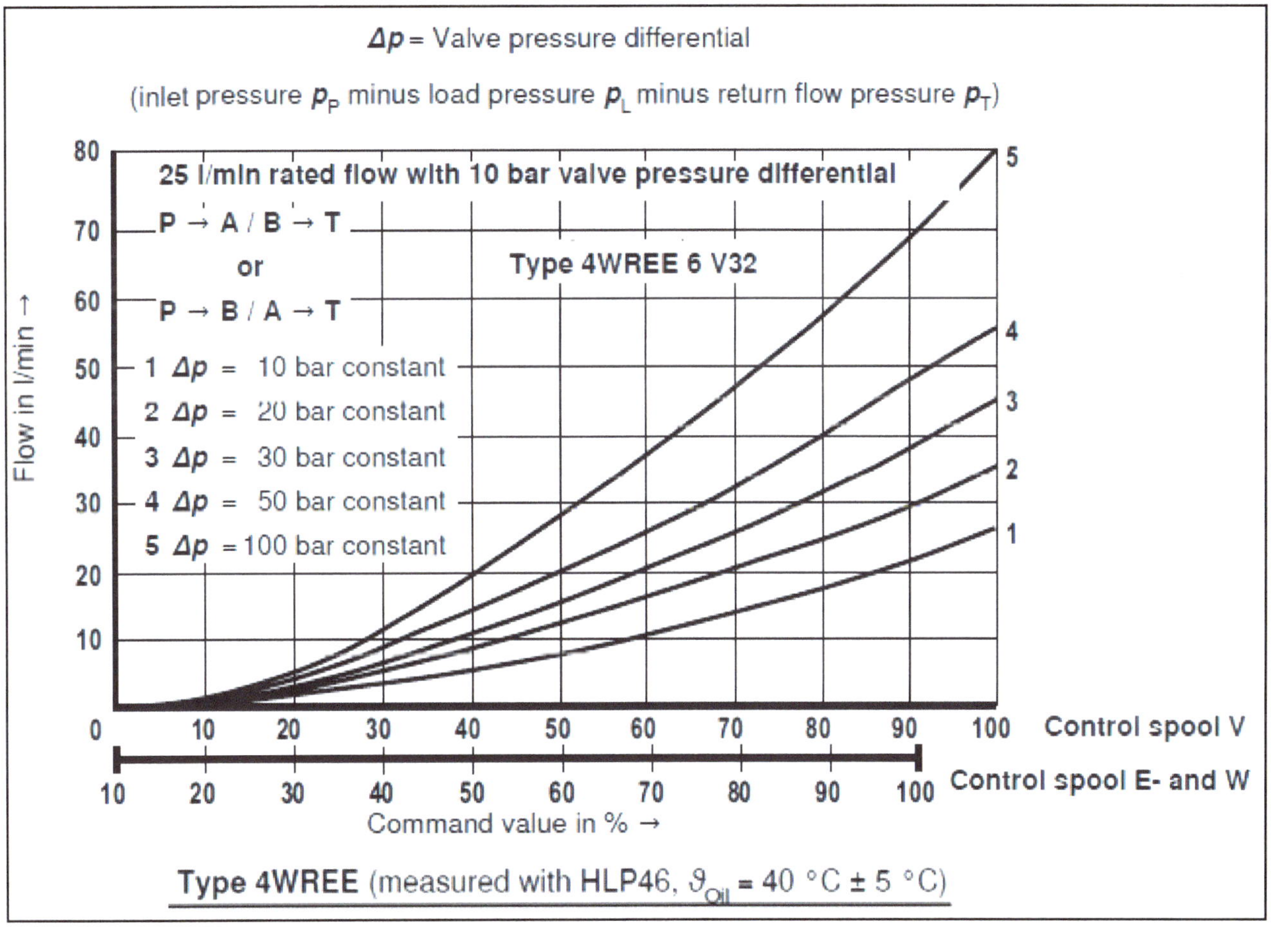

Fig. 5.19 - Example of PDV Flow Gain (Courtesy of Bosch Rexroth)

If a PDV is used to control pressure/load of an actuator, then the *Pressure Gain Characteristics* are needed. Fig. 5.20 shows that this particular valve can pass 100% of the supply pressure if it receives 3.7% of the maximum input signal. Figure 5.21 shows the *Null Leakage* characteristics of the valve. It shows the leakage through the valve when the spool is at rest. This test is also refereed as *Zero Flow*. It is an important test to identify the health condition of the valve.

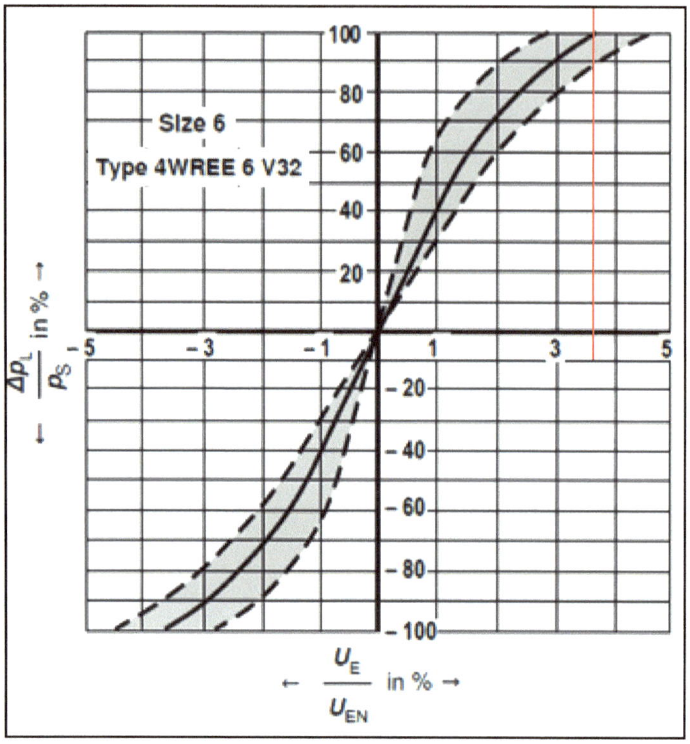

Fig. 5.20 - Example of PDV Pressure Gain (Courtesy of Bosch Rexroth)

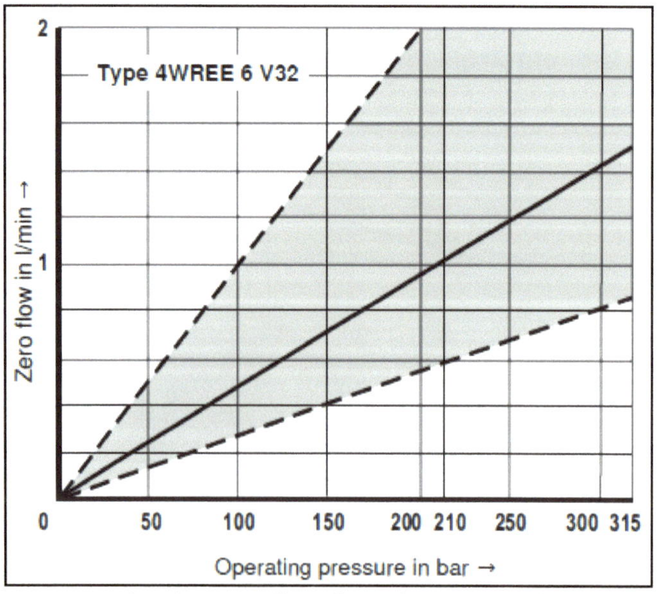

Fig. 5.21 - Example of PDV Null Leakage (Courtesy of Bosch Rexroth)

5.3.3- Direct-Operated Force-Controlled Proportional Directional Valve

Figure 5.22 shows the simplest form of a spool-type PDV in which the valve is direct-operated, and the proportional solenoid is force-controlled.

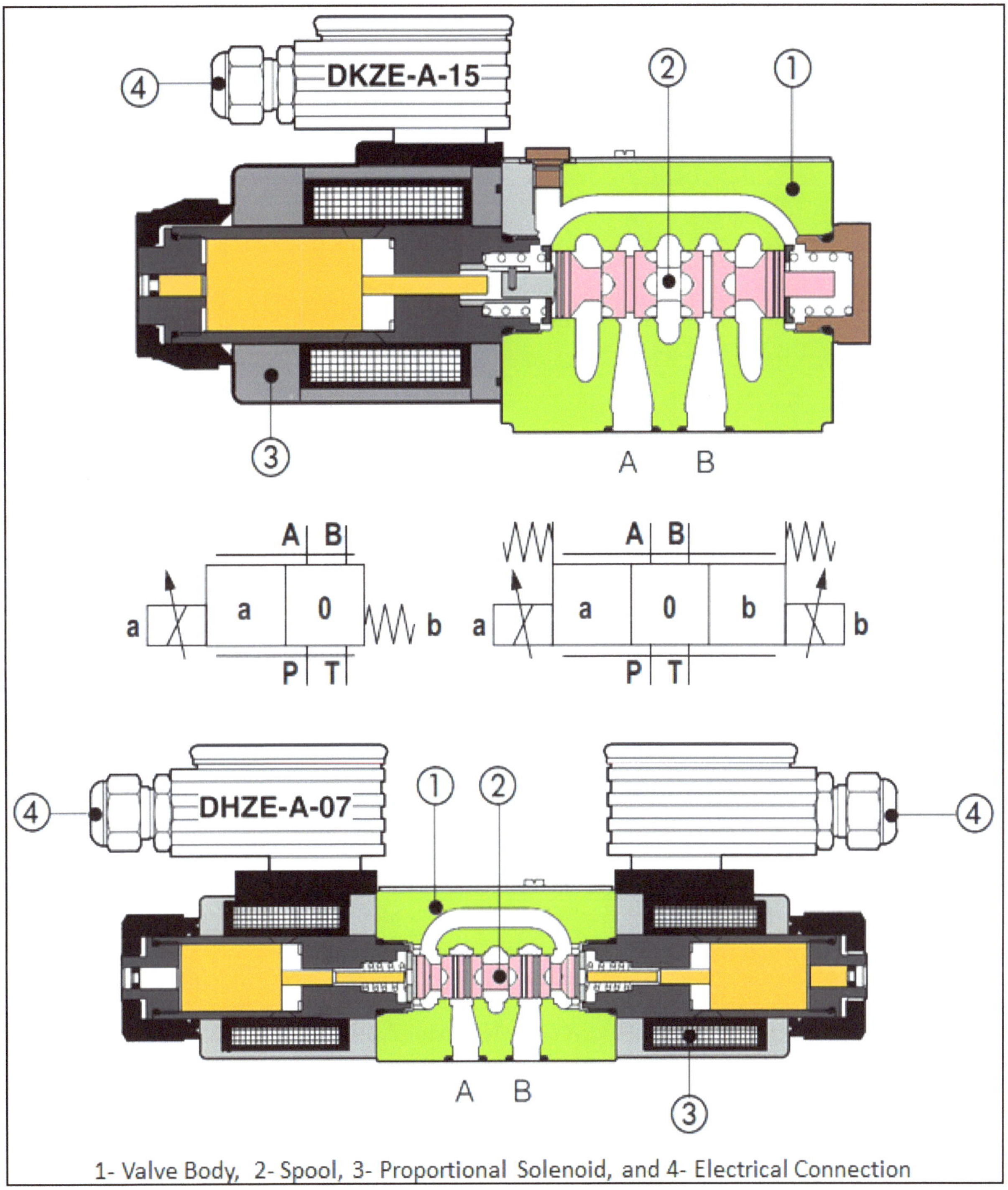

Fig. 5.22 - Direct-Operated Force-Controlled Spool-Type PDV with Separate ECU (Courtesy of Atos)

Figure 5.23 shows a 4/3 PDV direct-operated force-controlled spool-type PDV with the ECU On-Board. The use of an OBE versus a separate ECU will be discussed later. Center position could be any of the standard four-way types.

The main connector (7) is used to power the OBE and receives the input signal. The *Fieldbus* connector (6) is used for online communication and condition monitoring with the valve. The USB connector (5) is used for offline reset of the digital electronics to the factory defaults or to reprogram it to other values.

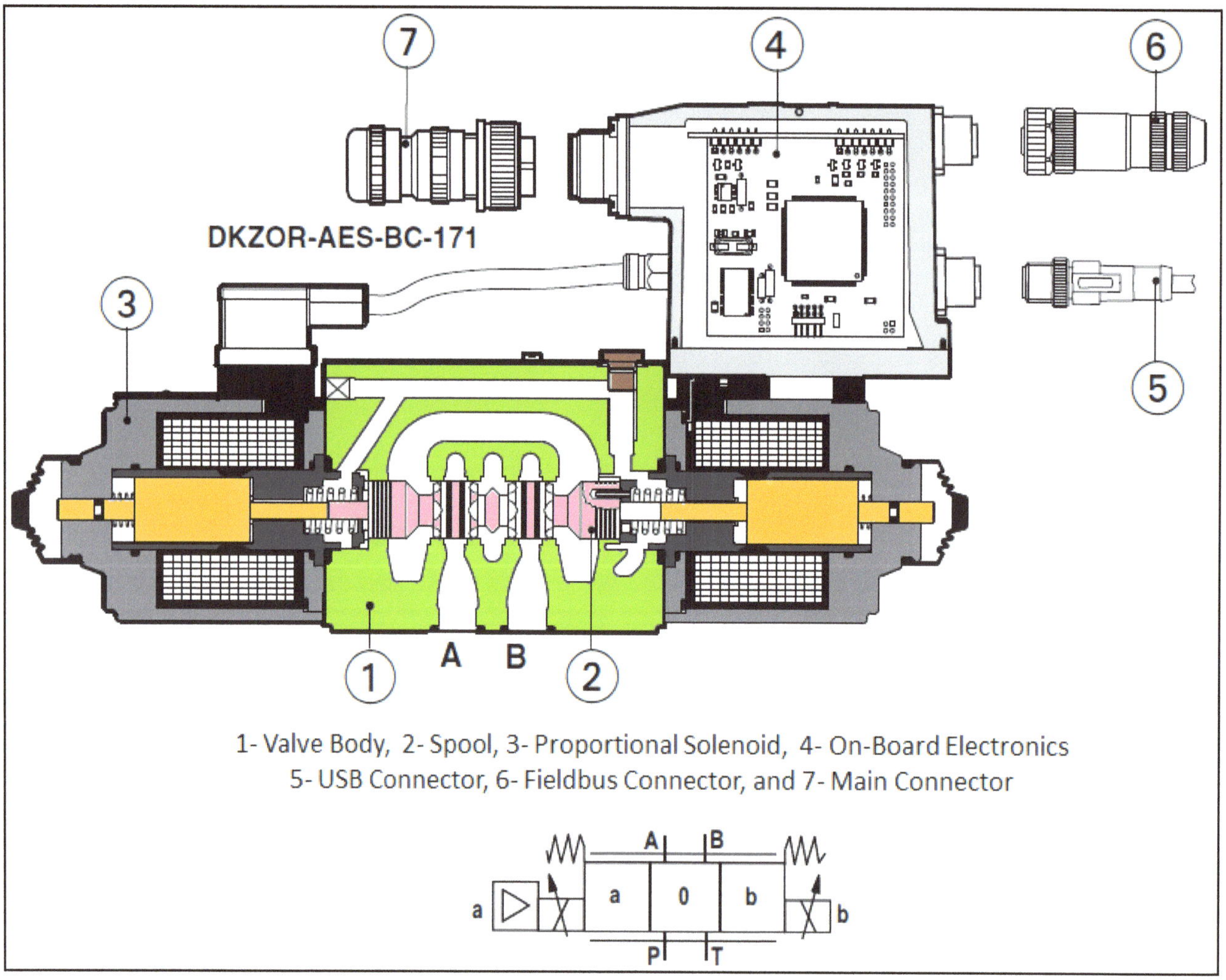

Fig. 5.23 - Direct-Operated Force-Controlled Spool-Type PDV with OBE (Courtesy of Atos)

Figure 5.24 shows a cartridge-type closed-center PDV. Cartridge valves are commonly used to build more compact manifold mounted control blocks.

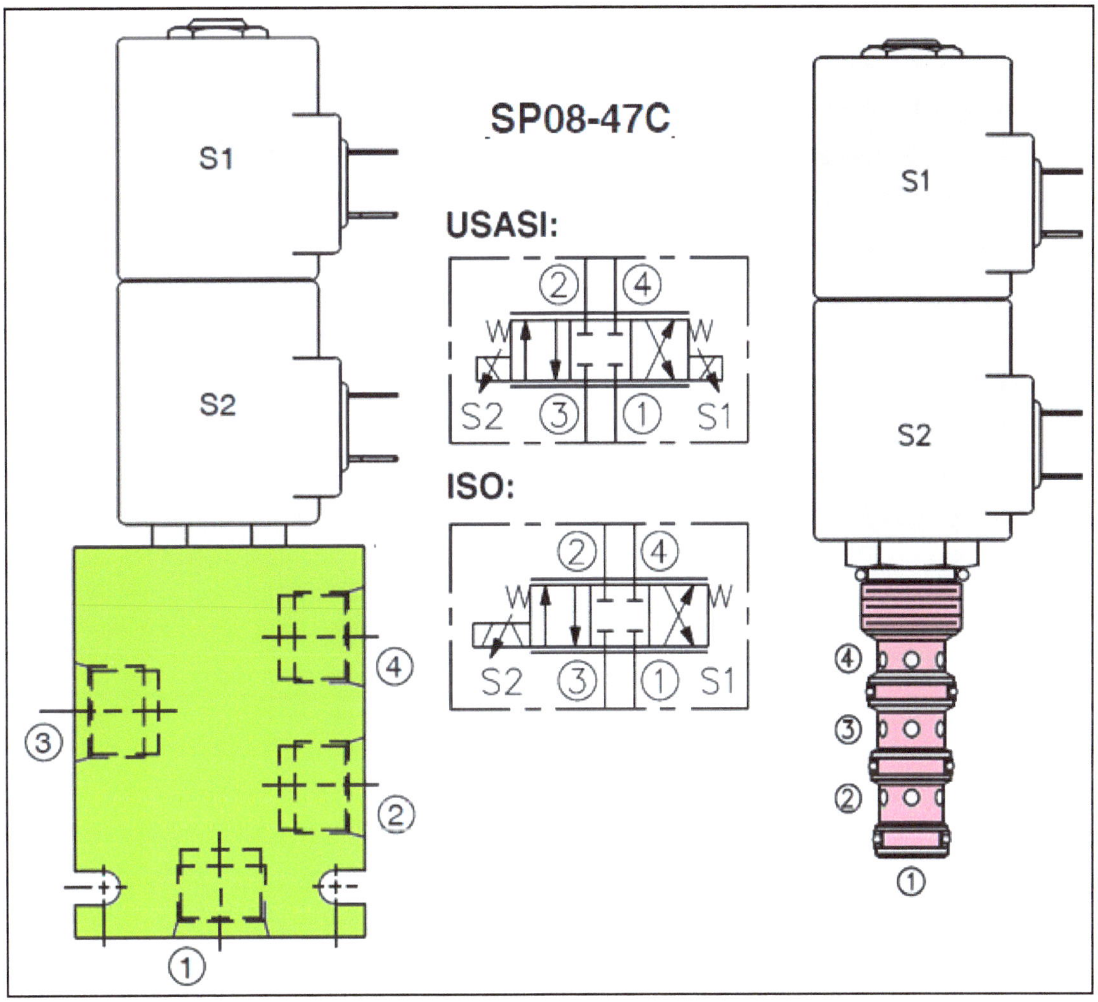

Fig. 5.24 - Direct-Operated Force-Controlled Cartridge-Type PDV with Separate ECU (Courtesy of Hydraforce)

5.3.4- Direct-Operated Stroke-Controlled Proportional Directional Valve

Figure 5.25 shows the outer shape of a direct-operated stroke-controlled spool-type PDV. The figure shows that such valves could be equipped with a separate ECU or with OBE. The table below the figure shows some of the electrical data reported by the manufacturer.

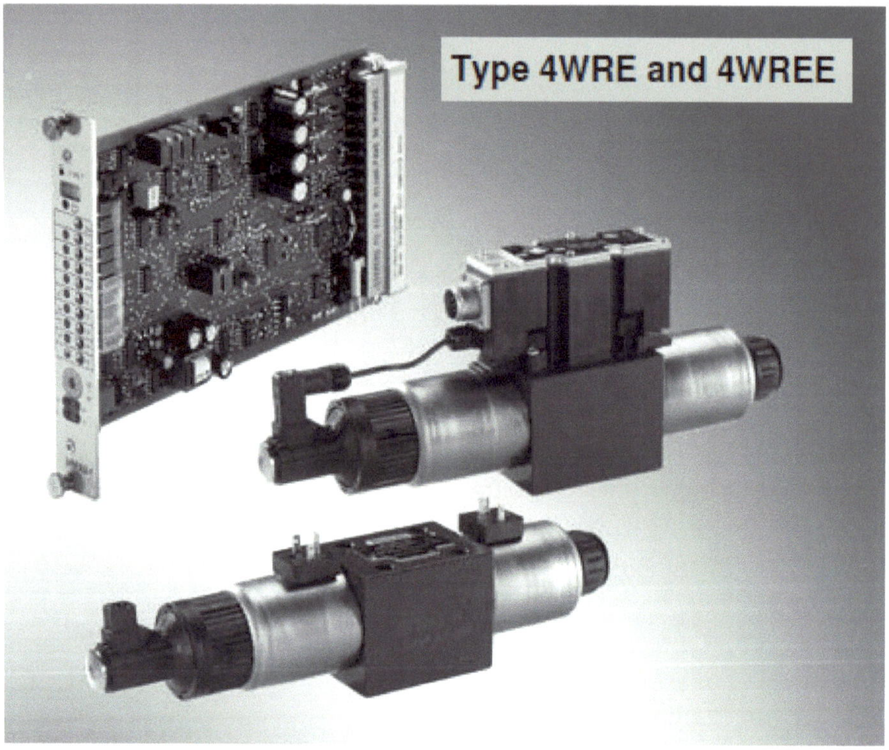

Duty cycle		%	100
Maximum coil temperature [1)]		°C	up to 150
Supply voltage	Nominal voltage	VDC	24
	lower limit value	V	19.4
	upper limit value	V	35
Current consumption of the amplifier	I_{max}	A	< 2
	Pulse current	A	3

Fig. 5.25 - Direct-Operated Stroke-Controlled Spool-Type PDV (Courtesy of Bosch Rexroth)

Figure 5.26 shows the internal construction of the valve that has a separate ECU. Figure 5.27 shows the same valve but with OBE.

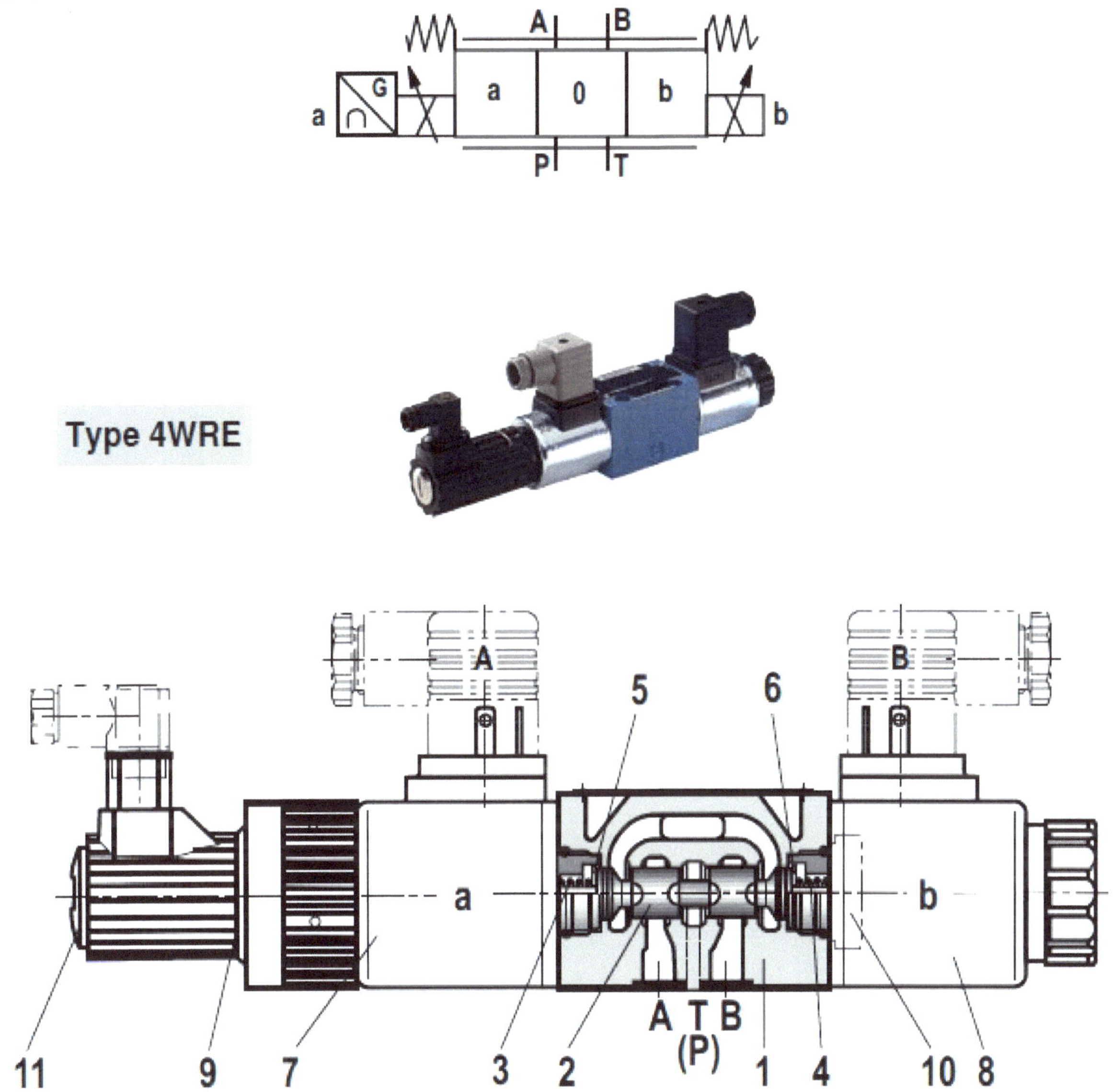

1- Housing, 2- Spool, 3 & 4- Compression Springs, 5 & 6- Spring Plates
7 & 8- Solenoid, 9- Position Transducer, 10- Plug Screw, 11-PG Fitting

Fig. 5.26 - Direct-Operated Stroke-Controlled Spool-Type PDV with Separate ECU (Courtesy of Bosch Rexroth)

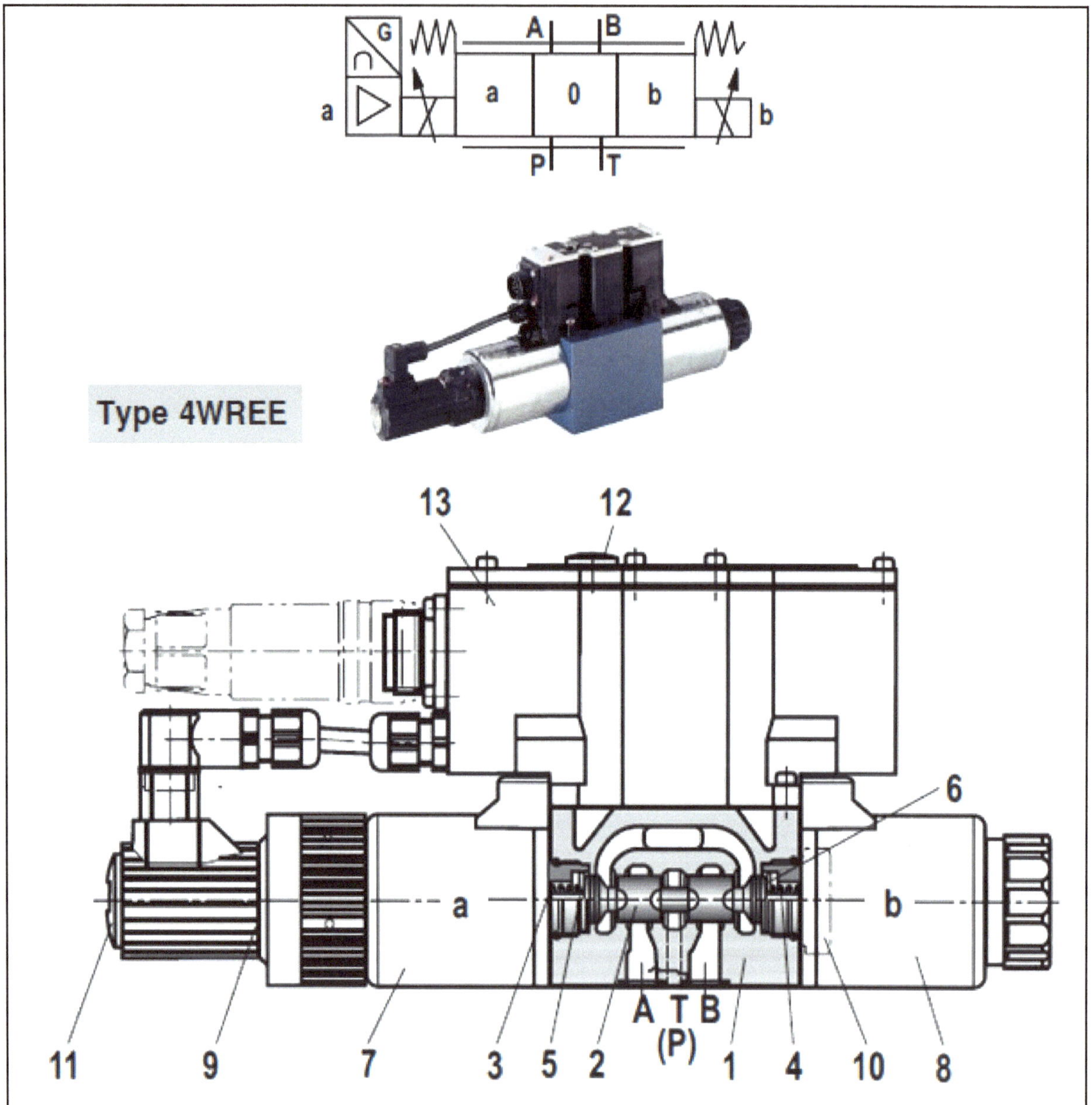

12- Electric Zero Point Adjustment, 13- OBE

Fig. 5.27 - Direct-Operated Stroke-Controlled Spool-Type PDV with OBE (Courtesy of Bosch Rexroth)

5.3.5- High Performance Proportional Directional Valve

Figure 5.28 shows an example of an upper end version of proportional directional valves. It is known as a *High Performance* or a *Servo Performance* PDV.

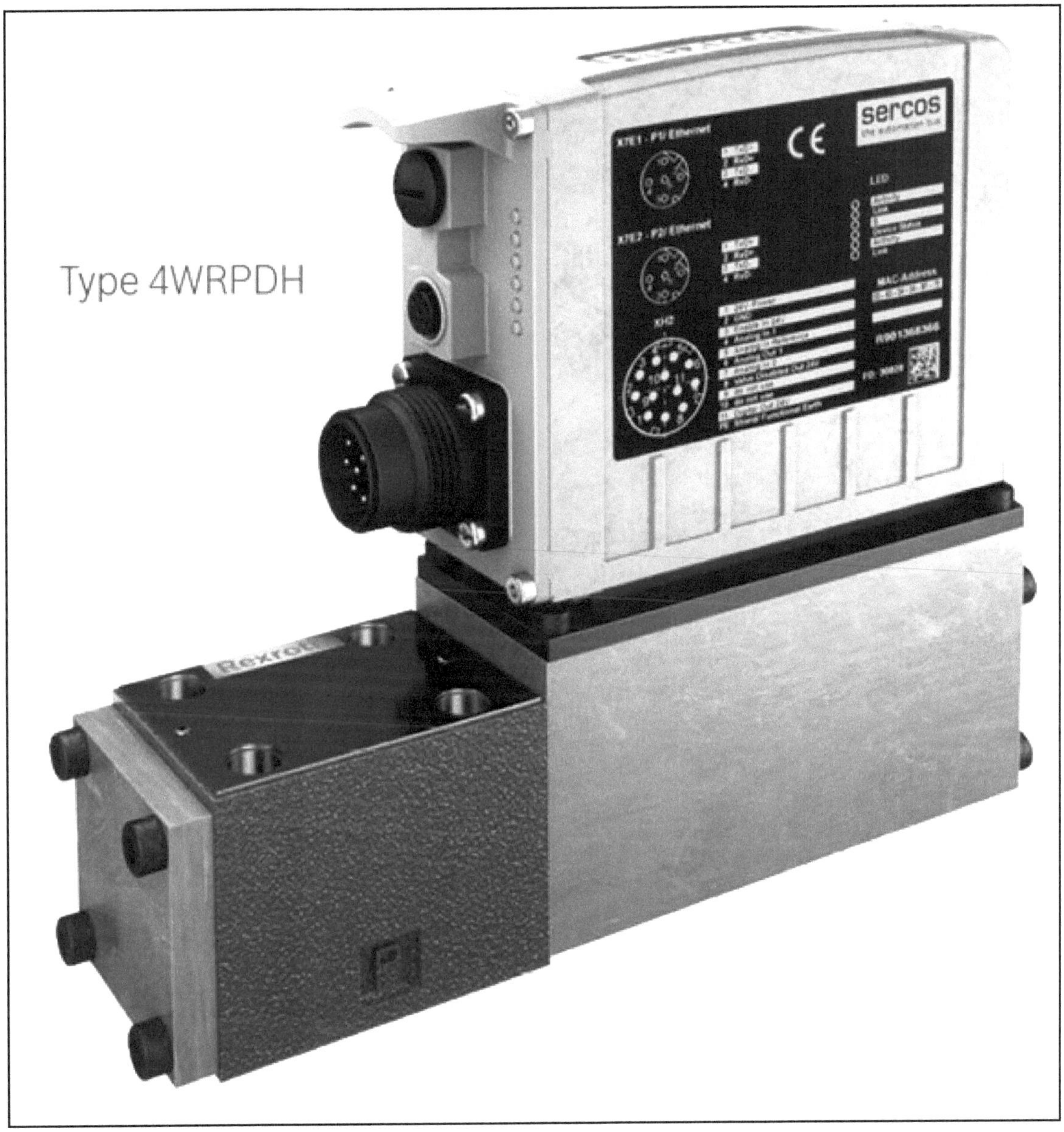

Fig. 5.28 - High-Performance Direct-Operated PDV (Courtesy of Bosch Rexroth)

A PDV is given such a name if it has some combination of the following features:

Electronics (Fig. 5.29):

- **OBE (1):** the valve is equipped with an OBE.
- **Analog Control (2):** the valve can send/receive analog signals for control and condition monitoring purposes. For example, the valve can receive analog input command and send analog output signal about the actual spool position.
- **Digital Control (3):** the OBE of the valve may contain digital control capability including a digital processor and the associated memory. The valve can run standalone programs without being connected to a computer.
- **Ethernet Interface (4):** the valve is equipped with Ethernet interface for bus connections so that it can be easily used to control an axis in industrial automated machines.

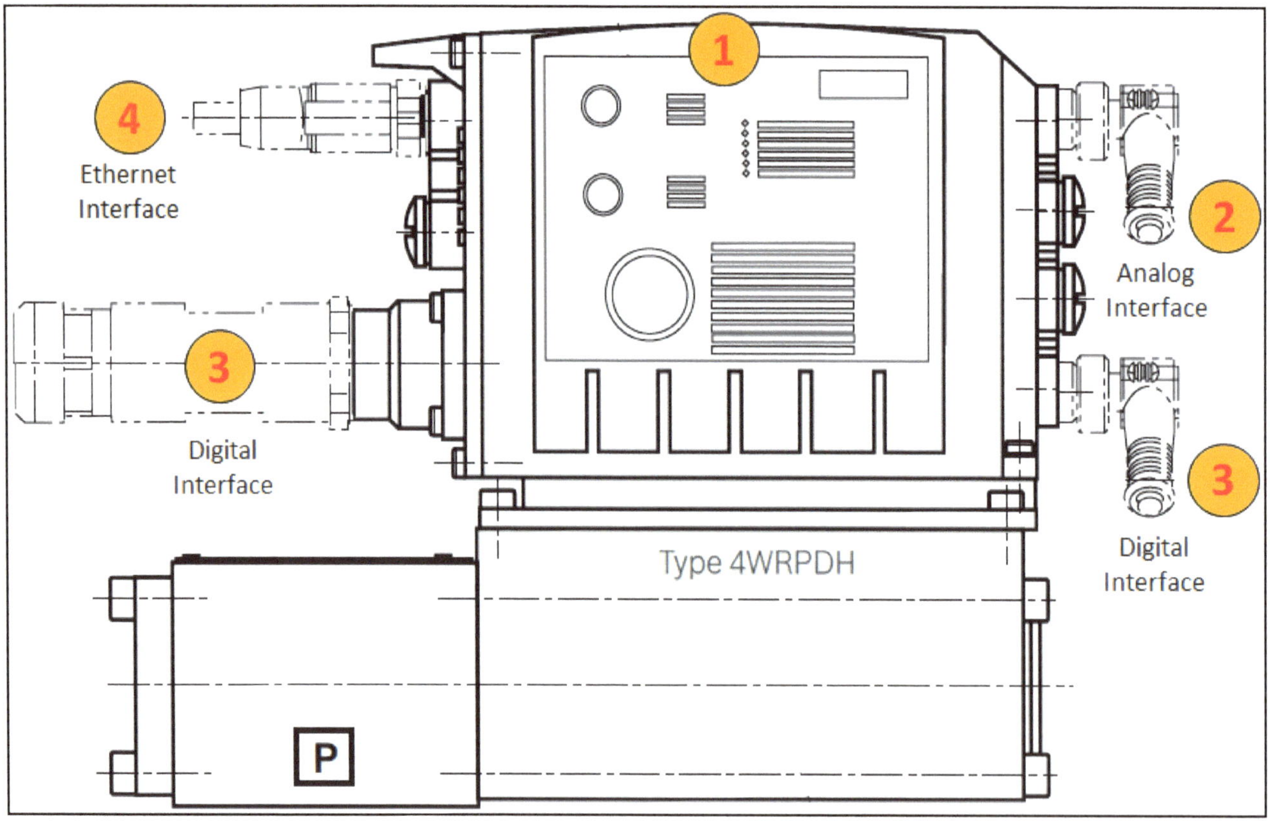

Fig. 5.29 - Electronic Features of High-Performance Direct-Operated PDV (Courtesy of Bosch Rexroth)

Solenoid (Fig. 5.30):

- **Push-Pull Proportional Solenoid (5):** for a three-position valve, a single solenoid has faster response than two solenoids.
- **Stroke-Controlled (6):** a valve with controlled spool in a closed-loop is more accurately positioned and compensates for aging conditions of the valve.

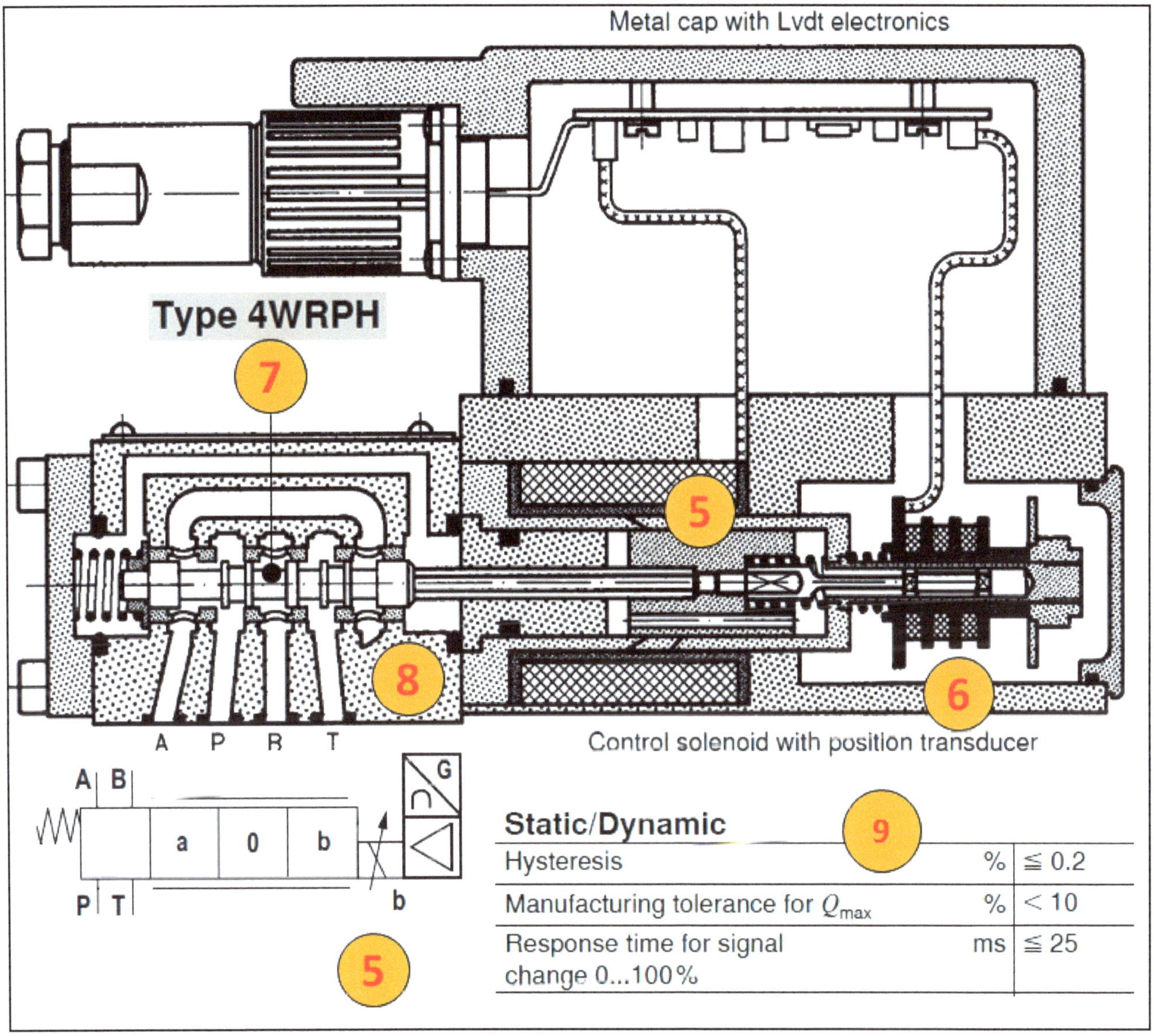

Fig. 5.30 - Solenoid and Hydraulic Features of High-Performance Direct-Operated PDV (Courtesy of Bosch Rexroth)

Hydraulics (Fig 5.30):

- **Servo-Spools (7):** zero-lapped spool configured to offer linear flow characteristics.
- **Anti-Rotation of Spool (8):** Large conventional spools tend to rotate when they are subjected to overflow. A high-performance PDV has an improved spool-sleeve assembly with *Anti-Rotation* design feature.
- **Improved Dynamics (9):** A high performance PDV has high response and little hysteresis which enable it to be used in a closed-loop control system.
- **Fail-Safe Fourth Position (10) (Fig. 5.31):** As shown in Fig. 6.31, a 4/4 high performance PDV secures a fourth position to protect the actuator when the input signal is lost due to any accidental reason.

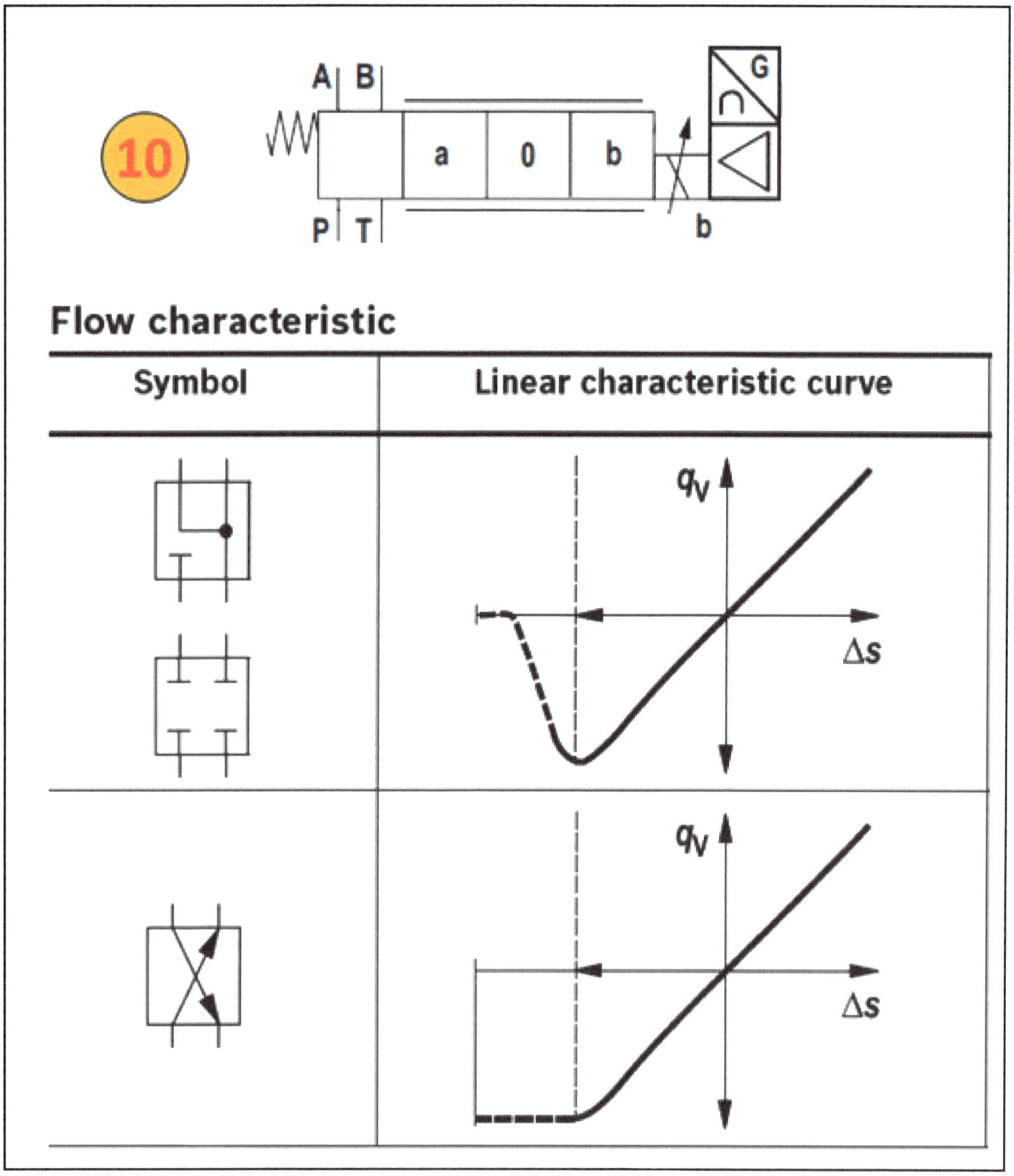

Fig. 5.31 - Fail-Safe Position in a High-Performance Direct-Operated PDV (Courtesy of Bosch Rexroth)

Figure 5.32 shows an example of using a high-performance PDV in closed-loop cylinder control within a system network. Such valves replace servo valves in many applications except where servo valve high dynamic features are needed.

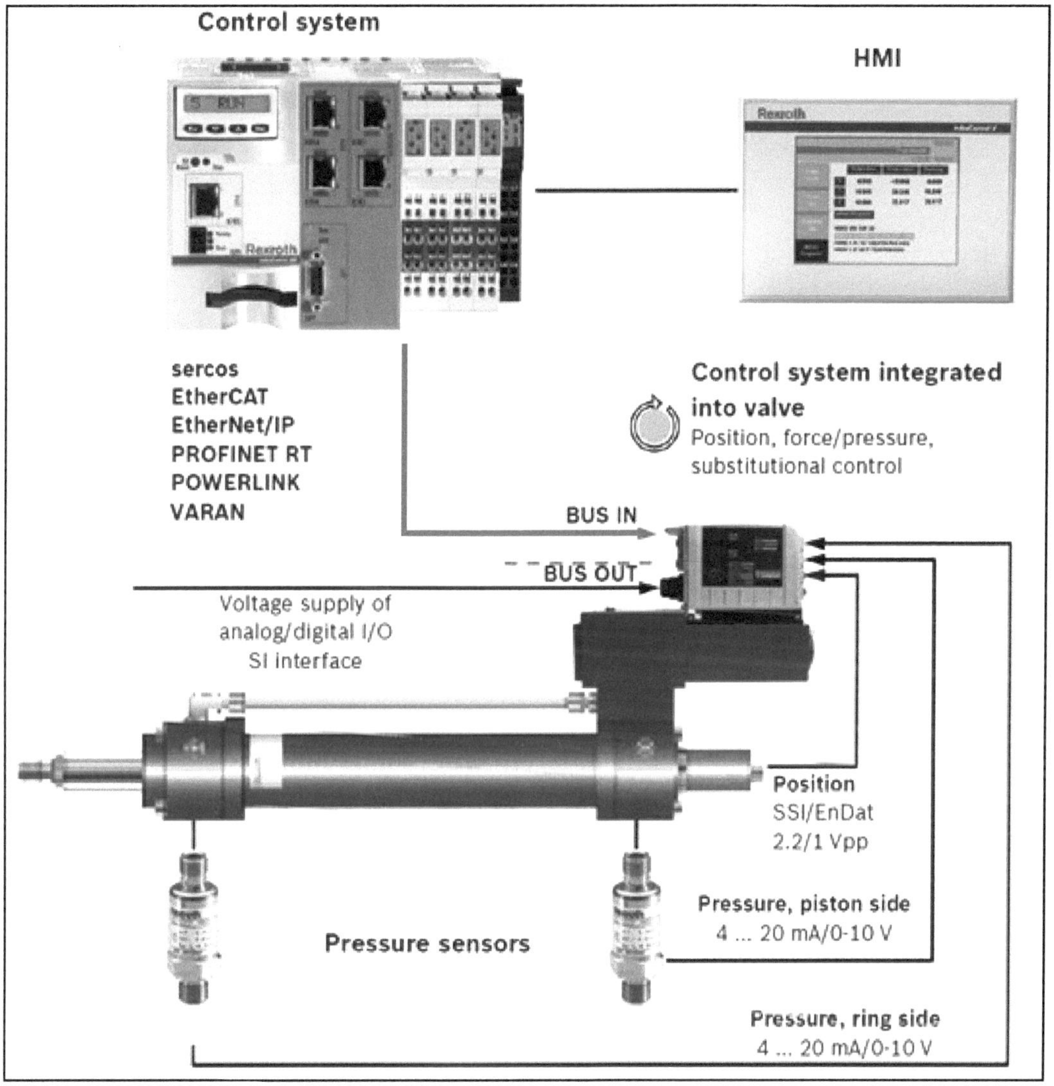

Fig. 5.32 - Example of using a High-Performance PDV in a Closed-Loop Application (Courtesy of Bosch Rexroth)

5.3.6- Pilot-Operated Force-Controlled Proportional Directional Valve

As it has been explained in Chapter 4, pilot-operated valves are used to overcome the flow forces so that it can handle large flow (approximately 100-120 lit/min or 20-25 GPM).

As shown in Fig. 5.33, similar to the pilot-operated switching directional valves:

- The pilot-operated proportional valve consists of two stages.
- The valve could be of two positions or three positions (dotted line in the figure).
- A pilot stage receives the control pressure **X** and uses it to control the main stage.
- The valve is called *Pilot-Controlled* if **X** is sourced externally; and is called *Direct-Controlled* if **X** is generated from the main pressure line **P**.
- Control pressure should not exceed a certain value (30 – 100 bar) which is specified by the valve manufacturer. Therefore, if needed and regardless the source of **X**, an optional sandwich-type pressure reducing valve (8) may be used to control the proper value of X to the pilot stage.
- It is easier to merge the pilot pressure return **Y** with the valve main tank line **T**. Line **Y** is drained externally if line **T** line is experiencing back pressure.
- If **X** is external, it is not necessary that **Y** to be external too.

As shown in Fig. 34, unlike the pilot-operated switching directional valves, positioning of the main spool in a pilot-operated force-controlled PDV is converted into an **open-loop pressure control** solution:

- The pilot stage here is a double three-way proportional pressure reducing valve.
- By actuating the proportional solenoids, the pilot valve creates a differential pressure as a %, proportional to the input signal, of the control pressure **X**. As a result, the main spool will be shifted "proportionally" and stops when the differential pressure force is balanced by the corresponding side spring force.
- For the main spool to move farther, the pilot valve must receive higher signal. The pilot valve must be kept energized so the main spool stays in its controlled position.
- The valve performance depends on the calibration between the control pressure and the supporting springs.
- In case the two proportional solenoids are de-energized, then both sides of the main spool will be vented and the centering springs will move the main spool to the center position.

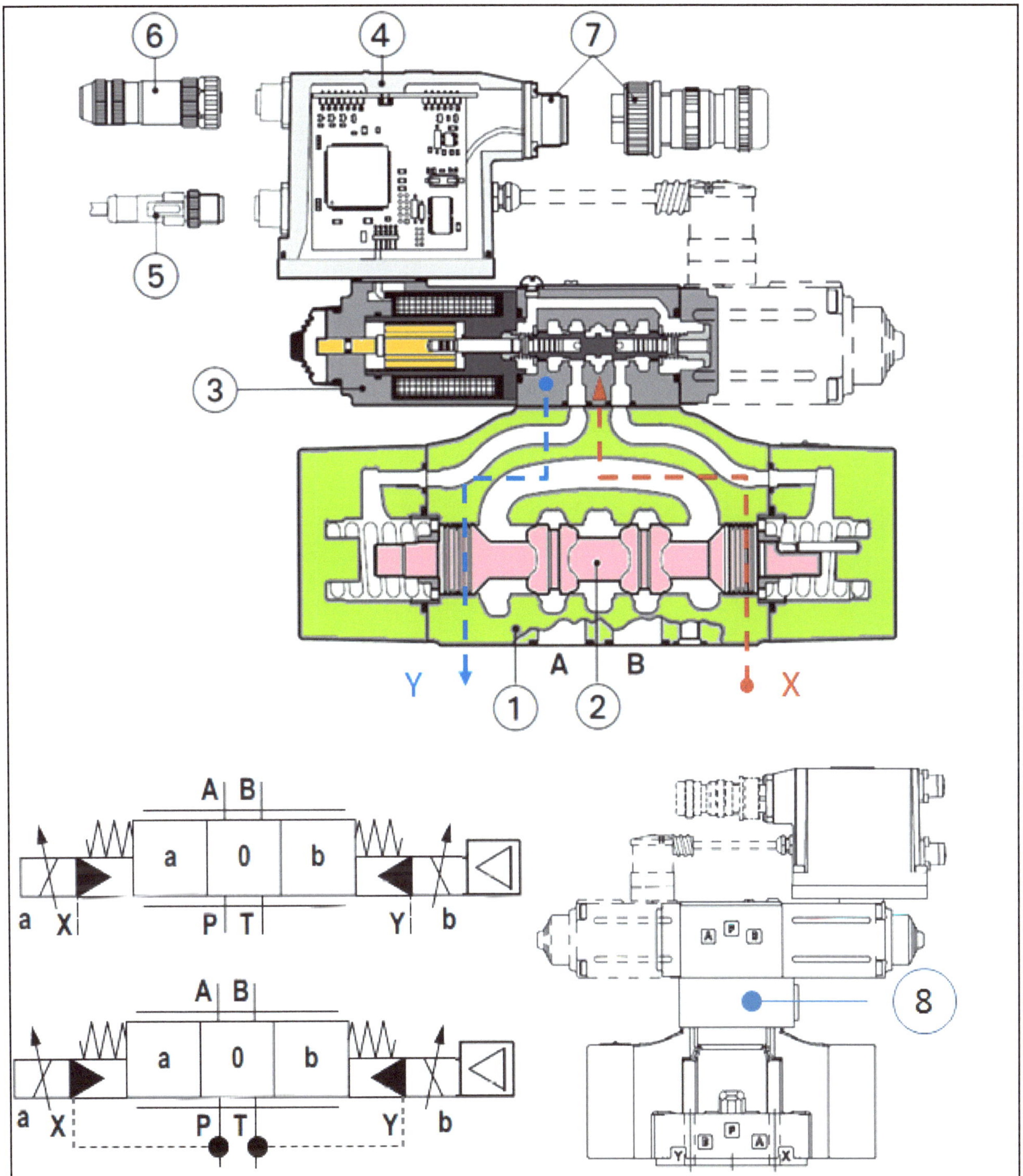

1- Main Valve Body, 2- Main Spool, 3- Pilot Valve (Pressure Reducing),
4- On-Board Electronics, 5- USB Connector,
6- Fieldbus Connector, 7- Main Connector, and 8- Optional Reducing Valve

Fig. 5.33 - Pilot-Operated Force-Controlled Spool-Type PDV with Two Balancing Springs (Courtesy of Atos)

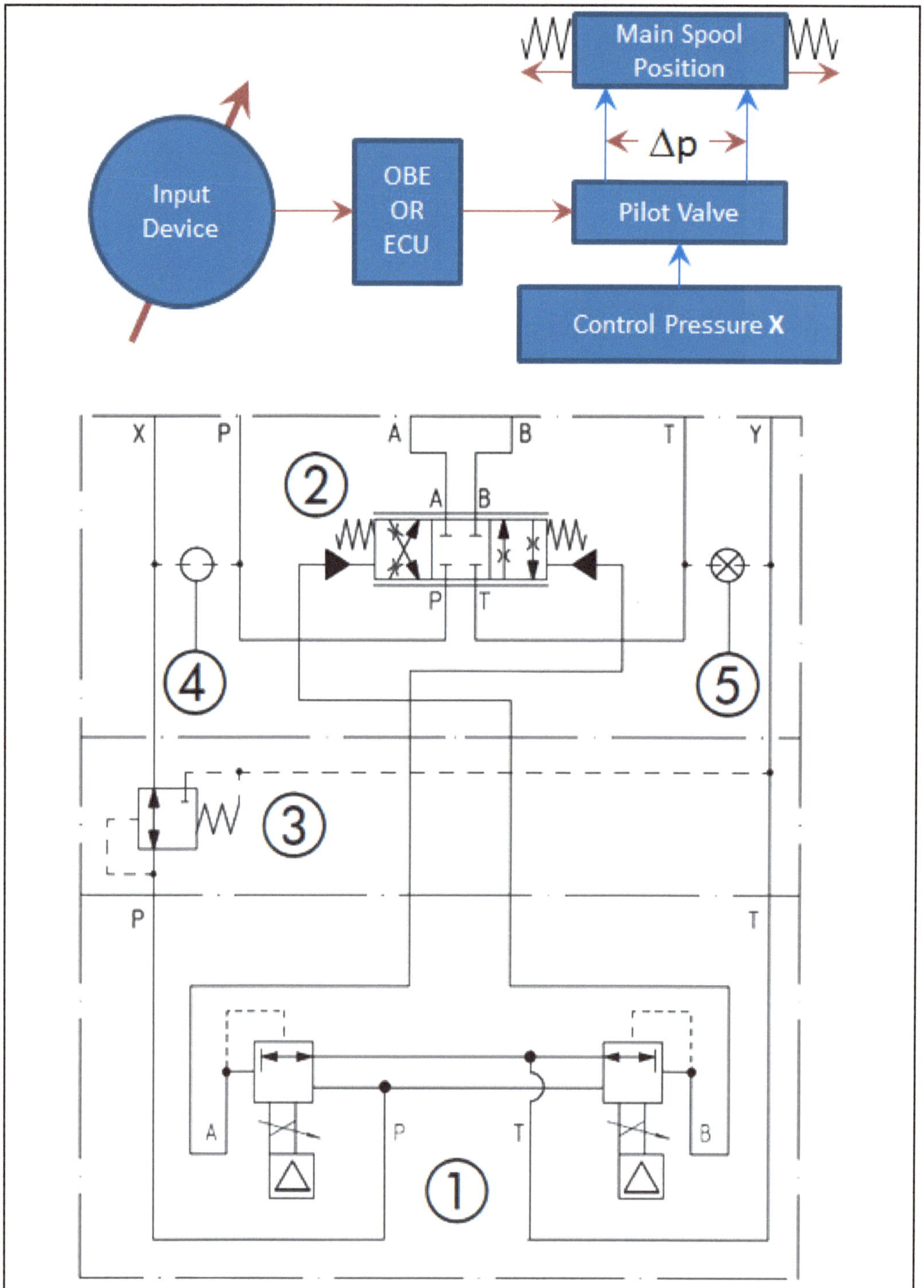

1- Pilot Valve (Double Three-Way Pressure Reducing Valve, 2- Main Stage,
3- Optional Sandwich-Type Pressure Reducing Valve,
4/5 – Plugs to be Added/Removed to Convert to External/Internal

Fig. 5.34 - Symbols for a Pilot-Operated Force-Controlled Spool-Type Two Balancing Springs (Courtesy of Atos)

Figure 5.35 shows the same previous valve but with one preloaded spring. The valve could be configured with OBE or with separate ECU.

Despite that the valve has one spring; the previous symbols are still applicable because the spring is pre-compressed to act on both sides.

As shown in Fig. 5.36, the advantages of having one preloaded spring are as follows:

- Ensures, an identical valve reaction in each direction to any given signal, and thus, equal deflection in each direction
- The change in the spring characterizes will be distributed evenly and will not affect the center position of the main spool. In case of two balancing springs, if the characteristics of springs are changed differently, this may result in load creeping when the two solenoids are de-energized.

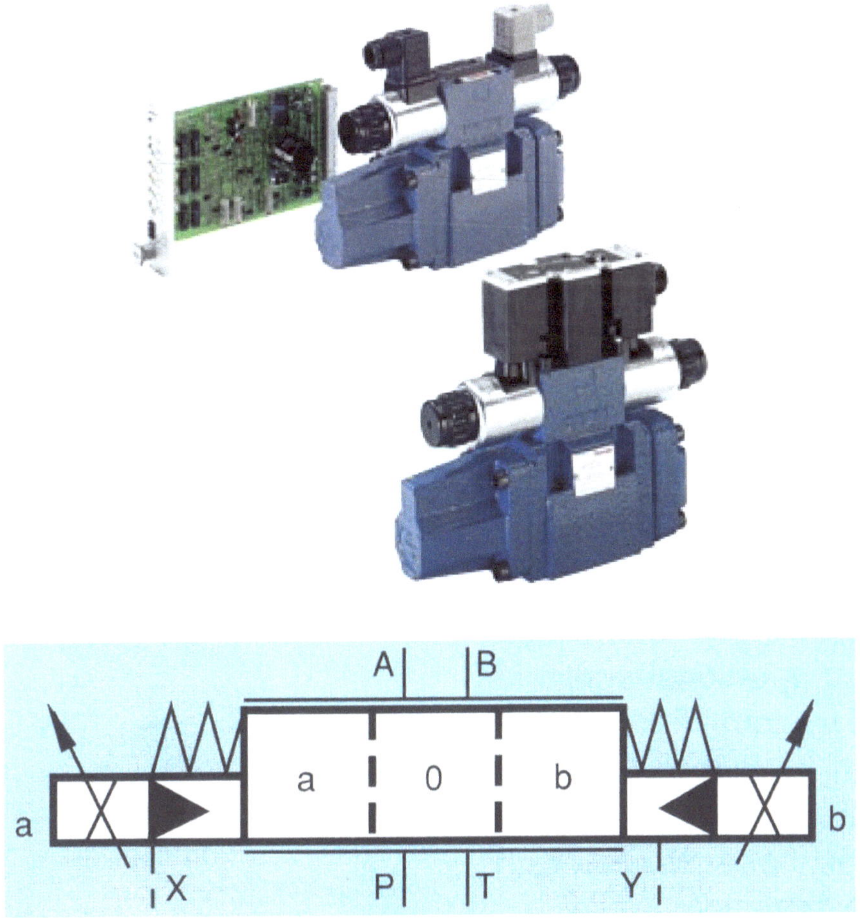

Fig. 5.35 - Symbol and Outer Shape for a Pilot-Operated Force-Controlled Spool-Type PDV with One Preloaded Spring (Courtesy of Bosch Rexroth)

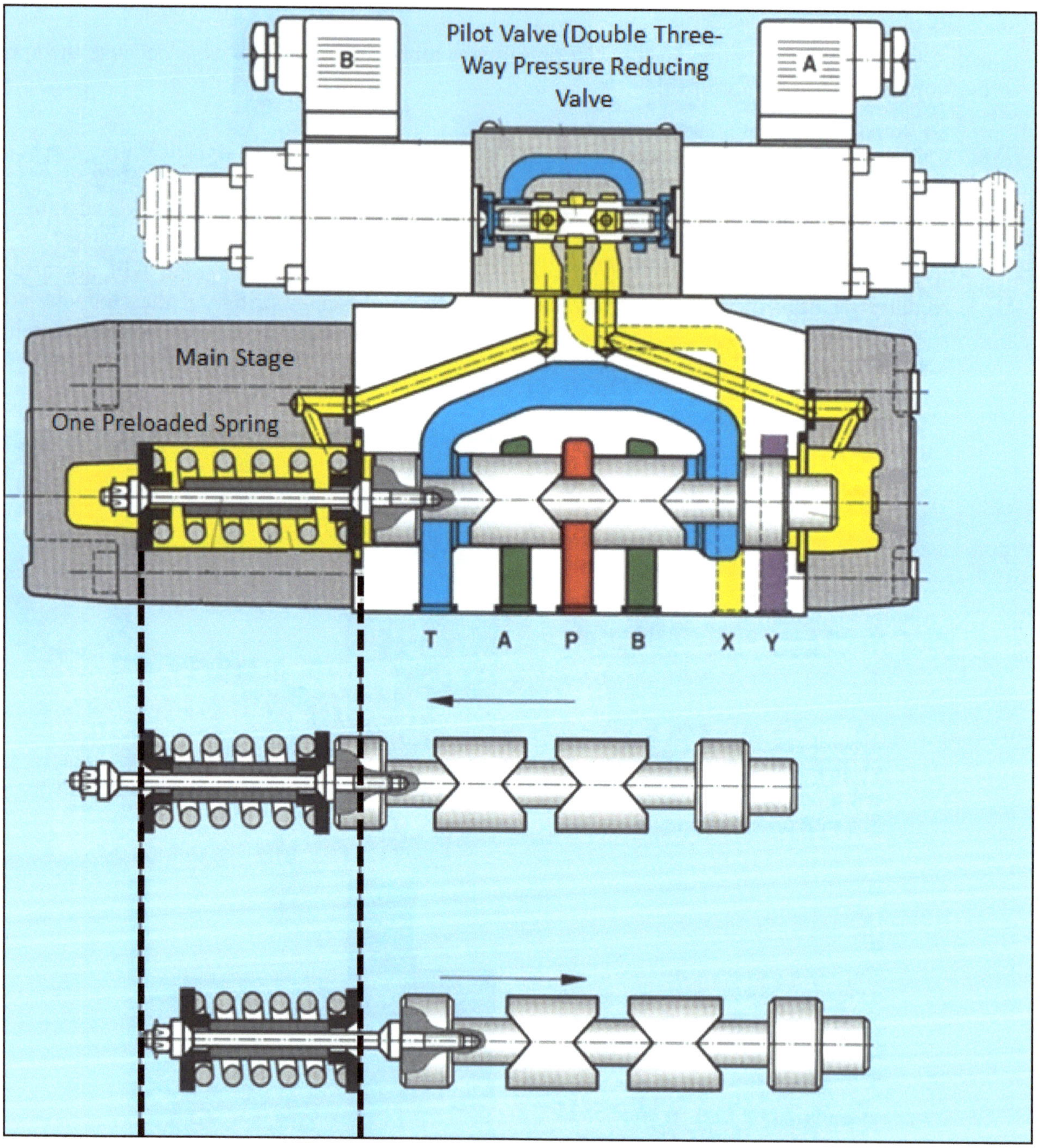

Fig. 5.36 - Pilot-Operated Force-Controlled Spool-Type PDV with One Preloaded Spring (Courtesy of Bosch Rexroth)

As shown in Fig. 5.37, pilot-operated, force-controlled, cartridge-type PDVs are also found. The symbol shows a pilot-operated and pilot-controlled valve.

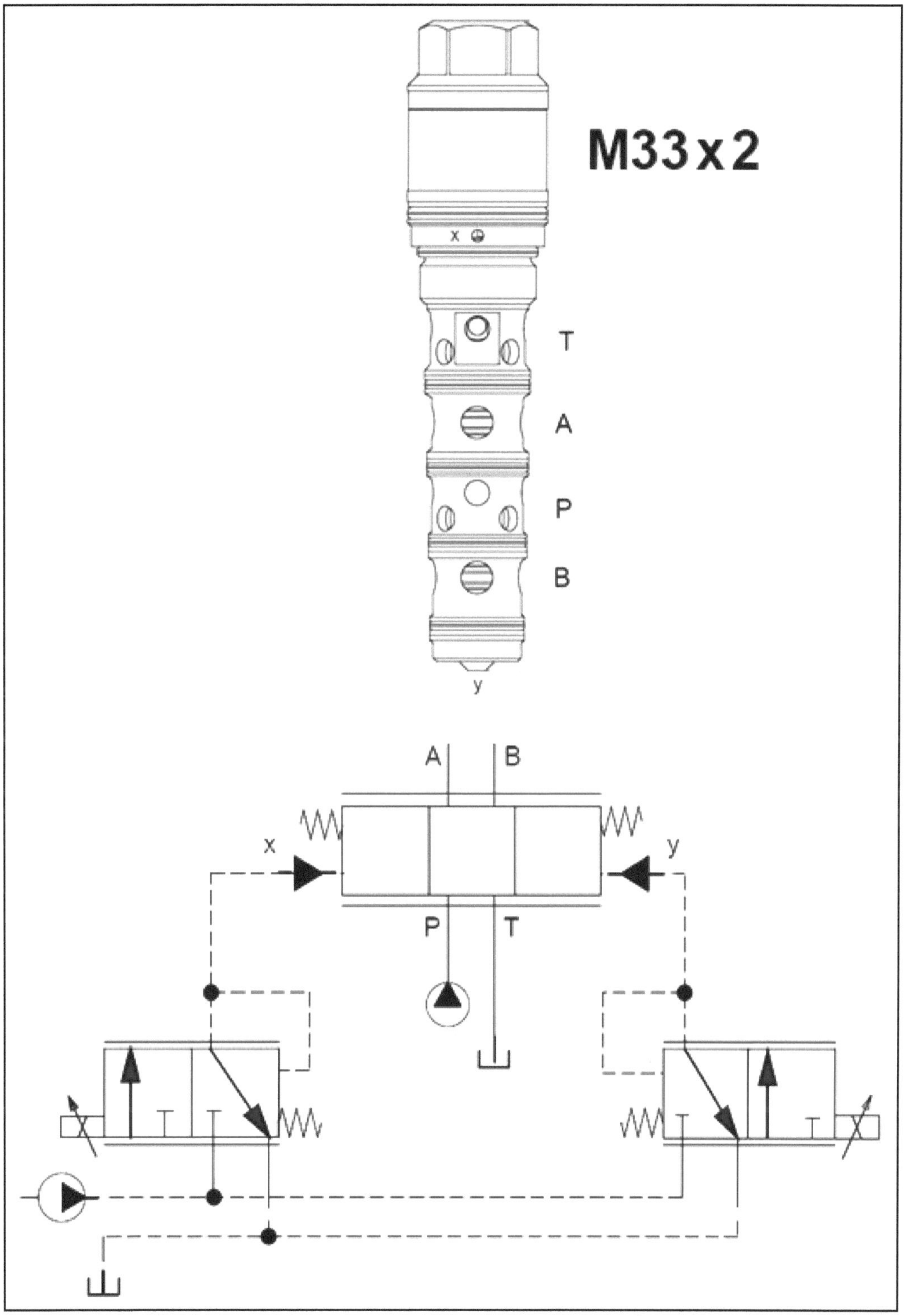

Fig. 5.37 - Pilot-Operated Force-Controlled Cartridge-Type (Courtesy of Wandfluh)

There is a complete range of proportional screw-in cartridge valves in the market. As shown in Fig. 5.38, directional control, pressure control as well as flow control valves are found. The valves are designed for maximum pressures of up to 400 bar. With volume flow of up to 400 l/min, they are ideal for controlling large consumers. Ambient temperatures of up to 70°C can be accepted without any loss of performance.

Fig. 5.38 - Cartridge-Type Proportional Valves (Courtesy of Wandfluh)

5.3.7- Pilot-Operated Stroke-Controlled Proportional Directional Valve

A two-stage stroke-controlled valve can be configured to have only the main stage is stroke-controlled or both the main and the pilot stage are stroke controlled. The block diagram shown in Fig. 5.39 explains the first case in which the position of the main spool is monitored and controlled in a closed-loop. The block diagram shown in Fig. 5.40 explains the second case in which the two stages are stroke-controlled. The second case is considered the upper end of pilot-operated proportional directional valves.

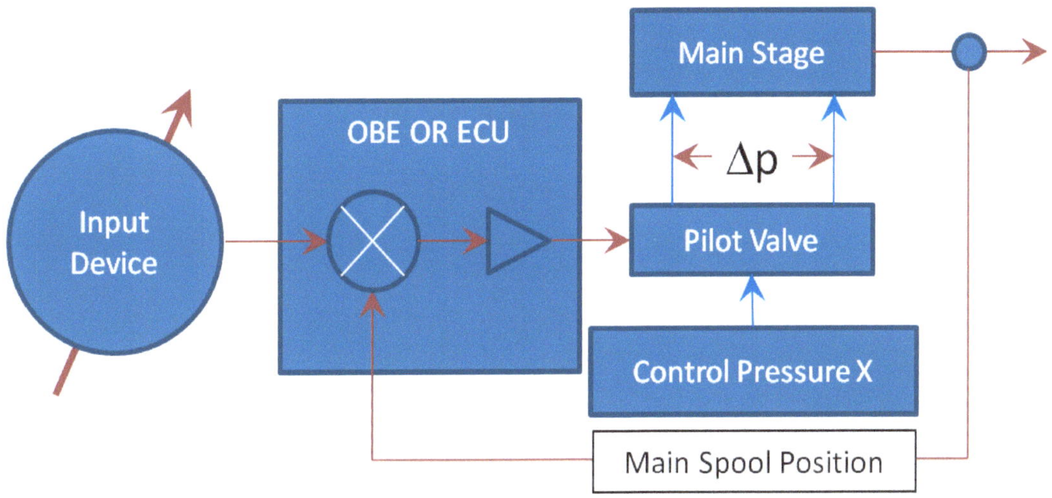

Figure 5.39 - Block Diagram of a Pilot-Operated Spool-Type PDV with Stroke-Controlled Main Stage

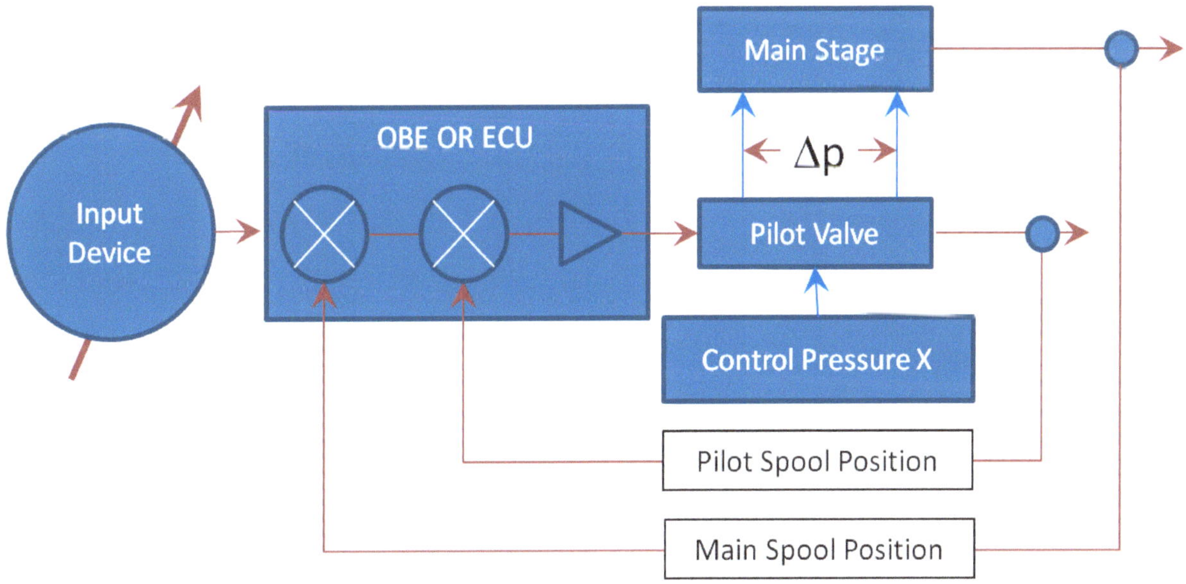

Fig. 5.40 - Block Diagram of a Pilot-Operated Spool-Type PDV with the two Stages Stroke-Controlled

Figure 5.41 shows an example of a typical valve of the first case in which the main stage is only stroke-controlled. The valve can be configured with or without OBE.

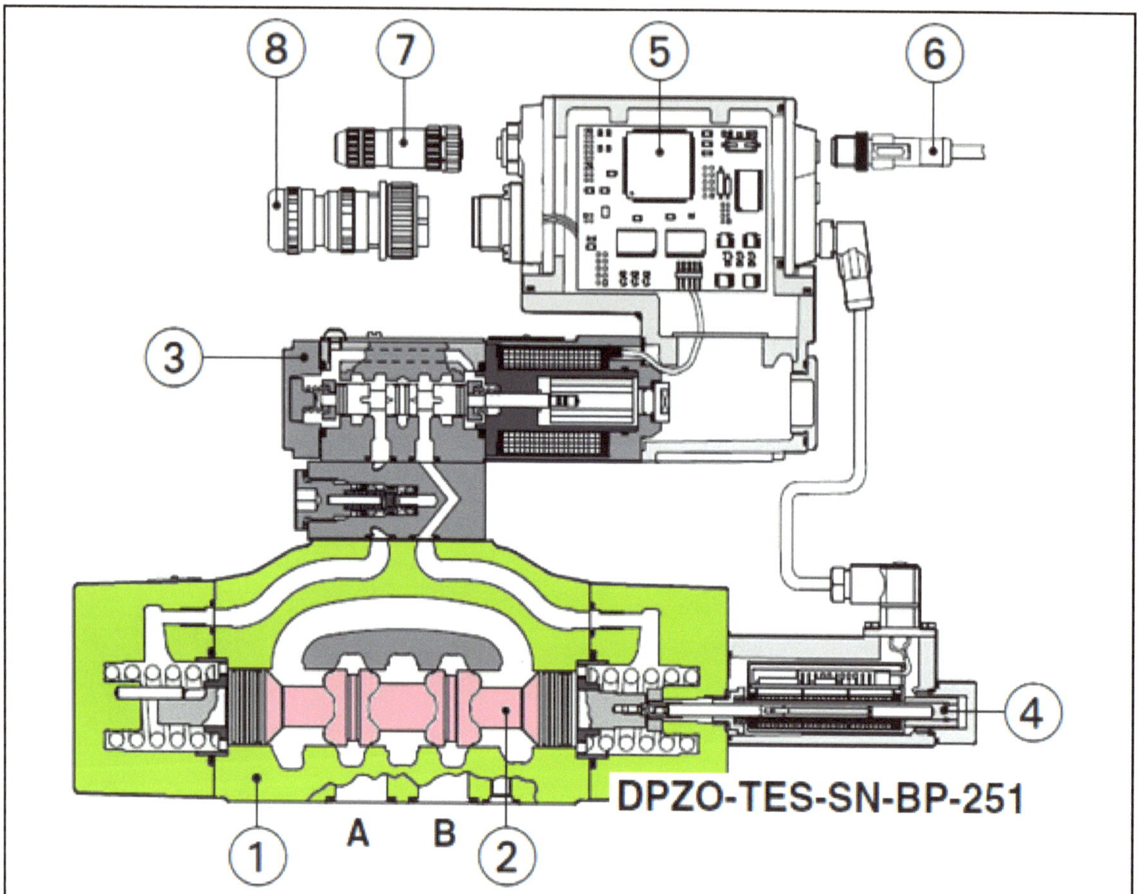

1- Main Valve Body, 2- Main Spool, 3- Pilot Valve, 4-Main Stage Position Transducers, 5- On-Board Electronics, 6- USB Connector, 7- Fieldbus Connector, 8- Main Connector,

Fig. 5.41 - Pilot-Operated Spool-Type PDV with Stroke-Controlled Main Stage (Courtesy of Atos)

Figure 5.42 shows the simplified and detailed symbols of the previous valve. It is to be reminded that the reduced symbol reflects the central position of the main stage spool. This stroke-controlled valve is configured as a C*ontrol System Type 0* (PID Controller is needed). The input signal to the pilot valve creates a differential pressure that drives the main spool towards its destination. Main spool moves against the return spring and its position is used as a feedback signal to the controller.

To move the main spool farther, the pilot valve must receive higher signal. The pilot valve must be kept energized so the main spool stays in its controlled position. If the pilot valve loses the signal, its centering springs will take the spool to its float center. Consequently, the two sides of the main spool will be vented and its centering springs will take the spool to its central position.

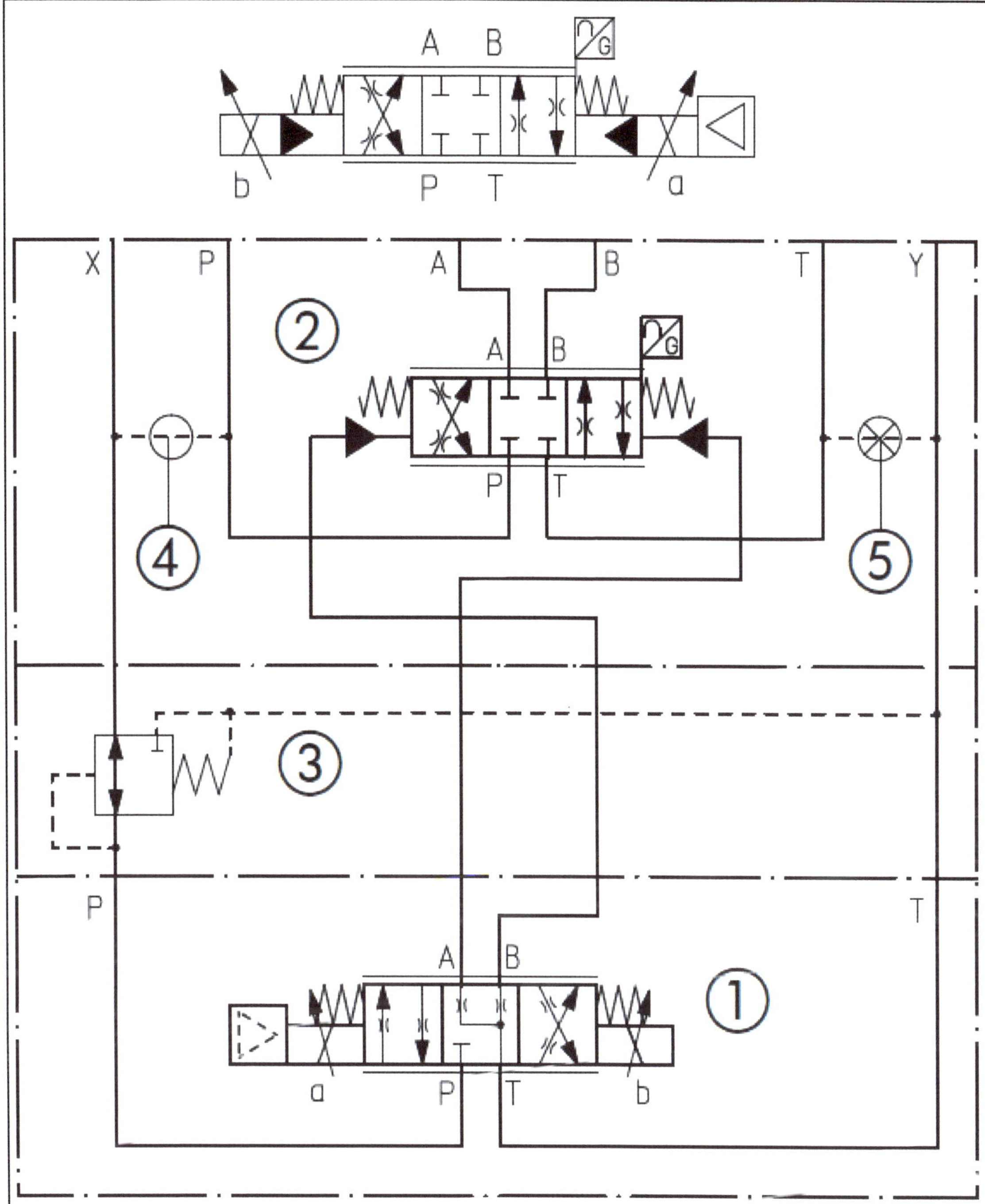

1- Pilot Valve, 2- Stroke-Controlled Main Stage, 3- Optional Sandwich-Type Pressure Reducing Valve, 4/5 – Plugs to be Added/Removed to Convert to External/Internal

Fig. 5.42 - Symbols for a Pilot-Operated Spool-Type PDV with Stroke-Controlled Main Stage (Courtesy of Atos)

Figure 5.43 shows a valve that has the same configuration as Fig. 5.41 but another manufacturer. The sectional view in the figure shows the OBE drives the proportional solenoids of the pilot stage and takes care of closing the loop on the position of the main stage spool.

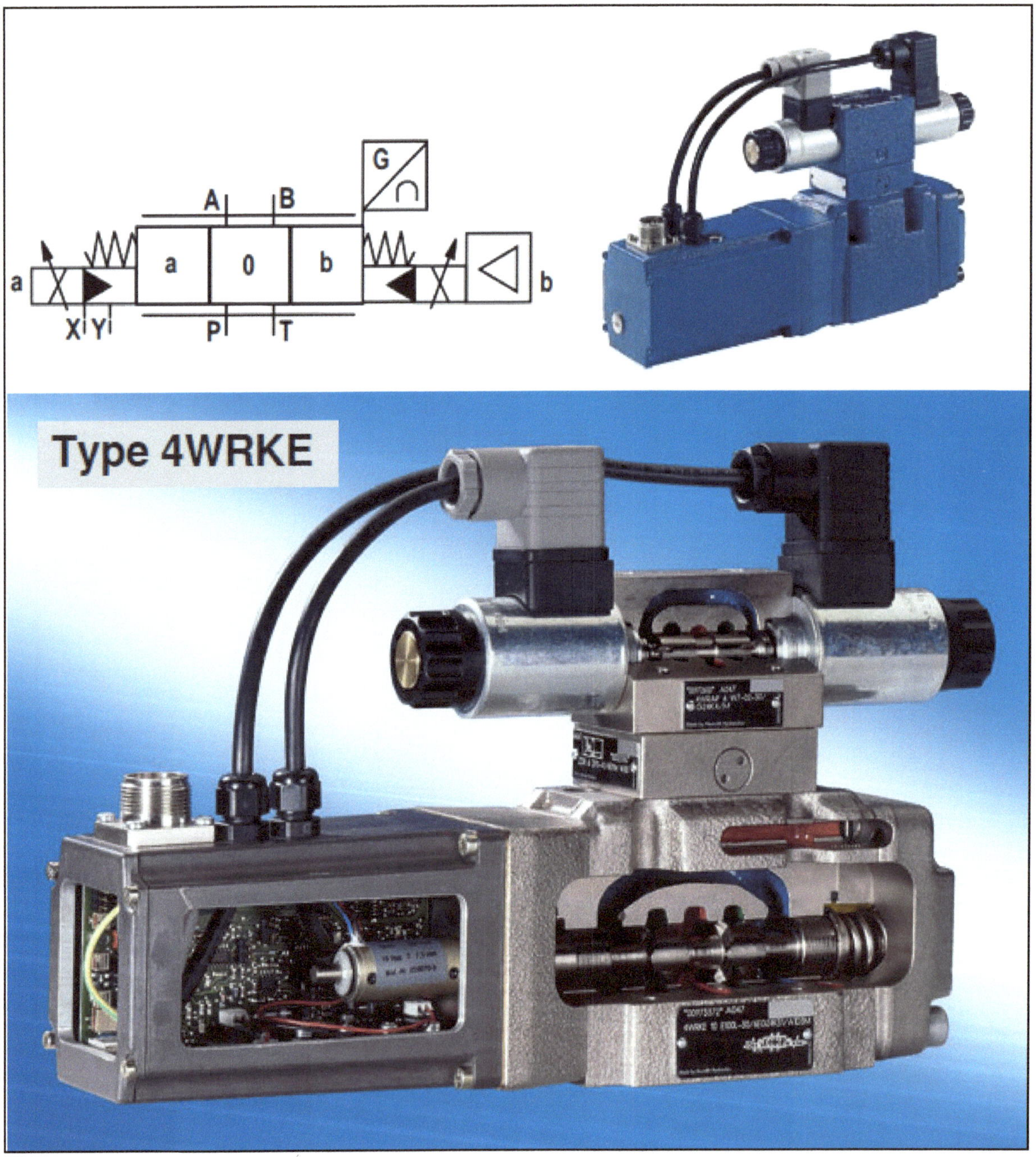

Fig. 5.43 - Pilot-Operated Spool-Type PDV with Stroke-Controlled Main Stage and with an OBE (Courtesy of Bosch Rexroth)

Figure 5.44 shows the upper end of pilot-operated, spool-type PDVs in which both the pilot stage and the main stage are stroke-controlled. The figure shows the simplified and detailed symbols of the valve. This stroke-controlled valve is configured as a *Control System Type 1* (only PD Controller is needed). Once the supply voltage to the valve is turned ON, the pilot spool takes its zero-lapped position holding the main spool in its center position. The spool position of the pilot valve is controlled proportional to the input signal by its own closed-loop. Movement of the pilot spool creates a differential pressure that drives the main spool towards its destination. Main spool position is used as a feedback signal in its closed-loop controller. Unlike the previous stroke-controlled valve, when the main spool reaches its desired destination, the error control signal given to the pilot valve goes down to zero. Therefore, the pilot spool goes to its zero-lapped position to hold the main spool at its new desired position. If the pilot valve lost the supplied power or in case of a cable break, its return spring returns it to its fail-safe fourth float position. Consequently, the two sides of the main spool will be vented, and its centering springs put it in its center position.

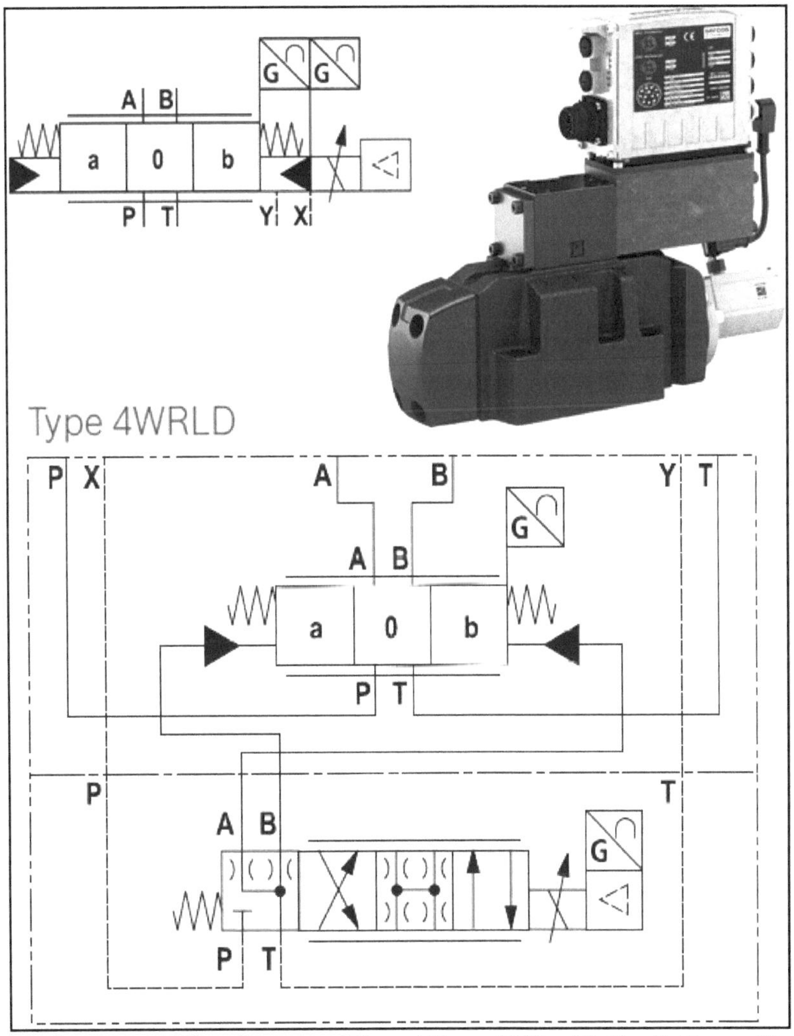

Fig. 5.44 - High-Performance Pilot-Operated Spool-Type PDV with the 2-Stages Stroke-Controlled (Courtesy of Bosch Rexroth)

Figure 5.45 shows the same configuration of the previous valve with possible special designs of the main spool. For certain machine operations, the main spool may be designed in a special way to achieve certain command-flow relationship.

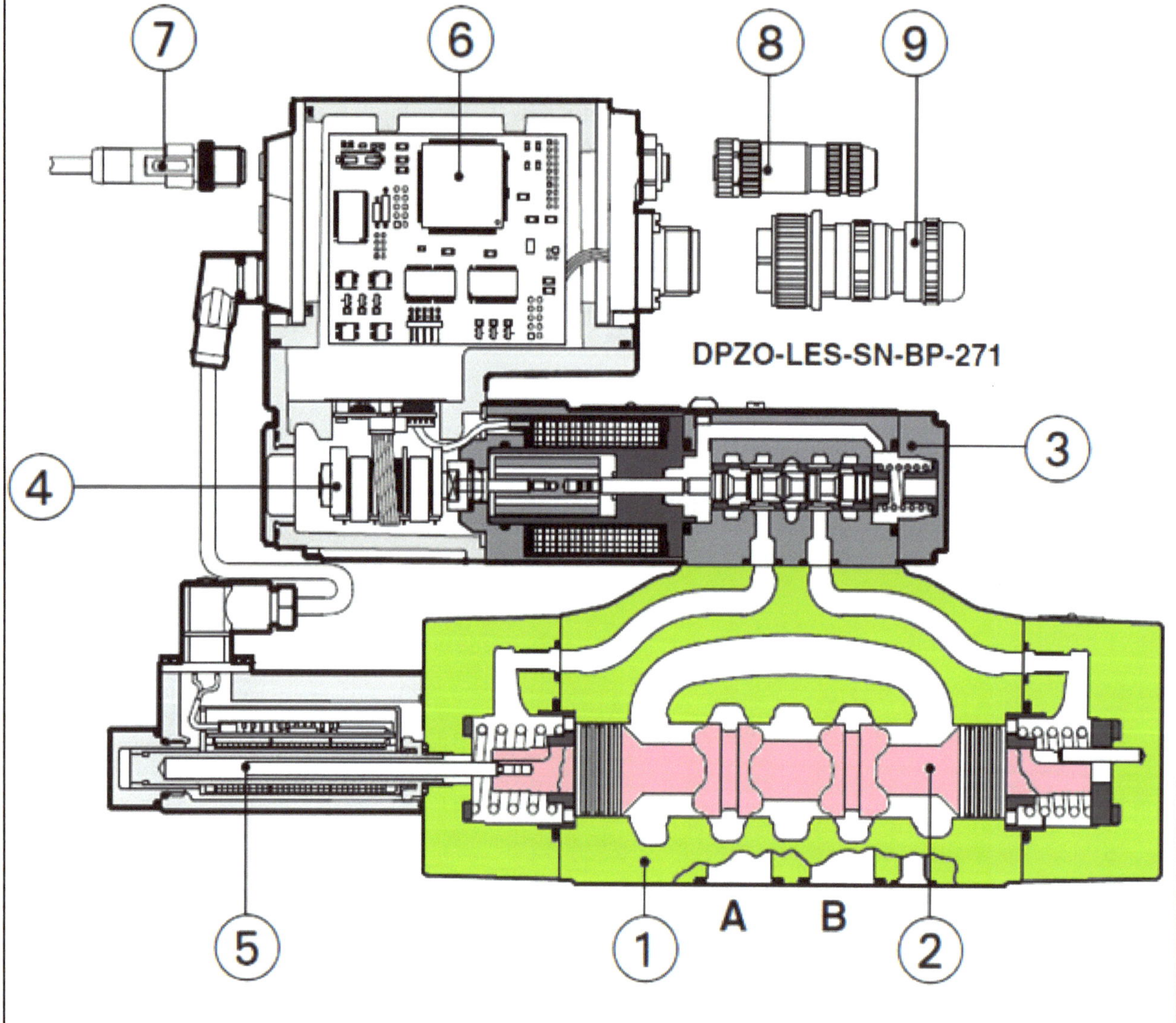

1- Main Valve Body, 2- Main Spool, 3- Pilot Valve, 4- Pilot Valve Position Transducers, 5-Main Stage Position Transducers, 6- On-Board Electronics, 7- USB Connector, 8- Fieldbus Connector, 9- Main Connector,

Fig. 5.45 - Pilot-Operated Spool-Type PDV with the 2-Stages Stroke-Controlled and a Special Design Proportional Spool (Courtesy of Atos)

As an example, Fig. 5.46 shows a main spool with four positions. With no signals on the pilot valve, the two sides of the main spool are vented, and the centering springs takes care of perfectly centering the main spool at its closed center which is the initial position.

The figure shows one position of the left side of the closed center. In that position the ports are connected as per the spool when it moves from 0 to -100% of its maximum stroke.

The figure shows two positions on the right side of the closed center. In fact, they are not two positions. The spool has been cut so that the ports are connected in two different ways depending on the signal. So, when the spool moves from 0 to 60 % of its maximum stroke, the ports are connected in the way shown in the curve. If the spool moves further, from 60 to 100, ports B and T are disconnected.

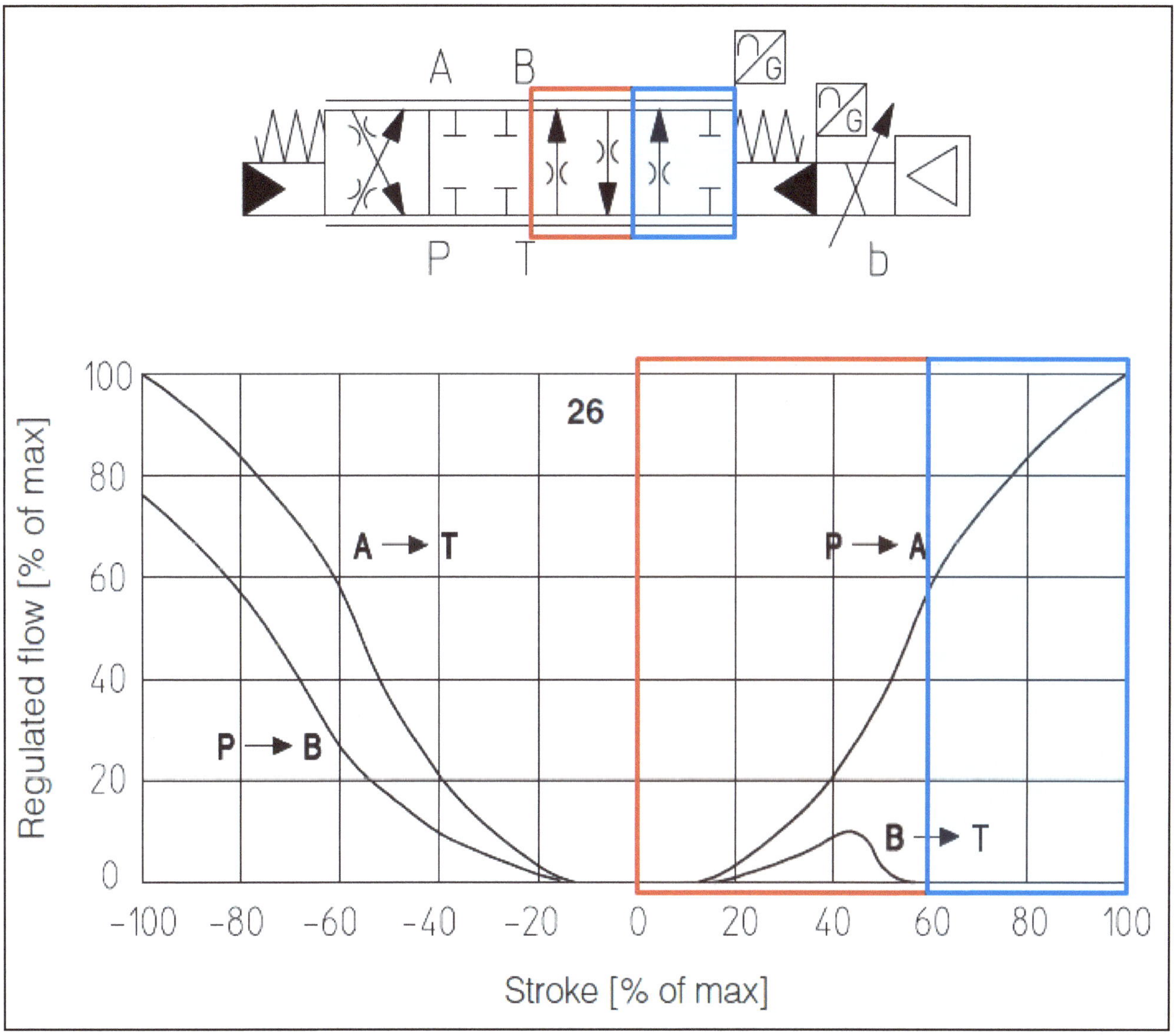

Fig. 5.46 - Performance of Proportional Directional Valve with Main Spool Special Design (Courtesy of Atos)

5.3.8- Proportional Directional Valves with Variable Load Compensator

Regardless the configuration of the PDV, as shown in Fig. 5.47, it is nothing but an electronically adjustable throttle valve. Therefore, as it has been comprehensively explained in Volume 1 of this series of textbooks, it won't maintain constant flow in cases where it drives a variable load in an open-loop. The characteristic curves shown in the figure indicate that the flow through a PDV depends on how far the valve is open and the differential pressure across the valve. For example, if the proportional valve has been set to 80% opening with 50 bar differential pressure, the motor will receive 40 lit/min and then spin at a constant speed as long as the torque on the motor is kept constant. At the same 80% valve opening, if the load increased so that the differential pressure reduced to 30 bar, the motor receives approximately 33 l/min and the motor slows down accordingly.

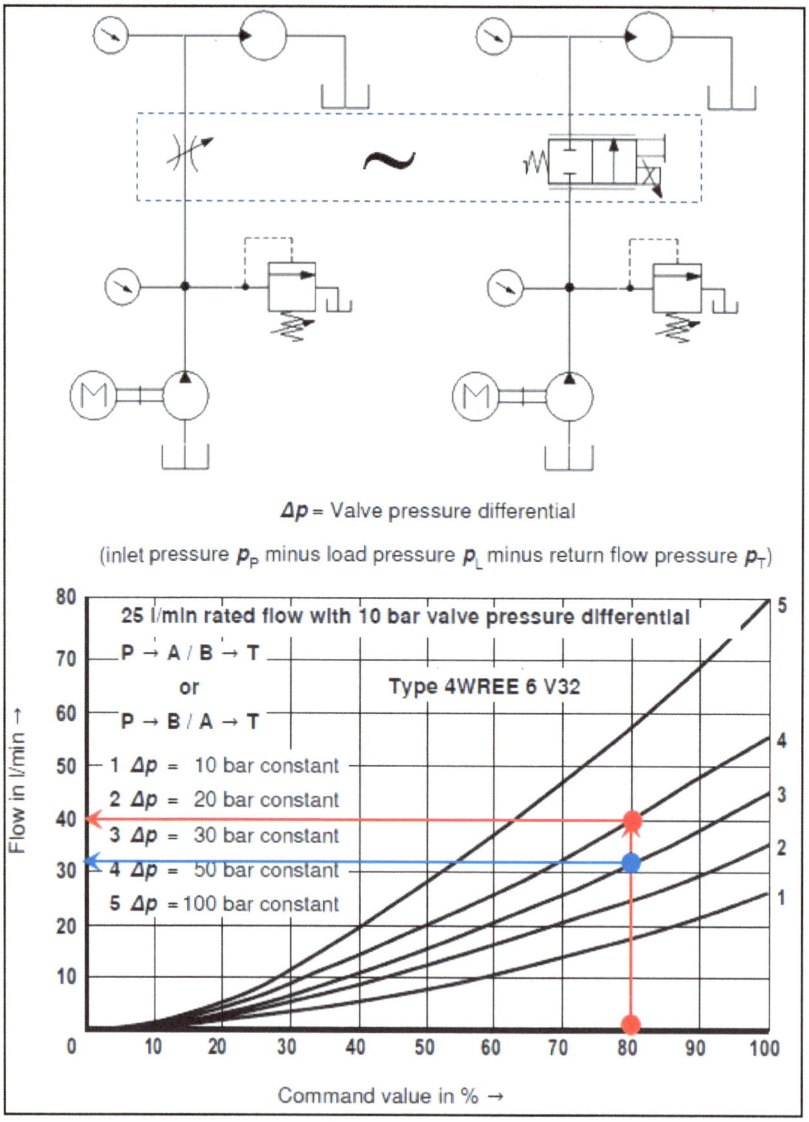

Fig. 5.47 - Need for Variable Load Compensator

As shown in Fig. 5.48, in order to maintain constant flow through the PDV (1), a *Variable Load Compensator* (*VLC*) (2) should be placed upstream of the PDV.

Within the active control zone of the VLC, the regulating pressure P_R acts on one side of the spool. The load pressure P_L acts on the other side of the spool. Because P_R is higher than P_L, a balancing spring is used to maintain the current balanced position of the VLC's spool.

If the torque on the motor increased, load pressure P_L increases and the motor would slow down. Alternatively, the spool of the VLC will be slightly shifted to the left side, the regulating restriction opens wider for the flow from the pump, and then the regulating pressure P_R is increased.

So, the VLC watches the load and accordingly takes a new balanced position to regulate the pressure P_R in order to maintain constant differential pressure across the PDV. Hence, constant flow passes across the PDV valve, and consequently a constant actuator speed is maintained regardless the load variation.

It is to be noted that the VLC is not needed if the PDV is used in a closed-loop because in such a case the control system will take care of regulating the spool of the PDV to maintain the desired speed of the motor in response to any external disturbances.

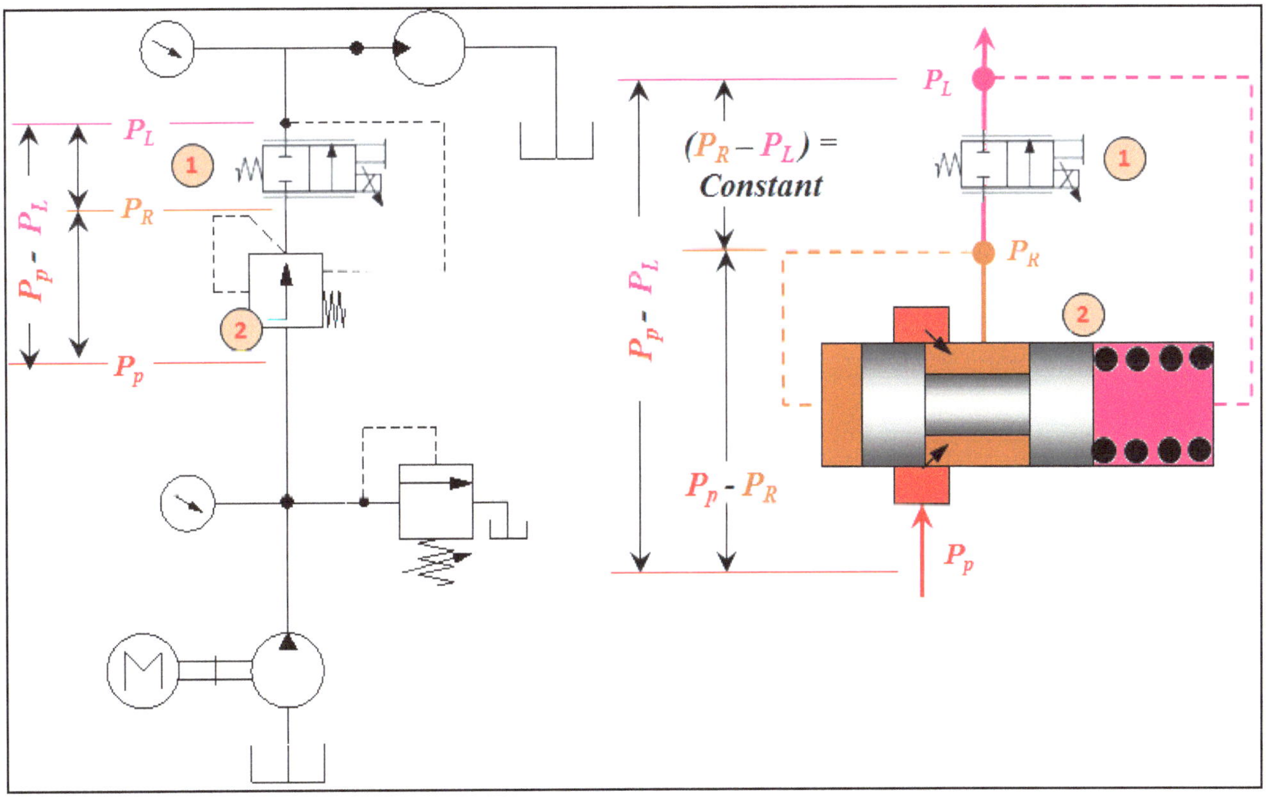

Fig. 5.48 - Placement of Load Compensator for Unidirectional Motors or Single Acting Cylinders

Figure 5.49 shows the placement of the load compensator when a double acting cylinder or a bidirectional motor is used. A shuttle valve is used to transmit the load pressure regardless the direction of motion of the actuator.

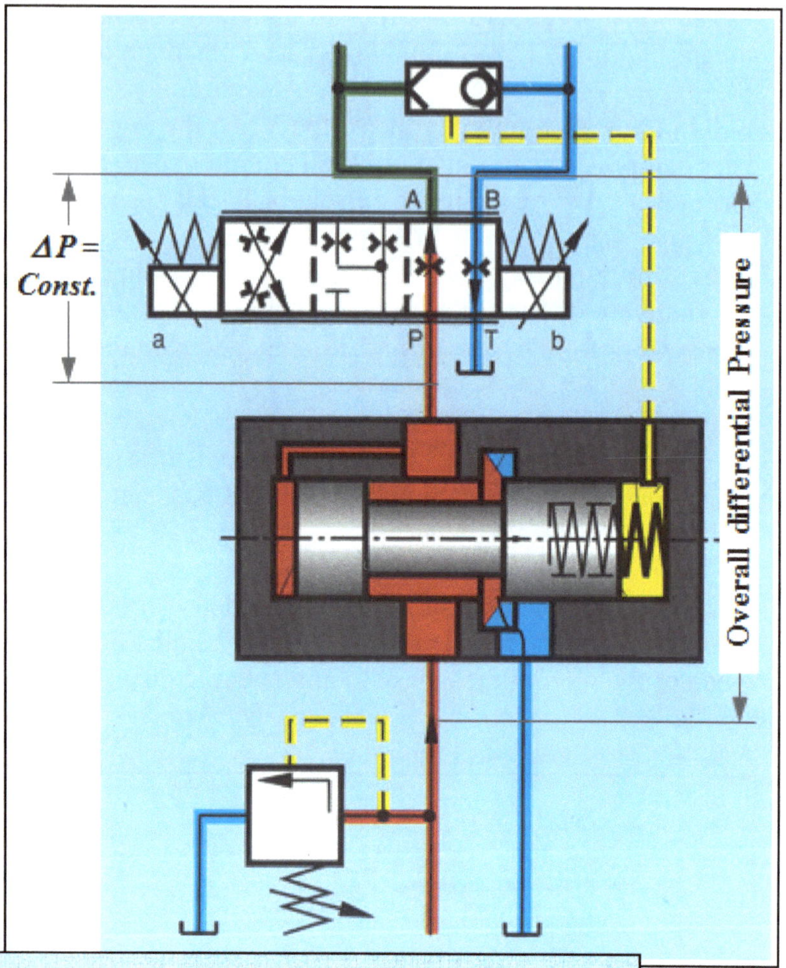

Fig. 5.49 - Placement of Load Compensator for Bidirectional Motors or Double Acting Cylinders (Courtesy of Bosch Rexroth)

Figure 5.50 shows the operation characteristics after adding the VLC. The figure shows that constant flow is maintained irrespective to the overall differential pressure, which is the pump (supply) pressure minus the load pressure.

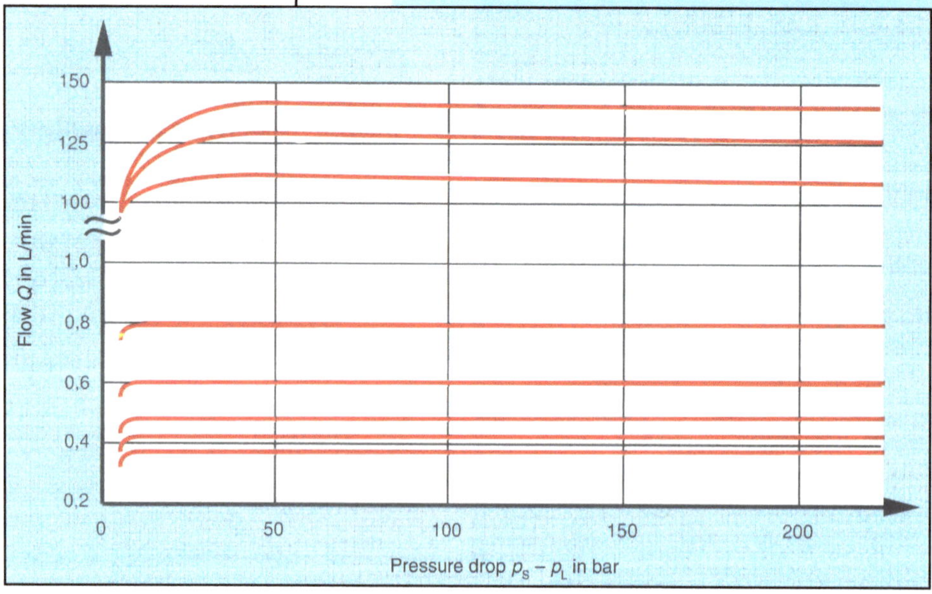

Fig. 5.50 - Operation Characteristics of a PDV after adding a VLC (Courtesy of Bosch Rexroth)

For the proper functioning of the VLC, a minimum overall differential pressure must be maintained otherwise the VLC will be saturated and acts like a fixed orifice. For example, if the load is blocked, then the load pressure increases to the limit that causes the spool to shift completely to the left side, then the system will act like the sum of two throttles in series.

Manufacturers of VLCs should publish characteristics similar to the one shown in Fig. 5.51. Such characteristics show, for every value of expected flow, what is the minimum overall differential pressure below which the compensator is saturated.

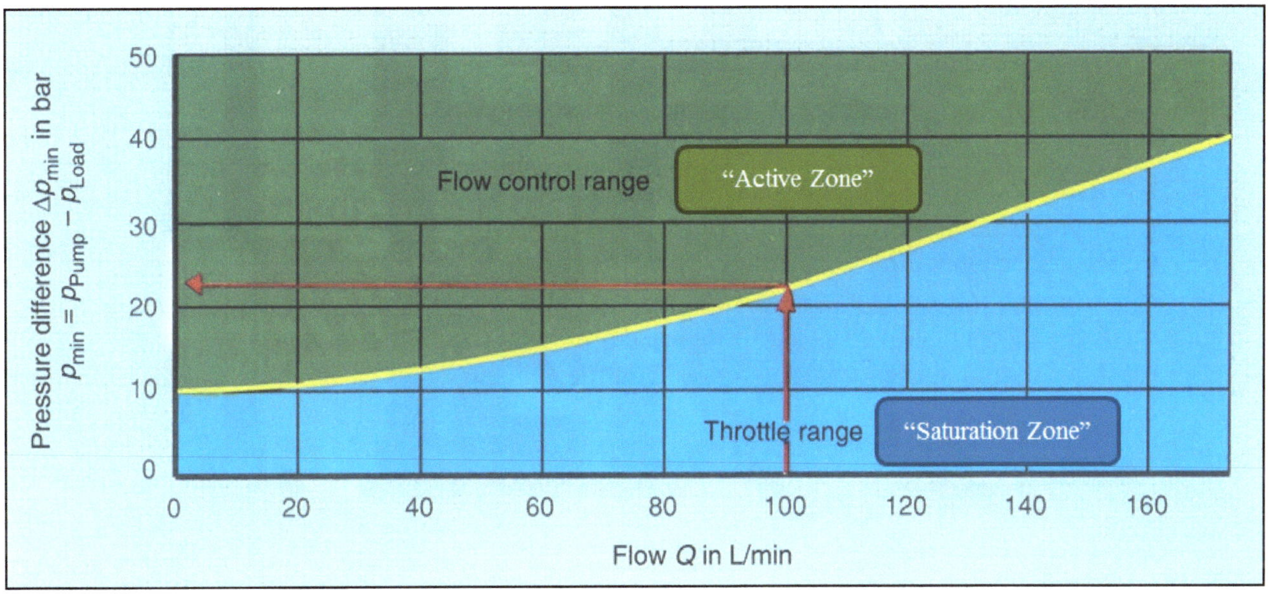

Fig. 5.51 - Performance Limits of a VLC (Courtesy of Bosch Rexroth)

5.4- Proportional Pressure Control Valves

A *Pressure Relief Valve,* (*PRV*), is the most important valve in hydraulic circuits because its prime job is safety. A proportional PRV functionally performs similar to the hydro-mechanical one except they are electronically adjustable.

5.4.1- Interpretation of Symbols for Proportional Pressure Relief Valves

Figure 5.52 explores symbols for various proportional PRVs. The symbols show the following features:

- The two parallel lines on top and bottom of the symbol reveal the infinite positioning of the proportional PRV spool.
- The arrow on the proportional solenoid indicates that the solenoid receives variable signal.
- A proportional PRV with a separate ECU versus with an OBE; (1 versus 2).
- A proportional PRV without versus with a built-in pressure transducer; (2 versus 3).
- A force-controlled versus stroke-controlled proportional PRV; (1 versus 4).
- A direct-operated versus a pilot-operated proportional PRV; (1 versus 5).
- A proportional PRV without versus with mechanical PRV; (5 versus 6).

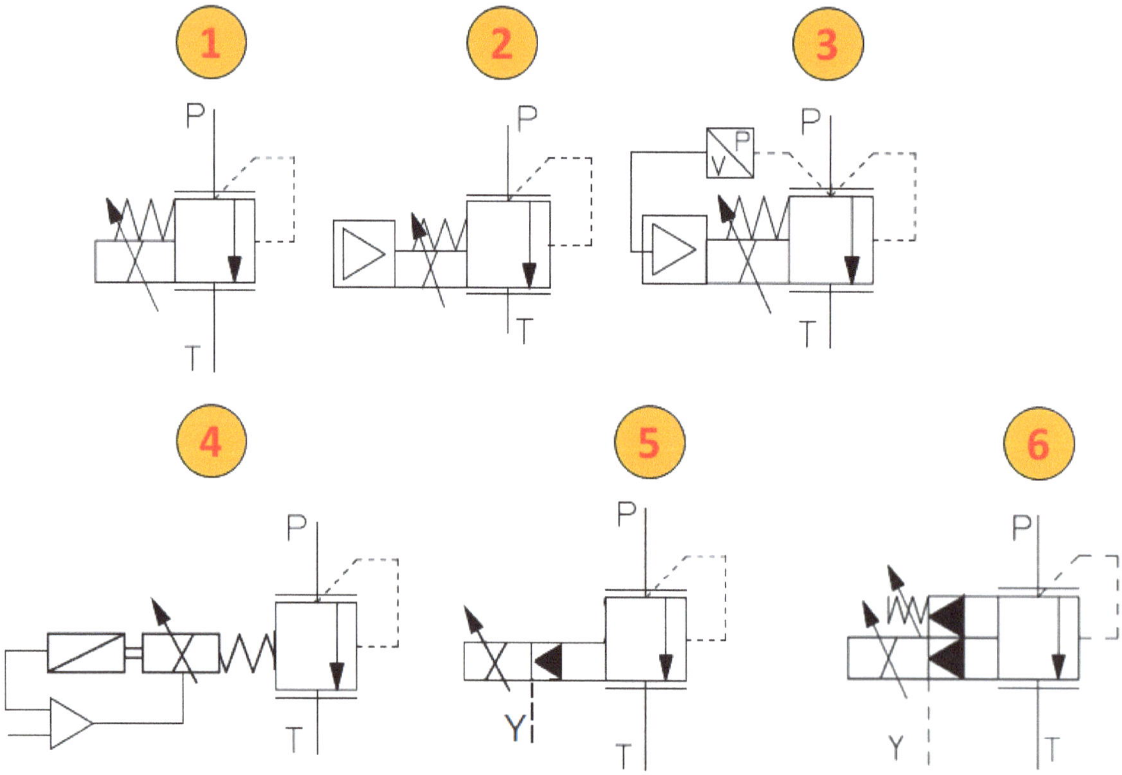

Fig. 5.52 - Symbols of Proportional Pressure Relief Valves

5.4.2- Hydraulic Static Characteristics of Proportional Pressure Relief Valves

Figures 5.53 shows typical characteristics of a proportional PRV as examples of what is published by the manufacturer to identify the valve performance. The upper curve shows the relation between the input signal and the % of the maximum pressure. The lower curve shows the ***p-Q*** characteristics for different maximum pressure settings.

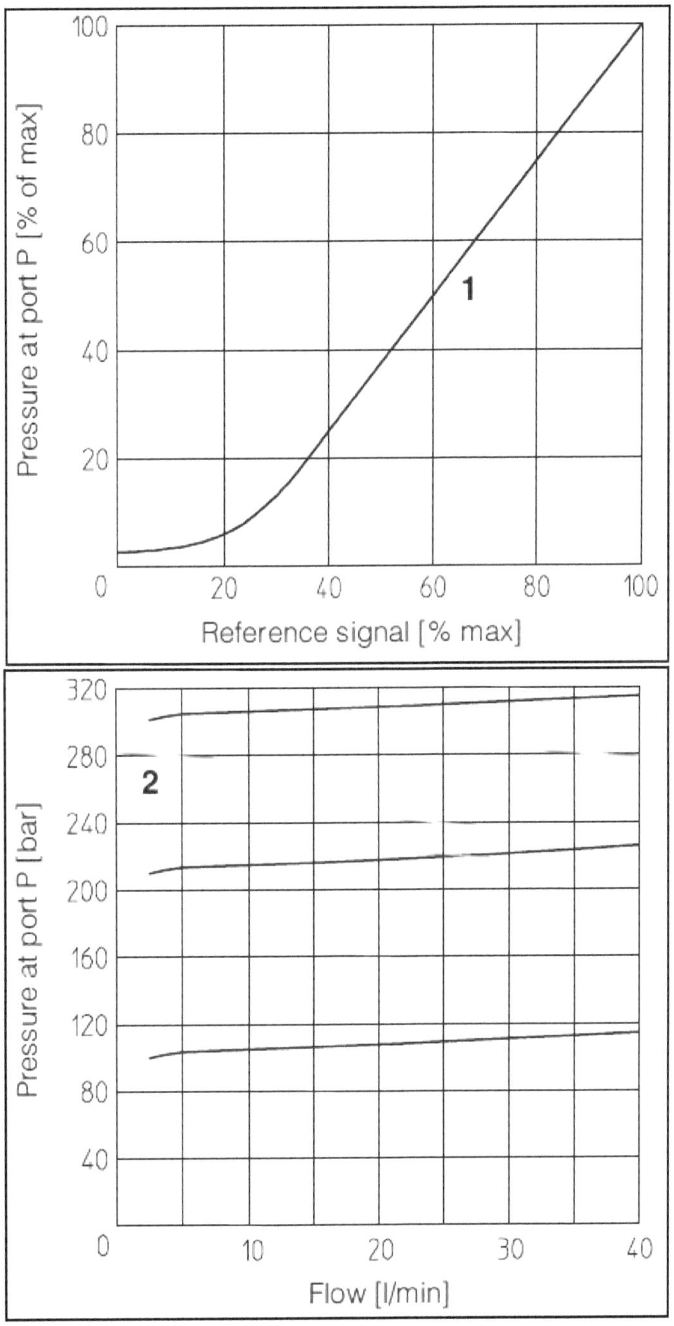

Fig. 5.53 - Example of PDV Performance (Courtesy of Atos)

5.4.3- Direct-Operated Force-Controlled Proportional Pressure Relief Valves

Figure 5.54 shows the conceptual construction of a direct-operated force-controlled proportional PRV. The proportional solenoid (6) replaces the knob of a conventional PRV that is used for manual setting of maximum pressure. With these valves, the system pressure that needs to be limited can be infinitely adjusted in relation to the position of the solenoid by means of external trigger electronics.

The front spring (3) is a fail-safe feature. If the power is turned off or the input signal is lost accidentally, the front spring (3) opens the valve so that the pump is automatically unloaded, and no residual pressure will be left in the system.

As is known, positive displacement pumps produce fluctuated flow that generates pressure ripples in the system. If the poppet (2) and the push rod (5) are made as one part, pressure ripples will cause this part to continuously vibrate. As a result, eddy current will be produced and overheat the coil. The rear spring (4), is to absorb the small pressure fluctuation and prevent transferring it back to the moving core of the proportional solenoid.

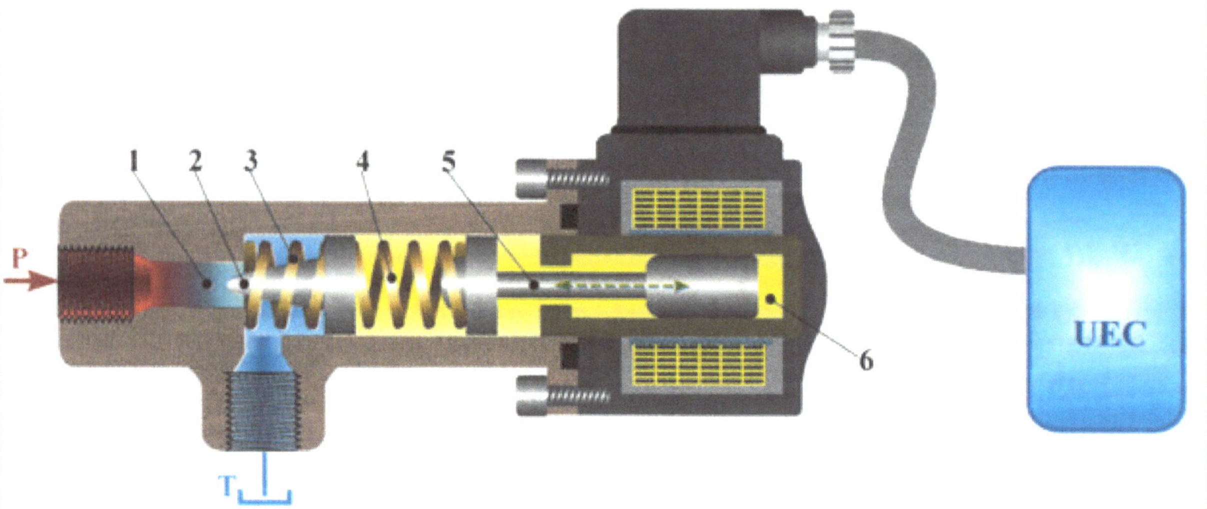

1- Poppet Seat, 2- Poppet, 3- Front Spring, 4-Back Spring, 5-Push Rod, and 6- Proportional Solenoid

Fig. 5.54 - Conceptual Construction of a Proportional PRV (Courtesy of ASSOFLUID)

Figure 5.55 shows an example of a direct-operated, force-controlled sub-plate mounted proportional PRV with a proportional solenoid. The symbols show that it could be with a separate ECU or with an OBE. At the first valve commissioning the air trapped inside the solenoid must be bled-off through the air bleed screw (9).

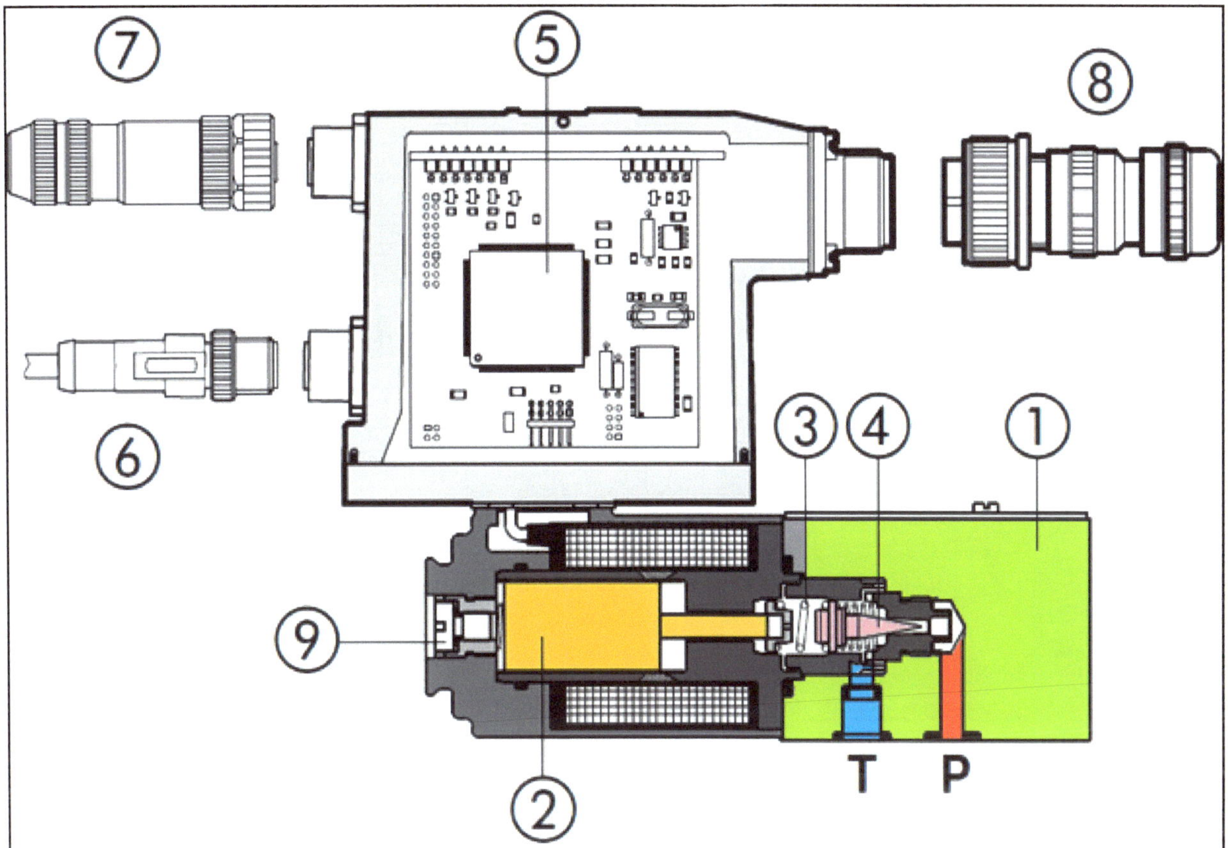

1- Main Valve Body, 2- Proportional Solenoid, 3 Spring, 4- Poppet, 5-On-Board Electronics, 6- USB Connector, 7- Fieldbus Connector, 8- Main Connector, and 9- Screw for Air Bleeding

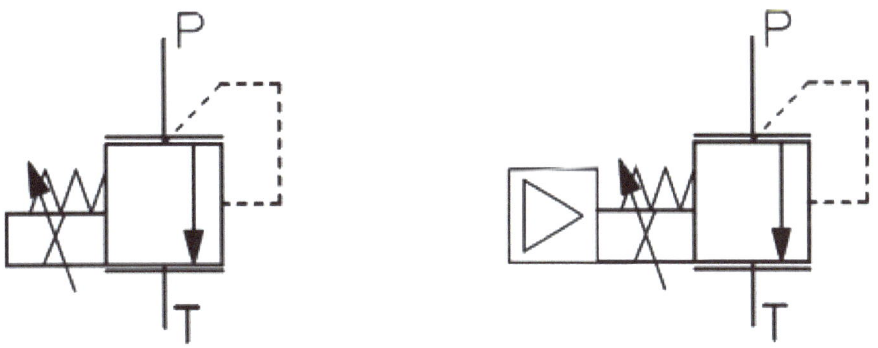

Fig. 5.55 - Example of a Direct-Operated, Force-Controlled Sub-Plate Mounted Proportional PRV (Courtesy of Atos)

Figure 5.56 shows the same previous valve but with a built-in pressure transducer. Despite the proportional solenoid still actuated in an open-loop, the operation of the valve is improved by controlling the maximum pressure in the system by a closed-loop. The pressure transducers sense the actual pressure and feeds it back to the associated electronics that takes care of positioning the proportional solenoid accordingly to assure that system pressure does not exceed the maximum value. The symbols show that it could be with a separate ECU or with an OBE.

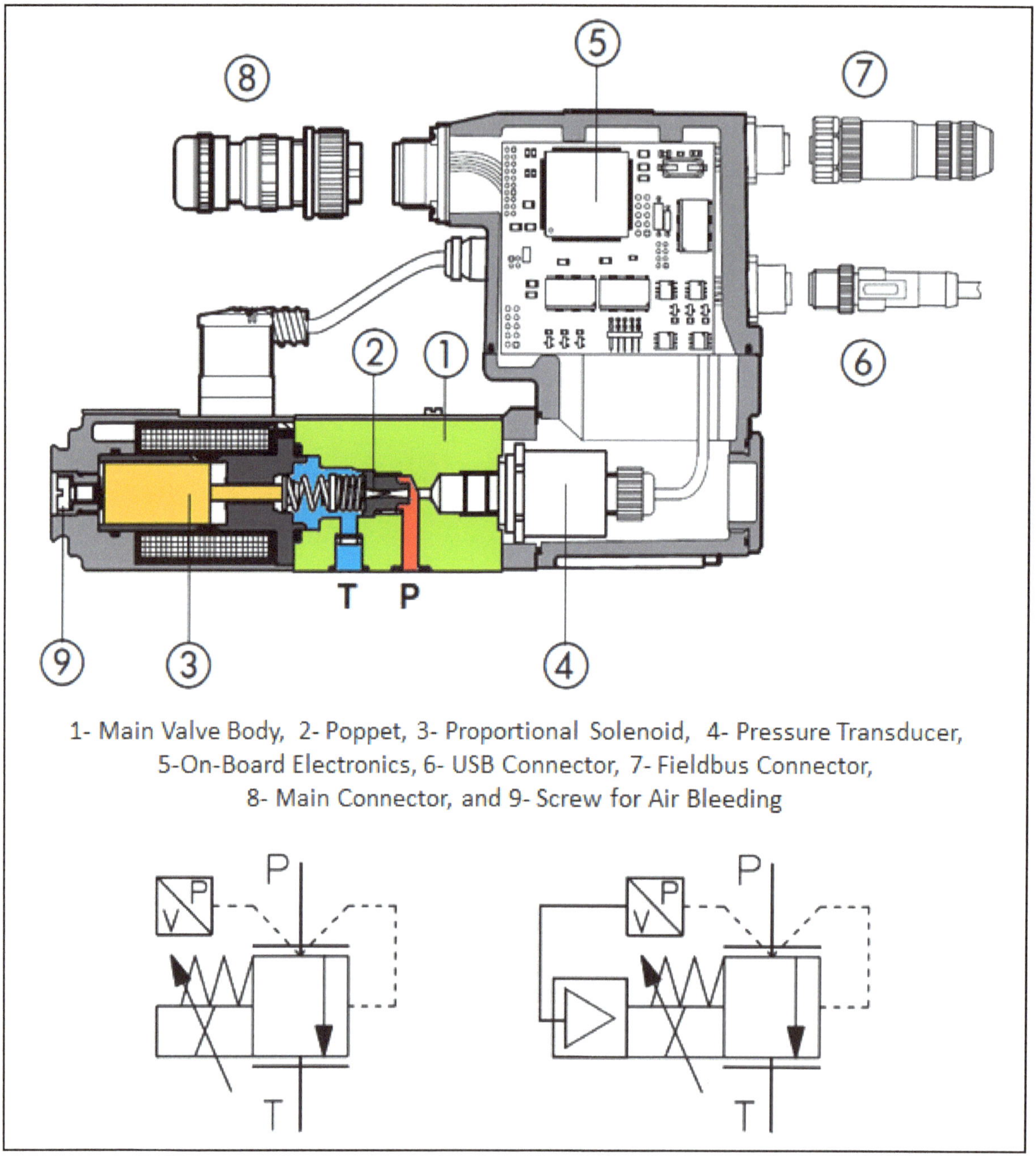

Fig. 5.56 - Example of a Proportional PRV with a built in Pressure Transducer (Courtesy of Atos)

Figure 5.57 shows an example of a direct-operated, force-controlled cartridge-type proportional PRV. The valve can be equipped with different standard types of electrical connectors.

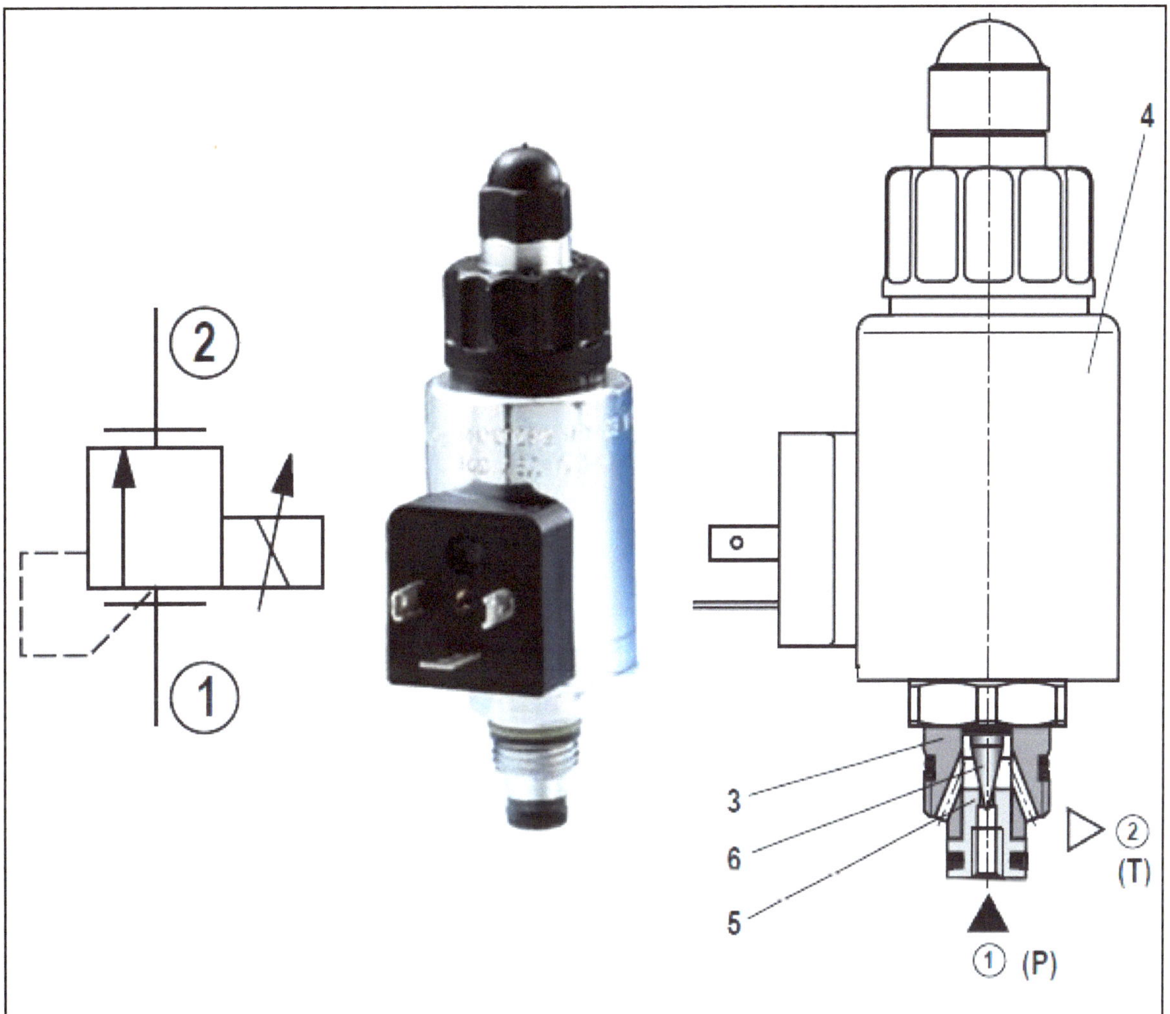

1- Pressure (Inlet) Port, 2- Tank (Outlet) Port, 3- Valve Body,
4- Proportional Solenoid, 5- Valve Seat, 6- Poppet

Fig. 5.57 - Example of a Direct-Operated Force-Controlled Cartridge-Type Proportional PRV (Courtesy of Bosch Rexroth)

5.4.4- Direct-Operated Stroke-Controlled Proportional Pressure Relief Valves

Figure 5.58 shows an example of a direct-operated, stroke-controlled sub-plate mounted proportional PRV. The valves are actuated by means of a position-controlled proportional solenoid.

To adjust the system pressure, a set-point is set in the trigger electronics. Based on this set-point, the electronics control the position of the armature on the compression spring by means of the signal from the position transducer.

A Stroke-controlled proportional PRV offers the following features over the force-controlled one:

- The position control ensures extremely low hysteresis.
- The position is maintained even in the event of external disturbances.

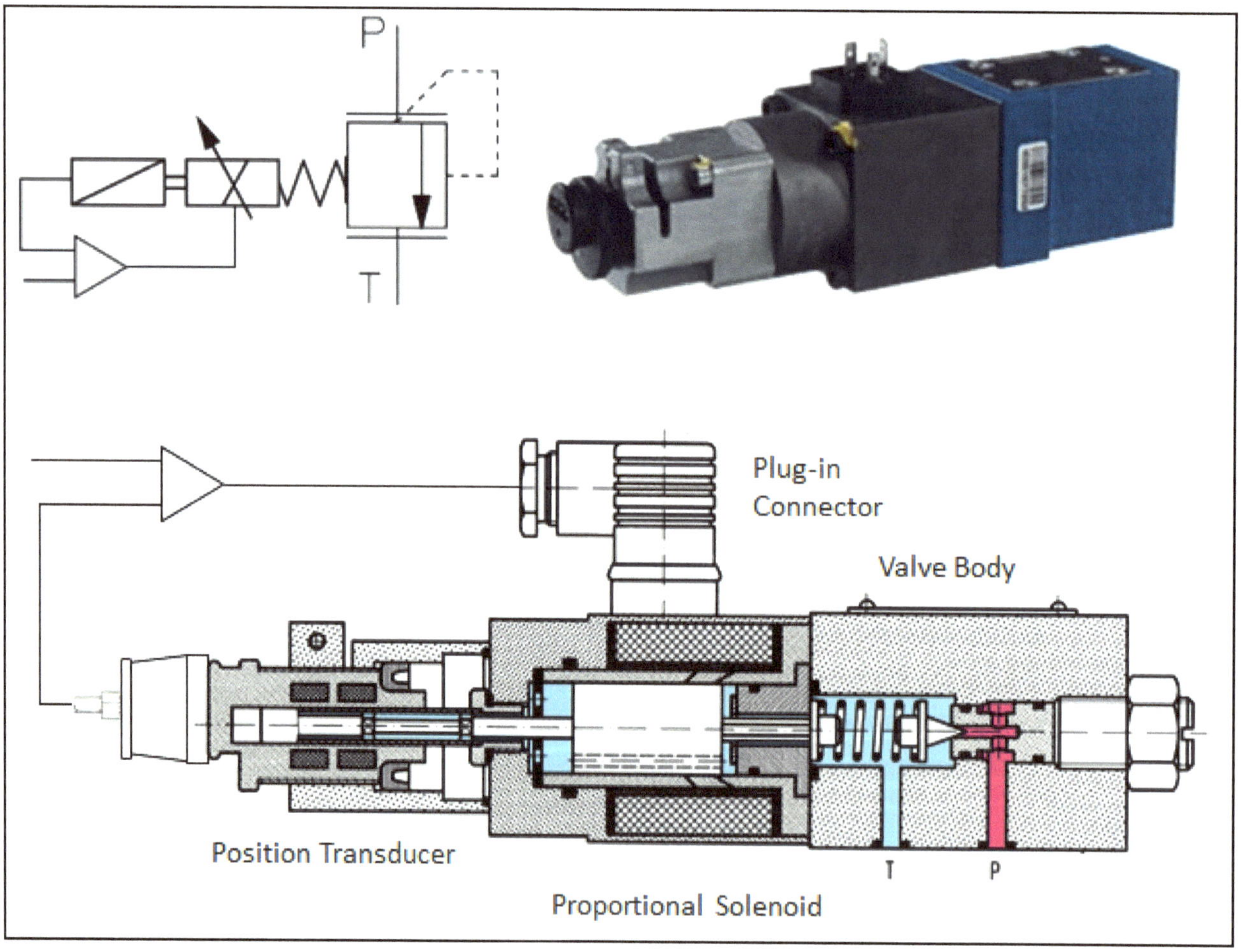

Fig. 5.58 - Example of a Direct-Operated, Stroke-Controlled Sub-Plate Mounted Proportional PRV with a Separate ECU (Courtesy of Bosch Rexroth)

Figure 5.59 shows the same previous valve but with an OBE. The figure shows that the position transducer is built in below the OBE that communicates with it and the proportional solenoid.

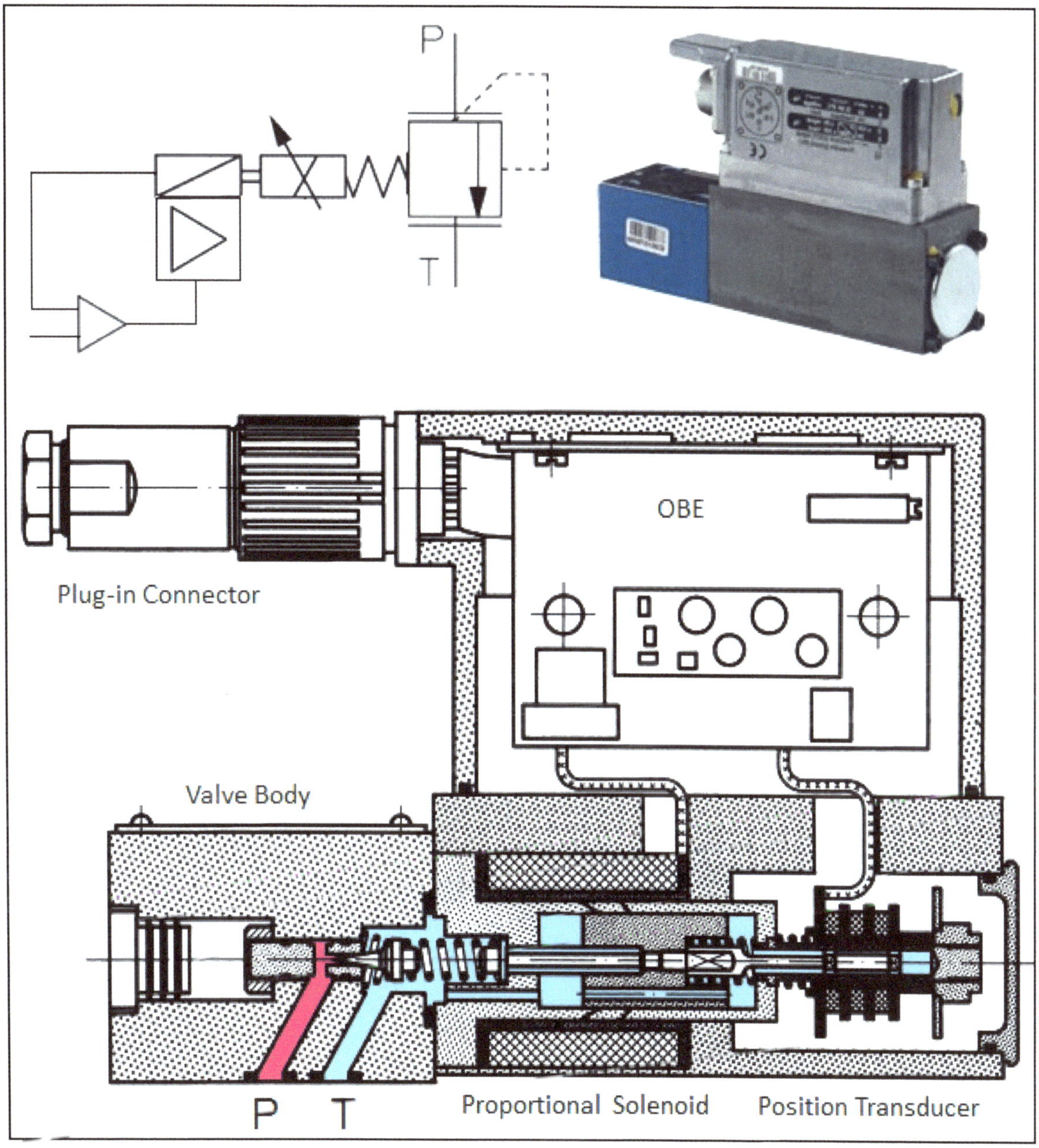

Fig. 5.59 - Example of a Direct-Operated, Stroke-Controlled Sub-Plate Mounted Proportional PRV with OBE (Courtesy of Bosch Rexroth)

5.4.5- Pilot-Operated Force-Controlled Proportional Pressure Relief Valves

The construction, concept of operation, and the purpose of using a pilot-operated PRV have been comprehensively explained in Volume 1 of this series of textbooks. Figure 5.60 shows the conceptual construction of a pilot-operated force-controlled proportional PRV. Briefly the valve consists of a pilot stage and a main stage. The pilot stage must open first at maximum system pressure in order for the main stage to open and to pass the large flow.

As compared to the hydro-mechanical pilot operated PRV, the mechanical knob is replaced by a proportional solenoid for remote and interactive adjustment of the maximum system pressure.

As compared to the pilot operated PRV that has a switching solenoid-actuated unloading directional valve, this valve can unload the pump by just sending a signal of zero value to the proportional solenoid. In such case the maximum pressure will be minimal, and the pump will be unloaded. It is also more compact since the unloading directional valve has been removed.

An additional feature is shown here, which is a mechanical relief valve has been added as a fail-safe feature. It is considered as a backup in case if a faulty signal is sent to the proportional solenoid.

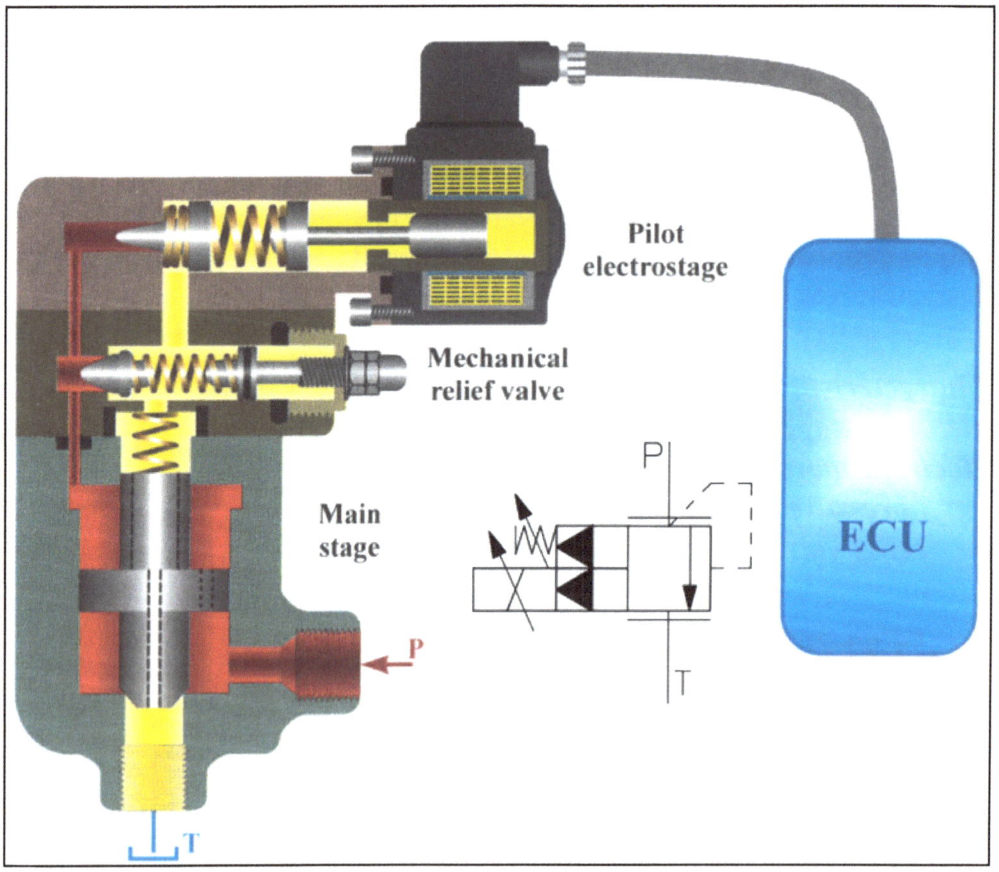

Fig. 5.60 - Conceptual Construction of a Pilot-Operated Proportional PRV (Courtesy of ASSOFLUID)

Figure 5.61 shows symbols for pilot-operated force-controlled sub-plate mounted proportional PRVs with a proportional solenoid. Figure 5.62 shows the detailed construction of the valve.

The port **X** is blocked on the sub-plate if the valve works as a PRV. This valve can perform proportional unloading function if the control pressure is supplied externally through the port **X**. In such a case, the valve does not control the main pressure at port **A**; it controls the value of the control pressure (at port **X**), at which port **A** is unloaded. Should the valve work as unloading valve, the port **X** is to be opened on the sub-plate and the blind plug must be installed to separate the port **X** from the port **P**.

The symbols show also that the control pressure is drained externally through port **Y**. The valve can be switched between internal and external drain of the control by either removing or installing the blind plug shown in the symbol. Draining the control pressure through port **Y** is mandatory if the main tank line **T** experiences back pressure from a filter or a cooler that are mounted in the main return line of the valve. Or if the main return line of the valve may experience pressure fluctuation due to frequent flow surges.

The symbols show that the valve can be equipped with a separate ECU or with an OBE.

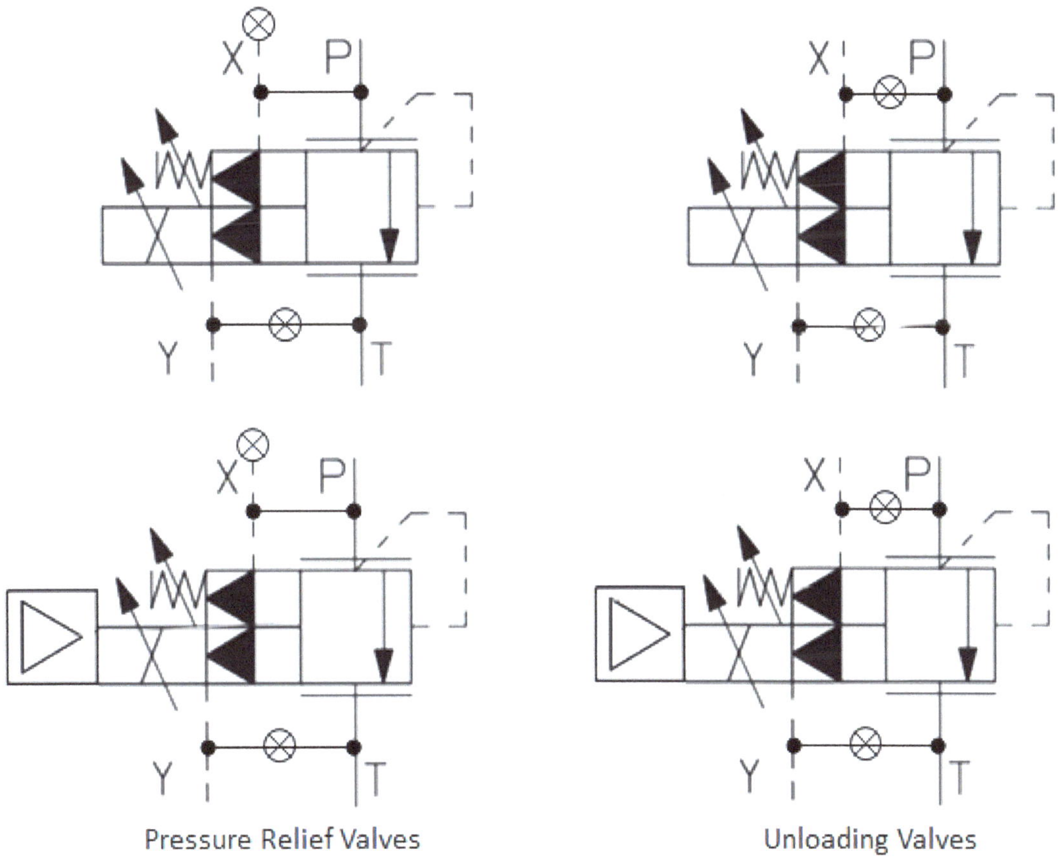

Fig. 5.61 - Symbols for Pilot-Operated Force-Controlled Proportional PRVs

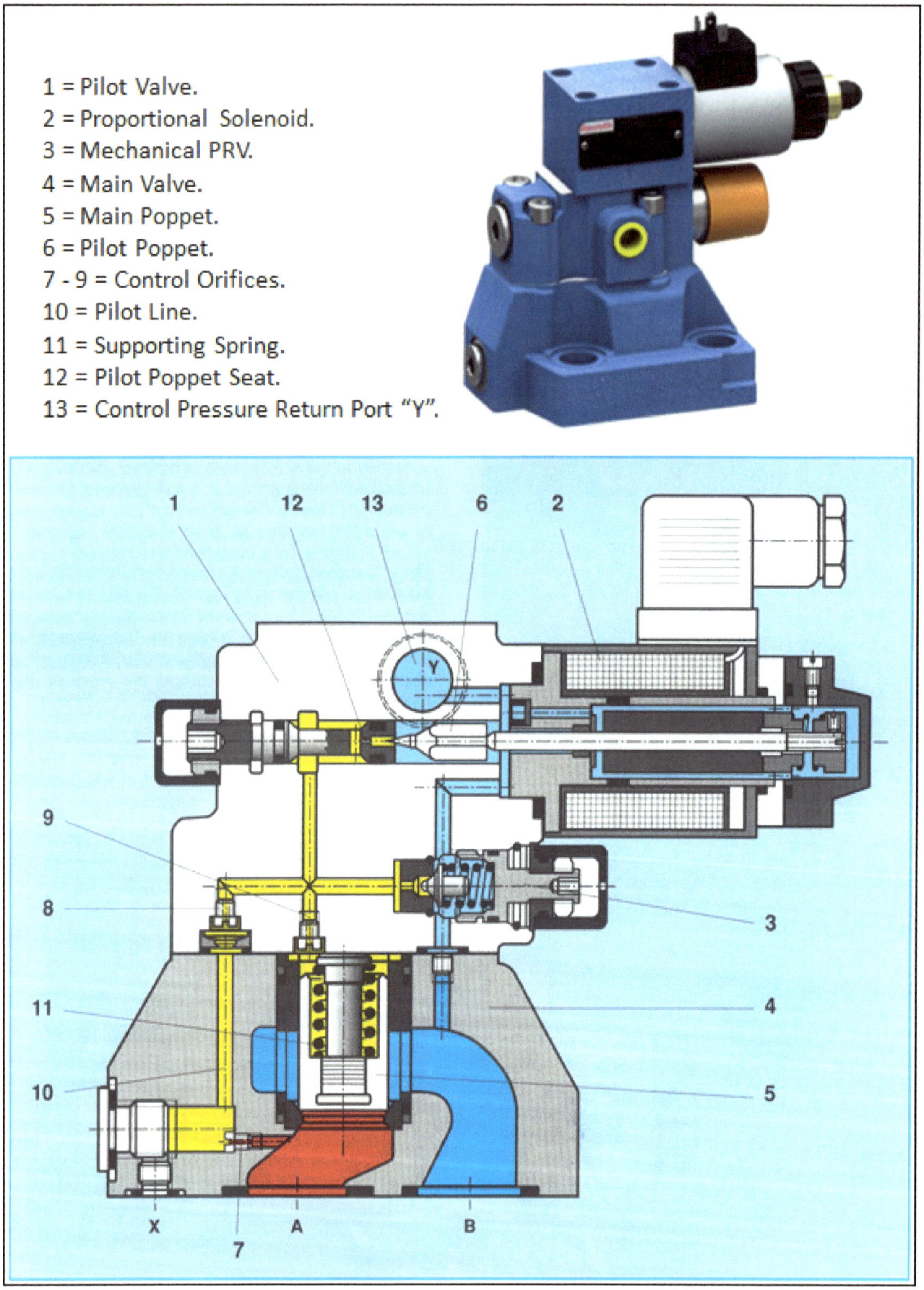

Fig. 5.62 - Example of Pilot-Operated Force-Controlled Sub-Plate Mounted Proportional PRV (Courtesy of Bosch Rexroth)

Figure 5.63 shows symbols for pilot-operated force-controlled sub-plate mounted proportional PRV with a proportional solenoid and a built-in pressure transducer. Similar to the previous valve, port **X** is for unloading valves. The symbols show that the valve can be equipped with a separate ECU or with an OBE.

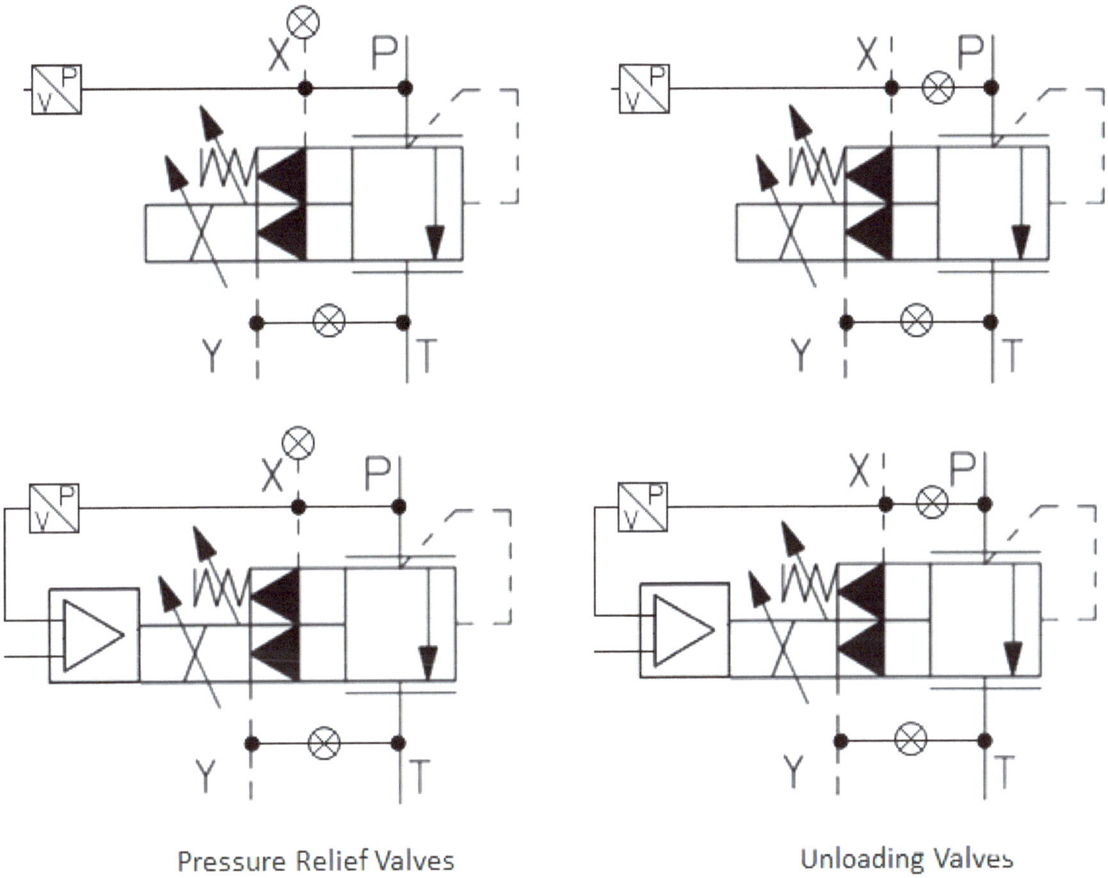

Fig. 5.63 - Symbols for Pilot-Operated Force-Controlled Proportional PRVs with a built in Pressure Transducer

Figure 5.64 shows the detailed construction of the valve. The valve can be switched between internal and external drain of the control by turning the screw (4) at port **Y**.

The port **X** is blocked on the sub-plate if the valve works as a PRV.

Should the valve work as unloading valve, the port **X** is to be opened on the sub-plate and the blind plug must be installed to separate the port **X** from the port **P**.

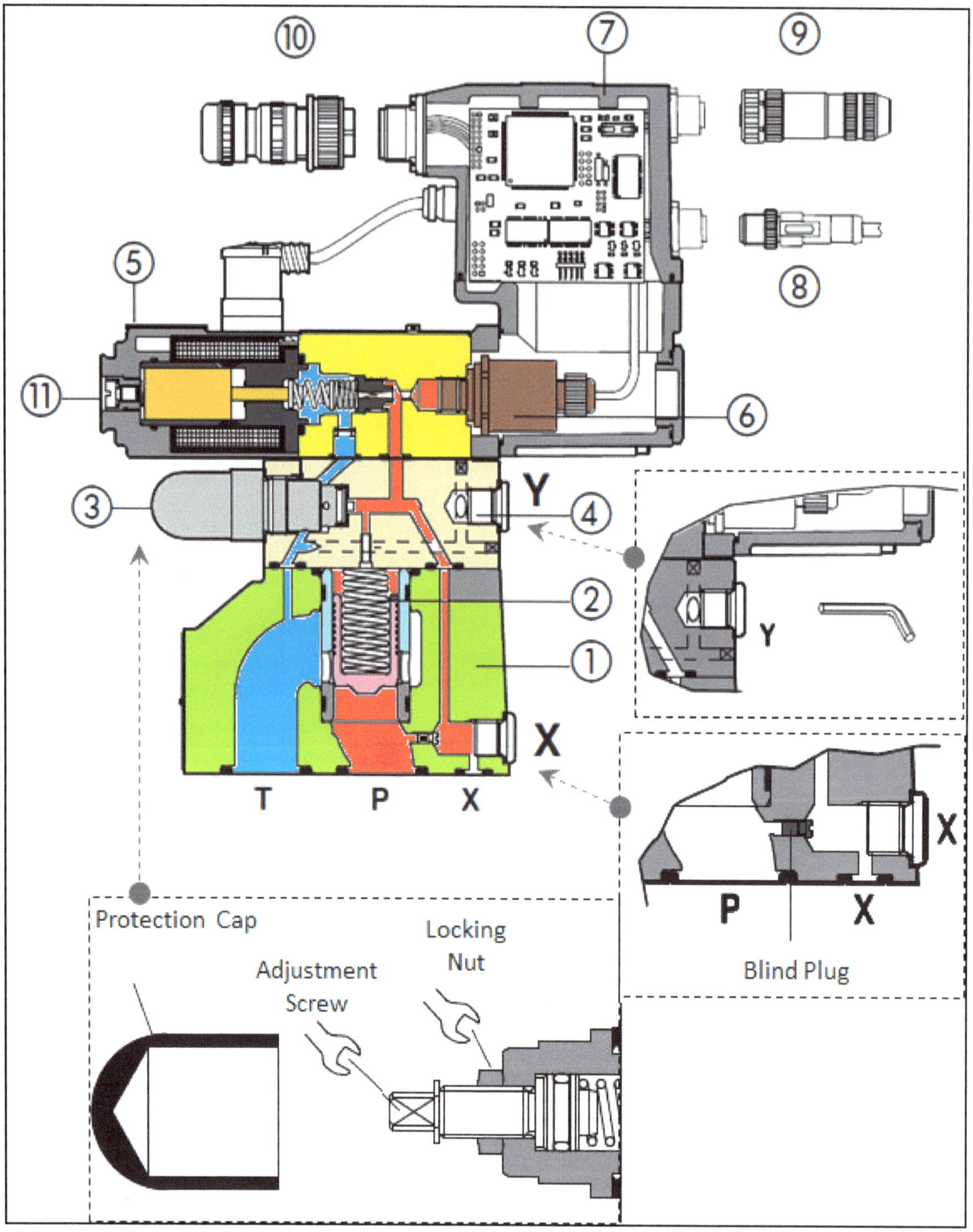

1- Main Valve Body, 2- Main Poppet, 3- Mechanical PRV, 4- Control Pressure Drain Port, 5-Proportional Solenoid, 6- Pressure Transducer, 7-On-Board Electronics, 8- USB Connector, 9- Fieldbus Connector, 10- Main Connector, and 11- Screw for Air Bleeding

Fig. 5.64 - Example of Pilot-Operated Force-Controlled Sub-Plate Mounted Proportional PRV With a built in Pressure Transducer (Courtesy of Atos)

Pressure at port P can be unloaded by sending a signal of zero value from the proportional control to the proportional solenoid. As shown in Fig. 5.65, an optional unloading directional valve can be added as a fail-safe feature in case if the proportional control fails. Within the functionality of PRV of the previous valve, the main pressure line **P** can be remotely unloaded by connecting the port **X** to a switching solenoid-actuated unloading directional valve. In such an option, the blind plug must be replaced by a metering orifice. As shown in the figure, the solenoid unloading directional valve is normally closed. Once the solenoid is energized, port X in vented allowing for a small pilot flow to pass to the tank. As a result, the pressure below the main poppet (at the upstream side of the metering orifice) becomes higher than the upper part of the main poppet (at the downstream side of the metering orifice). Then the main poppet opens to handle the large flow while the proportional control is by-passed.

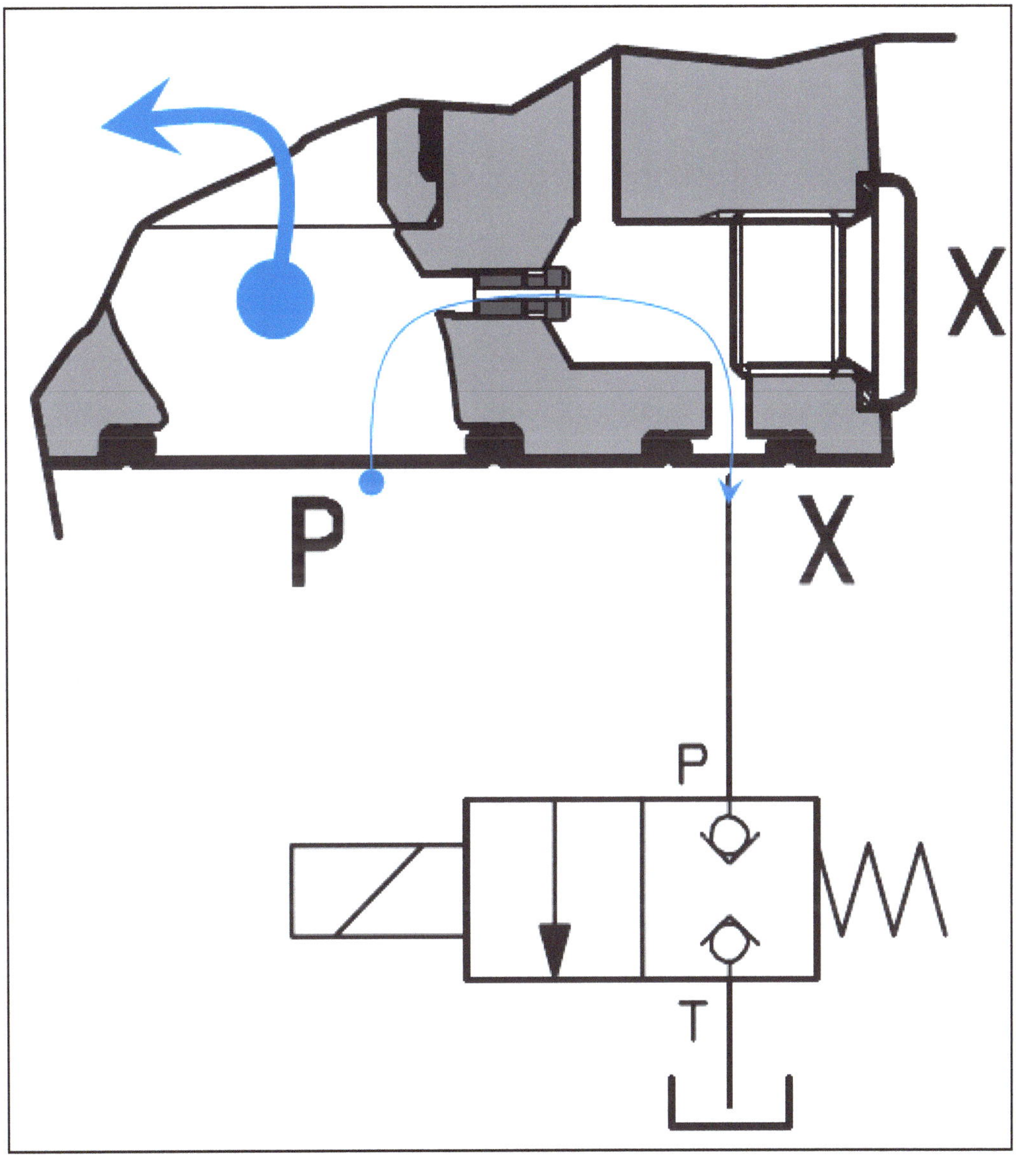

Fig. 5.65 - Pilot-Operated Proportional PRV with Unloading Feature

Pilot-operated proportional PRVs in cartridge configuration are also found as shown in Fig. 5.66. The symbol does not indicate whether it is sub-plate mounted or cartridge-type. The electrical connector can be any of the standard types.

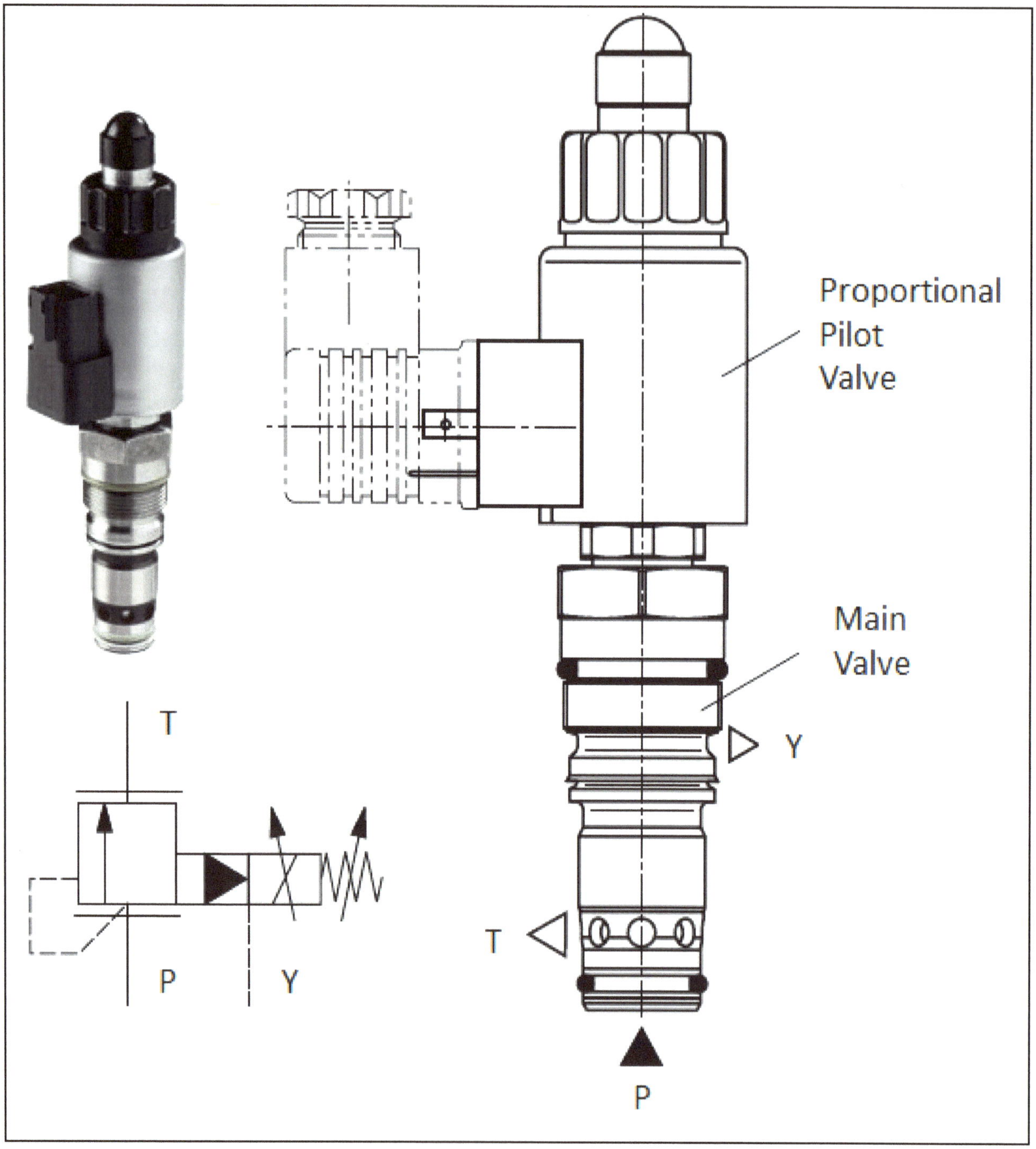

Fig. 5.66 - Example of Pilot-Operated Force-Controlled Cartridge-Type Proportional PRV (Courtesy of Bosch Rexroth)

5.4.6- Pilot-Operated Stroke-Controlled Proportional Pressure Relief Valves

Fig. 5.67 shows an example of pilot-operated, stroke-controlled sub-plate mounted proportional PRV. The figure shows the possibility of adding the remote unloading and the mechanical PRV features to the valve.

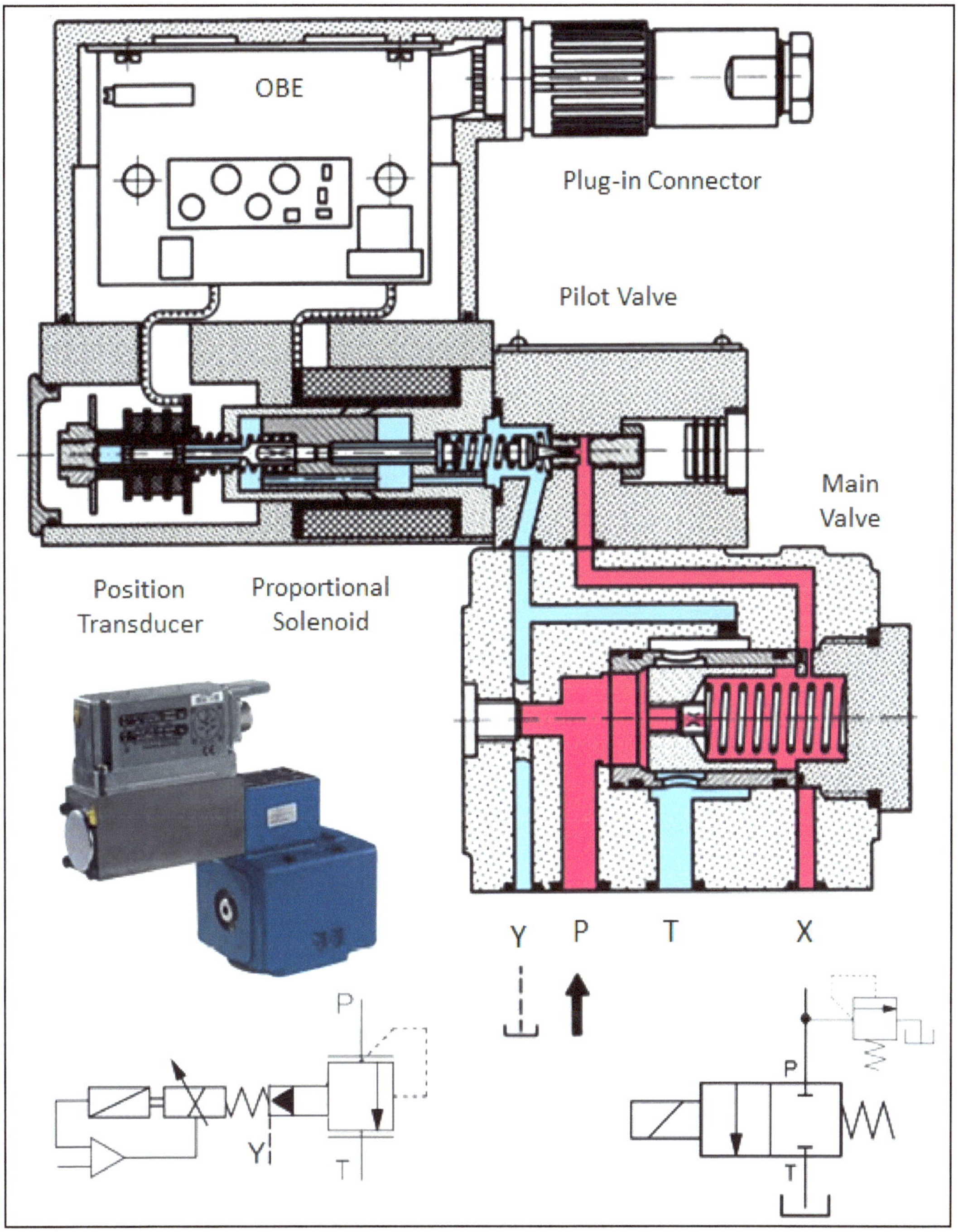

Fig. 5.67 - Example of Pilot-Operated Stroke-Controlled Sub-Plate Mounted Proportional PRV (Courtesy of Bosch Rexroth)

5.4.7- Proportional Pressure Reducing Valves

As shown in Fig. 5.68, proportional pressure reducing valves are eligible for all features that have been shown previously with the PRV, including: a direct-operated (1), a stroke-controlled valve (2), a built in pressure transducer (3) for closed-loop pressure control, a mechanical PRV (4) for safety, and OBE (5). Reducing valves can be also equipped optionally with a built in check valve (6) for reverse flow. It is to be noted that, in pilot-operated valves, the signal that operates the pilot stage is from the downstream side of the valve at port B.

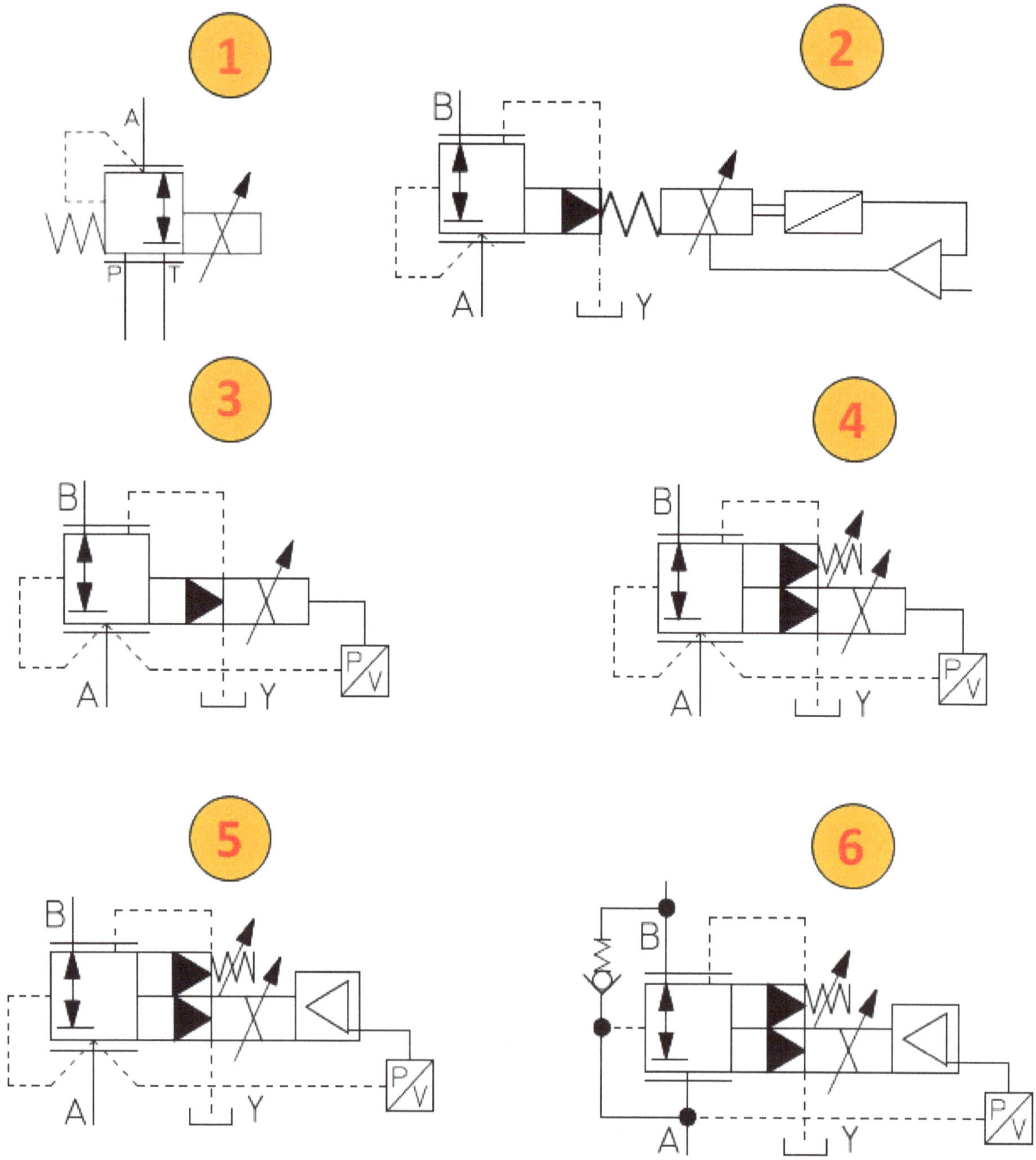

Fig. 5.68 - Symbols for Pilot-Operated Proportional Pressure Reducing Valves

The construction, concept of operation, and the purpose of using a pilot-operated pressure reducing valves have been comprehensively explained in Volume 1 of this series of textbooks. Figure 6.69 shows the conceptual construction of a pilot-operated force-controlled proportional pressure reducing valve. Briefly the poppet position of the pilot valve is set using the proportional control to pass a continuous pilot flow to the tank. As a result the spool of the main valve takes a balanced position to reduce the pressure based on the input signal.

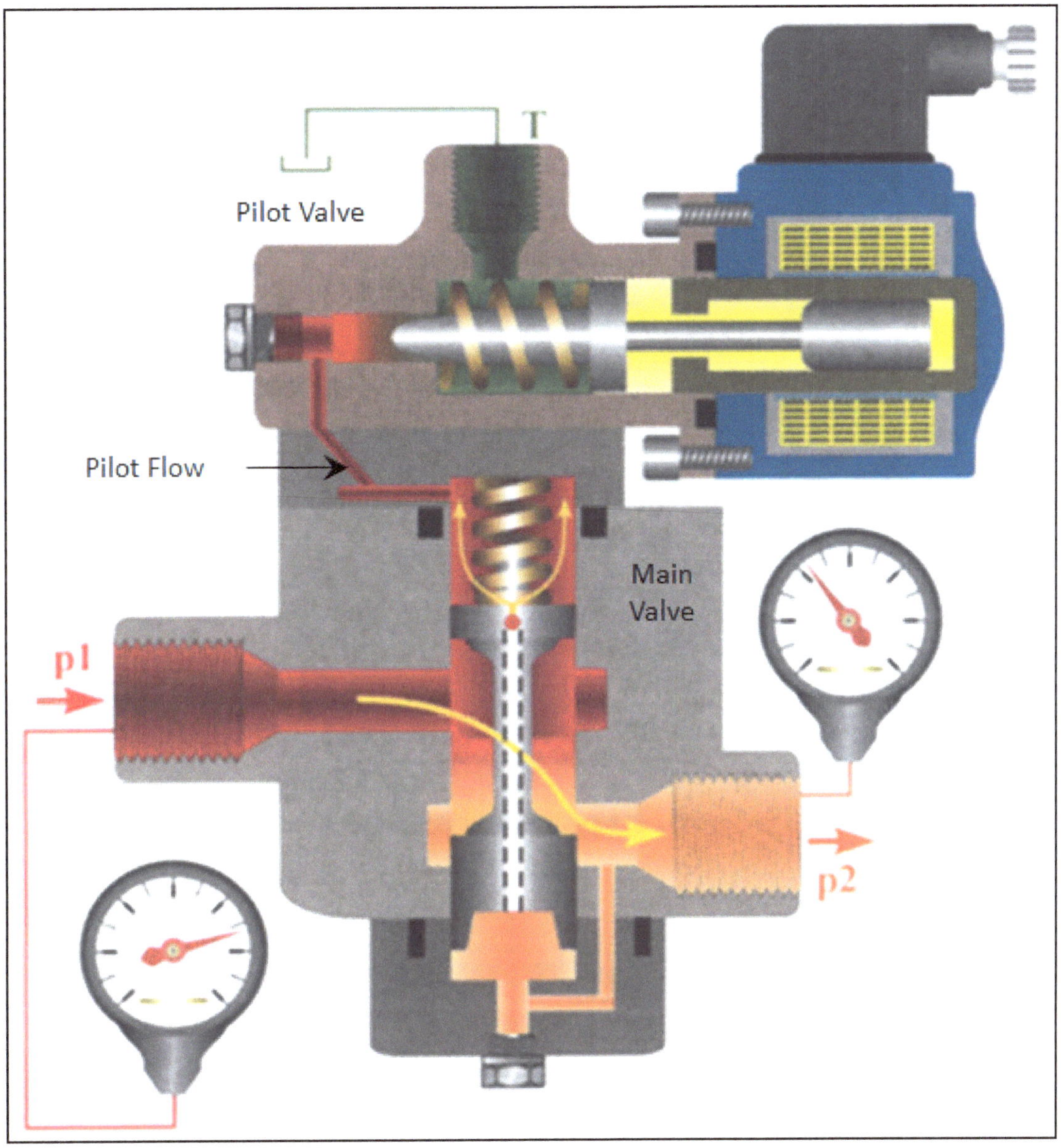

Fig. 5.69 - Conceptual Construction of a Pilot-Operated Proportional Pressure Reducing Valve (Courtesy of ASSOFLUID)

5.5- Proportional Flow Control Valves

As shown in Fig. 5.70, proportional flow control valves are eligible for all features that have shown previously, including: a force-controlled valve (1), stroke-controlled valve (2), a built in pressure compensator (3), a built in check valve (4), and a three-way flow control valve (5). All previous valves can be configured with separate ECU or with OBE.

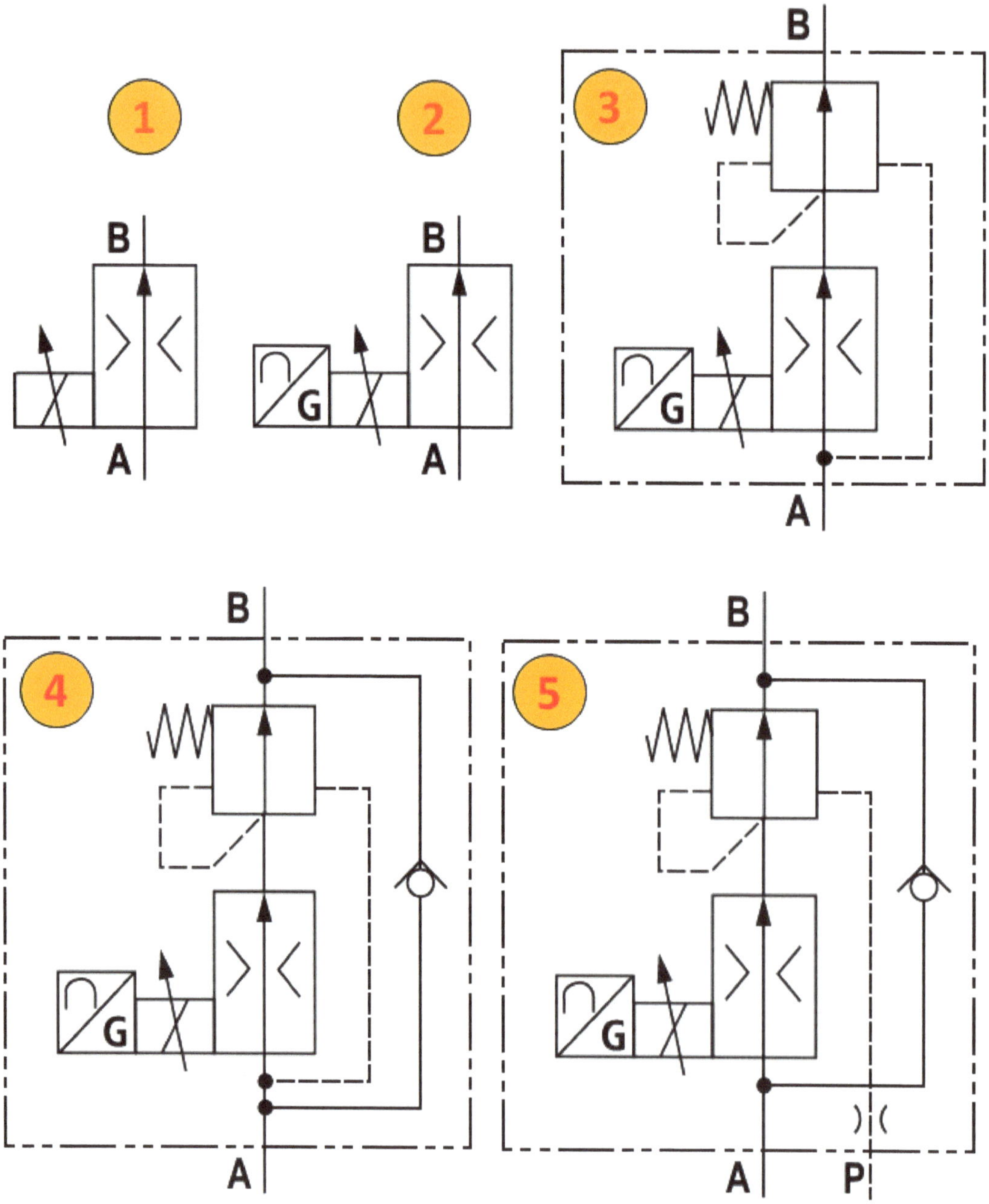

Fig. 5.70 - Symbols for Direct-Operated Proportional Flow Control Valves

Figure 5.71 shows the flow versus pressure drop through the non-compensated valve. It is to be noted that these characteristics are developed when the maximum signal is applied. Figure 5.72 shows the characteristics of the pressure compensated valve.

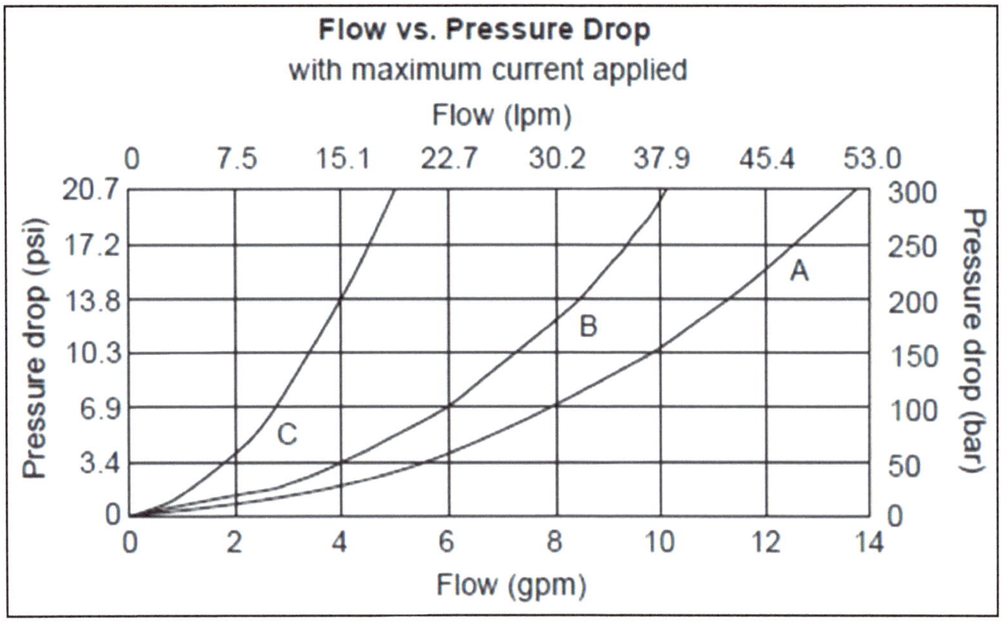

Fig. 5.71 - Characteristics of Non-Compensated Proportional Flow Control Valves

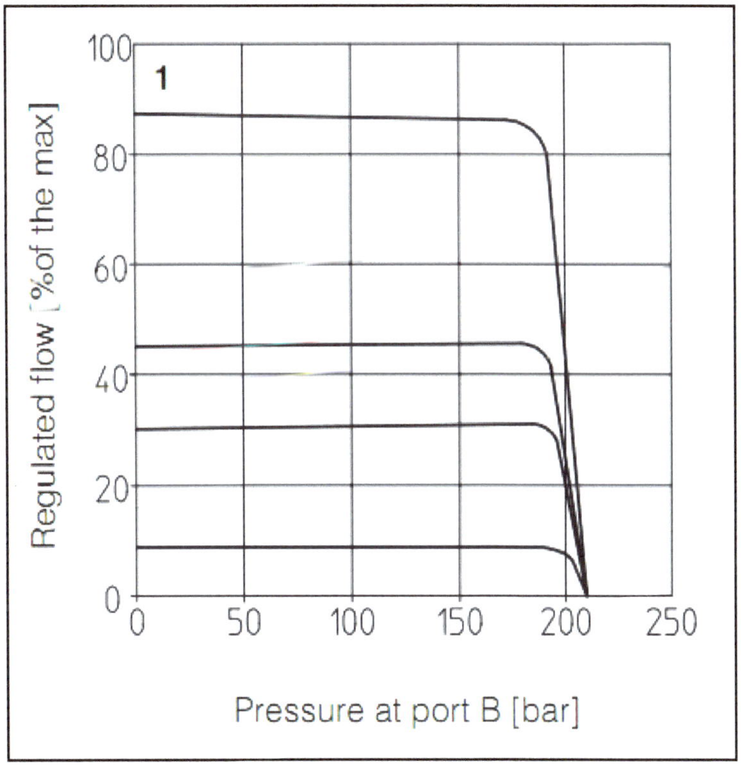

Fig. 5.72 - Characteristics of Compensated Proportional Flow Control Valves

Figure 5.73 shows the conceptual construction of direct-operated force-controlled proportional flow control valve. The proportional solenoid replaces the knob of a conventional throttle that is used for manual setting of the valve. With these valves, the flow can be infinitely adjusted in relation to the position of the solenoid by means of external trigger electronics.

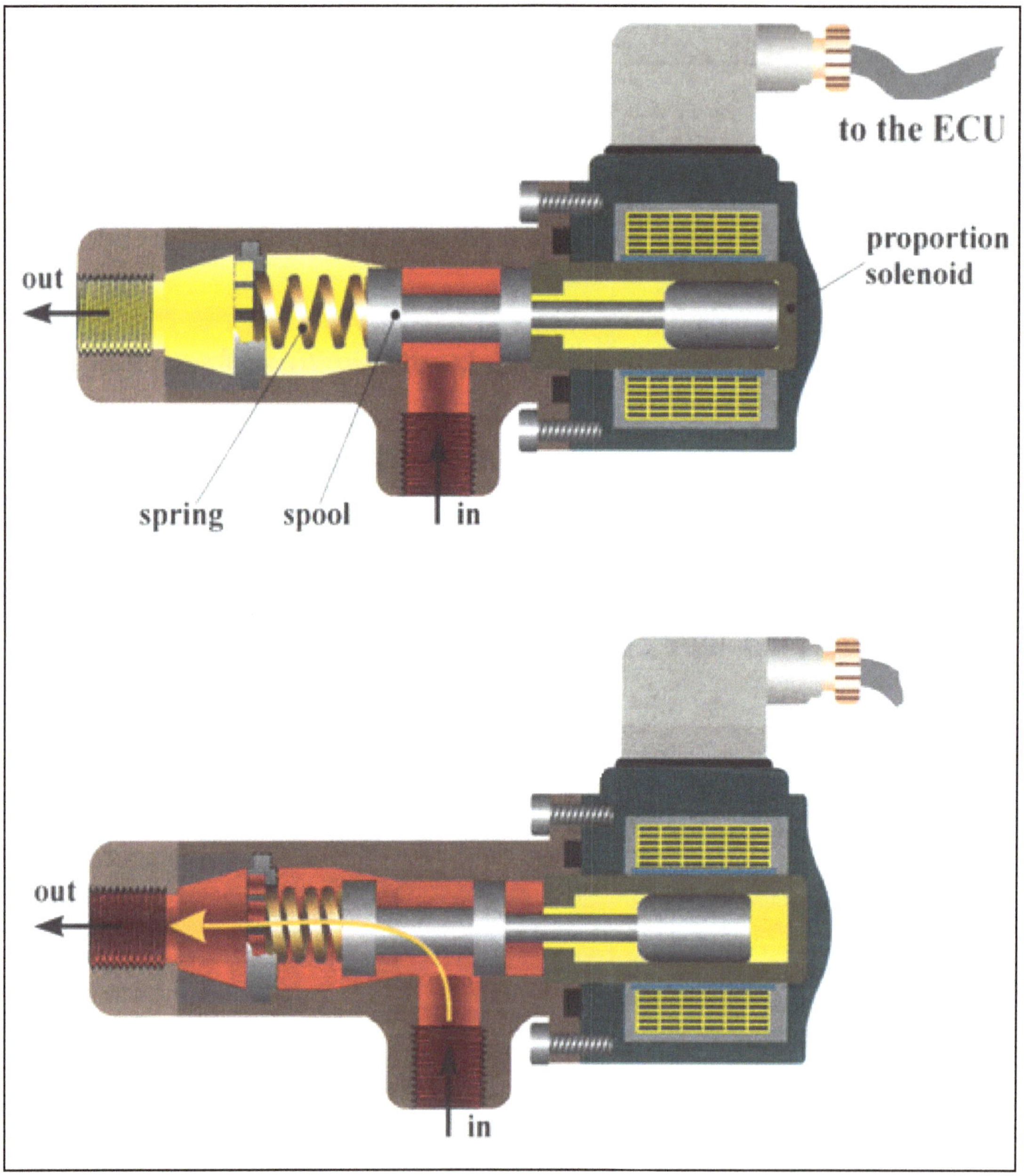

Fig. 5.73 - Conceptual Construction of a Direct-Operated Proportional Flow Control Valve (Courtesy of ASSOFLUID)

Figure 5.74 shows the conceptual construction of a direct-operated force-controlled proportional flow control valve with a built-in pressure compensator. As it has been explained in the directional proportional valves, the pressure compensator is to assure constant flow regardless of the load variation.

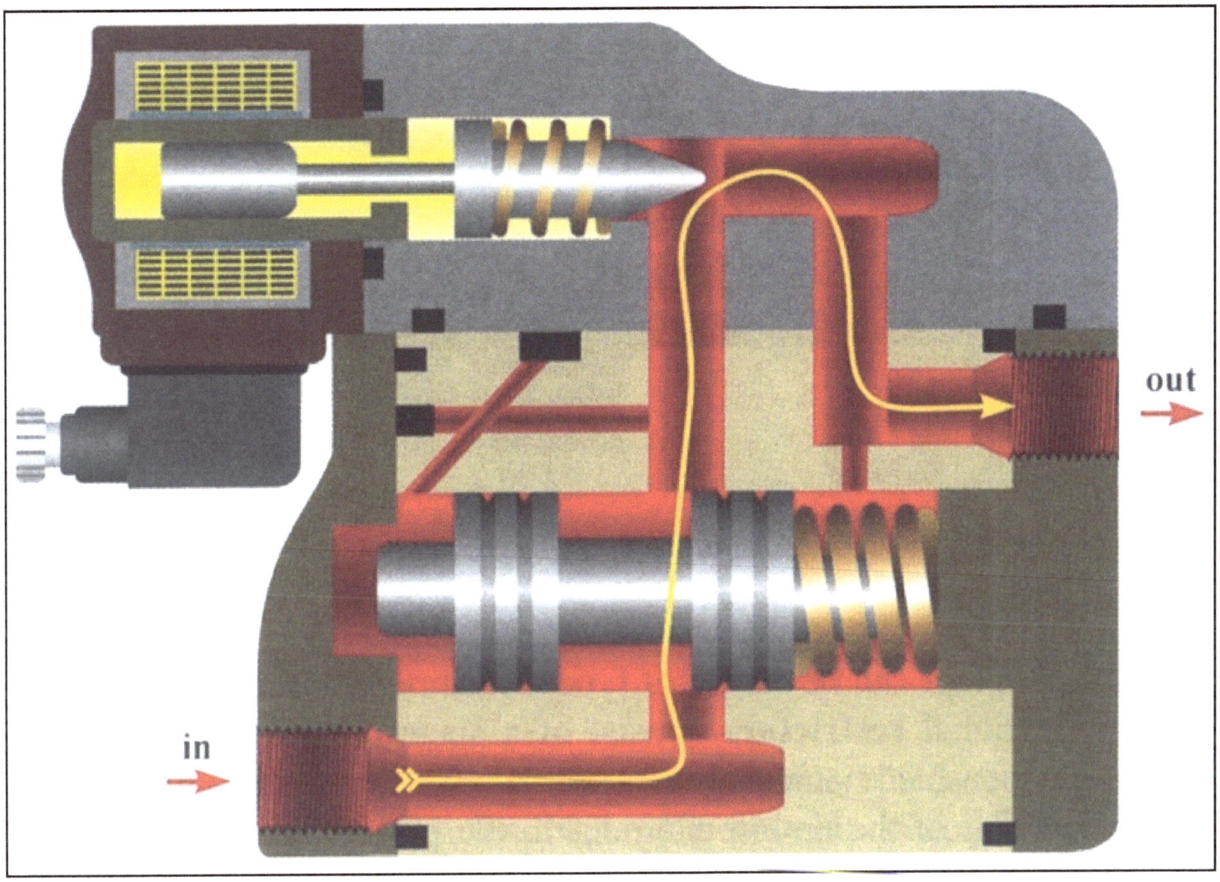

Fig. 5.74 - Conceptual Construction of a Direct-Operated Proportional Flow Control Valve with a built in Pressure Compensator (Courtesy of ASSOFLUID)

Slip-in cartridge proportional flow control valves are also available. As shown in Fig. 5.75, for large flow, a pilot-operated proportional flow control valve is used. The figure and the symbol show that both the pilot stage and the main stage are stroke-controlled, and they are driven by an OBE.

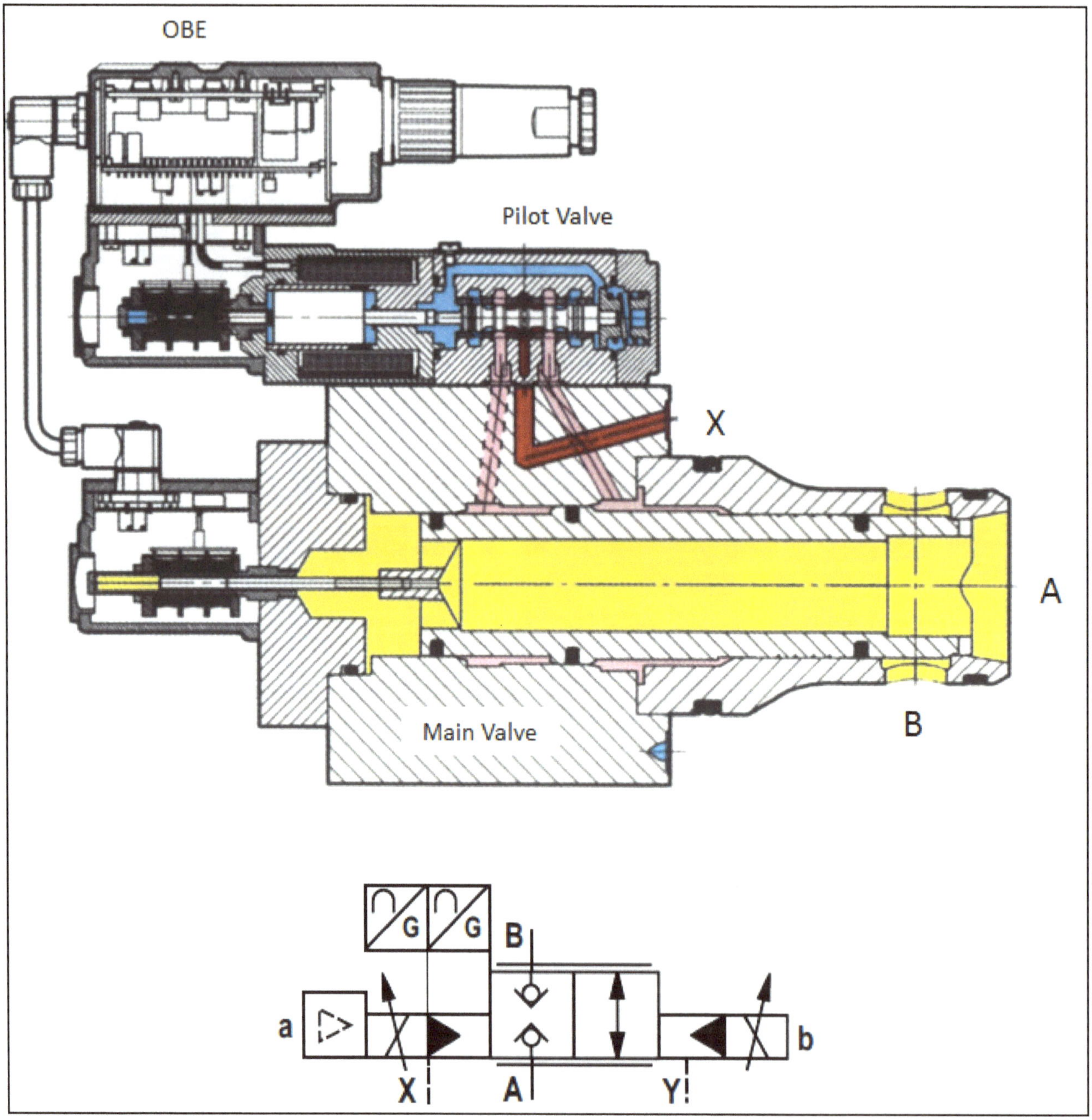

Fig. 5.75 - Pilot-Operated Stroke-Controlled Cartridge-Type Proportional Flow Control Valve (Courtesy of ASSOFLUID)

5.6- Proportional Valves for Mobile Applications

There are several manufacturers that focus on producing proportional control blocks for mobile machines. Figure 5.76 shows proportional control valves that cover simple to complex machinery needs in agriculture, forestry, earth-moving, paving, road-building, railway, marine, waste handling, crushers, shredders, or any application involving heavy mobile or industrial equipment.

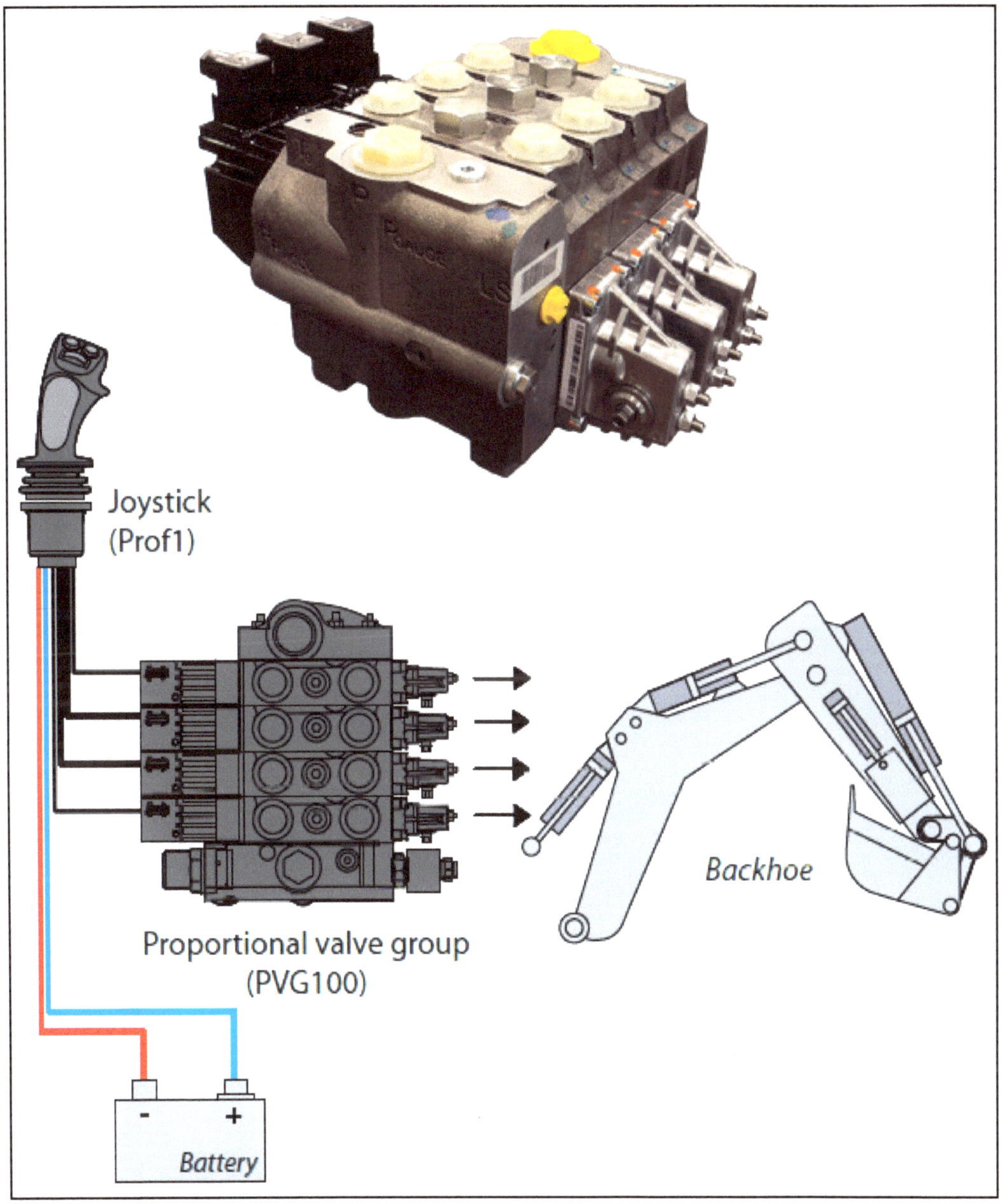

Fig. 5.76 – Proportional Control Block for Mobile Application (powersolutions.danfoss.com)

Chapter 6

Servo Valves

Objectives:

This chapter covers the construction and wiring methods of the main electric components in servo valves including torque motors. The chapter also covers the conceptual construction and the operation of flapper-nozzle and jet-pipe servo valves. The chapter introduces examples of typical valves with mechanical or electrical feedback.

Brief Contents:

Chapter 6 - Servo Valves

6.1- Introduction to Servo Hydraulics

6.1.1- Historical Background

Historically, prior to the World War II, the first servo valve configuration was based on the jet pipe concept. Following that, the half bridge flapper-nozzle was introduced and retrofitted later into the full bridge flapper-nozzle that still is being used today.

After the World War II, torque motors replaced traditional proportional solenoids. First a two-stage servo valve with a spool moving inside a sleeve was produced and upgraded into a stroke-controlled spool configuration. Pilot-operated, three-stage, servo valves were used to handle large power. After that, research and development efforts focus on improving the performance of servo valves through enhancing the spool-sleeve design. Starting in 1990, along with the vast developments on the side of electronics and software, on-board electronics with digital control capabilities were added to the design. Today, a servo valve can be completely standalone to control an actuator in a closed loop without additional accessories.

6.1.2- Servo Valve Applications

Figure 6.1 shows the arrangement of continuous valves in terms of dynamics and accuracy. As a matter of fact, a servo valve has faster dynamic response and higher accuracy than a proportional valve. Technically both valves can be used for open-loop or closed-loop applications. A proportional valve is more cost effective and convenient for open loop applications. Higher performance proportional valves can be used for closed-loop faster dynamics. There is no sense to use a servo valve unless its upper end capabilities are needed because there are other valves that can do the job for less cost.

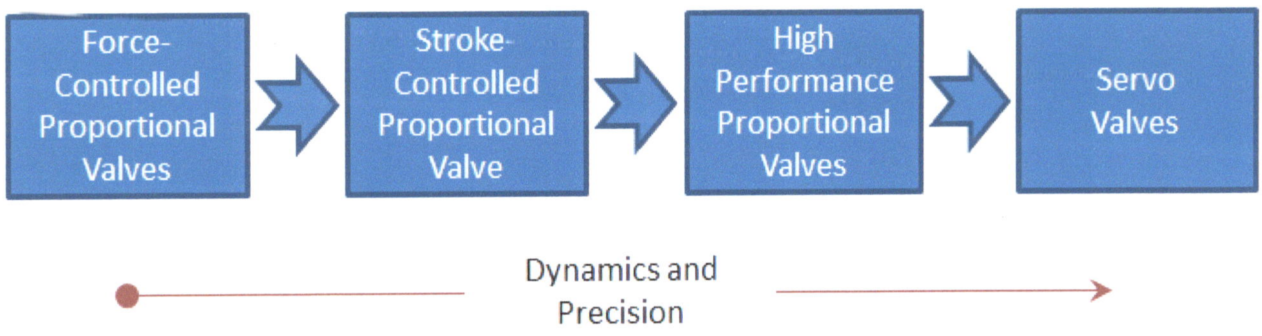

Fig. 6.1 - Servo Valve versus Proportional Valves

Figure 6.2 shows a servo valve that can be built-in with a hydraulic actuator to form a standalone single axis to control position, speed, and load in the two directions of motion of a hydraulic actuator.

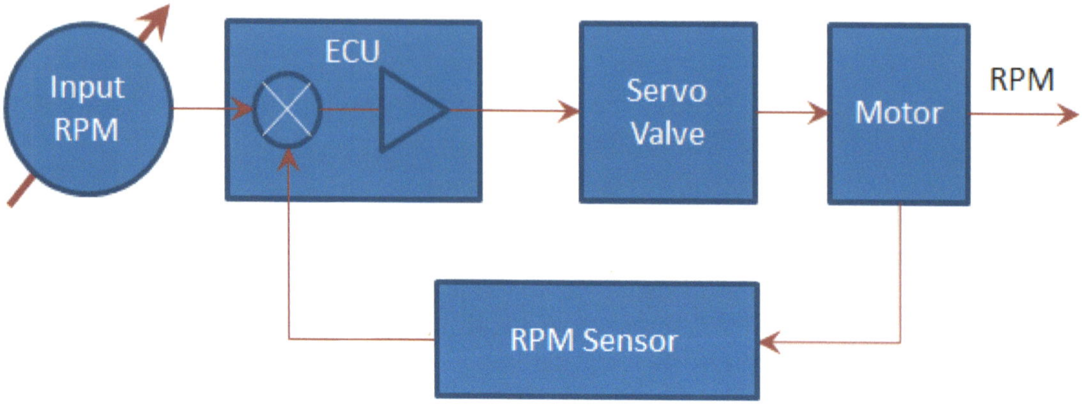

Fig. 6.2 - A Servo Valve in a Single-Axis System

As shown in Fig. 6.3, a servo valve can also be connected to a control system of a larger automated system.

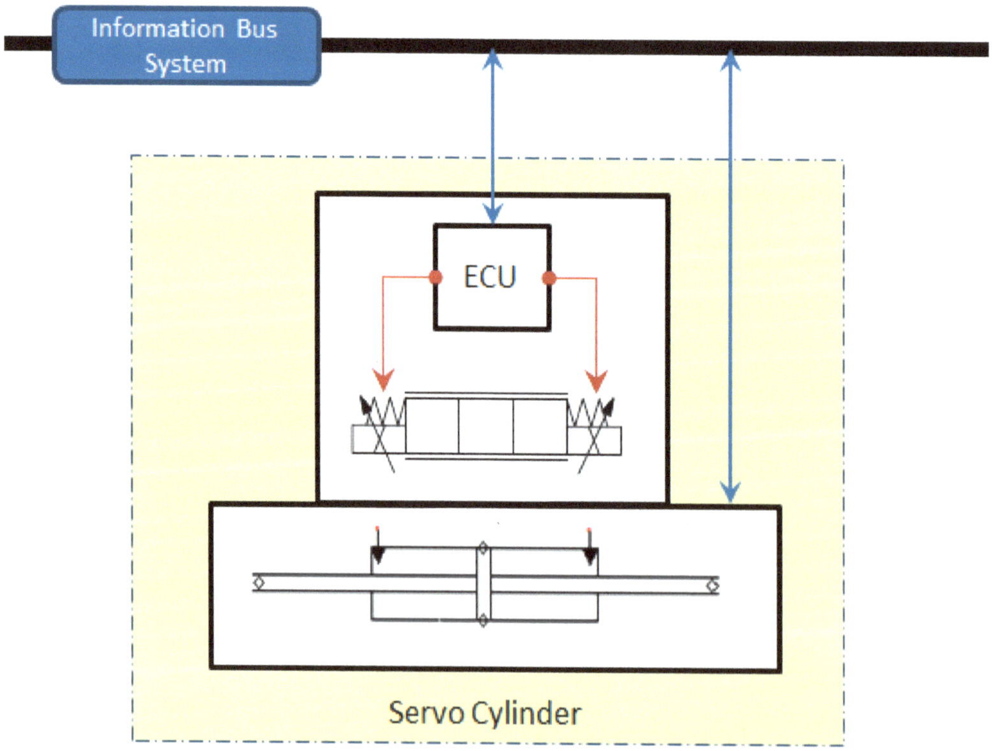

Fig. 6.3 - A Servo Valve in a Large Automated System

Table 6.1 shows, in approximate order of magnitude, a servo valve can be used for applications under the following conditions:

- Flow rates as low as 2 lit/min and up to 570 lit/min.
- Frequency ranging from 5-600 Hz.
- Hydraulic power of 1 to 250 kW.

It is noticeable that mobile applications are generally slower than industrial applications.

Field	Frequency (Hz)	Flow (l/min)	Power (kW)
Vibration Exciters	600	4	1.5
Missile Fin Control	400	4	1.5
Seekers Antenna	300	2	0.75
Oil Exploration Hammers	200	450	190
Fatigue Testing	100	115	40
Flight Simulators	50	190	75
Airplane Flight Controls	40	115	45
Metal Forming Rolling Mills	30	570	225
Agricultural Equipment	15	25	10
Cranes	7	75	30
Process Control	5	7.5	3

Table 6.1 - Servo Valve Application Examples

Three reasons make a servo valve faster than a proportional valve. These reasons will not be clearly understood before covering the construction and operation of servo valves. They can be briefly stated at this point as follows:

1-**Electrically:** A torque motor has faster response than a proportional solenoid and consumes much less power.

2-**Mechanically:** A Servo valve has a built-in internal feedback mechanism that assures the accurate and fast positioning of the main spool.

3-**Hydraulically:** The pilot stage in a servo valve creates the required differential pressure against the main stage spool much faster than in a pilot operated proportional valve.

6.2- Torque Motors

6.2.1- Torque Motor Construction and Operation

Function: A *Torque Motor* has an equivalent job to the proportional solenoid in the proportional valve. It is the part in the middle between the control portion and the hydraulic portion of a servo valve. The control portion may be separate or integrated with the valve.

Construction: As shown in Fig. 6.4, a torque motor consists of two identical coils framed by two pieces of permanent magnets. The figure shows that the upper piece is the *North Pole* of the magnet and the lower piece is the *South Pole*. A ferromagnetic *Core* is placed inside the coils and pivoted in the middle as shown in the figure. A vertical *Flapper* is integrated mechanically to the core and passes down through a *Flapper Tube*. The two coils in each servo valve are wound for equal turns with a normal production tolerance on coil resistance of ±12%. Copper magnet wire is used, resulting in a coil resistance that will vary significantly with temperature.

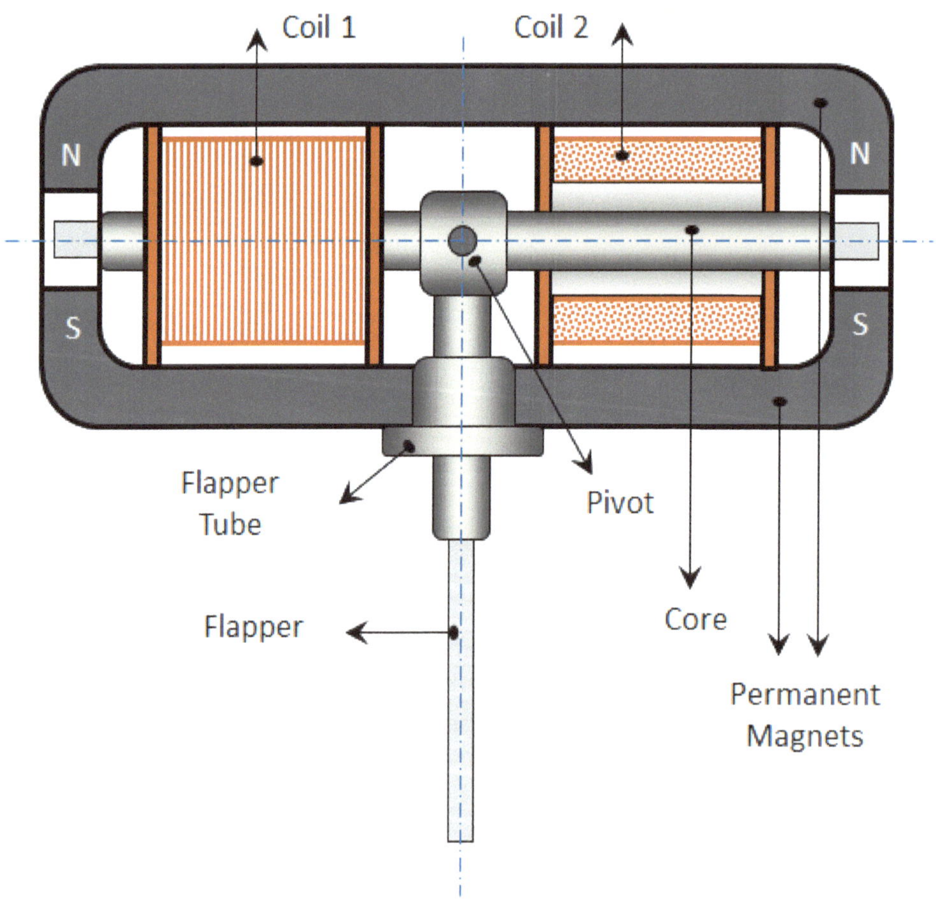

Fig. 6.4 - Torque Motor Construction

Operation: As shown in Fig. 6.5, once the coils are energized, the core will become magnetized, one end becomes North and the other becomes South. Accordingly, electromagnetic torque is generated that provides bi-directional, rotational motion. The direction of rotation depends of the polarity of the signal and the magnitude of the torque is proportional to the input signal.

It is to be noted that, if the torque motor is energized, the core will continue to rotate until the electromagnetic torque is balanced:

- Mechanically by a spring.
- Electrically due to the wiring style of the coils.

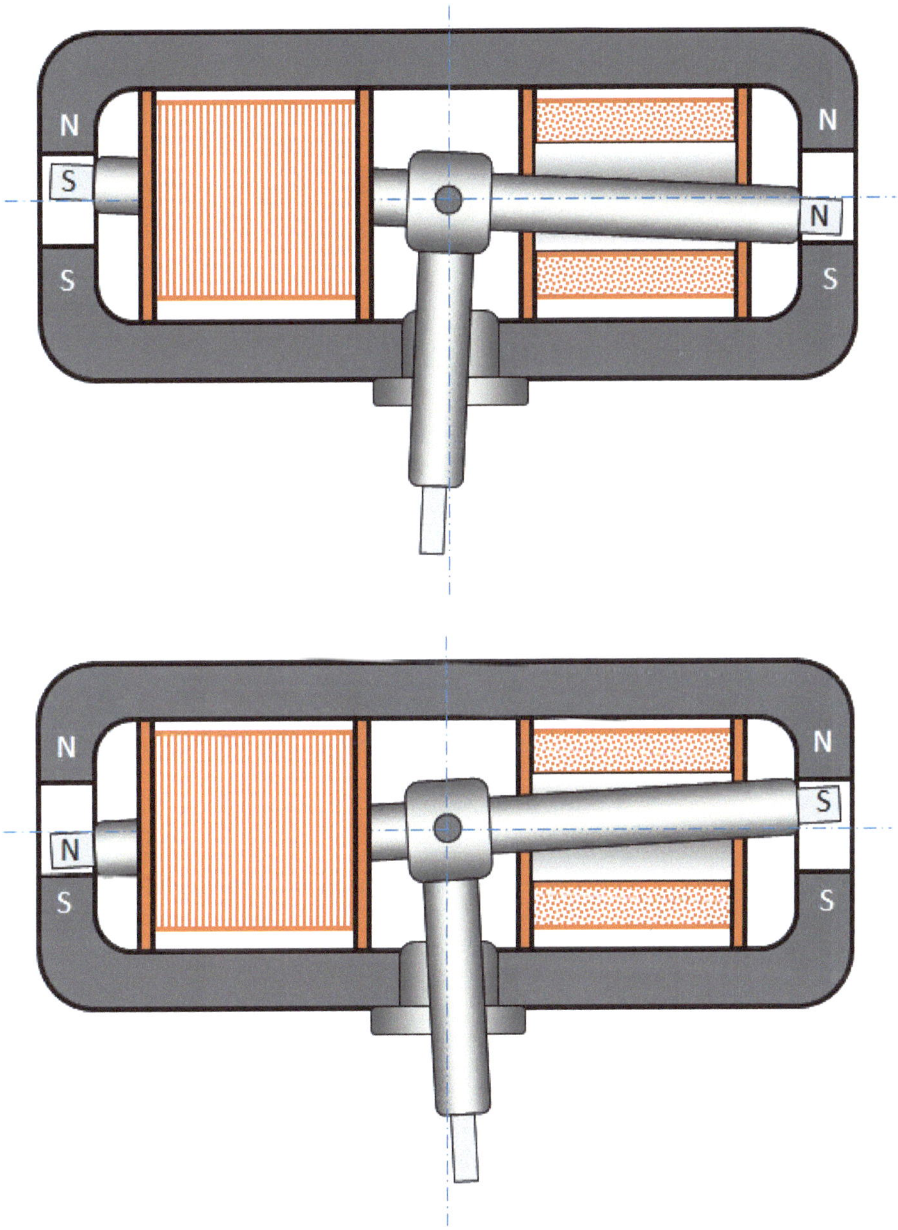

Fig. 6.5 - Torque Motor Operation

6.2.2- Torque Motor Wiring

Torque motors have the following types of wiring: parallel, series, and single. The following sections discuss the features of each type.

6.2.2.1- Parallel Wiring

As shown in a Fig. 6.6, in *Parallel Wiring*, the two coils assist each other in driving the core. The direction of rotation depends of the polarity of the signal and the magnitude of the torque is proportional to the input signal.

In parallel wiring, the core will continue to rotate until the electromagnetic torque is balanced with an external counter torque. The parallel wiring is preferred over the other types for reasons of safety. If one of the coil legs has been disconnected, the other coil will continue to work with half the power. The other advantage, as shown by Eq. 6.1, parallel wiring consumes less power than other types of wiring.

$$Power = I_P^2 \frac{R}{2} \qquad 6.1$$

Where:

Ip = Current per coil and R = Resistance per coil

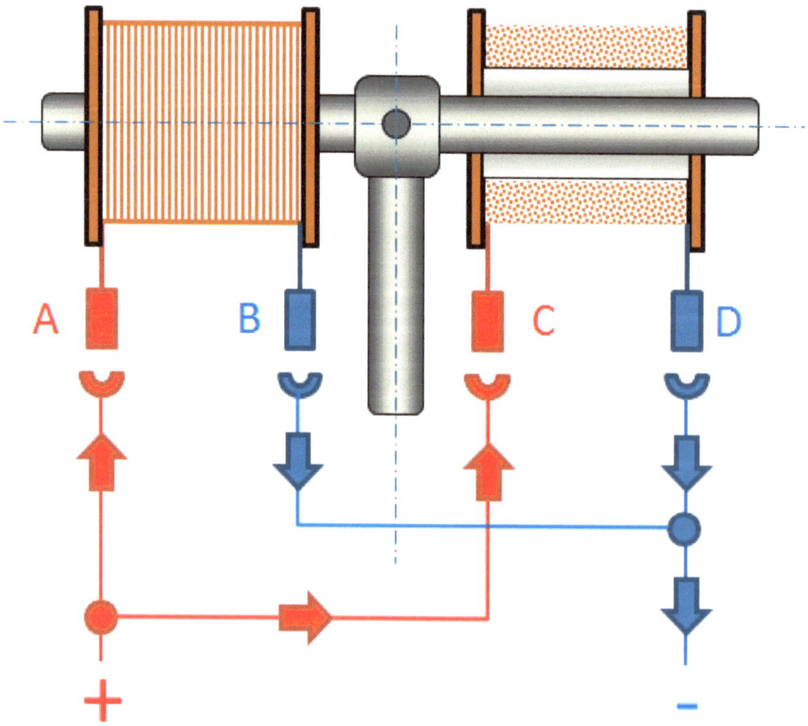

Fig. 6.6 - Parallel Wiring

6.2.2.2- Series Wiring

As shown in Fig. 6.7, in a *Series Wiring*, the end of the first coil is permanently connected to the beginning of the second coil. Therefore, only two ends are left to energize the torque motor. That could be considered as an advantage of simple wiring. But on the other hand, if one leg is disconnected, the torque motor will fail.

Like in parallel wiring:

- Coils assist each other and
- The direction of rotation depends of the polarity of the signal and the magnitude of the torque is proportional to the input signal.
- An external counter torque is needed to stop the core.

The only difference as compared to parallel wiring is that the series wiring works at double the voltage and half the current.

Equation 6.2 calculates the power consumption for series wiring.

$$Power = I_P^2\,(2R) \qquad 6.2$$

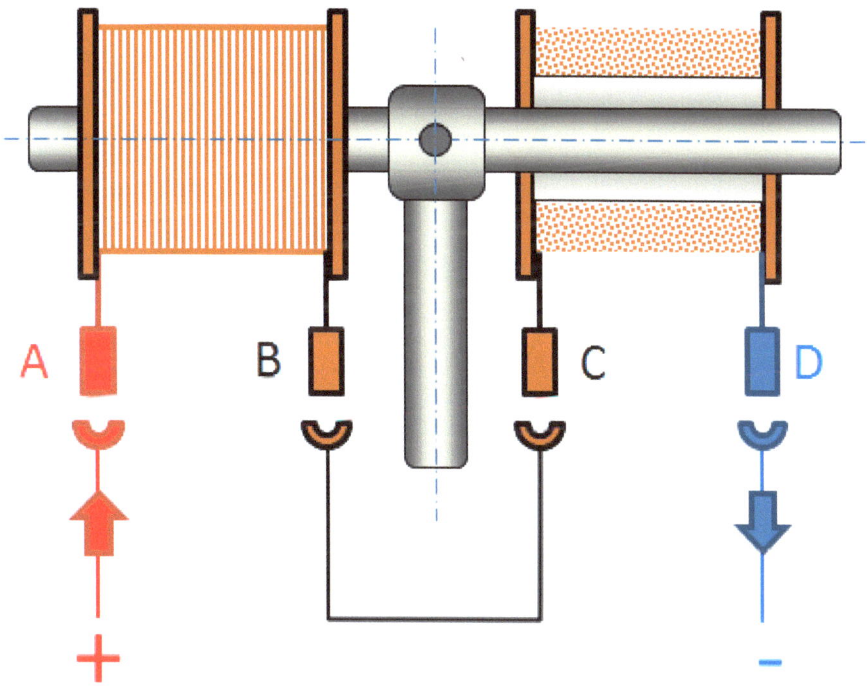

Fig. 6.7 - Series Wiring

6.2.2.3- Single Wiring

Single Wiring, shown in Fig. 6.8, works as described in following bullet points:

- Each coil is energized independently from 0 to maximum.
- The polarity of each coil is never reversed.
- Each coil has twice the strength of a coil used in parallel or in series.
- The coils resist each other.
- Direction of core rotation depends on the sign of the differential signal (I_1-I_2).
- Magnitude of the electromagnetic torque depends on the magnitude of differential signal.

The major advantage that single wiring offers is the ability to hold the angular position of the core without the need for an external counter torque. When the two signals are equal, the armature is held in position. Additionally, the single wiring method provides better stability of the core in an angular position by equal and opposite coils.

Equation 6.3 calculates the power consumption for single wiring.

$$\boldsymbol{Power} = (\Delta I)^2 R \qquad 6.3$$

Where: ΔI ***= Differential Current***

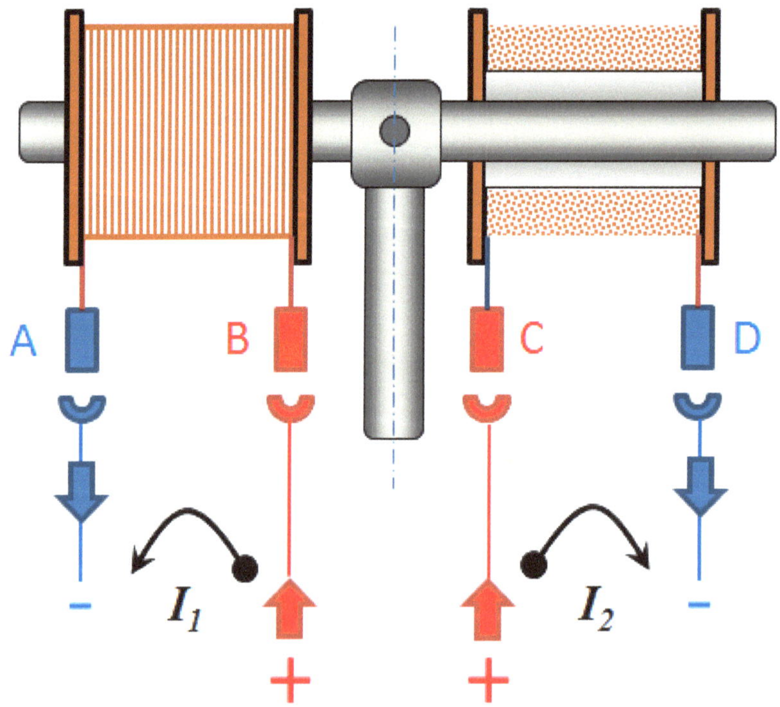

Fig. 6.8 - Single Wiring

6.3- Servo Valve Configurations

6.3.1- Power Assistance using Mechanical Servo Valve

Construction: Figure 6.9 shows a power assisted load drive using a mechanical servo valve. The valve is manually actuated by a hand lever that is integrated to the spool and pivoted to a frame of reference. The sleeve of the valve is attached to another frame of reference, which is the load. The load is driven by a hydraulic cylinder that is connected to the valve ports as shown in the figure.

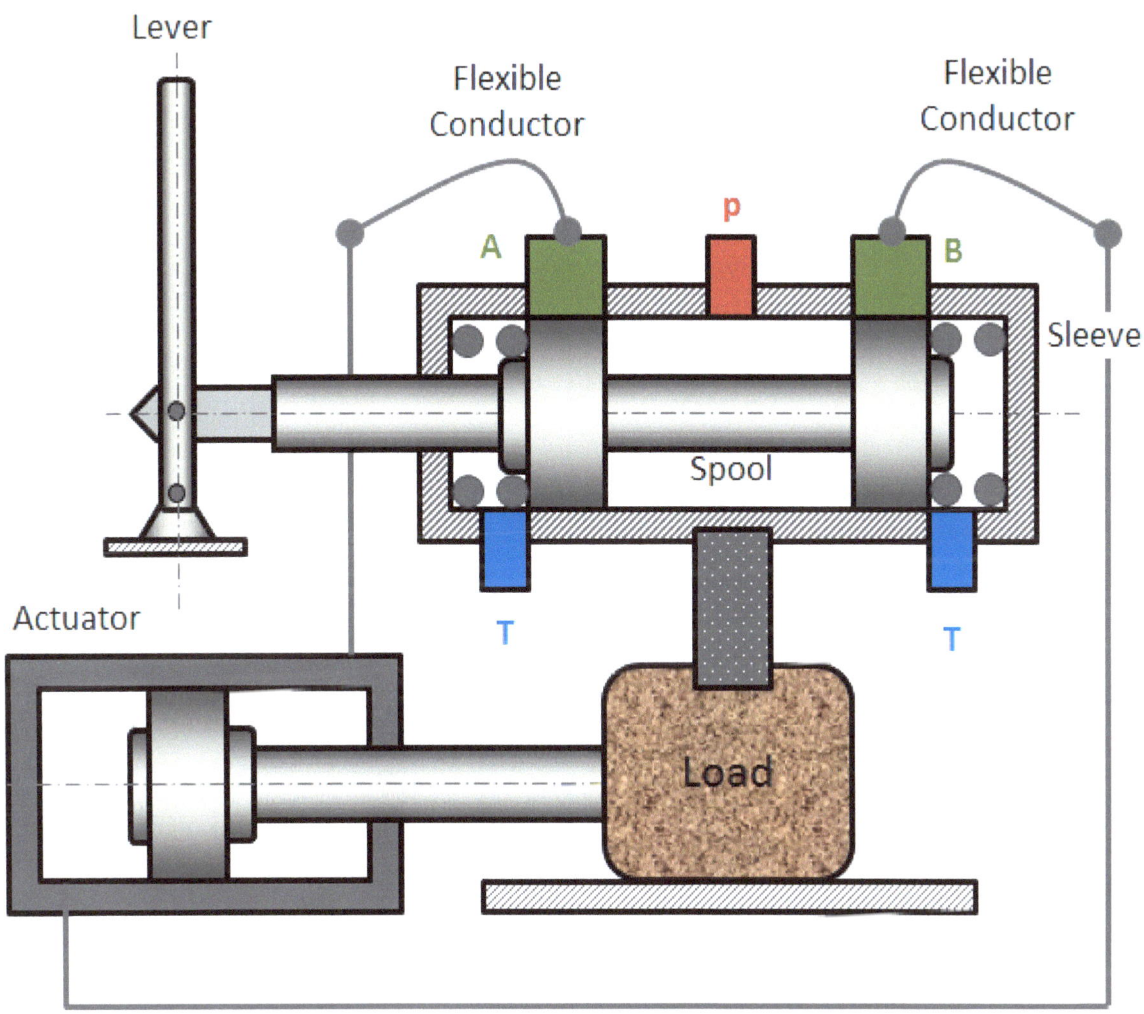

Fig. 6.9 - Power Assistance using Mechanical Servo Valve

Operation: Figure 6.10 assumes that the operator shifts the hand lever backward so that the spool has moved a distance "x". The pressure port **P** is connected to port **A** and consequently the rod side of the cylinder is pressurized. Simultaneously the piston side of the cylinder is vented. Consequently, the cylinder retracts pulling the valve sleeve. The cylinder will continue to retract until the sleeve moves backward the same distance "x", blocks the service ports **A** and **B**, and the cylinder is held in the new position. If the lever is shifted further, the load will move accordingly with the same distance the lever moves and in the same direction. So, it is a one-to-one relationship between the lever and the load. Use of such a valve is not very common and used for loads that move for relatively short distances.

This system is currently used to control the swash plate angle in variable-displacement piston pumps.

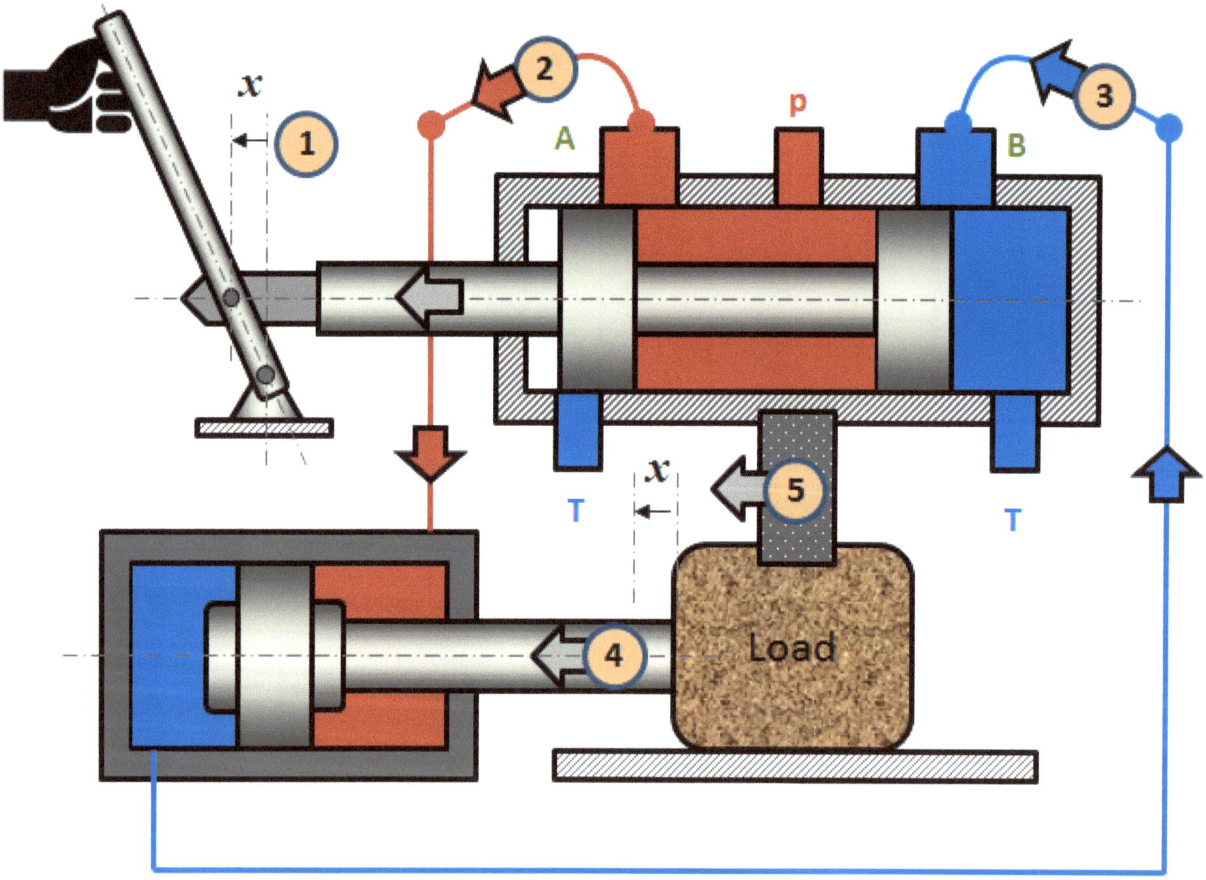

Fig. 6.10 - Concept of Operation of Mechanical Servo Valve

6.3.2- Single-Stage Servo Valve

Figure 6.11 shows the conceptual construction of an EH single-stage servo valve, in which a torque motor drives the spool directly against two balancing springs. So, it is considered a *Force Feedback*. This valve is functionally equivalent to a direct-operated force-controlled directional valve.

The torque motor must be sized to overcome spool friction, flow forces, and spring forces. Such a valve is good for small volume flow rates. Since the electromagnetic torque is countered by the balancing springs, then parallel or single methods of wiring are applicable and no need for single wiring.

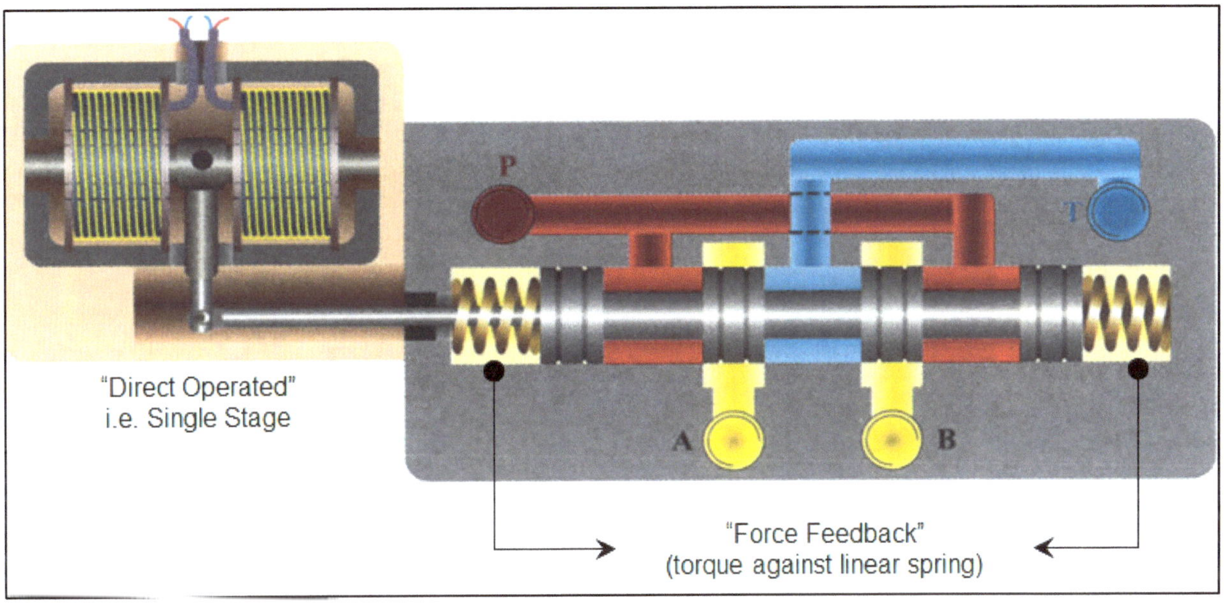

Fig. 6.11 - Conceptual Construction of a Single Stage Servo Valve (Courtesy of ASSOFLUID)

Figure 6.12 shows another example of a single stage servo valve, in which no balancing springs are used. Therefore, it became necessary to wire this torque motor based on using single wiring method. When input signal is received, the torque motor moves the tie rod (7) depends on the polarity of the input signal. The spool is then shifted until torque motor balances with one of the centering springs. Shifting the spool will cause the supply pressure to be routed to either port A or B to drive the attached load.

The hydraulic null point is adjusted by screw (8) which moves the control sleeve (10) in the housing (9) axially relative to the control spool (2).

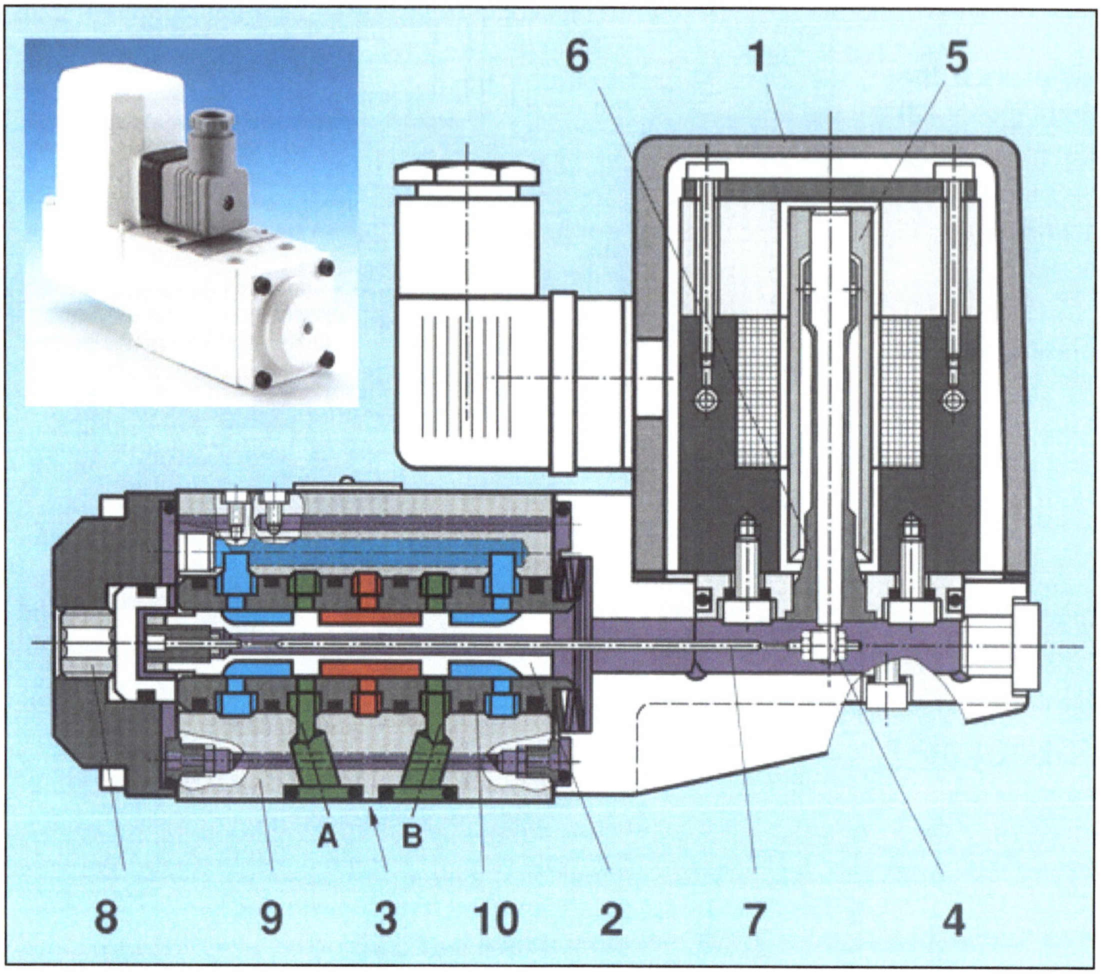

1. **Torque Motor**
2. **Control Spool**
3. **Spool Stage**
4. **Pin**
5. **Armature**
6. **Torque Tube**
7. **Tie Rod**
8. **Null Adjustment Screw**
9. **Housing**
10. **Sleeve**

Fig. 6.12 - Example of a Single Stage Servo Valve (Courtesy of Bosch Rexroth)

6.3.3- Two-Stage Servo Valve with Tracking Sleeve

Construction: Figure 6.13 shows the conceptual construction of a two-stage servo valve with a *Tracking Sleeve*. As a courtesy of ASSOFLUID, the valve consists of a small pilot spool connected directly to the torque motor. The pilot spool moves within a sliding sleeve that is mechanically linked to the main spool. The right-hand end of the main spool is permanently connected to the pilot pressure line, but because of the linkage rod, its area is reduced to an annulus of area **A**. Pressure at the left-hand end of the spool is controlled by the pilot valve. There is no area restriction at this end, and the valve is designed such that the spool has an area of **2A**.

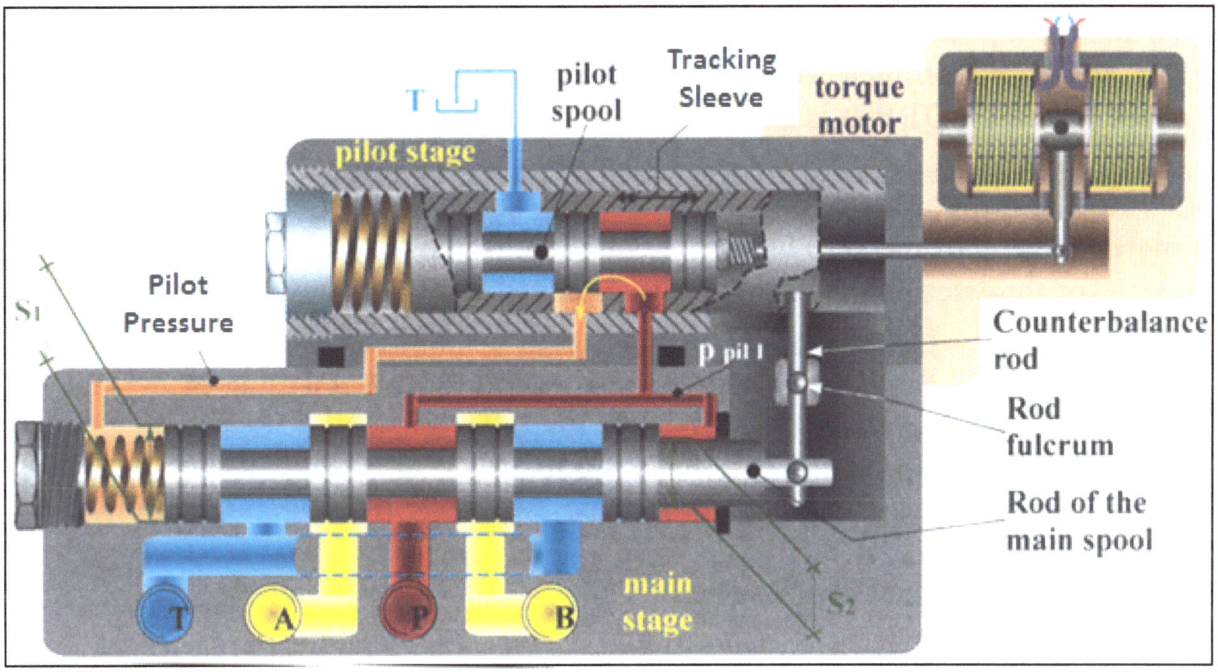

Fig. 6.13 - Conceptual Construction of a Servo Valve with Tracking Sleeve (Courtesy of ASSOFLUID)

Operation: Figure 6.14 explains the operation of the valve as follows. If the same pressure **P** is applied to both ends, the spool experiences a left force of (P x A) and a right force of (2P x A) causing a net force of (P x A) to the right, resulting in a shift of the main spool to the right.

If a pressure of **P** is applied to the right-hand end and **0.5P** is applied to the left-hand end, equal and opposite forces (P x A) result and the valve spool is stationary.

The pilot valve can thus move the main spool in either direction, in a controlled manner, by varying pressure at left-hand end of the main spool from zero to full pilot pressure.

The mechanical linkage between main spool and pilot sleeve controls the flow of fluid between the pilot valve and main valve, and hence controls pressure at the left-hand end of the main spool.

Suppose the electrical control signal (1) causes the pilot spool to shift left. This increases the pressure (2) causing the main valve to shift right (3) which in turn pushes the tracking sleeve left (4). The main spool stops moving when the control pressure port in the pilot sleeve exactly aligns with the land on the pilot spool. At that time, the control pressure on the left side (5) is restricted enough to balance the forces on the main spool. A change in electrical signal moving the pilot spool to the right reduces pressure at the left-hand end of the main spool by bleeding fluid back to the tank. This causes the main spool to move left until, again, the pilot sleeve and pilot lands are aligned. The main valve spool thus follows the pilot spool with equal, but opposite movements.

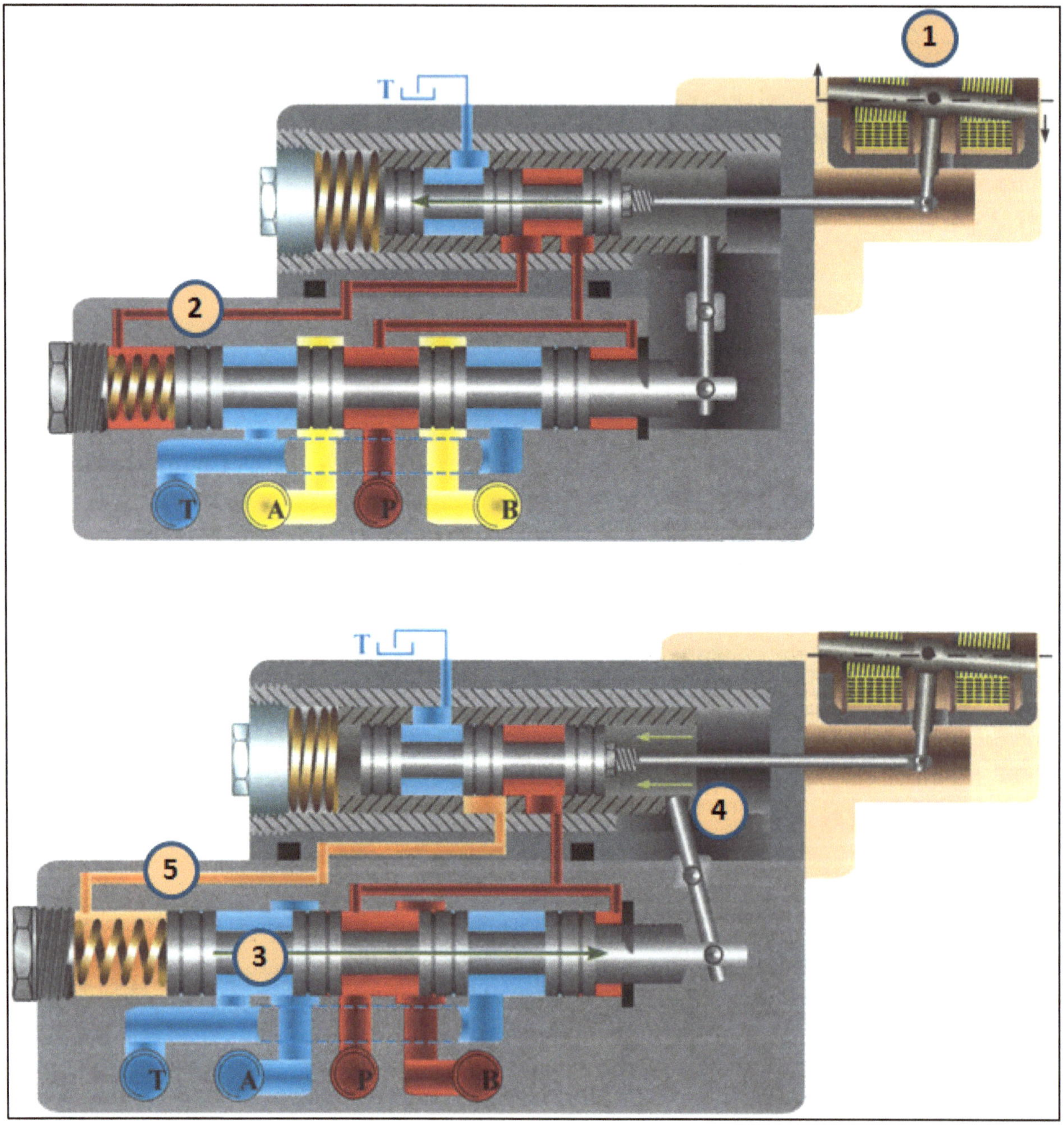

Fig. 6.14 - Concept of Operation of a Servo Valve with Tracking Sleeve (Courtesy of ASSOFLUID)

6.3.4- Flapper–Nozzle Concept of Operation

6.3.4.1- Half-Bridge Flapper-Nozzle Concept

Figure 6.15 shows conceptually the construction of a *Half-Bridge Flapper-Nozzle* stage. The figure shows that constant supply pressure p_s is connected to the nozzle via a fixed orifice. The nozzle is facing a movable flapper. The nozzle-flapper assembly, in the equivalent hydraulic schematic, is represented by a variable orifice. By moving the flapper towards the nozzle, the outlet pressure p_{out} at the port **C1** increases, and vice versa.

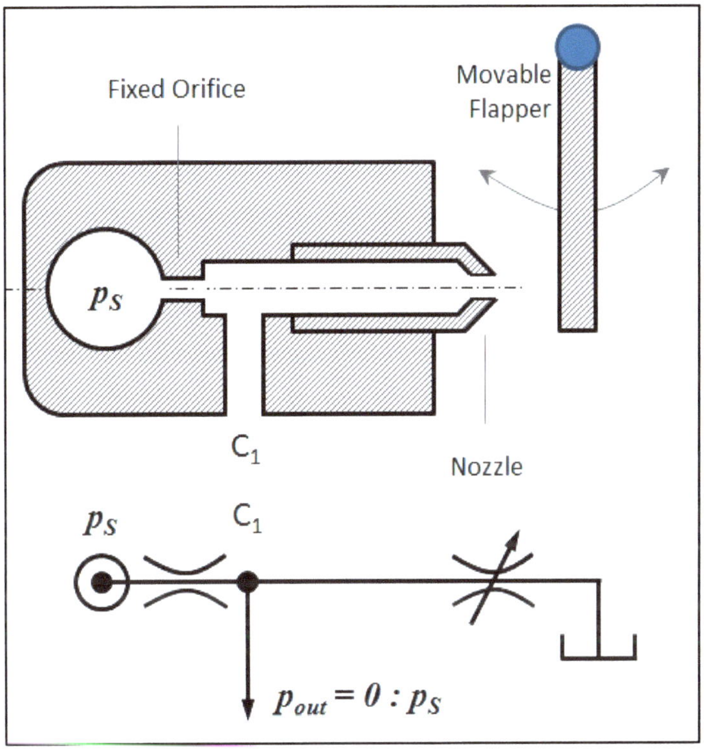

Fig. 6.15 - Half-Bridge Flapper-Nozzle Concept

Figure 6.16 shows a proof of the above-mentioned concept of operation. The figure shows that the flow goes from point 1 at the circular cross-sectional area of the nozzle to point 2 on the cylindrical curtain area as shown in the figure.

By applying Bernoulli's Equation 6.4 between points 1 and 2, the flapper moves closer to the nozzle, curtain area decreases, fluid speed increases, and pressure decreases at point 2. Consequently, the outlet pressure at point 1 increases. If the flapper closes the nozzle completely, then outlet pressure p_{out} equal supply pressure p_s.

Equation 6.5 indicates that, to achieve this controllability, the curtain area must be kept much less the than the nozzle area. In order of magnitude, the flapper travel x_f apart from the nozzle shouldn't exceed one tenth of the nozzle diameter d_n. Otherwise, the outlet pressure will fall to tank pressure

The above explanation is based on the assumption of ignoring the losses and the potential head between point 1 and 2.

$$\left(\frac{p}{\rho g} + \frac{v^2}{2g}\right)_1 = \left(\frac{p}{\rho g} + \frac{v^2}{2g}\right)_2 \qquad 6.4$$

$$(A\uparrow \;\rightarrow v\downarrow \rightarrow p\uparrow)_1 \qquad (A\downarrow \;\rightarrow v\uparrow \rightarrow p\downarrow)_2$$

$$x_f \cdot \pi d_n \ll \frac{\pi}{4}\, d_n^2 \qquad 6.5$$

$$x_f = 0.1 d_n$$

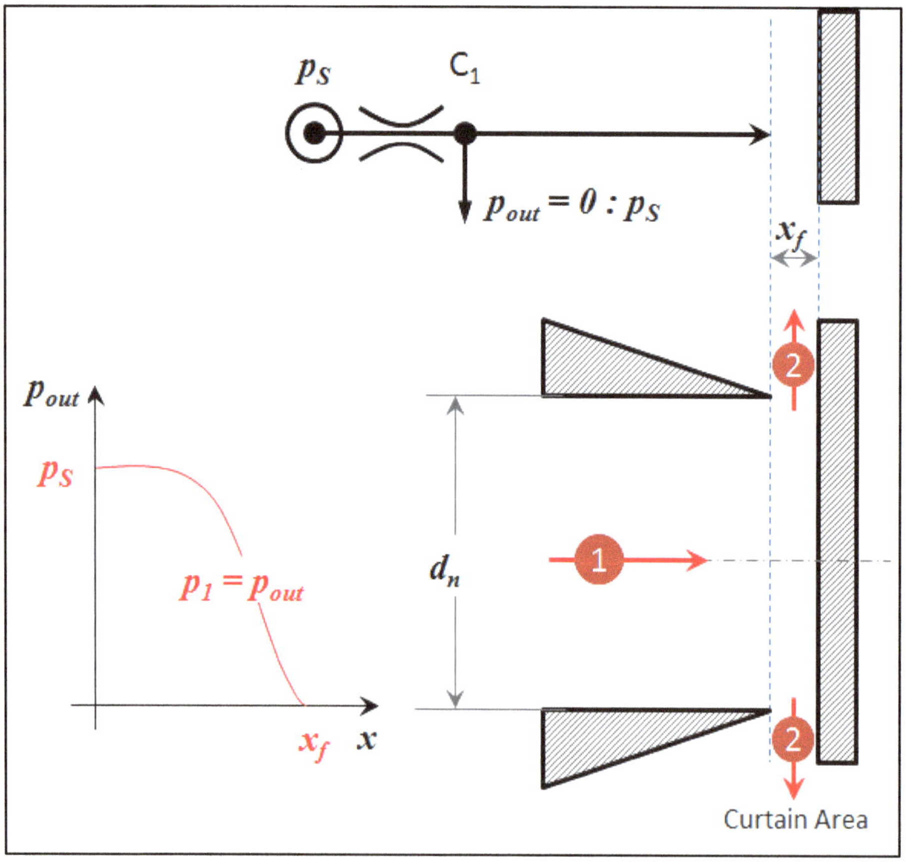

Fig. 6.16 - Bernoulli's Equation as Applied on Flapper-Nozzle Concept

6.3.4.2- Full-Bridge Flapper-Nozzle Concept

Figure 6.17 shows conceptually the construction of a *Full-Bridge Flapper-Nozzle* stage and the equivalent hydraulic diagram. The figure shows that constant supply pressure p_s is connected to the two nozzles simultaneously via two fixed orifices. A movable flapper has been placed accurately in the middle between the two nozzles. The nozzle-flapper assembly, in the equivalent hydraulic schematic, is represented by two variable orifices. When the flapper is kept equidistant from the two nozzles, the outlet pressures p_1 and p_2 are equal and the differential pressure $\Delta p = 0$. By moving the flapper towards one of the nozzles, the pressure behind this nozzle increases and the pressure behind the other nozzle deceases. Hence, the full-bridge flapper-nozzle assembly works like a pressure divider that creates an output differential pressure that is proportional to the flapper movement. When the flapper blocks one of the nozzles, the differential pressure equals the supply pressure. The full-bridge flapper-nozzle assembly can be used as a single stage so that the created differential pressure drives small loads directly. For large valves, it is used as a pilot stage so that the differential pressure drives the main stage's spool.

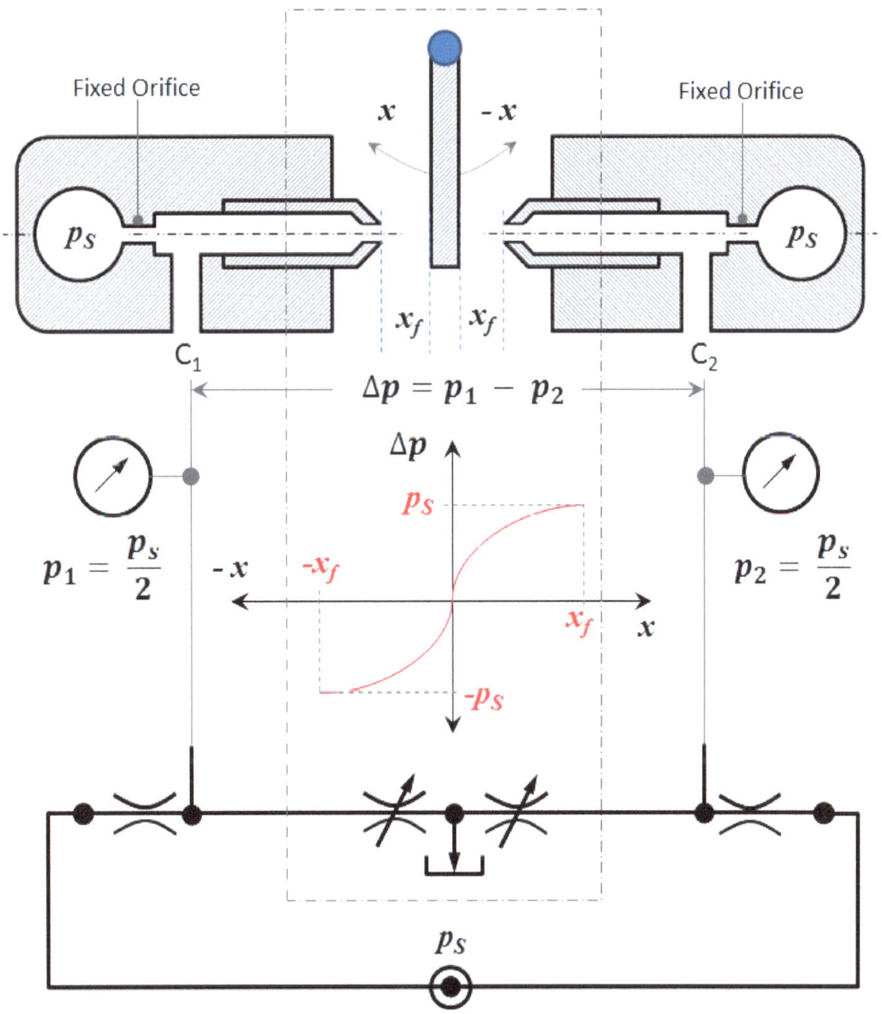

Fig. 6.17 - Full Bridge Flapper-Nozzle Concept

Figure 6.18 shows typical operating curves for a servo valve in which the flapper-nozzle is a pilot stage. The valve can be designed to generate linear flow gain versus the spool displacement (input command) with certain tolerance.

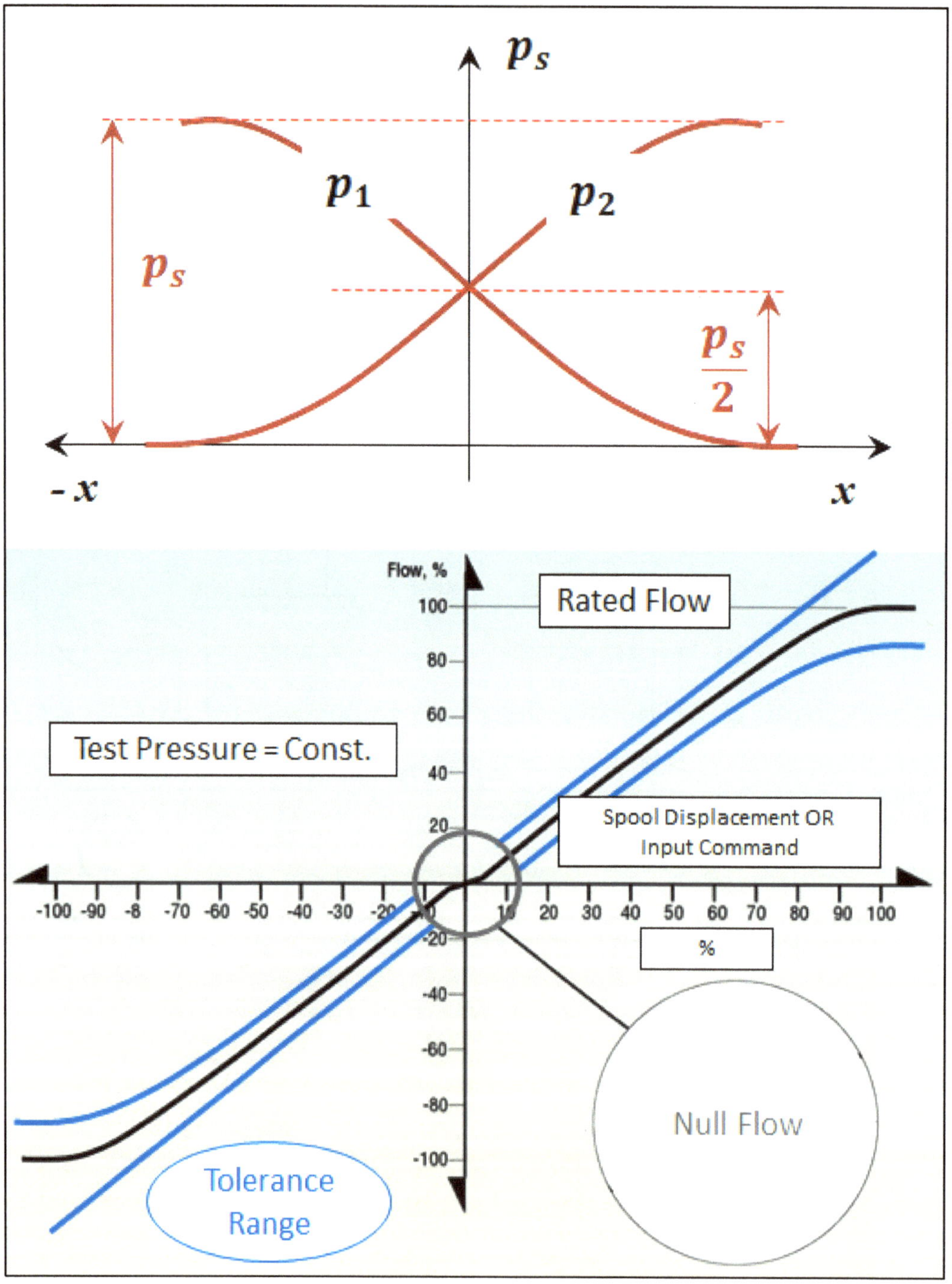

Fig. 6.18 - Typical Operating Curves for Full-Bridge Flapper-Nozzle Stage

6.3.4.3- Power Gain of Servo Valve

Figure 6.19 explains the concept of *Power Gain* of a servo valve. An example of this assumes a torque motor uses 0.1 Watts of electric power to control 100-Watt hydraulic power at the flapper-nozzle stage. If the flapper-nozzle stage controls 1000 Watts of hydraulic power at the main stage, then the total power gain of the valve is ten thousand. It is to be noted that this power gain of the valve shouldn't be understood as power amplification since the power at the flapper-nozzle stage and at the main stage are already sourced externally. The power gain of the valve should be understood as how input small power can be used to control large output power.

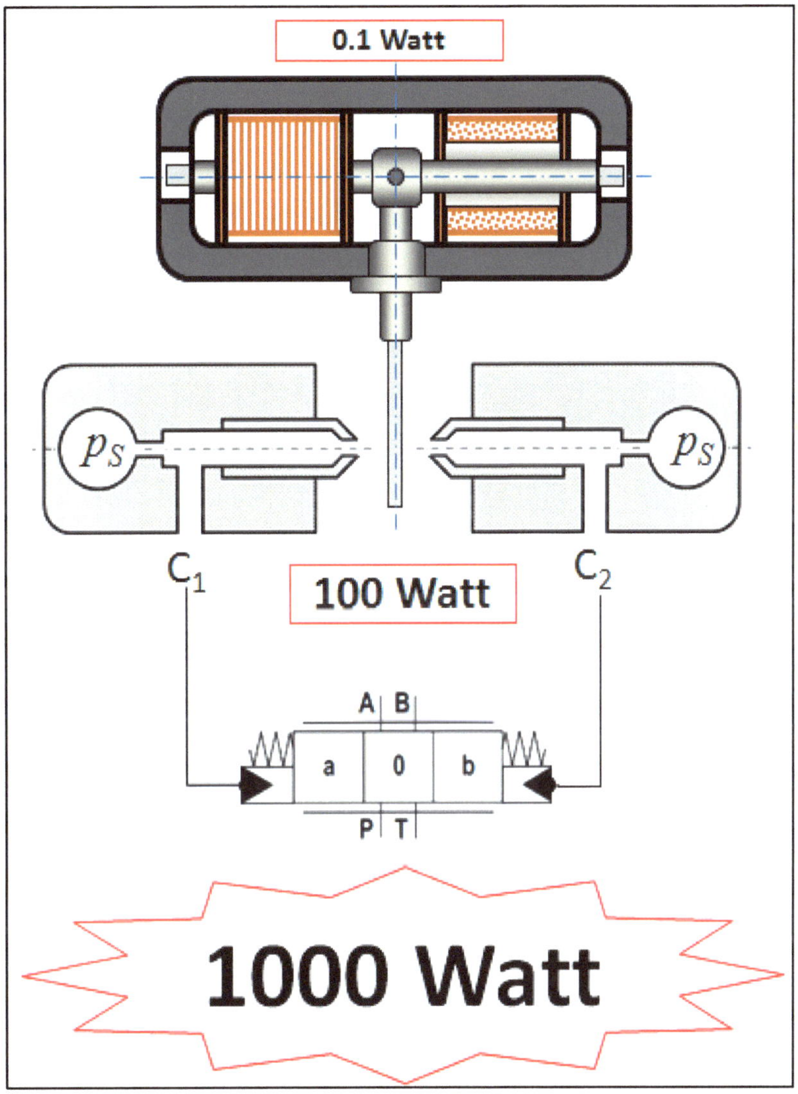

Fig. 6.19 - Power Gain of Full-Bridge Flapper-Nozzle Stage

6.3.5- Two-Stage Flapper–Nozzle Servo Valve with Barometric Feedback

Figure 6.20 shows a two-stage flapper-nozzle valve with *Barometric Feedback.* This valve configuration has the equivalent functionality of a force-controlled proportional valve, in which the movement of the main spool is proportional to the input command without feedback on the spool position. When the torque motor receives an input signal, it drives the flapper towards one of the two nozzles so that the control spool moves accordingly. The differential pressure created by the flapper movement is balanced by the control springs. In order to continue to hold the spool in the desired position, the torque motor must continue to receive the input signal, and the flapper must be biased towards one of the nozzles. In this valve, the torque motor must be wired in a single mode in order to hold the angular position of the flapper.

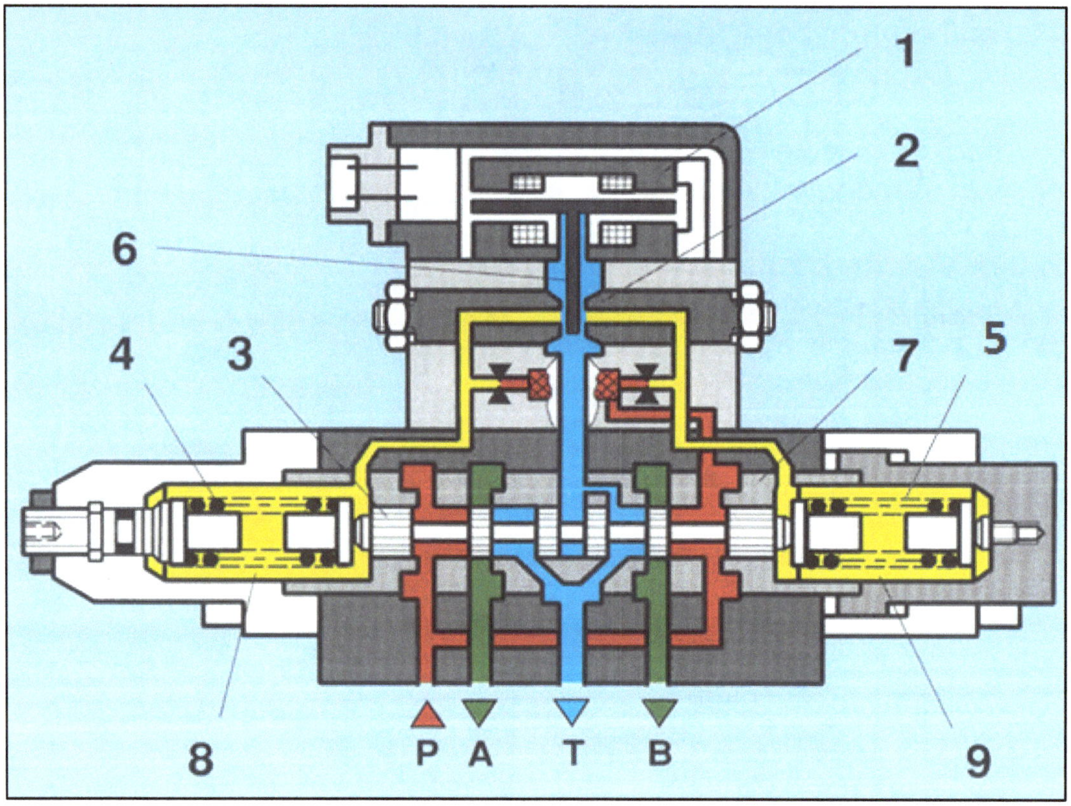

1. **Torque Motor**
2. **Two Nozzles**
3. **Control Spool**
4. **Control Spring**
5. **Control Spring**
6. **Flapper**
7. **Interchangeable Control Sleeve**
8. **Control Chamber**
9. **Control Chamber**

Fig. 6.20 - Two-Stage Flapper–Nozzle Servo Valve with Barometric Feedback (Courtesy of Bosch Rexroth)

As shown in Fig. 6.21, the control spool and the interchangeable control sleeve are manufactured as a pair that can be changed together.

Fig. 6.21 - Servo Valve Interchangeable Spool and Sleeve (Courtesy of Moog)

6.3.6- Two-Stage Flapper–Nozzle Servo Valve with Mechanical Feedback

Construction: Figure 6.22A shows the conceptual construction of the most common and usable two-stage flapper-nozzle valve. The valve is considered has Mechanical Feedback or in other words it is equivalent to proportional stroke-controlled but mechanically.

In such a valve, the mechanical feedback is through a wire spring that connects the end of the flapper to the main spool. As shown in the figure, the wire spring forms a ball-in-socket joint with the spool. The figure also shows that, whether the valve is direct-controlled (pilot pressure is sourced internally) or pilot-controlled (pilot pressure is sourced externally), there should be last chance filters to protect the nozzles.

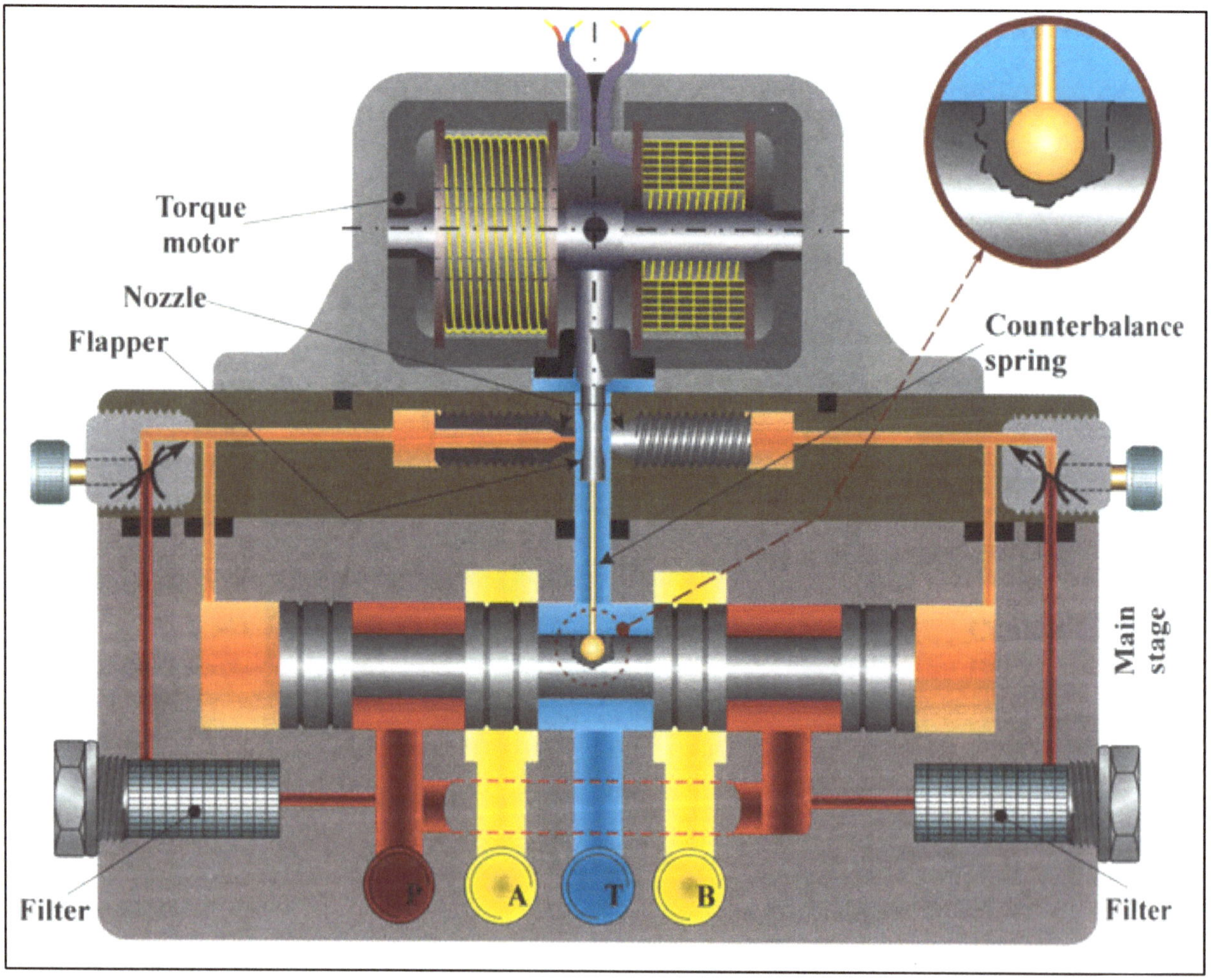

Fig. 6.22A - Conceptual Construction of a 2-Stage Flapper-Nozzle Servo Valve with Mechanical Feedback (Courtesy of ASSOFLUID)

Operation: Figure 6.22B shows the valve operation once the torque motor receives an input signal. The flapper moves accordingly creating differential pressure that drives the spool in the opposite direction of the torque motor.

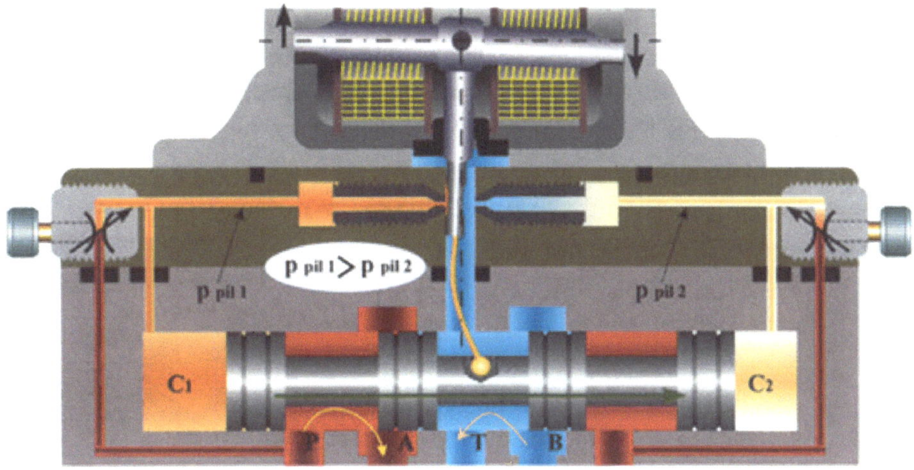

Fig. 6.22B - Torque Motor Receives Signal (Courtesy of ASSOFLUID)

Figure 6.22C shows that the spool continues to move, opposite to the torque motor, pulling back the flapper via the feedback spring towards the central position between the two nozzles. Once the flapper is perfectly centered between the two nozzles, the differential pressure becomes zero and the spool stops at a new balanced position. To move the spool further, the input signal must be increased and vice versa. Since the feedback in this valve is mechanical, the single wiring method is not necessary, all wiring methods are applicable. At any spool position, once the torque motor is de-energized, the current flapper position will create a differential pressure that will automatically drive the spool to the center position and the valve will reset.

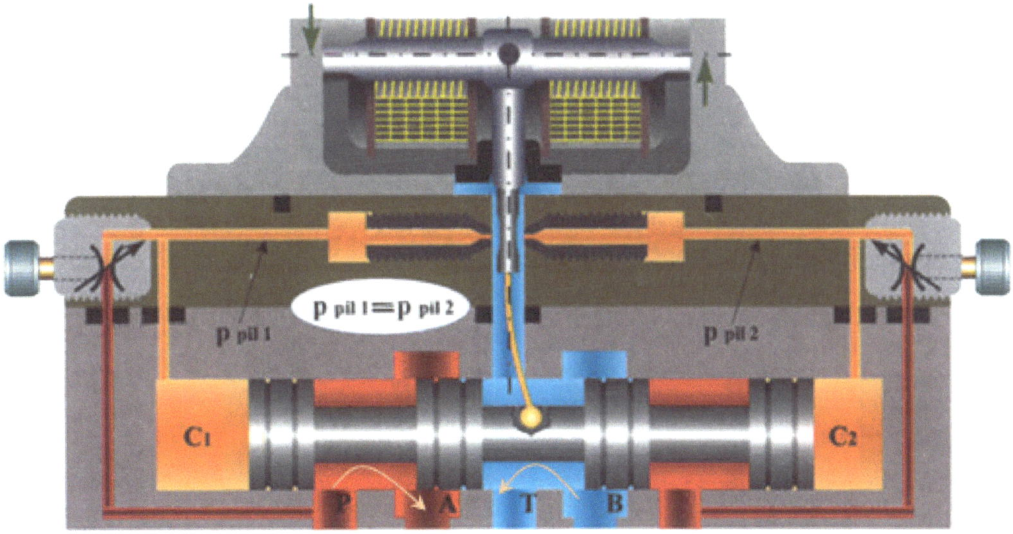

Fig. 6.22C - Spool Balanced (Courtesy of ASSOFLUID)

Figure 6.23 shows the flow distribution within the valve during operation. The associated figure 6.24 describes the step-by-step operation of the valve as follows:

1. A torque motor receives an input signal (control current i).
2. The torque motor generates a torque (+T) proportional to the input signal.
3. The torque (+T) drives the flapper towards one of the nozzles an angle (+θ).
4. A differential pressure (ΔP_S) is created across the main spool.
5. The main spool moves a distance X pulling back the wire spring.
6. The wire spring generates a feedback torque (-T).
7. The feedback torque (-T) pulls back the flapper an angle distance (-θ).
8. When the error in the flapper angular movement (θ_{Error}) becomes zero, the differential pressure across the spool (ΔP_S) becomes zero and spool stops moving and is held in its current balanced position.

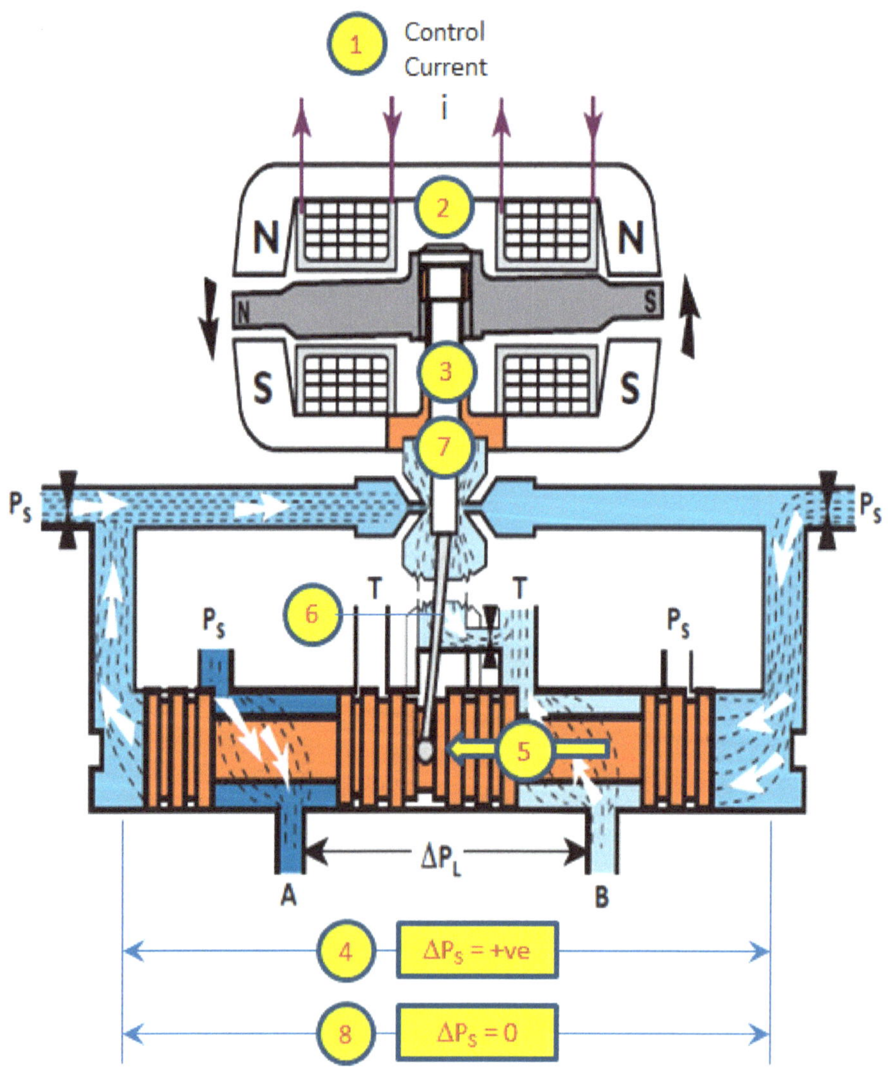

Fig. 6.23 - Flow Distribution in Flapper-Nozzle Servo Valve

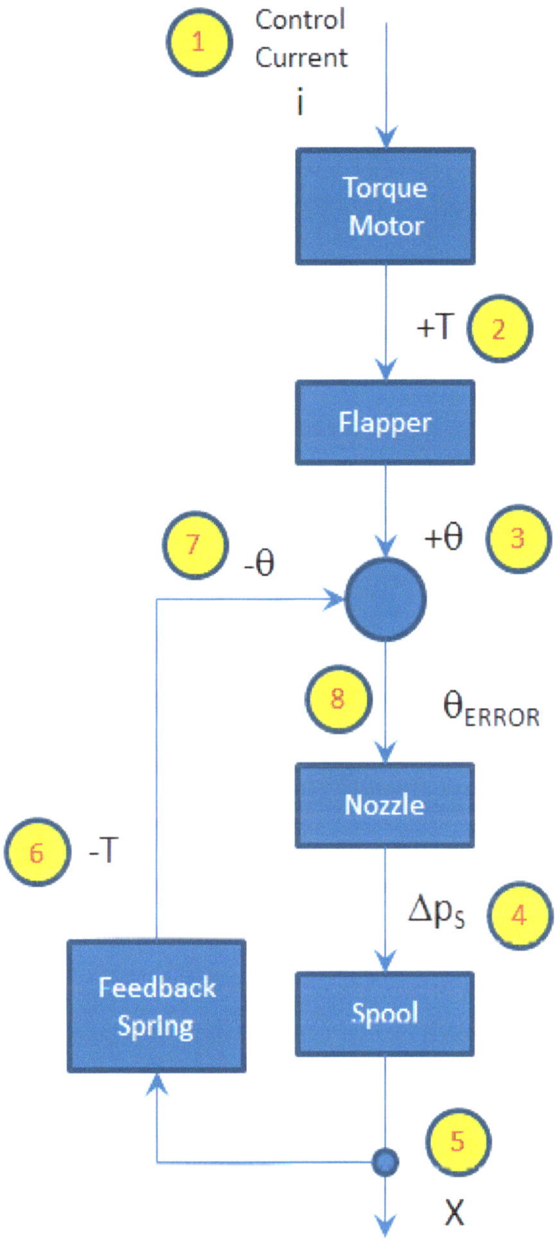

Fig. 6.24 - Concept of Mechanical feedback

Figure 6.25 shows a small size "mini" flapper-nozzle servo valves. As shown in the figure, the valve has flying ends to connect the valve to the control electronics.

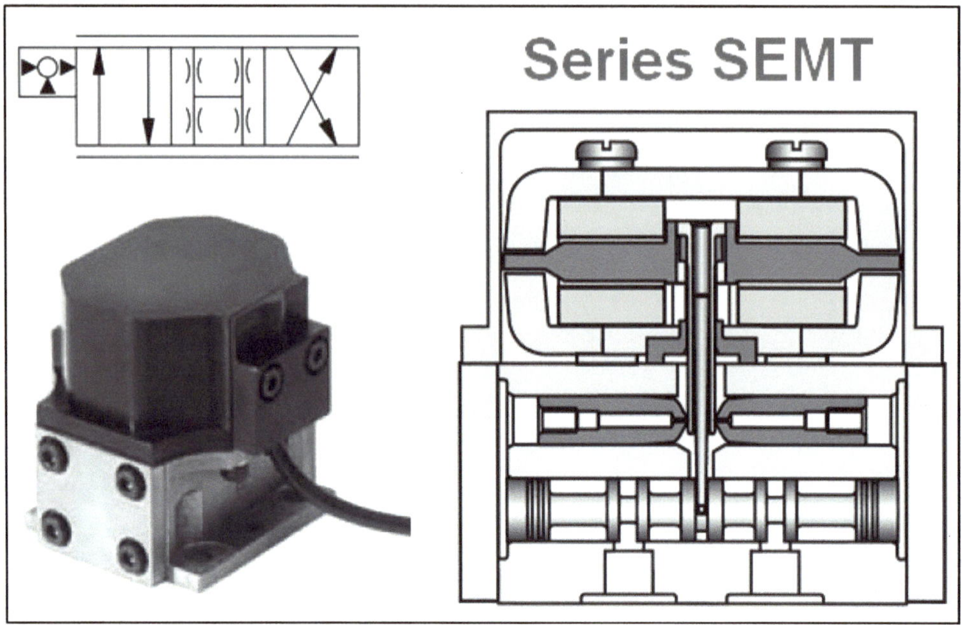

General Description

The Parker mini-valve, SEMT, is a two stage, 4-way, flapper and nozzle style servovalve. Its remarkably small size makes it optimal for Remotely Operated Vehicles (ROV), motorsport suspension control, or any application requiring a compact, and light weight, high performance servovalve.

A special jewel feedback design enhances durability and prevents ball glitch problems, which can occur in other types of servovalves. This valve is rated for 210 bar service. Higher pressure capability is available upon request.

Fig. 6.25 - Mini Flapper-Nozzle Servo Valve (Courtesy of Parker)

Figure 6.26 shows the technical specifications of the previous valve.

Specifications

Parameter	Unit	Value
Flow Rating ± 10% (at 70 bar)	[l/min]	2, 4, 7
Supply pressure	[bar]	15 - 210
Tank port pressure	[bar]	max. 210 < 10 for best performance
Pilot and null leakage flow (at 140 bar)	[l/min]	0.4 - 0.7
Input command	[mA]	±10 std.
Frequency response (at 90° phase shift)	[Hz]	> 170 (see Bode plots)
Non-linearity	[%]	≤ 10
Hysteresis	[%]	≤ 3
Threshold	[%]	≤ 0.5
Null shift with temperature	[%]	≤ 2 per 55°C
Null shift with pressure	[%]	≤ 2 per 70 bar
Pressure gain % change in pressure per 1% change in input command		60% typical
Step response		0 - 100%, < 4 ms
Fluid		Petroleum based mineral oil 10 to 110 cSt at 38 °C
Fluid cleanliness		ISO 4406 15/12 or better
Operating temp.	[°C]	-30 to +130
Protection class		NEMA 4, IP65

Fig. 6.26 - Technical Specifications of a Mini Flapper-Nozzle Servo Valve (Courtesy of Parker)

Figure 6.27A shows a typical two-stage flapper-nozzle servo valve with mechanical feedback. As shown in the valve features, it is driven by a separate ECU and the last chance filter can be externally replaced.

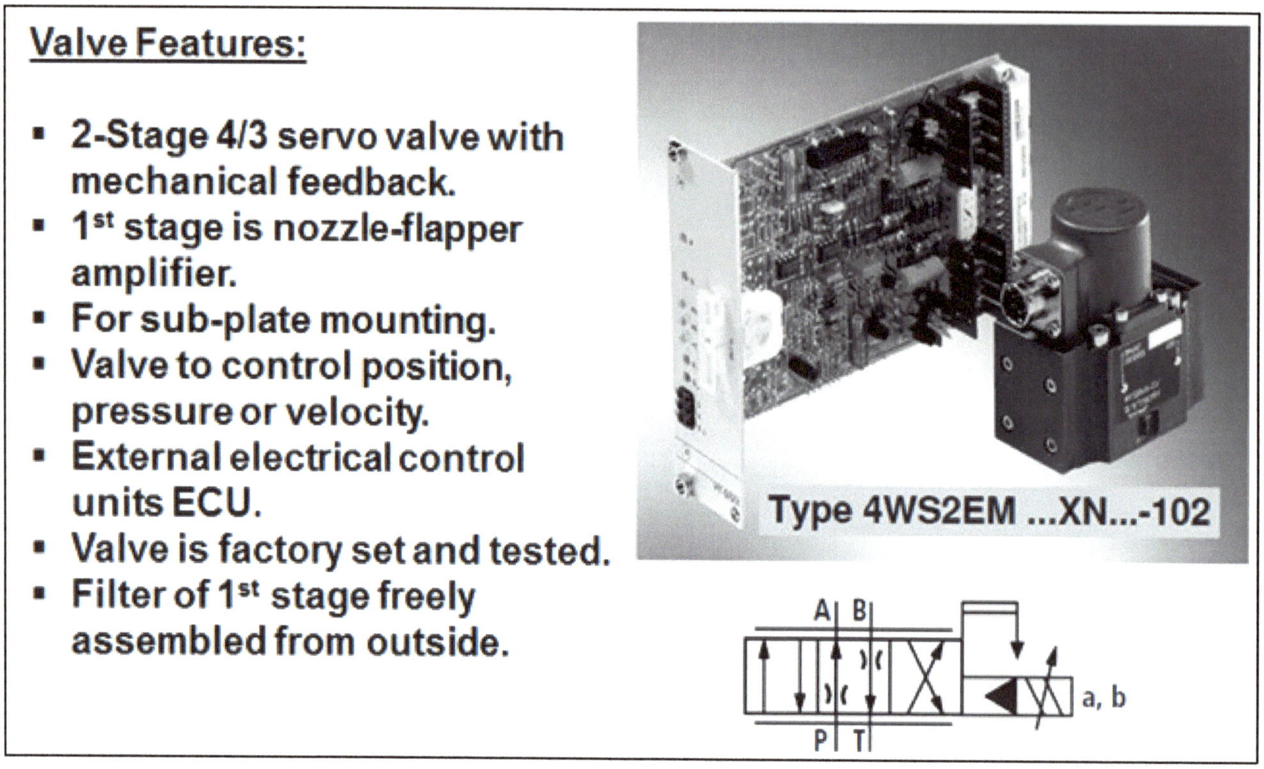

Fig. 6.27A - Example of a 2-Stage Flapper–Nozzle Servo Valve with Mechanical Feedback (Courtesy of Bosch Rexroth)

Figure 6.27B shows the main elements of the previous valve. The two adjusting screws (8) are used to adjust the null position of the valve.

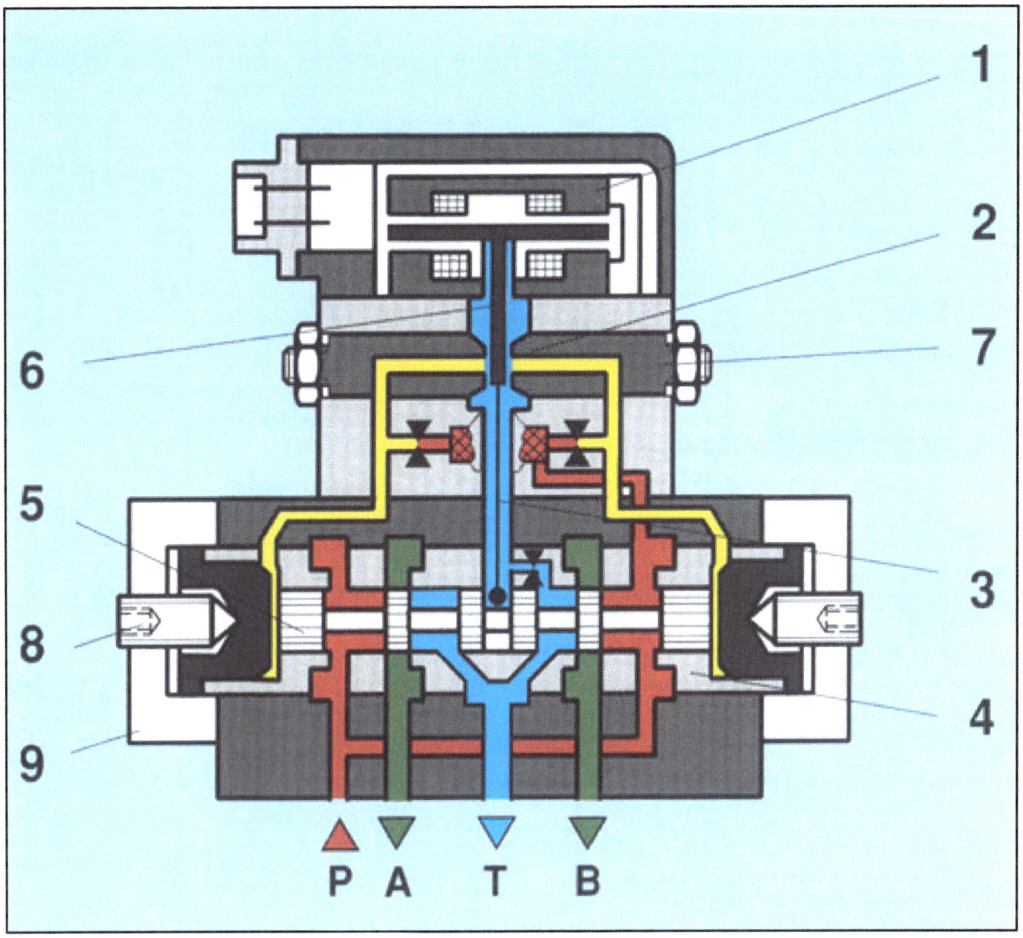

1. **Torque Motor**
2. **Control Chamber**
3. **Feedback Spring**
4. **Interchangeable Control Sleeve**
5. **Control Spool**
6. **Flapper**
7. **Two Nozzles**
8. **Two Adjusting Screws**
9. **Two Covers**

Fig. 6.27B - Detailed Valve Construction (Courtesy of Bosch Rexroth)

Figure 6.27C shows typical hydraulic characteristics of the valve that are published by the valve manufacturer.

hydraulic (measured with HLP32, ϑ_{Oil} = 40 °C ± 5 °C)

Operating pressure	Ports P, A, B	bar	10 … 210 or 10 … 315
Return flow pressure	Port T	bar	Pressure peaks < 100 static < 10
Hydraulic fluid			Mineral oil (HL, HLP) according to DIN 51524 Ignition temperature > 150 °C
Hydraulic fluid temperature range		°C	–15 ... +70; preferably +40 ... +50
Viscosity range		mm²/s	15 ... 380; preferably 30 ... 45
Cleanliness class according to ISO 4406 (c)			Class 18/16/13 [1)]
Rated flows $q_{v\,nom}$, tolerance ±10 % with valve pressure differential Δp = 70 bar		l/min	2; 5; 10; 15; 20; 25

Fig. 6.27C - Valve Hydraulic Specifications (Courtesy of Bosch Rexroth)

Figure 6.27D shows typical electrical characteristics of the valve that are published by the valve manufacturer. The figure shows that the torque motor can be wired in series or in parallel. The figure indicates also the direction of flow versus the polarity of wiring.

electric

Type of signal			Analog
Rated current per coil		mA	30
Resistance per coil		Ω	85
Inductivity with 60 Hz and 100% rated current	Serial connection	H	1.0
	Parallel connection	H	0.25

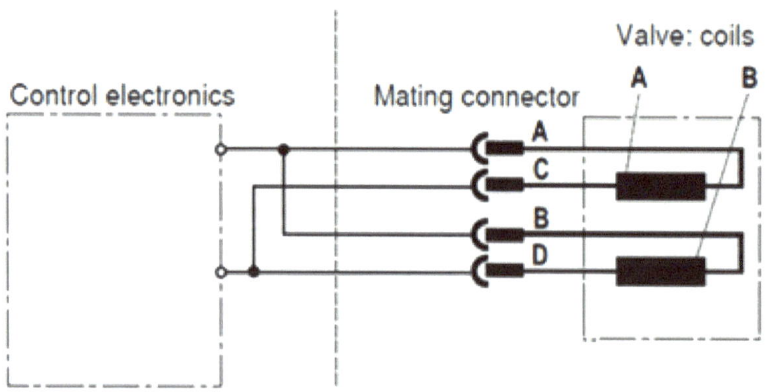

- **Electrical control from A (+) to D (-) provides direction of flow P→A and B → T.**
- **Reverse electrical control provides flow P→ B and A → T.**

Fig. 6.27D - Valve Electrical Specifications (Courtesy of Bosch Rexroth)

Figures 28A through 28C shows another example of the same type valve but from a different manufacturer and is being driven by an OBE unit.

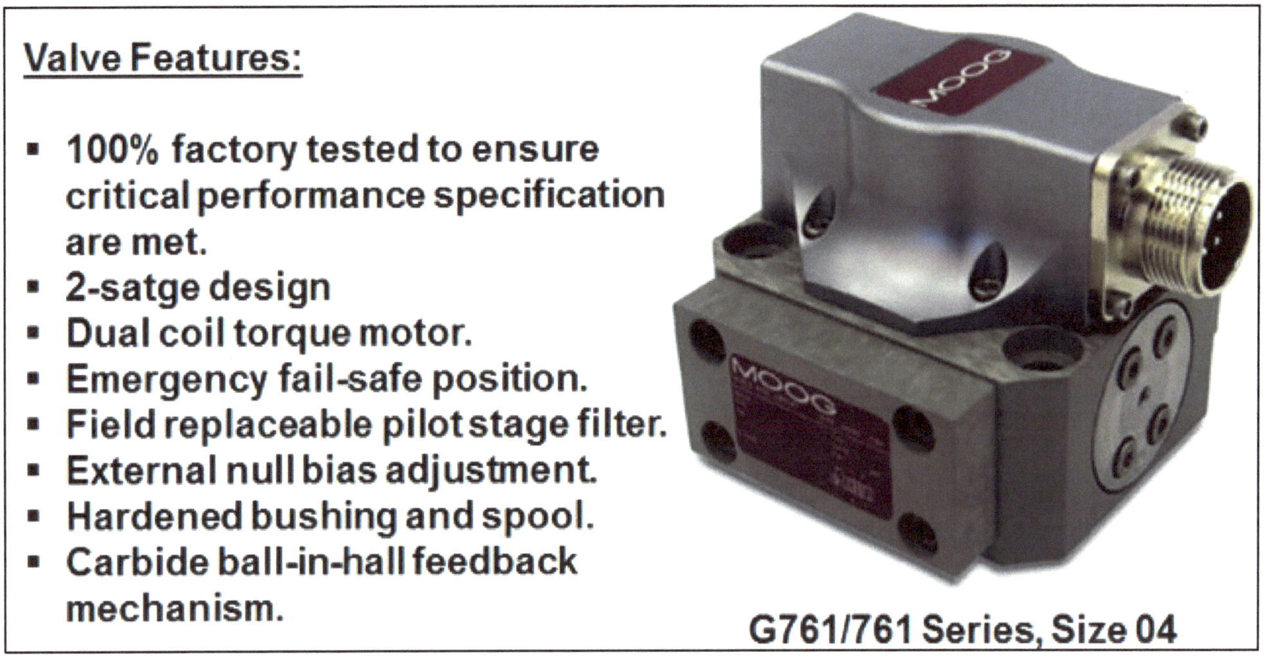

Fig. 6.28A - Example of a 2-Stage Flapper–Nozzle Servo Valve with Mechanical Feedback (Courtesy of Moog)

General Technical Data

Valve design	2-stage, with spool and busing and dry torque motor
Pilot stage	Nozzle Flapper
Mounting pattern	ISO 10372-04-04-0-92
Installation position	Any orientation, fixed or movable
Weight	1.08 kg (2.4 lb)
Storage temperature range	-40 to +60 °C (-40 to +140 °F)
Ambient temperature range	-40 to +135 °C (-40 to +275 °F)
Vibration resistance	30 g, 3 axis, 10 Hz to 2 kHz
Shock resistance	30 g, 3 axis
Seal material	Fluorocarbon (FKM) 85 SHORE D Others upon request

Fig. 6.28B - Valve General Specifications (Courtesy of Moog)

Hydraulic Data	
Maximum operating pressure to ports P, T, A, B, X	315 bar (4,500 psi)
Rated flow at p_N 35 bar/spool land (500 psi/spool land)	4/10/19/38/63 l/min (1/2.5/5/10/16.5 gpm)
Maximum main stage leakage flow rate (zero lap)	2.3 l/min (0.60 gpm)
Null adjust authority	Greater than 10% of rated flow
Hydraulic fluid	Hydraulic oil as per DIN 51524 parts 1 to 3 and ISO 11158. Other fluids on request.
Temperature range	-40 to +60 °C (-40 to +140 °F)
Recommended viscosity range	10 to 97 mm²/s (cSt)
Maximum permissible viscosity range	5 to 1,250 mm²/s (cSt)
Recommended cleanliness class as per ISO 4406	
For functional safety	17/14/11
For longer life	15/13/10
Recommended filter rating	
For functional safety	$\beta_{10} \leq 75$ (10µm absolute)
For longer life	$\beta_5 \leq 75$ (5 µm absolute)

Fig. 6.28C - Valve Hydraulic Specifications (Courtesy of Moog)

6.3.7- Two-Stage Flapper–Nozzle Servo Valve with Electrical Feedback

Figure 6.29 shows a two-stage flapper-nozzle valve with Electrical Feedback. This valve configuration has the equivalent functionality of a stroke-controlled proportional valve, in which the movement of the main spool is proportional to the input command with an electrical feedback on the spool position.

This valve is configured as a system type 1 that is being used with a PD controller to perform spool position control. The wire spring has been removed and replaced by electrical feedback on the spool position.

When the torque motor receives a signal, it drives the flapper towards one of the two nozzles so that the spool moves accordingly. Simultaneously, an inductive position transducer feeds back the spool position to the control electronics.

When the spool reaches the desired position, the torque motor receives zero error signal, the flapper is centered, and the control pressure across the spool is equalized.

Single wiring method is required in order to control the angular position of the flapper with the movement of the spool in a closed loop.

1. **Torque Motor**
2. **Feedback Output**
3. **Interchangeable Control Sleeve**
4. **inductive position transducer**
5. **Core**
6. **Control Spool**
7. **Flapper**
8. **Two Nozzles**
9. **Two Control Chambers**

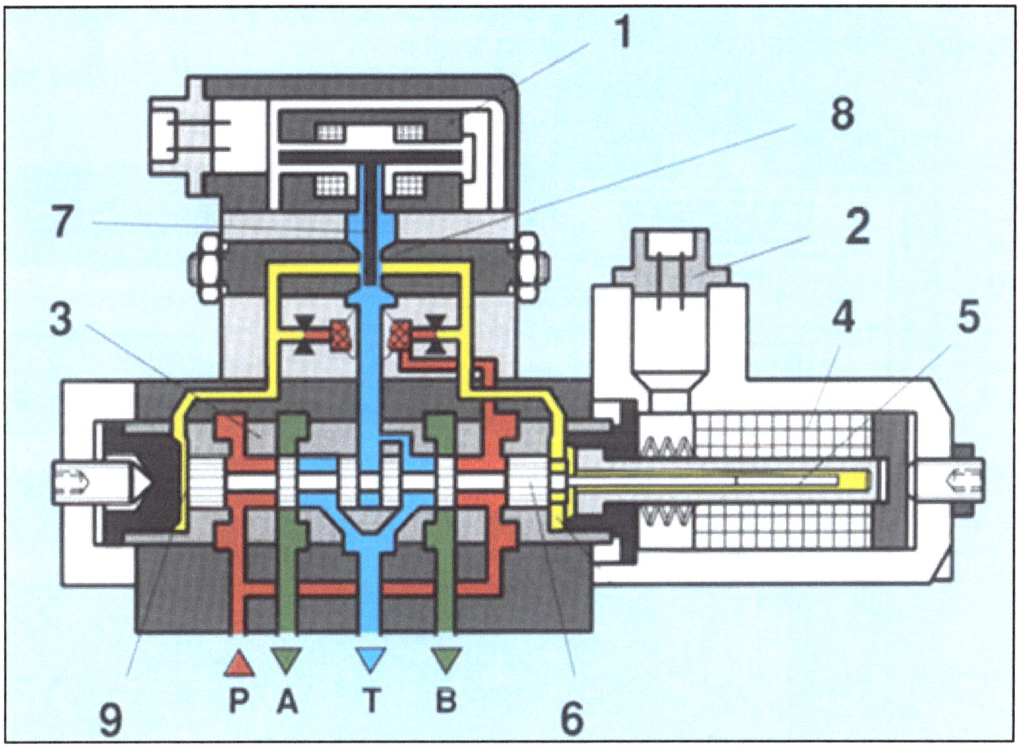

Fig. 6.29 - Example of a 2-Stage Flapper–Nozzle Servo Valve with Electrical Feedback (Courtesy of Bosch Rexroth)

6.3.8- Two-Stage Flapper–Nozzle Servo Valve with Mechanical and Electrical Feedback

For a faster response and a higher level of accuracy, the valve shown in Fig. 6.30A has been configured to contain both mechanical and electrical feedback. Figure 6.30B shows a sectional view of the valve.

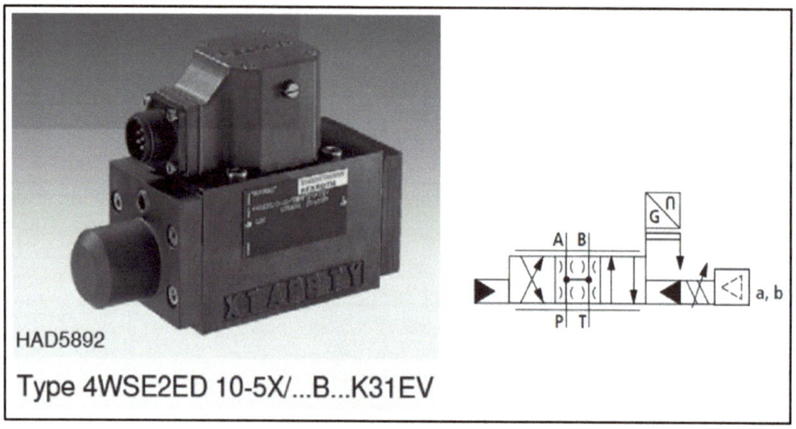

Fig. 6.30A - Example of a 2-Stage Flapper–Nozzle Servo Valve with Mechanical and Electrical Feedback (Courtesy of Bosch Rexroth)

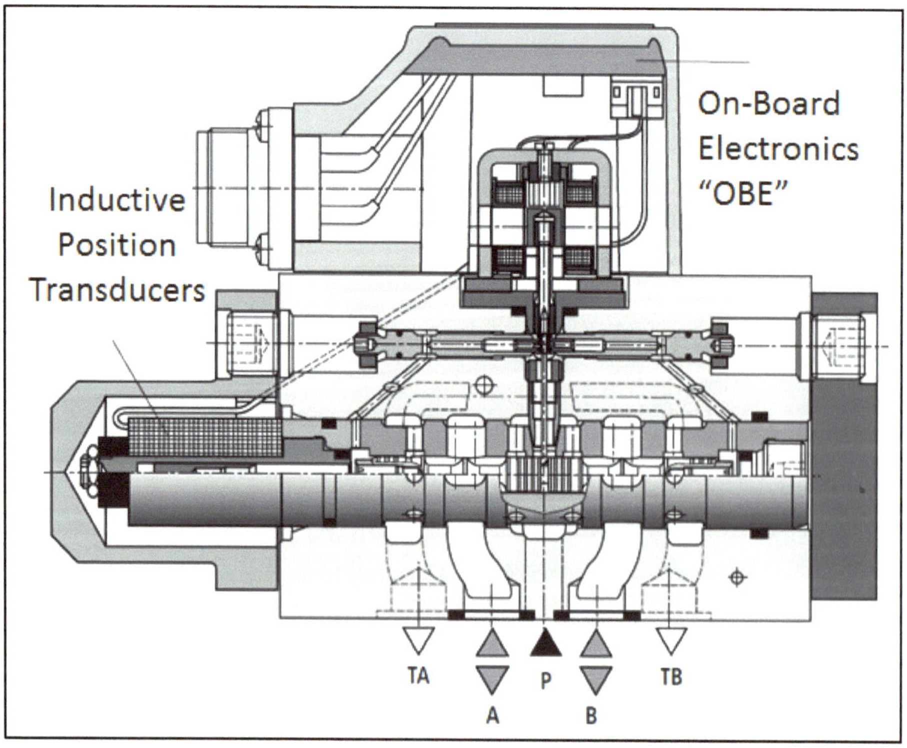

Fig. 6.30B - Valve Sectional View (Courtesy of Bosch Rexroth)

6.3.9- Three-Stage Flapper–Nozzle Servo Valve

Figure 6.31 shows the conceptual construction of a 3-Stage flapper–nozzle servo valve with an electrical feedback on the main stage. In this valve a complete 2-stage flapper-nozzle servo valve with a mechanical feedback is used as a pilot stage to control a large third stage.

1. **Main Spool**
2. **Main Sleeve**
3. **Two Control Chambers**
4. **inductive position transducer**
5. **Core**

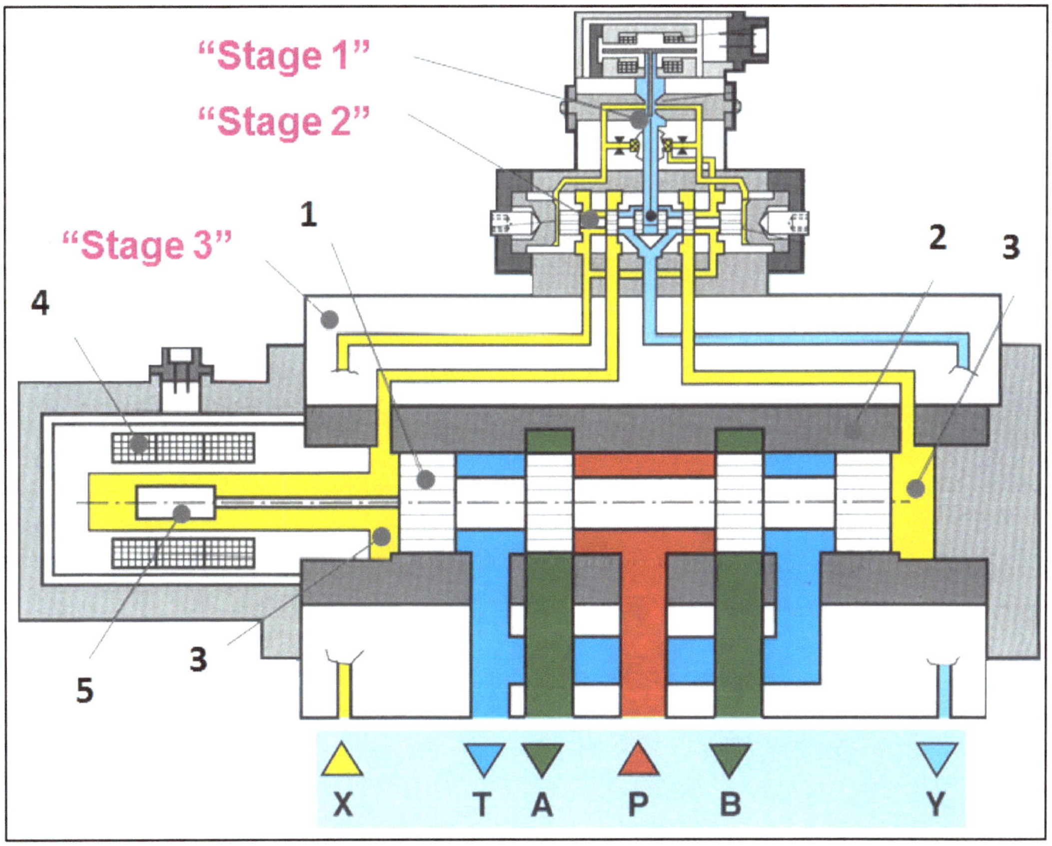

Fig. 6.31 - Conceptual Construction of a 3-Stage Flapper–Nozzle Servo Valve with Electrical Feedback (Courtesy of Bosch Rexroth)

Figure 6.32A shows a typical example of the previous valve and the valve features that are published by the valve manufacturer. The symbol of the valve has two arrows that indicate the two pilot stages.

Valve Features:

- 3-Stage 4/3 servo valve.
- High dynamics 2-stage pilot control valve of size 6
- Main stage with electrical feedback.
- For sub-plate mounting.
- Valve to control position, pressure or velocity.
- High response sensitivity, very low hysteresis and zero point drift
- External electrical control units ECU.
- Valve is factory set and tested.
- Filter of 1st stage freely assembled from outside.
- Internal or external pilot oil supply and return

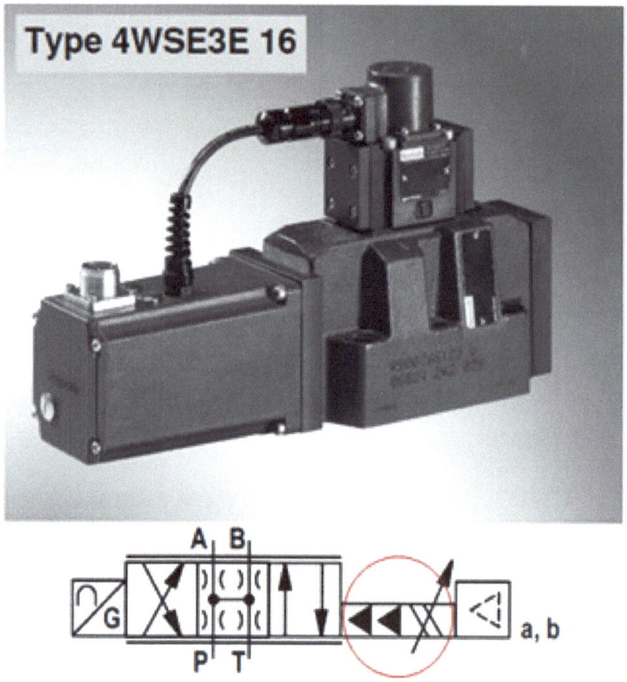

Fig. 6.32A - Example of a 3-Stage Flapper–Nozzle Servo Valve with Electrical Feedback (Courtesy of Bosch Rexroth)

Figure 6.32B shows an outer shape of the valve and some optional infield tuning features of the valve. This high-tech type of valve is usually tuned and maintained by experienced and sometimes certified technicians. As shown in the figure, if needed, special instructions are given by the valve manufacturer.

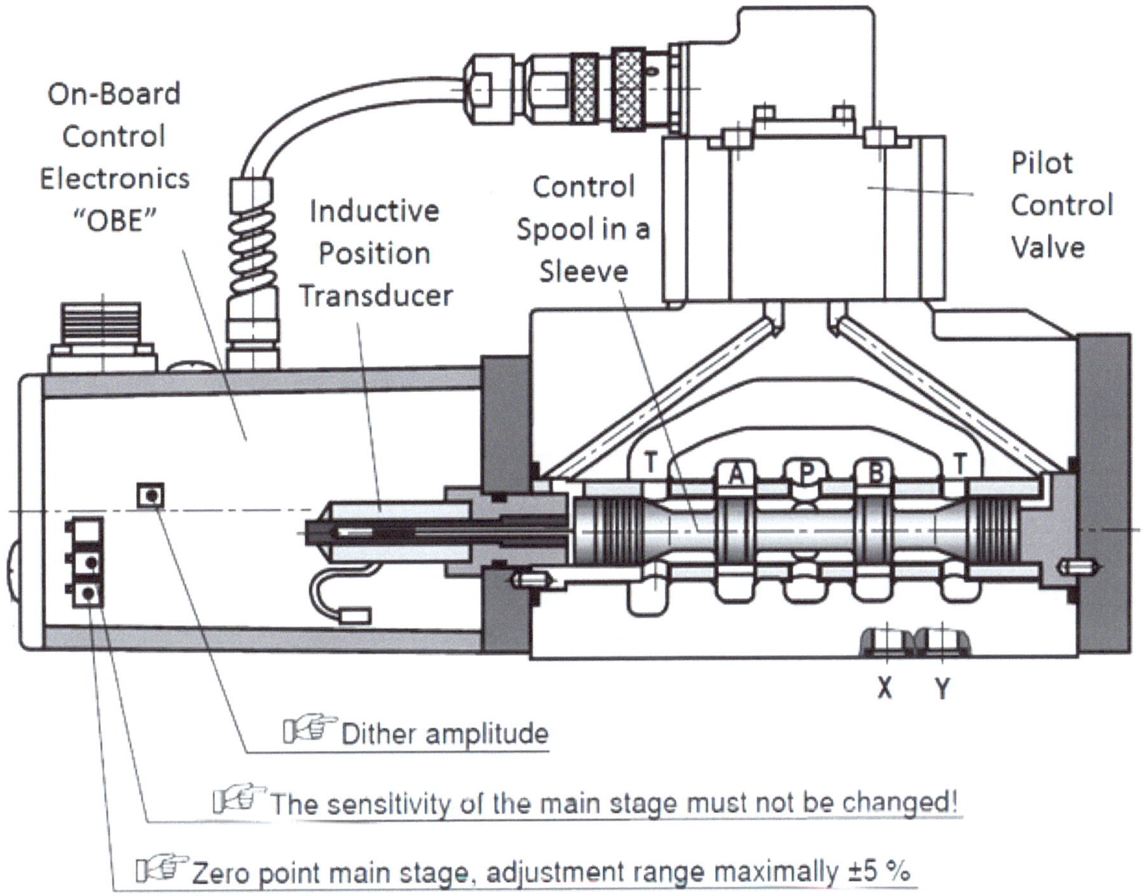

Notice!

Changes in the zero point and/or the dither amplitude may result in damage to the system and may only be implemented by instructed specialists.

The pilot control valve may only be maintained by Bosch Rexroth employees. An exception to this is the replacement of the filter element – see data sheet 29564.

Fig. 6.32B - Valve Sectional View (Courtesy of Bosch Rexroth)

In a further upgrade, the valve shown in Fig. 6.33A is the same as the previous valve except that the pilot valve has both mechanical and electrical feedback. Figure 6.33B shows a sectional view of the valve.

Valve Features:

- Electrical position feedback with pressure isolated position transducer, no wear.
- Integrated electronics with false polarity protection.
- Optional external pilot supply and return connections via fifth and sixth port in valve body.
- Low hysteresis.
- Excellent null stability.
- Pre-adjusted at factory.

Servo Valves with OBE D791 and D792 Series

Fig. 6.33A - Example of a 3-Stage Flapper–Nozzle Servo Valve with Electrical Feedback (Courtesy of Moog)

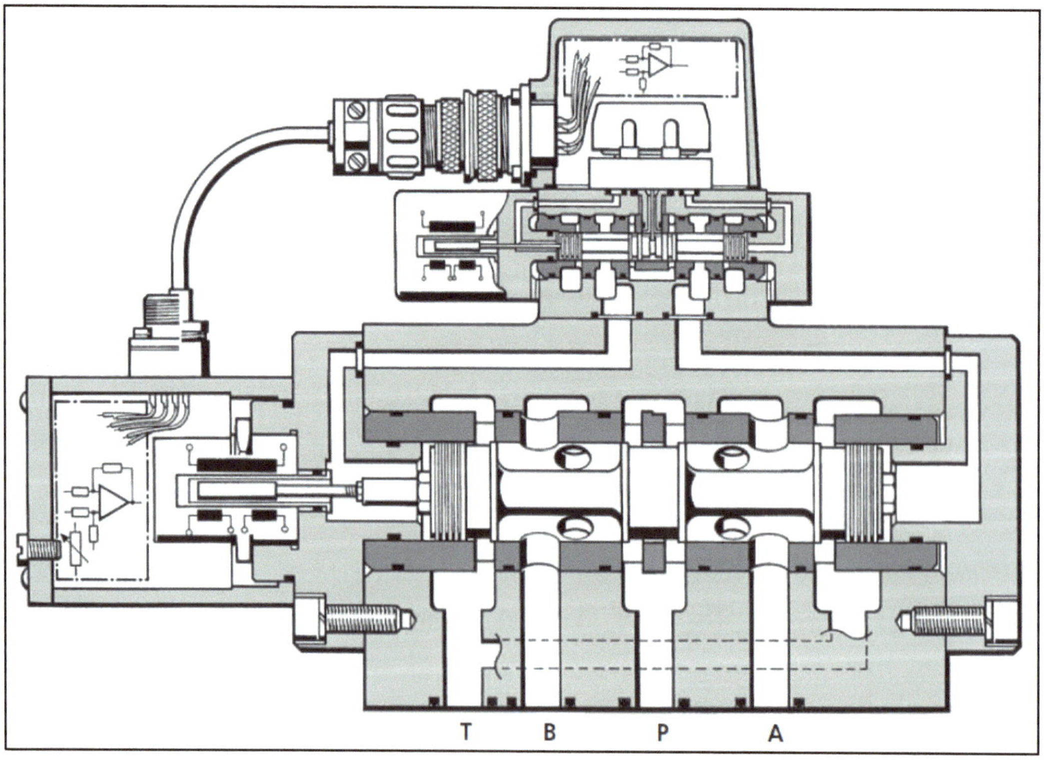

Fig. 6.33B - Valve Sectional View (Courtesy of Moog)

6.3.10- Jet Pipe Concept of Operation

Figure 6.34 explains the concept of *Jet Pipe* servo valves. Constant supply pressure $\boldsymbol{p_s}$ is fed into a movable jet pipe against two adjacent receptacles (receiving) holes. As the jet pipe moves angularity, it goes near to one of the holes and simultaneously away from the other. As a result, a differential pressure $\boldsymbol{\Delta p}$ is generated between the two holes. The magnitude of the differential pressure depends on the supply pressure and the pipe angular displacement. The direction of the differential pressure depends on the direction of the jet pipe angular movement.

In order to achieve such controllability, the clearance $\boldsymbol{X_{jn}}$ between the plane of the jet pipe and the plane of the receiving holes must be at least twice as the jet pipe diameter $\boldsymbol{d_j}$.

Servo valves with jet pipe stages are referred as "Servo Jets".

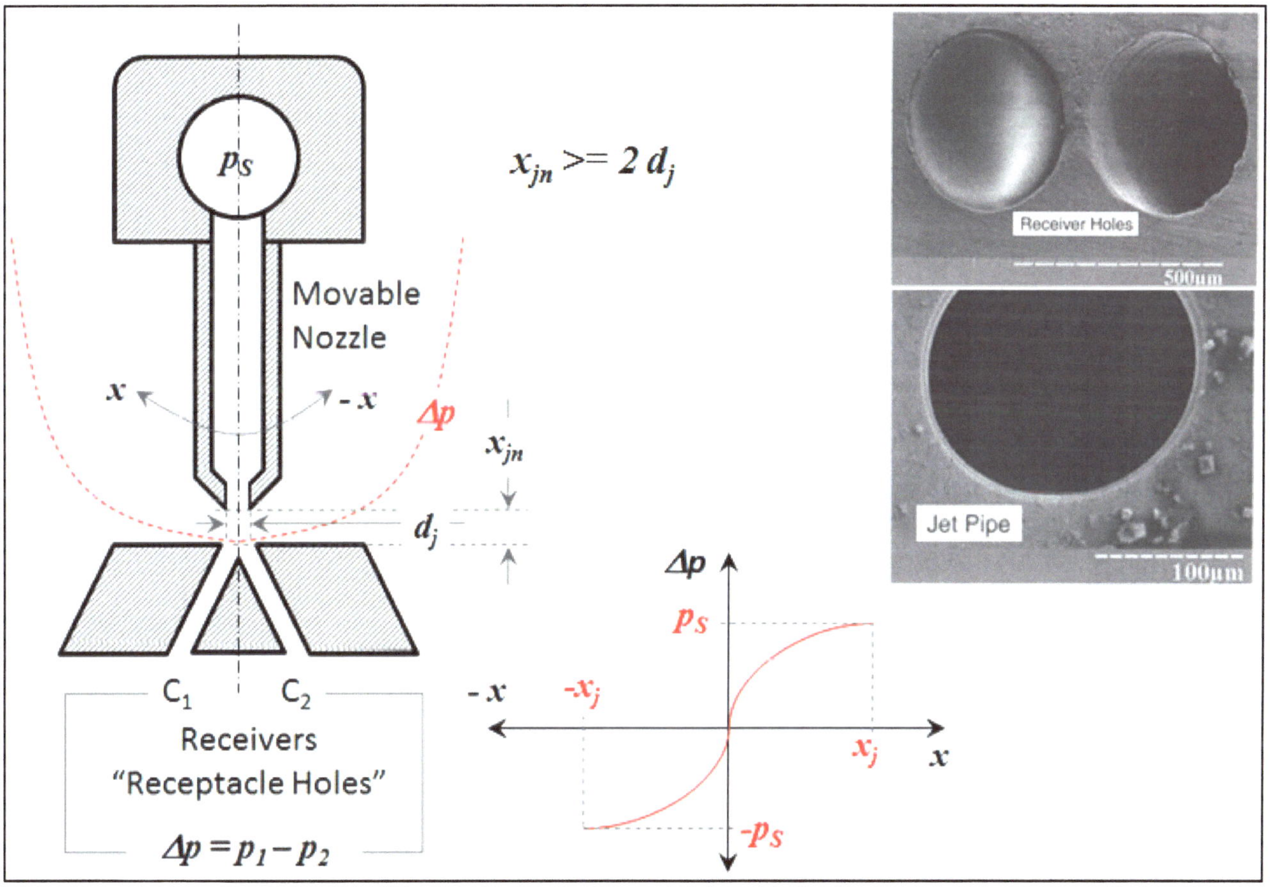

Fig. 6.34 - Jet-Pipe Concept

6.3.11- Two-Stage Jet Pipe Servo Valve with Mechanical Feedback

Construction: Figure 6.35 shows the conceptual construction of a two-stage Jet pipe servo valve with mechanical feedback. The figure shows that this valve has the same construction as the flapper-nozzle one except using a jet pipe stage rather than a flapper nozzle stage. A feedback wire spring connects the jet pipe to the main spool. The pilot pressure feeds the jet pipe through a last chance filter to remove any possible contamination that may block the jet pipe.

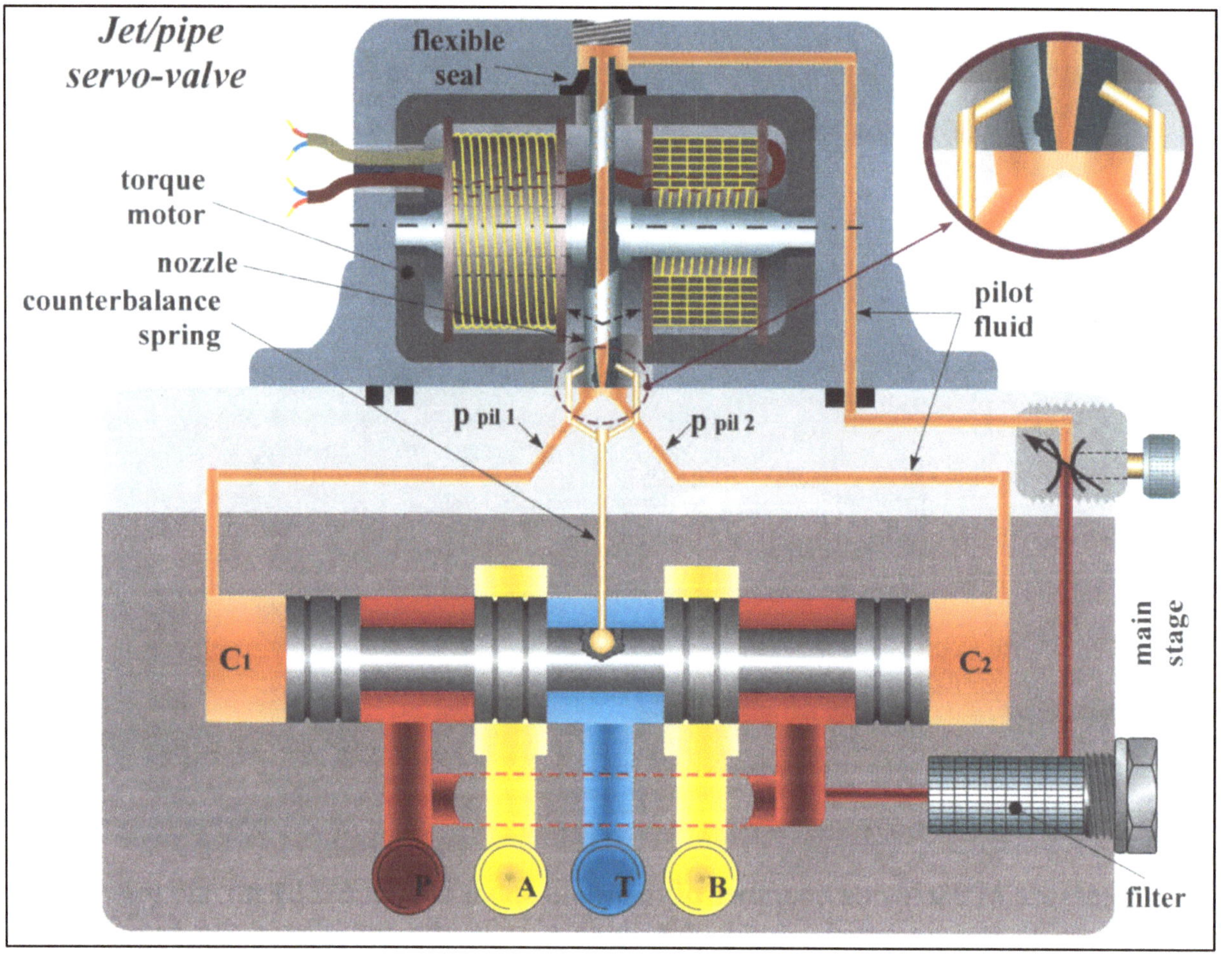

Fig. 6.35 - Conceptual Construction of a 2-Stage Jet-Pipe Servo Valve (Courtesy of ASSOFLUID)

Operation: Figure 6.36A shows the valve at rest where the torque motor is de-energized. As shown in Fig. 6.36B, when the torque motor is energized, the jet pipe moves accordingly, a differential pressure is created across the main spool, and the spool is shifted pulling back the jet pipe via the feedback wire spring.

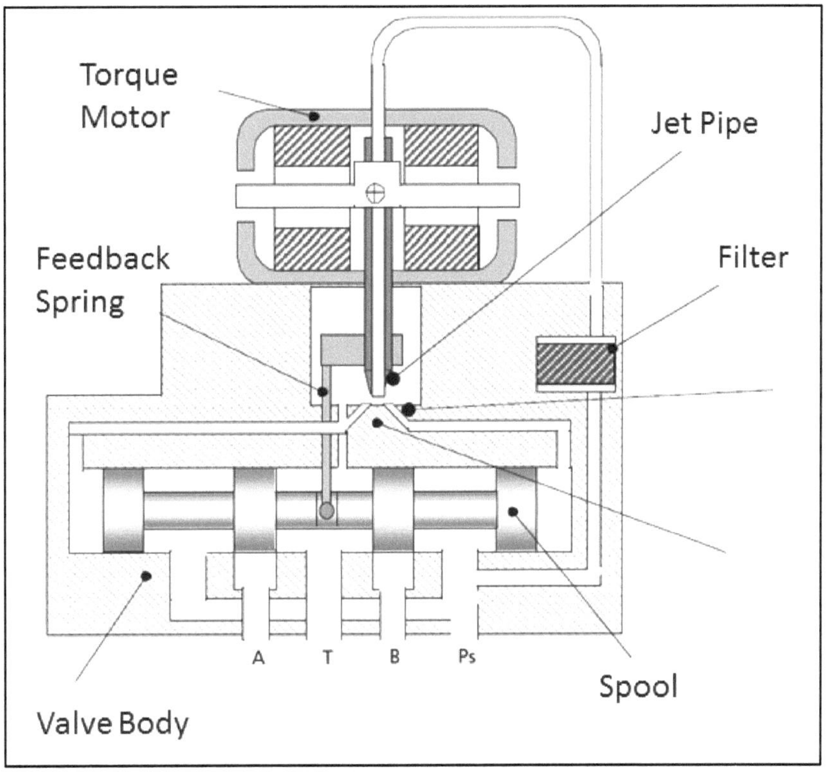

Fig. 6.36A - Valve at Rest (Courtesy of Moog)

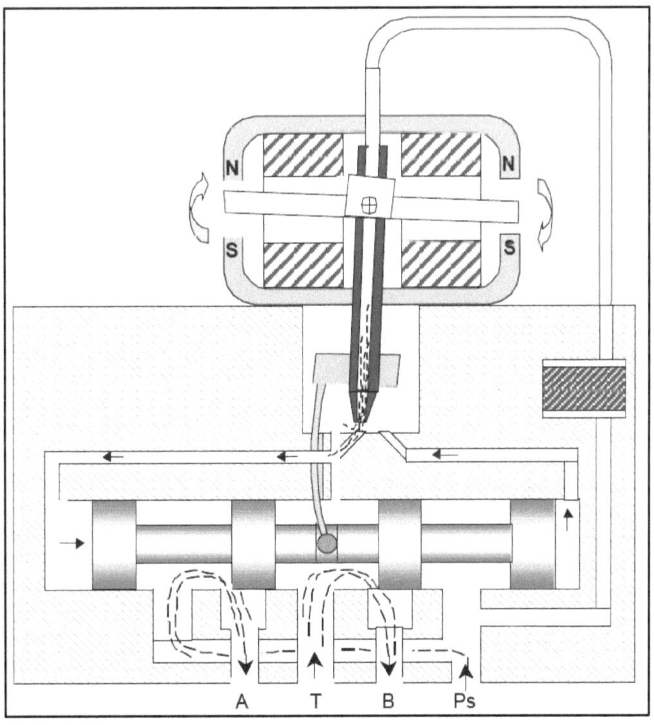

Fig. 6.36B- Torque Motor Receives Signal (Courtesy of Moog)

Figure 6.36C shows when the jet pipe gets back to the central position between the two receptacle holes, the spool stops moving. So the valve is considered a mechanically stroke-controlled valve.

As shown in the figure, in the new balanced position of the spool, the main pressure connects to one of the ports while the other port is vented simultaneously. The valve is designed so that the spool moves in a direction depending on the sign of the input signal; and with a distance that is proportional to the magnitude of the input signal. For a linear valve, the percentage of the spool movement equals the percentage of the input signal.

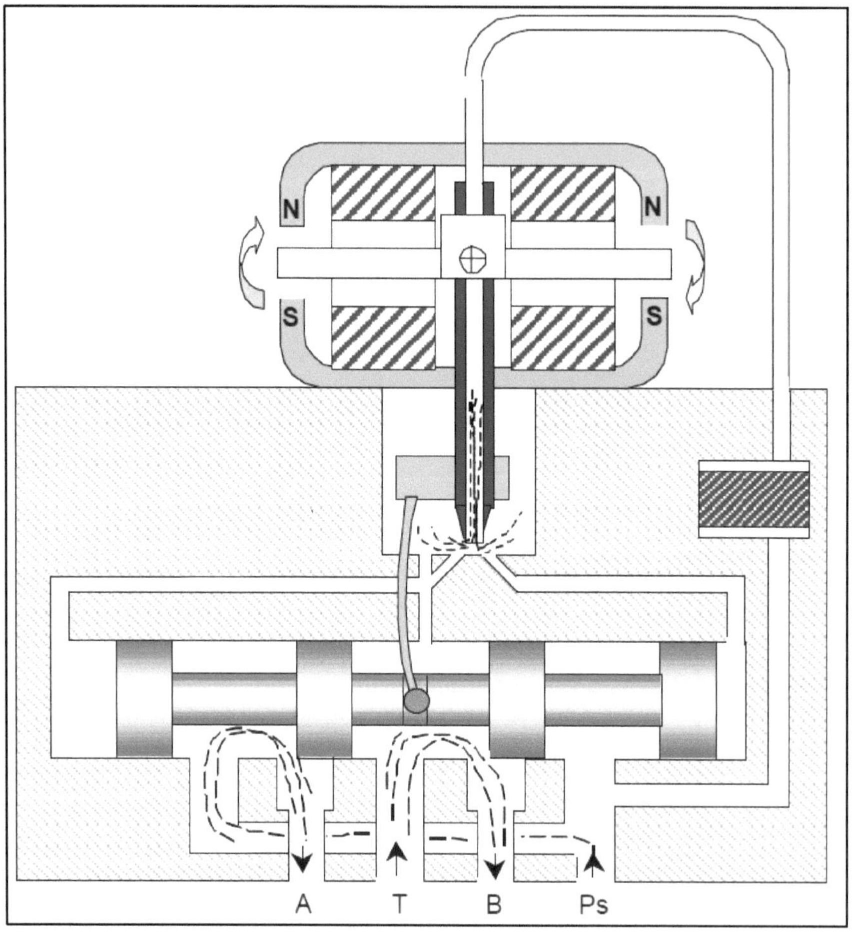

Fig. 6.36C- Spool Balanced (Courtesy of Moog)

Figure 6.37 introduces an example of a 2-Stage Jet Pipe servo valve with mechanical feedback. The figure shows also some of the technical features published by the valve manufacturer.

Specifications	
Hydraulic Data	
Rated Flow	0 to 225 l/min (0 to 60 gpm) @ Δp 70 bar (1000 psi)
Maximum Operating Pressure	350 bar (5,000 psi)
Mounting Pattern	ISO 10372
100% Step Response	4 to 32 ms @ 210 bar (3,000 psi)
Electrical Data	
Supply Voltage	N/A
Rated Signal Options	+/-10, +/-20 and +/-50 mA

Fig. 6.37-Example of a 2-Stage Jet-Pipe Servo Valve with Mechanical Feedback (Courtesy of Moog)

6.3.12- Two-Stage Jet-Pipe Servo Valve with Electrical Feedback

In this valve, as shown in Fig. 6.38, the feedback wire spring is removed and replaced by an electrical feedback configuration. The valve is then considered as an electrically stroke-controlled valve.

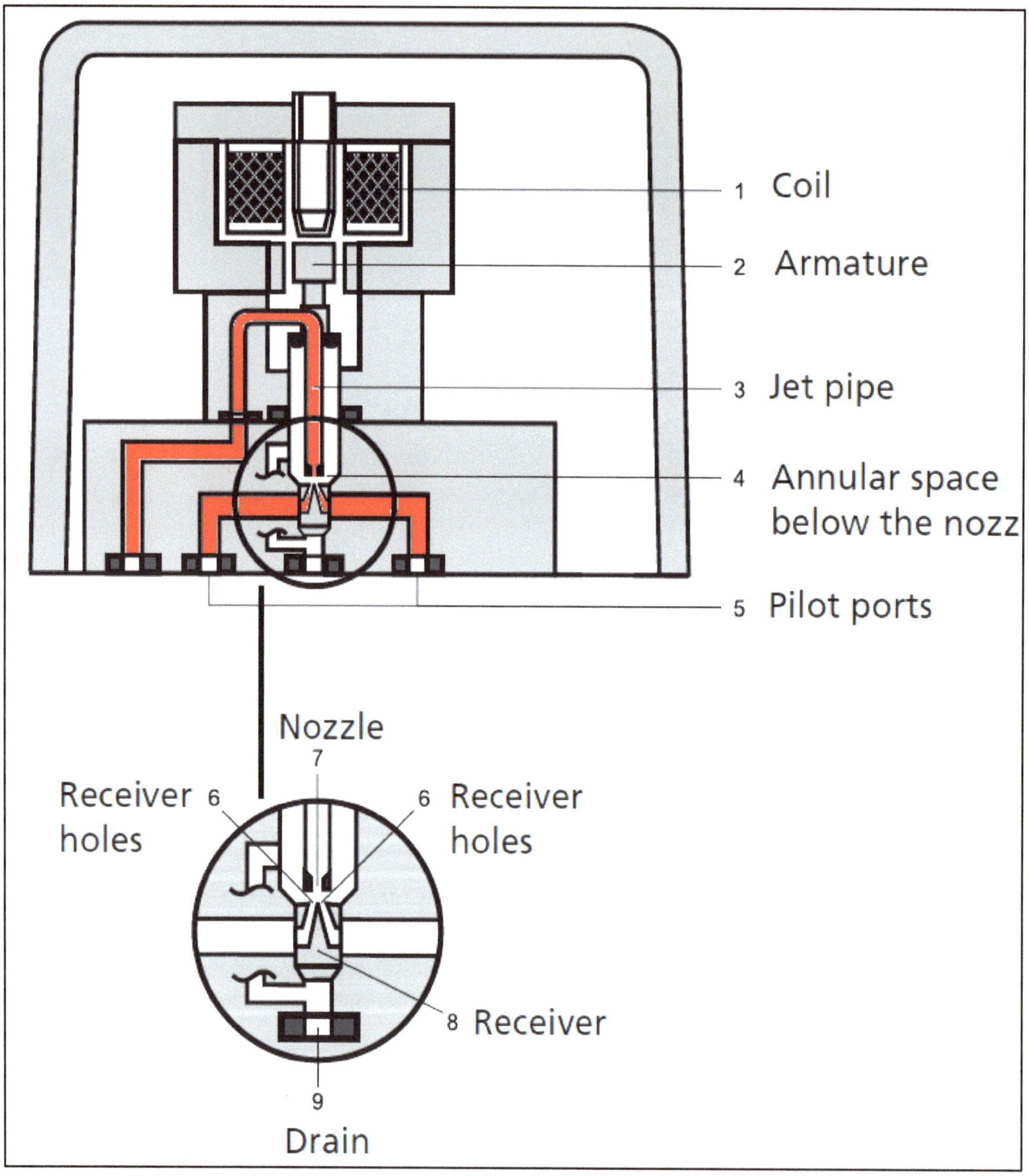

Fig. 6.38- Jet-Pipe Stage for Servo Valve with Electrical Feedback (Courtesy of Moog)

Figure 6.39A introduces an example of the previous valve with a sectional view shown in Fig. 6.39B.

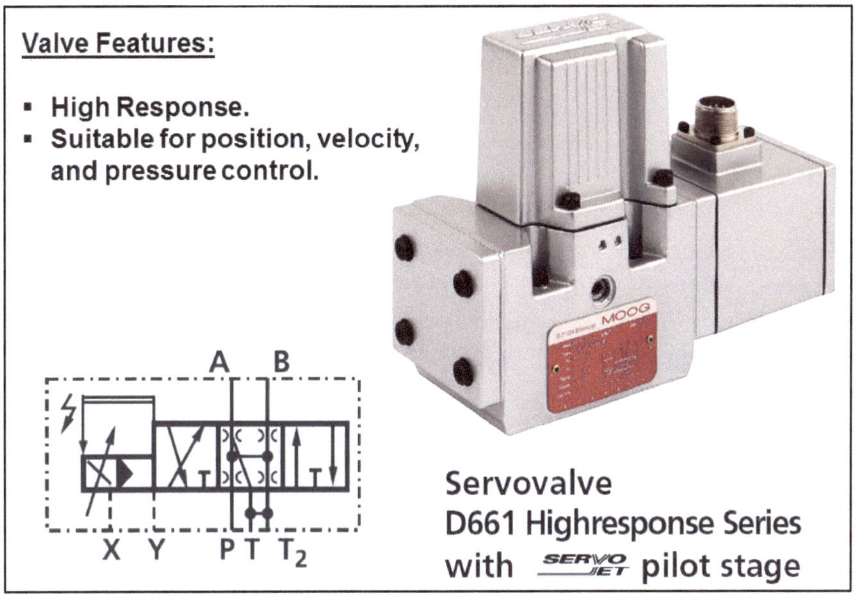

Fig. 6.39A-Example of a 2-Stage Jet-Pipe Servo Valve with Electrical Feedback (Courtesy of Moog)

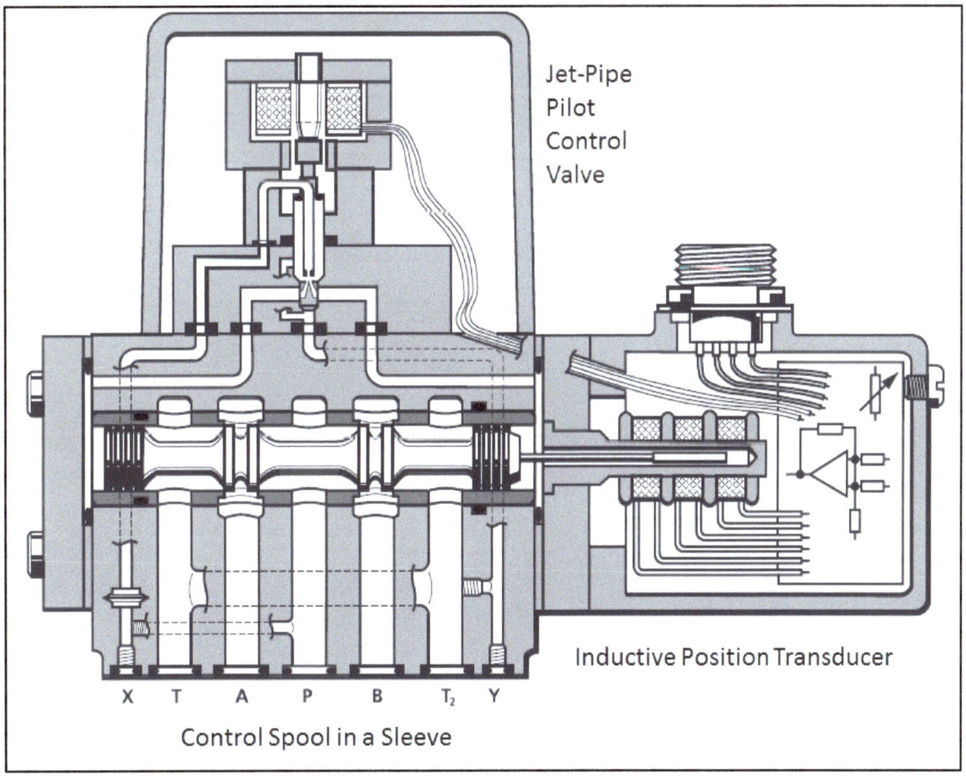

Fig. 6.39B-Valve Sectional View (Courtesy of Moog)

6.3.13- Jet Pipe versus Flapper-Nozzle Pilot Stage

As shown in Fig. 6.40, a jet-pipe servo valve has the following advantages:

- Relatively larger clearances so that it is less sensitive to contamination.
- Fails passively or safely as compared to the flapper nozzle. To explain that, assume a piece of contamination blocked the jet pipe, none of the receiving holes will receive any power. Therefore, the main spool stays in neutral position and the load as well. On the other hand, if one of the nozzles is blocked by contamination, this is equivalent to a situation where the torque motor is energized with maximum signal. In such a case the spool will suddenly shifted to full stroke and the load will suddenly receive maximum hydraulic power.
- The erosion in a jet pipe stage happens almost in a symmetric way between the two receiving holes. That distributes the deviation in the valve characteristics evenly. In a flapper-nozzle, if the erosion in one side of the flapper is different from the other side by 30 microns, this will result in a 50% change in the valve flow gain and doubles the valve leakage.

A jet-pipe servo valve has the following disadvantages:

- Difficult null adjustments because of the complex fluid flow characteristics.
- Requires relatively larger power consumption to move the jet pipe.
- Has relatively slower dynamics because moving jet pipe that contains fluid.

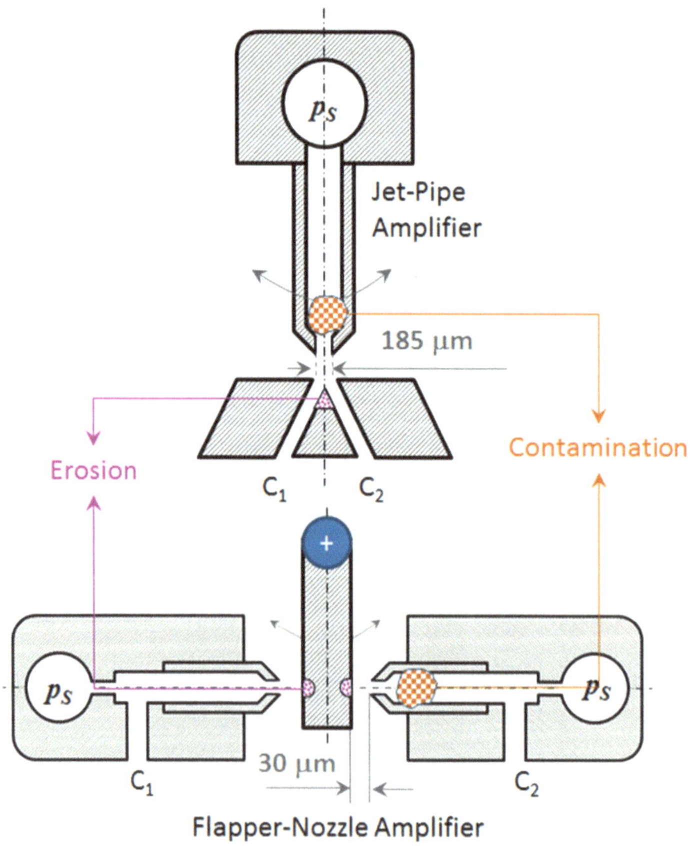

Fig. 6.40- Jet-Pipe versus Flapper-Nozzle Pilot Stage

6.3.14- Two-Stage Jet-Deflector Servo Valve with Mechanical Feedback

Figure 6.41A shows another configuration named *Jet Deflector* that is mainly developed to overcome the shortcoming of the conventional jet pipe configuration. As shown in the figure, a jet deflector is used to transmit the pilot pressure between a fixed jet pipe and the receiving holes. Obviously moving the jet deflector is easier than the jet pipe, thus improving the null adjustment, valve dynamics, and power consumption.

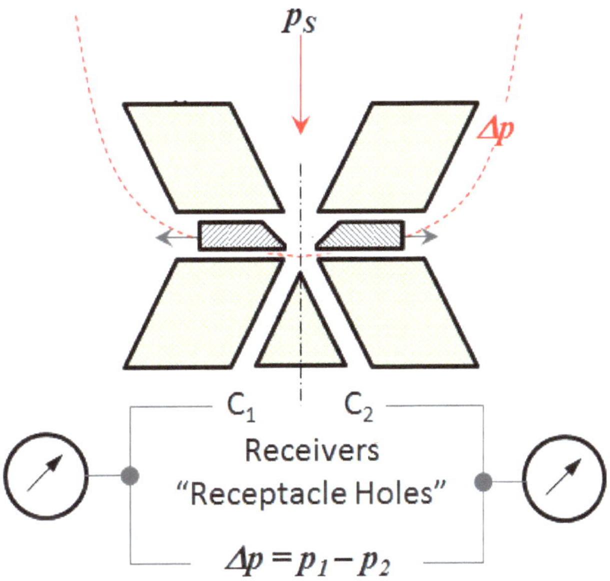

Fig. 6.41B- Conceptual Construction of a 2-Stage Jet-Deflector Servo Valve with Mechanical Feedback (Excerption of FD Norvelle)

Fig. 6.41B shows the concept and the construction of a 2-stage jet deflector servo valve with mechanical feedback. The figure shows that the torque motor moves the jet deflector while it is connected to the spool by the feedback wire spring.

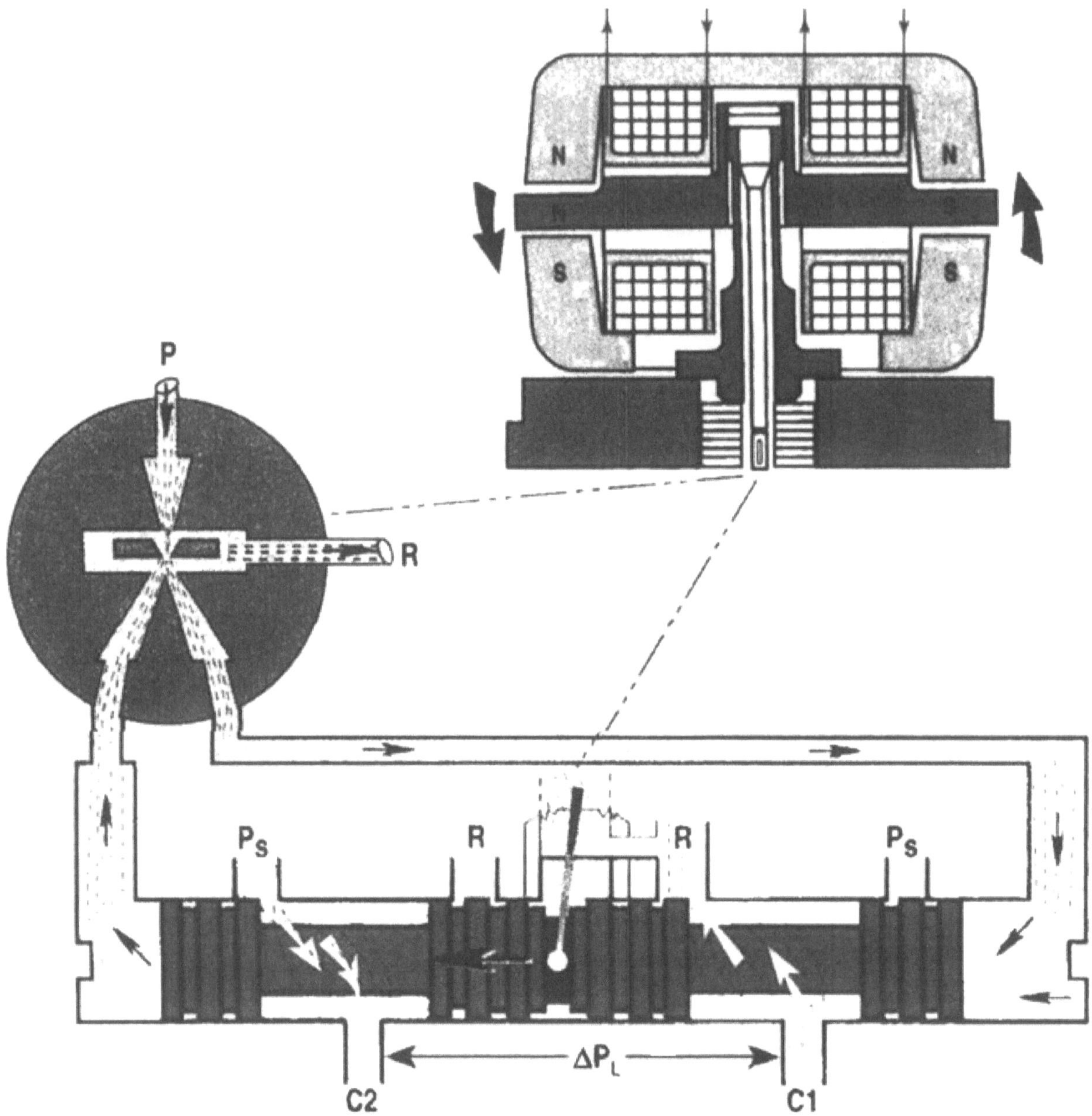

Fig. 6.41B- Conceptual Construction of a 2-Stage Jet-Deflector Servo Valve with Mechanical Feedback (Excerption of FD Norvelle)

Chapter 7
Electro-Hydraulic Valve Selection Criteria

Objectives:

This chapter presents the various criteria to select a specific valve for an application. These criteria include, valve type, spool design, operating conditions, static and dynamic characteristics. The chapter also provides examples of the current valves that are produced by existing manufacturers.

Brief Contents:

Chapter 7 – Electro-Hydraulic Valve Selection Criteria

7.1- Importance of Proper Selection of an EH Valve

Proper valve selection is the most critical action item in designing an electro-hydraulic closed-loop control system, based on which the entire system performance is determined.

In critical applications, as shown in Fig. 7.1, electro-hydraulic valves should have high accuracy and high speed response.

Proper valve selection will result in:

- Optimized closed-loop overall gain to maintain a stable and responsive system.
- Better control resolution.
- Reduced energy loss and heat generation in the system.

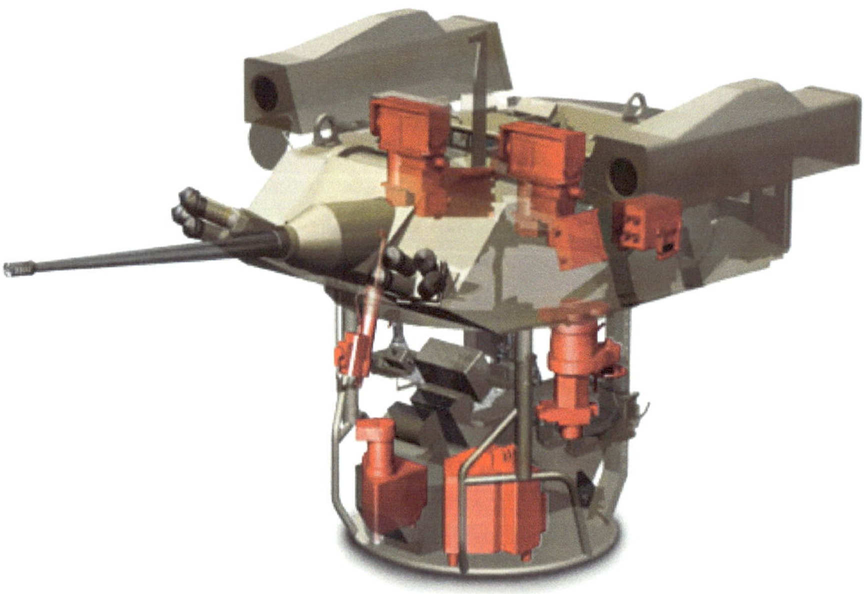

Fig. 7.1- Ammunition Handling and Control Systems

7.2- Electro-Hydraulic Valve Selection Criteria

In order to select an EH valve for an application, there are many specifications that should be reviewed to ensure suitability of the valve for the application. Figure 7.2 shows a detailed chart for EH valve specifications based on which a valve is selected.

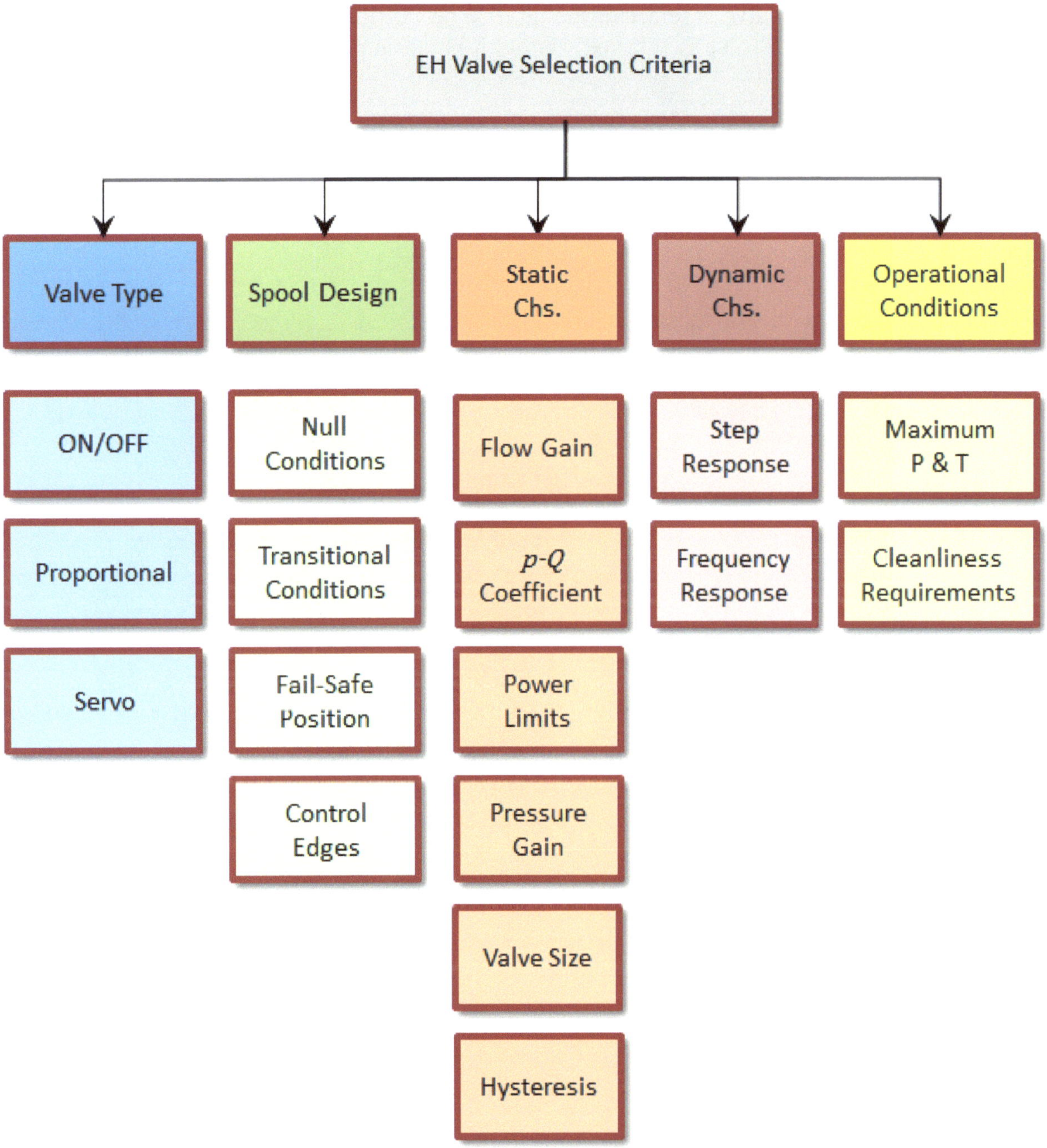

Fig. 7.2- Electro-Hydraulic Valve Selection Criteria

7.3-Valve Type

As shown in Fig. 7.3, the first set of questions to be answered are as follows:

- Which valve type should be selected?" Or should the valve be of an ON/OFF type or continuous type?
- If a continuous valve is selected, should it be proportional or servo?
- If a proportional valve is selected, should it be basic proportional valve or high performance one?

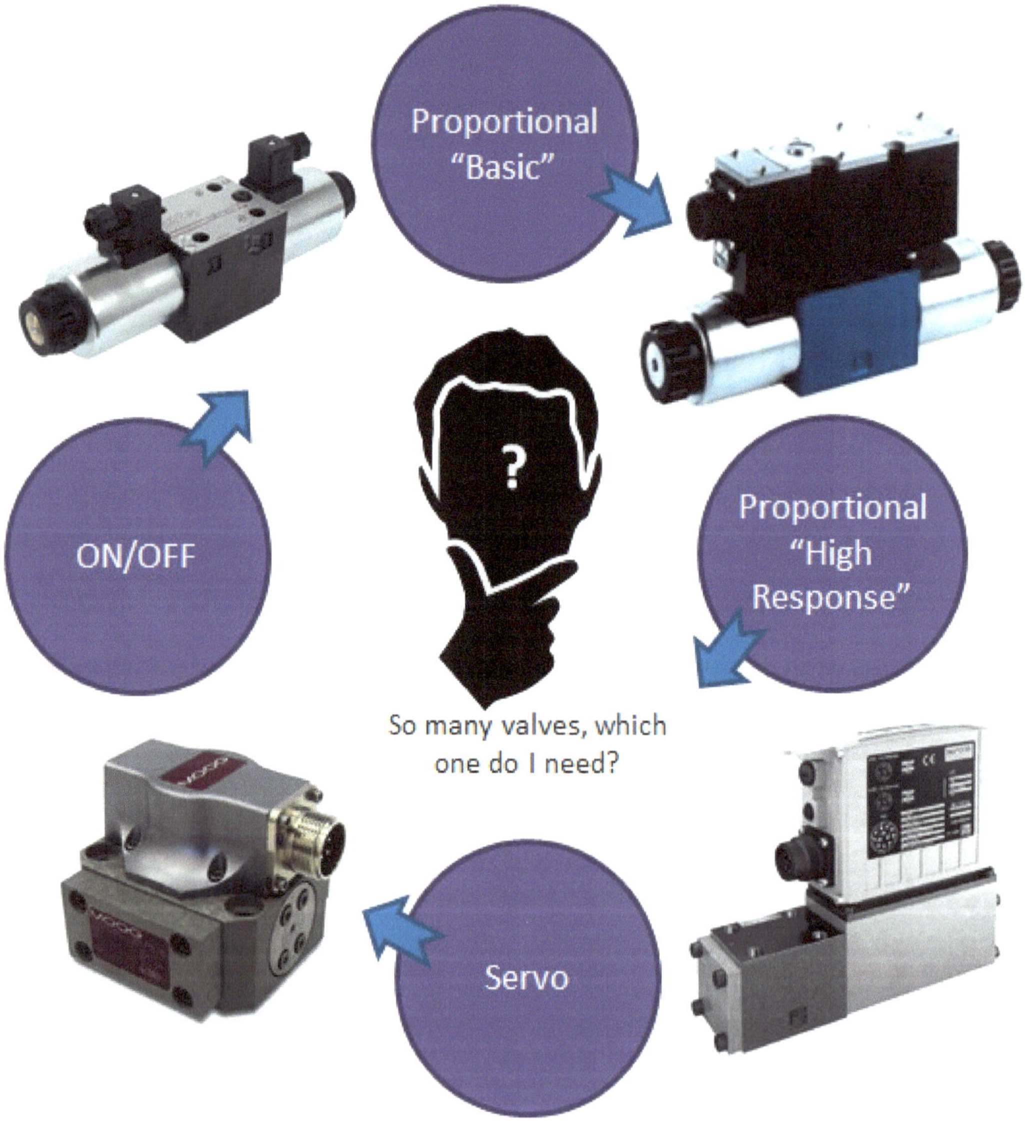

Fig. 7.3- Valve Type Selection

As shown in Fig. 7.4, the answer to the above questions depends on the control mode and the accuracy level of the application.

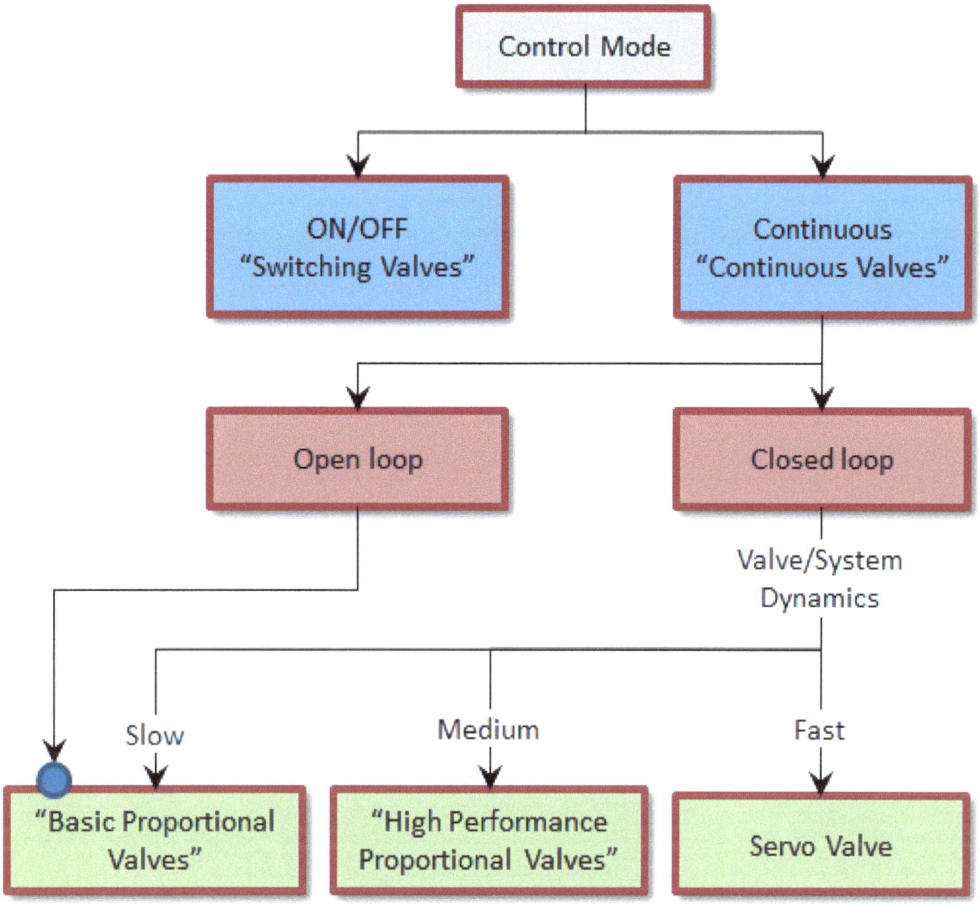

Fig. 7.4- Qualitative Methodology for Selecting an EH Valve

ON/OFF Valves: For applications that require a slow change in the controlled parameter, an ON/OFF control is more cost effective and convenient. Figure 7.5 shows an example of a cooling fan speed control using an ON/OFF valve.

Basic Proportional Valves: Proportional valves were known for long time as devices that are good only for open loop because of its limited dynamic characteristics. Nowadays the gap between proportional and servo valves has narrowed because of the high-performance proportional valves characteristics.

High Performance Proportional Valves: Such valves are used for closed-loop control systems as well as servo valves.

Servo Valves: Technically servo valves can be used to perform low dynamics control applications, but there is no sense to do that if the job can be done using a less expensive valve. Servo valves should only be used when their high capabilities are required.

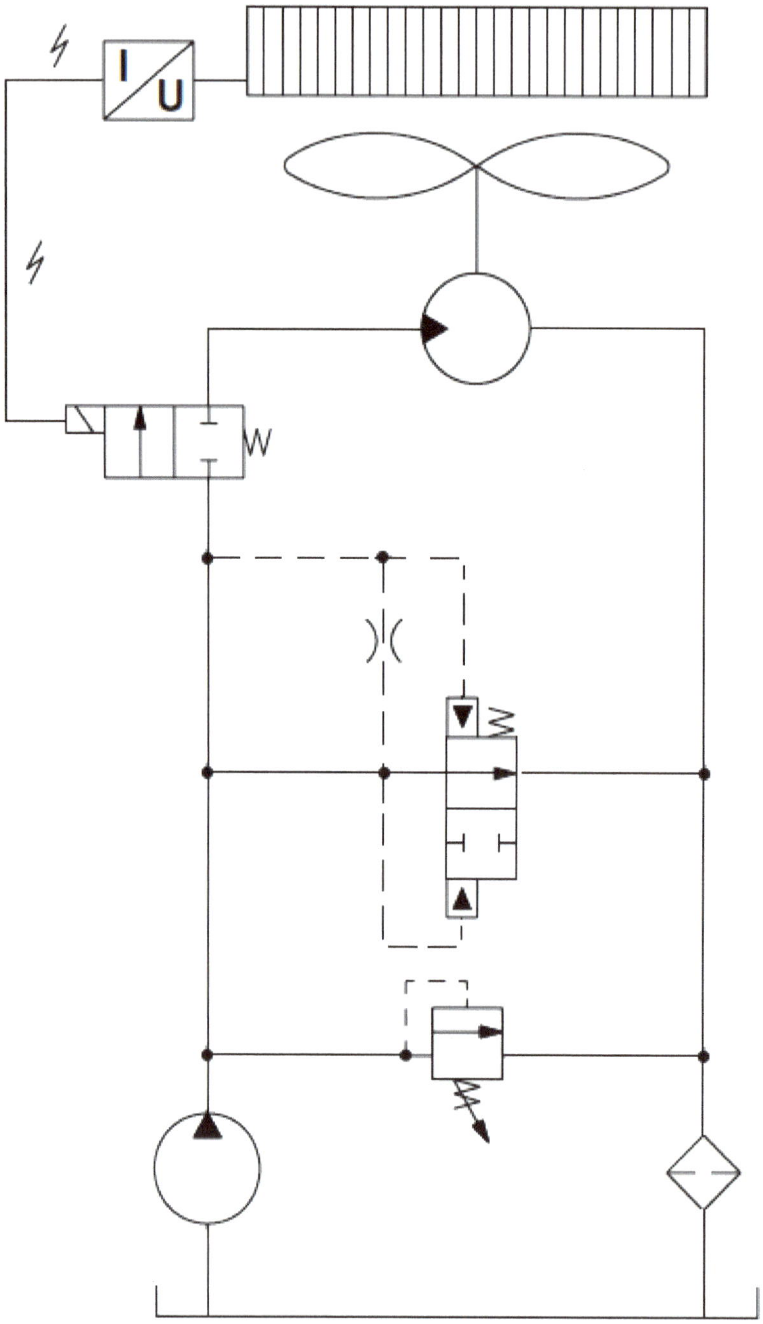

Fig. 7.5- ON/OFF Cooling Fan Speed Control

Table 7.1 shows qualitative guidelines to select EH valves based on the controlled parameter in a closed-loop control system. The cells in the table that are marked with green buttons indicate the suitability of the corresponding valve for the assigned operation.

The other cells that are marked with yellow buttons indicate that the valve can technically do the job but there are some concerns about the suitability of the valve.

The concern in the first yellow cell is that the accuracy level of basic proportional valves may not be suitable to perform position control with good resolution. Additionally, the majority of basic proportional valves have overlapped spools that result in steady state error in position control. Such error requires careful controller gain adjustment and/or enabling dead band elimination function.

The concerns on the second and third yellow cells are that it is not recommended to use a servo valve for unidirectional operation. For example, if a servo valve is used to control the speed of a unidirectional hydraulic motor, the spool of the valve will be used only in one direction and that accelerates valve failure.

The concern on the fourth yellow cell is that controlling pressure requires an underlapped valve with very high sensitivity. Controlling pressure is performed by moving an underlapped valve within ±5% of the spool stroke. Therefore, basic proportional valves are not suitable for such systems unless their specifications meet the requirements.

	Basic Proportional Valves	High Response Proportional Valves	Servo Valves
Position	1 🟡	🟢	🟢
Flow/Speed (one direction)	🟢	🟢	2 🟡
Flow/Speed (two directions)	🟢	🟢	🟢
Pressure/Load (one direction)	🟢	🟢	3 🟡
Pressure/Load (two directions)	4 🟡	🟢	🟢

Table 7.1- Valve Selection Based on the Controlled Parameter in a Closed Loop

Some manufacturers provide guideline charts to promote selecting their valve brands. An example of this is shown in Table 7.2.

The group of valves marked (1) are either excluded or yielded from selection because they can't technically do the job. The group of valves marked (2) are either excluded or yielded from selection because there are other cost-effective valves that can do the job.

Matrix of proportional directional valves

	Valve model	Nominal flow (l/min)	Nominal Δp (bar)	Data sheet RE..	Overlap compensation (with E, W spool)	Valve dynamics (natural frequency) [1]	Typical application [2] Open control loop	Closed-loop position control With low precision	Closed-loop position control With high precision	Closed-loop pressure control
Direct operated	4WRA(E)	Size 6: 7, 15, 30 Size 10: 30, 60	10	29055	Yes	Very low	✓	▽	STOP	STOP
	4WRP(E)	Size 6: 8, 18, 32 Size 10: 50, 80	10	29022 29025	Yes	Low	✓	✓	STOP	STOP
	4WRE(E)	Size 6: 4, 8, 16, 32 Size 10: 25, 50, 75	10	29061	No	Medium	✓	✓	✓	✓
	4WRSE	Size 6: 4, 10, 20, 35 Size 10: 25, 50, 80	10	29067	No	High	STOP	▽	✓	✓
	4WRPE(H)	Size 6: 2, 4, 12, 24, 40 Size 10: 50, 100	70	29035 29037 29028 29032	No	High	STOP	▽	✓	✓
	4WRREH	Size 6: 4, 8, 12, 24, 40	70	29041	No	Very high	STOP	STOP	✓	✓
	4WS(E)2E	Size 6: 2, 5, 10, 15, 20 Size 10: 10, 20, 30, 45, 60, 75, 90	70	29564 29583	No	Very high	STOP	STOP	✓	✓
	4WRZ(E)	Size 10: 25, 50, 85 Size 16: 100, 150 Size 25: 220, 325 Size 32: 360, 520 Size 52: 1000	10	29115	No	Very low	✓	▽	STOP	STOP

Table 7.2- Valve Selection Chart (Courtesy of Bosch Rexroth

7.4- Valve Spool Design

7.4.1- Spool Null Conditions

The term *Null Condition* of a spool is used to describe the dimensional relationship of the spool land and the ports on the sleeve when the spool is in its center position.

The term *Null Leakage* describes the land-to-land (typically p to T when A & B are blocked) leakage flow when the spool is in its center position, i.e. when the valve is de-energized.

The term *Flow Gain* describe the flow from the pressure port to either of the service ports based on the input signal or the spool displacement and at constant differential pressure across the valve.

Null leakage and flow gain are standard tests that should be conducted frequently to check the health conditions of the valve.

The following sections explore the various types of null conditions in EH valves and their suitability to specific applications.

7.4.1.1-Overlapped (Closed-Center) Null Conditions

As shown in Fig. 7.6, *Overlapped* condition means that the spool land dimension is larger than the dimension of the port on the sleeve by an overlap distance X_O. Overlapped valves are also referred to as *Closed-Center* valves.

Figure 7.7 shows a detailed description of a generic flow gain of an overlapped valve. The figure shows that a typical dead zone range is ±(5-20)%. Within the active zones, flow through the valve is proportional to the spool displacement and the differential pressure across the valve. The valve is called *Saturated* when the spool is displaced at full stroke and the flow becomes flat.

Features:
Null Leakage: Reduced null leakage and offers better degree of safety.

Flow Gain: Nonlinear with dead zone. No flow through the valve unless the spool moves a distance greater than X_O in either direction.

Cost: Inexpensive to produce overlapped valve with some tolerance.

Applications:
Position Control (Not Recommended): Overlapped valve results in steady state error.

Flow/Speed Control (Recommended): This valve is recommended for *Speed Control* because:
- Since the valve will be opened most of the time, it does not really matter what the valve null condition is.
- A valve with overlap null conditions stops the actuator completely before changing its direction of motion. So, if the actuator is integrated to inertial load, using an overlapped reduces pressure spikes and mechanical stresses when the motion of the actuator is reversed.

Pressure/Load Control (Not Applicable): Pressure control requires an underlapped valve.

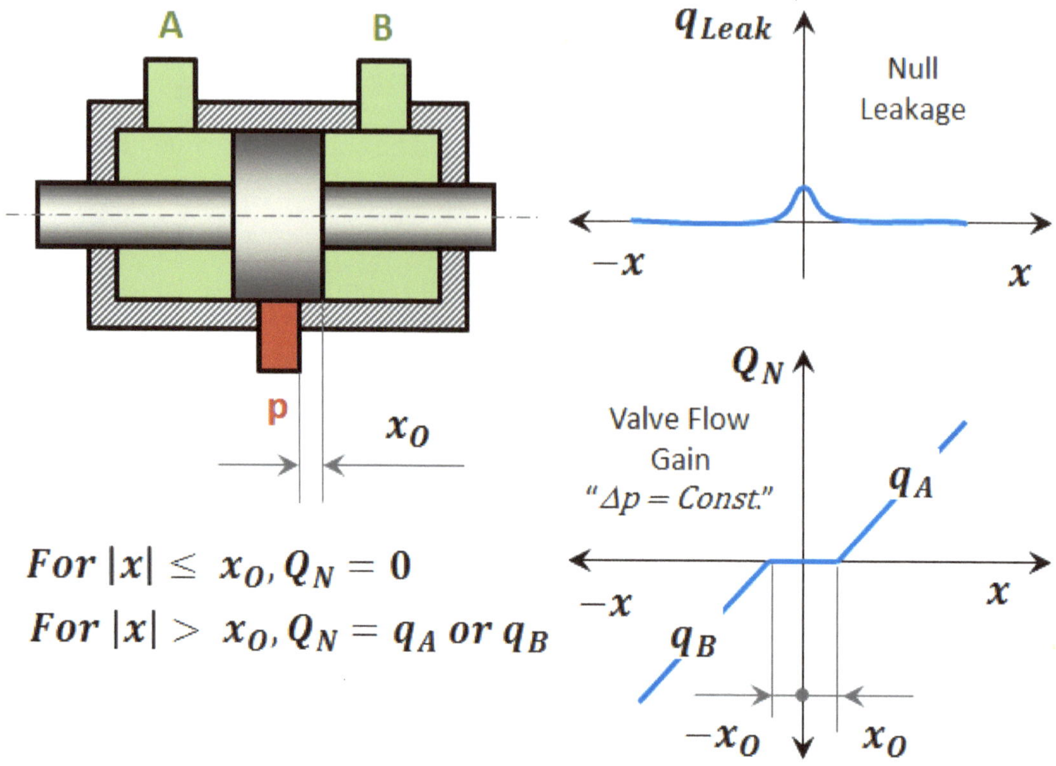

Fig. 7.6- Overlapped (Closed-Center) Null Conditions

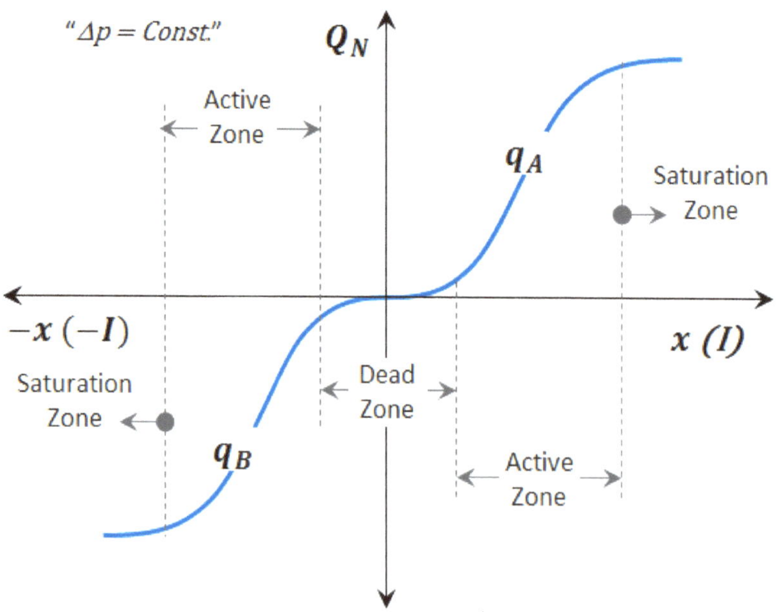

Typical Overlap Value = ± (5-20)%

Fig. 7.7- Generic Flow Gain for an Overlapped Valve

7.4.1.2- Zero-Lapped (Critical-Center) Null Conditions

As shown in Fig. 7.8, *Zero-Lapped* valves means that the spool land dimension is exactly equal to the dimension of the port on the sleeve. They are also referred to as *Critical-Center* valves.

Features:
Null Leakage: The valve has medium null leakage.
Flow Gain: Flow developed linearly through the valve once the spool moves from the null position.
Cost: Expensive to produce exact dimensions with no tolerances.
Applications:
Position Control (Recommended): This valve is perfect for *Position Control* application because:

- No steady state error.
- No need to enable/tune dead band eliminator.
- Easy to design a control loop using linear control theory.

Flow/Speed Control (Not Recommended): This valve is not recommended because:

- The job can be done by a cost-effective overlapped valve.
- Possible load creep when the valve is positioned in its null position.
- Possible pressure spikes during motion reversal because no pause period to stop the actuator before reversing its motion.

Pressure Control (Not Applicable): since pressure control requires an underlapped valve.

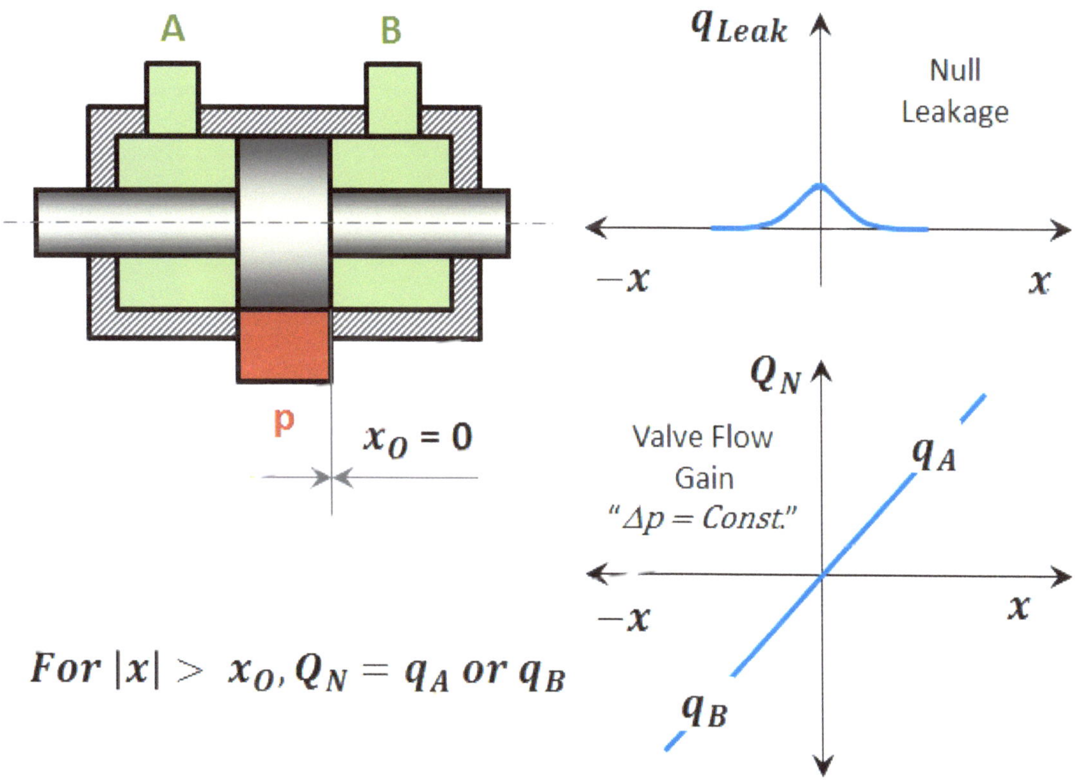

Fig. 7.8- Zero-Lapped (Critical-Center) Null Conditions

7.4.1.3- Underlapped (Open-Center) Null Conditions

As shown in Fig. 7.9, *Under-Lapped* valves means that the spool land dimension is smaller than the dimension of the port on the sleeve by an underlap distance X_O. Underlapped valves are also referred to as *Open-Center* valves.

Features:
Null Leakage: An underlapped valve has continuous leakage from the pressure port to the tank port and high null leakage between the service ports.
Flow Gain: Nonlinear with discontinuity around the null position.
Cost: the valve isn't inexpensive to produce
Applications:
Position Control (Not Applicable): Such underlapped valves are not applicable for position control because the load will float at null position.

Flow/Speed Control (Conditionally Recommended): The valve can be used for speed control with the following cautions:

- Actuator should be float at null position, e.g. cooling fan control.
- Cycling slow actuator or small loads that can be reversed without pressure spikes.

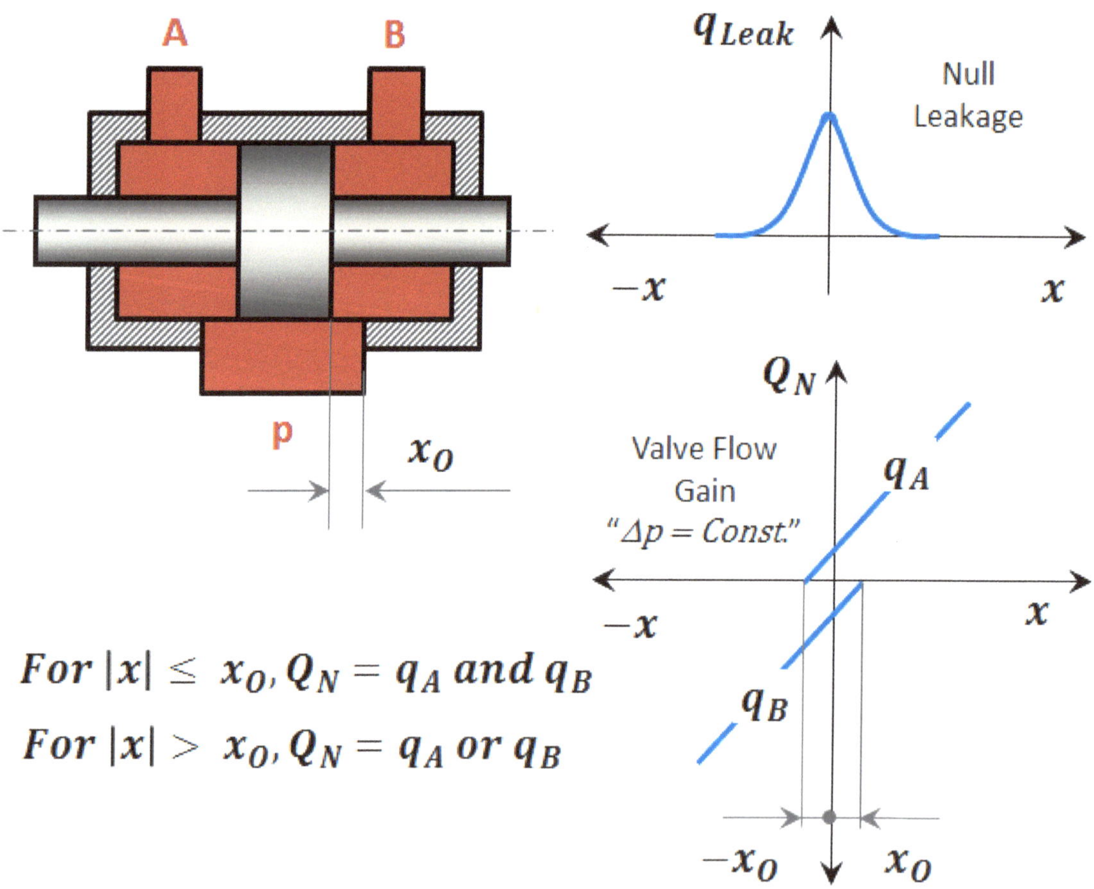

$$For\ |x| \leq x_O, Q_N = q_A\ and\ q_B$$
$$For\ |x| > x_O, Q_N = q_A\ or\ q_B$$

Fig. 7.9- Underlapped (Open-Center) Null Conditions

Pressure/Load Control:

Such an underlapped spool is the only one applicable for *Pressure/Load Control*. As shown in Fig. 7.10, control pressure is performed around the null position.

The figure shows that, when the valve is in its center position, the supply pressure will split into two equal halves that act on the ports A and B.

Within the underlap dimension, the spool movement will be accompanied by flow from the pressure port to both the service ports. Assuming the spool moves to the right side, the pressure port P is accordingly restricted from the port B and connected more to the port A. As a result, differential pressure is created.

Once the spool moves for a distance greater than or equal the underlap distance, the pressure port will be connected only to one of the service ports while terminated from the other one and the differential pressure across the actuator will equal the supply pressure.

If an overlapped or a zero lapped valve is used, differential pressure won't be created because one land is always closed.

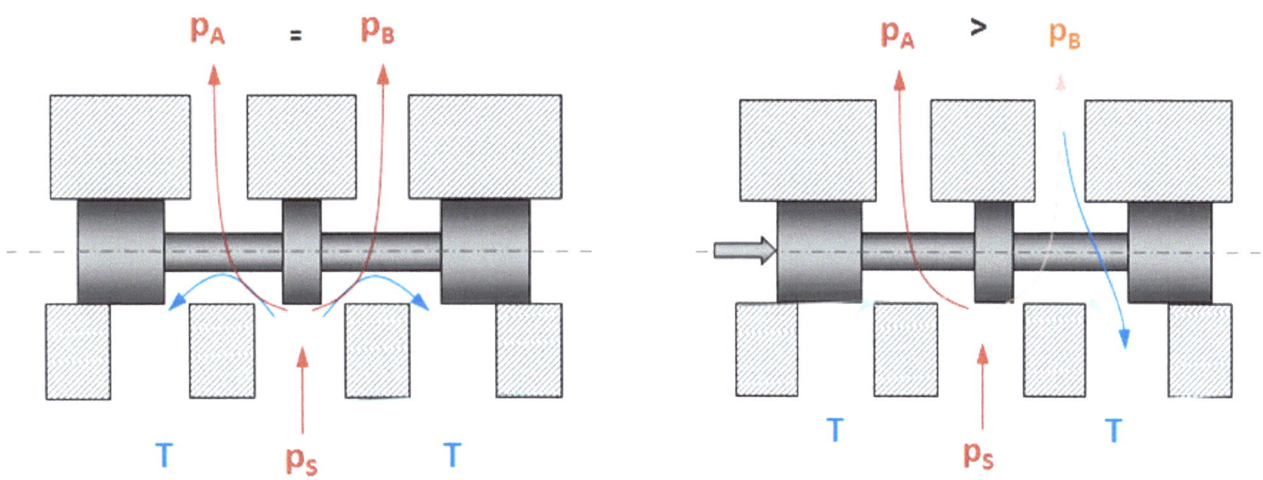

Fig. 7.10- Underlapped Valve in Pressure Control

Table 7.3 concludes qualitatively the selection criteria for valve null position based on control system.

	Overlapped "Closed-Center"	Zero Lapped "Critical-Center"	Underlapped "Open-Center"
Position control			
	Steady state error	Recommended	Float actuator around valve null position
Speed/Flow Control			
	Recommended	Expensive valve is used for no reason	Float actuator around valve null position. Good for slow actuators and small loads
Load/Pressure Control			
	One land is always closed	One land is always closed	Recommended

Table 7.3 - Valve Spool Selection Based on Control Solutions

7.4.2- Spool Transitional Conditions

Spool *Transitional Condition* means how the ports are connected during the transitional movement of the spool. Figure 7.11 shows examples of spool transitional conditions. The importance of reviewing such specifications is that it affects the operational performance. For example, if a valve with original closed-center transitional condition is replaced by an open-center transitional condition, the load will experience some jerky motion during valve shifting.

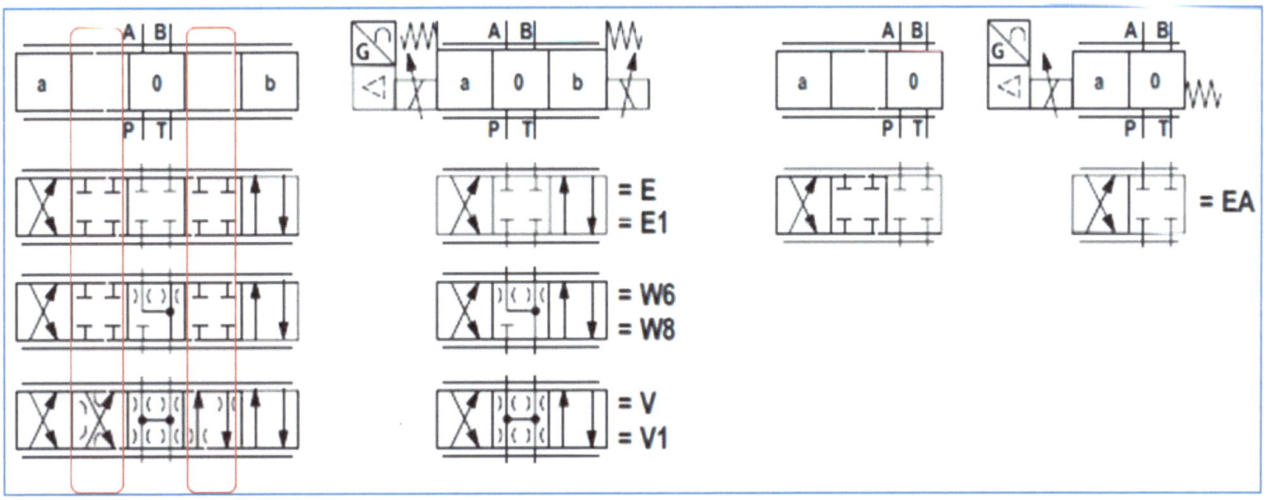

Fig. 7.11- Transitional Conditions of Selected Spools (Courtesy of Bosch Rexroth)

7.4.3- Spool Fail-Safe Position

Fail Safe Position in an EH valve is a position to which the spool moves when the valve is de-energized. This could happen in case of emergency, e.g., power outage or loss of signal.

The purpose of this position is to secure certain safe situation for the actuator in case of emergency.

Figure 7.12 shows examples of valves that have fail-safe positions.

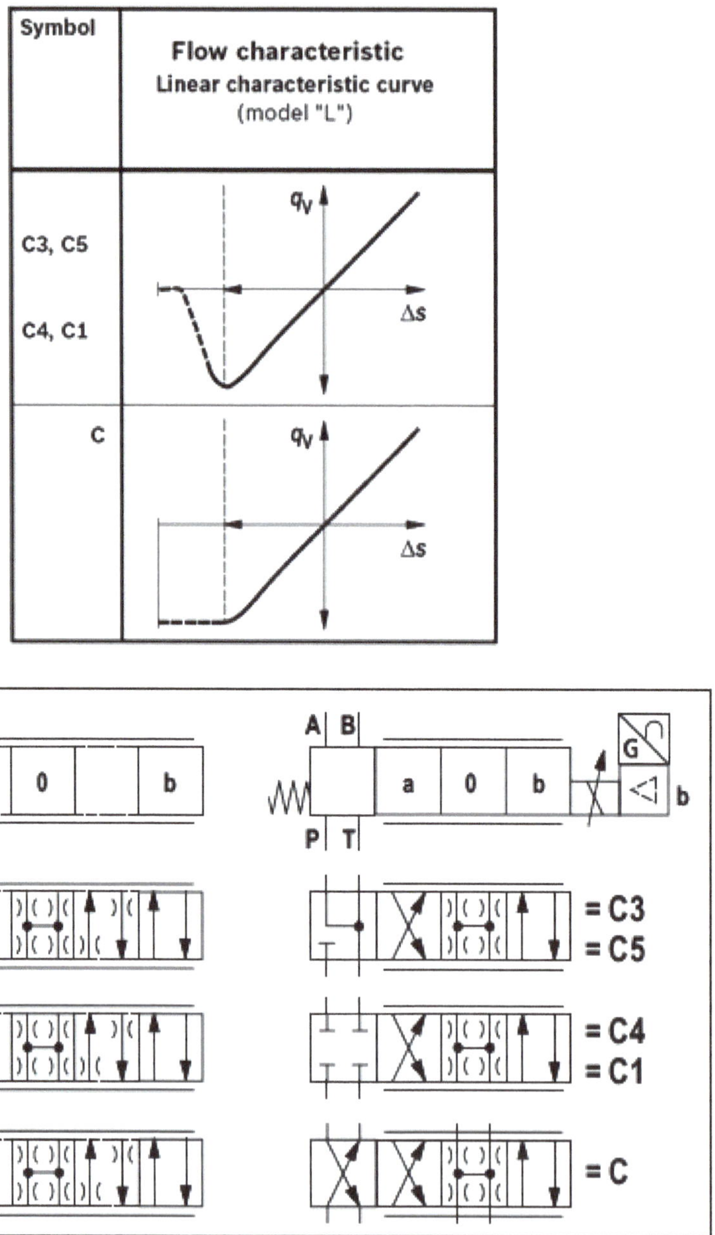

Fig. 7.12- Fail-Safe Position of Selected Spools (Courtesy of Bosch Rexroth)

7.4.4- Spool Control Edges

Spools of proportional and servo valves are designed with *Control Edges* to obtain certain flow gain that improves the machine operation. Spool control edges are also referred to as "*Metering Orifices*" or "*Control Notches*".

7.4.4.1- Holes versus Control Edges

As shown in Fig. 7.13, ports on the sleeve are connected via drilled holes or control edges on the spool. The drilled holes are standard and cost-effective method. But it does not offer flexibility in designing the flow gain of the valve.

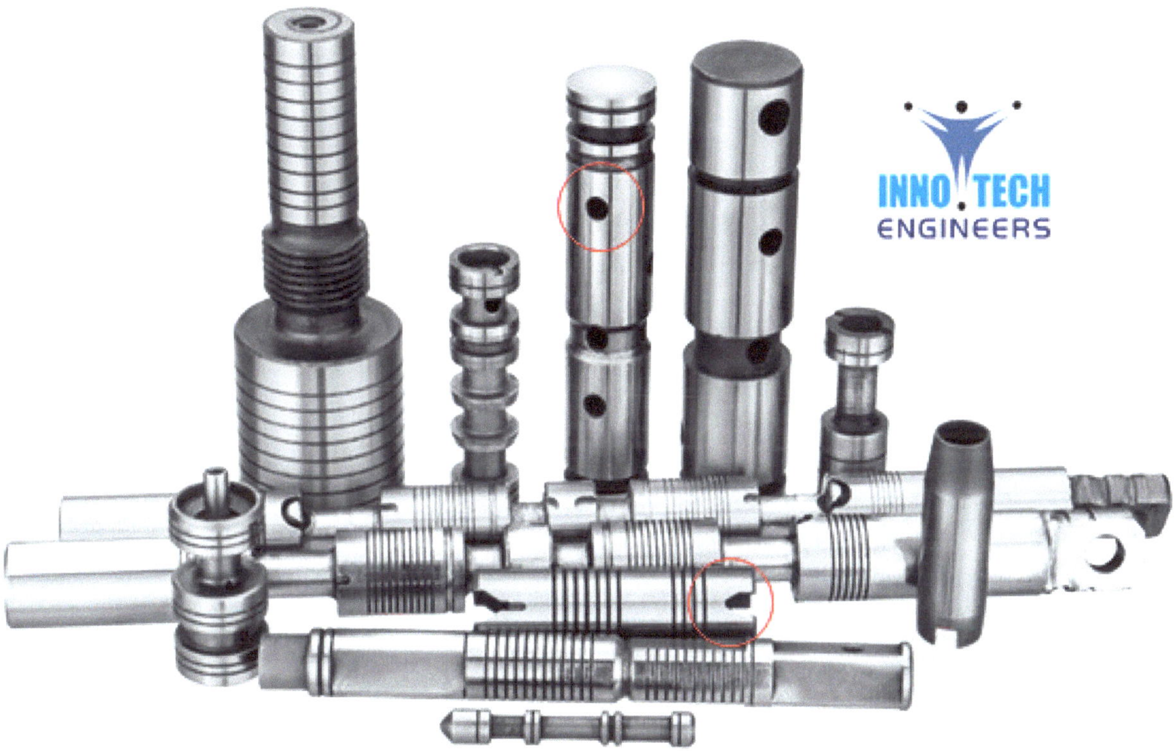

Fig. 7.13- Holes versus Control Edges

7.4.4.2- Custom Control Edges

Figure 7.14 explains the procedure of designing custom control edges for a continuous valve.

- The design procedure starts by defining the machine motion profile for a specific dynamic job or duty cycle.
- Based on the machine motion profile, the actuator speed is obtained.
- The flow gain of the valve, to achieve the predefined actuator speed, is then calculated.
- The flow gain of the valve is function of the opening area in the valve and the differential pressure across the valve. Then, based on the predefined flow gain, the control edges are designed.

$$v = f(Q) \rightarrow Q = f(A) = C_D A \sqrt{\frac{2\Delta p}{\rho}}$$

$$\rightarrow A = f(x)$$

Fig. 7.14- Design Methodology for Custom Spool Control Edges

7.4.4.3- Standard Control Edges

Square Notches:

Figure 7.15 shows a spool designed with *Square Notches*. When the spool moves a displacement ***X***, the valve opens an area equals to the spool displacement times the vertical dimension of the notch, which is constant. As a result, the flow gain is linearly proportional to the spool displacement.

For such a linear flow gain, the flow starts with sharp acceleration and accompanied by backward flow forces. Therefore, such a valve is good for small flow rate.

Figure 7.16, shows a typical characteristic of a valve that has linear flow gain. The figure indicates that the valve is overlapped (closed-center). The figure also shows flow gain (1) for a small valve and flow gain (2) for a large valve. Note that, under constant differential pressure, flow through the valve is linearly proportional to the opening area. Therefore, if the opening area is linearly function of the spool displacement, the resulted flow gain is linear as shown in the figure.

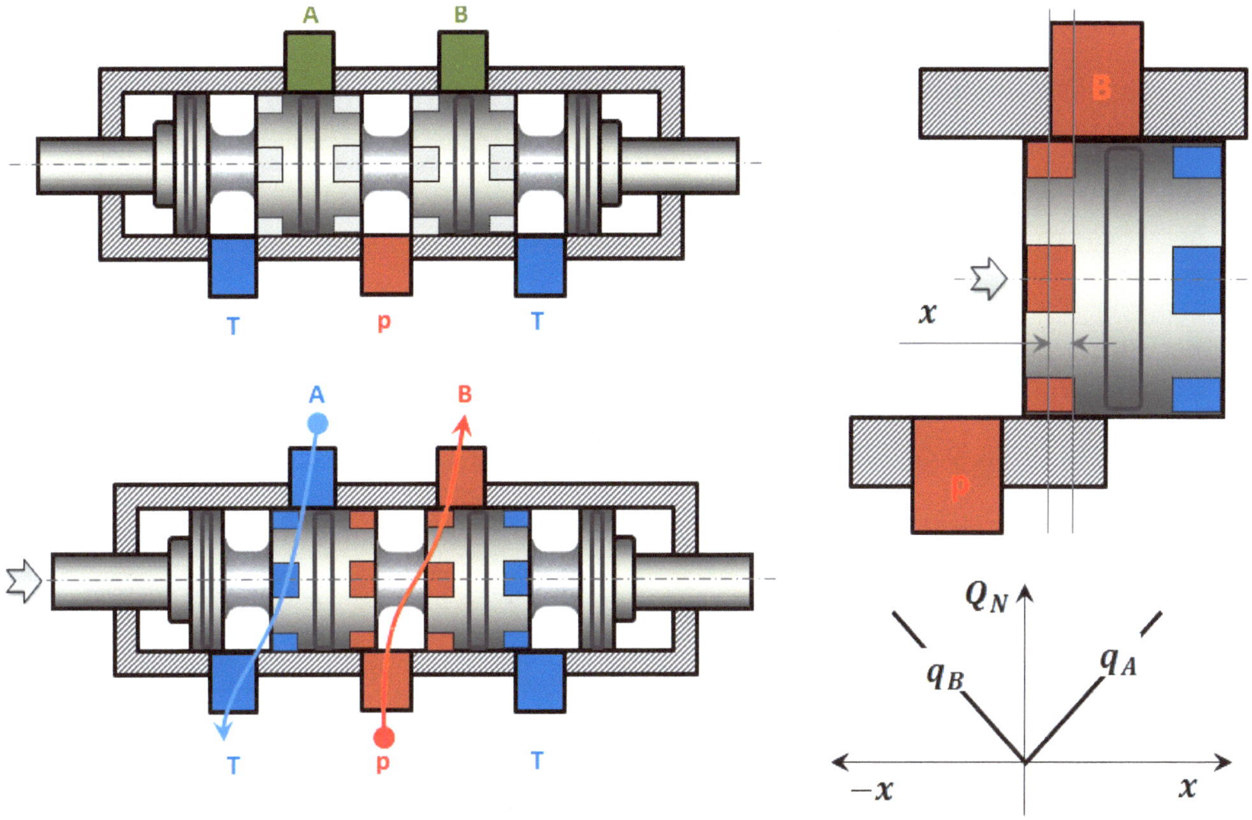

Fig. 7.15- Square Notches

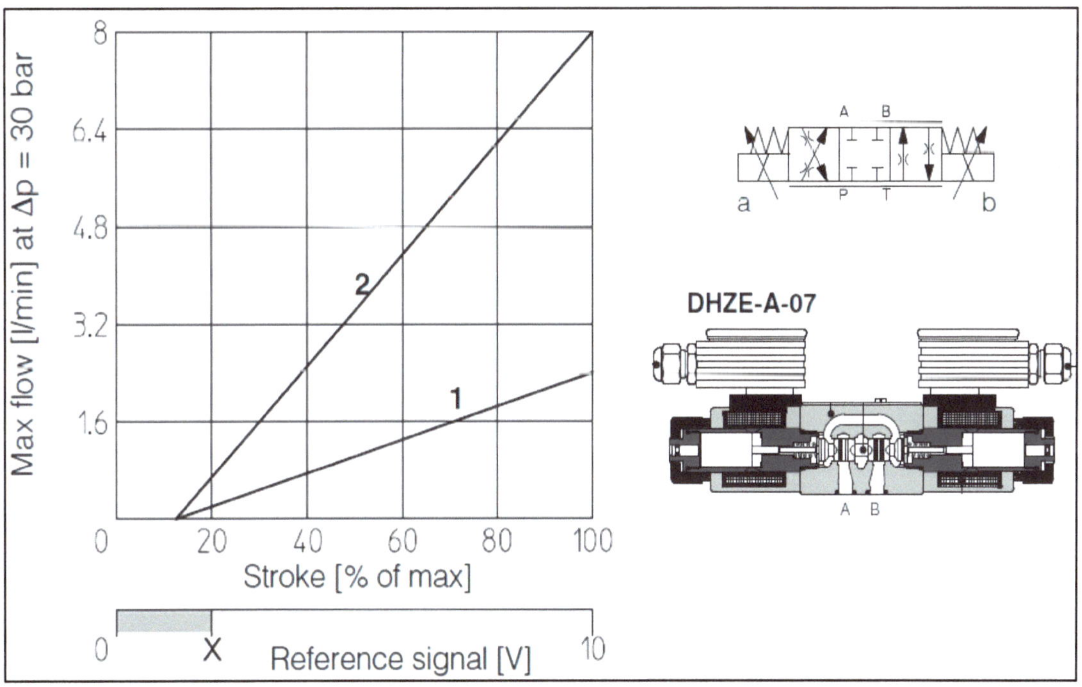

Fig. 7.16- Example of Linear Flow Characteristics (Courtesy of Atos)

Equilateral Triangular Notches:

Figure 7.17 shows a spool designed with *Equilateral Triangular Notches*. Such a type of notches is referred to as "*V Notches*" or "*Delta Notches*".

When the spool moves a displacement ***X***, the valve opens a triangular area, in which the triangle height equals the spool displacement. As a result, parabolic flow gain versus spool displacement is created.

For such a parabolic flow gain, the flow starts with easy acceleration. Therefore, such a valve is good for large flow rate.

Figure 7.18 shows a typical overlapped valve that has parabolic flow gain.

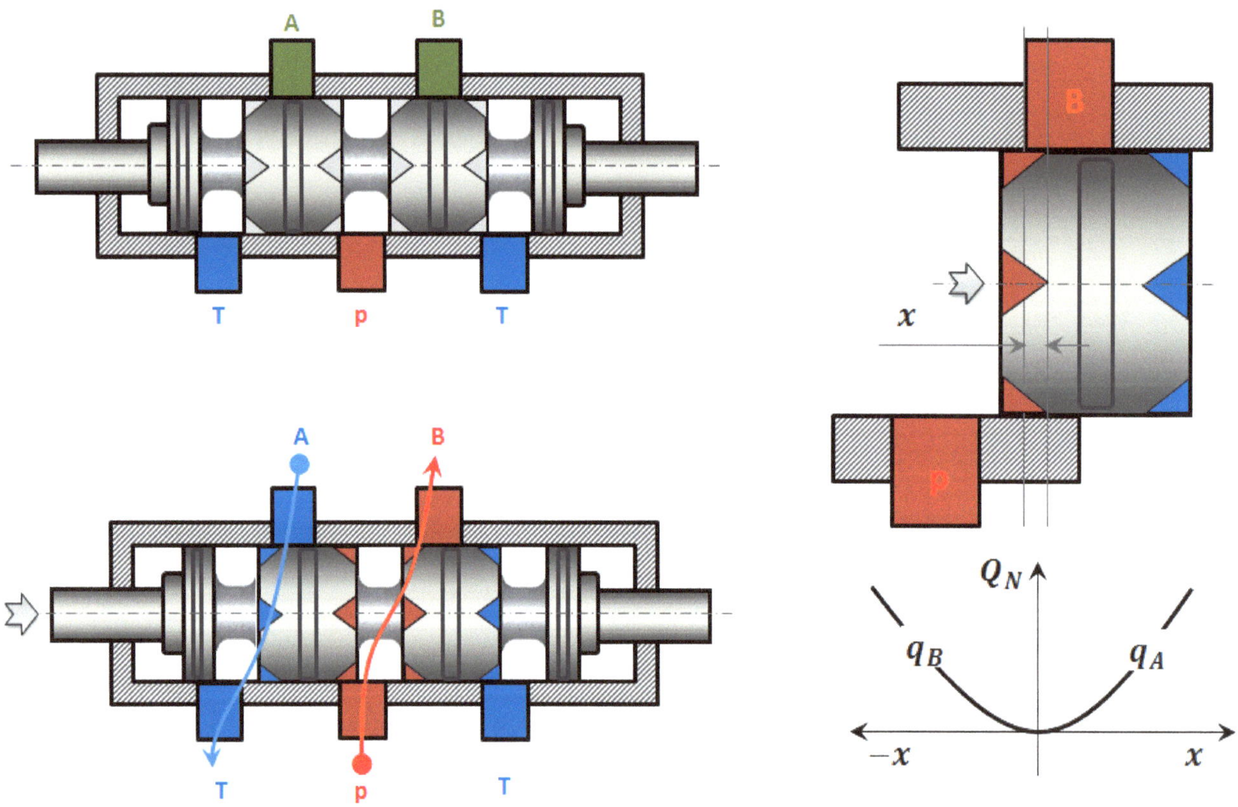

Fig. 7.17- Equilateral Triangular Notches

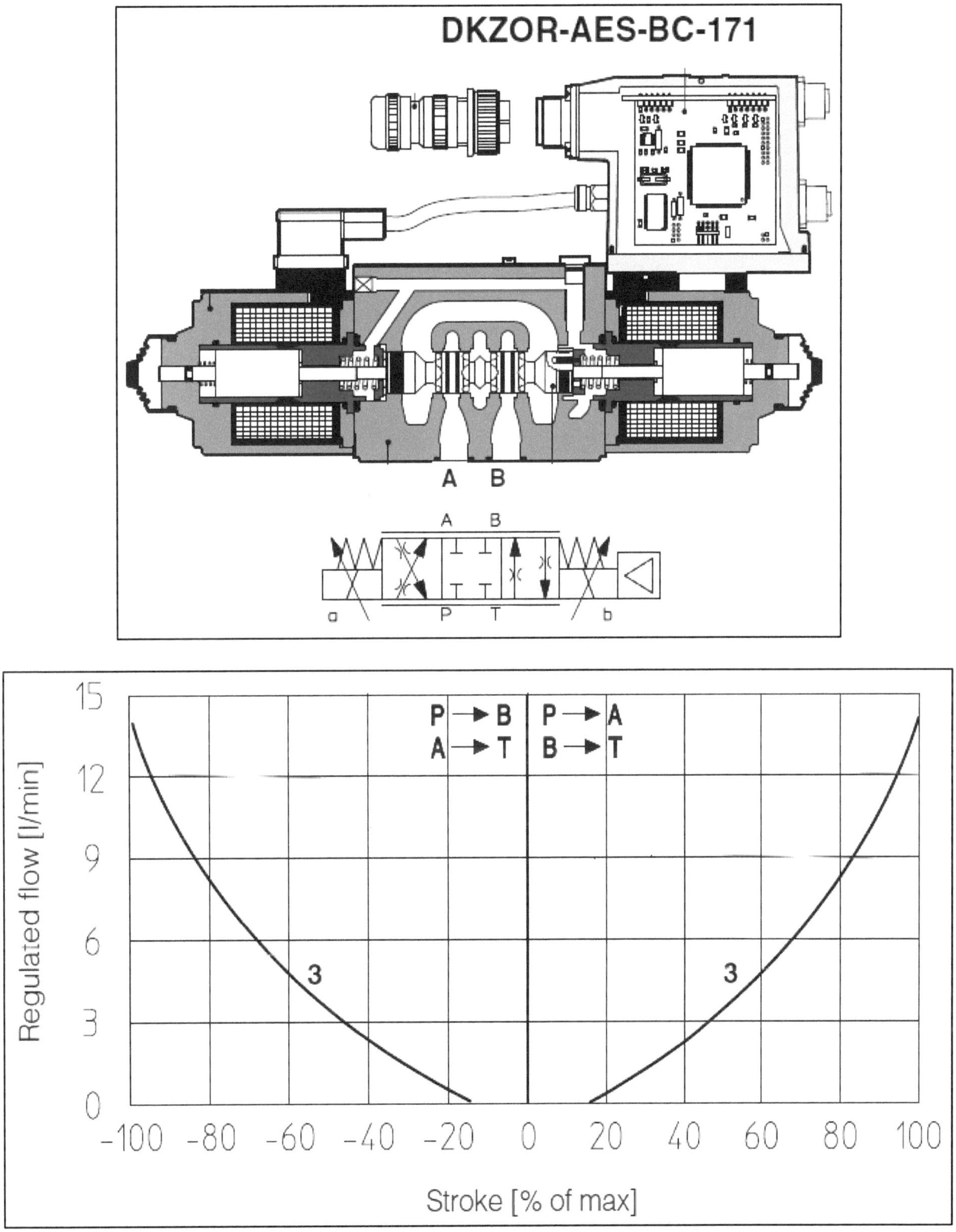

Fig. 7.18- Example of Parabolic Flow Characteristics (Courtesy of Atos)

Semicircular Notches:

Figure 7.19 shows a spool designed with *Semicircular Notches*. Such type of notches is referred to as *"D Notches"*.

When the spool moves a displacement ***X***, the valve opens a portion of a circle. As a result, *S-Shaped* flow gain versus spool displacement is created.

For such S-shaped flow gain, the flow starts with easy acceleration ends with easy deceleration. Therefore, such a valve can produce easy motion starts and stops, which is good for position control applications.

Figure 7.20 shows two typical overlapped valves with S-shaped flow gain.

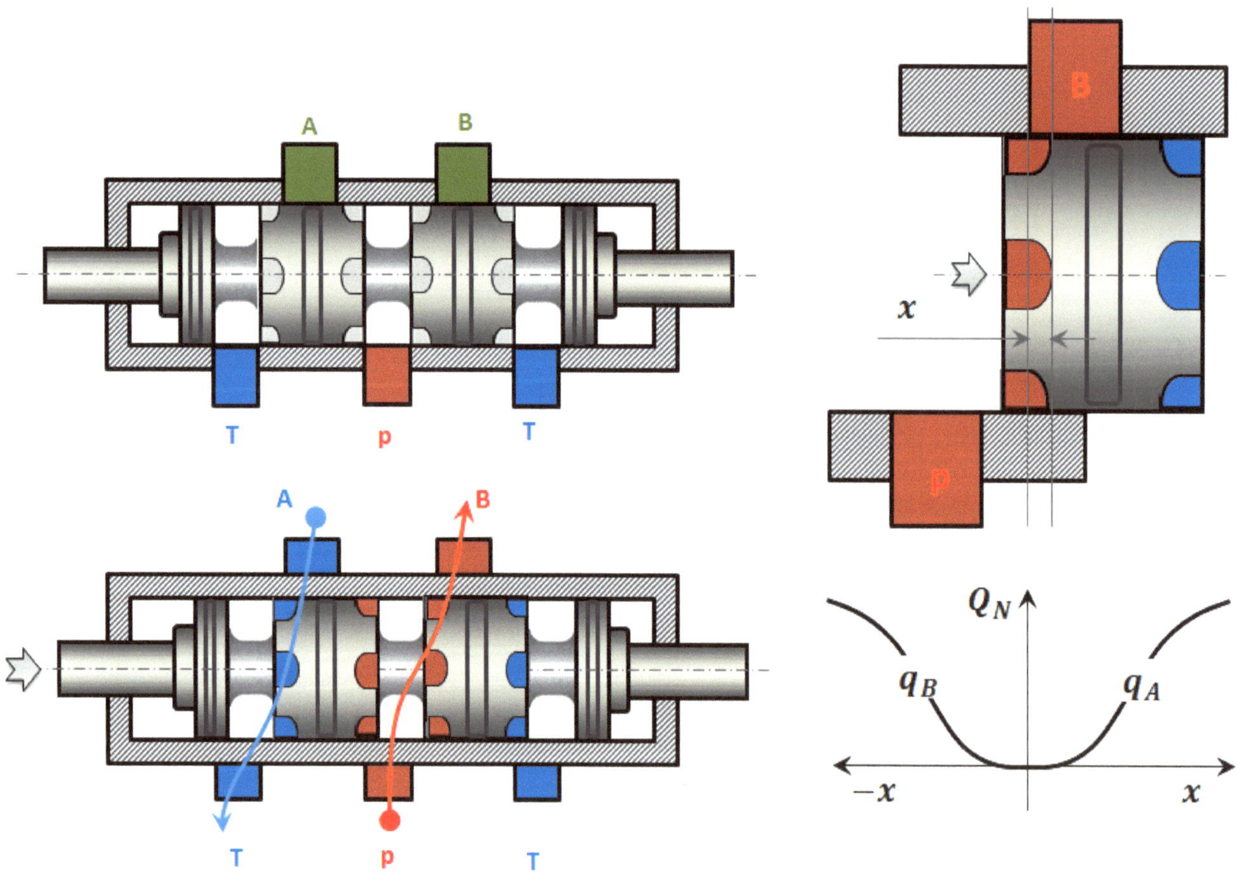

Fig. 7.19- Semicircular Notches

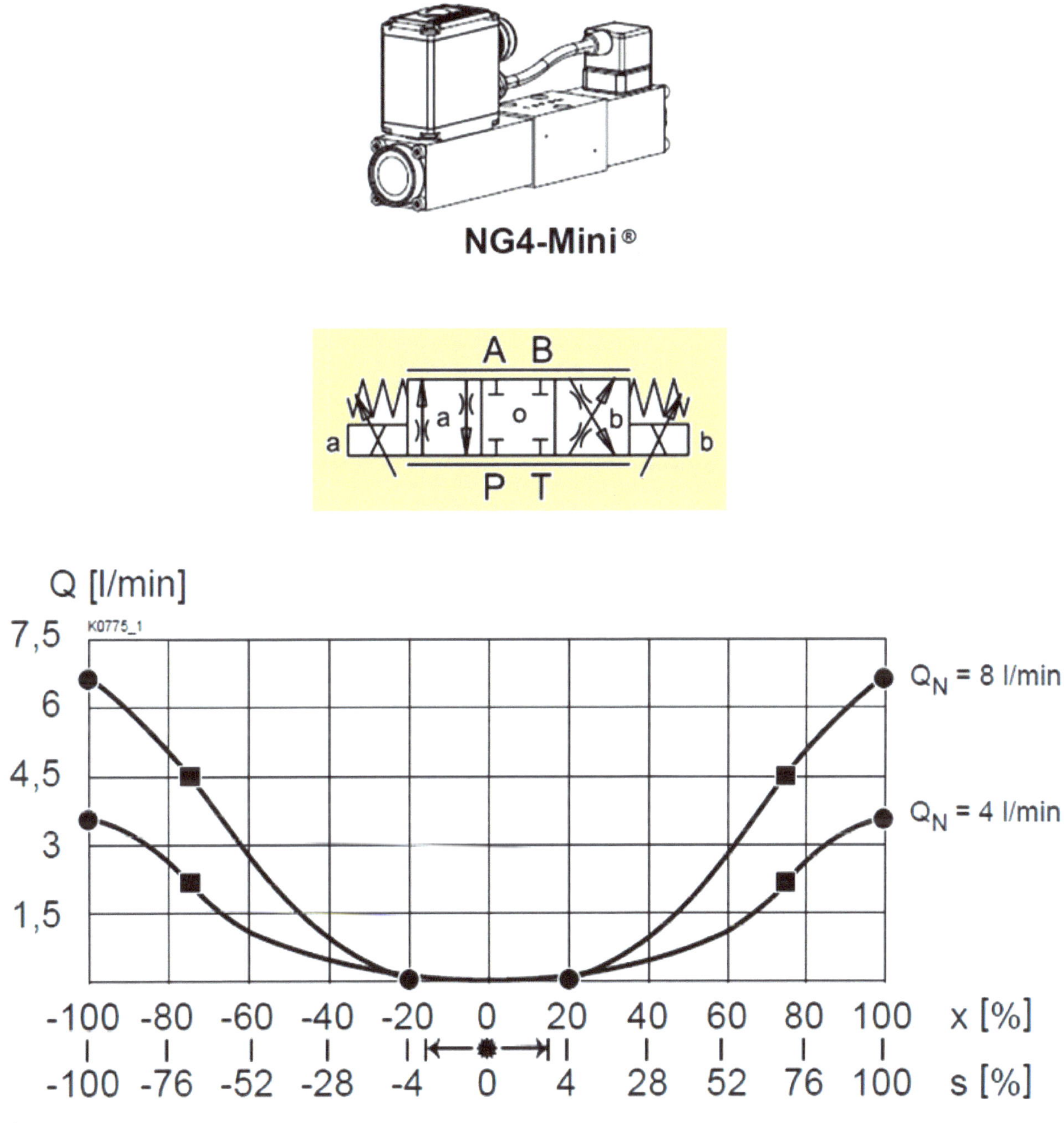

Fig. 7.20- Example of S-Shaped Flow Characteristics (Courtesy of Wandfluh)

7.4.4.4- Distribution of Control Edges

As shown in Fig. 7.21, for hydraulic motors and synchronous cylinders, valves with symmetric spools are used. Using such symmetric spools, the actuator is driven in either direction with same speed.

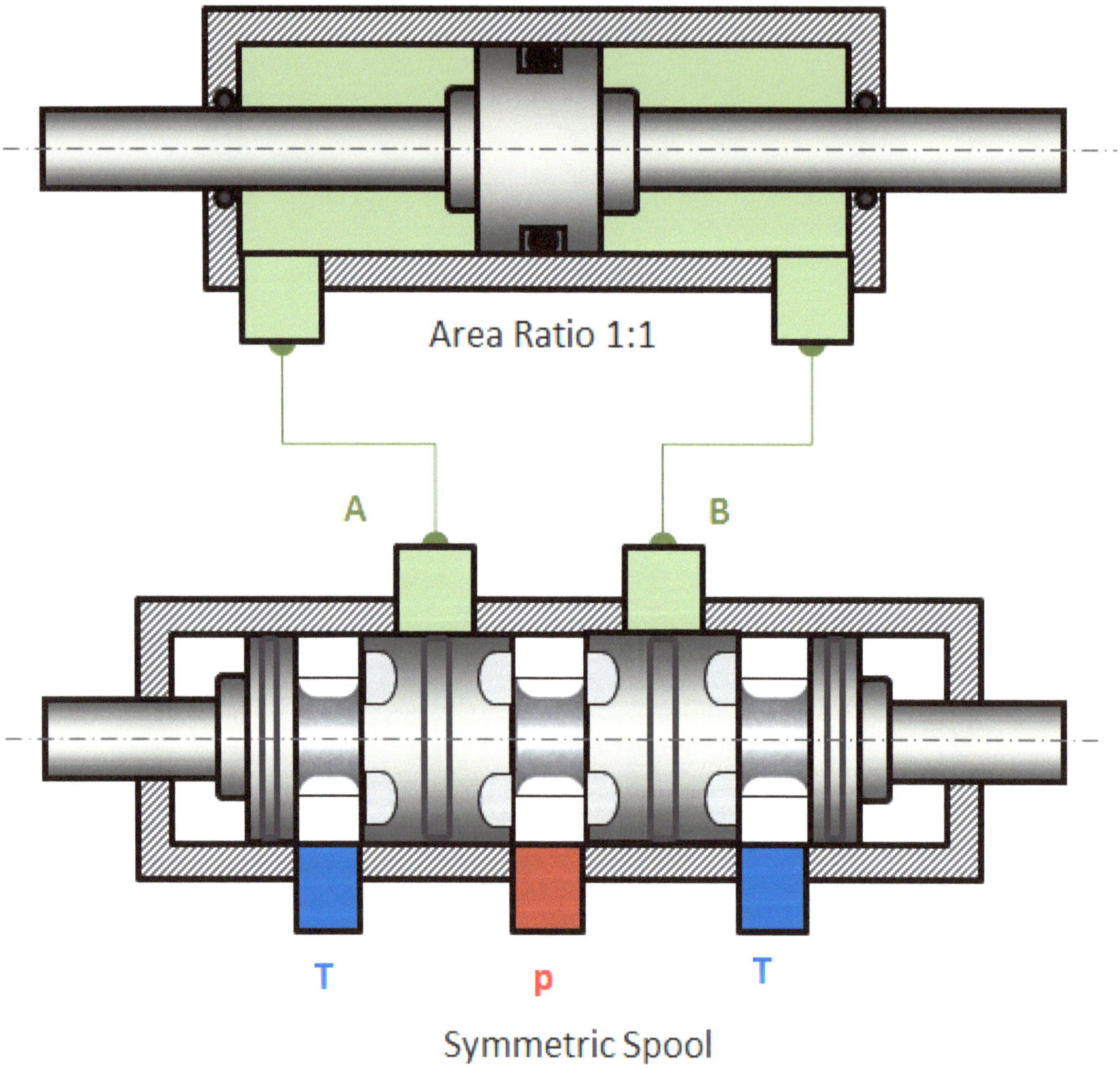

Fig. 7.21- Symmetric Distribution for Control Edges

For a differential cylinder to retract, it requires less flow than what is needed to extend. Therefore, if a symmetric spool is used to drive a differential cylinder, the spool should be shifted differently to extend and retract the cylinder with the same speed. For example, to extend a cylinder that has area ratio 2:1, the spool should be displaced a distance twice the distance that is required to retract the cylinder with the same speed.

To resolve this issue, as shown in Fig. 7.22, dissymmetric spool is recommended for differential cylinder. The operator of the valve, in open loop, will shift the spool equally to cycle the differential cylinder with constant speed.

In a closed loop, the controller can take care of cycling a differential cylinder with constant speed; but the spool will still be shifted differently. Therefore, it is still recommended to use the dissymmetric spool even with a closed loop.

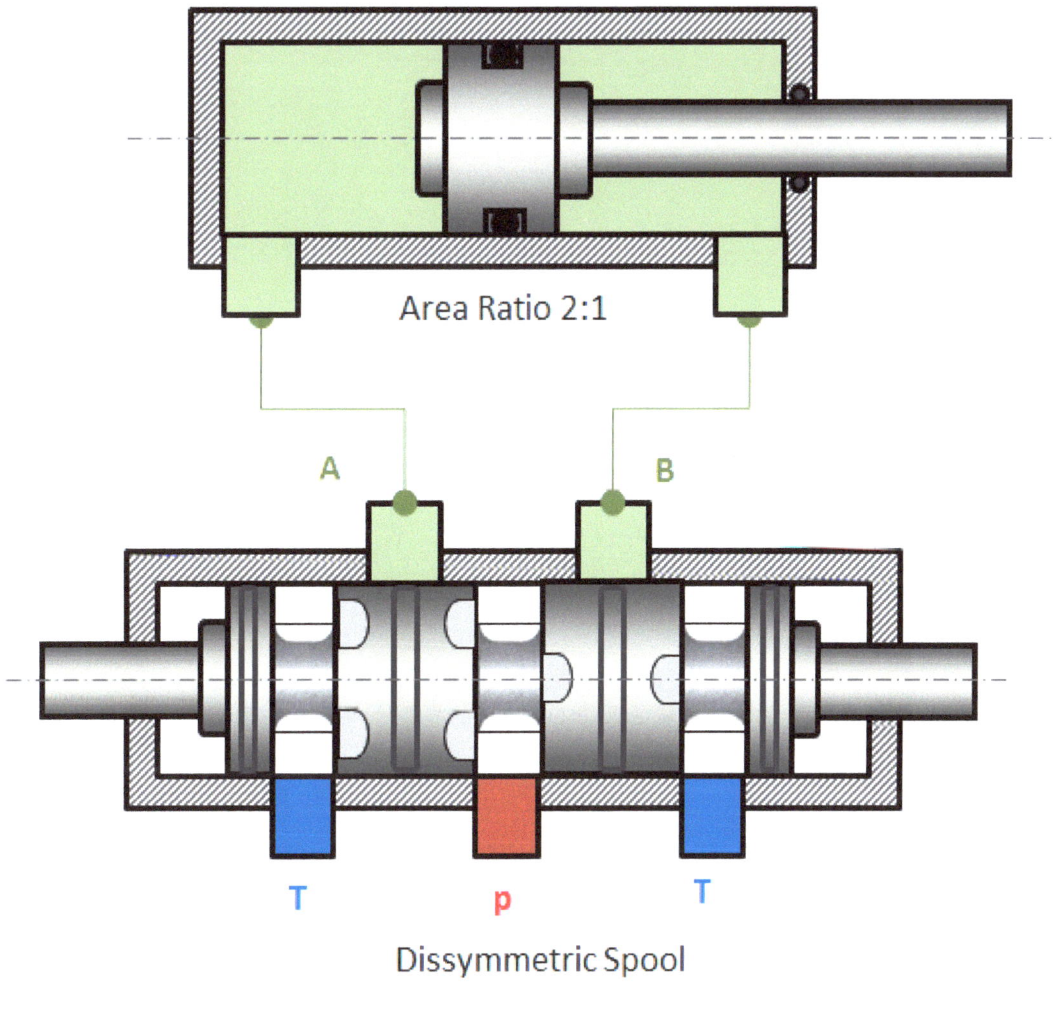

Fig. 7.22- Dissymmetric Distribution for Control Edges

7.5- Valve Static Characteristics

Figure 7.23 shows that the flow ***Q*** through a continuous valve depends on the differential pressure ***Δp*** across the valve and the spool displacement, which is function of the input signal ***I***. Based on that, three main static characteristics can be developed by testing the valve.

- **Flow Gain:** required for position control, flow control, and speed control.
- **Flow-Pressure Coefficient: required** similar to flow gain and additionally for power losses and heat calculations.
- **Pressure Gain:** required for pressure and load control.

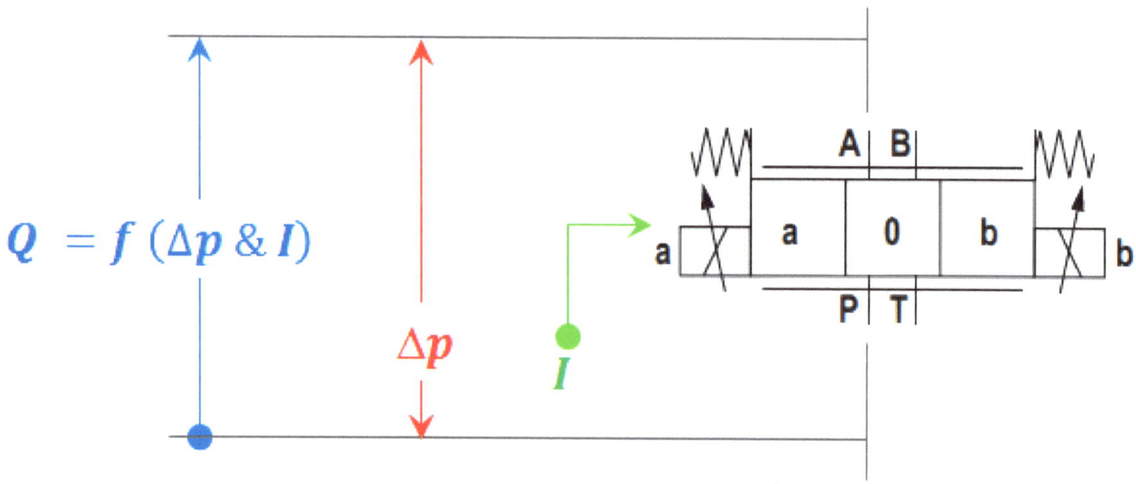

Fig. 7.23- Flow Dependency through a Continuous Valve

7.5.1-Valve Flow Gain

As shown in Fig. 7.24, *Flow Gain* is defined as the flow through the valve as function of the input signal at constant differential pressure. Equation 7.1 uses the partial derivative expression to calculate the flow gain because the flow through the valve is function of more than one parameter (input signal and the differential pressure).

$$\textit{Flow Gain} = k_{QI} = \left[\frac{\partial Q_N}{\partial I_N}\right]_{\Delta p_{N=Constant}} \qquad 7.1$$

- ❑ Q_N = Output Nominal "Rated" Flow (developed by testing the valve).
- ❑ Δp_N = Constant Nominal "Rated" Differential Pressure (used to test the valve).
- ❑ I_N = Variable Nominal Input Command to the valve.

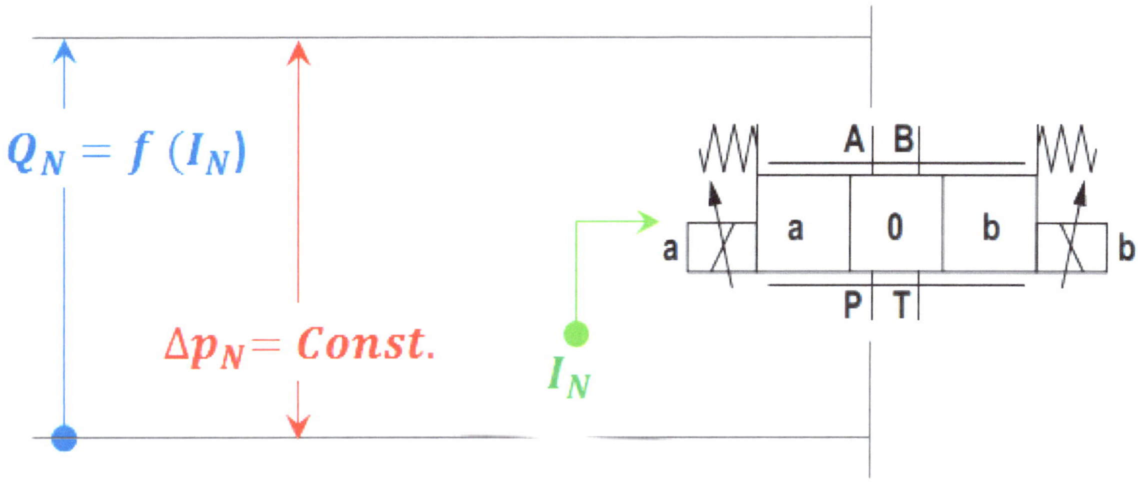

Fig. 7.24- Interpretation of Valve Flow Gain

Figure 7.25 shows a generic family of curves developed by testing a valve under three different differential pressures. The figure shows that, at a spool displacement X_S (that is function of the input signal I_N), more flow is driven through the valve at higher differential pressure. It is to be noted that such a test must be performed using predefined fluid viscosity and working temperature.

On the same plot, a family of curves could be developed to show the valve flow gain through the different passages inside the valve, e.g., P→A, P→B, A→T, BT, and P→T.

In the standard test protocols, 35 bar (500 psi) is a nominal pressure per pass (leg). Hopefully all valve manufactures follow the standard valve test protocol that unifies the test conditions, so that the end users can make apple-to-apple comparison. Unfortunately, this isn't the case!

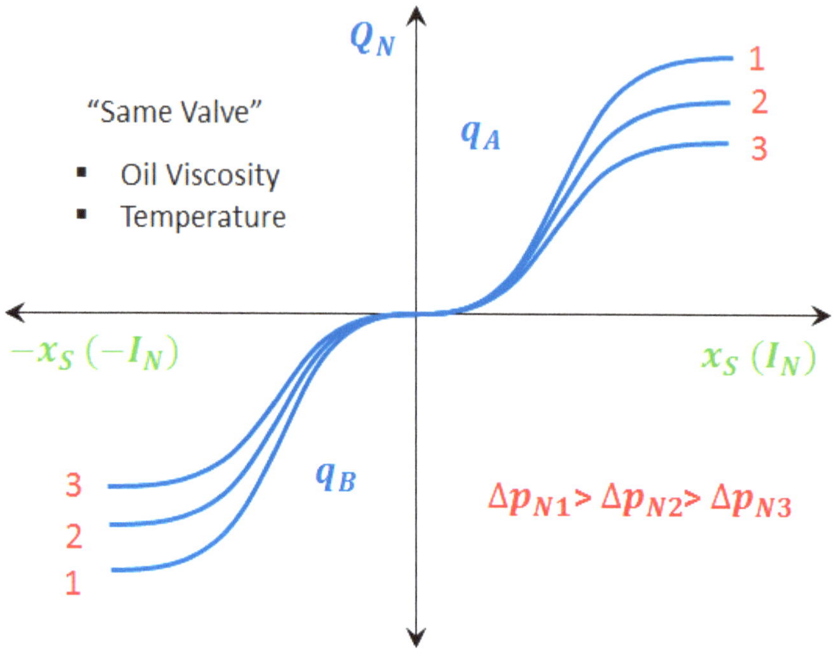

Fig. 7.25- Generic Flow Gains for a Continuous Valve at Several Constant Differential Pressures

For design purposes of some applications, valve flow gain can be considered linear within a certain range of input command. That makes it easy for a system designer to design system controllers and to calibrate the input command versus the output control parameter.

Figure 7.26 shows an example in which flow gain range of (5-95) % has been considered linear within (15-85)% range of the input command. To avoid the valve nonlinearities, a system designer should avoid operating the valve out of these ranges.

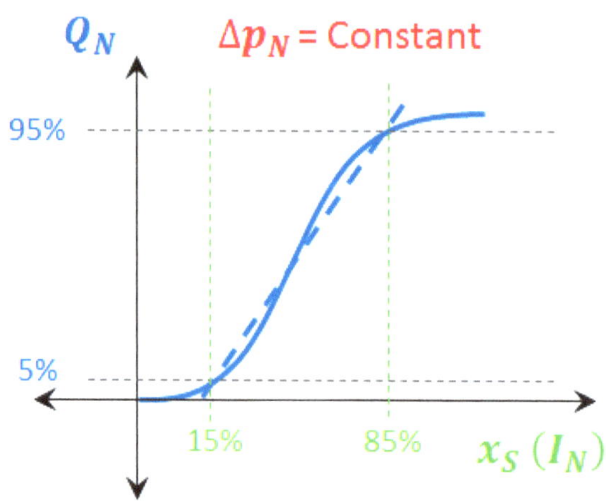

Fig. 7.26- Flow Gain Linearization

The following section introduces several flow gain examples of actual valves. It is common in all of the introduced examples that the manufacturer should state the conditions under which the valve is tested, particularly the nominal differential pressure, fluid viscosity, and the temperature. Examples 1 through 6 shows flow gain of one valve under different nominal pressures.

Example 1, Fig. 7.27, shows flow gain for a three-position zero-lapped proportional directional valve. The curve shows that, if the valve receives maximum input signal and subjected to 35 bar differential pressure, the valve will pass maximum flow of 40 liter/min.

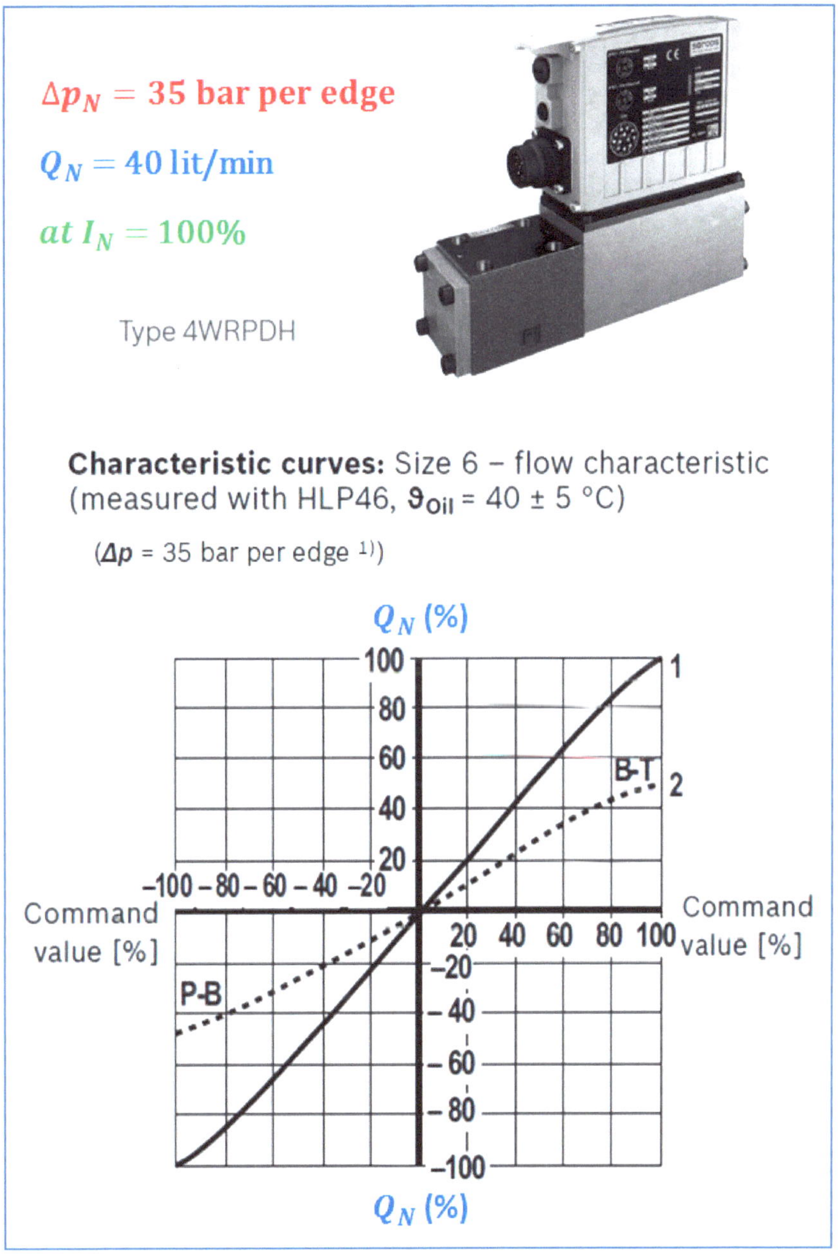

Fig. 7.27- Typical Flow Gain for a Proportional Valve - Example 1, (Courtesy of Bosch Rexroth)

Example 2, Fig. 7.28, shows flow gain at different nominal differential pressures for a three-position overlapped proportional valve.

The figure shows nonlinear flow gain. The valve requires 15% minimum input command and the valve is saturated after 85% spool stroke. If only one quadrant is drawn for a three-position valve, this means that the valve is symmetric.

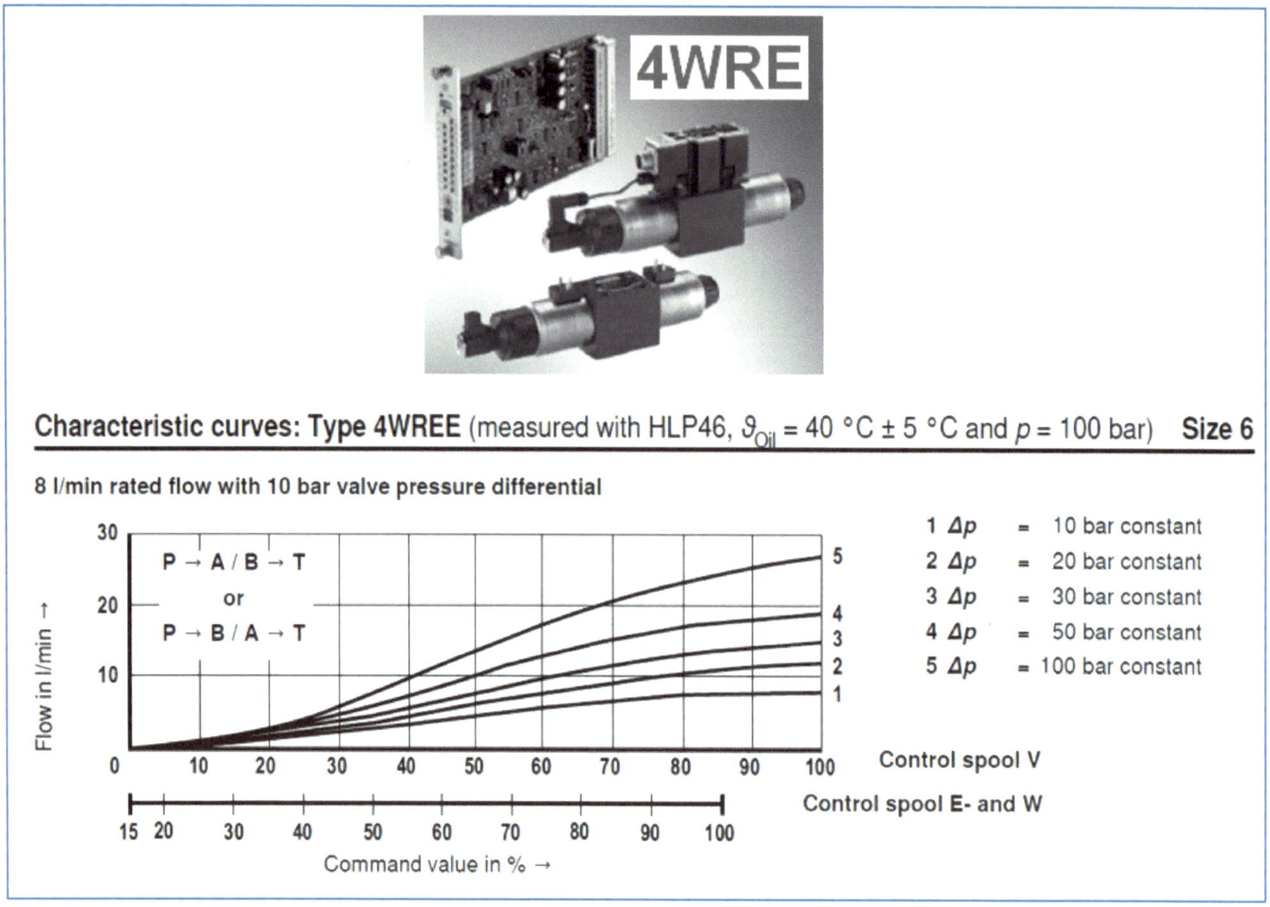

Fig. 7.28- Typical Flow Gain for a Proportional Valve - Example 2, (Courtesy of Bosch Rexroth)

Example 3, Fig. 7.29, shows flow gain at three different nominal pressures for a small size two-position overlapped proportional valve. The figure shows approximately 22% overlap condition.

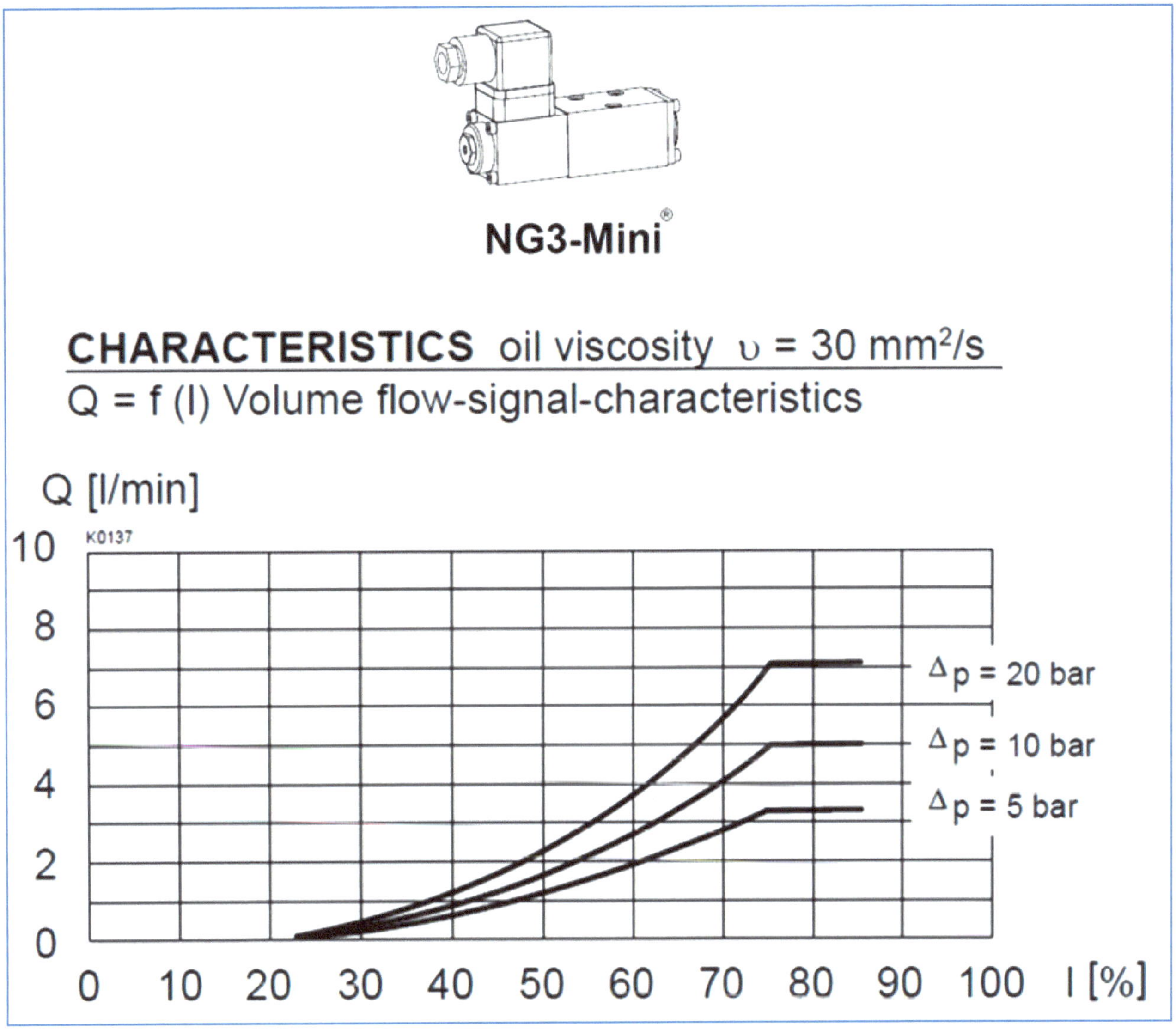

Fig. 7.29- Typical Flow Gain for a Proportional Valve - Example 3, (Courtesy of Wandfluh)

Example 4, Fig. 7.30, shows flow gain for an overlapped cartridge proportional valve at two nominal differential pressures. The figure shows the tolerance for each flow gain as a shaded area.

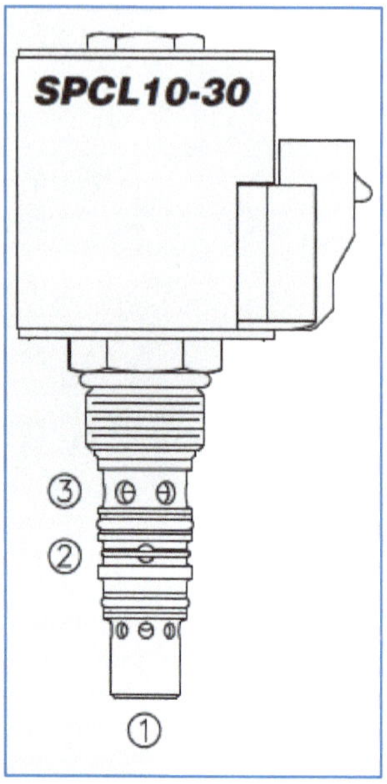

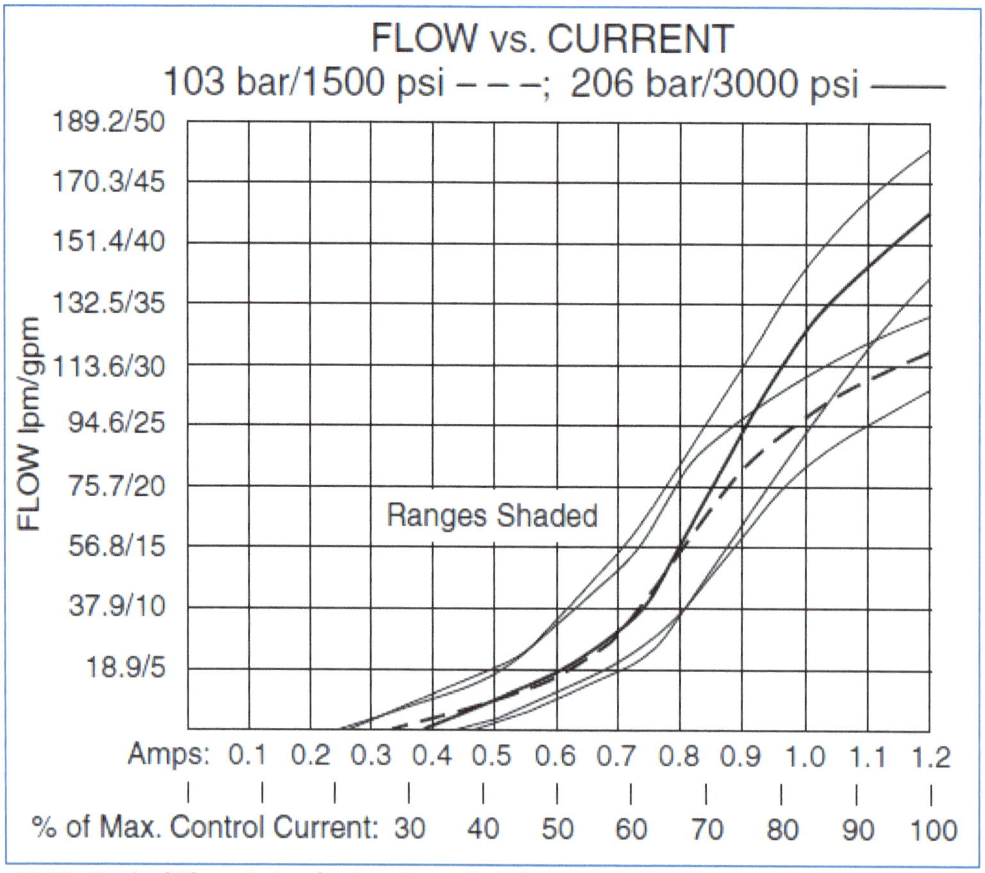

Fig. 7.30- Typical Flow Gain of a Continuous Valve – Example 4, (Courtesy of Hydraforce)

Example 5, Fig. 7.31, shows flow gain for a direct-operated overlapped directional proportional valve.

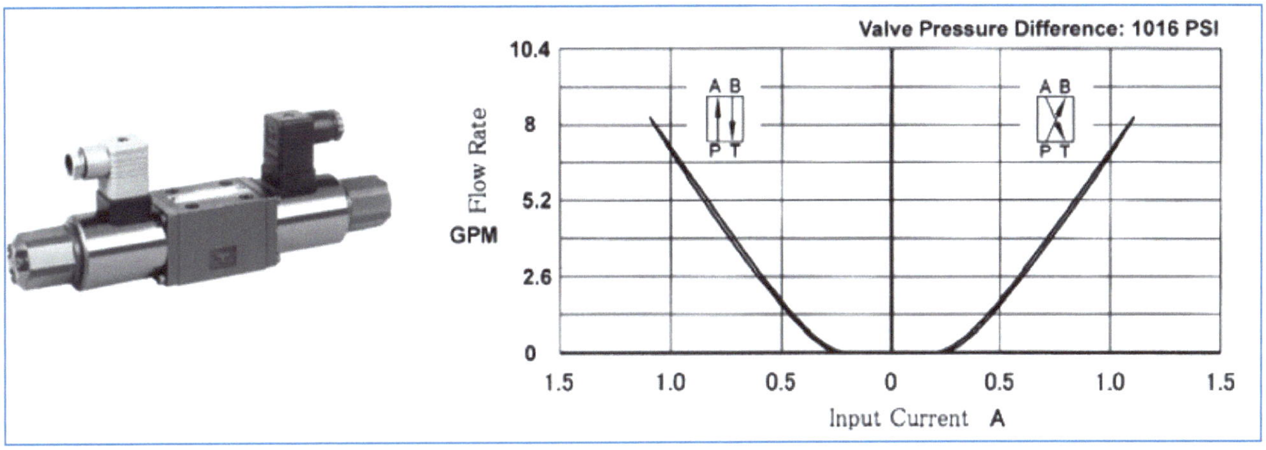

Fig. 7.31- Typical Flow Gain for a Proportional Valve - Example 5, (Excerption from Yuken)

Example 6, Fig. 7.32, shows flow gain for an overlapped pilot-operated valve at various nominal pressures.

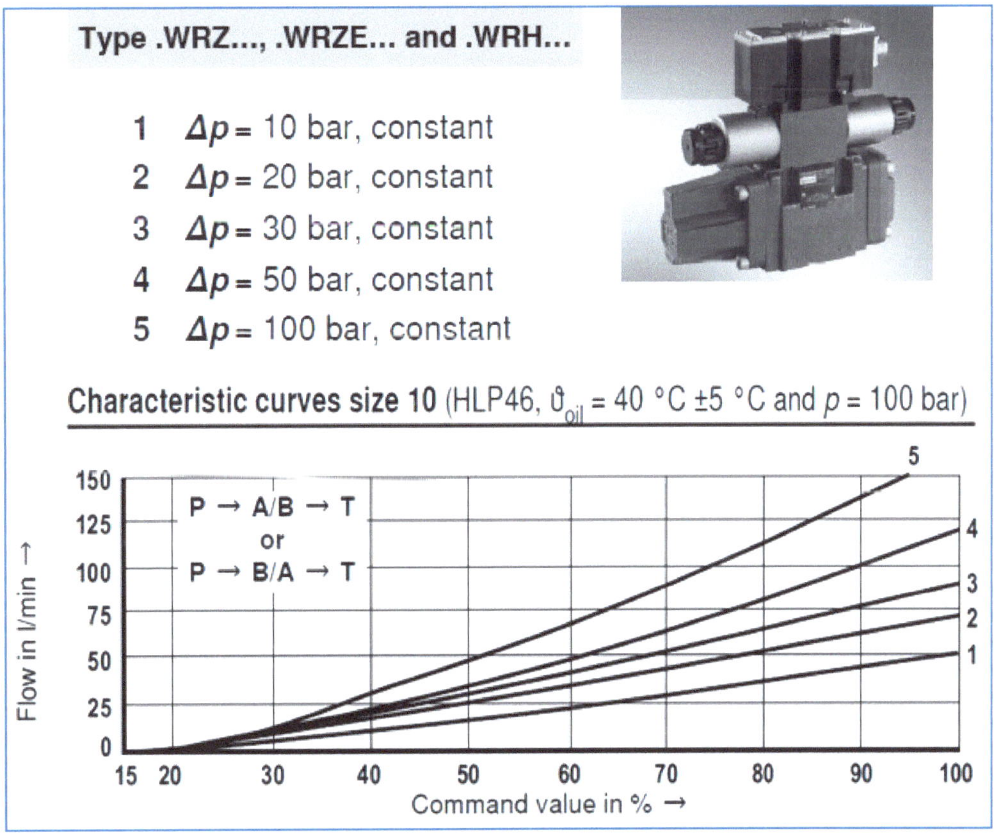

Fig. 7.32- Typical Flow Gain of a Proportional Valve - Example 6, (Courtesy of Bosch Rexroth)

As shown generically on Fig. 7.33, the following section introduces examples of flow gains for different valves were developed on the same graph.

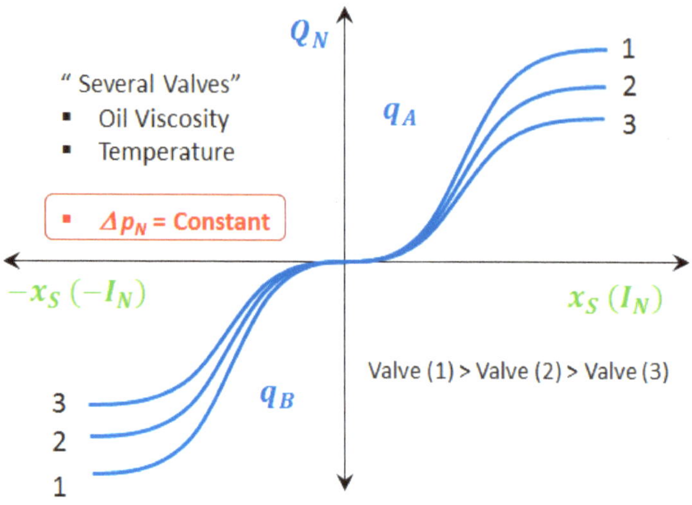

Fig. 7.33- Generic Flow Gains for Multiple Continuous Valves at a Constant Differential Pressure

Example 7, Fig. 7.34, shows flow gains for two valves under a maximum pressure of 25 MPa.

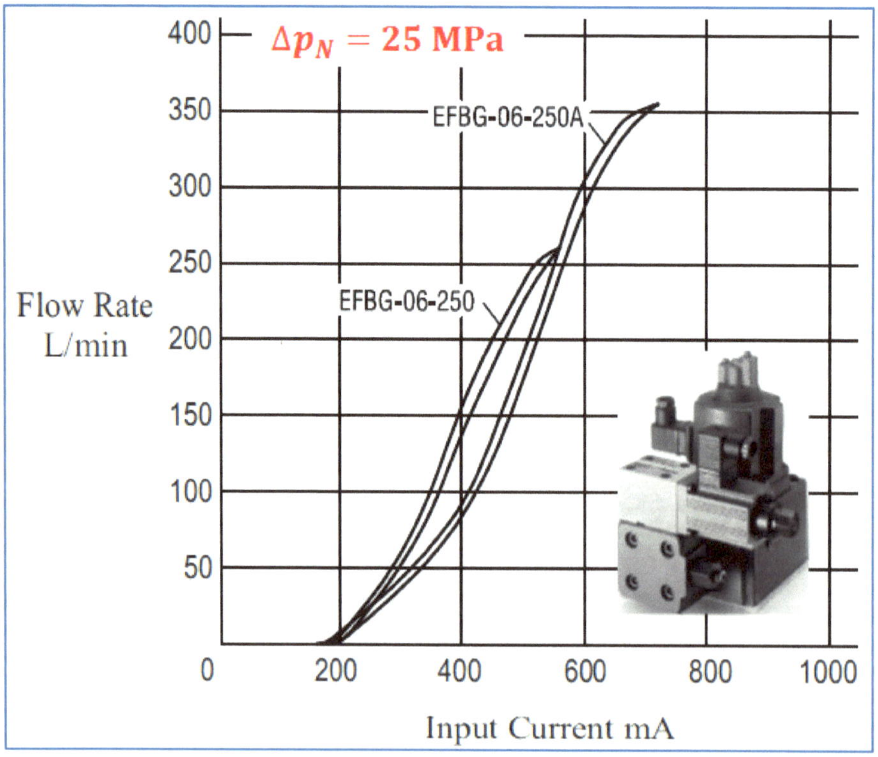

Fig. 7.34- Typical Flow Gains for Proportional Valves - Example 7, (Excerption from Eaton)

Example 8, Fig. 7.35, shows flow gains for six valves under a nominal pressure of 30 bar. The figure shows that, despite curves 3 and 4 have the same maximum flow, curve 3 is for a spool with square notches and curve 4 is for a spool with delta notches. Similarly curves 5 and 6.

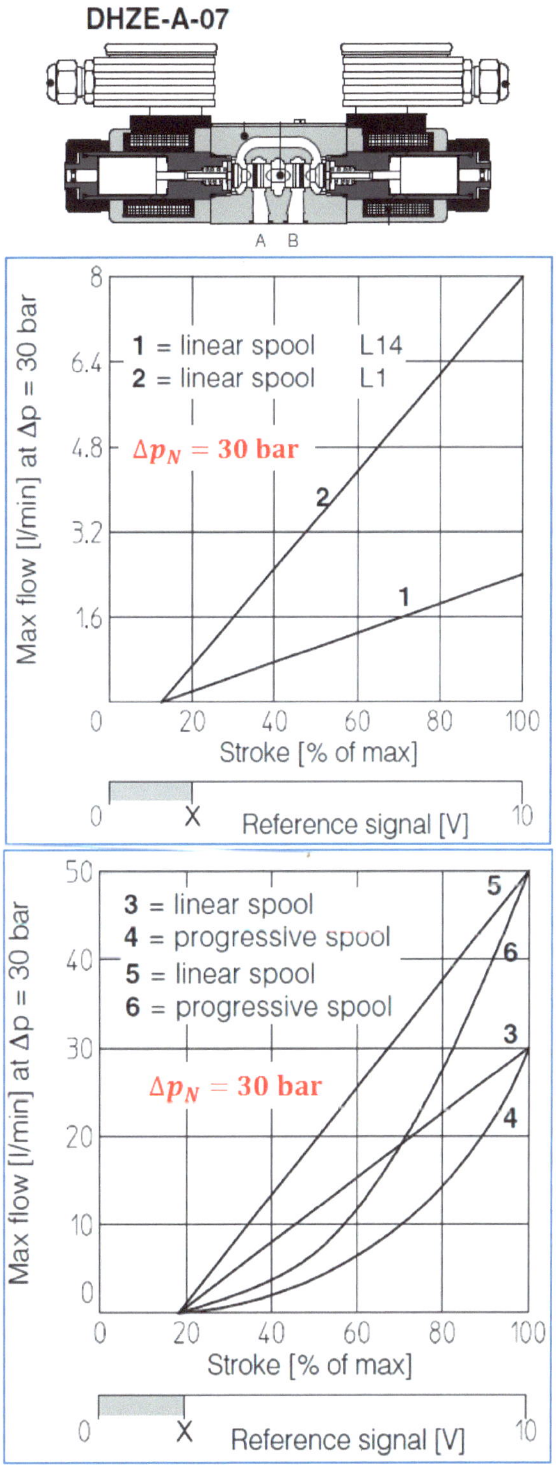

Fig. 7.35- Typical Flow Gains for Proportional Valves - Example 8, (Courtesy of Atos)

Example 9, Fig. 7.36, shows flow gains for five pilot-operated overlapped valves under a nominal pressure of 10 bar.

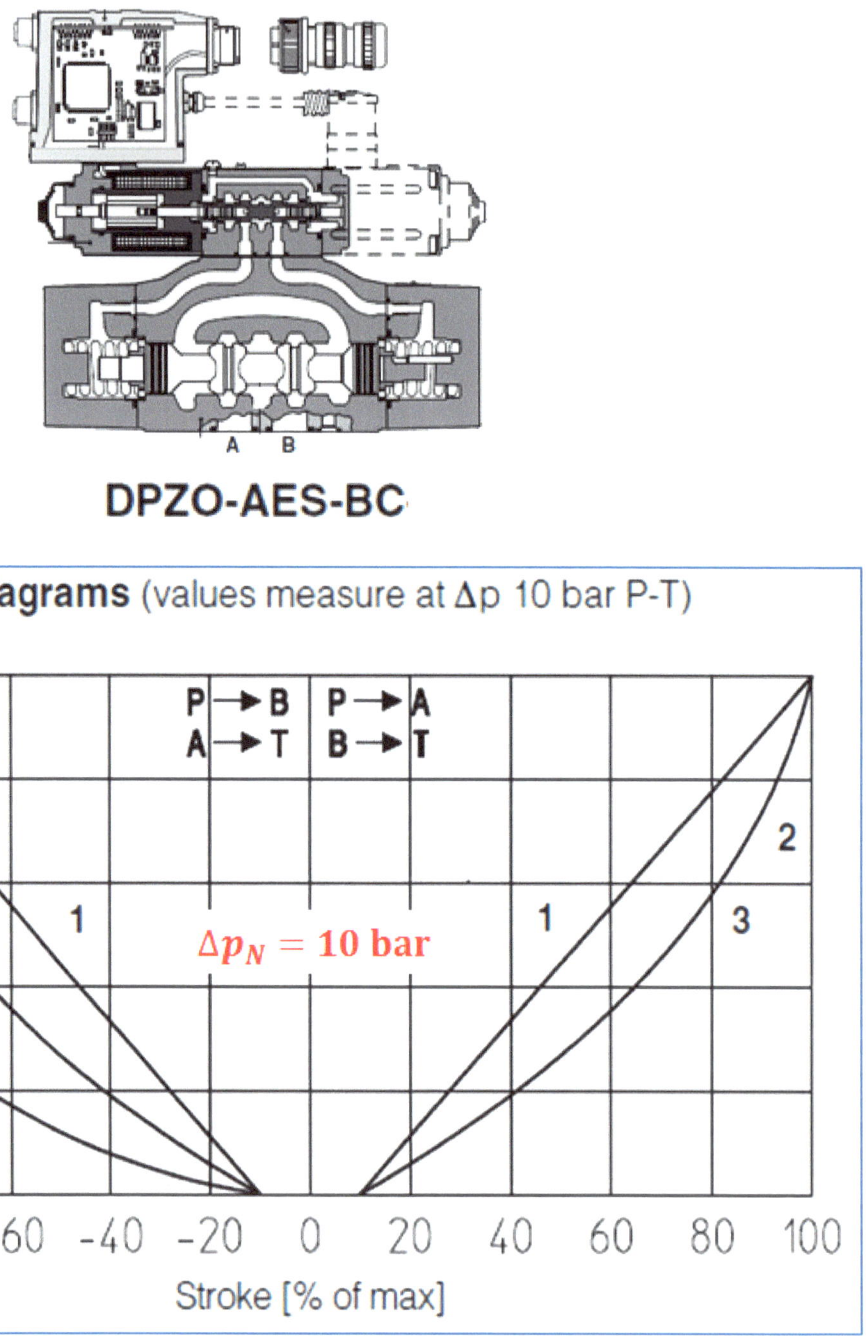

Fig. 7.36- Typical Flow Gains for Proportional Valves - Example 9, (Courtesy of Atos)

7.5.2- Valve Flow-Pressure Coefficient

As shown in Fig. 7.37, Flow-Pressure Coefficient is defined as the flow through the valve as function of the differential pressure at a constant input command.

Equation 7.2 uses the partial derivative expression to calculate the pressure-coefficient because the flow through the valve is function of more than one parameter (input signal and the differential pressure).

$$(\boldsymbol{Flow - Pressure})\ \boldsymbol{Coeifficient} = \boldsymbol{k_{Qp}} = \left[\frac{\partial Q_N}{\partial p_N}\right]_{I_{N=\,Constant}} \qquad \mathbf{7.2}$$

- ❑ Q_N = Output Nominal "Rated" Flow (developed by testing the valve).
- ❑ I_N = Constant Nominal "Rated" Input Signal (used to test the valve).
- ❑ P_N = Variable Differential Pressure (across the valve).

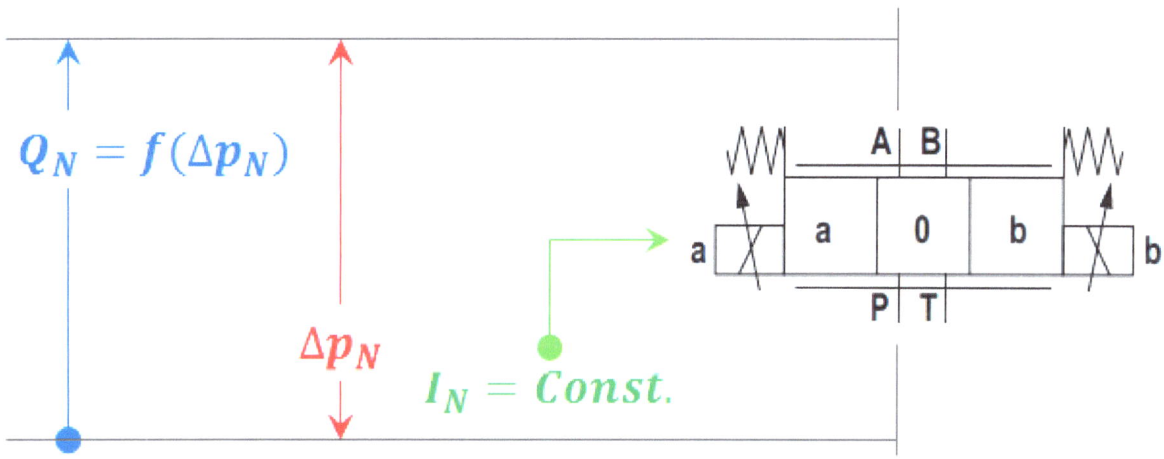

Fig. 7.37- Interpretation of Valve Flow-Pressure Coefficient

Figure 7.38 shows a generic flow-pressure coefficient for a continuous valve where the valve opens with increment of 25% of the spool stroke. The valve should be tested with varying pressure from no load conditions to blocked load conditions. As shown in the figure, a throttle valve is used to simulate the load.

At a constant input command, the test starts with a fully opened throttle (no load condition). The valve is gradually closed, and the corresponding flow is recorded. When the throttle valve is fully closed (blocked load condition) load pressure p_A equals the supply pressure p_S and no flow passes through the valve. If the valve is symmetric, one quadrant is enough.

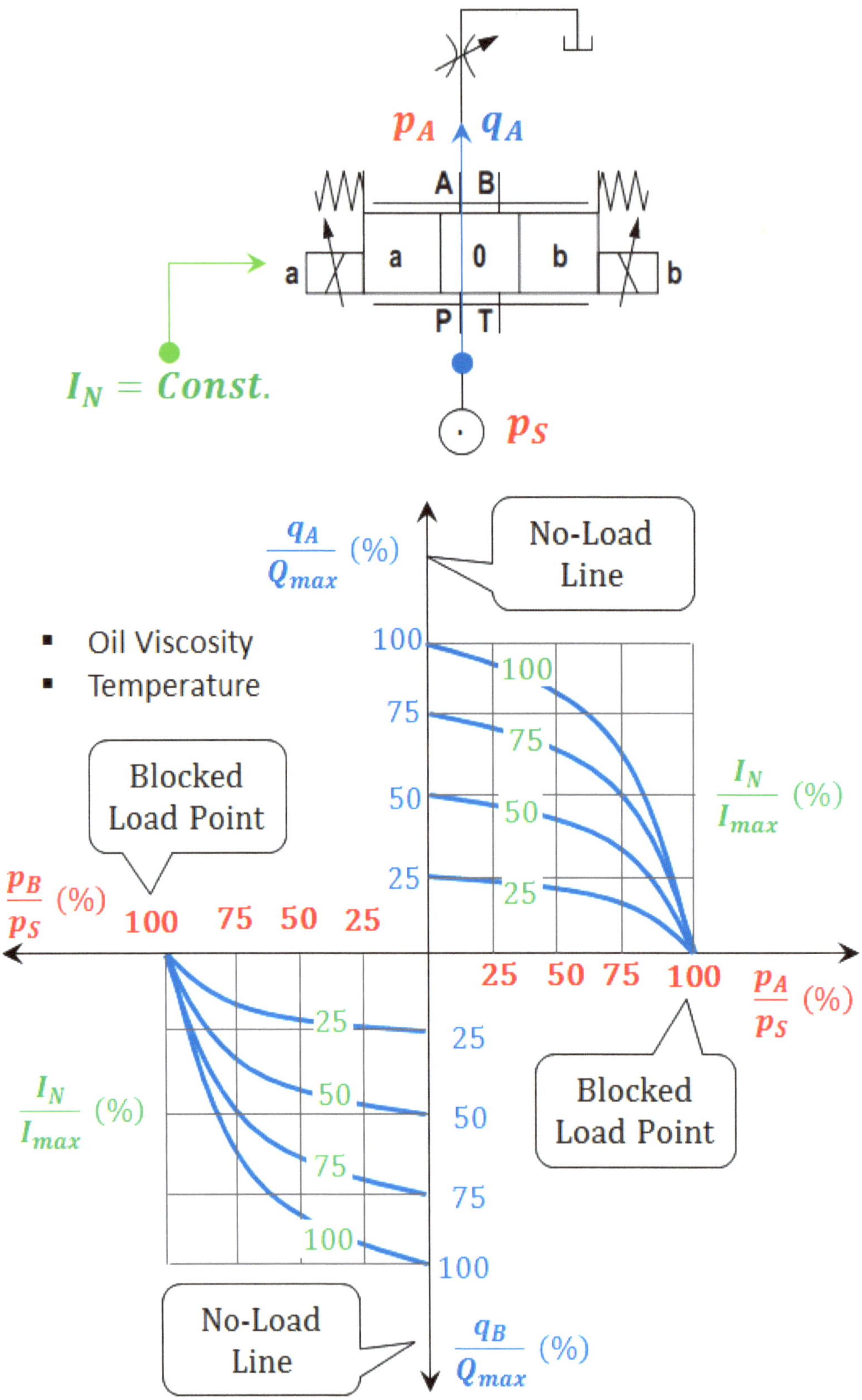

Fig. 7.38- Generic Flow-Pressure Coefficient for a Continuous Valve

The following section shows typical flow-pressure examples of actual valves.

Example 1, Fig. 7.39, shows the flow-pressure coefficient for a servo valve.

Rated Flow @ p 1000 PSI = 3.78-151 LPM (1 – 40 GPM)

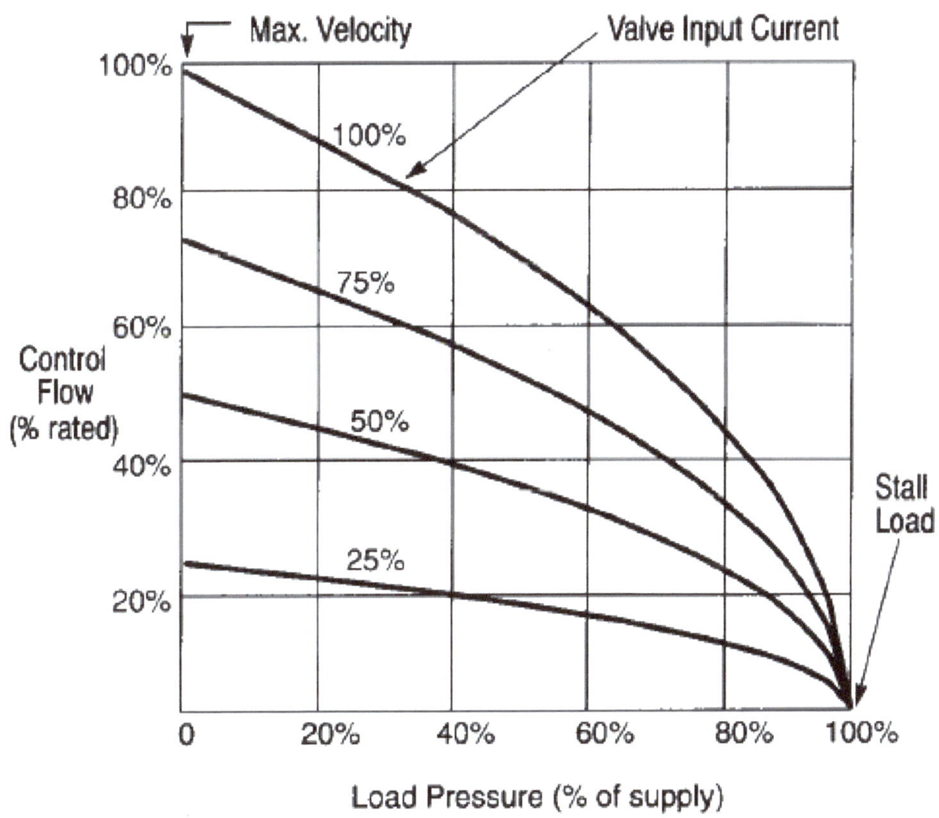

Fig. 7.39- Typical Flow-Pressure Coefficient for a Servo Valve - Example 1 (Courtesy of Parker)

Example 2, Fig. 7.40, shows the flow-pressure coefficient for six sizes of servo valves at 100% command. The relationship is nonlinear, but the graph uses logarithmic scale to present the data in linear format. Bolded vertical line at 70 bar, which is the standard nominal (test) differential pressure for two valve passages as indicated on the graph (p_P - p_A - p_T).

Ordering code	Rated flow	Curve
2	2 l/min	1
5	5 l/min	2
10	10 l/min	3
15	15 l/min	4
20	20 l/min	5
25	25 l/min	6

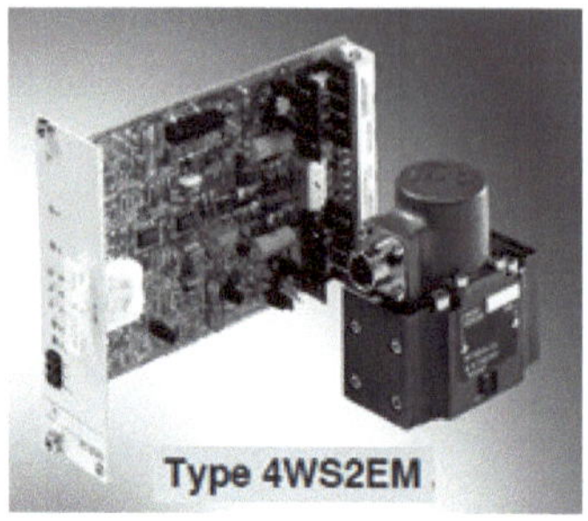

Δp = Valve pressure differential (inlet pressure p_P minus load pressure p_L minus return flow pressure p_T)

Characteristic curves (measured with HLP32, ϑ_{Oil} = 40 °C ± 5 °C)

Flow/load function (tolerance ±10 %) with 100 % command value signal

Notice: Observe the flow values in max. command value range

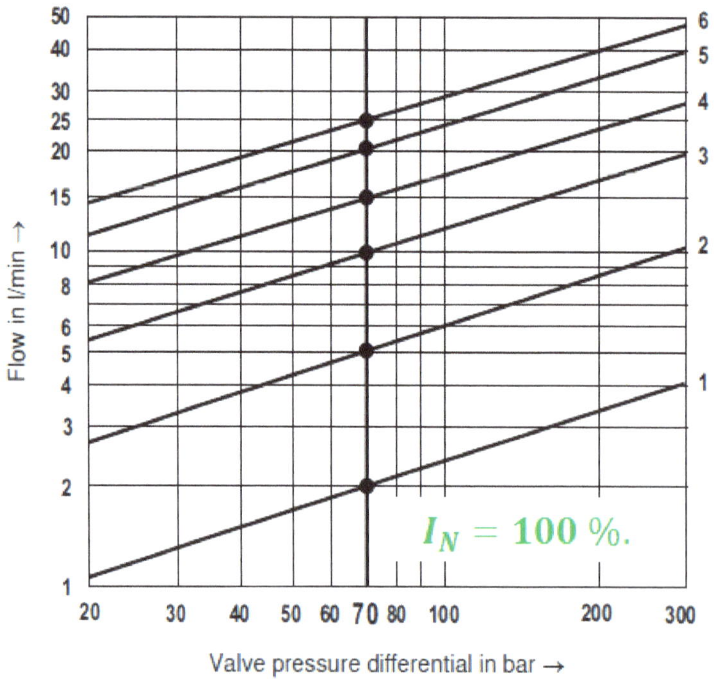

Fig. 7.40- Typical Flow-Pressure Coefficients for Servo Valves - Example 2 (Courtesy of Bosh Rexroth)

Example 3, Fig. 7.41, shows same as in the previous example but for a different valve brand.

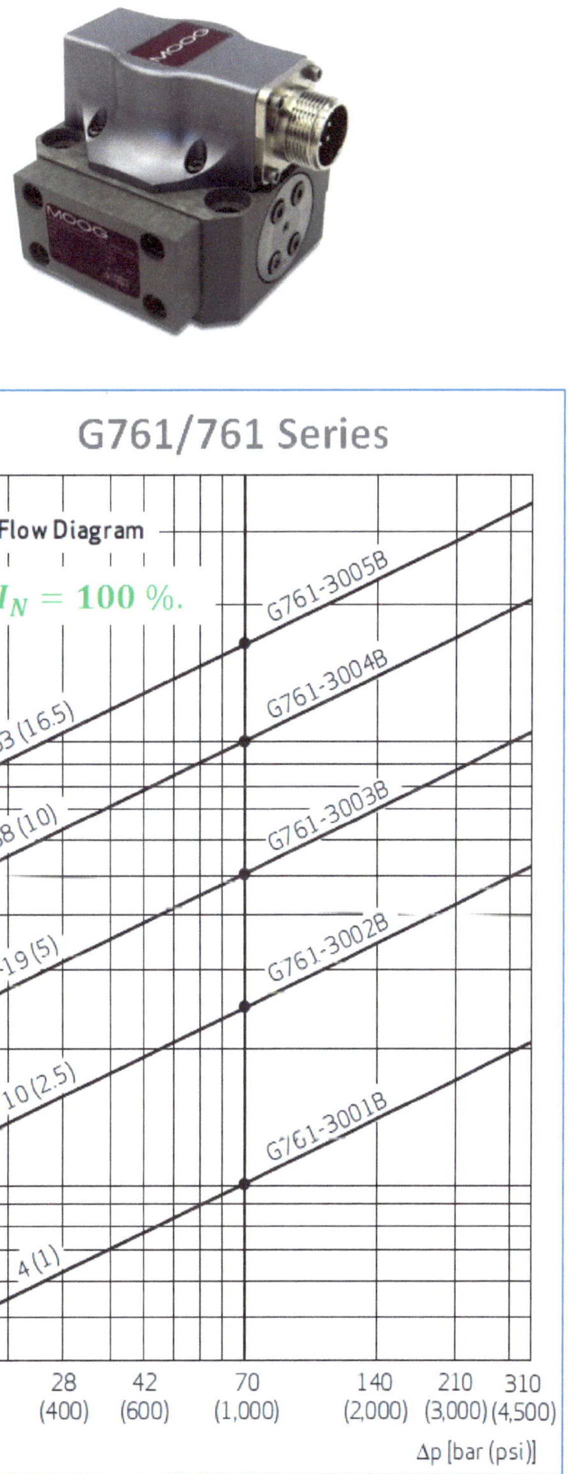

Fig. 7.41- Typical Flow-Pressure Coefficients for Servo Valves – Example 3 (Courtesy of Moog)

7.5.3-Power Limits

As explained in Fig. 7.42, flow through the valve is function of the differential pressure. The power is quantified as the product of the flow times the differential pressure.

Within the normal operating range, increasing the differential pressure increases the flow and consequently the amount of power through the valve.

Beyond a certain power limit, increasing the differential pressure will result in decreasing the flow and consequently the power. That happens because of:

- Increasing flow forces trying to close the valve.
- Significant increase of the fluid speed results in cavitation and air lock.

Therefore, a system designer should not just rely on the mathematics to predict the flow versus the differential pressure across the valve. *Power Limit* of the valve must be checked.

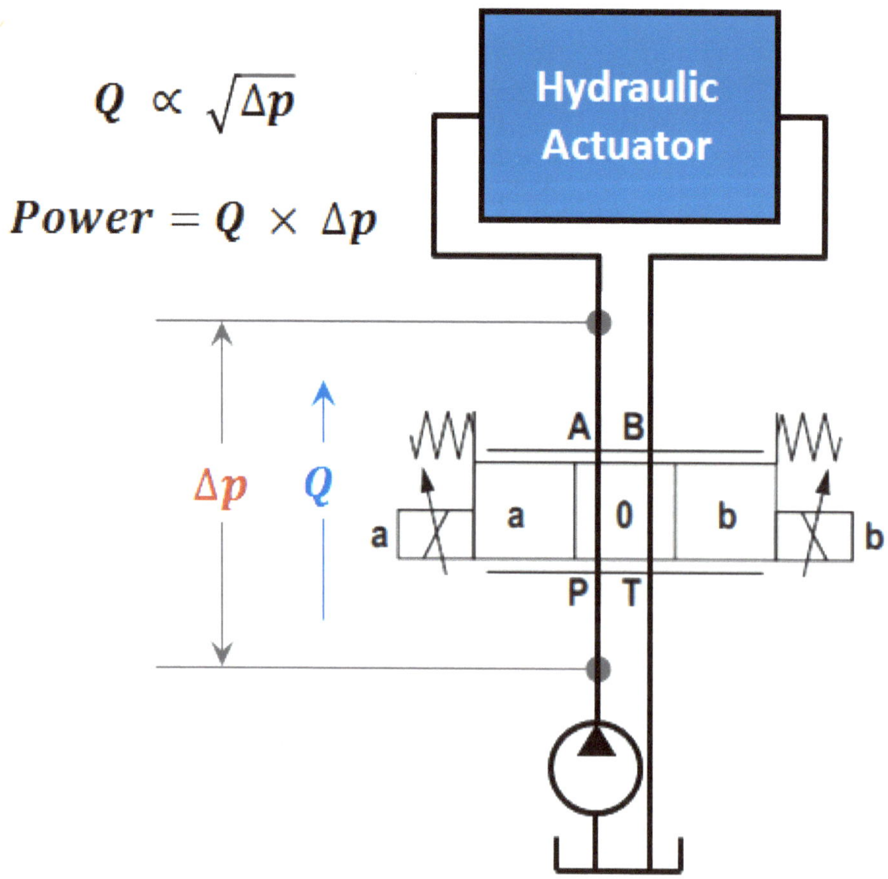

Fig. 7.42- Valve Power Limit Interpretation

Example 1, Fig. 7.43, shows the power limit for a small proportional valve where the valve is fully opened. A system designer has the choice of determining the working differential pressure range with the consideration of the power limits.

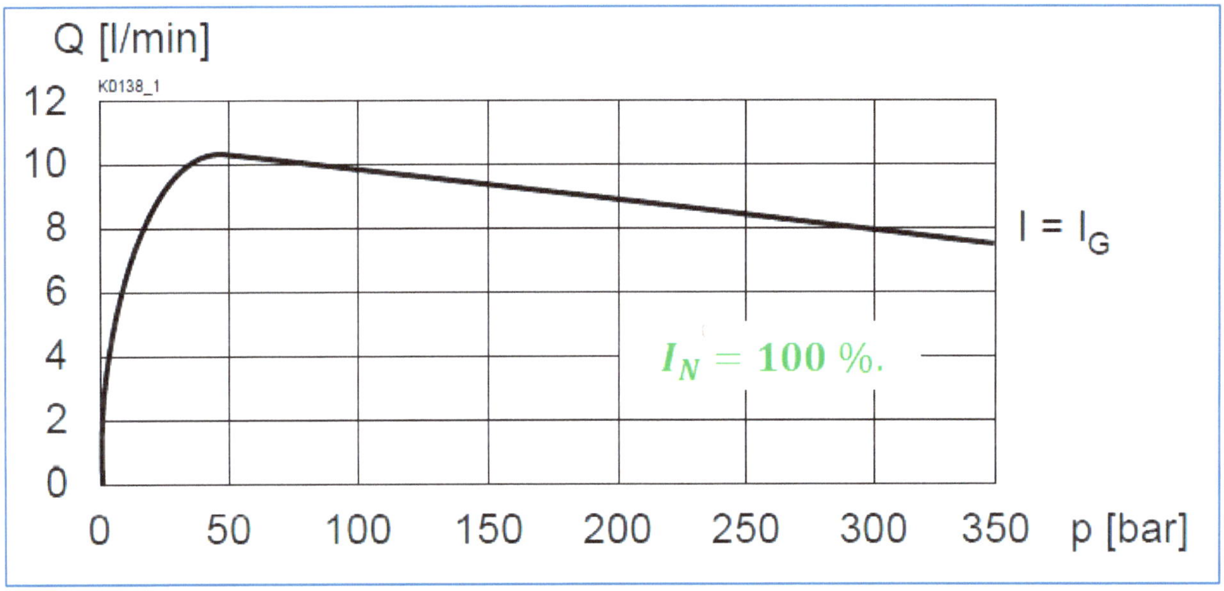

Fig. 7.43- Typical Performance Limits for a Proportional Valve - Example 1, (Courtesy of Wandfluh)

Example 2, Fig. 7.44, shows other presentation of the power limits for a proportional valve where the axis swapped. The figure shows that above 70 bar differential pressure, the flow through the valve starts to decrease.

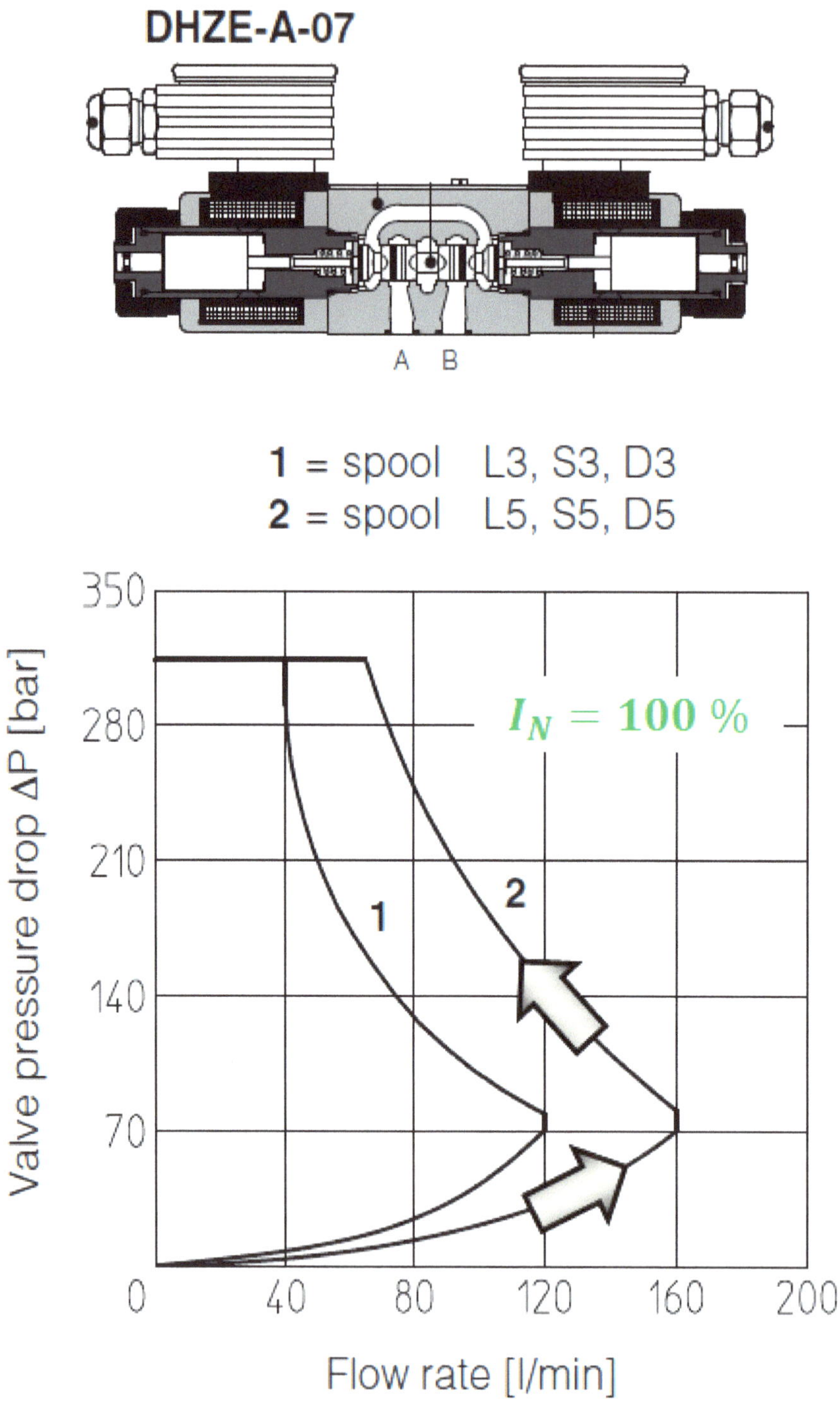

Fig. 7.44- Typical Performance Limits for a Proportional Valve - Example 2, (Courtesy of Atos)

7.5.4-Pressure Gain

Valve *Pressure Gain* is required for designing an open-loop or closed-loop pressure control or load control systems.

With the aid of Fig. 7.45, Eq. 7.3 defines the pressure gain as the differential pressure across the valve versus the input command at constant flow through the valve.

$$\textit{Pressure Gain} = k_{pI} = \left[\frac{\partial p_N}{\partial I_N}\right]_{Q_N=Constant} = \frac{k_{QI}}{k_{Qp}} \qquad 7.3$$

- ❑ p_N = Output Nominal "Rated" Differential Pressure (developed by testing the valve).
- ❑ Q_N = Constant Nominal "Rated" Flow (used to test the valve).
- ❑ I_N = Variable Input Command to the valve.

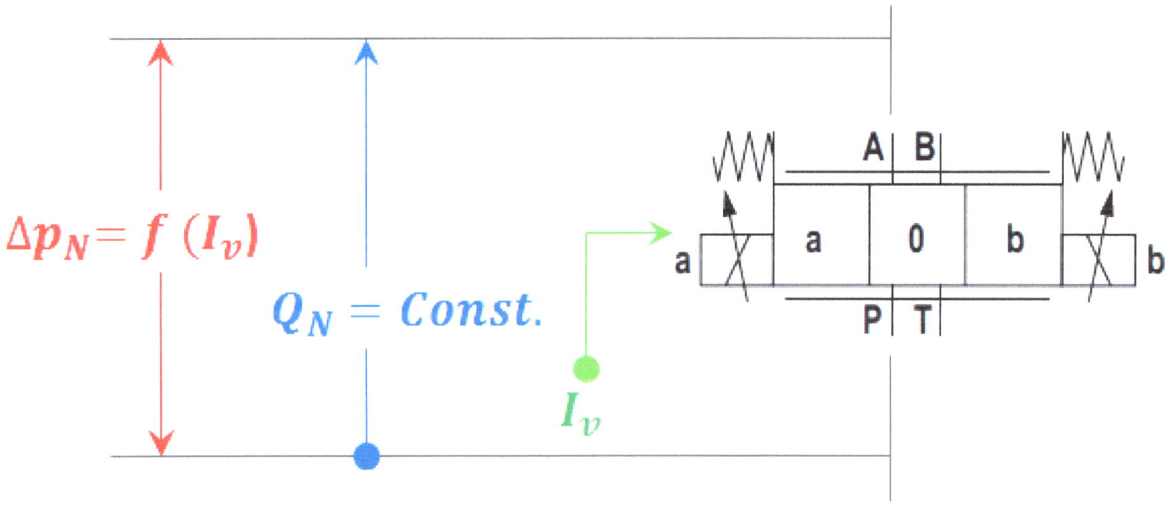

Fig. 7.45- Interpretation of Pressure Gain

Figure 7.46 shows a generic pressure gain plot. The curve has been developed in normalized form with respect to the maximum values. The blocked load point is when the load pressure (p_A or p_B) equals the supply pressure p_S. It is to be remembered that, as it has been explained earlier, the valve should be underlapped. The input command should be within the underlapped percentage of the spool stroke.

Figure 7.47 shows pressure gain for a typical valve. Maximum supply pressure and maximum input command must be reviewed from the data sheet of the valve.

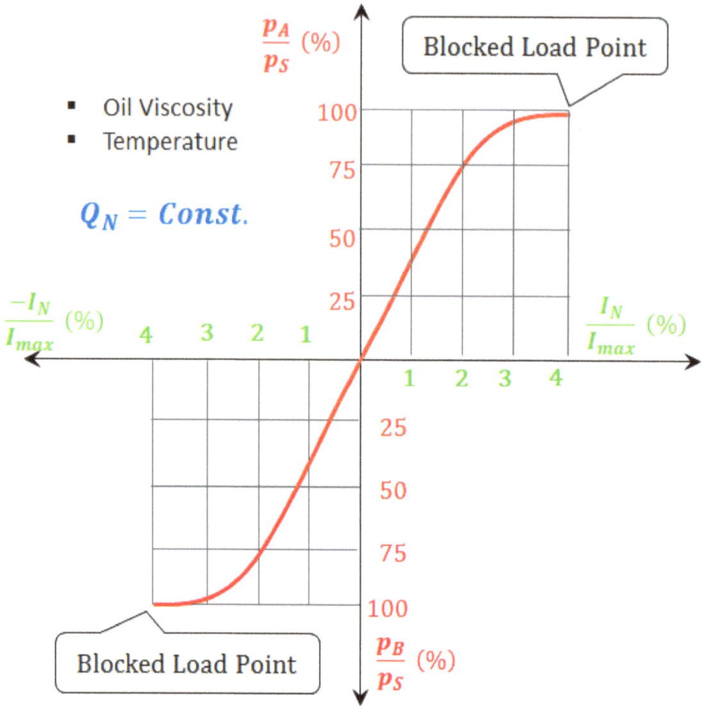

Fig. 7.46- Generic Pressure Gain for a Continuous Valve

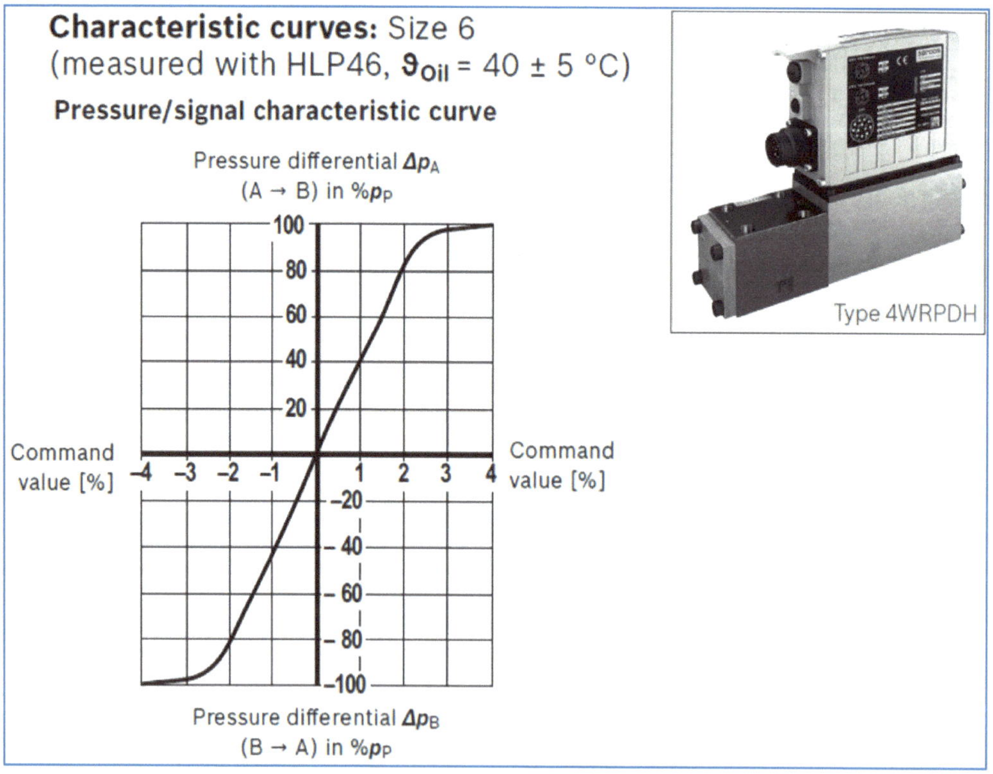

Fig. 7.47- Typical Pressure Gain (Courtesy of Bosch Rexroth)

7.5.5-Valve Sizing

Selecting a valve size is a crucial step in designing an electro-hydraulic control system. All hydraulic components, including valves are sized primarily based on flow (NOT pressure). Under sizing a valve will make it act as a throttle valve no matter what the original function of the valve is. Oversizing a valve will reduce the control resolution or its ability to perform the control function.

Direct-operated (single-stage) valves can handle up to approximately 100 liter/min (20 gpm). Above this flow range, valves are configured as pilot-operated (two or three-stages).

The following section, with the aid of Fig. 7.48, presents a step-by-step procedure of sizing a valve for a specific application:

Step 1-Load Flow (Q_L) Analysis:
Based on the motion profile or the duty cycle of the actuator, load flow to meet the speed requirements is determined considering the actuator's two directions of motion.

Step 2-Load Pressure (p_L) Analysis:
Load pressure based on load analysis considering the actuator's two directions of motion.

Step 3-Constant Supply Pressure (p_S):
Define the constant supply pressure at the pump outlet. There are various techniques to secure constant supply pressure, but this topic is out of scope of this book. However, to ease the valve selection, it is recommended to select p_S so that the resulted load differential pressure ($\Delta p_L = p_S - p_L$) is equal to the nominal differential pressure Δp_N used by the manufacturer to test the valve.

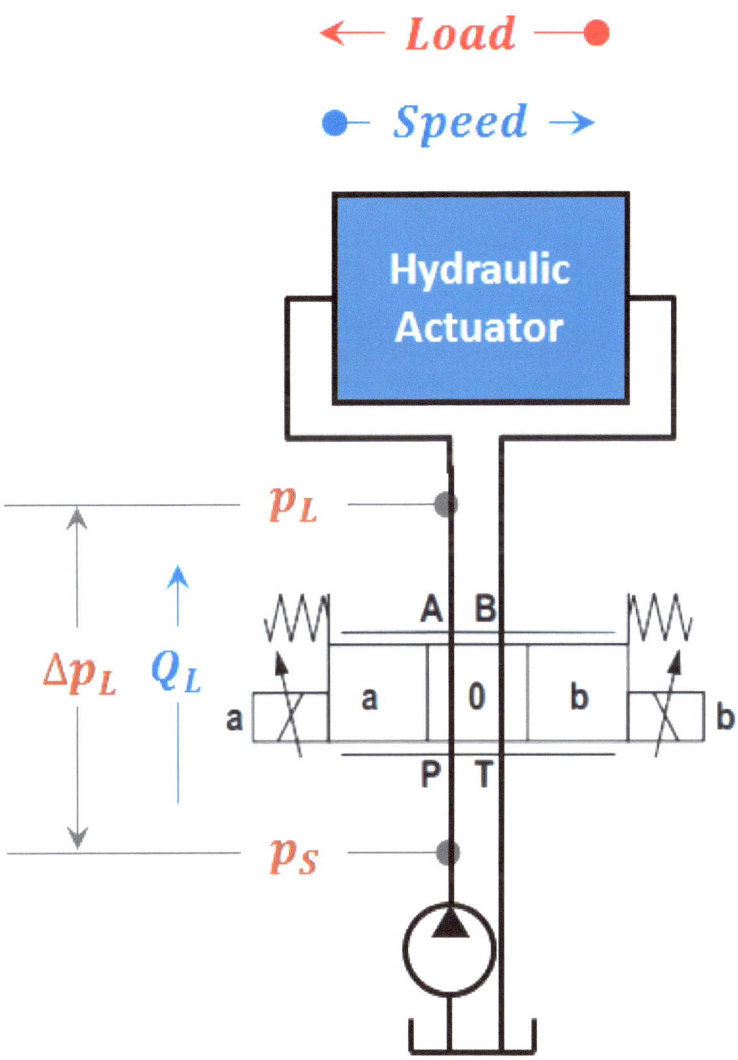

Fig. 7.48- Valve Sizing Calculations

Step 4-Selecting a Valve from a Catalogue:

Choice A: based on (Flow-Pressure) Coefficient:

This is the easiest way because the suitability of a valve to the application can be directly checked by locating the operating point on the curve. As best practices, valve could be sized based on (80-90) % of input signal to avoid the nonlinear portion of the characteristics when the valve is saturated.

The following section introduces some examples based on an assumption of:

$$\Delta p_L = 100 \text{ bar and } Q_L = 30 \text{ lit/min.}$$

Example 1, Fig. 7.49. By locating the operating point, the figure shows that the valve can meet the application at approximately 94% of the input command, which is Ok.

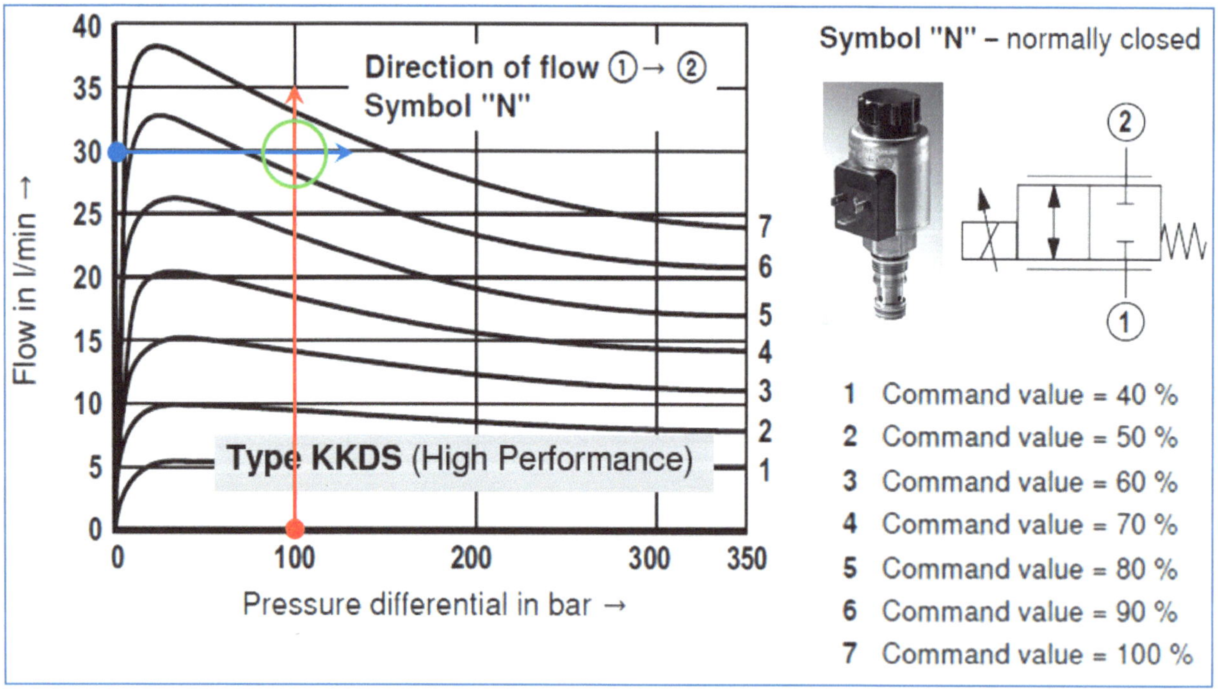

Fig. 7.49- Valve Selection, Example 1 (Courtesy of Bosch Rexroth)

Example 2, Fig. 7.50. The figure shows that valve with ordering code 25 (curve # 6) is ok for the job. It is to be noted here that this condition is under an input command of 100%.

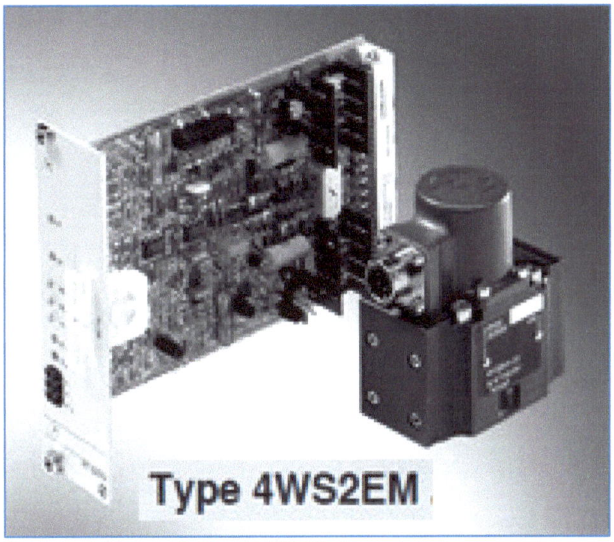

Characteristic curves (measured with HLP32, ϑ_{Oil} = 40 °C ± 5 °C)

Flow/load function (tolerance ±10 %) with 100 % command value signal

Notice: Observe the flow values in max. command value range

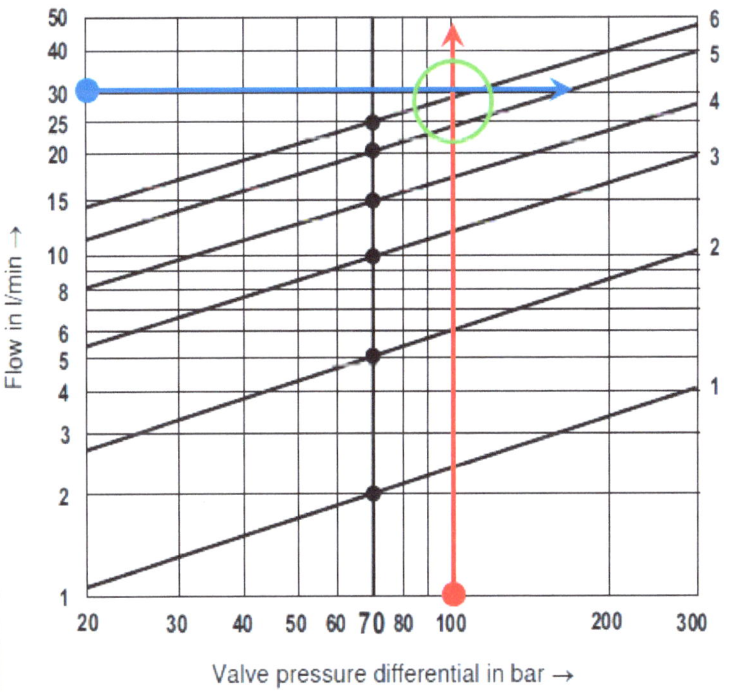

Ordering code	Rated flow	Curve
2	2 l/min	1
5	5 l/min	2
10	10 l/min	3
15	15 l/min	4
20	20 l/min	5
25	25 l/min	6

Δp = Valve pressure differential (inlet pressure p_P minus load pressure p_L minus return flow pressure p_T)

Fig. 7.50- Valve Selection, Example 2 (Courtesy of Bosch Rexroth)

Example 3, Fig. 7.51. The figure shows that the operating point is out of the valve operating range. The required load flow through this valve can be met at a very low pressure, which means the valve is oversized.

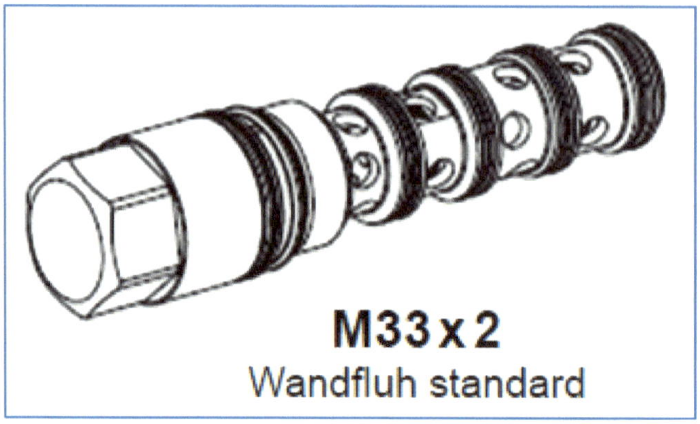

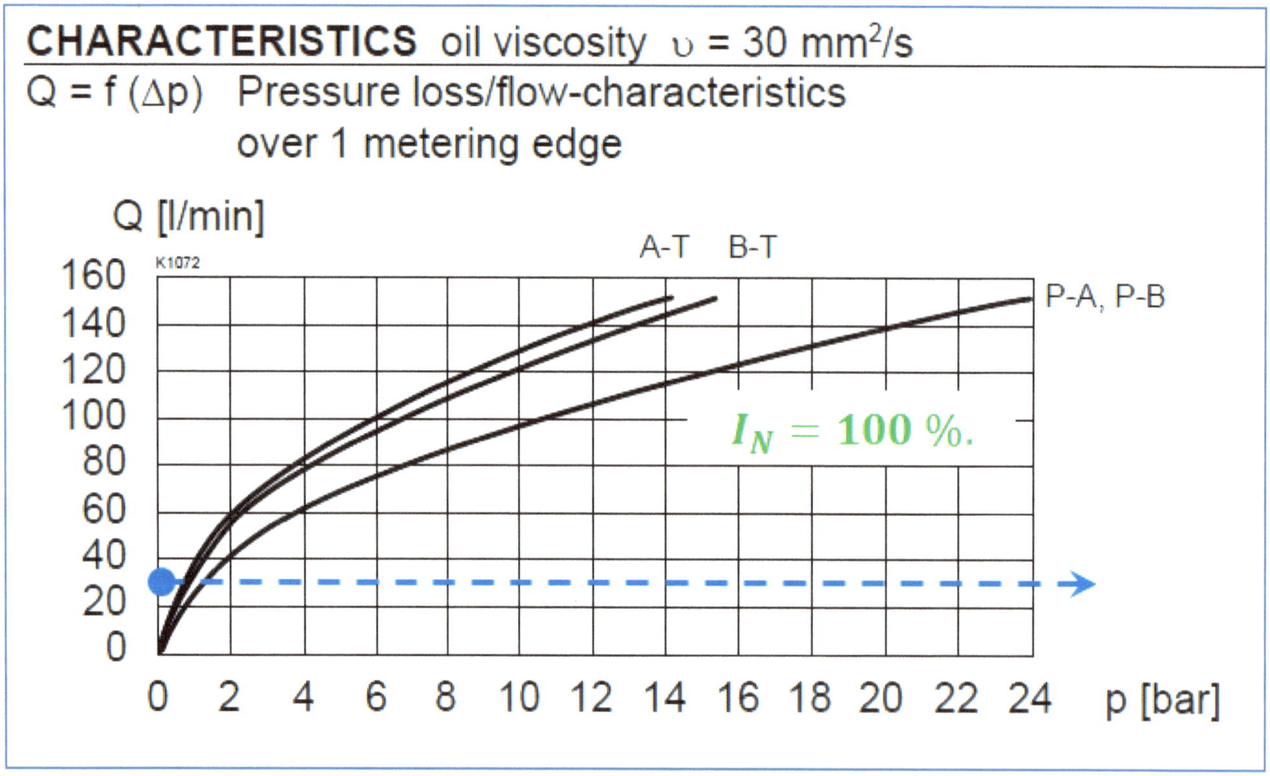

Fig. 7.51- Valve Selection, Example 3 (Courtesy of Wandfluh)

Example 4, Fig. 7.52. The figure shows that the operating point is out of the valve operating range. By applying the load differential pressure, the valve passes flow lower than needed, which means the valve is undersized.

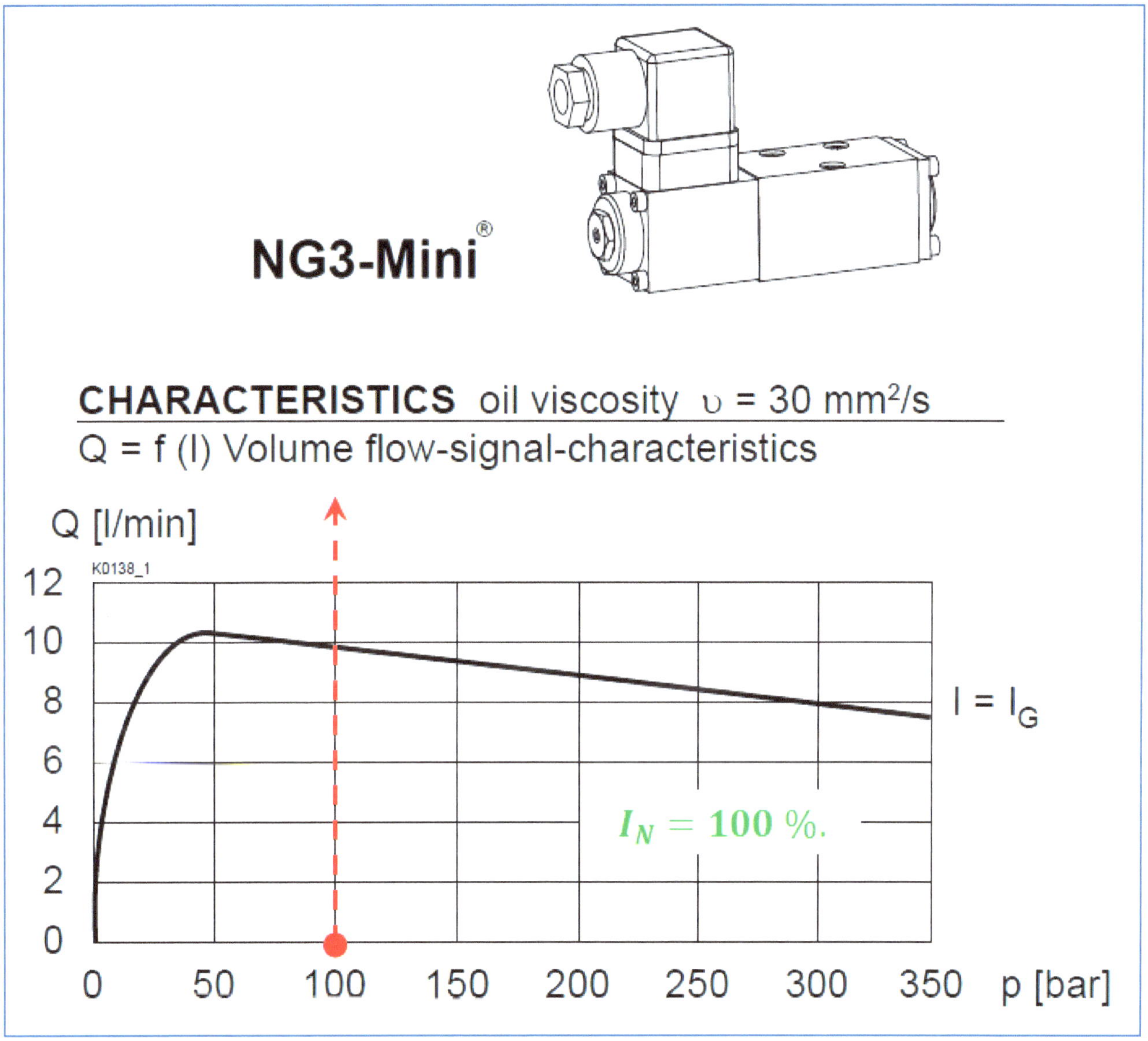

Fig. 7.52- Valve Selection, Example 4 (Courtesy of Wandfluh)

Choice B: Selecting a valve based on Flow Gain:

If the flow-pressure coefficient isn't found, there is no choice but using the flow gain of the valve. In this case there are one of two possibilities:

Case # 1- Direct Selection if $\Delta p_L = \Delta p_N$:

If the system designer were able to tweak his design to make $\Delta p_L = \Delta p_N$. The only step needed here is to check if the valve can pass the required load flow within (80-100) % of the input command.

Example 5, Fig. 7.53. In this example, one of the nominal test differential pressures (curve #5) equals the application's load differential pressure. By checking the flow at 100% command, the valve is found little small, so it can barely do the job.

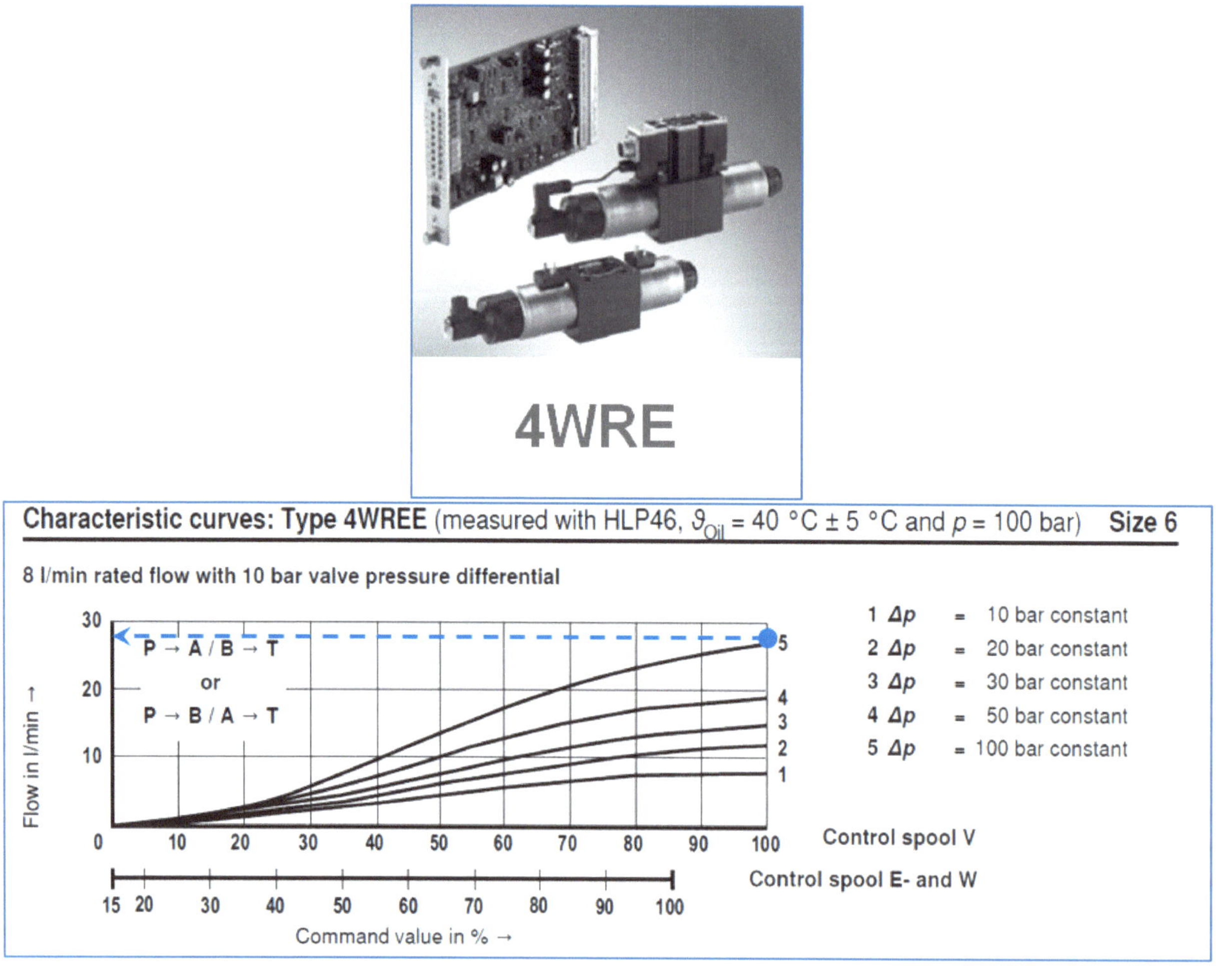

Fig. 7.53- Valve Selection, Example 5 (Courtesy of Bosch Rexroth)

Example 6, Fig. 7.54. It is obvious that the valve in this example is larger than needed. If this valve is selected, the spool will move to a maximum of approximately 42% of its full stroke to meet the required load flow.

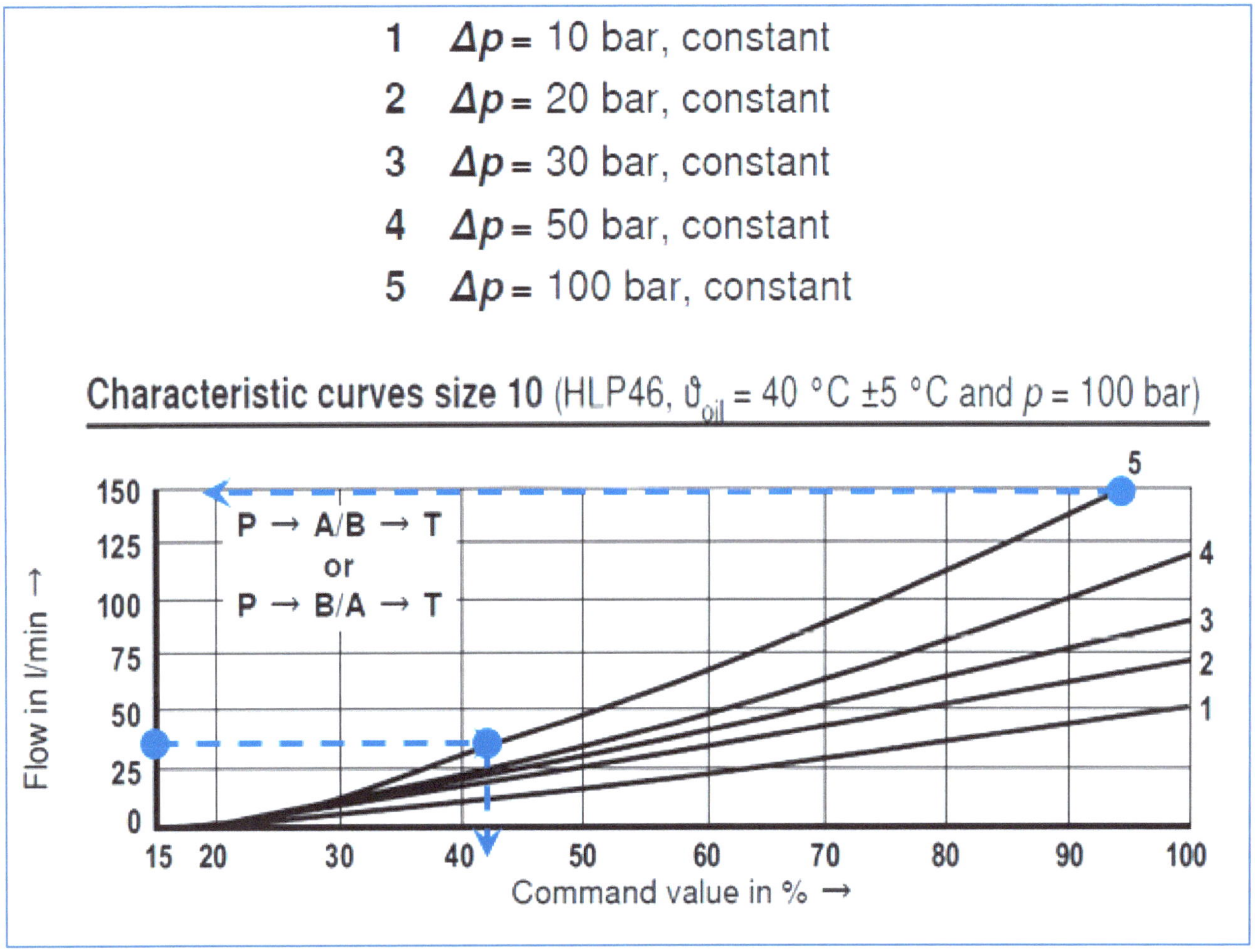

Fig. 7.54- Valve Selection, Example 6 (Courtesy of Bosch Rexroth)

Example 7, Fig. 7.55. This valve is also oversized and can do the job within 60% of the spool stroke.

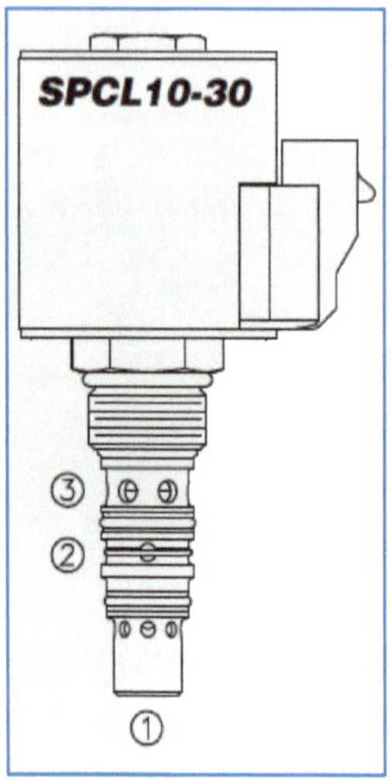

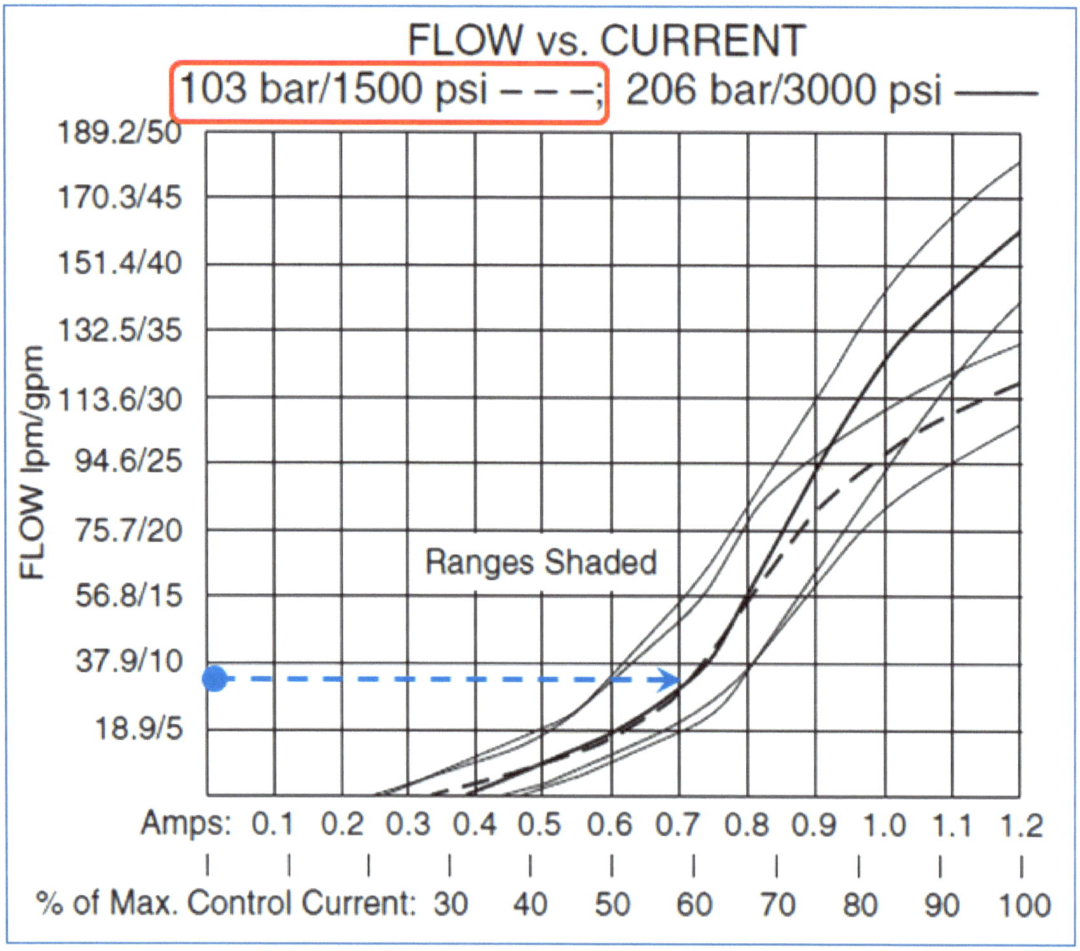

Fig. 7.55- Valve Selection, Example 7 (Courtesy of Hydraforce)

Case # 2- Indirect Selection if $\Delta p_L \neq \Delta p_N$::

If there is no flow gain developed at the load differential pressure and the system design can't be tweaked to meet the published flow gains. This case is very possible because it should be understood that the manufacturer can't develop flow gains at an infinite number of nominal differential pressures to cover all possible actual applications.

In such a case, an answer to the following question is required.

If the system design requires Q_L at Δp_L, what is the equivalent Q_N if the valve is tested at Δp_N?

To answer this question, equation 7.4 should be solved. The equation has been developed based on the matter of fact that Q is proportional to Δp.

$$Q \propto \sqrt{\Delta p} \rightarrow \frac{Q_N}{Q_L} = \sqrt{\frac{\Delta p_N}{\Delta p_L}} \quad \textit{at a constnat input signal}$$

$$\rightarrow Q_N = Q_L \sqrt{\frac{\Delta p_N}{\Delta p_L}} \qquad 7.4$$

Then the valve should be checked if it can pass Q_N at Δp_N. It is to be remembered here that this step must be followed by checking the valve power limits.

Example 8, Fig. 7.56. In this example, a group of valves have been tested at Δp_N = 30 bar. Then by applying Eq. 7.4, the size of the nominal valve is 16.4 lit/min. The best selection then is valve number 4 because it passes the flow around 80% of the spool stroke.

$$Q_N\left(\frac{lit}{min}\right) = 30\left(\frac{lit}{min}\right)\sqrt{\frac{30\ bar}{100\ bar}} = 16.4$$

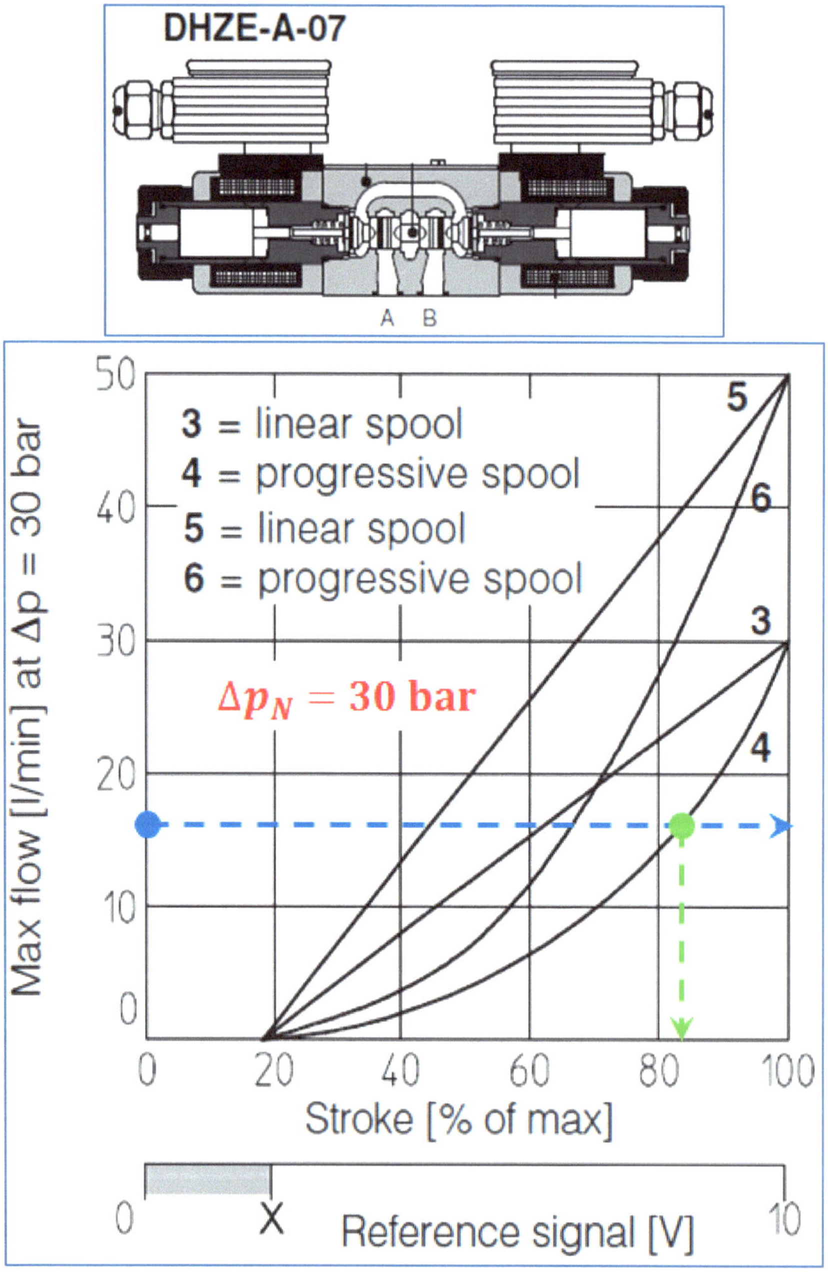

Fig. 7.56- Valve Selection, Example 8 (Courtesy of Atos)

Example 9, Fig. 7.57. The valve in this example has been tested at 30 bar nominal differential pressure as the valves in the previous example, so the same calculation is valid. By checking the nominal flow rate, the valve has less than 16.4 lit/min even at maximum spool stroke. It can be used by increasing the supply pressure.

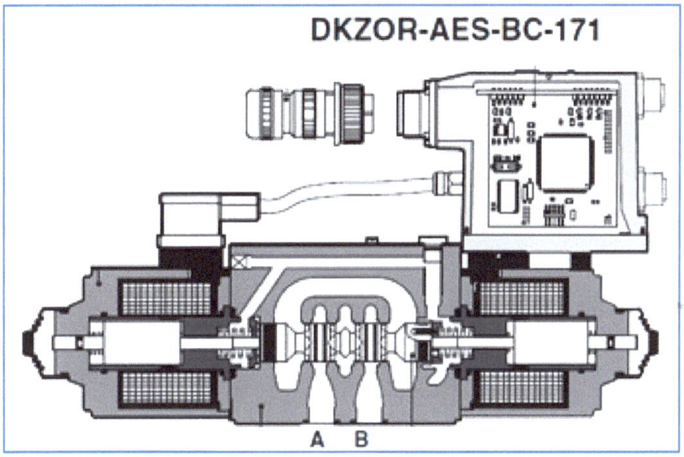

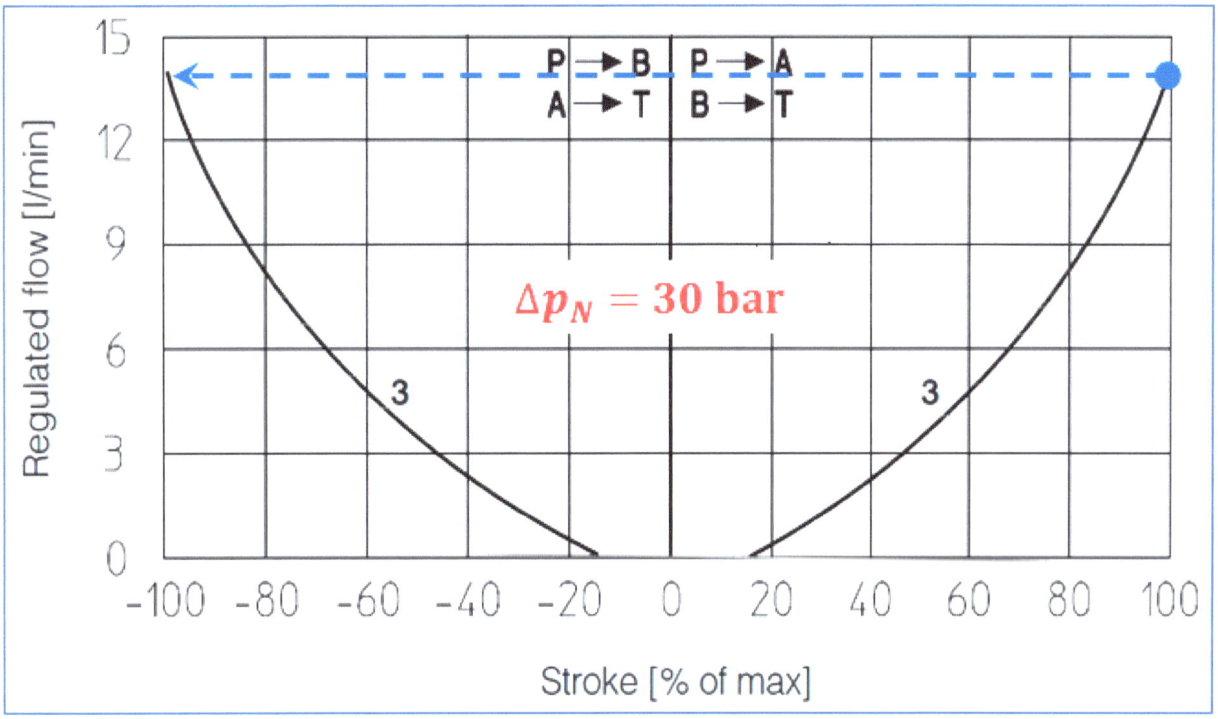

Fig. 7.57- Valve Selection, Example 9 (Courtesy of Atos)

Example 10, Fig. 7.58. The valve in this example, obviously without calculation, is oversized. The reason is that the valve passes more flow than needed at less differential pressure.

$\Delta p_N = 35$ **bar per edge**

$Q_N = 40$ lit/min

$at\ I_v = 100\%$

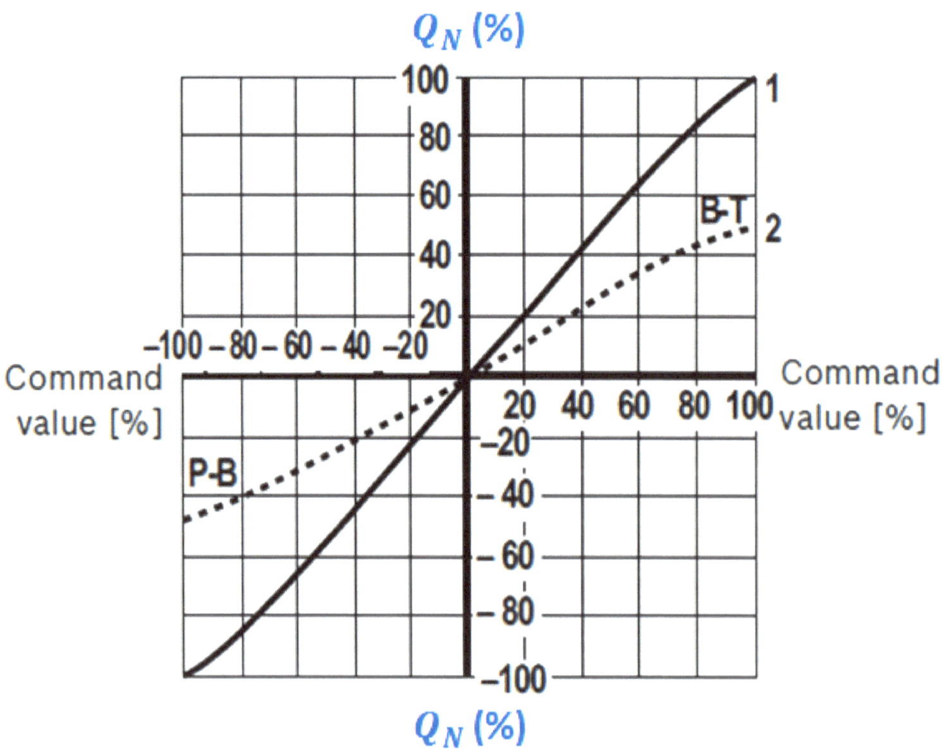

Fig. 7.58- Valve Selection, Example 10 (Courtesy of Bosch Rexroth)

7.5.6-Valve Hysteresis

Definition:

Valve *Hysteresis* is defined as a measure of the valve response deviation from the original (design) values. Hysteresis values are provided as a curve versus the input command. Figures 7.59, 7.60, and 7.61 show generic hysteresis curves for zero-lapped, overlapped, and underlapped valves; respectively.

Hysteresis Measures:

From hysteresis point of view, valves are compared based on either of the following measures:

- Absolute maximum hysteresis no matter where along the spool stroke.
- Hysteresis around the Null (almost maximum because of high friction around Null).
- Best practice is to check the hysteresis at a specific spool position around which the valve works in the application.

Order of magnitude:

- Force-controlled Proportional Valves ≤ 5%.
- Stroke-controlled Proportional Valves ≤1%
- Servo Valves ≤ 0.5%

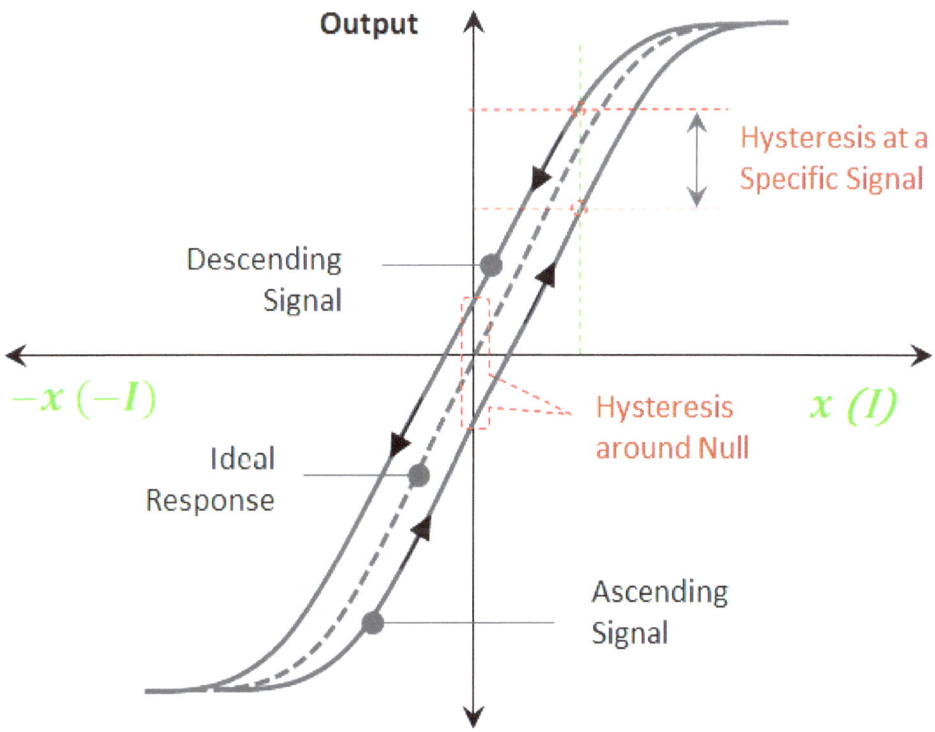

Fig. 7.59- Generic Hysteresis Curve for a Zero-Lapped Valve

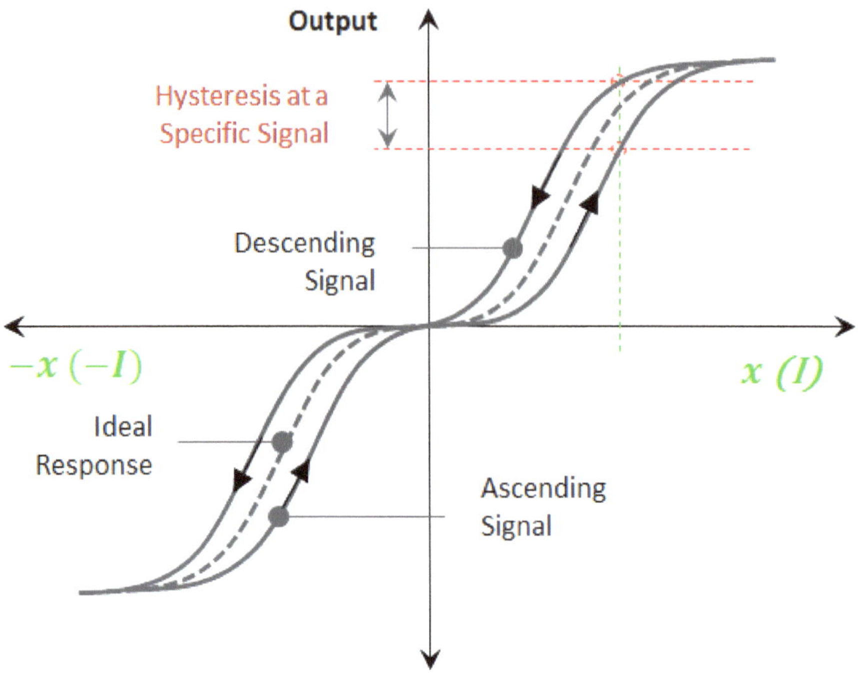

Fig. 7.60- Generic Hysteresis Curve for an Overlapped Valve

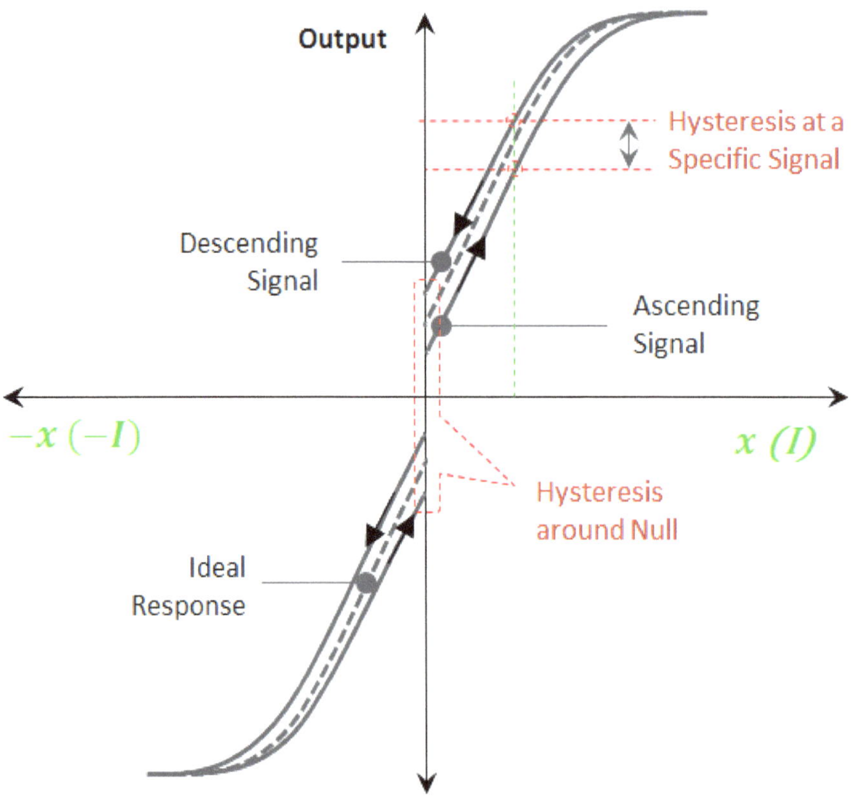

Fig. 7.61- Generic Hysteresis Curve for an Underlapped Valve

Effect of Hysteresis:
It affects the accuracy and the repeatability of control systems, particularly the position control.

Working Conditions that Increase Hysteresis:
- High working temperature.
- High viscous fluid.
- Contamination.
- Increased spool-sleeve friction.

Threshold (Response Sensitivity):
It is another measure of valve performance deviation. Assume a valve spool has moved to a certain position within the spool stroke and then stay there for a certain dwell time. If the spool is ordered to move forward, as shown in Fig. 7.62, the deviation of the curve is known as the *Threshold* or Response Sensitivity.

Reversal Error:
After the spool stays at rest for a certain dwell time, if the spool is ordered to move backward as shown in Fig 7.63, the deviation of the curve is known as the *Reversal Error* or Inversion Range.

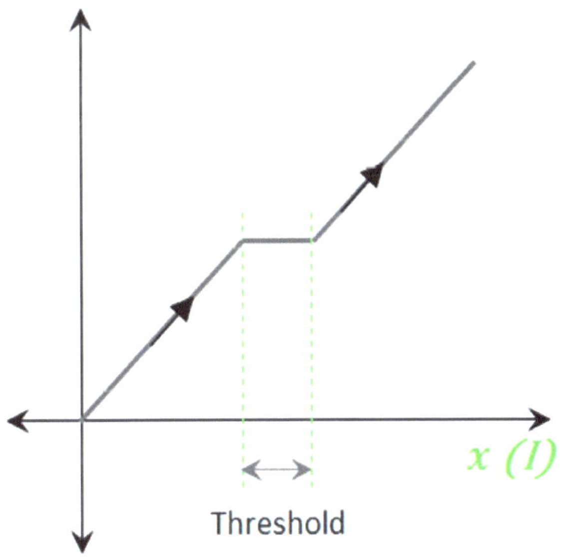

Fig. 7.62- Threshold for a Continuous Valve

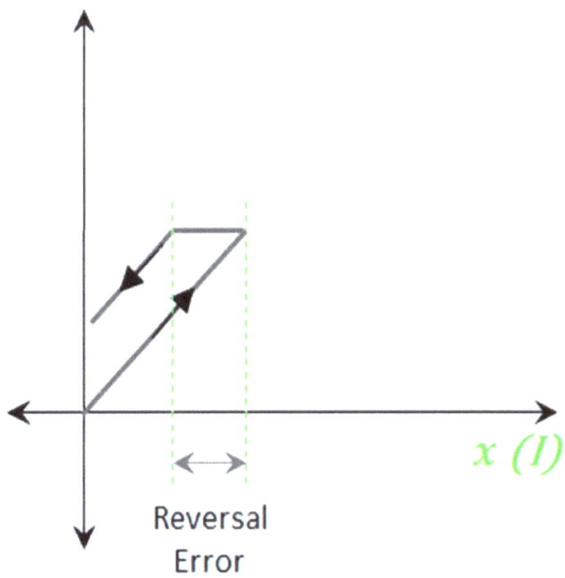

Fig. 7.63- Reversal Error for a Continuous Valve

7.6- Valve Dynamic Characteristics

Every dynamic system, including hydraulic valves, must be tested to check if it can work under severe dynamic loads.

7.6.1- Step Response

Step Response test develops a valve response when the valve is requested to move suddenly and as fast as possible to certain position of the spool stroke. Figure 7.64 shows a generic step response of a valve. The figure shows that the valve response is affected by the damping factor in the valve design. Typical dynamic system design considers undamped system with the following measures:

***Rise Time*:** time required for the spool to move from 10% to 90% of the maximum spool stroke.

***Peak Time*:** time at the first (highest) peak.

***Settling Time*:** settling time is defined as the time the spool takes to settle around a predefined *Steady State Error (SSE).* Typically (3-5) %.

Overshoot: spool position beyond the commanded value.

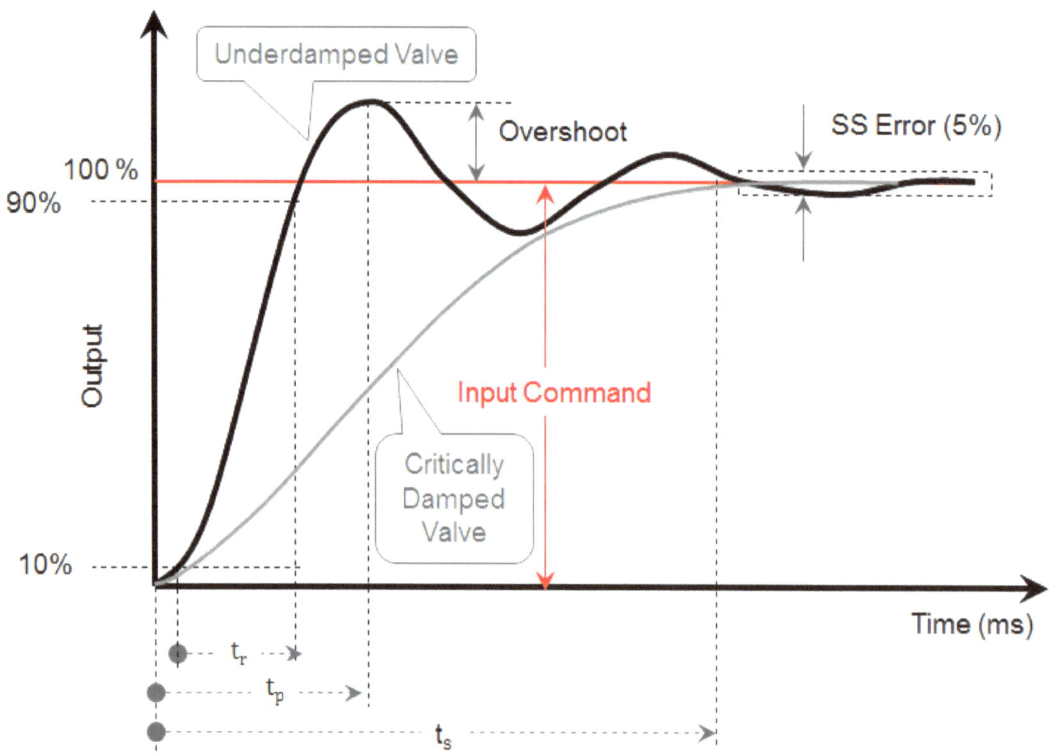

Fig. 7.64- Generic Step Response for a Continuous Valve

Valves are typically tested for spool positions 25, 50, 75, and 100% of the full spool stroke and in both directions, i.e. from 0-max and back to 0. The following section presents some examples of step response results of actual valves.

Example 1, Fig.7.65. The valve in this example is a direct operated proportional valve and requires 40 ms for full stroke travel in both directions.

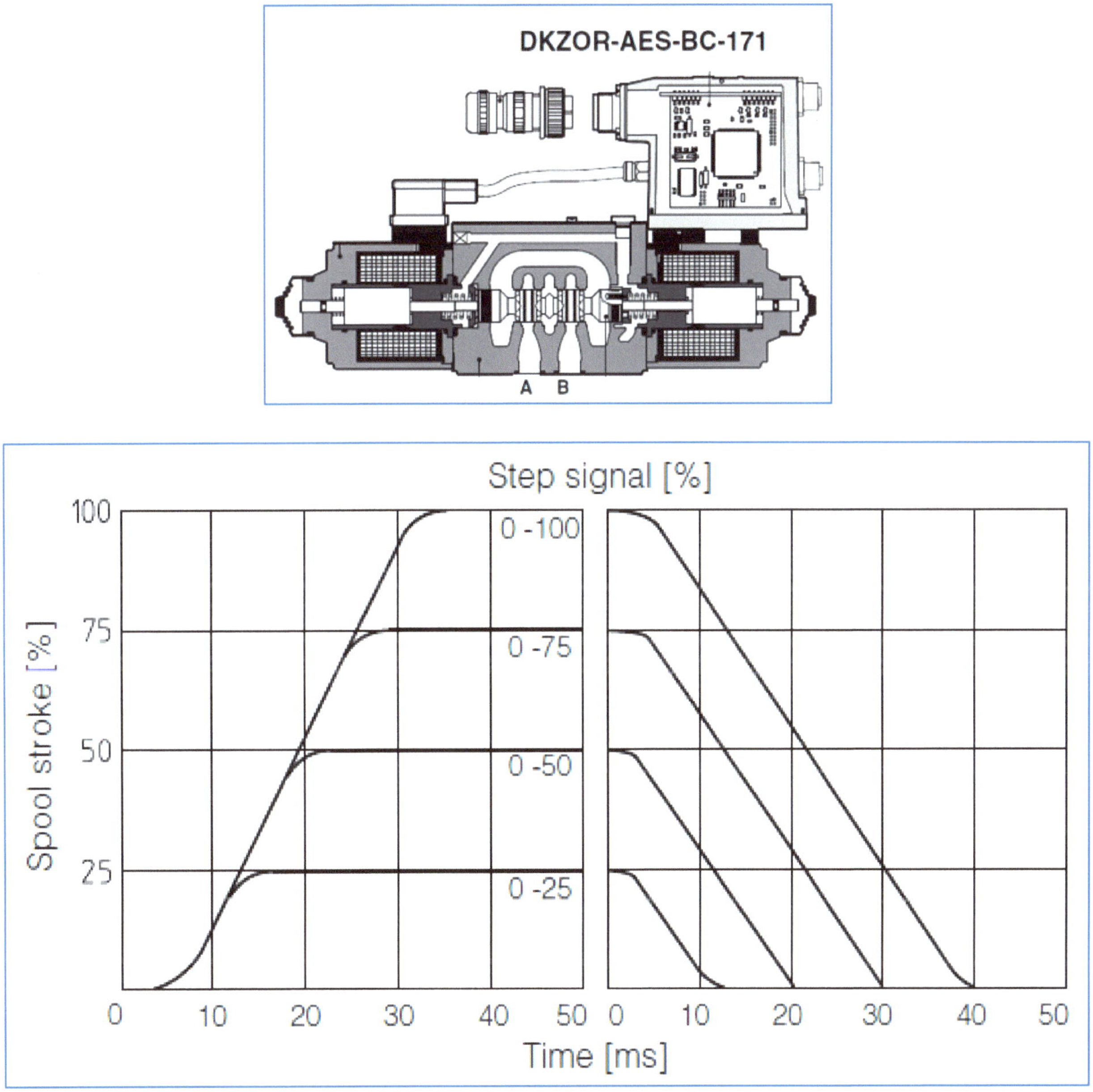

Fig. 7.65- Typical Step Response for a Proportional Valve - Example 1 (Courtesy of Atos)

Example 2, Fig. 7.66. The valve in this example is a pilot operated proportional valve. Despite that the valve is larger in size, it requires the same travel time. This valve experiences little overshot.

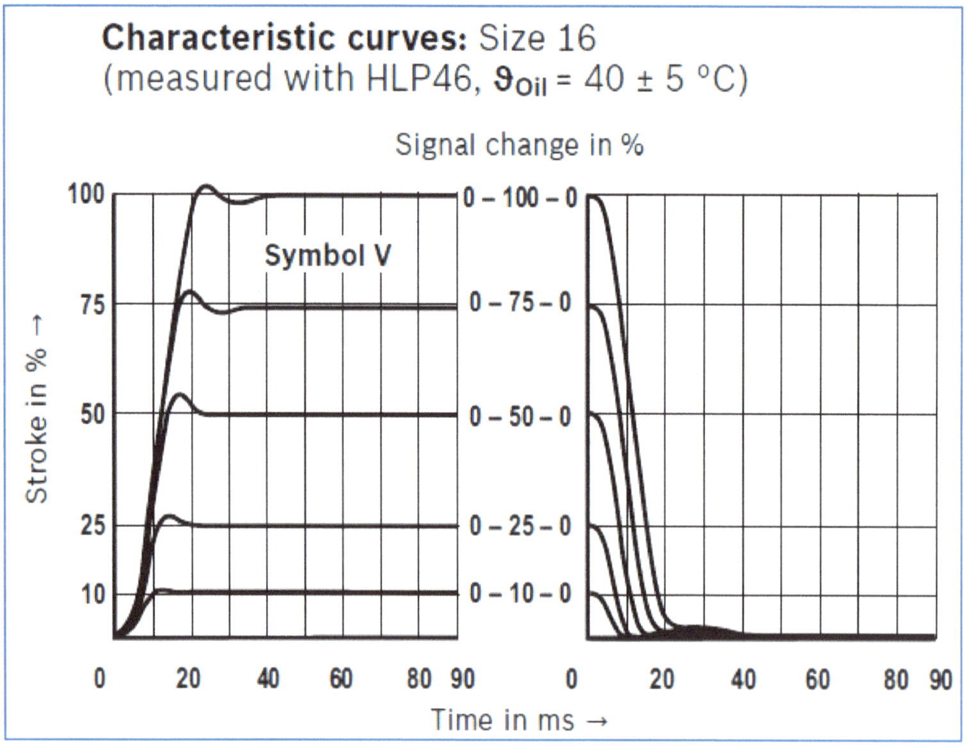

Fig. 7.66- Typical Step Response for a Proportional Valve - Example 2, (Courtesy of Bosch Rexroth)

Example 3, Fig. 7.67. This example explores how fast a servo valve is as compared to proportional valves. The figure also shows that the step response is affected by the test pressure. The valve responds faster at higher pressure.

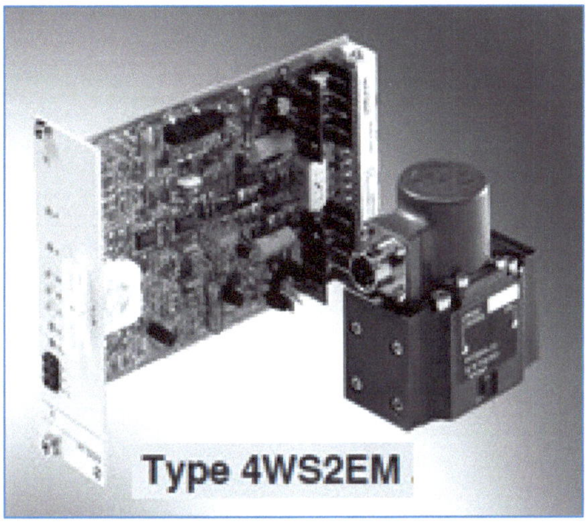

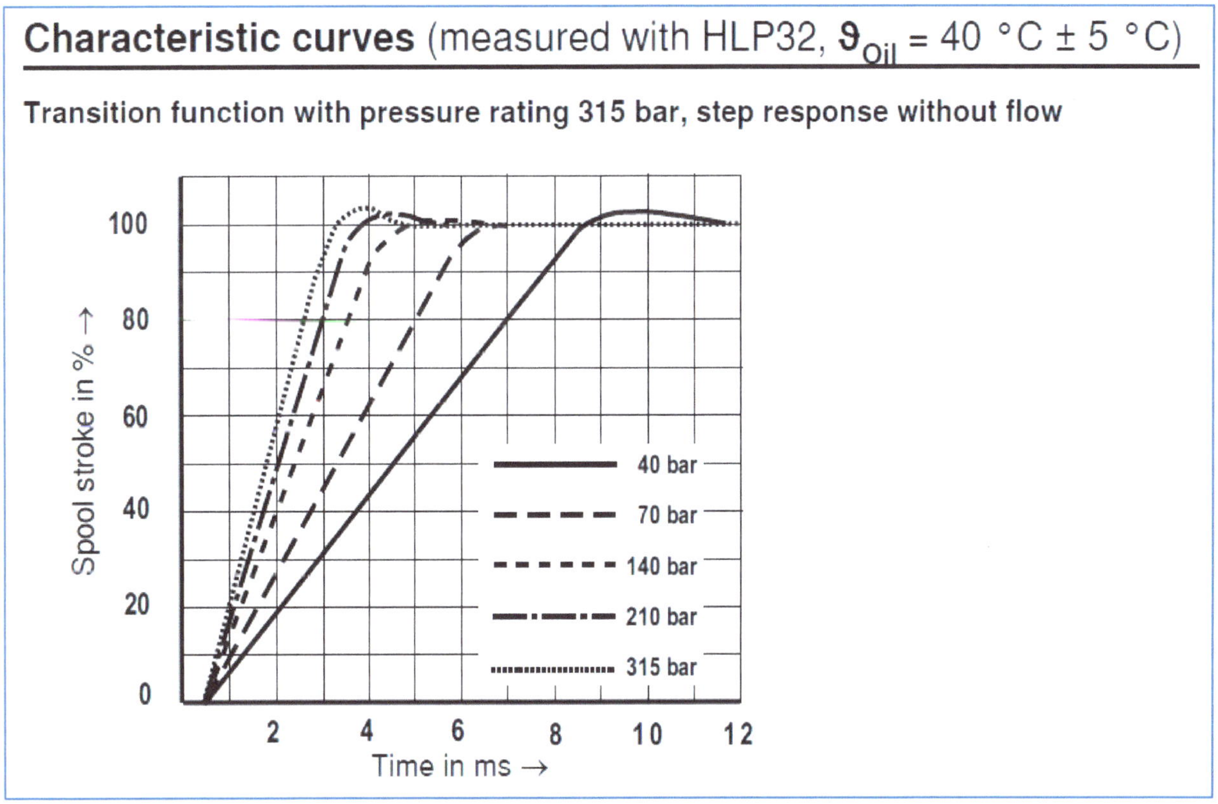

Fig. 7.67- Typical Step Response for a Servo Valve - Example 3, (Courtesy of Bosch Rexroth)

Example 4, Fig. 7.68. This example shows step response of a servo valve that requires just 6 ms for full stroke travel.

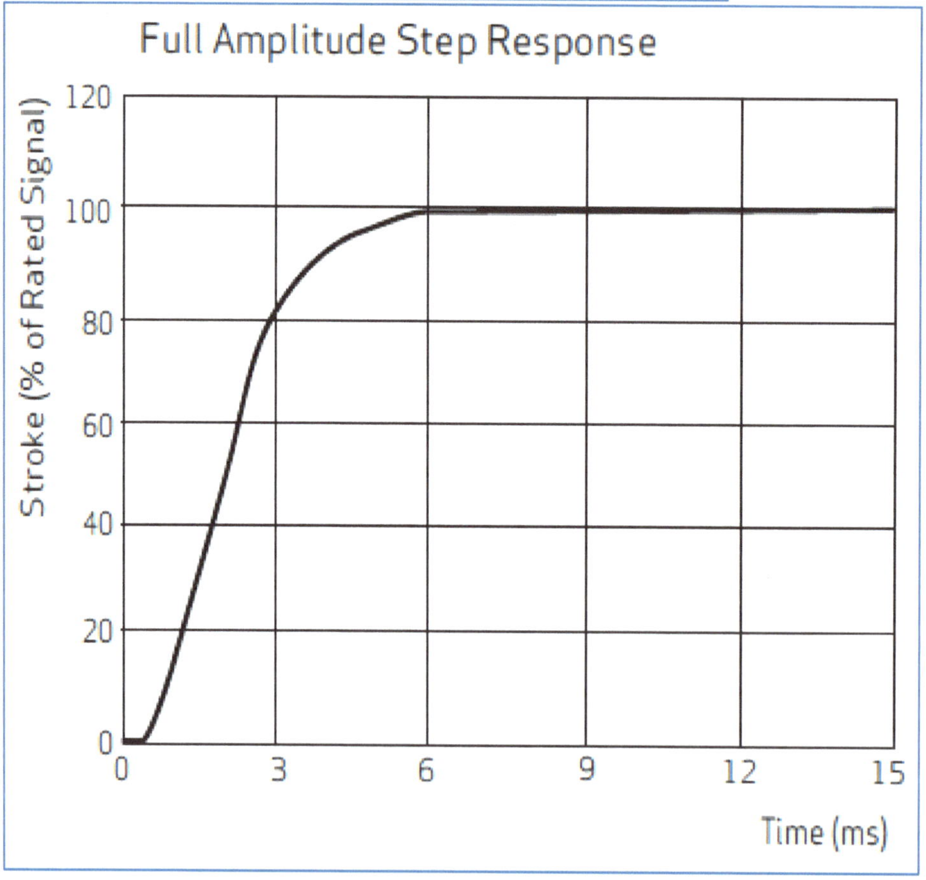

Fig. 7.68- Typical Step Response for a Servo Valve - Example 4, (Courtesy of Moog)

Example 5, Fig. 7.69. This example shows step response for a servo valve with electrical feedback.

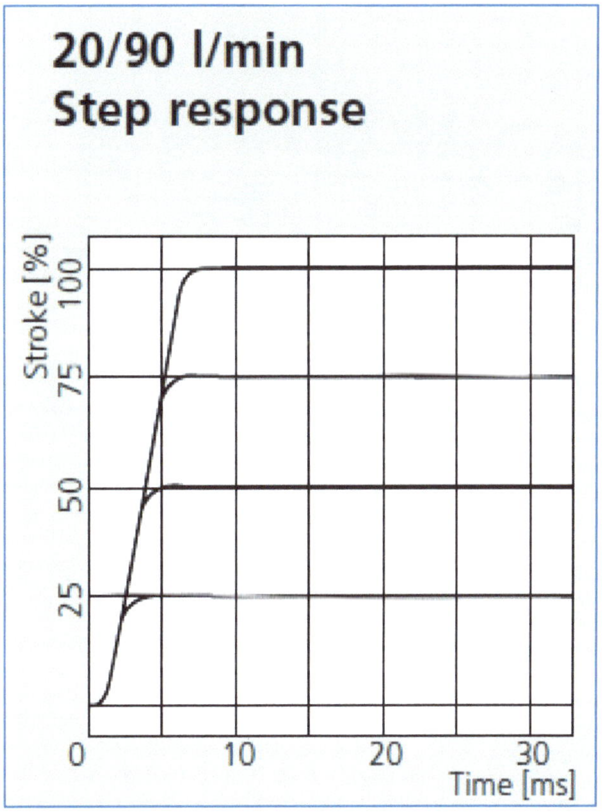

Fig. 7.69- Typical Step Response for a Servo Valve - Example 5 (Courtesy of Moog)

7.6.2-Frequency Response

Frequency Response test develops a valve response to a harmonic load. A proportional or servo valve may be used to stabilize an actuator against periodic load, e.g. stabilizing control surfaces of an airplane or a rocket against aerodynamic forces. They can also be used to cycle an actuator at high frequency, e.g., as in vibration exciters.

Applying Fourier analysis shows that any periodic load contains number of harmonics. A harmonic load is the most harmful on a dynamic system among periodic loads. Therefore, frequency response is a test that develops a valve response to a harmonic load. If the valve passes the frequency response test, it should be able to work against a periodic load that contains such a frequency range.

As shown in Fig. 7.70, if a valve is excited by a harmonic input signal of certain frequency, the valve responds with certain *Amplitude Attenuation* and *Phase Lag*.

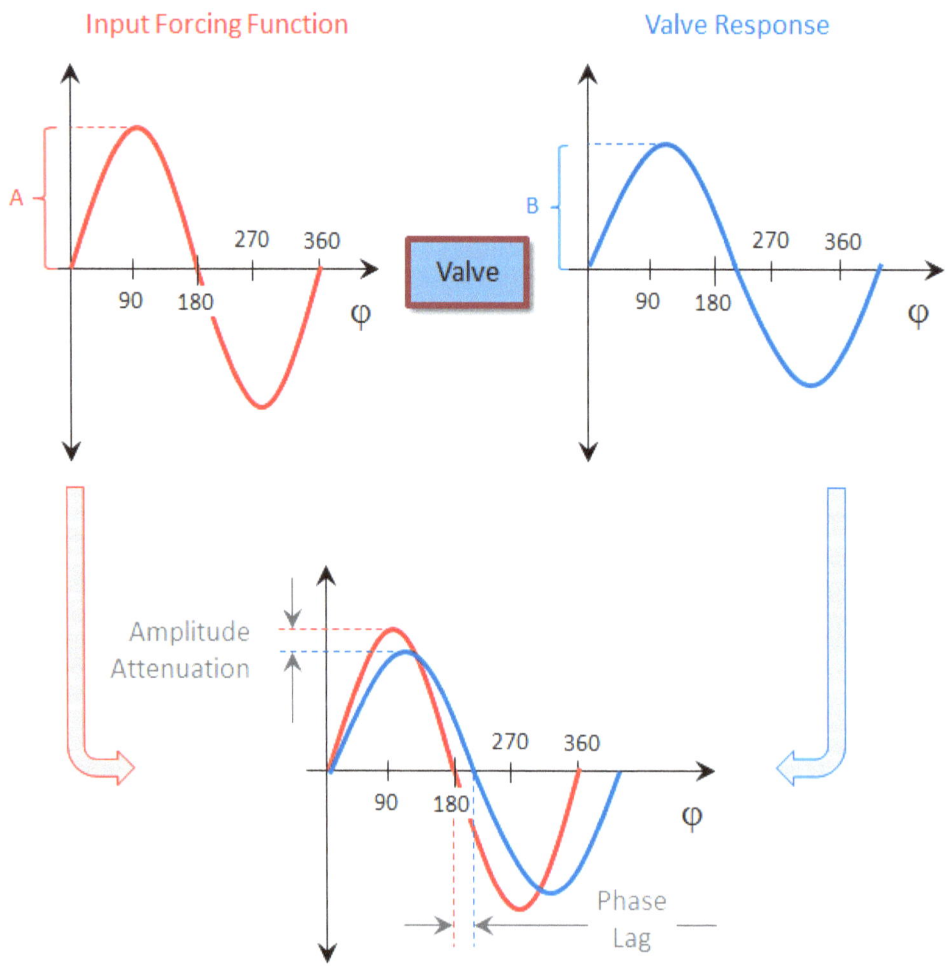

Fig. 7.70- Generic Frequency Response for a Continuous Valve

As shown generically in Fig. 7.71, the frequency response is presented in a *Bode Plot*. The figure shows two curves named Phase Curve and Gain Curve.

***Phase Curve*:**
The phase curve plots the phase lag (delay) in degrees versus the frequency. The figure shows that at low frequencies, the valve is able to catch the exciting frequency without delay. Beyond certain frequency, the valve response is delayed behind the exciting frequency. The frequency at which the valve is lagged by 90 degrees is called the *Natural Frequency*. It is a crucial parameter in designing an EH system and it defines how fast the valve is.

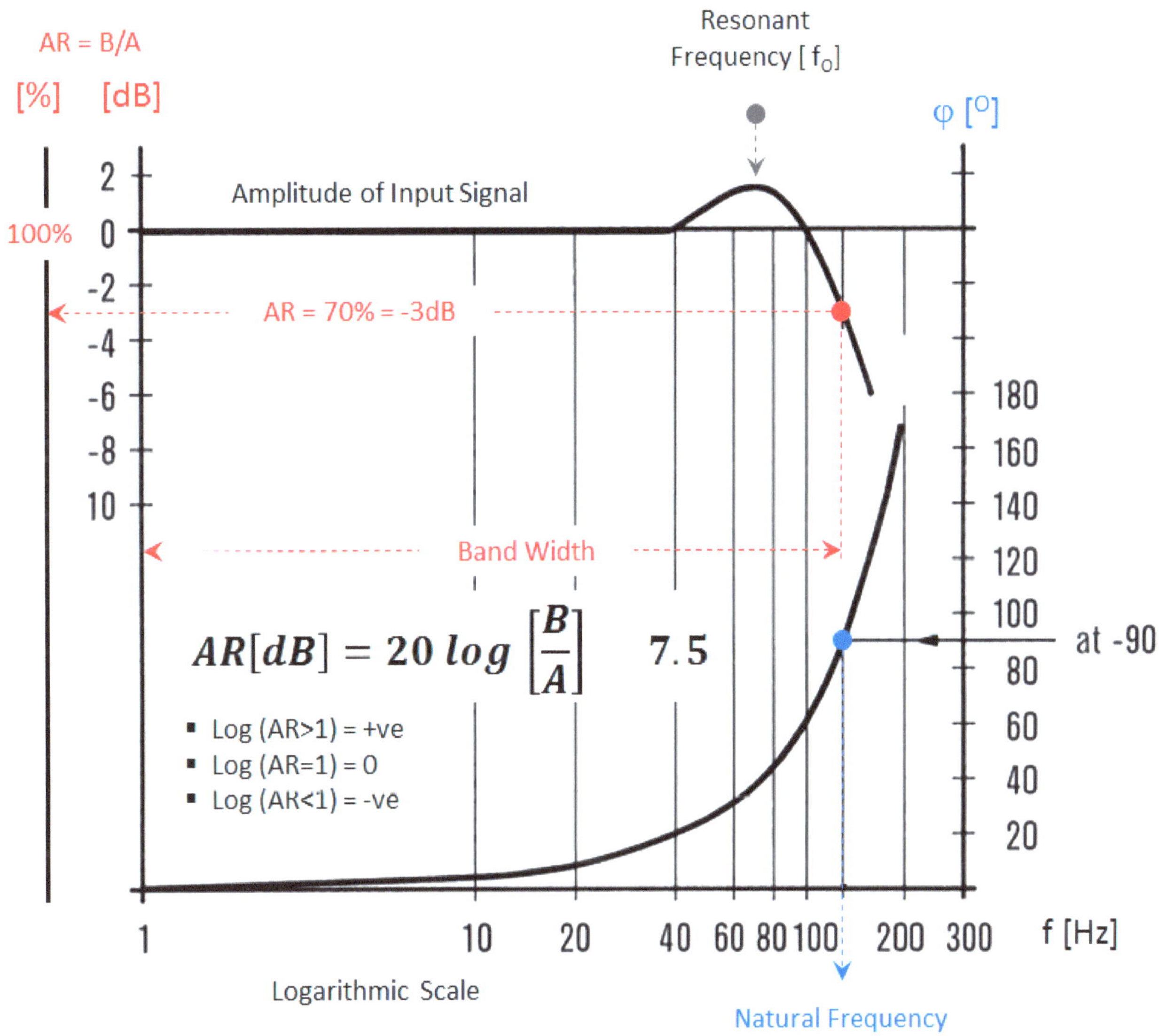

Fig. 7.71- Generic Bode Plot for a Continuous Valve

***Gain Curve*:**

The gain curve plots the amplitude attenuation versus the frequency. Equation 7.5 shows the calculation of the gain value, which is the amplitude ratio AR in dB units where **B** is the output amplitude and **A** is the amplitude of the exciting frequency.

$$AR[dB] = 20 \log \left[\frac{B}{A}\right] \qquad 7.5$$

The figure shows that at low frequencies, the valve responds exactly similar to the exciting frequency without amplitude attenuation. During thus frequency range, the amplitude ratio equals 0 dB, which is equivalent to 100% on a linear scale.

Beyond a certain frequency, if the valve is underdamped, there is a possibility that the spool will overshoot. Then the amplitude ratio will be positive dB value, which is equivalent to greater than 100%. The frequency at the maximum overshoot is called the *Resonant Frequency*. Critically damped and over damped valves do not experience such overshoot.

Increasing the frequency further will reduce the amplitude ratio. The frequency at which the amplitude ratio equal -3dB (70%) is called the *Bandwidth* of the valve.

It is not necessary that the natural frequency equal the bandwidth of the valve. It is due to the system designer to compare the valves based on any of the mentioned critical frequencies. Best practices are to compare the valves based on whichever lower with the consideration of avoiding driving the valve near the overshoot frequencies.

The following section presents some examples of frequency response results for real valves.

Example 1, Figure 7.72. In this example the valve has been tested with different amplitudes. When the valve is excited with the full stroke, bandwidth was found equal 40 Hz and the natural frequency us 70 Hz. Obviously when the valve is excited with smaller amplitude, it delays less and bandwidth becomes larger.

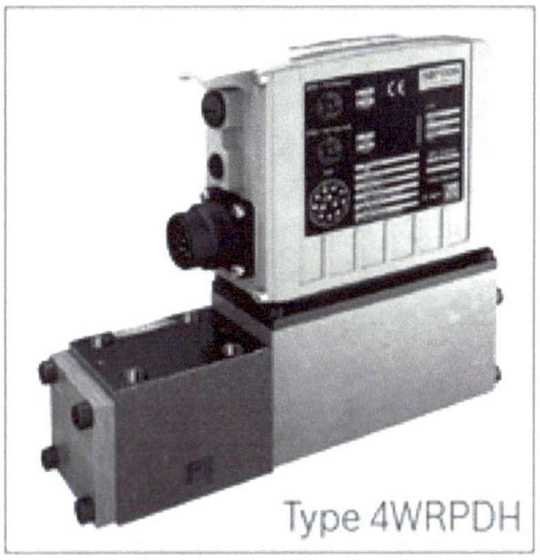

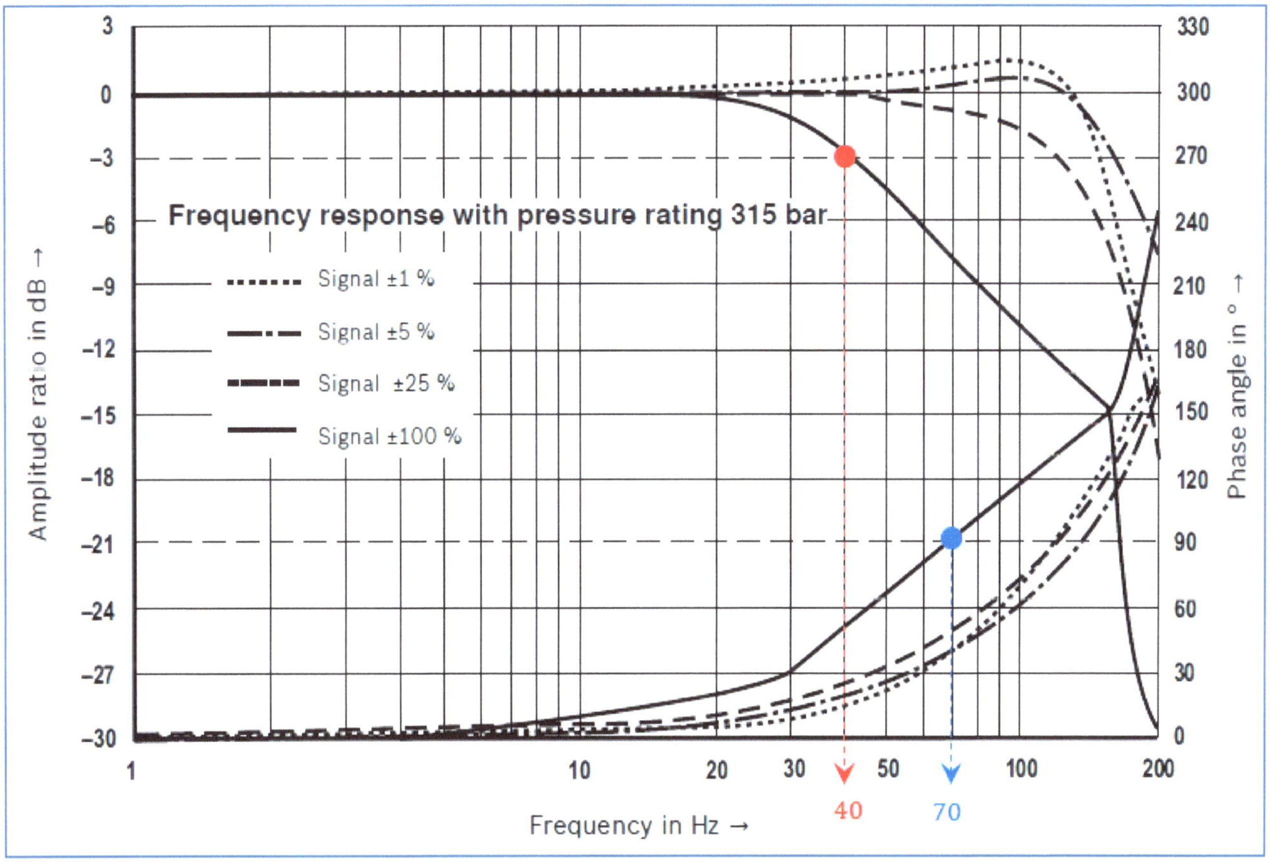

Fig. 7.72- Typical Frequency Response for a Proportional Valve - Example 1 (Courtesy of Bosch Rexroth)

It is to be noted that such tests have been developed under certain working pressure. If the valve in the application works at different pressure, correction factor should be considered.

As shown in Fig. 7.73, assuming the application pressure is 50% of the test pressure, the correction factor is 0.83. The valve bandwidth and natural frequency under the application pressure are calculated as shown in the figure.

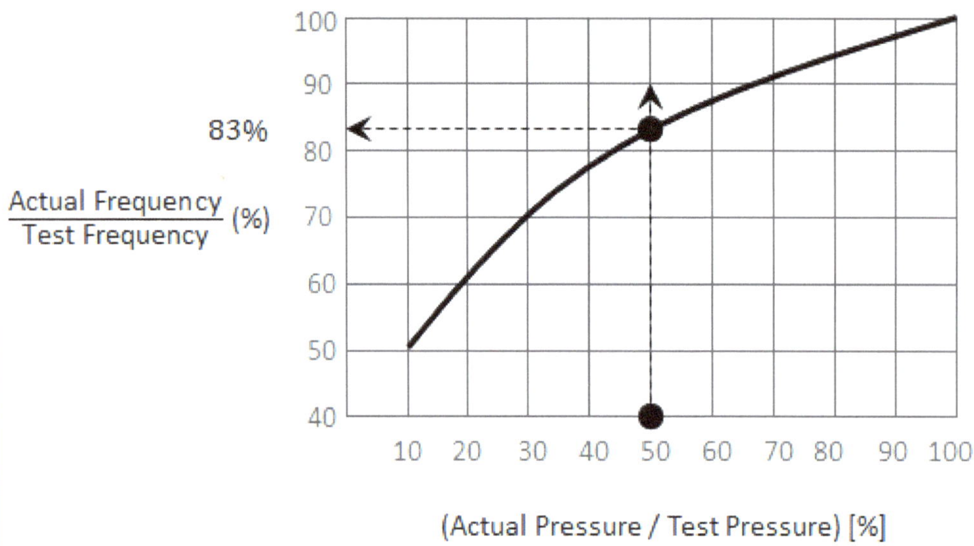

$$Band\ Width\ (at\ 50\%\ pressure) = 0.83 \times 40 = 33.2\ \text{Hz}$$

$$Natural\ Frequency\ (at\ 50\%\ pressure) = 0.83 \times 70 = 58.1\ \text{Hz}$$

Fig. 7.73- Frequency Response Correction Based on Working Pressure

Example 2, Fig. 7.74. This example proves that the dynamics of a servo valve are much faster than a proportional valve.

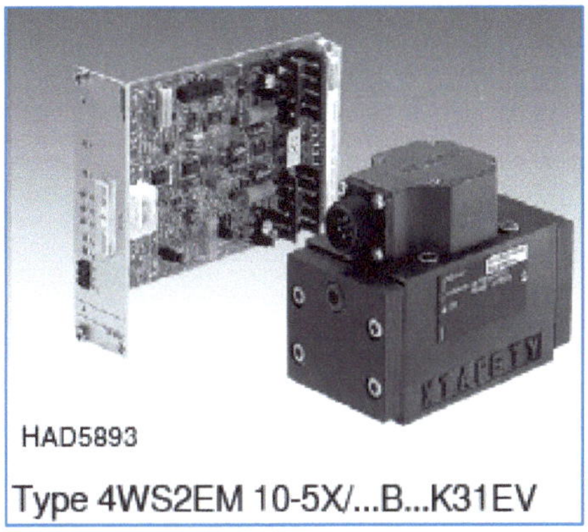

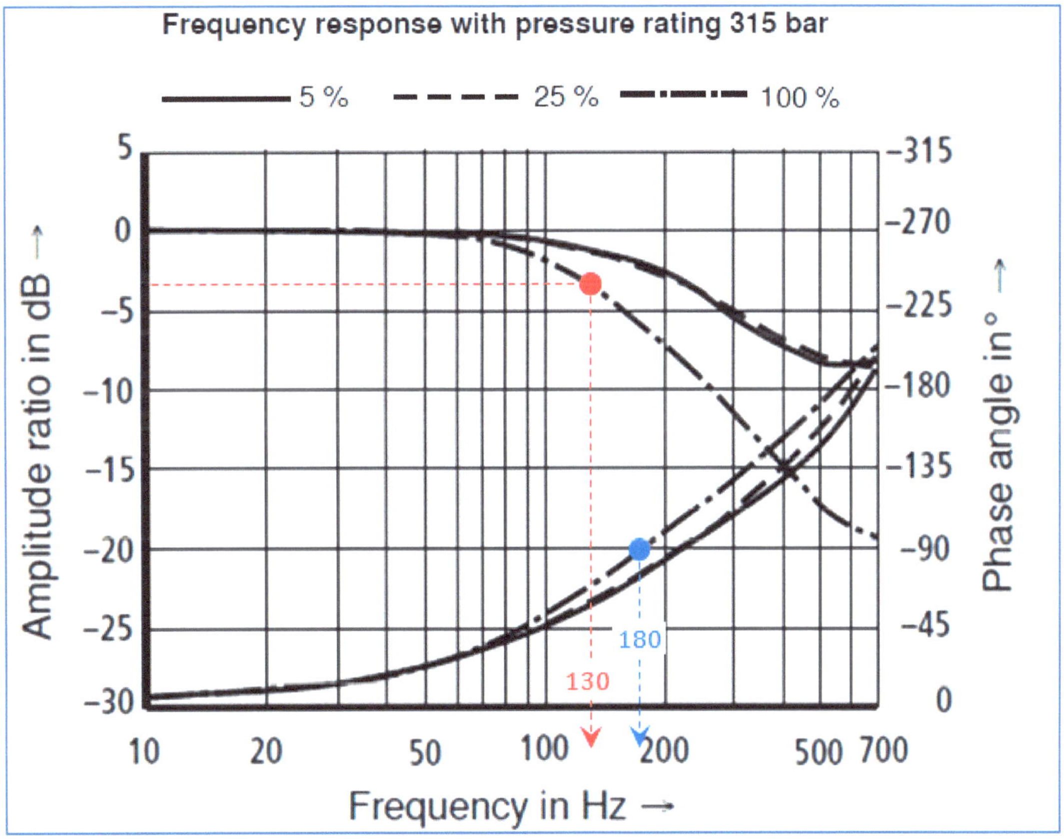

Fig. 7.74- Typical Frequency Response for a Servo Valve - Example 2 (Courtesy of Bosch Rexroth

Example 3, Fig. 4.75. This example presents frequency response for a servo valve from a different manufacturer.

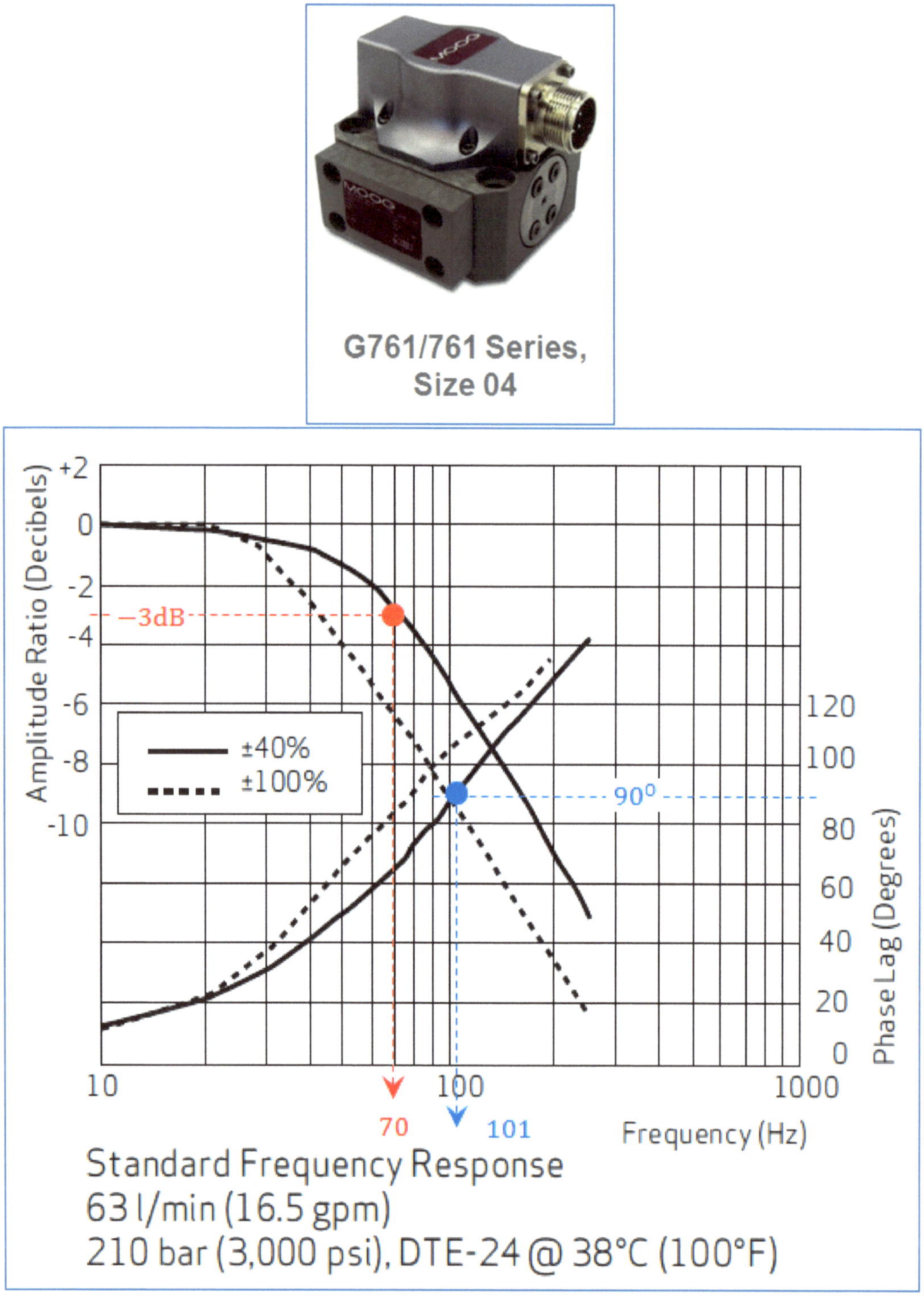

Fig. 7.75- Typical Frequency Response for a Servo Valve - Example 3 (Courtesy of Moog)

7.7-Valve Operating Conditions

Valve operating conditions are reported by the valve manufacturer. Valve operating conditions should be reviewed by a system designer to make sure the valve will work safely and properly. The following bullets form a checklist to be reviewed and considered in EH system design:

- Maximum working temperature.
- Maximum working pressure.
- Hydraulic fluids: recommended types and viscosities.
- Cleanliness requirements. It may affect the filtration system design. A system designer should be aware of the valve manufacturer's recommendations about:
    - Location of the main filter (pressure filter or return filter).
    - Mesh size.
    - Filter beta ratio (or efficiency).

7.8-Examples of Published Data

Usually valve manufacturers publish the valve specifications in sections such as General, Hydraulic, and Electrical. The following specifications are the common ones that are published by most of the valve manufacturers:

- Maximum temperature and pressure.
- Rated flow, tolerance flow, and leakage flow.
- Hydraulic fluids, recommended viscosity, cleanliness requirements, and filter rating.
- Sealing recommendation.
- Step response, frequency response, threshold, hysteresis, and repeatability.
- Weight, assembly position, sub plate surface finish.
- Coil resistance, supply voltage, current, power consumption, duty factor.
- Other custom data.

Figures 7.76 through 7.83 present examples of actual valves. With all the presented knowledge base in this chapter, it is up to the reader to go line by line to familiarize himself with the published data on each valve.

DHZE-A-07

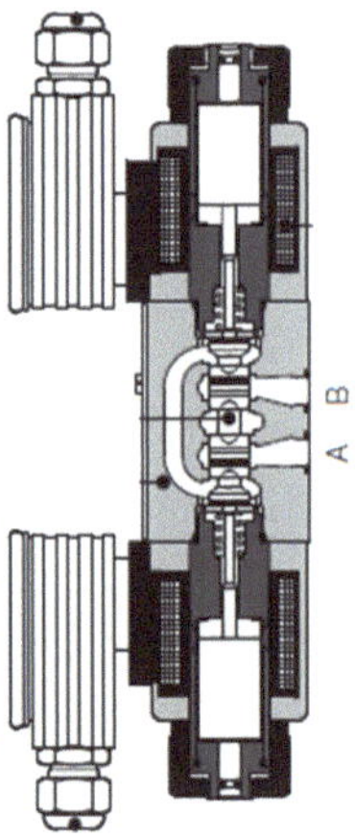

HYDRAULIC CHARACTERISTICS (based on mineral oil ISO VG 46 at 50 °C)

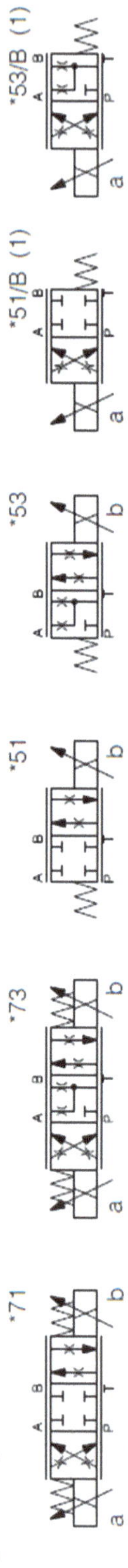

Hydraulic symbols	*71	*73	*51	*53	*51/B (1)	*53/B (1)
Valve model	DHZE				DKZE	
Spool overlapping	**1, 3**	**1**	**1, 3**	**1, 3**	**1, 3**	**1, 3**
Spool type and size (2)	**L14**	**L1**	**S3, L3, D3**	**S5, L5, D5**	**S3, L3, D3**	**S5, L5, D5**
Pressure limits [bar]	ports P, A, B = 350; T = 210				ports P, A, B = 315; T = 210	
Max flow (3) [l/min]						
at Δp = 10 bar (P-T)	1	4,5	17	28	45	60
at Δp = 30 bar (P-T)	2	8	30	50	80	105
at Δp = 70 bar (P-T)	3	12	45	70	120	160
Response time (4) [ms]	< 30				< 40	
Hysteresis [%]	≤ 5%				≤ 5%	
Repeatability	± 1%				± 1%	

Fig. 7.76- Published Data for a Proportional Valve - Example 1 – Part 1 (Courtesy of Atos)

MAIN CHARACTERISTICS, SEALS AND HYDRAULIC FLUID

Assembly position / location	Any position					
Subplate surface finishing	Roughness index Ra 0,4 - flatness ratio 0,01/100 (ISO 1101)					
Ambient temperature	Standard execution = -30°C ÷ +70°C; /PE option = -20°C ÷ +70°C; /BT option = -40°C ÷ +70°C					
Seals, recommended fluid temperature	NBR seals (standard) = -20°C ÷ +60°C, with HFC hydraulic fluids = -20°C ÷ +50°C FKM seals (/PE option)= -20°C ÷ +80°C HNBR seals (/BT option)= -40°C ÷ +60°C, with HFC hydraulic fluids = -40°C ÷ +50°C					
Recommended viscosity	15÷100 mm²/s - max allowed range 2.8 ÷ 500 mm²/s					
Fluid contamination class	ISO 4406 class 21/19/16 NAS 1638 class 10, in line filters of 25 μm (β10 ≥75 recommended)					
Hydraulic fluid	**Suitable seals type**		**Classification**		**Ref. Standard**	
Mineral oils	NBR, FKM, HNBR		HL, HLP, HLPD, HVLP, HVLPD		DIN 51524	
Flame resistant without water	FKM		HFDU, HFDR		ISO 12922	
Flame resistant with water	NBR, HNBR		HFC			
Flow direction	As shown in the symbols of table 3					
Coil code	DHZE-A*			DKZE-A*		
	standard	option /6 (1)	option /18 (2)	standard	option /6 (1)	option /18 (2)
Coil resistance R at 20°C	3 ÷ 3,3 Ω	2 ÷ 2,2 Ω	13 ÷ 13,4 Ω	3,8 ÷ 4,1 Ω	2,2 ÷ 2,4 Ω	12 ÷ 12,5 Ω
Max. solenoid current	2,2 A	2,75 A	1 A	2,6 A	3,25 A	1,2 A
Max. power	30 Watt			35 Watt		
Protection degree (CEI EN-60529)	IP65					
Duty factor	Continuous rating (ED=100%)					

Fig. 7.77- Published Data for a Proportional Valve - Example 1 – Part 2, (Courtesy of Atos)

G761/761 Series

Hydraulic Data

Maximum operating pressure to ports P, T, A, B, X	315 bar (4,500 psi)
Rated flow at p_N 35 bar/spool land (500 psi/spool land)	4/10/19/38/63 l/min (1/2.5/5/10/16.5 gpm)
Maximum main stage leakage flow rate (zero lap)	2.3 l/min (0.60 gpm)
Null adjust authority	Greater than 10% of rated flow
Hydraulic fluid	Hydraulic oil as per DIN 51524 parts 1 to 3 and ISO 11158. Other fluids on request.
Temperature range	-40 to +60 °C (-40 to +140 °F)
Recommended viscosity range	10 to 97 mm^2/s (cSt)
Maximum permissible viscosity range	5 to 1,250 mm^2/s (cSt)
Recommended cleanliness class as per ISO 4406	
For functional safety	17/14/11
For longer life	15/13/10
Recommended filter rating	
For functional safety	$ß_{10} \leq 75$ (10µm absolute)
For longer life	$ß_5 \leq 75$ (5 µm absolute)

Fig. 7.78- Published Data for a Servo Valve - Example 2 – Part 1 (Courtesy of Moog)

Static and Dynamic Data					
Rated Flow	4 l/min (1 gpm)	10 l/min (2.5 gpm)	19 l/min (5 gpm)	38 l/min (10 gpm)	63 l/min (16.5 gpm)
Tolerance of rated flow	± 10% of rated flow				
Step response time for 0 to 100% stroke	5 ms	5 ms	5 ms	7 ms	16 ms
Threshold	≤ 0.5% of rated signal				
Hysteresis	≤ 3.0% of rated signal				
Null shift at $\Delta T = 38°C$ (100°C)	≤ 2.0% of rated signal				

Fig. 7.79- Published Data for a Servo Valve - Example 2 – Part 2 (Courtesy of Moog)

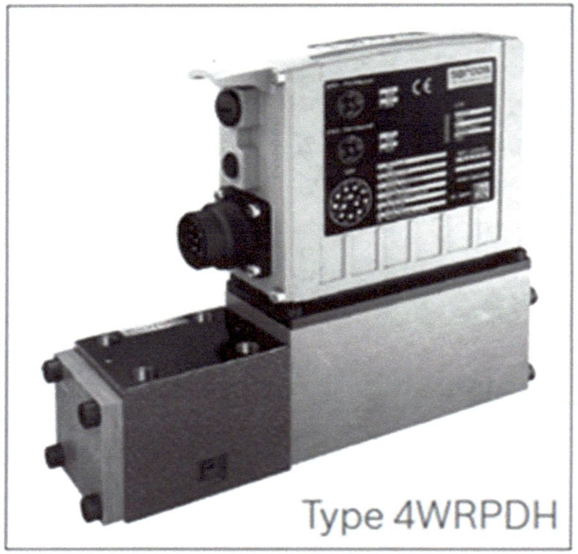

general			
Size	Size	6	10
Design		Directional spool valve, direct operated, with steel sleeve	
Actuation		Proportional solenoid with position control, OBE	
Type of connection		Plate connection, porting pattern according to ISO 4401	
Installation position		Any	
Ambient temperature range	°C	–20 ... +60	
Storage temperature range	°C	+5 ... +40	
Sine test according to DIN EN 60068-2-6		10 ... 2000 Hz / maximum of 10 g / 10 cycles / 3 axes	
Noise test according to DIN EN 60068-2-64		20 ... 2000 Hz / 10 g_{RMS} / 30 g peak / 30 min / 3 axes	
Transport shock according to DIN EN 60068-2-27		15g / 11 ms / 3 axes	
Weight	kg	3.2	7.2
Maximum relative humidity (no condensation)	%	95	
Maximum surface temperature	°C	150	

Fig. 7.80- Published Data for a Proportional Valve - Example 3 – Part 1 (Courtesy of Bosch Rexroth)

hydraulic										
Maximum operating pressure	▸ Port A, B, P	bar	350 (size 6); 315 (size 10)							
	▸ Port T	bar	250							
Rated flow (Δp = 35 bar per edge [1])		l/min	2	4	12	15	24/25	40	50	100
Limitation of use (transition in fail safe position)	▸ Symbol C3, C5	bar	350	350	350	350	350	160	315	160
	▸ Symbol C4, C1	bar	350	350	350	280	250	100	250	100
Leakage flow (at 100 bar)	▸ Linear characteristic curve "L"	cm³/min	< 150	< 180	< 300	–	< 500	< 900	< 1200	< 1500
	▸ Inflected characteristic curve "P"	cm³/min	–	–	–	< 180	< 300	< 450	< 600 (1:1) < 500 (2:1)	< 600
Hydraulic fluid			See table page 7							
Viscosity range	▸ Recommended	mm²/s	20 ... 100							
	▸ Maximum admissible	mm²/s	10 ... 800							
Hydraulic fluid temperature range (flown-through)		°C	–20 ... +60							
Maximum admissible degree of contamination of the hydraulic fluid, cleanliness class according to ISO 4406 (c)			Class 18/16/13 [2]							

Fig. 7.81- Published Data for a Proportional Valve - Example 3 - Part 2 (Courtesy of Bosch Rexroth)

static / dynamic		
Hysteresis	%	≤ 0.2
Manufacturing tolerance q_{Vmax}	%	≤ 10
Temperature drift	%/10 K	Zero shift < 0.25
Pressure drift	%/100 bar	Zero shift < 0.15
Zero compensation		Ex plant ±1 %

Technical data
(For applications outside these values, please consult us!)

Hydraulic fluid		**Classification**	**Suitable sealing materials**
Mineral oils		HL, HLP, HLPD, HVLP, HVLPD	NBR, FKM
Bio-degradable	▶ Insoluble in water	HETG	NBR, FKM
		HEES	FKM
	▶ Soluble in water	HEPG	FKM

Fig. 7.82- Published Data for a Proportional Valve - Example 3 - Part 3 (Courtesy of Bosch Rexroth)

electrical, integrated electronics (OBE)			
Relative duty cycle		%	100 (continuous operation)
Protection class according to EN 60529			IP 65 with mounted and locked plug-in connectors
Supply voltage [3; 4]	▶ Nominal voltage	VDC	24
	▶ Lower limit value	VDC	18
	▶ Upper limit value	VDC	36
	▶ Maximum admissible residual ripple	Vpp	2.5 (Comply with absolute supply voltage limit value)
Current consumption	▶ Maximum [5]	A	2.5
	▶ Impulse current	A	4
Maximum power consumption	▶ Size 6	W	40
	▶ Size 10	W	60
AD/DA resolution	▶ Analog inputs		12 bit
	▶ Analog output		10 bit

Fig. 7.83- Published Data for a Proportional Valve - Example 3 - Part 4 (Courtesy of Bosch Rexroth)

Chapter 8

Open-Loop versus Closed-Loop EH Applications

Objectives:

This chapter explores the differences between electro-hydraulic open-loop and closed-loop systems. Several examples have been presented to discuss various ways to control load, speed and position of a hydraulic actuator. The chapter concludes by introducing examples from industrial and mobile applications.

Brief Contents:

Chapter 8 – Open-Loop versus Closed-Loop EH Applications

8.1- Electro-Hydraulic Open-Loop Control Systems

8.1.1- Structure of Electro-Hydraulic Open-Loop Control Systems

Figure 8.1 shows a block diagram of an *Open-Loop* electro-hydraulic system. The system starts with an input device that generates an Input *(Reference) Signal* **U_R** that is proportional to the desired value of the controlled parameter. Following the input device is an Electric Control Unit (ECU) that has the basic function of generating a *Control Signal* **U_C**, which is proportional to the input signal **U_R** and powerful enough to drive the following *Continuous Valve*. The latter drives an *Actuator* to achieve the desired *Output Value* **O** of the *Controlled Parameter*.

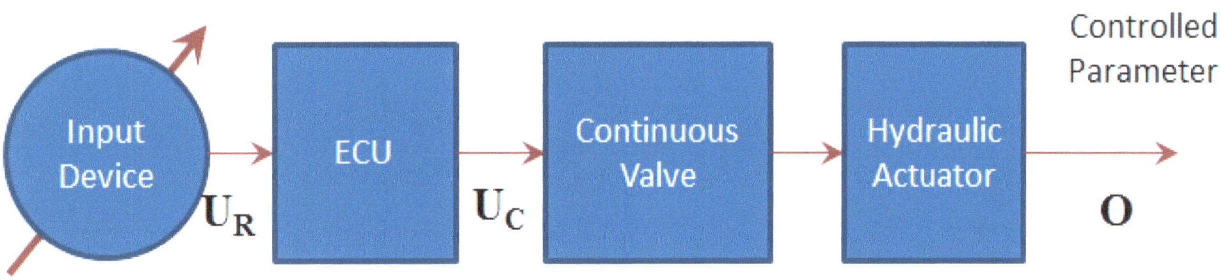

Fig. 8.1 - Block Diagram of an EH Open-Loop Control System using a Force-Controlled Continuous Valve

In the open-loop control system shown in Fig. 8.2, for a better accuracy, a *Stroke-Controlled* proportional valve is used. Such a valve has a built-in closed-loop to assure the accurate positioning of the valve spool based on the control signal **U_C**. The additional advantage of using a stroke-controlled valve is eliminating the errors from the valve. The closed-loop of the stroke-controlled valve compensates the error of the spool position and maintains the required spool position automatically.

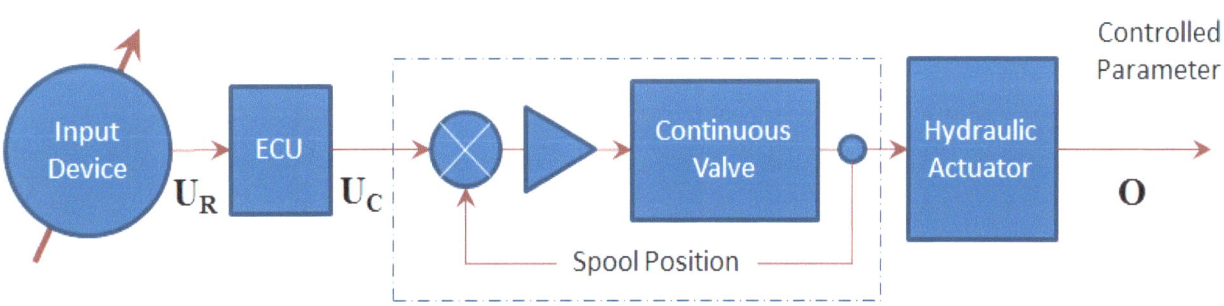

Fig. 8.2 - Block Diagram of an EH Open-Loop Control System using a Stroke-Controlled Continuous Valve

8.1.2- Features of Electro-Hydraulic Open-Loop Control Systems

Figure 8.3 shows the main features and the operating conditions of an open-loop control system.

Simplicity & Cost Effectiveness: Since there is no need for feedback, processor, and experienced tuning, an open-loop system is simpler and cheaper to construct than a closed-loop system.

Continuous Valves: A *Proportional Valve* is recommended because of the limited dynamics of the open-loop system. Although a servo valve technically can be used in an open-loop system, there is no sense to use such an expensive and highly dynamic valve in an open-loop system.

Response to External Disturbances and Loading Conditions: If the actuator experiences external disturbances, the output signal **O** of the controlled parameter will deviate from the desired input signal **U_R** that is set by the input device. The operator must manually correct the input signal to maintain the desired value of the controlled parameter. The fact is that the operator builds a virtual closed-loop by using his eyes as a sensor, his brain as a processor, and his arm as part of the actuation mechanisms. So, the operator is part of the loop.

The conclusion is, an open-loop system can't automatically maintain the desired value of the controlled parameter and to be used for a system that works under a constant load condition.

Calibration & Tuning: U_R-O signals in an open-loop system must be calibrated in the design stage, e.g. the potentiometer signal 0-10 Volts corresponds to a controlled parameter 0-max. An open-loop system requires frequent recalibration and in-field tuning to overcome sources of errors.

Accuracy and Source of Errors: The accuracy of an open-loop system depends on the accuracy of the components used. The sources of errors in an open-loop system are due to the change in the conditions of the continuous valve and the hydraulic actuator, e.g. valve spool sticking and leakage in the valve and the hydraulic actuator.

Controlled Parameters: The control signal **U_C,** which is supplied to the valve, results in a flow that passes through the valve based on the *Valve Flow Gain* or a pressure based on the *Valve Pressure Gain*. Therefore, an open-loop EH system can be used to perform flow (speed) control and pressure (load) control by direct calibration between the input signal **U_R** and the output signal **O** during the design stage of the system.

An open-loop EH system can't be used for position control because an actuator position can't be directly calibrated with the input signal.

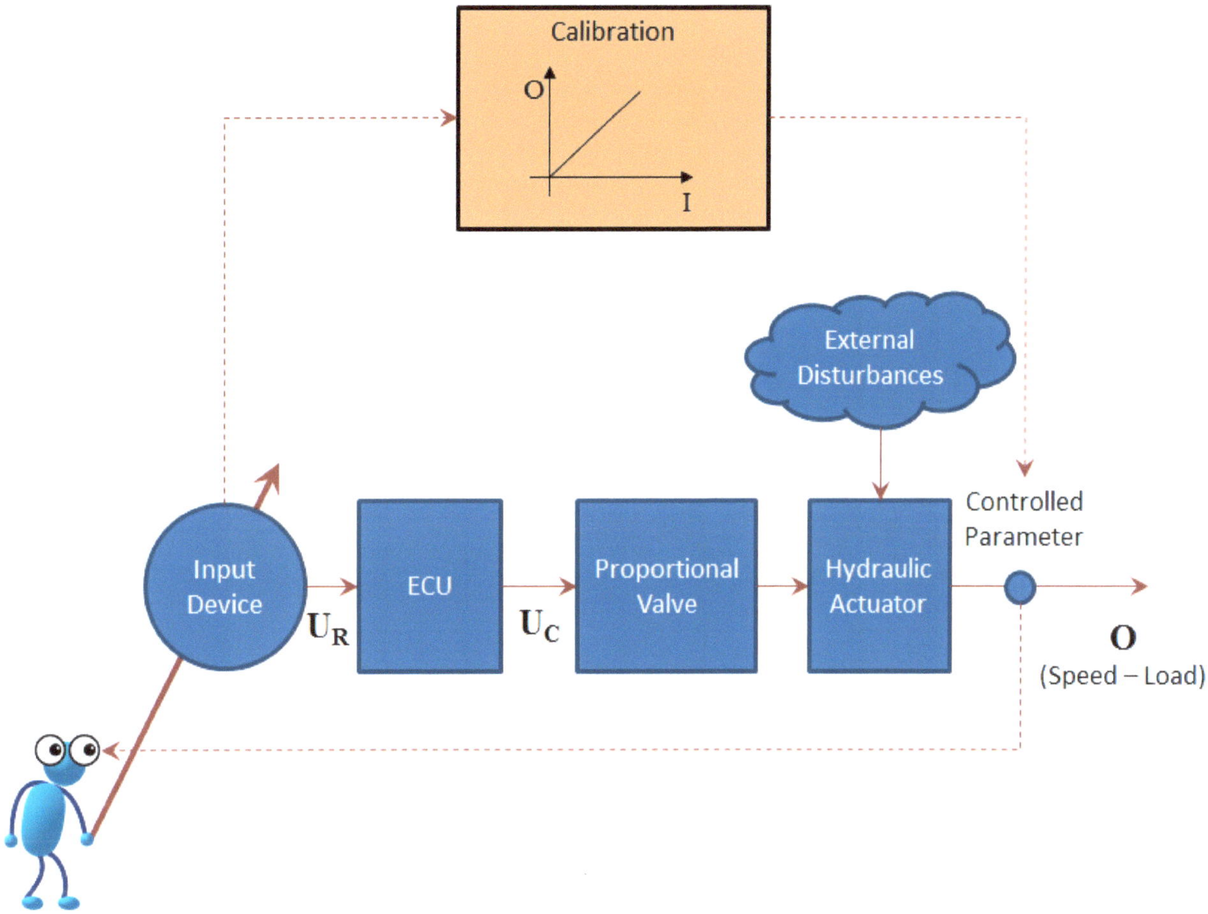

Fig. 8.3 - Features of Electro-Hydraulic Open-Loop Control Systems

8.1.3- Electro-Hydraulic Open-Loop Speed Control Systems

The following discussion is also applicable for a cylinder speed control. However, the example introduced considers the application of motor speed control.

Figure 8.4 shows a block diagram of an open-loop electro-hydraulic motor speed control system. A potentiometer is used as an input device that produces the input signal $\mathbf{U_R}$. The ECU conditions and amplifies the input signal $\mathbf{U_R}$ to produce the control signal $\mathbf{U_C}$. A 4/3 closed-center proportional directional valve is used to control both the direction and the speed of the motor. The valve opens proportional to the control signal $\mathbf{U_C}$. As a result, fluid flow that passes through the valve based on the *Valve Flow Gain*. Then the motor spins with a speed that is proportional to the flow through the valve. The proportionality factor is the motor size.

Figure 8.5 shows the operating conditions of the hydraulic circuit diagram of the previous system. The figure indicates that some tuning is required because the actual RPM is a bit off from the input signal. A flow control valve replaces the directional valve for a unidirectional valve.

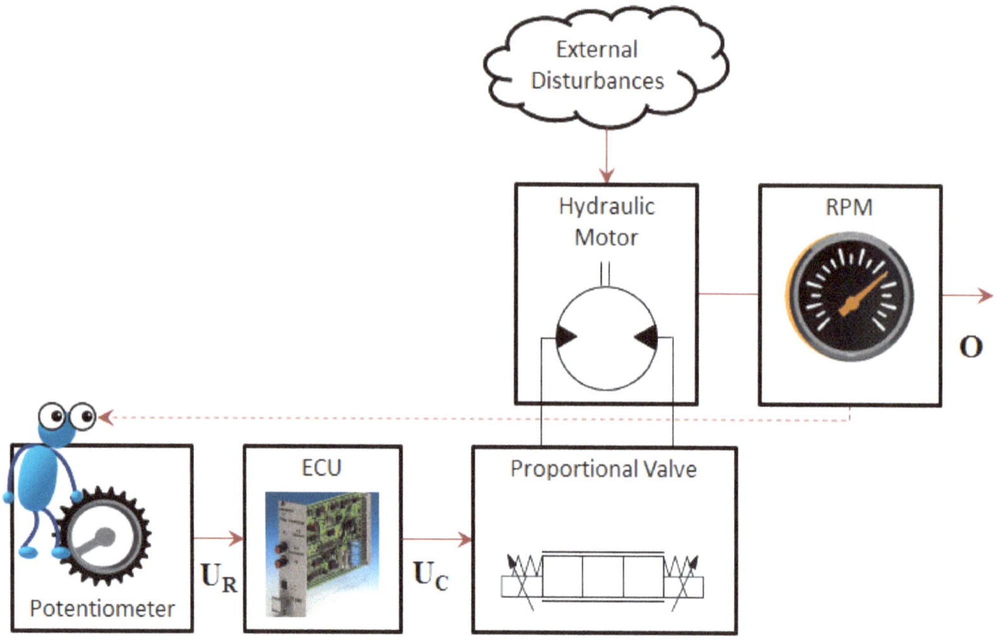

Fig. 8.4 - Block Diagram of an EH Open-Loop Motor Speed Control System

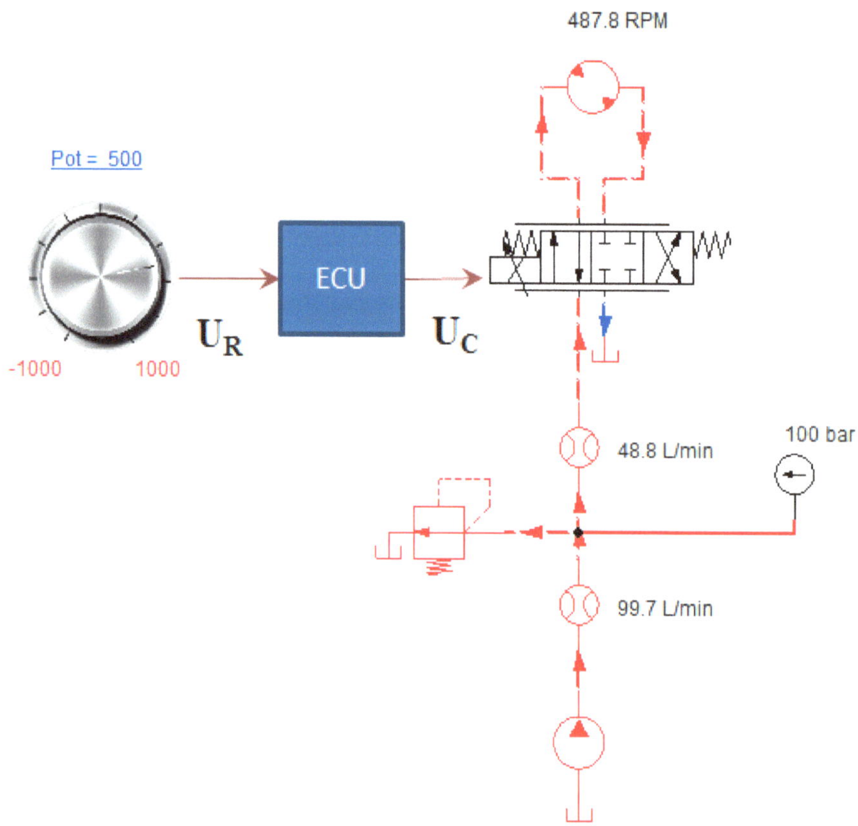

Fig. 8.5 - Hydraulic Circuit Diagram of an EH Open-Loop Motor Speed Control System

8.1.4- Electro-Hydraulic Open-Loop Load Control Systems

The following discussion is also applicable for a motor torque control or a cylinder force control. However, the example presented considers the application of a cylinder force control.

Single-Acting Cylinder:

Figure 8.6 shows a block diagram of an electro-hydraulic open-loop single-acting cylinder force control system. A potentiometer is used as an input device that produces the input signal $\mathbf{U_R}$. The ECU conditions and amplifies the input signal $\mathbf{U_R}$ to produce the control signal $\mathbf{U_C}$. A proportional pressure relief valve is used to maintain the required system pressure so that the force produced by a cylinder is controlled. The valve responds proportional to the control signal $\mathbf{U_C}$. As a result, pressure that acts on the cylinder is controlled. Consequently, the cylinder produces a force that is proportional to the pressure. The proportionality factor is the cylinder size.

Figure 8.7 shows the hydraulic circuit diagram of the previous system. The figure indicates that the system is well calibrated and the offset between the actual maximum pressure and the input signal is within the acceptable range.

Double-Acting Cylinder:

Figure 8.8 shows a block diagram of an electro-hydraulic open-loop double-acting cylinder force control system. A potentiometer is used as an input device that produces the input signal $\mathbf{U_R}$. The ECU conditions and amplifies the input signal $\mathbf{U_R}$ to produce the control signal $\mathbf{U_C}$. A 4/3 open-center proportional directional valve is used to control both the direction and the force of the cylinder. The valve opens proportional to the control signal $\mathbf{U_C}$. As a result, the pressure is distributed across the cylinder's piston based on the *Valve Pressure Gain*. Consequently, the cylinder produces a force that is proportional to the *differential pressure*. The proportionality factor is the cylinder size. Direction of the force is based on the sign of the input signal.

Figure 8.9 shows the hydraulic circuit diagram of the same previous system. The figure indicates that the relation between the input signal and the resulted differential pressure in nonlinear. Therefore, a point-to-point calibration curve must be developed and added as a lockup table inside the ECU to relate the input signal to the differential pressure.

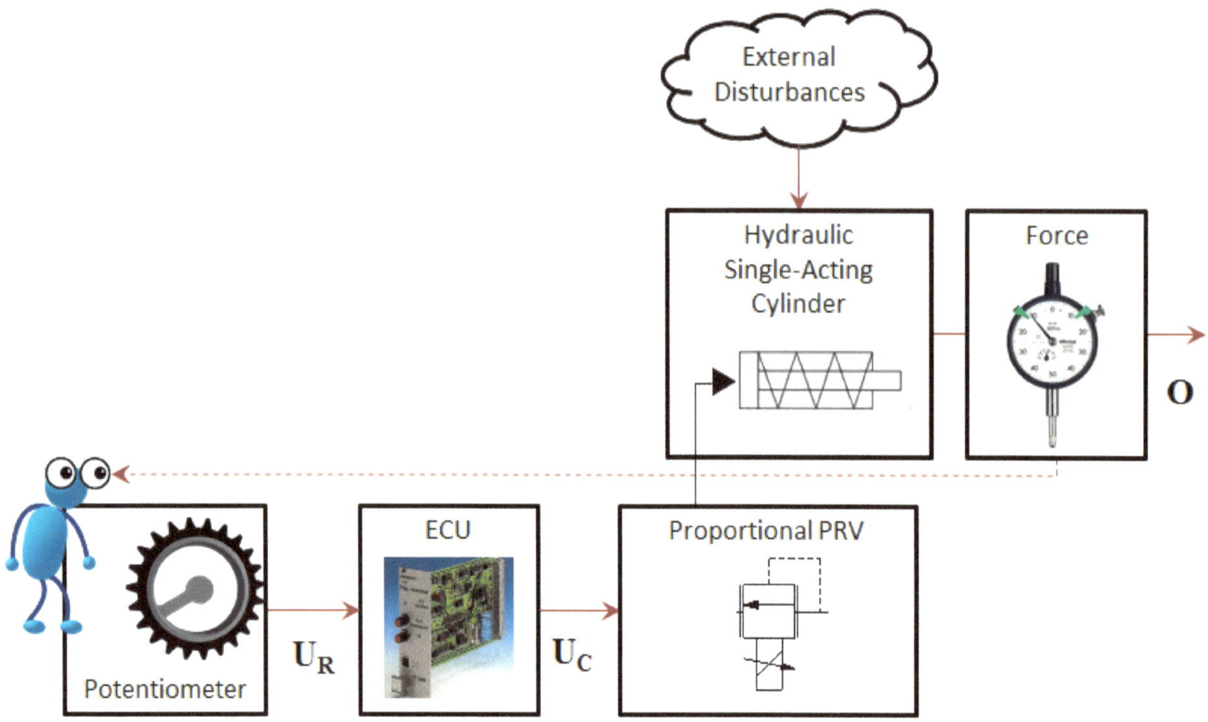

Fig. 8.6 - Block Diagram of an EH Open-Loop Single-Acting Cylinder Force Control System

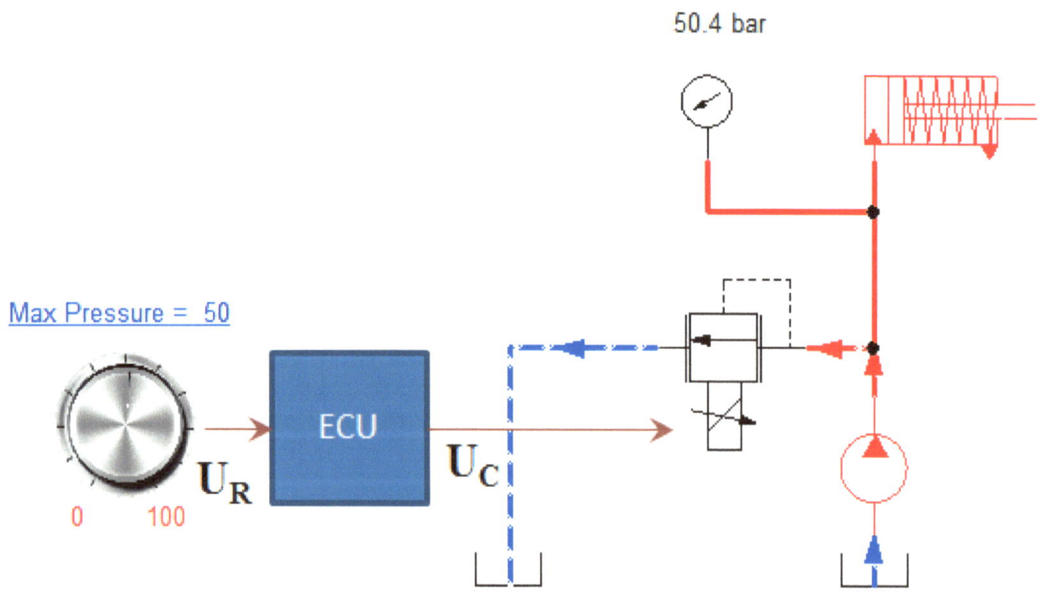

Fig. 8.7 - Hydraulic Circuit Diagram of an EH Open-Loop Single-Acting Cylinder Force Control System

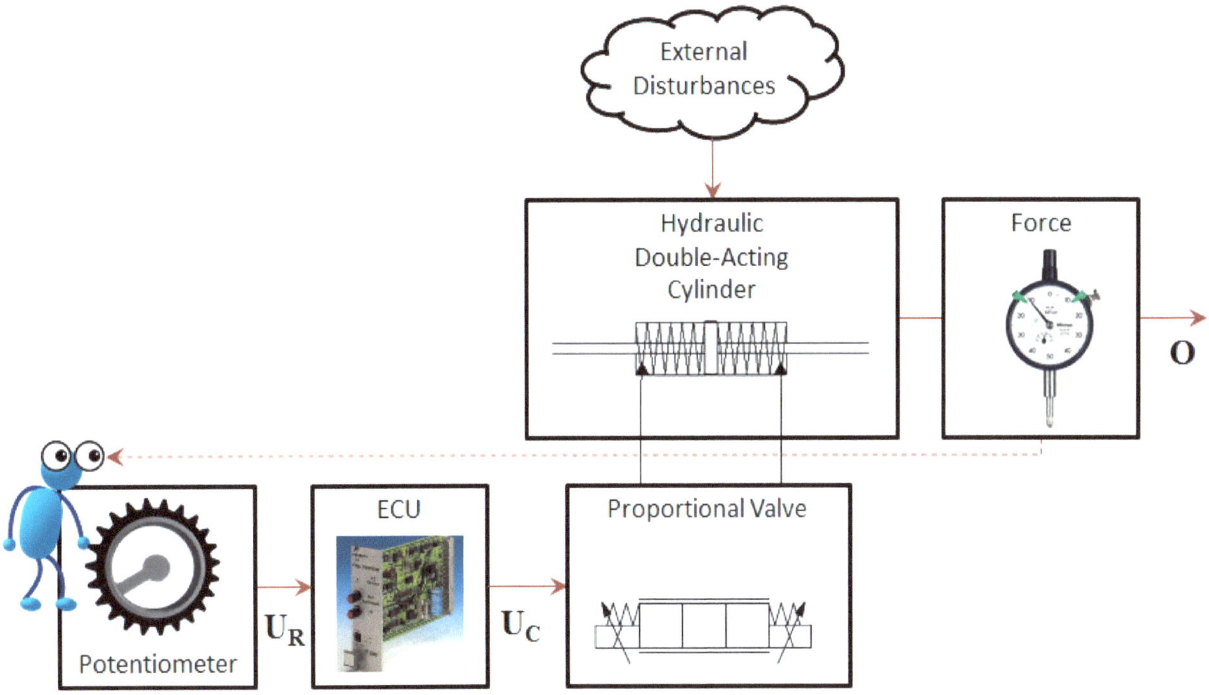

Fig. 8.8 - Block Diagram of an EH Open-Loop Double-Acting Cylinder Force Control System

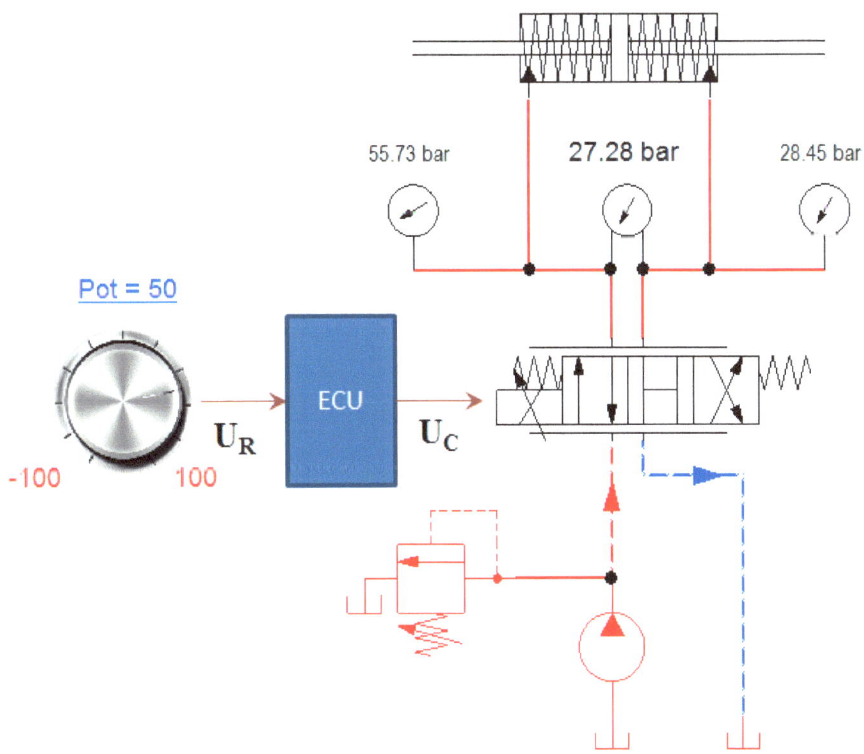

Fig. 8.9 - Hydraulic Circuit Diagram of an EH Open-Loop Double-Acting Cylinder Force Control System

8.2- Electro-Hydraulic Closed-Loop Control Systems

8.2.1- Structure of Electro-Hydraulic Closed-Loop Control Systems

Figure 8.10 shows a block diagram of an electro-hydraulic *Closed-Loop* system. The system starts with an input device that generates an input signal $\mathbf{U_R}$ that is proportional to the desired value of the controlled parameter. In addition to many functions to improve the system performance, the ECU in a closed-loop system has at least the following basic components: a *Summing Point, Controller,* and an *Amplifier*. The summing point calculates the instantaneous Error Signal $\mathbf{U_E}$ as a result of continuous comparing of the input signal $\mathbf{U_R}$ with the *Feedback Signal* $\mathbf{U_F}$. The *Error Signal* $\mathbf{U_E}$ is manipulated by the controller. Different types of controllers ranging from classical to advanced and intelligent controllers could be used. The amplifier finally produces the control signal **C**, which is proportional to the input signal $\mathbf{U_R}$ and powerful enough to drive the continuous valve. The control valve drives an actuator to achieve the desired output controlled parameter value **O**. The sensor receives the output signal **O** and feeds it back to the summing point inside the ECU. Figure 8.11 shows a closed-loop that uses a stroke-controlled valve. Such a structure represents the upper end of closed-loop electro-hydraulic system and is refereed as *Cascade Control System*.

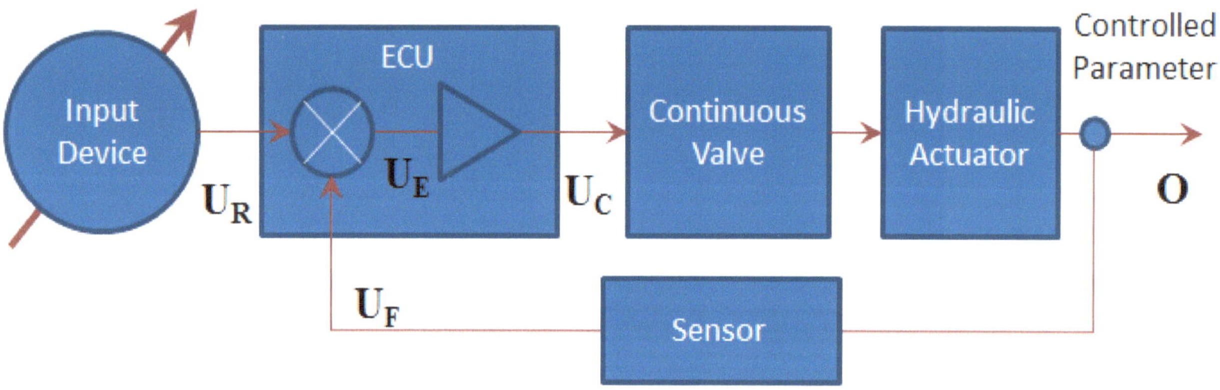

Fig. 8.10 - Block Diagram of an EH Closed-Loop Control System using a Force-Controlled Continuous Valve

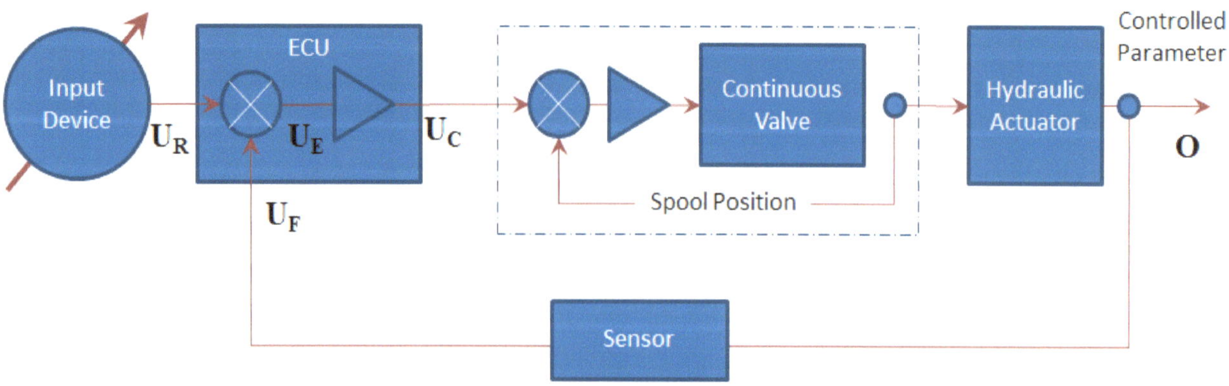

Fig. 8.11 - Block Diagram of an EH Closed-Loop Control System using a Stroke-Controlled Continuous Valve

8.2.2- Features of Electro-Hydraulic Closed-Loop Control Systems

Figure 8.12 shows the main features and the operating conditions of a closed-loop control system.

Simplicity & Cost Effectiveness: Obviously a closed-loop system is more complex and expensive than an open-loop system because it contains more elements and requires an experienced designer.

Continuous Valves: For best accuracy and highest speed of response, a *Servo Valve* is recommended for a closed-loop system. From a cost point of view, advanced generations of proportional valves can be used as long as they meet the dynamic requirements of the system they are used for.

Response to External Disturbances and Loading Conditions: A closed-loop system is primarily used to maintain automatically the desired value of the controlled parameter. If the actuator experienced external disturbances, the output signal **O** of the controlled parameter will be deviated from the desired input signal $\mathbf{U_R}$ that is set by the input device. Once the controller receives an error signal $\mathbf{U_E}$, it will automatically correct the valve opening to drive the actuator back to the desired value. It is to be noted that this automatic correction will continue to happen as long as the valve is not saturated, i.e. not fully opened.

Calibration & Tuning: A closed-loop system requires less frequent recalibration checks because the closed-loop, to some extent and within the saturation limits of the control valve, can compensate for the sources of errors in the system. To ease the process of tuning the controller, Input signal **I** and the Feedback Signal $\mathbf{U_F}$ should be unified in magnitude. This can be accomplished by calibrating the sensor.

The ECU of a closed-loop system contains more built in functions such as a ramp generator, dead band eliminator, and dither signal. Enabling and tuning these functions improves the system response. These functions will be discussed in a later chapter.

Accuracy and Source of Errors: Use of a closed-loop control obviously results in better accuracy of the controlled parameter. A closed-loop, to some extend within the saturation limit of the control valve, can compensate for the error that may result due to the change in the valve and the actuator conditions. However, other sources of errors are still there, e.g. controller adjustment and errors from the sensor.

Controlled Parameters: Since an electrical sensor is used to monitor the controlled parameter, an electro-hydraulic closed-loop control system can be used to perform any of the following control functions: position control, flow (speed) control, and pressure (load) control.

Figure 8.13 shows a generic wiring diagram of a closed-loop control system. Both the sensor and the continuous valve must be powered separately or through the signal cable. The figure does not show the shielding pins.

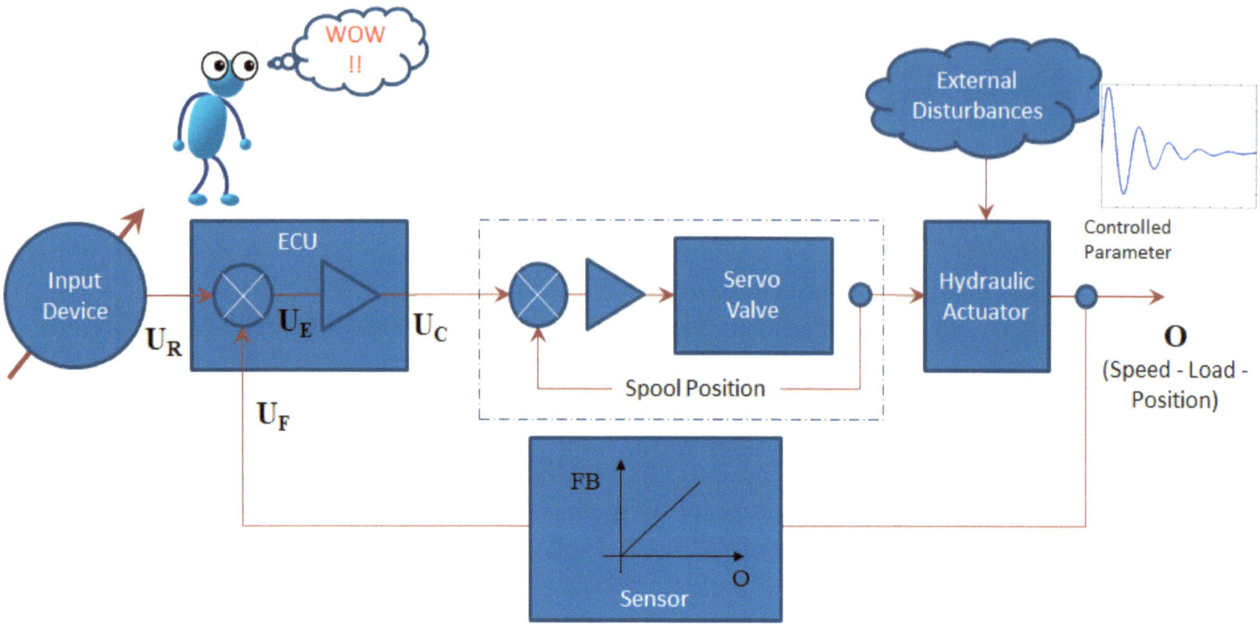

Fig. 8.12 - Features of Electro-Hydraulic Closed-Loop Control Systems

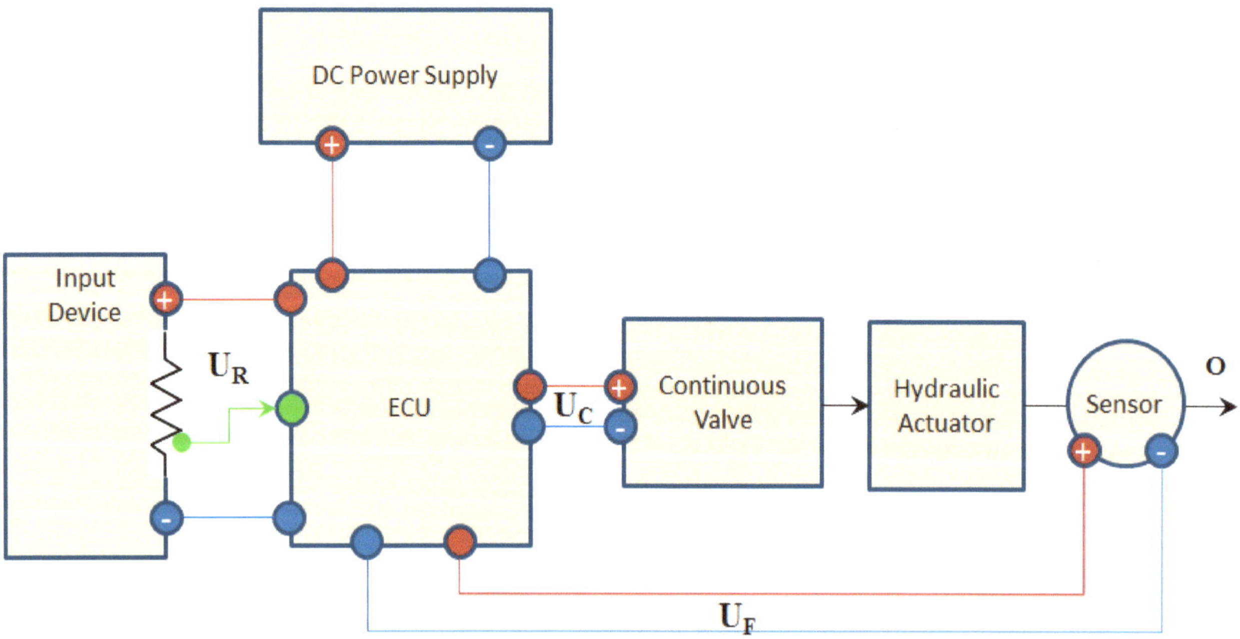

Fig. 8.13 - Generic Wiring Diagram of an EH Closed-Loop Control System

8.2.3- Electro-Hydraulic Closed-Loop Speed Control Systems

The following discussion is applicable for a motor speed control or a cylinder speed control. However, the examples presented consider the application of a motor speed control. Depending on where feedback is generated from, a motor speed can be controlled by closing the loop on the flow or on the motor RPM.

8.2.3.1- Electro-Hydraulic Closed-Loop Flow Control System

Figure 8.14 shows a block diagram of an electro-hydraulic closed-loop flow control system.

Input Device: A potentiometer is used as an input device that produces the input signal $\mathbf{U_R}$.

Sensor: Since the motor speed is a function of the flow, then the output of the flow sensor can be calibrated with the motor speed and then the loop is closed on the flow. A *Flow Sensor* feeds backward the actual magnitude of the flow rate.

Calibration and Feedback: To ease the tuning process of the controller, a calibration should be done to unify the range of the flow sensor and the input signal. As shown in the figure, if the input device is used to adjust the motor RPM, then the calibration should be done between the range of the flow sensor and the range of the potentiometer. In such a case, assuming no leakage in the motor and linear characteristics of the flow sensor, the reciprocal of the motor size is used as a calibration factor. Units must be balanced in the calibration equation. An actual point-to-point calibration is a more typical calibration practice. The output of the calibration stage continues to go backward to the summing point inside the ECU in the form of a feedback signal $\mathbf{U_F}$.

ECU: The ECU calculates the error signal $\mathbf{U_E}$, then conditions and amplifies it to produces the control signal $\mathbf{U_C}$. Controller tuning is a crucial step in closed-loop system design.

Control Valve: A 4/3 *Closed-Center (Overlapped) Proportional Directional Valve* is recommended to control both the direction and the speed of the motor. The valve opens proportional to the control signal **C**. As a result, fluid flow passes through the valve based on the *Valve Flow Gain*. Then the motor spins with a speed that is proportional to the flow through the valve. The proportionality factor is the motor size.

Energy Efficiency of the Operation: The control valve is kept open as long as the motor rotates and closes only when the motor stops. Therefore, the energy wasted in this system is larger than in load or position control.

System Operation and Sources of Error: If the actual fluid flow that is received by the motor is subjected to change, it will be seen and automatically corrected by the system. However, the effect of the external disturbances on the motor speed won't be seen by the system as long as the flow stays unchanged. Therefore, such a system is also good only for constant loading conditions. Additional sources of errors are the sensor and the adjustment of the controller.

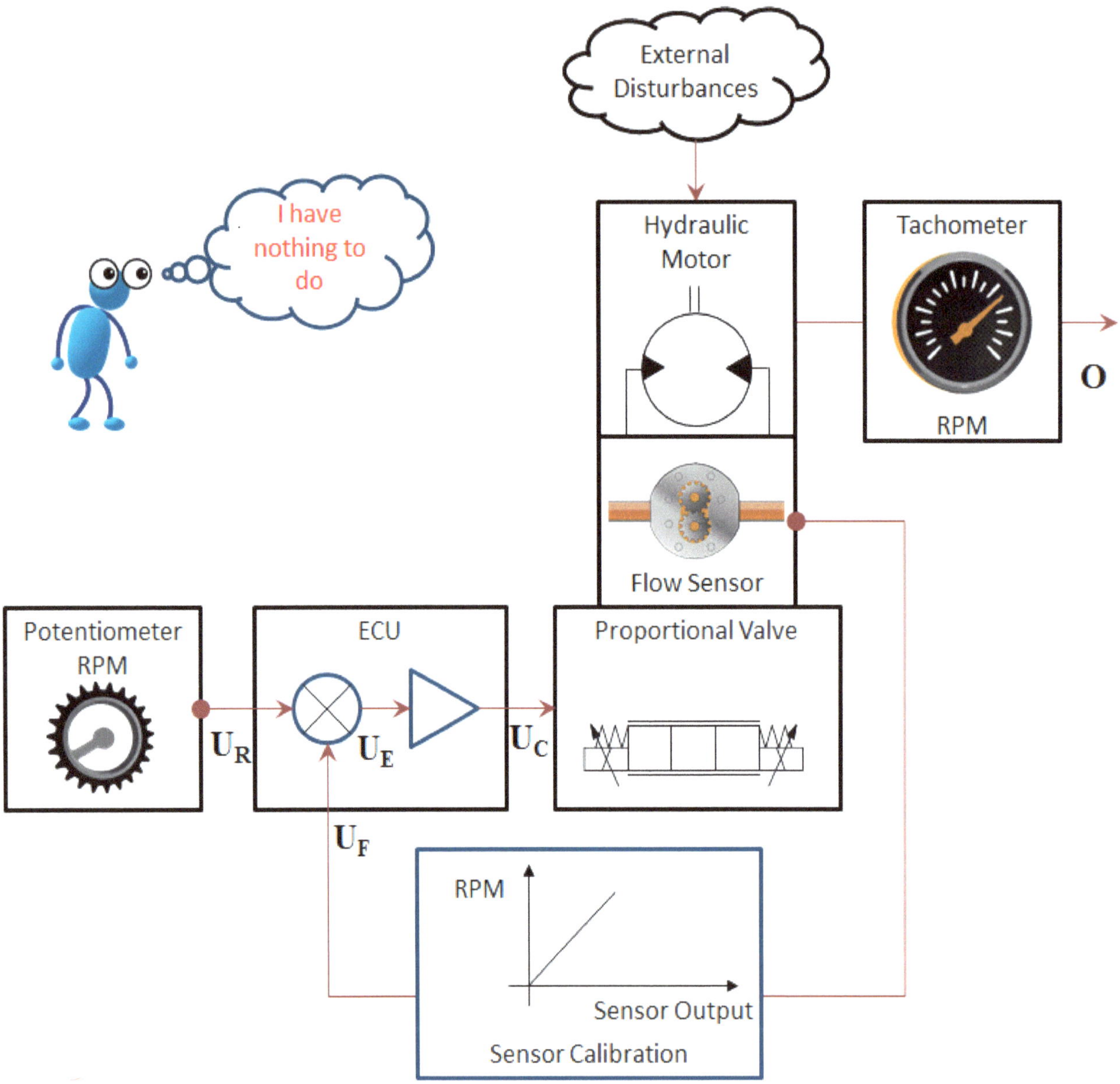

Fig. 8.14 - Block Diagram of a Closed-Loop EH Flow Control System

Figure 8.15 shows the operating conditions of the hydraulic circuit diagram of the previous system. A calibration factor equal to 10 has been used to convert the flow (lit/min) to an RPM based on a motor of 100 cc/rev size.

The figure indicates that the system is well calibrated, the controller is well tuned, and the offset between the input and the actual motor RPM is within the acceptable range (0.02%).

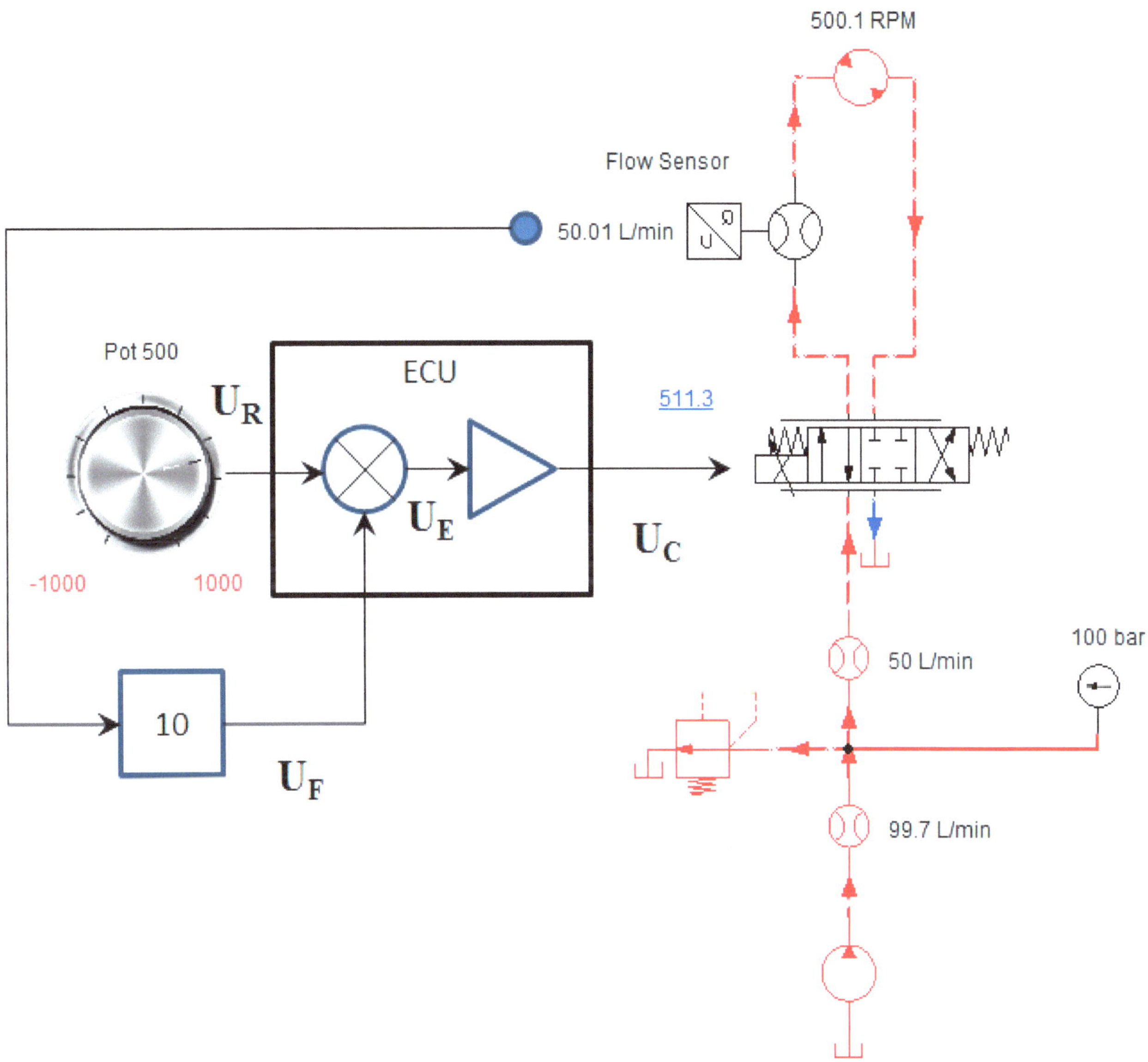

Fig. 8.15 - Hydraulic Circuit Diagram of an EH Closed-Loop Flow Control System

8.2.3.2- Electro-Hydraulic Closed-Loop Motor Speed Control System

Figure 8.16 shows an electro-hydraulic block diagram of a closed-loop motor speed control system. This system works the same way as the one shown in Fig. 8.14 except that the loop is closed on the actual motor RPM using an RPM sensor.

In such a system, the closed-loop will compensate for the changes in the valve and the motor conditions and also for the effect of the external disturbances on the motor speed. The only remaining sources of errors are the sensor and the adjustment of the controller.

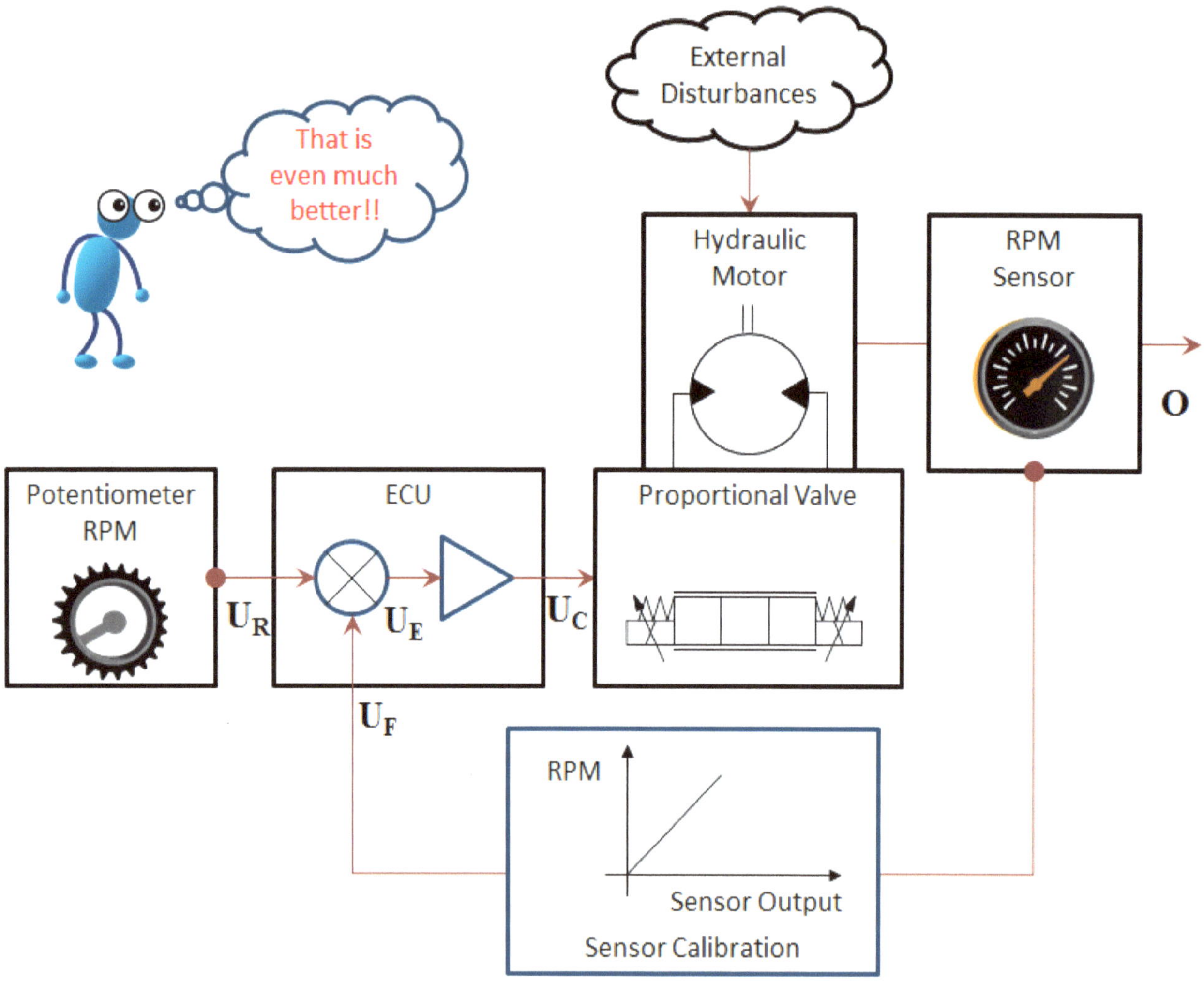

Fig. 8.16 - Block Diagram of an EH Closed-Loop Motor Speed Control System

Figure 8.17 shows the operating conditions of the hydraulic circuit diagram of the previous system. The figure indicates that the system is well calibrated, the controller is well tuned, and the offset between the input and the actual motor RPM is within the acceptable range (0.02%).

Although the % steady state error in this system is larger than the previous one, when the loop is closed on the flow, this system compensates for disturbances on the motor. Careful controller tuning can eliminate the steady state error. The calibration factor equals one because both the input device and the RPM sensor have the same range.

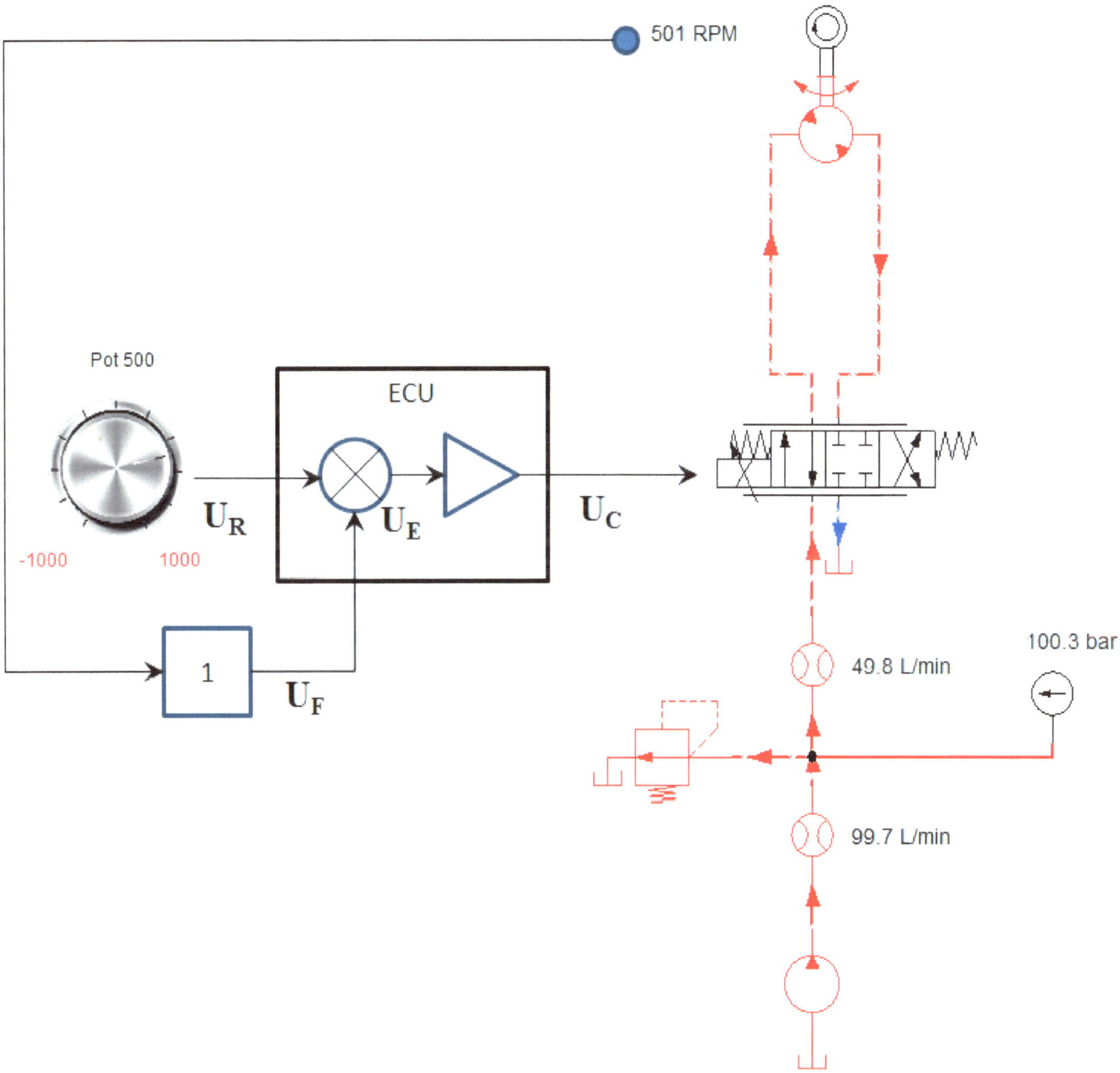

Fig. 8.17 - Hydraulic Circuit Diagram of an EH Closed-Loop Motor Speed Control System

8.2.4- Electro-Hydraulic Closed-Loop Load Control Systems

The following discussion is applicable for a motor torque control or a cylinder force control. However, the examples presented consider the application of a cylinder force control. Depending on where the feedback is generated from, a cylinder force can be controlled by closing the loop on the pressure or on the cylinder force.

8.2.4.1- Electro-Hydraulic Closed-Loop Single-Acting Cylinder Pressure Control System

Figure 8.18 shows a block diagram of an electro-hydraulic closed-loop pressure control system for a single-acting cylinder.

Input Device: A potentiometer is used as an input device that produces the input signal $\mathbf{U_R}$.

Sensor: Since the cylinder force is a function of the pressure, the output of the pressure sensor can be calibrated with the cylinder force and then the loop is closed on the pressure. A *pressure Sensor* feeds backward the actual magnitude of the pressure.

Calibration and Feedback: To ease the tuning process of the controller, a calibration should be done to unify the range of the pressure sensor and the input signal. As shown in the figure, if the input device is used to assign the pressure value, then the calibration should be done between the range of the pressure sensor and the range of the potentiometer. In such a case, assuming no leakage in the cylinder and linear characteristics of the pressure sensor, the cylinder effective area is used as a calibration factor with the units be balanced in the calibration equation. An actual point-to-point calibration is a more typical calibration practice. The output of the calibration stage continues to go backward to the summing point inside the ECU in the form of a feedback signal $\mathbf{U_F}$.

ECU: The ECU calculates the error signal $\mathbf{U_E}$, then conditions and amplifies it to produce the control signal **C.** Controller tuning is a crucial step in closed-loop system design.

Control Valve: A *Proportional Pressure Relief Valve* is used to maintain the required system pressure so that the force produced by a cylinder is controlled. The valve responds proportional to the control signal **C**. As a result, the pressure that acts on the cylinder is controlled. Consequently, the cylinder produces a force that is proportional to the pressure.

Energy Efficiency of the Operation: The pump forces the control valve (PRV) all the time. Therefore, the energy wasted in this system is considerable.

System Operation and Sources of Error: If the actual pressure that acts on the cylinder is subjected to change, it will be seen and automatically corrected by the system. However, the effect of the external disturbances on the cylinder won't be seen by the system as long as the pressure stays unchanged. Therefore, such systems are good only for constant loading condition. Additional sources of errors are the sensor and the adjustment of the controller.

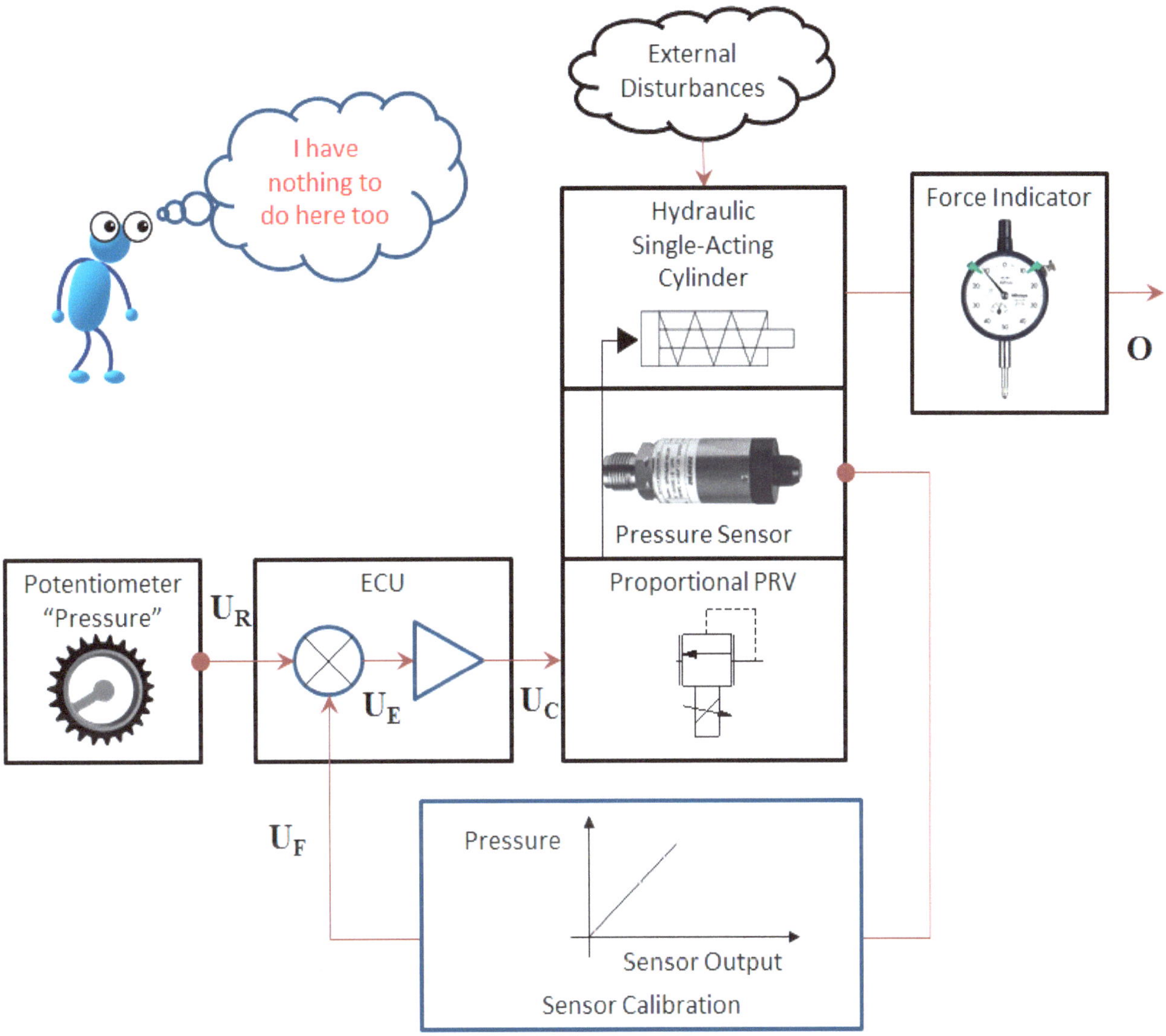

Fig. 8.18 - Block Diagram of an EH Closed-Loop Single-Acting Cylinder Pressure Control System

Figure 8.19 shows the operating conditions of the hydraulic circuit diagram of the previous system. Since the potentiometer is set to assign the pressure value, and the loop is closed on the pressure (bar), then a calibration factor equal to one is used. The cylinder then produces a force (kg) based on the cylinder's effective area (cm^2).

For example, as shown in the figure, for a pressure of 50 (bar), the cylinder that has an effective area of 10 cm^2 produces 500 kg force.

The figure indicates that the system is well calibrated, the controller is well tuned, and there is no steady state error.

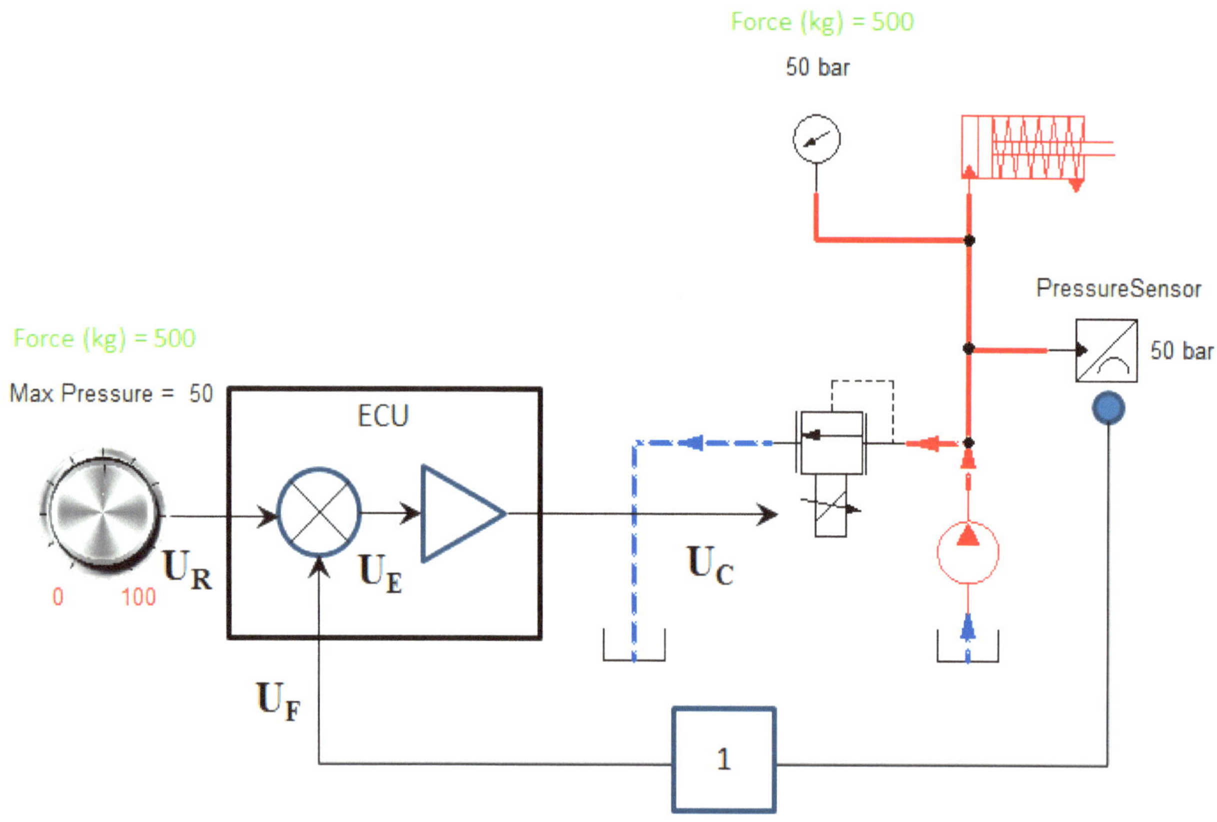

Fig. 8.19 - Hydraulic Circuit Diagram of an EH Closed-Loop Single-Acting Cylinder Pressure Control System

8.2.4.2- Electro-Hydraulic Closed-Loop Single-Acting Cylinder Force Control System

Figure 8.20 shows a block diagram of an electro-hydraulic closed-loop single-acting cylinder force control system. This system works the same way as the one shown in Fig. 8.18 except that the loop is closed on the actual cylinder force using a load cell.

In such a system, the closed-loop will compensate for the changes in the valve and the cylinder conditions and also the effect of the external disturbances on the cylinder force. The only remaining sources of errors are the sensor and the adjustment of the controller.

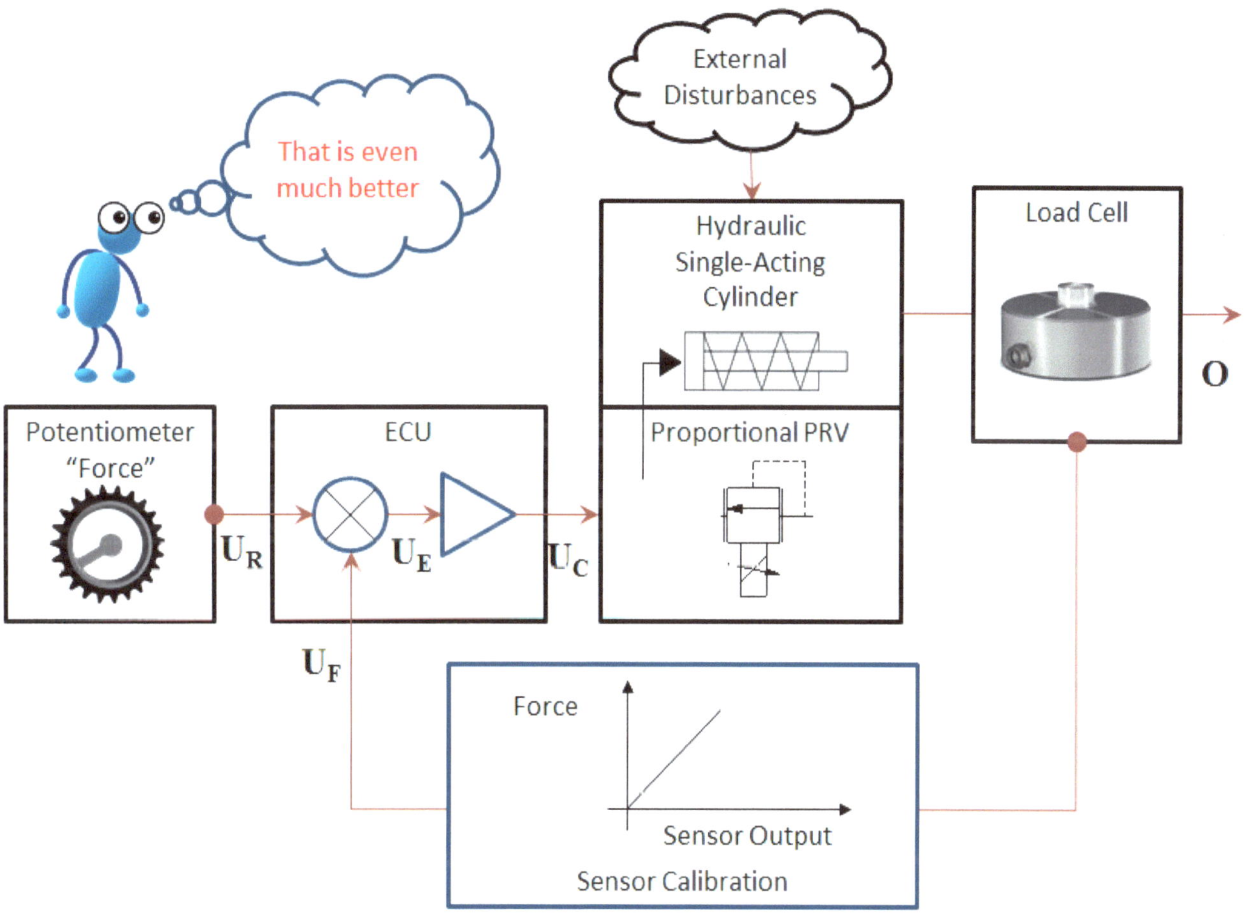

Fig. 8.20 - Block Diagram of an EH Closed-Loop Single-Acting Cylinder Force Control System

8.2.4.3- Electro-Hydraulic Closed-Loop Double-Acting Cylinder Pressure Control System

Figure 8.21 shows a block diagram of an electro-hydraulic pressure closed-loop control system for a double-acting cylinder.

Input Device: A potentiometer is used as an input device that produces the input signal $\mathbf{U_R}$.

Sensor: Since the cylinder force is a function of the pressure, the output of the pressure sensor can be calibrated with the cylinder force and then the loop is closed on the pressure. A *Differential Pressure Sensor* feeds backward the actual magnitude of the pressure. The sign of the signal indicates the direction of the force.

Calibration and Feedback: To ease the tuning process of the controller, a calibration should be done to unify the range of the differential pressure sensor and the input signal. As shown in the figure, if the input device is used to assign the pressure value, then the calibration should be done between the range of the differential pressure sensor and the range of the potentiometer. In such a case, assuming no leakage in the cylinder and linear characteristics of the pressure sensor, the cylinder effective area is used as a calibration factor with the units which must be balanced in the calibration equation. An actual point-to-point calibration is a more typical calibration practice. The output of the calibration stage continues to go backward to the summing point inside the ECU in the form of a feedback signal $\mathbf{U_F}$.

ECU: The ECU calculates the error signal $\mathbf{U_E}$, then conditions and amplifies it to produce the control signal $\mathbf{U_C}$. Controller tuning is a crucial step in closed-loop system design.

Control Valve: A 4/3 *Open-Center (Underlapped) Proportional Directional Valve* is recommended to control both the direction and the force of the cylinder. The valve opens proportional to the control signal U_C. As a result, the pressure is distributed across the cylinder's piston based on the *Valve Pressure Gain*. Consequently, the cylinder produces a force that is proportional to the *differential pressure*.

Energy Efficiency of the Operation: The control valve is an open-center valve that is leaking the pump discharge continuously to the tank. Therefore, the energy wasted in this system is considerable.

System Operation and Sources of Error: If the actual differential pressure that acts on the cylinder is subjected to change, it will be seen and automatically corrected by the system. However, the effect of the external disturbances on the cylinder won't be seen by the system as long as the differential pressure stays unchanged. Therefore, such system is also good only for constant loading conditions. Additional sources of errors are the sensor and the adjustment of the controller.

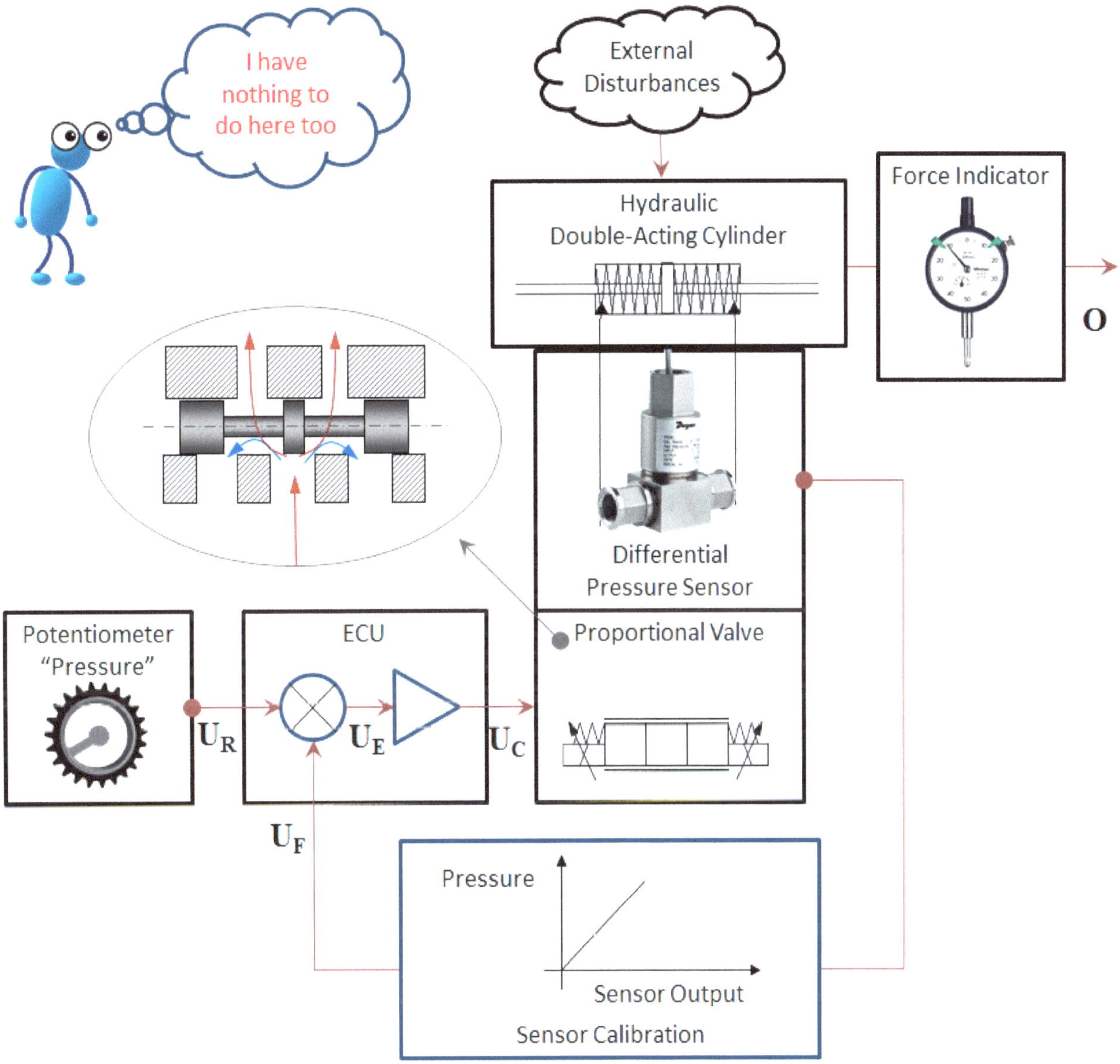

Fig. 8.21 - Block Diagram of an EH Closed-Loop Double-Acting Cylinder Pressure Control System

Figure 8.22 shows the operating conditions of the hydraulic circuit diagram of the previous system. Since the potentiometer is set to place the pressure, and the loop is closed on the pressure (bar), then a calibration factor equal one is used. The cylinder then produces force (kg) based on it's effective area (cm^2).

For example, as shown in the figure for a differential pressure of 50 (bar), the cylinder that has an effective area of 10 cm^2 produces a 500 kg force.

The figure indicates that the system is well calibrated, the controller is well tuned, and the offset between the input and the actual differential pressure is within the acceptable range.

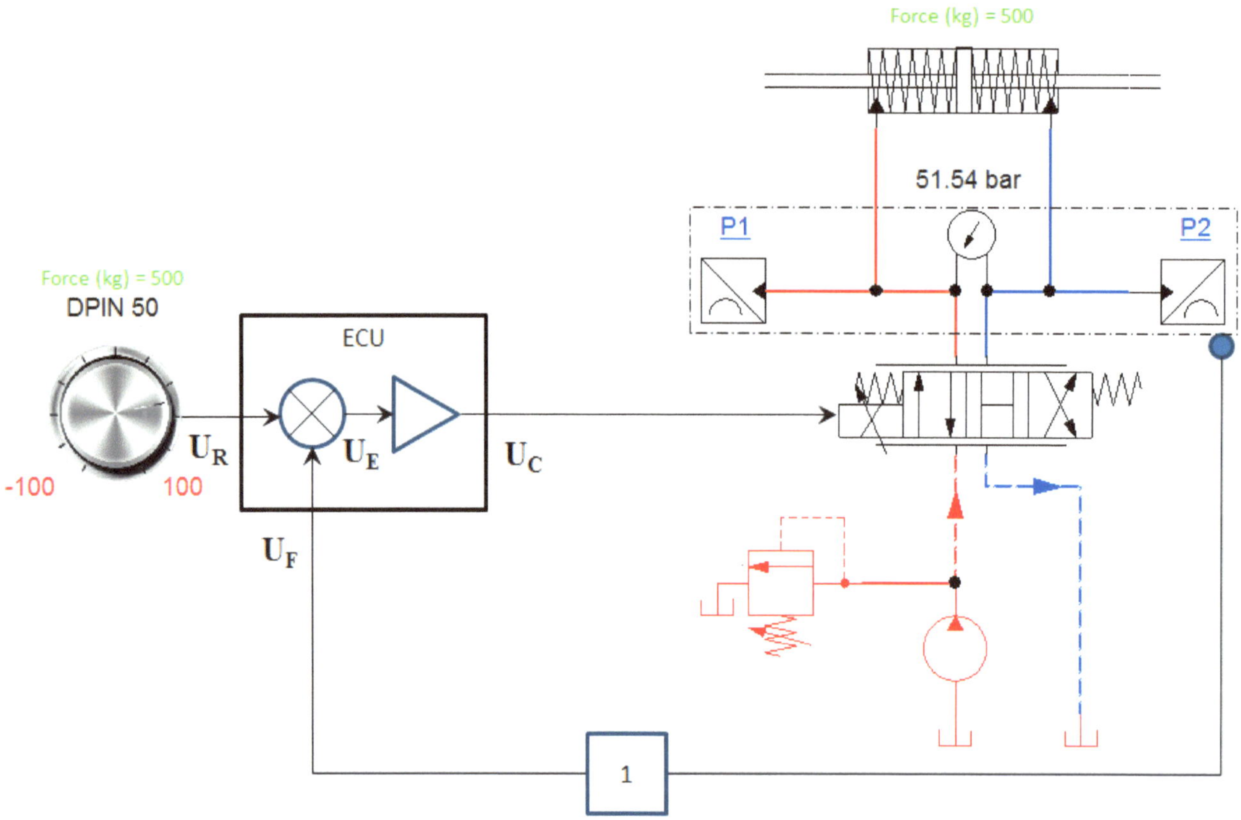

Fig. 8.22 - Hydraulic Circuit Diagram of an EH Closed-Loop Double-Acting Cylinder Pressure Control System

8.2.4.4- Electro-Hydraulic Closed-Loop Double-Acting Cylinder Force Control System

Figure 8.23 shows a block diagram of an electro-hydraulic closed-loop double-acting cylinder force control system. This system works the same way as the one shown in Fig. 8.21 except that the loop is closed on the actual cylinder force using a load cell.

In such a system, the closed-loop will compensate for the changes in the valve and the cylinder conditions and also the effect of the external disturbances on the cylinder force. The only remaining sources of errors are the sensor and the adjustment of the controller.

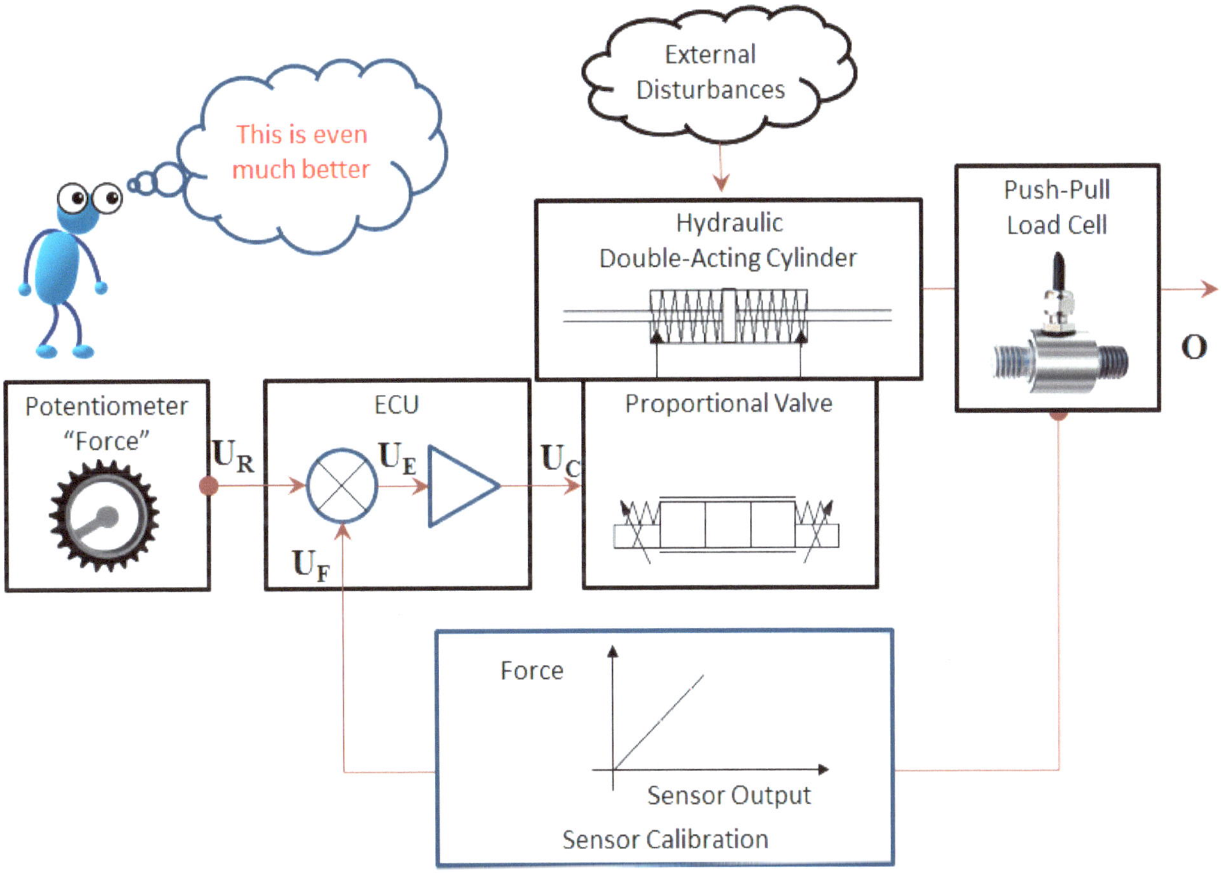

Fig. 8.23 - Block Diagram of an EH Closed-Loop Double-Acting Cylinder Force Control System

8.2.5- Electro-Hydraulic Closed-Loop Position Control Systems

The following discussion is applicable for a cylinder position control or a rotary actuator angular position control. However, the example presented considers the application of a position control. For a position control system there is no choice but to close the loop on the position.

Figure 8.24 shows a block diagram of an electro-hydraulic closed-loop cylinder position control system.

Input Device: A potentiometer is used as an input device that produces the input signal $\mathbf{U_R}$.

Sensor: In the system in the figure, the valve drives the cylinder with a speed proportional to the flow rate that passes through the valve and the cylinder size. There is no way then to calibrate the valve opening or the flow rate with the cylinder position because the factor of time is missing. Then there in no way to build position control system but to close the loop on cylinder position. A *Position Sensor* feeds backward the actual position of the cylinder.

Calibration and Feedback: To ease the tuning process of the controller, a calibration should be done to unify the range of the position sensor and the input signal. As shown in the figure, if the input device is used to place the cylinder position, then the calibration should be done between the range of the position sensor and the range of the potentiometer. The output of the calibration stage continues to go backward to the summing point inside the ECU in form of a feedback signal $\mathbf{U_F}$.

ECU: The ECU calculates the error signal $\mathbf{U_E}$, then conditions and amplifies it to produces the control signal $\mathbf{U_C}$. Controller tuning is a crucial step in closed-loop system design.

Control Valve: A 4/3 *Critical-Center (Zero-Lapped) Servo Valve* is recommended to control both the direction and the position of the cylinder. The valve opens proportional to the control signal $\mathbf{U_C}$. As a result, fluid flow passes through the valve and the cylinder is displaced towards the desired destination. As the cylinder approaches the final destination, the valve closes gradually to decelerate the cylinder. The valve closes at the final destination to maintain the desired position of the cylinder.

Energy Efficiency of the Operation: Such a system wastes the least amount of energy, as compared to a speed control system and a load control system. The valve is open only temporarily while the cylinder is moving till it reaches the desired position.

System Operation and Source of Errors: The effect of the external disturbances on the cylinder will be seen and automatically corrected by the system. Sources of errors are the sensor and the adjustment of the controller. If an overlapped valve is used instead of a critical-lapped one, it will result in a steady state error. This issue and resolving it will be discussed later.

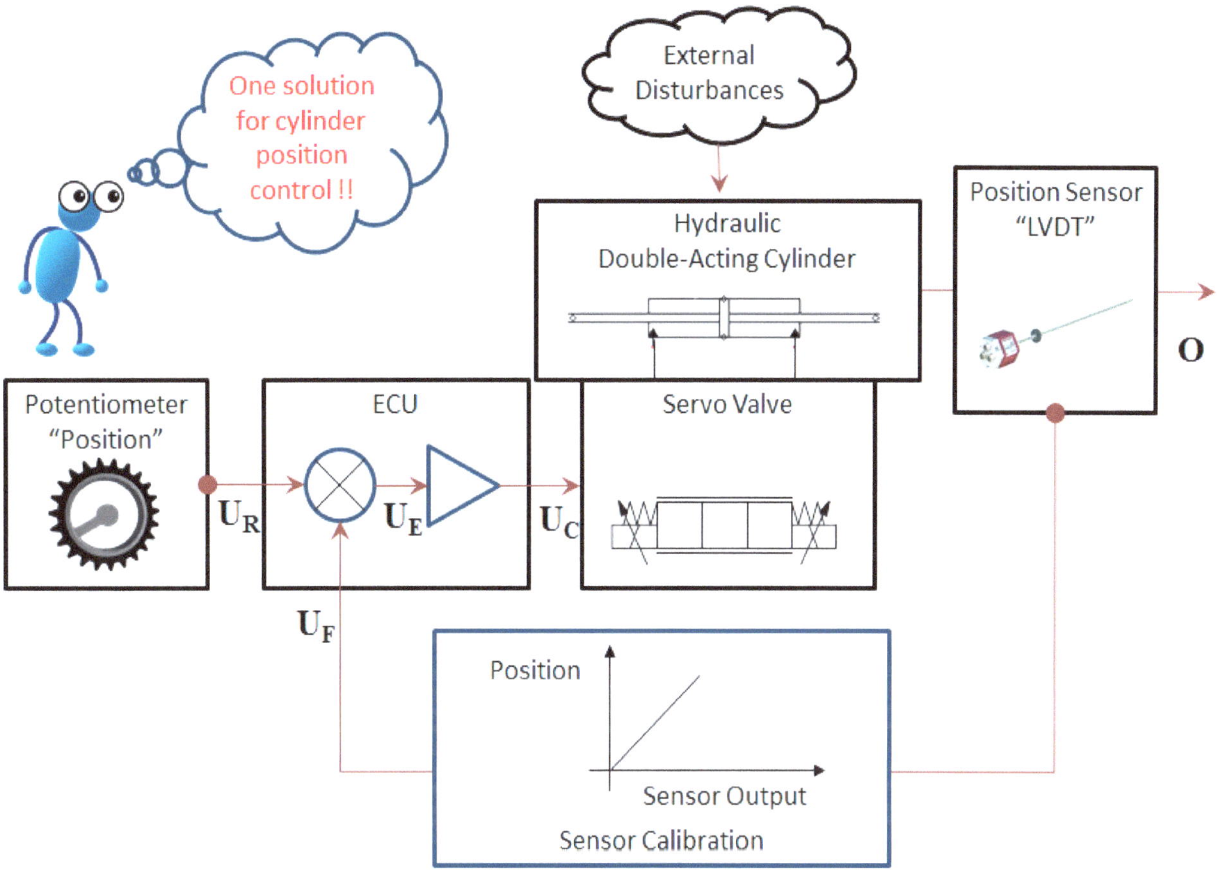

Fig. 8.24 - Block Diagram of an EH Closed-Loop Cylinder Position Control System

Figure 8.25 shows the operating conditions of the hydraulic circuit diagram of the previous system. Since the potentiometer is set to place the cylinder position, and the loop is closed on the cylinder position, then a calibration factor equal to one is used. The figure indicates that the system is well calibrated, the controller is well tuned, and there is no steady state error.

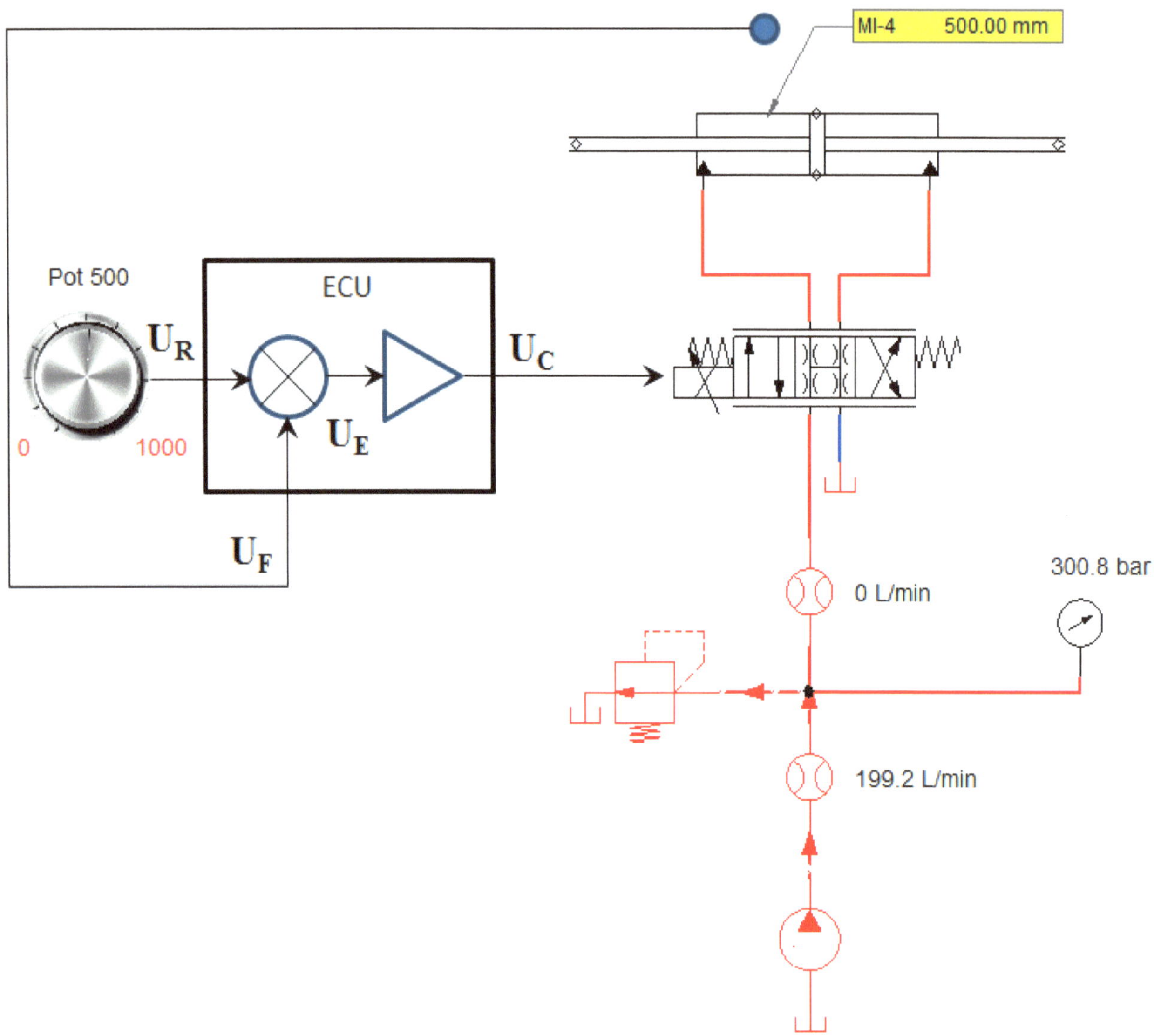

Fig. 8.25 - Hydraulic Circuit Diagram of an EH Closed-Loop Cylinder Position Control System

Figure 8.26 shows a turn-key cylinder position control. It is referred to as *"Servo Cylinder"*. The cylinder has the control valve built in and is equipped with all the necessary accessories. The end user just sends the input signal to the valve. Such a cylinder, as shown in Fig. 8.27 can be inserted as a subsystem in a larger automated process. The valve cables are hooked to an information bus system and shares its information with the main process.

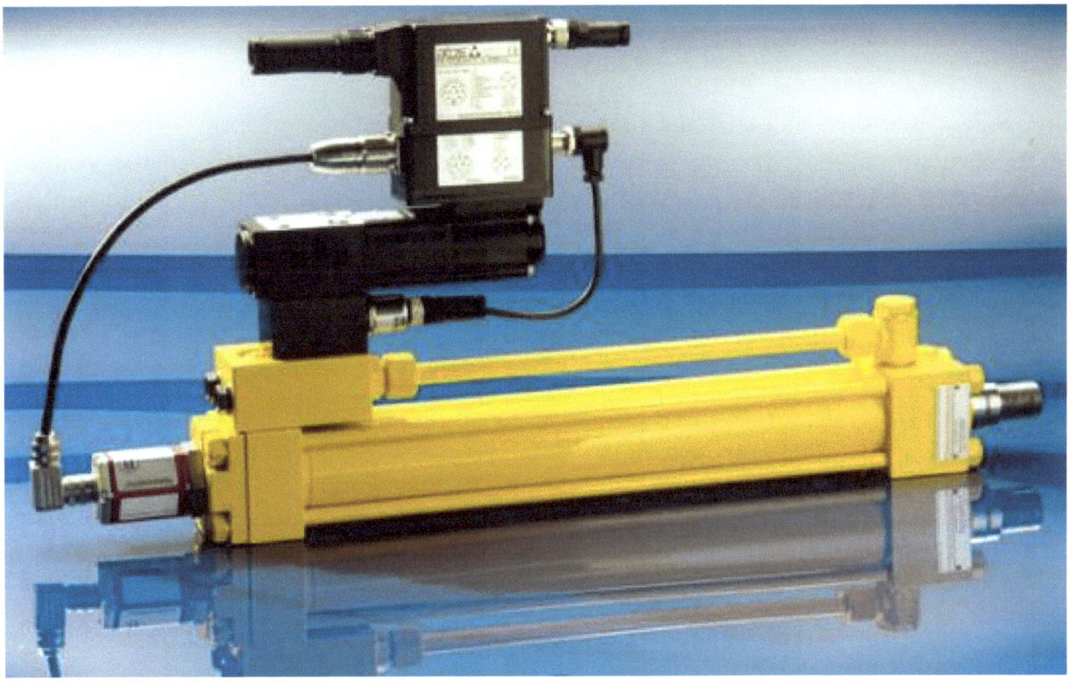

Fig. 8.26 - Servo Cylinder (Courtesy of Atos)

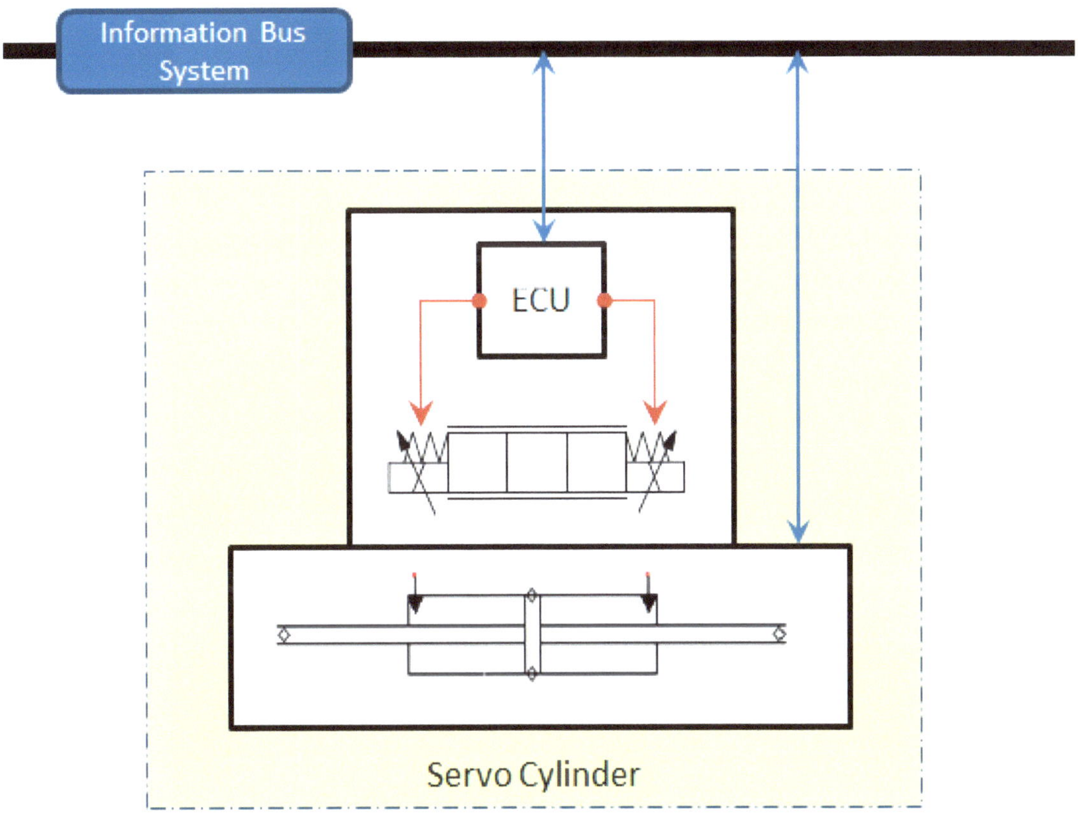

Fig. 8.27 - Servo Cylinder in Industrial Automation

8.3- Closed-Loop Performance Analysis

Figure 8.28 shows a block diagram for a generic EH closed-loop control system with the components gain shown. As shown in the figure, each of the elements in the system has its own *Gain*. The term gain means the ratio between the output and the input of each element as follows:

- G_{Amp} = **Amplifier gain:** The amplifier produces an output control signal **U_C** in "Current" that is proportional to the input error signal **U_E** in "Voltage". Amplifier gain is the only tunable gain infield.

- G_{Valve} = **Valve Gain:** The valve produces an outlet "Flow" or "Pressure" that is proportional to the input control signal C in "Current" of the amplifier.

- $G_{Actuator}$ = **Hydraulic Actuator Gain:** The hydraulic actuator transforms the hydraulic fluid flow or pressure into output controlled parameter such as load, speed or position.

- G_{Sensor} = **Sensor:** The feedback transducer converts the output to a feedback signal in "Voltage".

The expression "*Loop Gain*", K_V, refers to the non-dimensional product of each component gain in a system. Equation 3.1

$$K_v = G_{Amp} \times G_{Valve} \times G_{Actuator} \times G_{Sensor} \qquad \textbf{8.1}$$

Figure 8.29 shows the response of a generic electro-hydraulic closed-loop control system to a step input. As shown in the figure, the higher K_V is, the faster the system performances are to some extent since an excessive loop gain causes the instability of the system. As a rule of thumb, electro-hydraulic systems are designed so that the loop gain equals one third of the system's *Natural Frequency* "ω".

The calculation of the system's natural frequency is out of the scope of this textbook. It will be explained in other textbooks that will cover design strategies of an EH system.

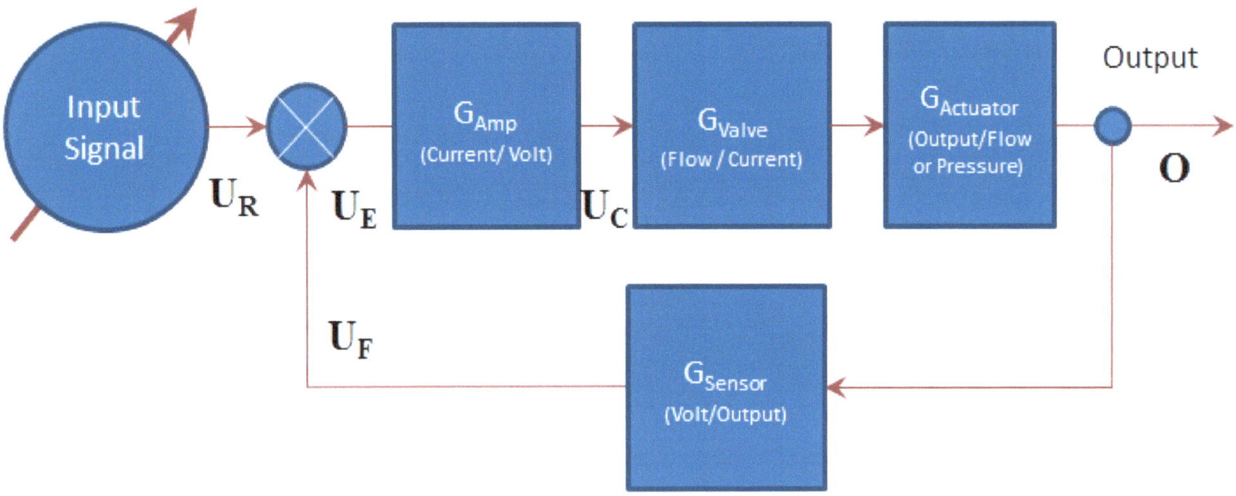

Fig. 8.28 - Components Gain in a Generic EH Closed-Loop Control System

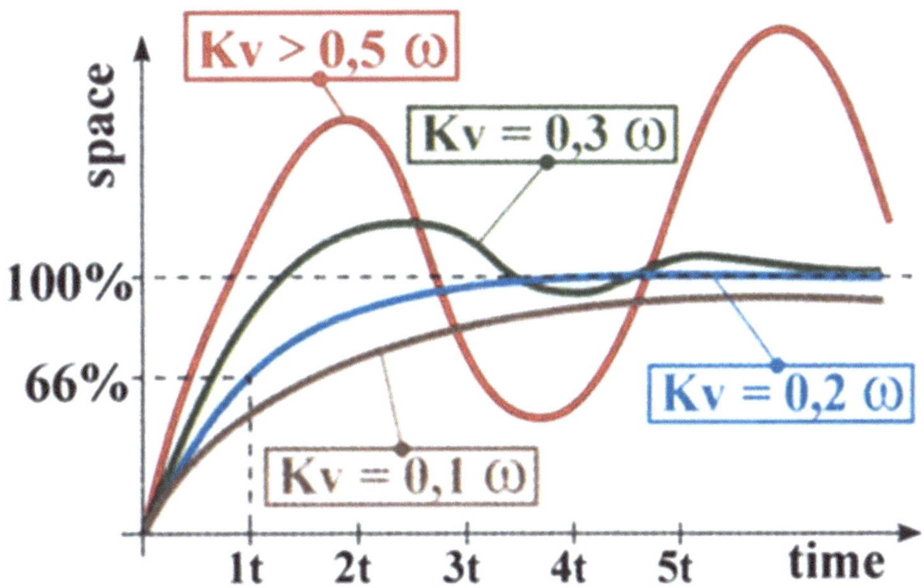

Fig. 8.29 – Closed-Loop Control System Performance (Courtesy of ASSOFLUID)

8.4- Electro-Hydraulic Control Systems Applications Examples

8.4.1- Industrial Applications

Figure 8.30 shows an electro-hydraulic closed-loop pump displacement control. Actual pump pressure and displacement are fed back to the controller that takes care of changing the pump displacement based on the predefined control logics and the control parameters preset there.

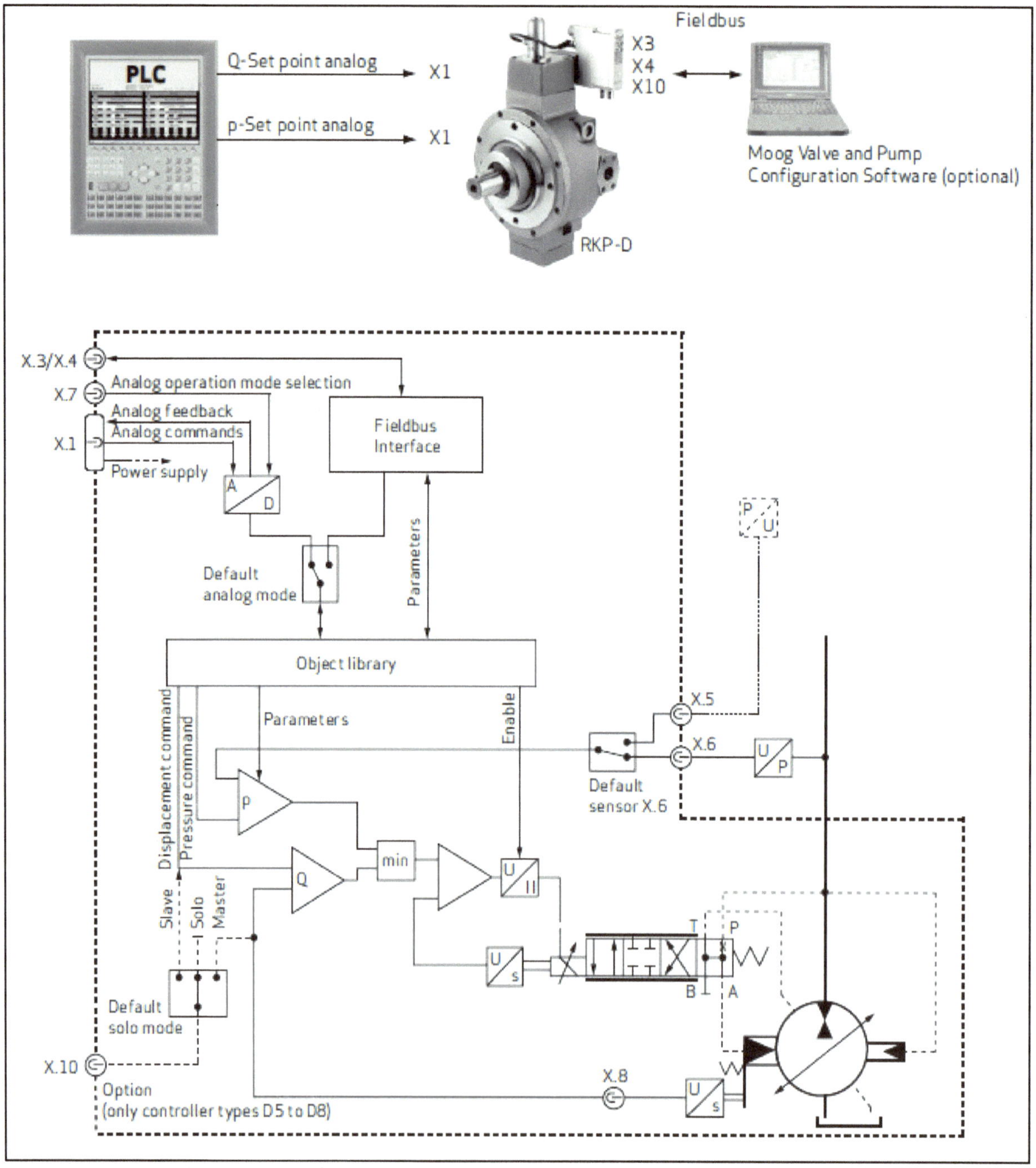

Fig. 8.30- EH Closed-Loop Pump Power Control (Courtesy of Moog)

The hydraulic circuit diagram shown in Fig. 8.31 is a portion of a complex automated drilling machine. This portion shows that a stroke controlled proportional valve is used to control the position of the drilling spindle so that it can control how deep the hole is drilled.

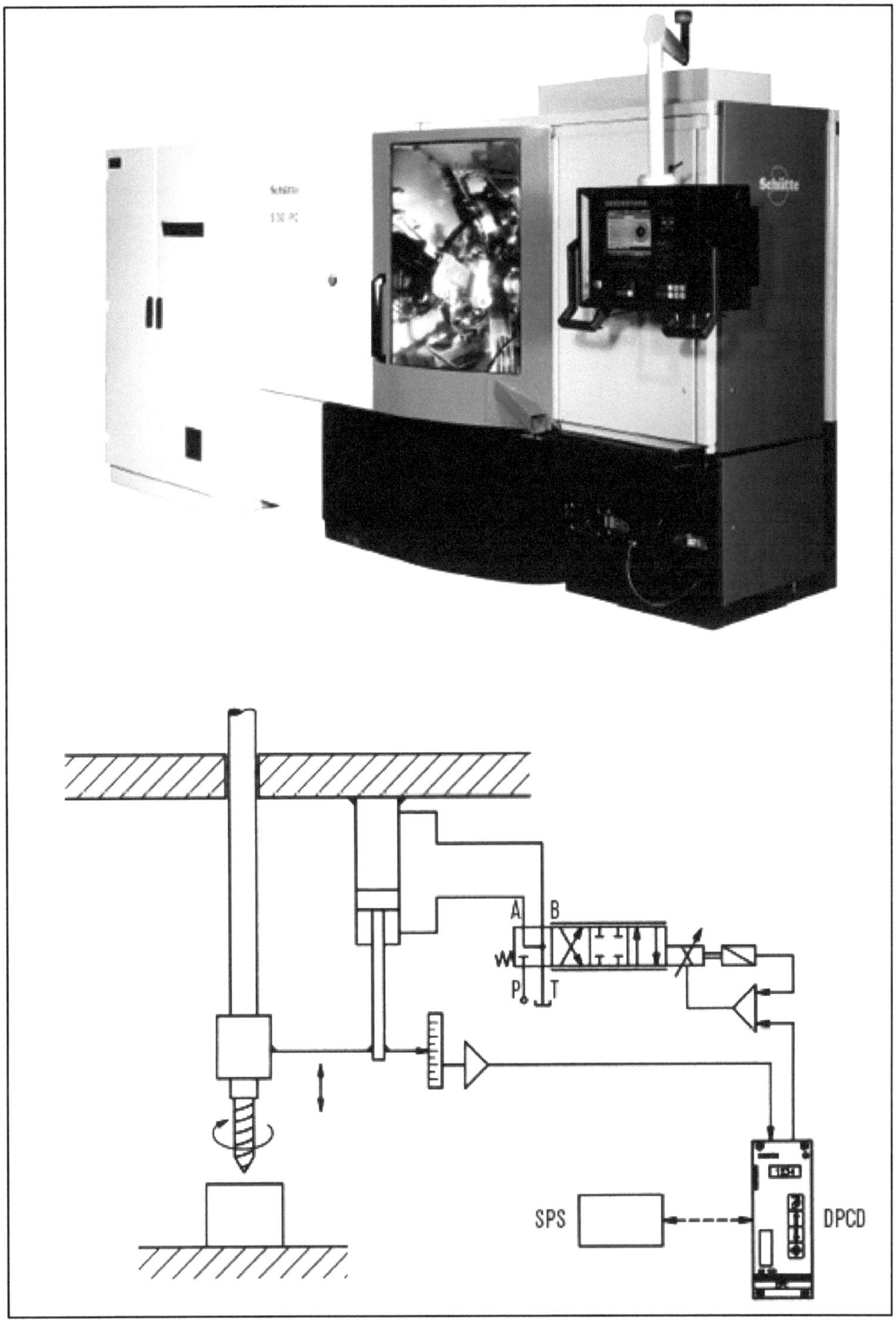

Fig. 8.31- EH Closed-Loop Single-Axis Cylinder Position Control in a Drilling Machine (Courtesy of Bosh Rexroth)

The CNC (Computerized Numerical Controlled) machine shown in Fig. 8.32 has two axes, each of which is an electrohydraulic closed-loop position control system. The area of application is unlimited. However, one famous example is milling machine tool.

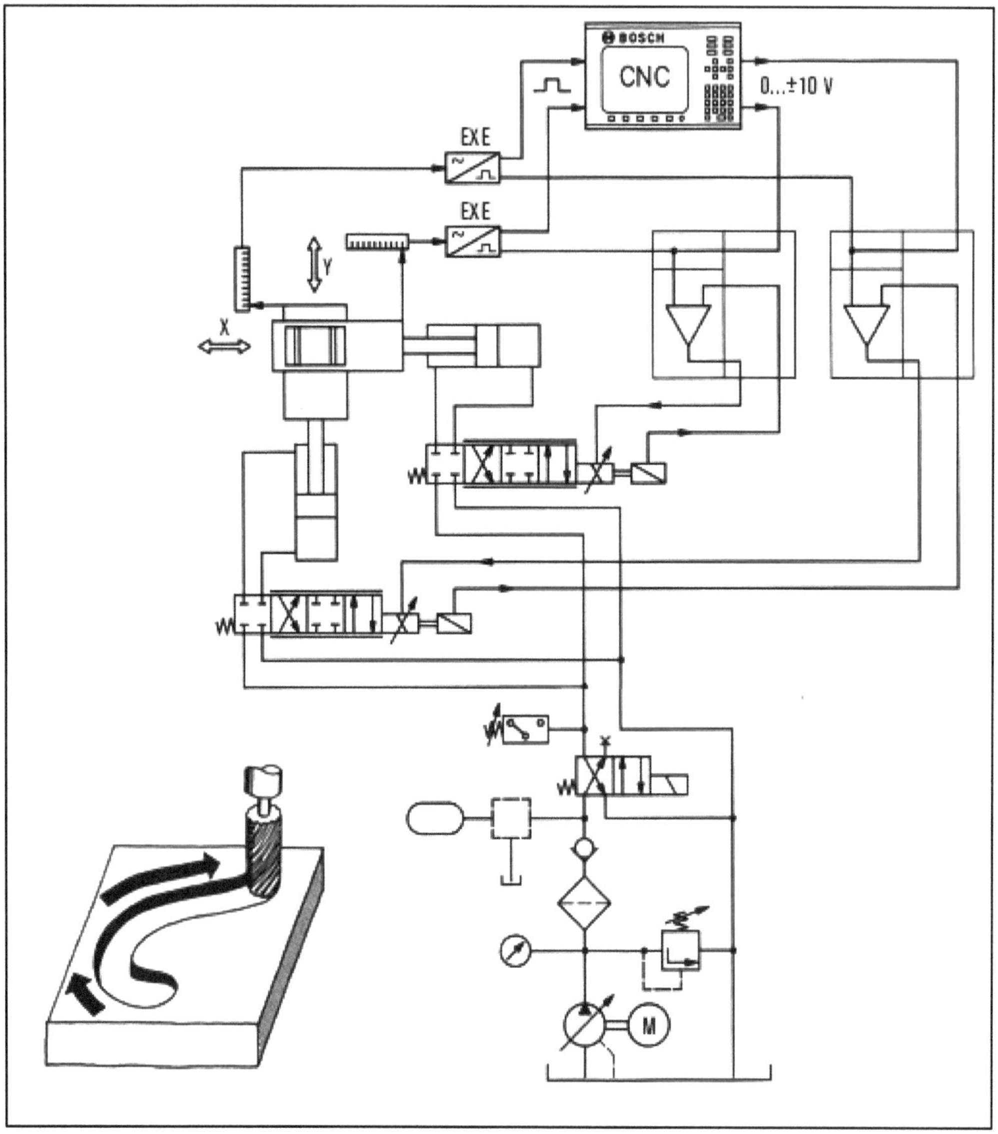

Fig. 8.32- EH Closed-Loop Double-Axis Cylinder Position Control in a CNC Machine (Courtesy of Bosh Rexroth)

The high-level control of a wind power plant continuously processes parameters like wind speed, direction of the wind and rotational speed of the rotor. From these quantities, the set point for the rotor blade angle is calculated and supplied to a separate position controller module. Figure 8.33 shows the operating principle of pitch adjustment of the rotor blade.

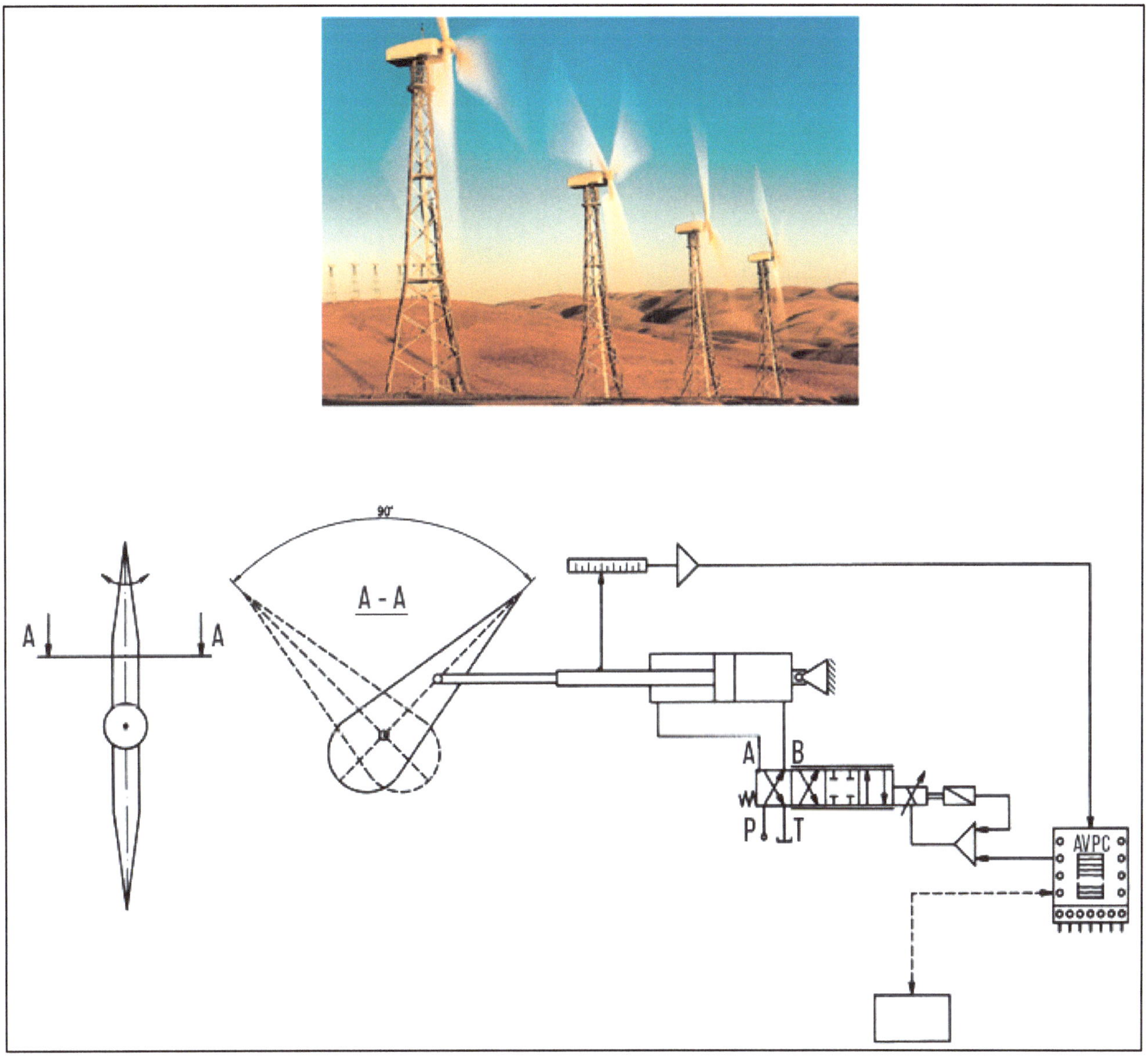

Fig. 8.33- EH Closed-Loop Cylinder Position Control in a Wind Turbine (Courtesy of Bosh Rexroth)

Figure 8.34 shows an electro-hydraulic open-loop speed control for a circular saw for round wood. As a courtesy of Bosch Rexroth, in the lumber industry, the length of the stem which is to be processed is sawn in less than a second. The disk of the circular saw has a diameter of approx. 2 m and a thickness of up to 10 mm. The advance cylinder with a piston diameter of approx. 80 mm is driven at a pressure of 180 bar with a flow of up to approx. 200 l/min. The control of the cylinder movement takes place by a stroke-controlled proportional directional control valve.

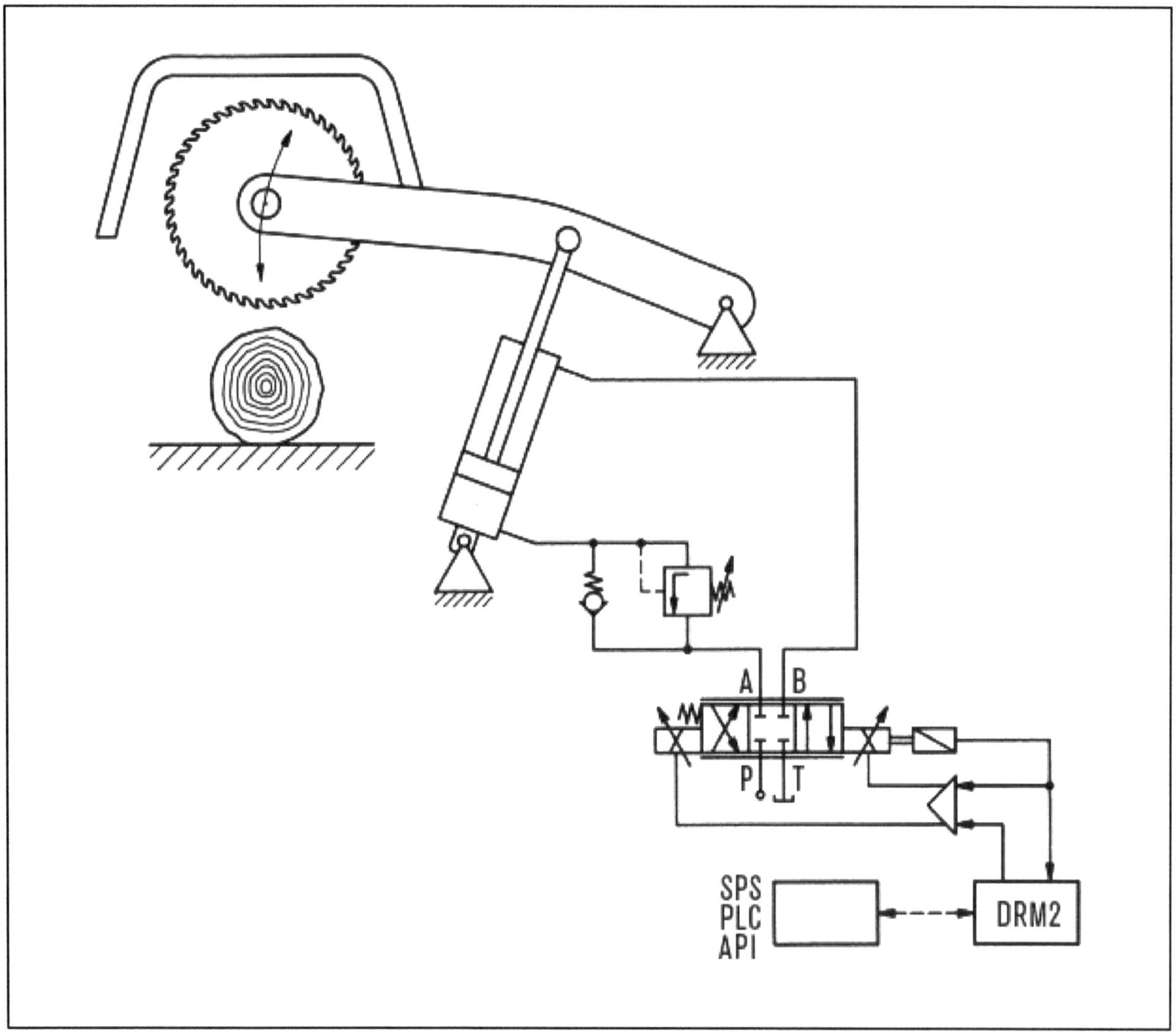

Fig. 8.34- EH Open-Loop Cylinder Speed Control for a Circular Saw for Round Wood (Courtesy of Bosh Rexroth)

Figure 8.35 shows a typical EH closed-loop motor speed control system that can be used in various applications. Fig. 8.36 shows an electro-hydraulic circuit for an injection molding machine. A stroke-controlled proportional throttle valve (1) and a stroke-controlled proportional pressure control valve (2) are used to ensure precision control of spindle rotation and force.

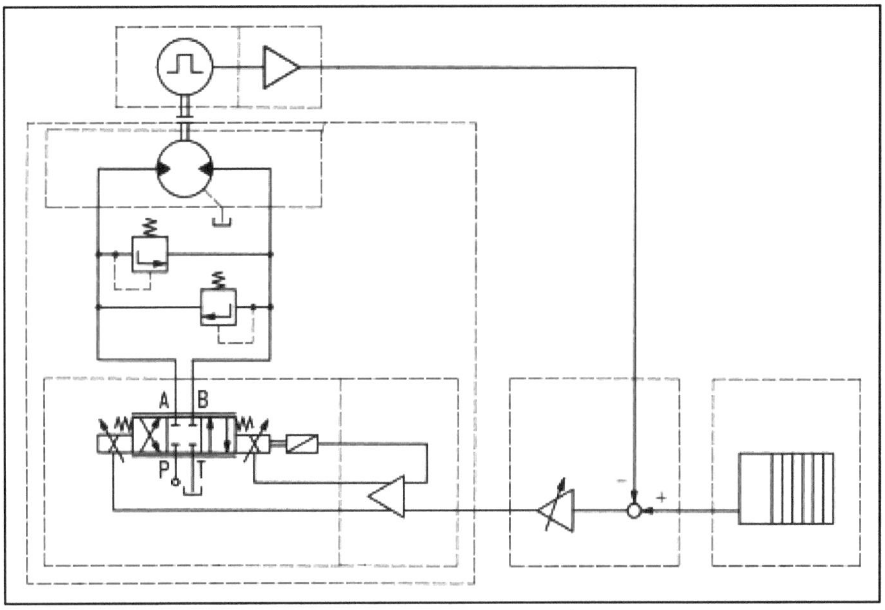

Fig. 8.35- EH Closed-Loop Motor Speed Control (Courtesy of Bosh Rexroth)

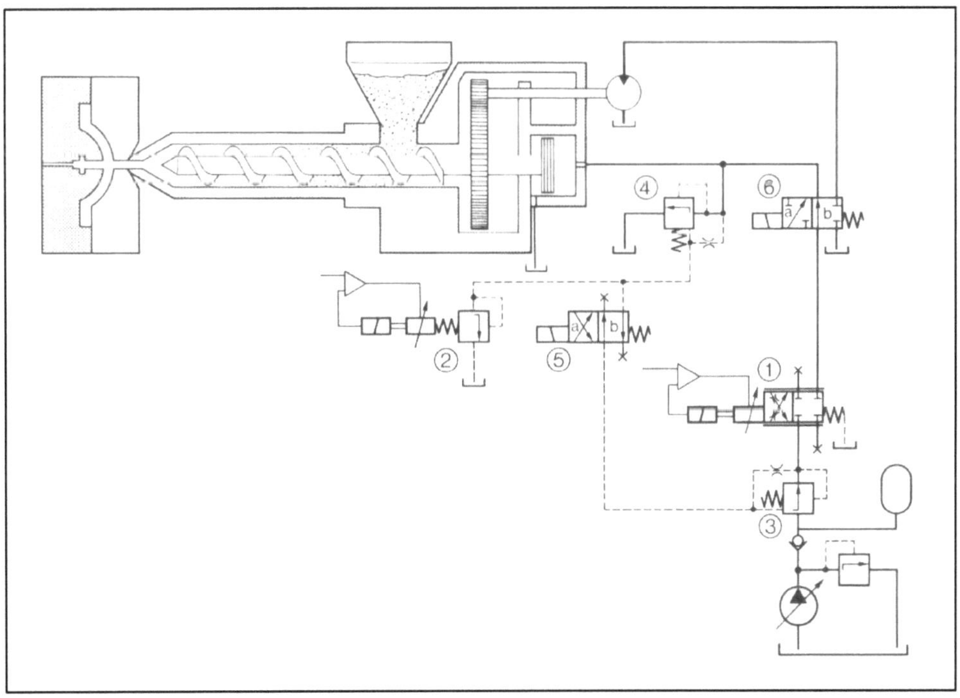

Fig. 8.36- EH Open-Loop Speed Control and Pressure Control in an Injection Molding Machine (Courtesy of Bosh Rexroth)

A welding machine, as shown in Fig. 8.37, the application of a differential pressure control ensures controlling the force at the weld seam.

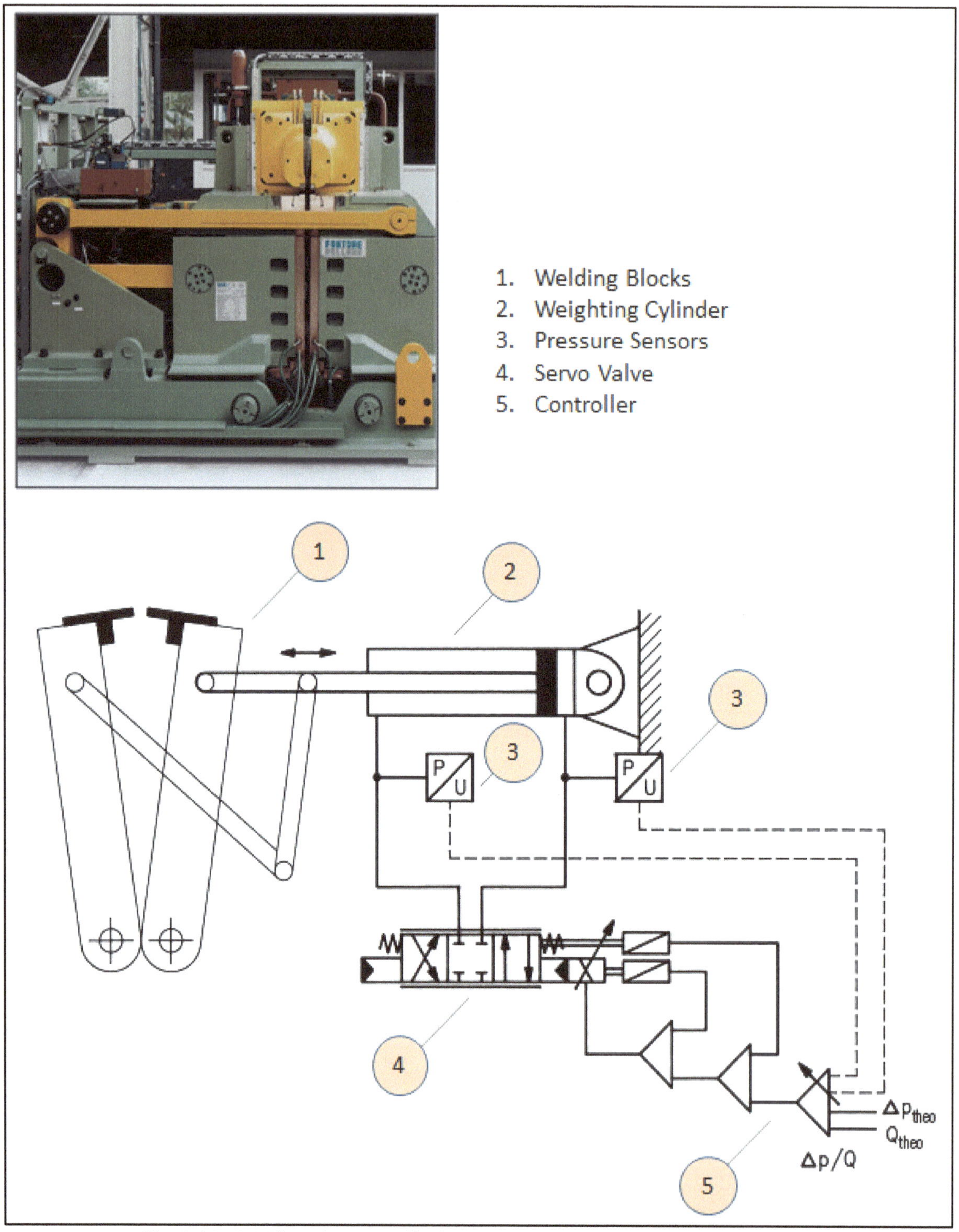

Fig. 8.37- EH Closed-Loop Cylinder Force Control in a Welding Machine (Courtesy of Bosh Rexroth)

Figure 8.38 shows an EH Speed Control Systems for Concrete Saw. To increase the efficiency of the cutting process, a variable rate hydraulic-driven saw-feed mechanism has been used. The system is designed to maintain constant speed electric drive motor at its full power no matter how thick or thin the slab might be.

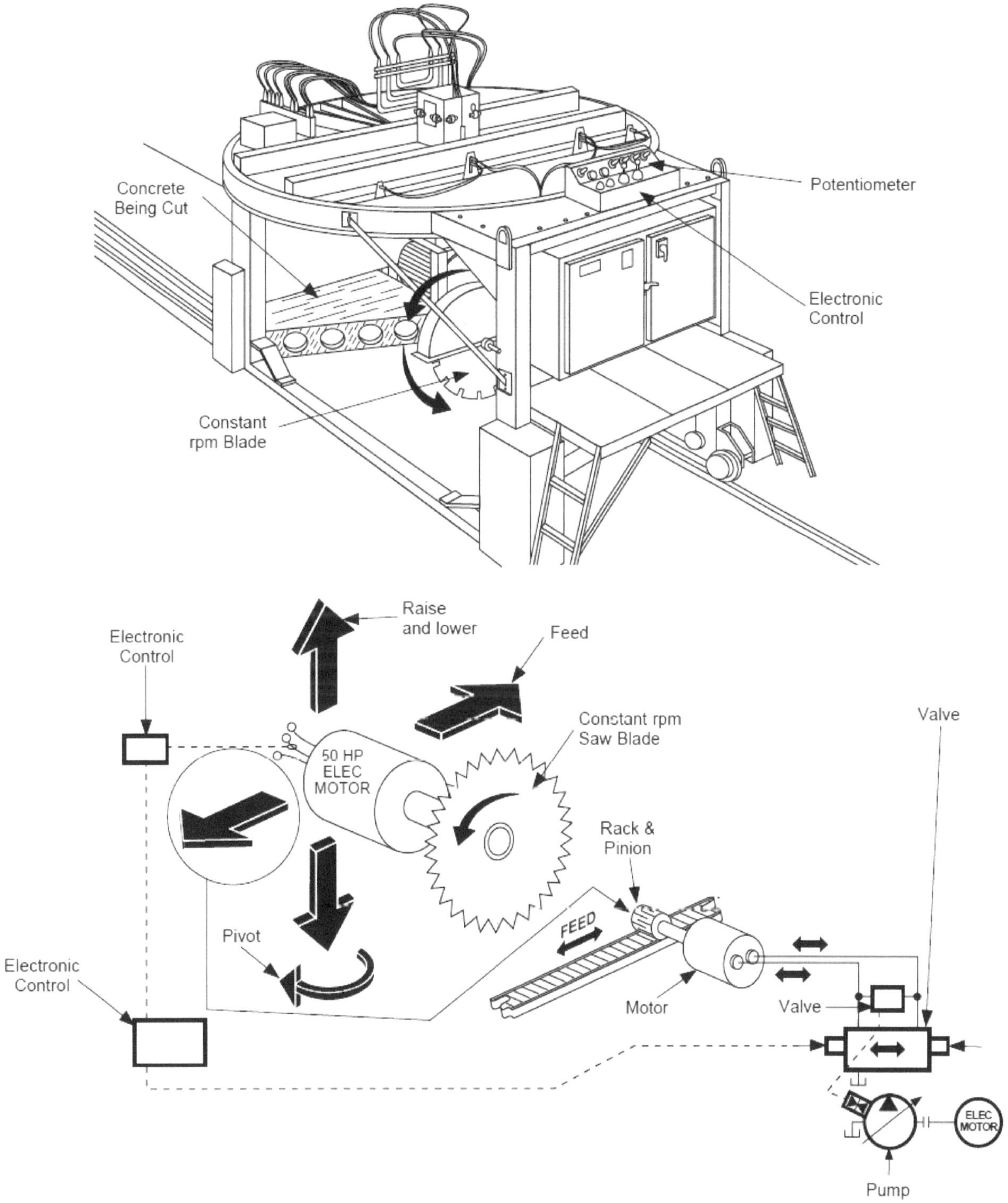

Fig. 8.38- EH Closed-Loop Speed Control Systems for Concrete Saw (Courtesy of NFPA)

8.4.2- Mobile Applications

In the machine shown in Fig. 8.39, the speeds of two motors are controlled in open-loop systems. Despite the open-loop mode, the control logic offers some level of dependency between the speeds of the two motors so that the salt spreading width and quantity is interactively controlled.

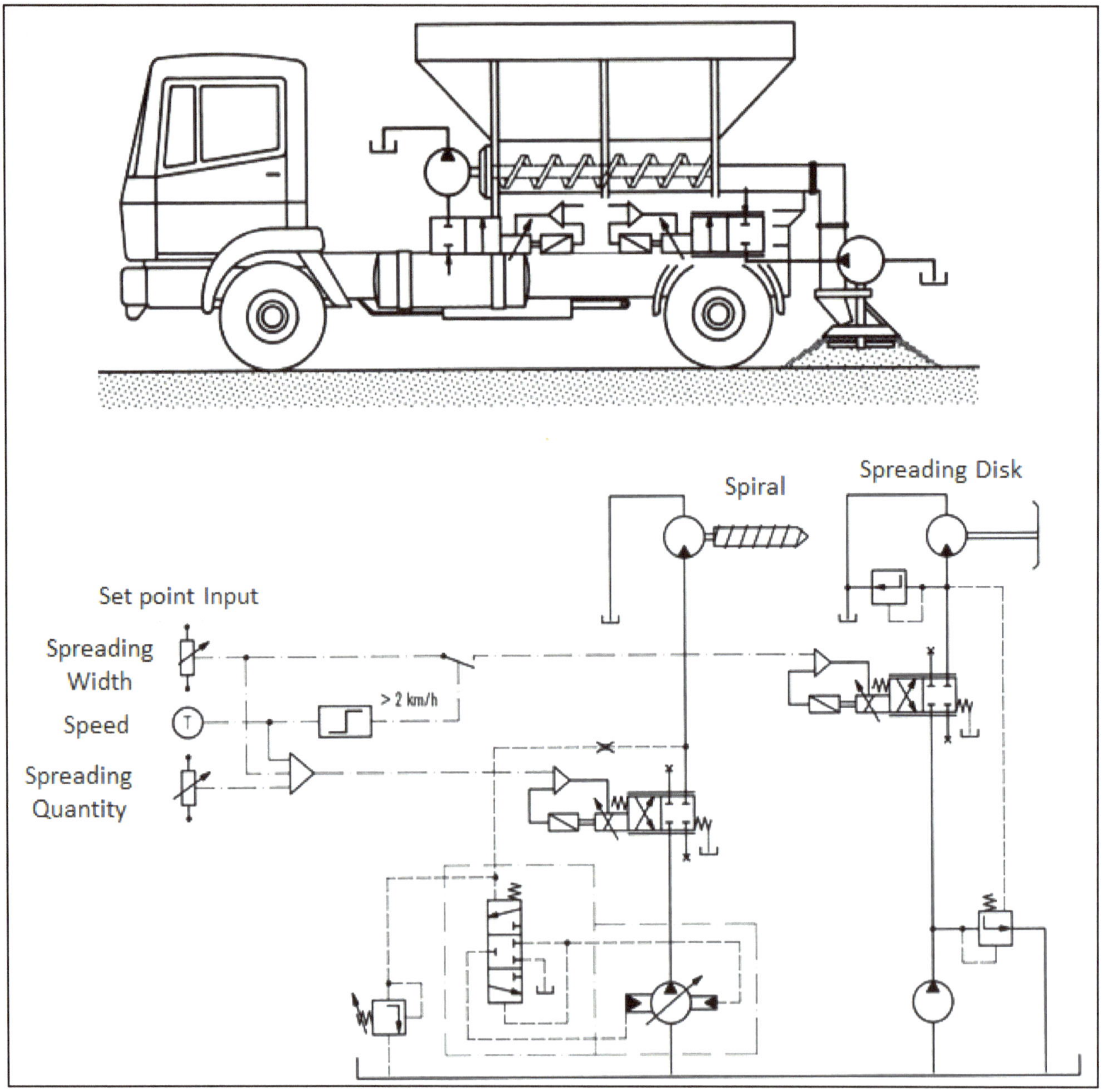

Fig. 8.39- EH Open-Loop Double-Axes Motor Speed Control in Snow Fighting Vehicles (Courtesy of Bosh Rexroth)

Figure 8.40 shows an electrohydraulic closed-loop steering control system. A stroke controlled proportional valve is used to control the position of a cylinder so that the angular position of the tire is controlled. An angular position sensor feeds the actual angular position back to the controller.

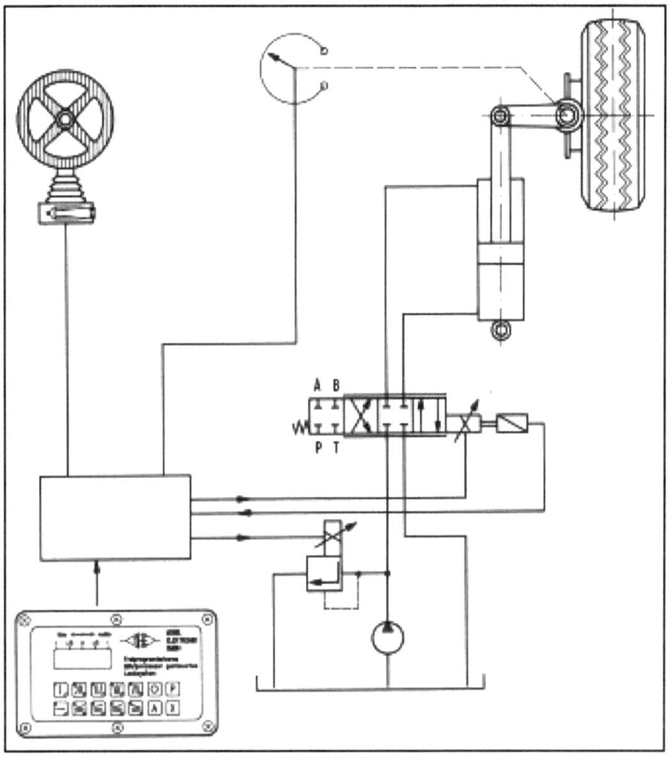

Fig. 8.40- EH Closed-Loop Cylinder Position Control in a Steering System (Courtesy of Bosh Rexroth)

Mobile equipment are the most applications for hydrostatic drives. However, the drives are so powerful that unless automatic torque and speed control are included in the system design, the internal combustion engine prime mover will stall. Figure 8.47 shows an automatic electronic control that senses the losses in engine speed, overrides the operator's manual control, and decrease the swashplate angle of the pump in accordance with the load increase.

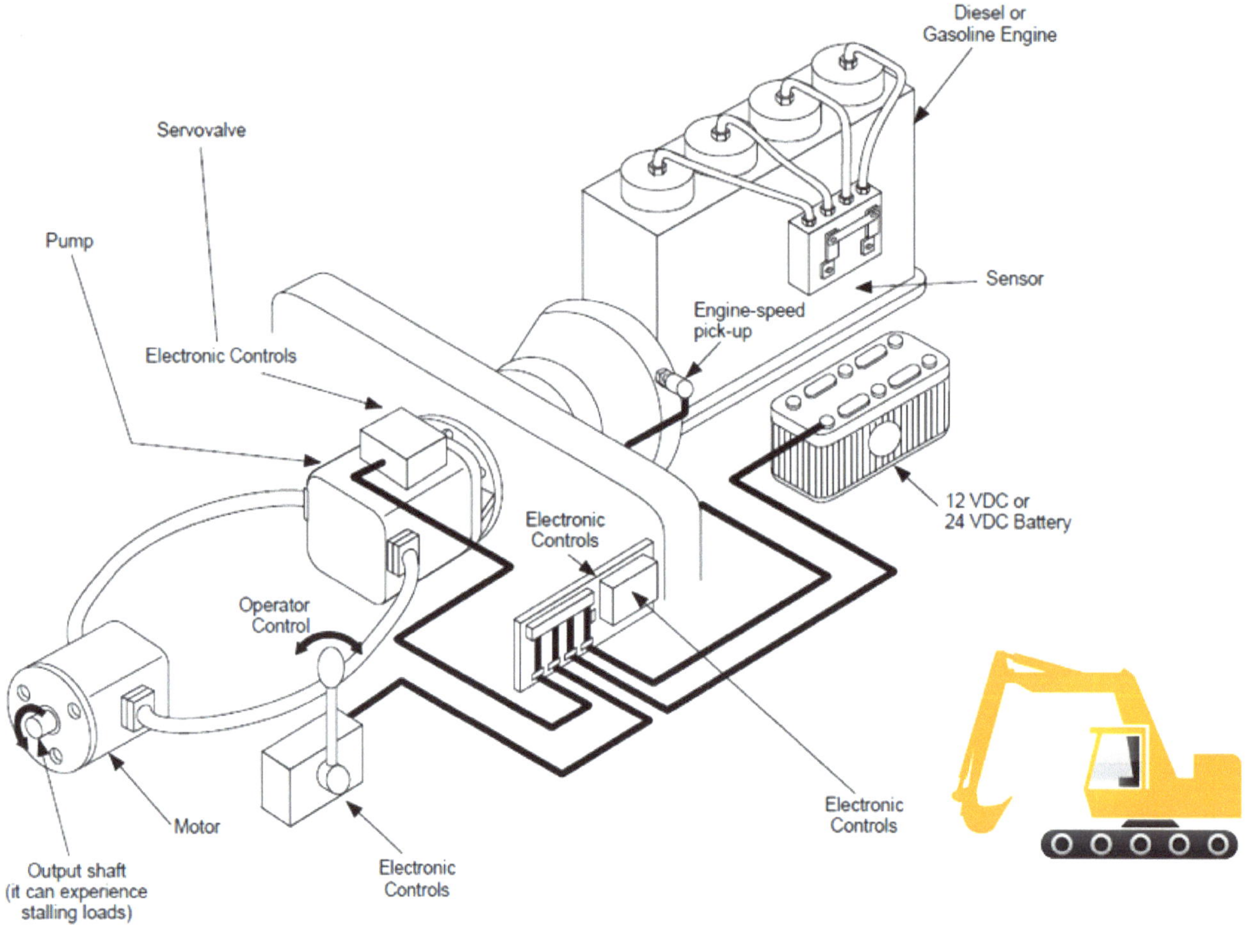

Fig. 8.41 - Anti-Stall Control for Hydrostatic Drive (Courtesy of NFPA)

On the rough ground around building construction, operating a fork lift truck with high requires skill and care. One of the most difficult operation in this setting involves inserting loads horizontally. Figure 8.42 shows an EH automatic leveling and insertion of the fork.

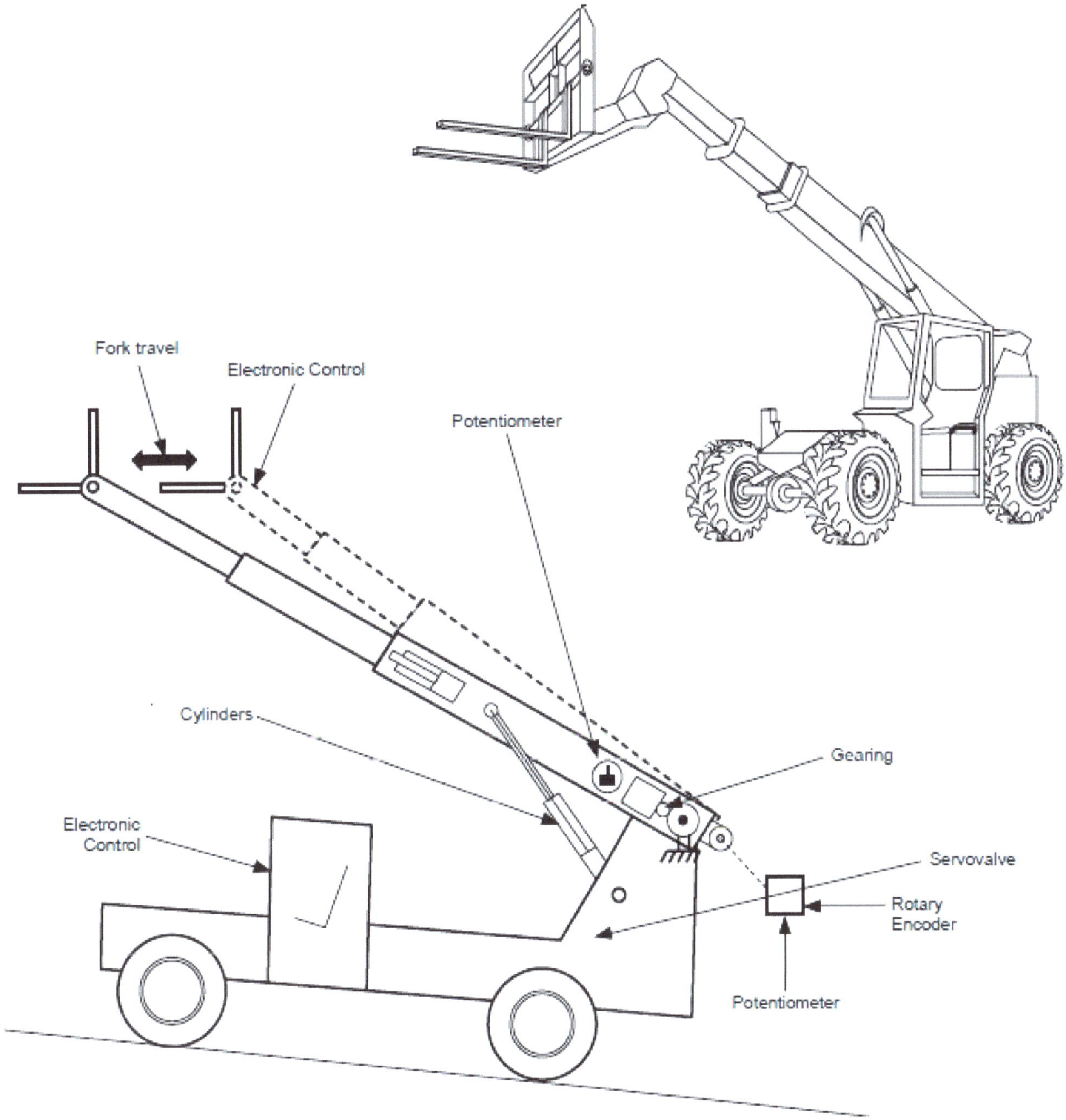

Fig. 8.42- EH Automatic Leveling and Insertion for Forklifts (Courtesy of NFPA)

Figure 8.43 shows the hydraulic circuit of a fan control system. The design of the circuit offers the following features: fail safe, proportional pressure control, automatic fan drive within temperature closed-loop control.

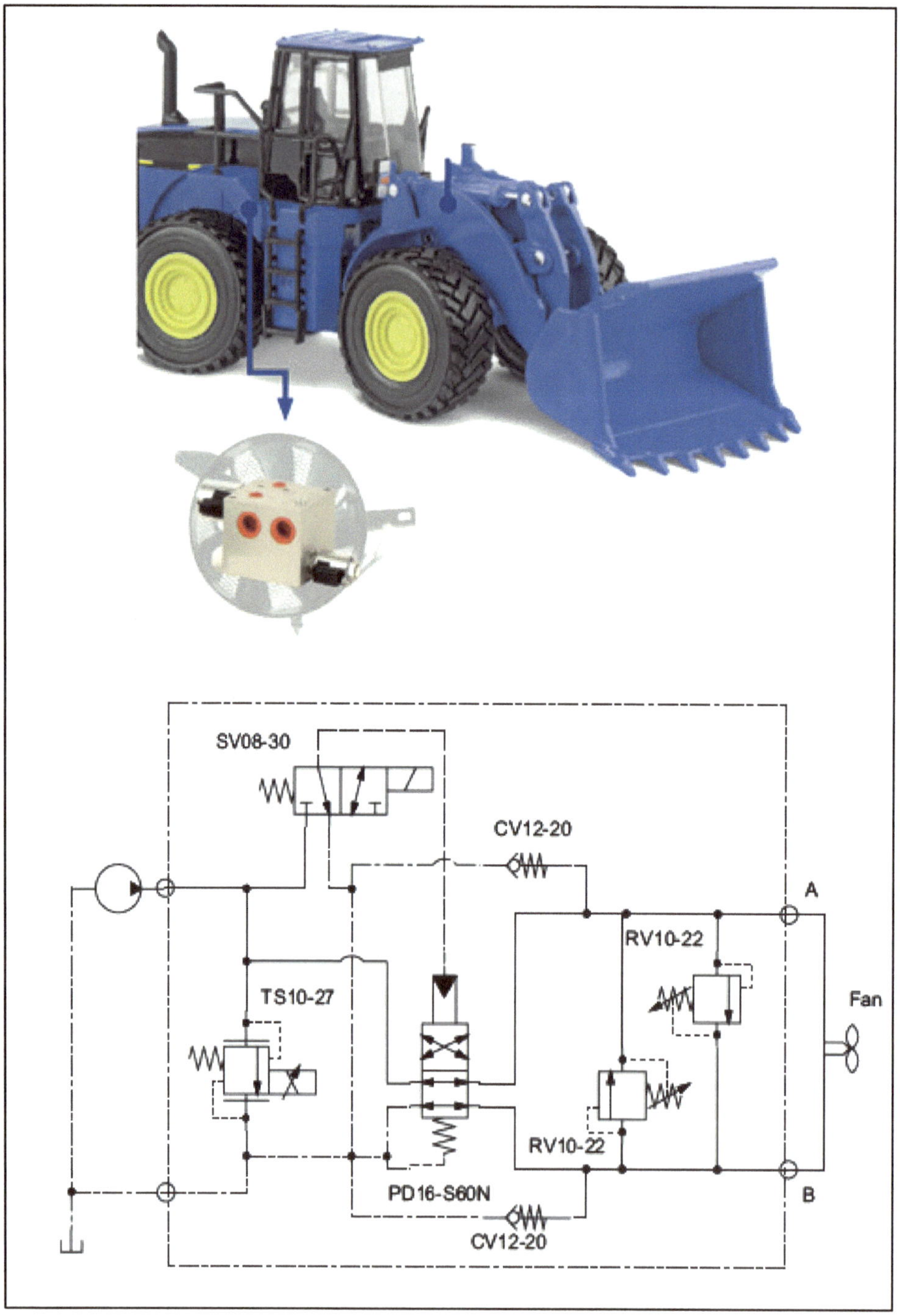

Fig. 8.43- EH Cooling Fan Control System for a Wheel Loader (Courtesy of Hydraforce)

Figure 8.44A shows the distribution of hydraulic components in a tractor. In the agricultural industry, as shown in Fig. 844B, a digital display proportional valve driver is used to control the height of the booms on crop sprayers.

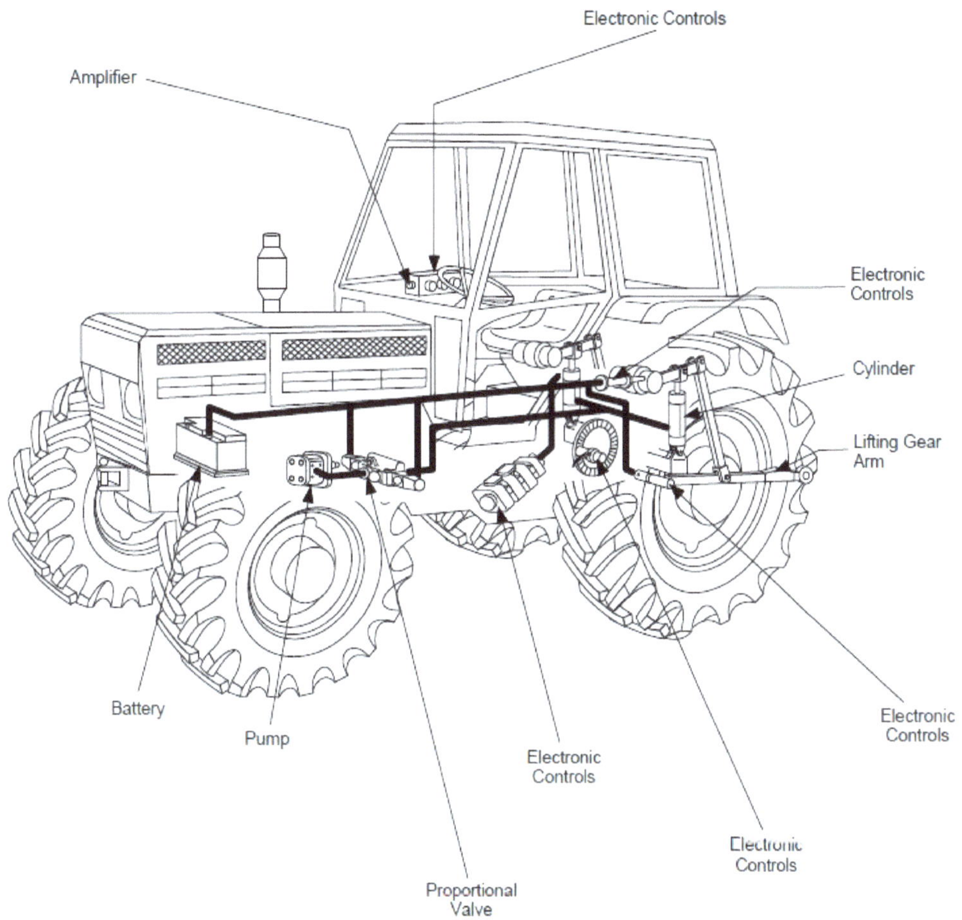

Fig. 8.44A- Force and Position Control in Tractors (Courtesy of NFPA)

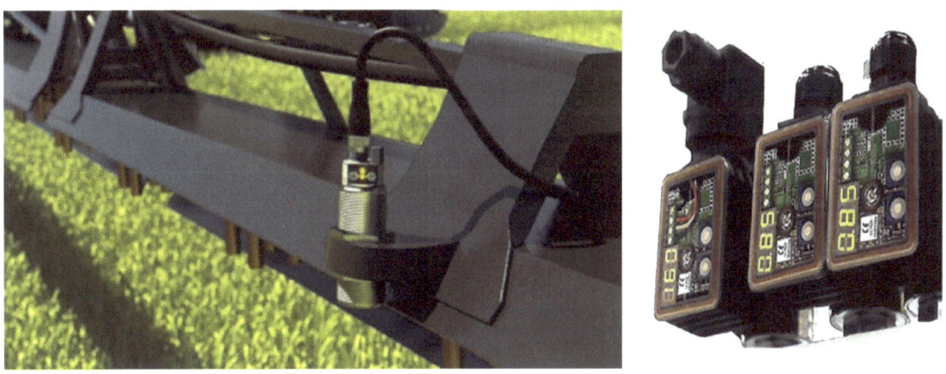

Fig. 8.44B- Level Control in Agricultural Machines (Excerption from Lynch)

Conventional car-transport are difficult to load and keep secure, and the cars they carry are under the mercy of the environment. Figure 8.45 shows a multi-axis force, position, and velocity control for a modern car loader. Only electro-hydraulic can do the job accurately and safely. Eight hydraulic cylinders power the linkages and commanded by a microprocessor.

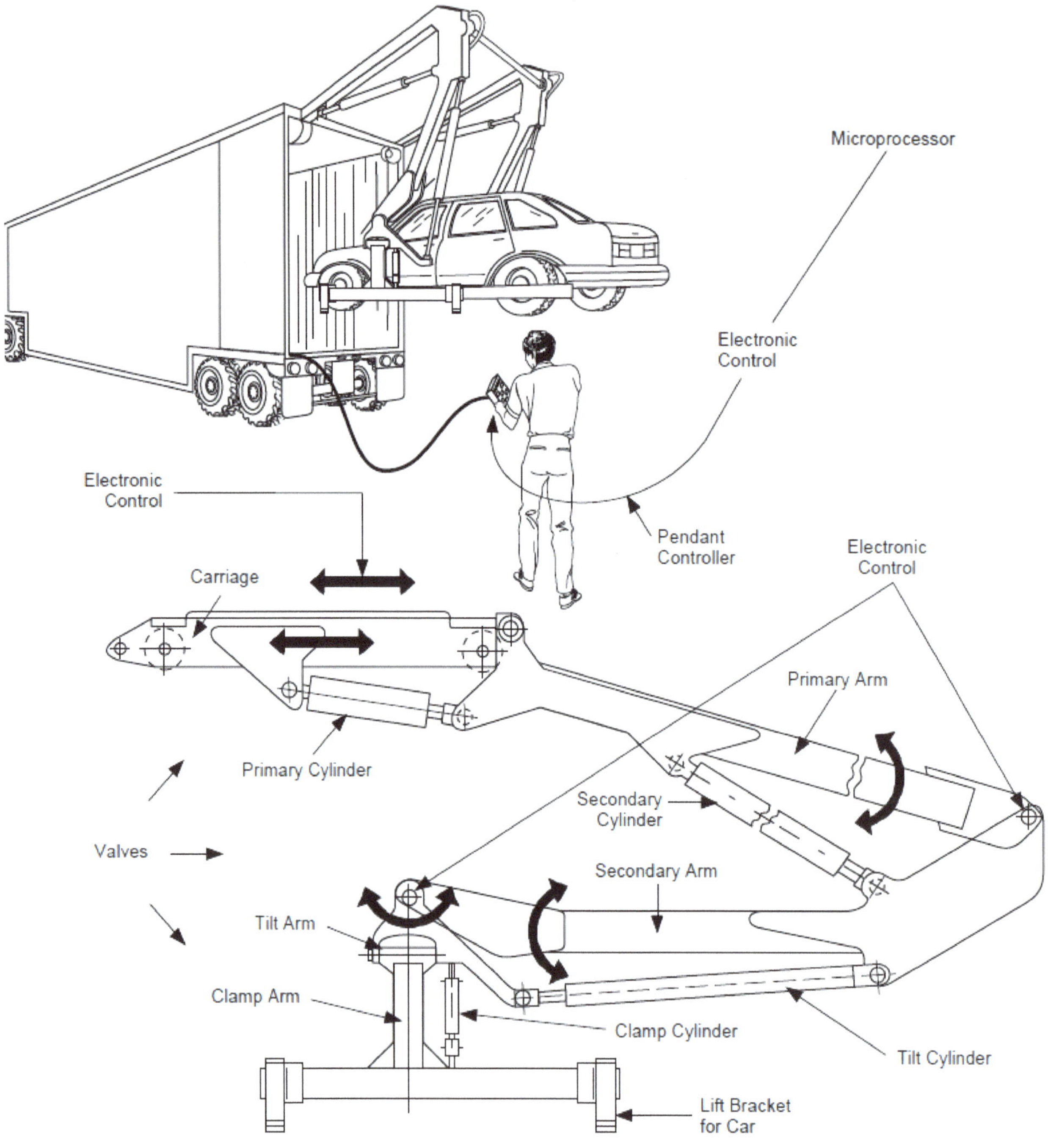

Fig. 8.45- Force, Position, and Velocity Control (Courtesy of NFPA)

Chapter 9

Control Electronics for Electro-Hydraulic Valves

Objectives:

This chapter presents various configurations and basic functions contained in control electronics for EH valves. The chapter discusses the changes in the system performance when these functions are enabled, including gain adjustor, dead band eliminator, ramp, and dither.

Brief Contents:

Chapter 9 - Control Electronics for Electro-Hydraulic Valves

9.1- Control Electronics Basic Functions

Electric control units that drive EH valves contain several elements that perform multiple functions. These functions, as shown in Fig. 9.1, can be categorized in the following basic functions

Power Functions:

- Supplying the required DC power by an external or built-in DC power supplier.
- Amplifying the low power input signal to the valve to an output current signal proportional to the input signal and powerful enough to drive a proportional solenoid or a torque motor.
- Conditioning the final output signal to improve the valve response (e.g. dither signal) or to reduce power consumption (e.g. pulse width modulation).

Regulation Functions:

- Regulate the signal received by the proportional solenoid or the torque motor to assure proper and safe operation (e.g. null adjustor and saturation adjustor)
- Overcome undesirable valve characteristics (e.g. ramp generator and dead band eliminator).

Control Functions:

- Apply the desired control algorithm within the entire system operation including gain adjustor and feedback signal calibration.

Figure 9.2 shows the distribution of these functions as the control electronics are used to form a closed-loop control system.

The following sections present detailed descriptions of these functions and their effects on system performance.

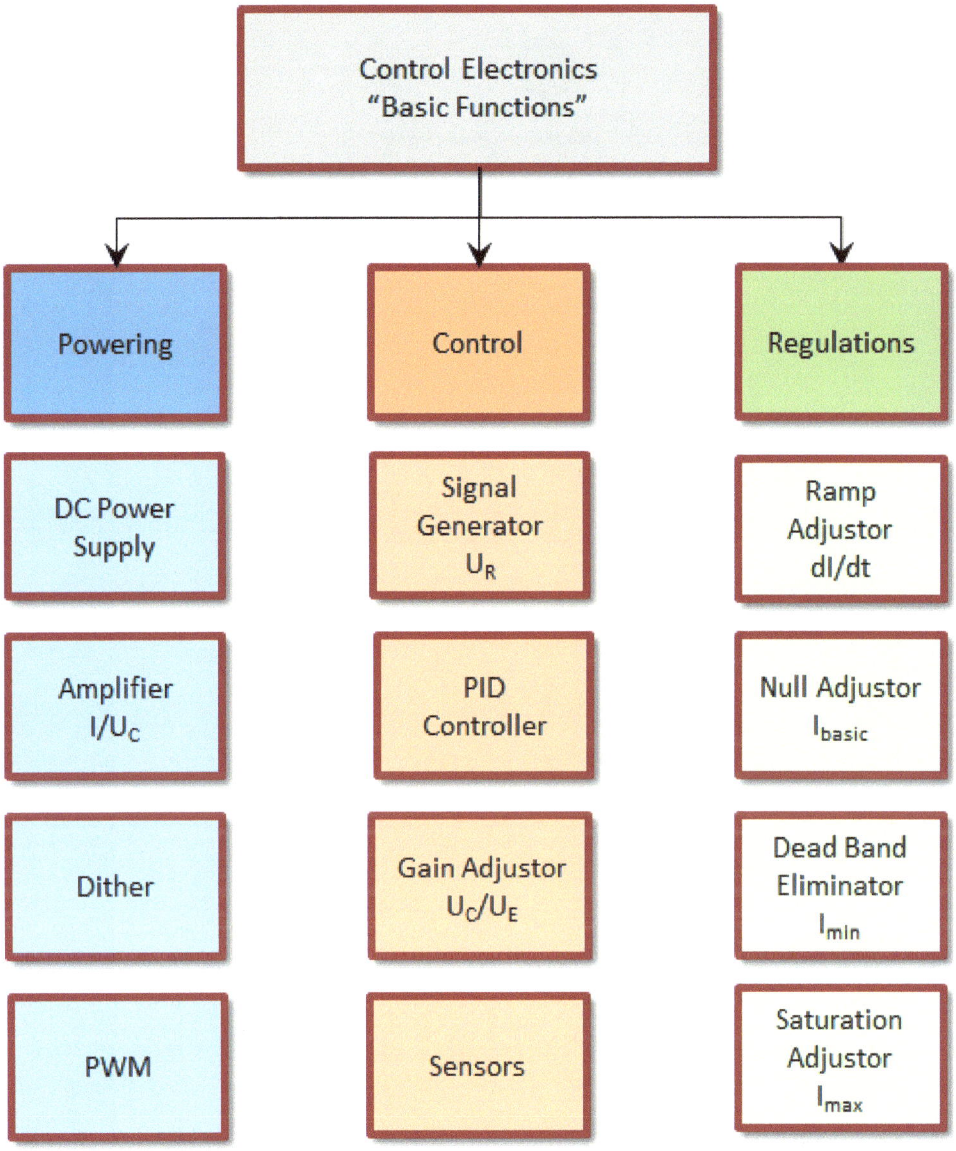

Fig. 9.1 - Control Electronics Basic Functions for EH Valves

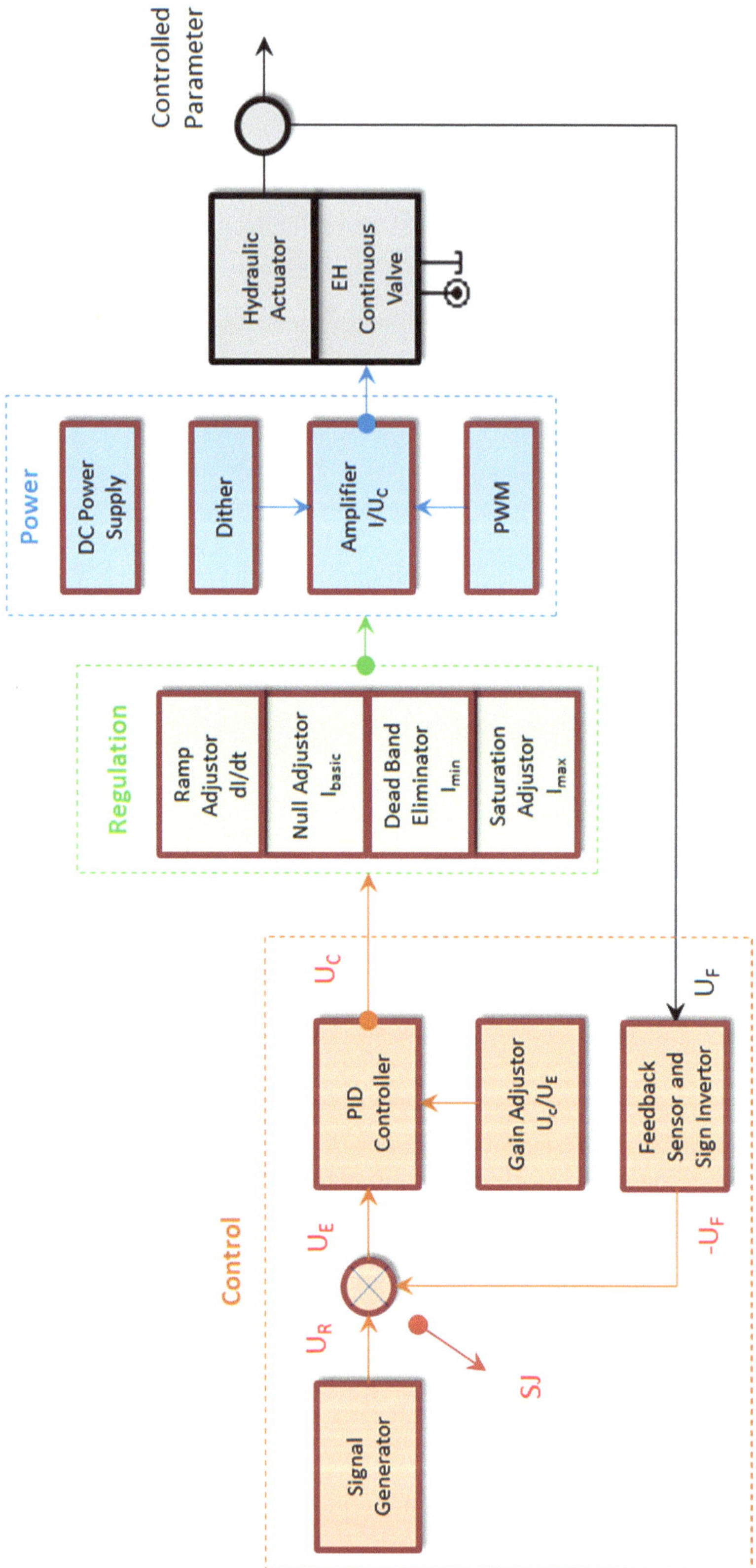

Fig. 9.2 - Control Electronics Basic Functions Distribution

Power Functions

9.2- DC Power Supply

Control electronics for proportional and servo valves work based on DC power only. Therefore, there should be an external or built-in source of DC power. AC power is more difficult to control and associated with a number of undesirable phenomena such as Inrush current, Eddy Current, and AC Hum. A DC power supply, as shown in Fig, 9.3, consists of the following basic components: Transformer, Rectifier, Condenser, and Stabilizer. As shown in Fig. 9.4, power supply pins must be assigned on the control electronics by numbers and/or color codes.

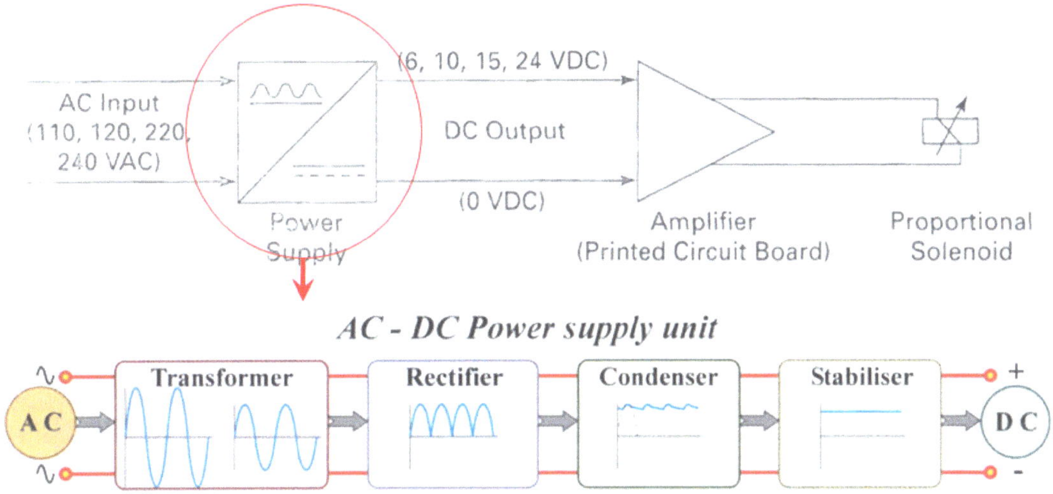

Fig. 9.3 - Basic Structure of AC to DC Power Convertor (Courtesy of ASSOFLUID)

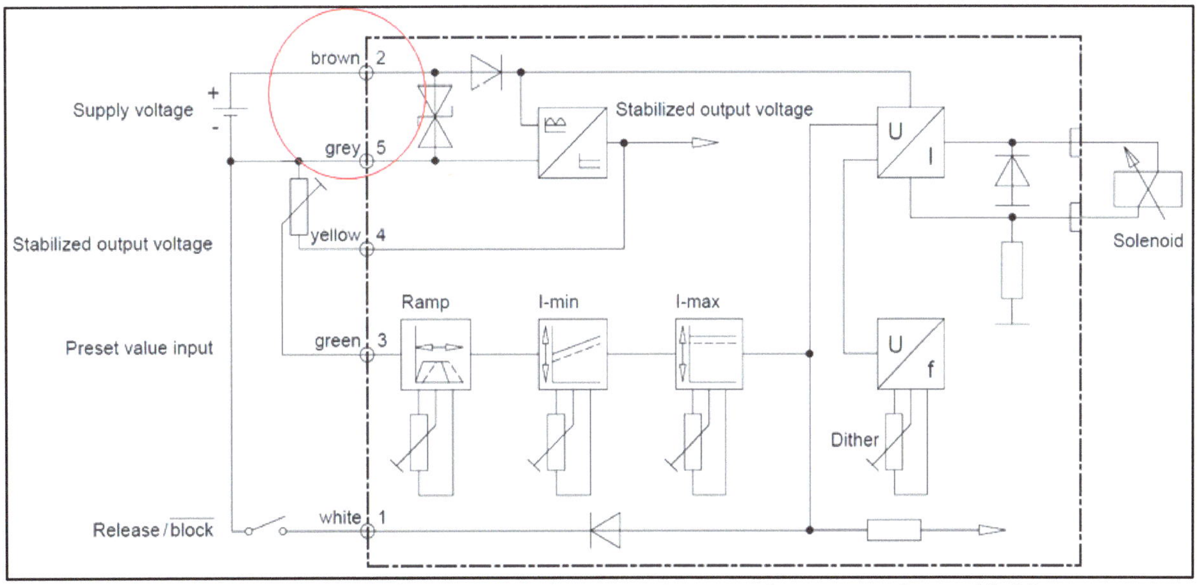

Fig. 9.4 - Example of Power Supply Pin Assignment (Courtesy of Wandfluh)

9.3- Signal Amplifier

The key element in control electronics is the amplifier. That is why the control electronics are referred to as the *Amplifier* (shortened to 'amp').

Its job is to amplify an input voltage low power signal to an output current signal that is powerful enough to drive the solenoid. The input signal could be a result of:

- In an open loop: signal generated from a potentiometer or any signal generator.
- In a closed loop: signal generated by the controller.

Figure 9.5 shows the symbols of the power amplifier the power amplifier physically consists of a number of transistor stages.

As shown in the figure, the power amplifier is placed just before the solenoid. It receives the low power voltage signal after it has been manipulated by all the control and regulation functions.

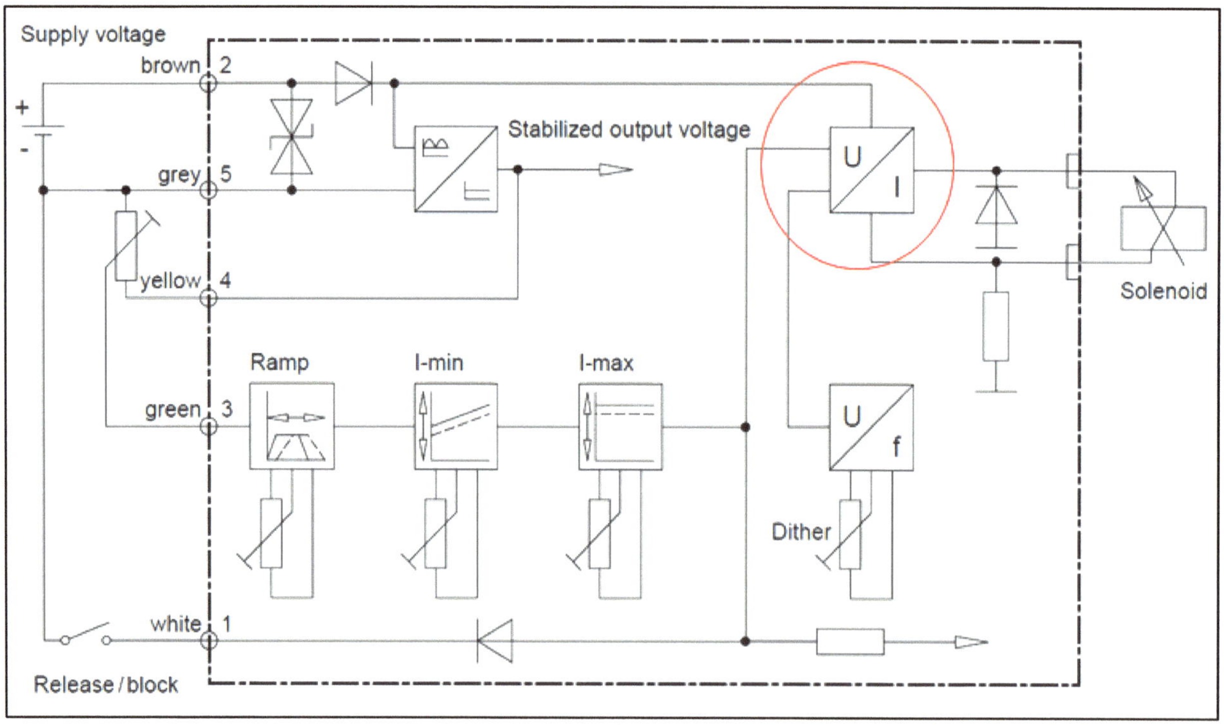

Fig. 9.5 - Example of a Signal Amplifier (Courtesy of Wandfluh)

9.4- Dither Signal

As shown in Fig. 9.6, *Dither* signal is a micro-vibrating signal produced and superimposed on the final signal to the solenoid. Dither signal also vibrates the spool when it is at rest. The purpose of the dither signal is to keep the spool performs micro vibration all the time waiting for the next input signal. Hence, applying a dither signal considerably reduces spool adhesion and thereby improves linearity and hysteresis performance characteristics of the valve. Dither signal frequency should be within the bandwidth of the valve in order that the valve can respond to it. Dither signal amplitude should be on micro level so that it does not affect the null conditions of the valve. Best estimate is the spool vibrates with amplitude of 10 microns. Both the dither frequency and amplitude may possibly be adjustable in field.

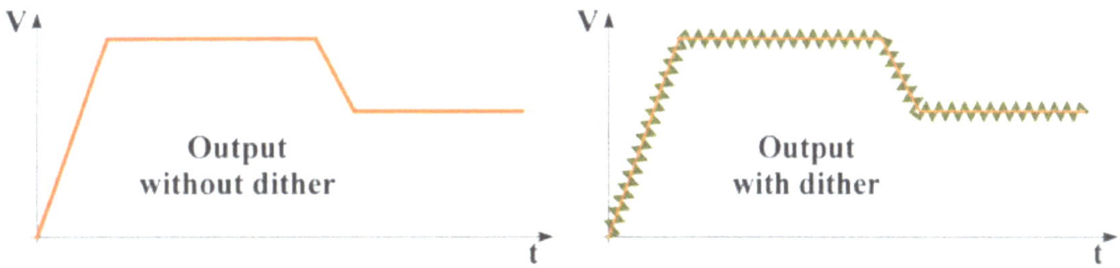

Fig. 9.6 - Dither Signal (Courtesy of ASSOFLUID)

9.5- Pulse Width Modulation

What is PWM? *Pulse Width Modulation (PWM)* is a technique used originally to encode electronic message and information. It was found useful to drive electrohydraulic valves, DC motors, etc. As shown in Fig. 9.7, it is a technique to supply a continuous output signal via a pulsated input signal.

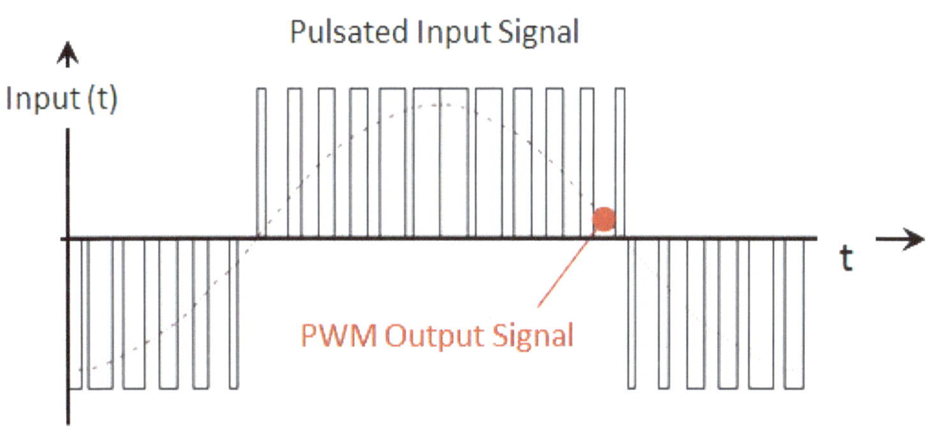

Fig. 9.7 - Pulse Width Modulated Signal

Why PWM? The main purpose of using PWM is to reduce power consumption. Applying DC power to a solenoid on 100% duty cycle warms the solenoid more than its ability to dissipate the heat. Hence a heat sink is needed. Using PWM reduces power consumption and may make the control electronics compact by getting rid of the heat sink. Additionally, as shown in Fig. 9.8, PWM may act like a dither by improving the hysteresis performance of the valve. PWM became a standard feature for EH valves.

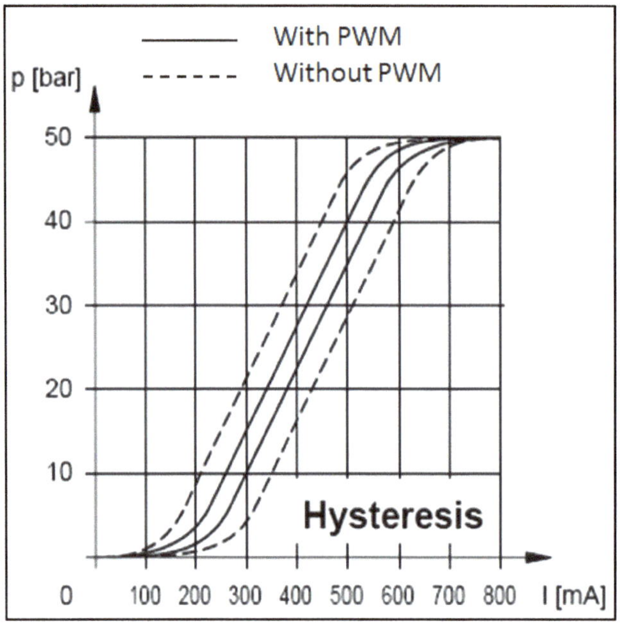

Fig. 9.8 - Effect of Pulse Width Modulation on Hysteresis

How to Apply PWM? There are no modifications required to the valve solenoid in order to use PWM. As shown in Fig. 9.9, the input signal is generated by regulator structured within the ECU. A built-in pulse generator is used to switch the input signal ON and OFF at a certain frequency.

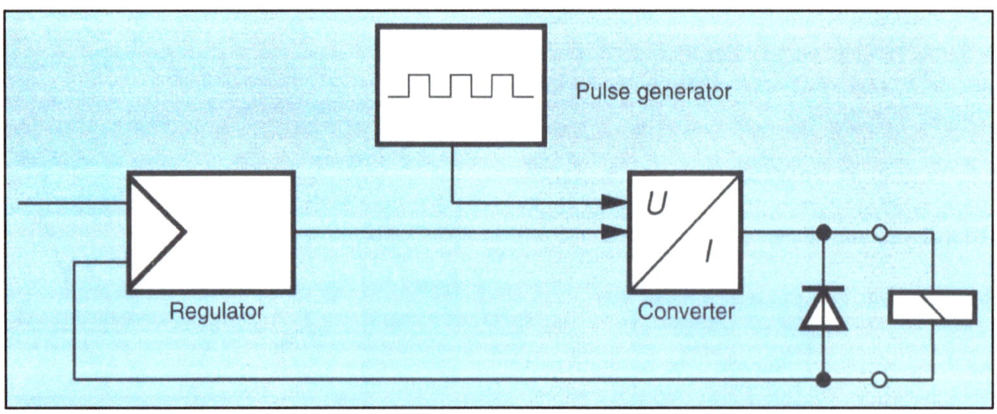

Fig. 9.9 - Placement of PWM and Dither Signals (Courtesy of Bosch Rexroth)

How Conceptually PWM Works? As shown in Fig. 9.10, a pulse generator generates pulsated signal at steady frequency and amplitude equal to the maximum signal for the solenoid. The signal after all regulation functions is used to adjust the duty cycle.

The term *Duty Cycle* describes the proportion of 'ON' time compared to the regular cycle time. Duty cycle is expressed in percent, 100% being fully ON. The longer the switch is ON compared to OFF periods, the higher the total power supplied to the load.

Figure 9.11 presents examples of applying PWM to an input signal. The figure shows the average output for duty cycles 25%, 50%, and 75%.

PWM Frequency: The PWM switching frequency has to be much higher than what would affect the load (the device that uses the power). In case of an EH valve, PWM frequency must be much higher than the valve bandwidth, typically 1 kHz. Higher frequency improves the continuity of the average output signal.

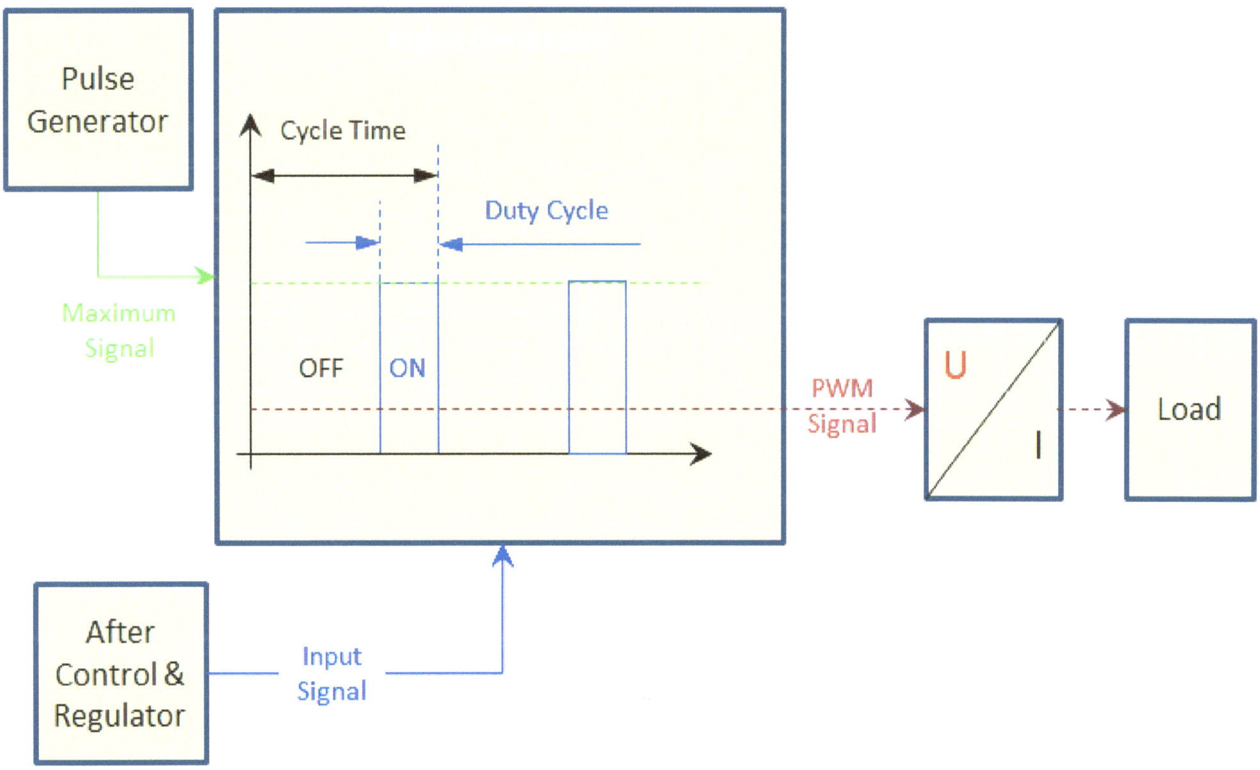

Fig. 9.10 - Pulse Width Modulation Concept

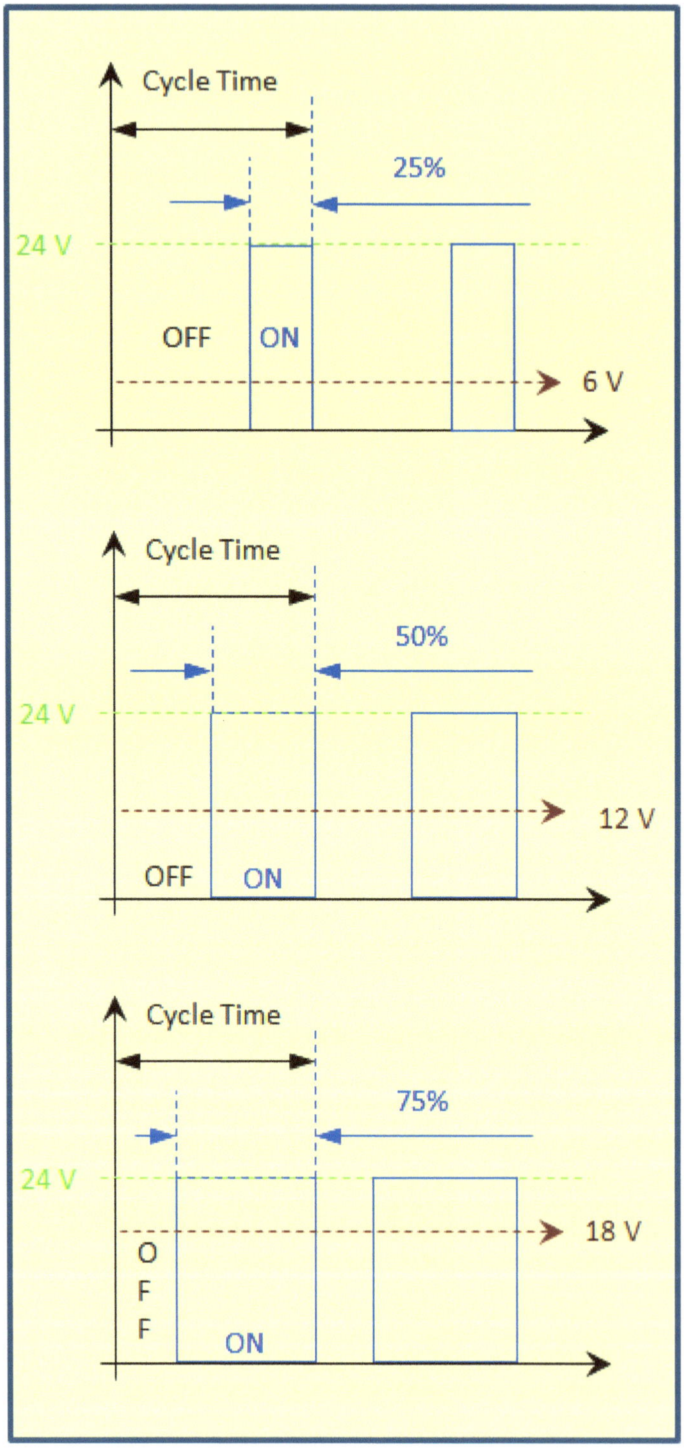

Fig. 9.11 - Examples of Applying PWM to an Input Signal

Control Functions

9.6- Input Signal Generator

Figure 9.12 shows that the input signal is generated differently depending on the format, the source, and the mode.

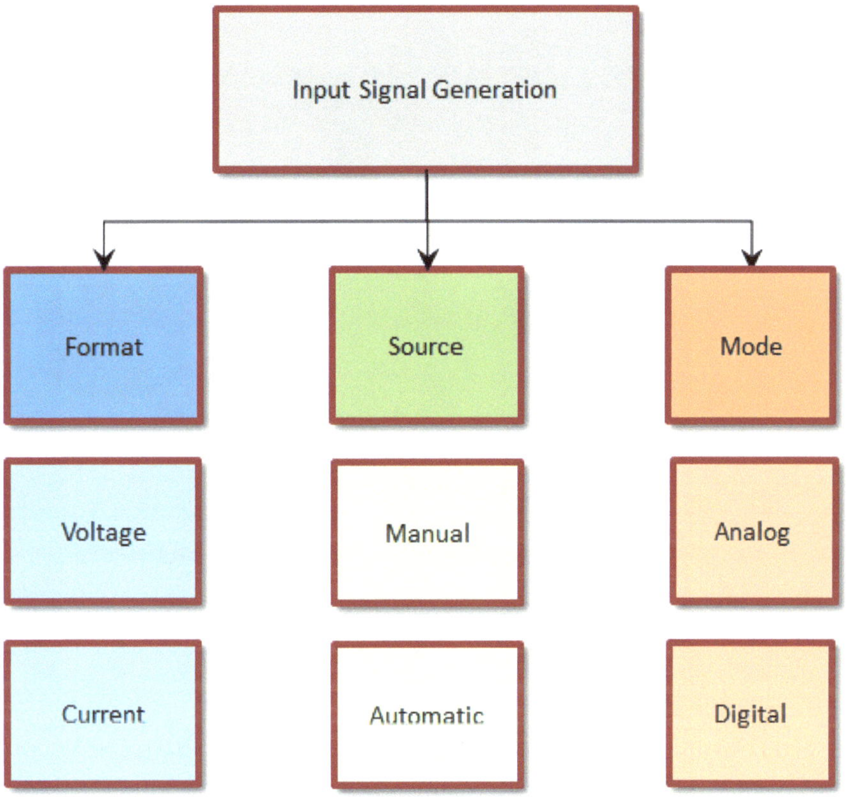

Fig. 9.12 - Choices of Input Signal Generation

9.6.1- Format of Input Signal

Input signals can be a voltage or current. Voltage input signals are common for all industrial controllers because they are easy to produce and to calibrate. Voltage signals are subject to voltage drop in long transmission cables and affected by noise. Values of voltage signals are typically (0-10) V for unidirectional parameter control, e.g. cylinder position. It could also be of ±10 V for bidirectional parameter control, e.g. bidirectional hydraulic motor speed.

Current input signals resist noise and losses. They are preferable for transmission through long cables. It is also recommended for troubleshooting purposes because 4 mA means zero signal but 0 mA means loss of signal. Current input signals are typically of (4-20) mA or (0-20) mA.

9.6.2- Source and Mode of Input Signal

Input signals are sourced either manually or automatically. Manual input signals can be generated using analog or digital devices. Figure 9.13 shows examples of manual analog devices. Potentiometers generally have linear characteristics as shown in Fig 9.14.

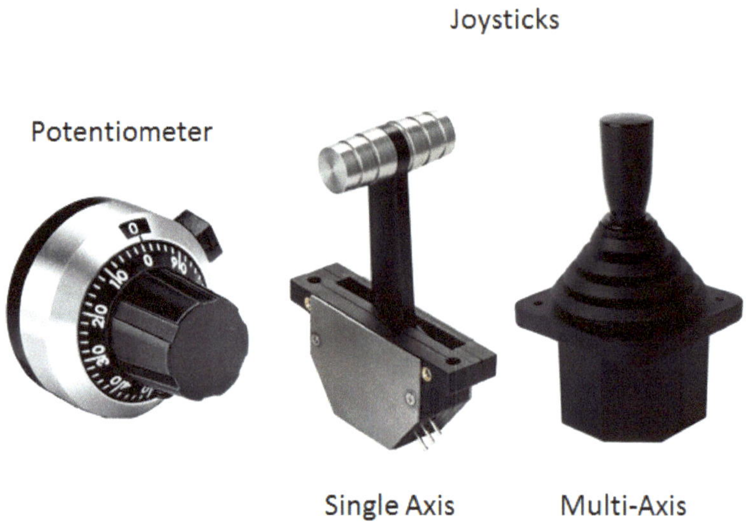

Fig. 9.13 - Examples of Manual Input Signal Devices

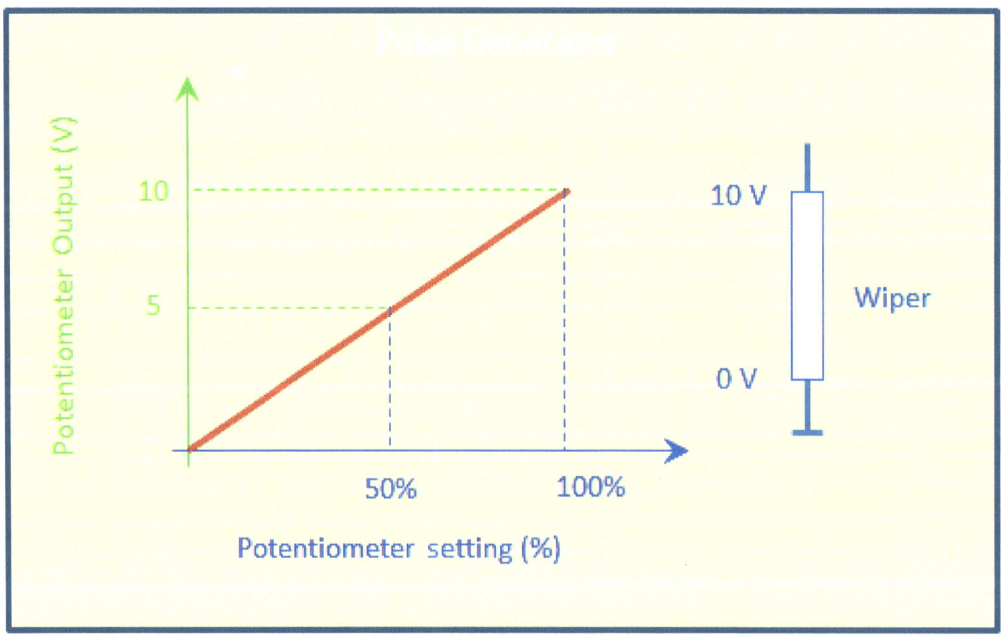

Fig. 9.14 - Performance Curve for Linear Analog Potentiometers

Input signals can still be generated manually but digitally. One way to do that, as shown in Fig. 9.15, is to use a Human Machine Interface (HMI) or any type of industrial PC with a touch screen.

As mentioned earlier, input signals can be generated automatically using an external function generator. An example of standalone function generator is shown in Fig. 9.16. As shown in Fig. 9.17, an input signal can be generated also by a PLC, PC, or whatever controller is used in the system. The host controller could be built-in on the control electronics itself.

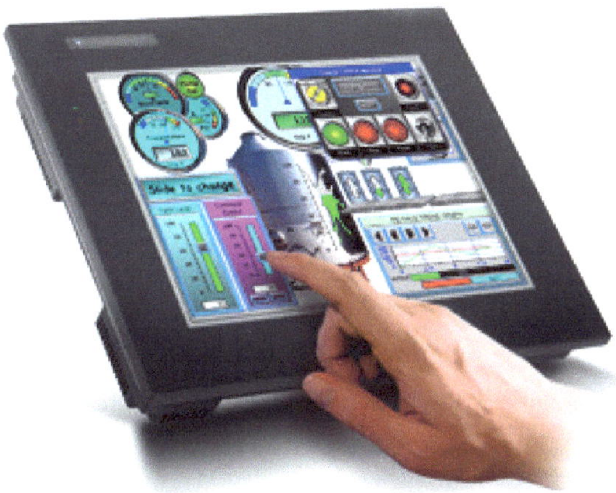

Fig. 9.15 - Touch Screen Operator Interface

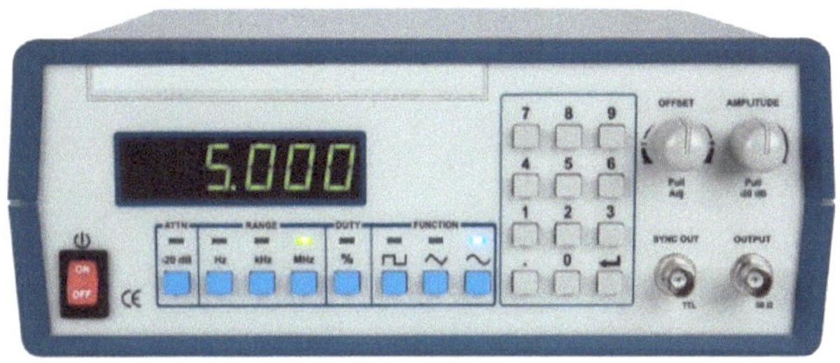

Fig.9.16 - Example of an Analog Function Generator

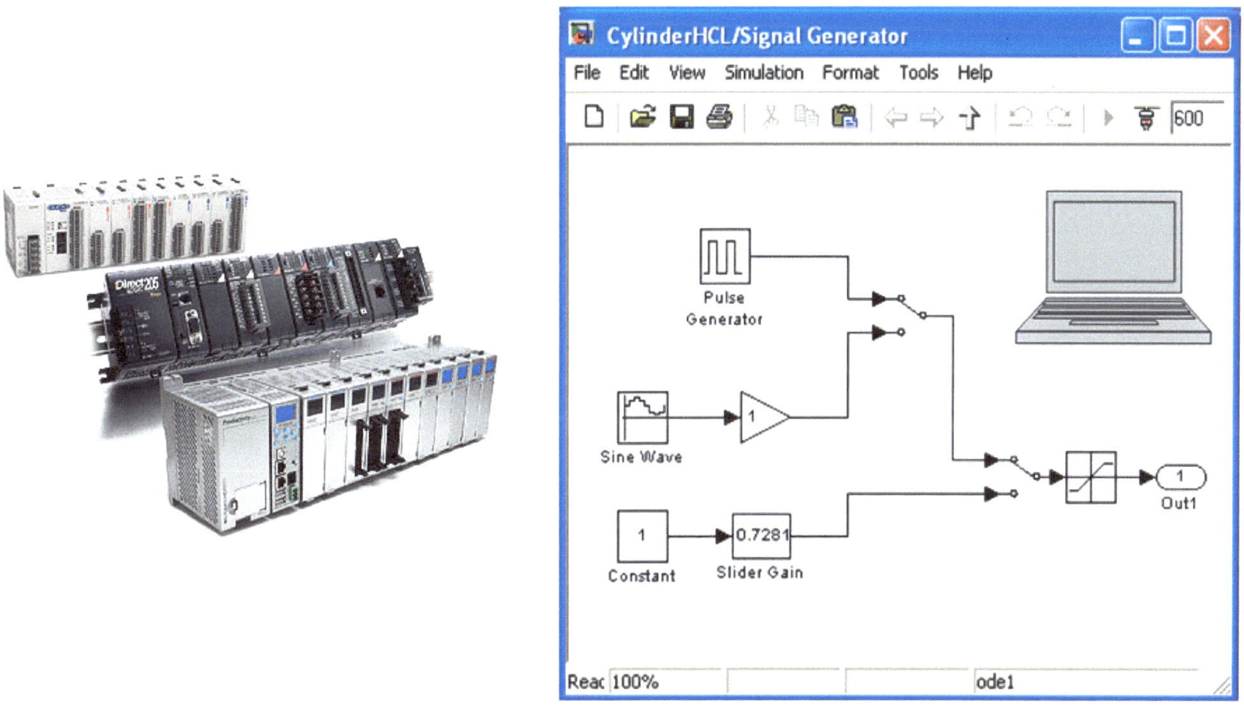

Fig. 9.17 - Automatic and Digital Input Signal Generation by PLC or PC

9.7- Feedback Sensors

Chapter 4 presents a brief idea about the different types of switches being used with ON/OFF valves. Sensors are elements that produce analog signals that are used with continuous valves, proportional or servo, to form open-loop and closed-loop systems. Figure 9.18 lists the most common sensors used in EH systems.

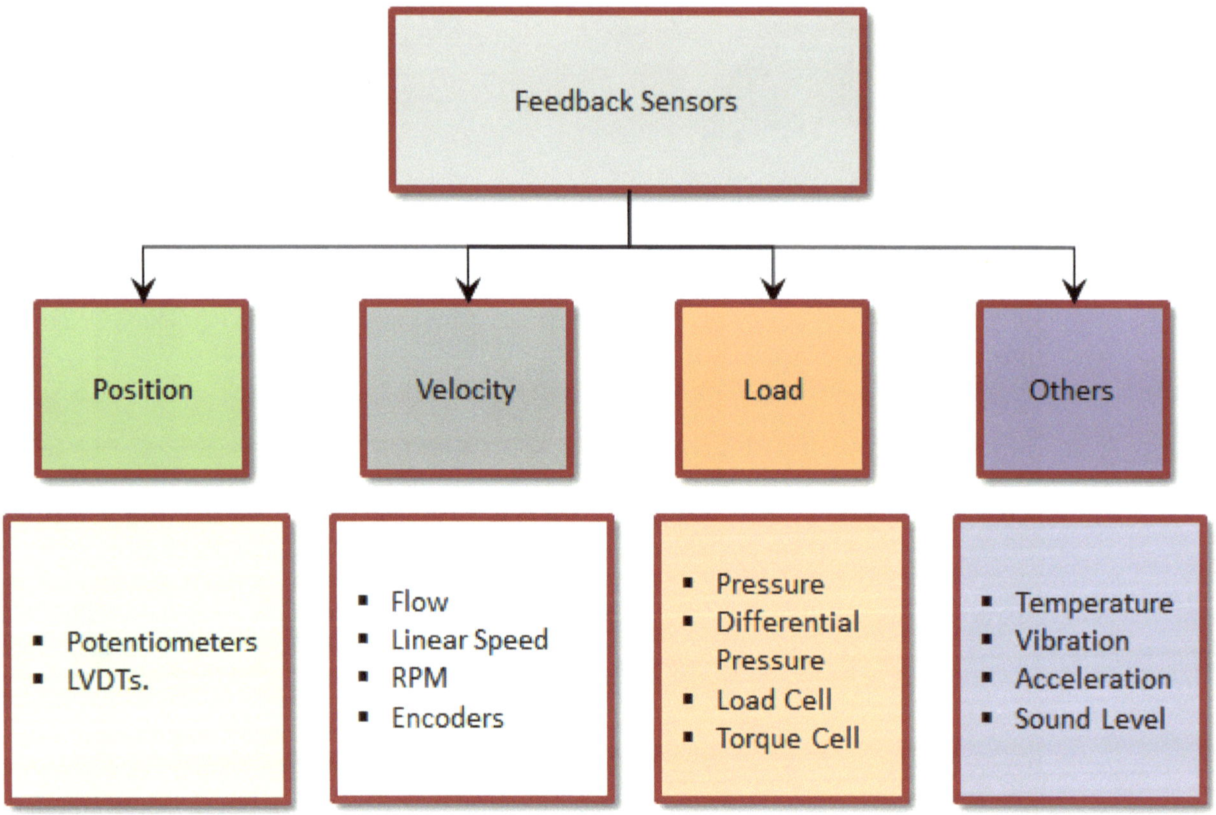

Fig. 9.18 - Classifications of Sensors commonly used in EH Systems

9.8- PID Controller

Despite the legacy of the *PID Controller*, it still is the most common controller and the core of most of updated intelligent controllers. For a PID controller in a closed-loop, feedback signal $\mathbf{U_F}$ must be subtracted from the reference signal $\mathbf{U_R}$.

Figure 9.19 shows the symbols for *Sign Invertor* used by schematics for control electronics. The figure shows also, in one symbol, *Summing Junction* and sign invertor. The resulting Error Signal $\mathbf{U_E}$ will be manipulated by the PID controller.

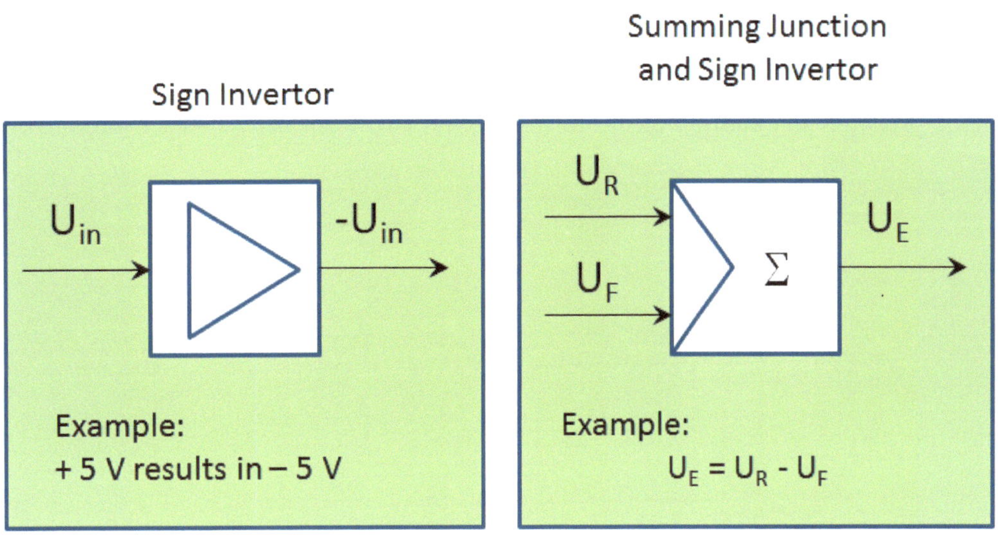

Fig. 9.19 - Symbols for Sign Inventor and Summing Junction

Figure 9.20 shows the symbol for a PID controller that receives the error signal and produces the control signal **Uc**. Equation 9.1 shows how mathematically the control signal is calculated.

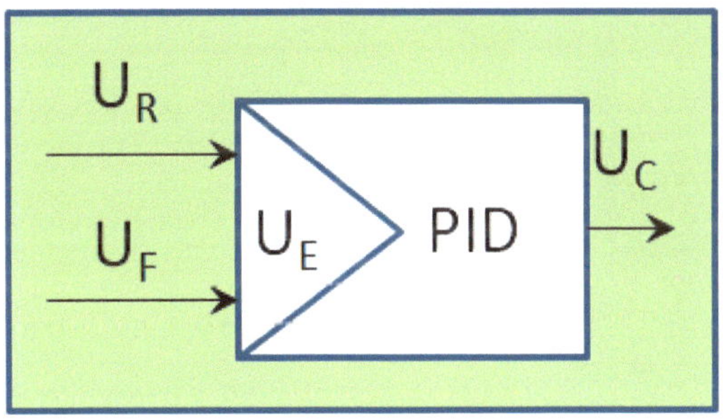

Fig. 9.20 - Symbol for PID Controller

$$U_C = k_P U_E + k_I \int U_E + k_D \frac{dU_E}{dt} \qquad 9.1$$

The parameters contained in the equation can be briefly explained follows:

k_P = *Proportional Gain*:
It is responsible for the valve shifting displacement and consequently the speed of the actuator.

k_I = *Integral Gain*:
It is responsible for eliminating the steady state error.

k_D = *Derivative Gain*:
It is responsible for removing oscillations and instability of the controlled parameters.

PID controller can be built-in the control electronics and constructed as an analog device. Figure 9.21 shows a closed-loop system to control position of a cylinder. The shown system contains an analog PID controller that includes an amplifier as well. It is known as an *Operational Amplifier*. Figure 9.22 presents an example of building a digital PID controller using Matlab-Simulink.

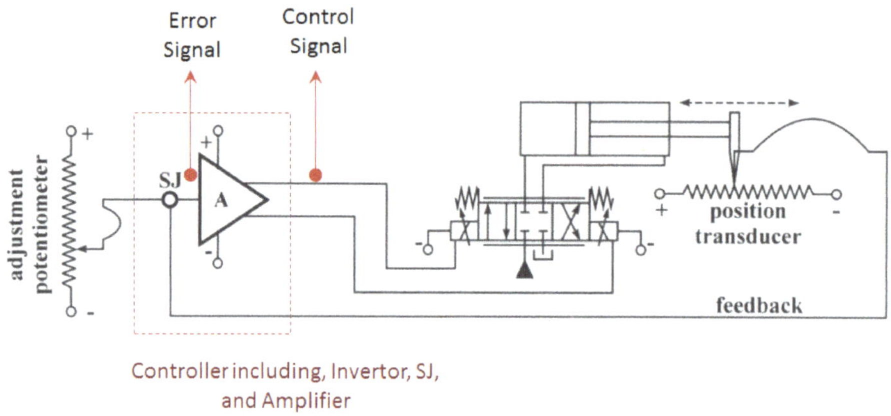

Fig. 9.21 - Example of using an Analog Controller in a Cylinder Position Control System (Courtesy of ASSOFLUID)

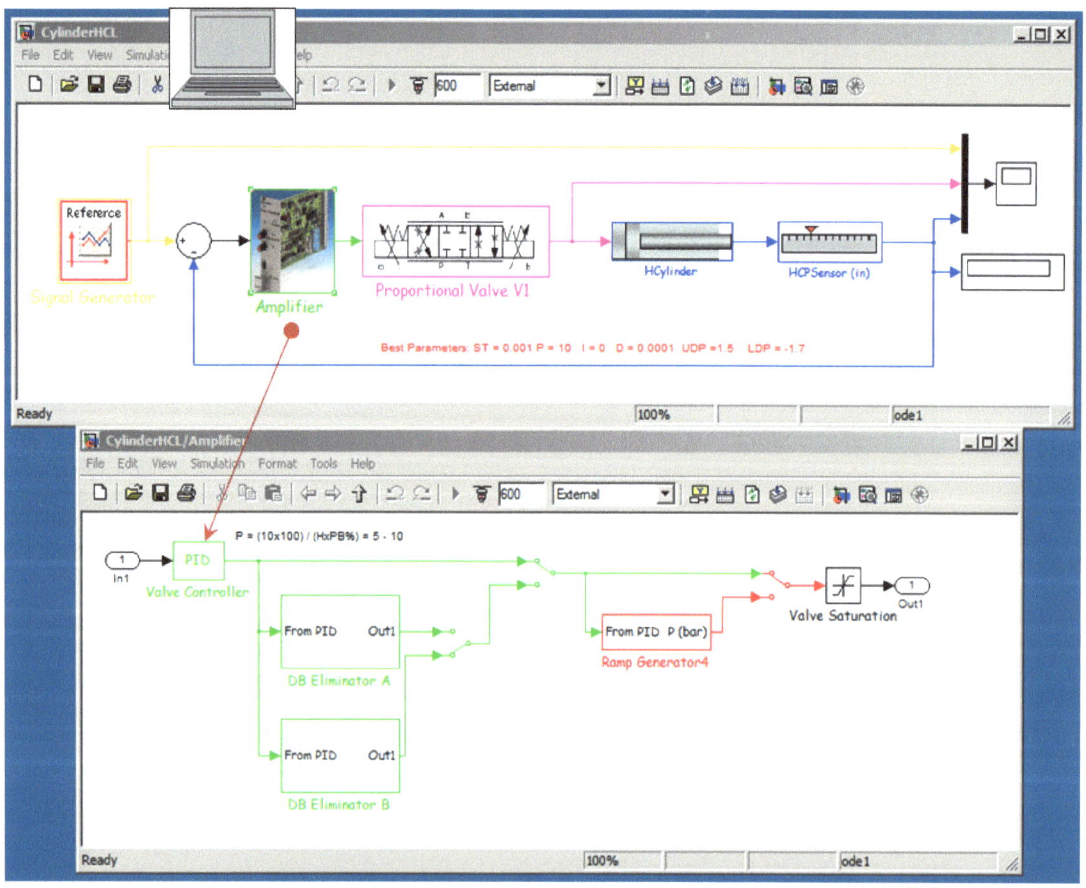

Fig. 9.22 - Example of using a Digital Controller in a Cylinder Position Control System

9.9- Gain Adjustor

As shown previously in Eq. 9.1, three gain factors are used to calculate the control signal. Integral and derivative gains are optional depending on the system performance. Proportional gain is the essential part of the equation. That is why, if the word "gain" is mentioned, by default it is understood as the proportional gain.

Figure 9.23 shows that for the same error signal, the higher the gain the wider the valve opens and consequently the faster the driven actuator.

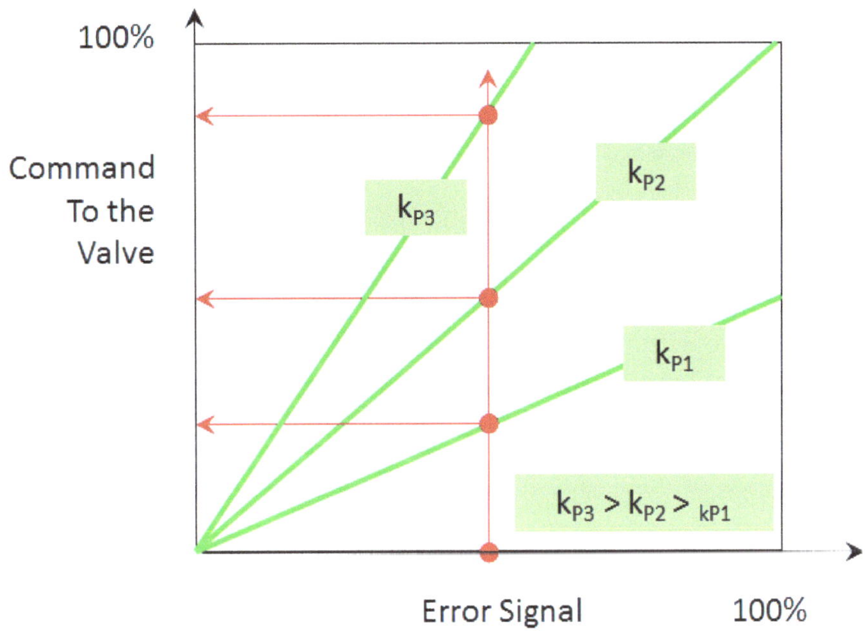

Fig. 9.23 - Effect of Proportional Gain on the Valve Opening

Figure 9.24 shows a block diagram for a closed-loop cylinder position control system to discuss the effect of the proportional gain on the system performance.

The presented example was based on the following assumptions:

- Input Signal U_R = 0 - 10 V.
- Cylinder Stroke = 10 Inches.
- Cylinder is fully retracted → U_F = 0 V.
- Cylinder is required to fully extend → U_R = 10 V

Then, once the cylinder is turned ON, error signal U_R = 10 V:

Case 1 when $k_P = 1 \rightarrow U_C = 10$ Volt (at the start):

Valve starts fully opened and cylinder starts at maximum speed. Once the cylinder moves, feedback signal U_F reduces, valve gradually closes, and cylinder decelerates accordingly. The valve forms one-to-one relationship with the error signal.

Case 2 when $k_P = 0.5 \rightarrow U_C = 5$ Volt (at the start).

Valve starts half open and cylinder at with medium speed. Once the cylinder moves, feedback signal U_F reduces, valve gradually closes, and cylinder decelerates accordingly. The system is sluggish as compared to case 1.

Case 3 when $k_P = 2 \rightarrow U_C = 20$ Volt (at the start).

If the control signal exceeds the maximum signal to the valve, the control electronics takes care of protecting the valve by limiting the input signal to the assigned maximum allowable value. The in this case, the valve starts fully open and cylinder starts at maximum speed. Cylinder will continue moving at maximum speed till U_C drops below 10. This happen when U_E becomes less than 5, i.e. when the cylinder passes 50% of its stroke. After that, the valve starts to close.

Conclusion: assume that $k_p = 3$, then cylinder starts to decelerate after 66.6% of its stroke, and so on. Then increasing the proportional gain, increases the valve shifting displacement and the time the valve is kept opened, and consequently the system speed of response.

Caution: increasing proportional gain may drive the actuator to overshoot that my result in system instability.

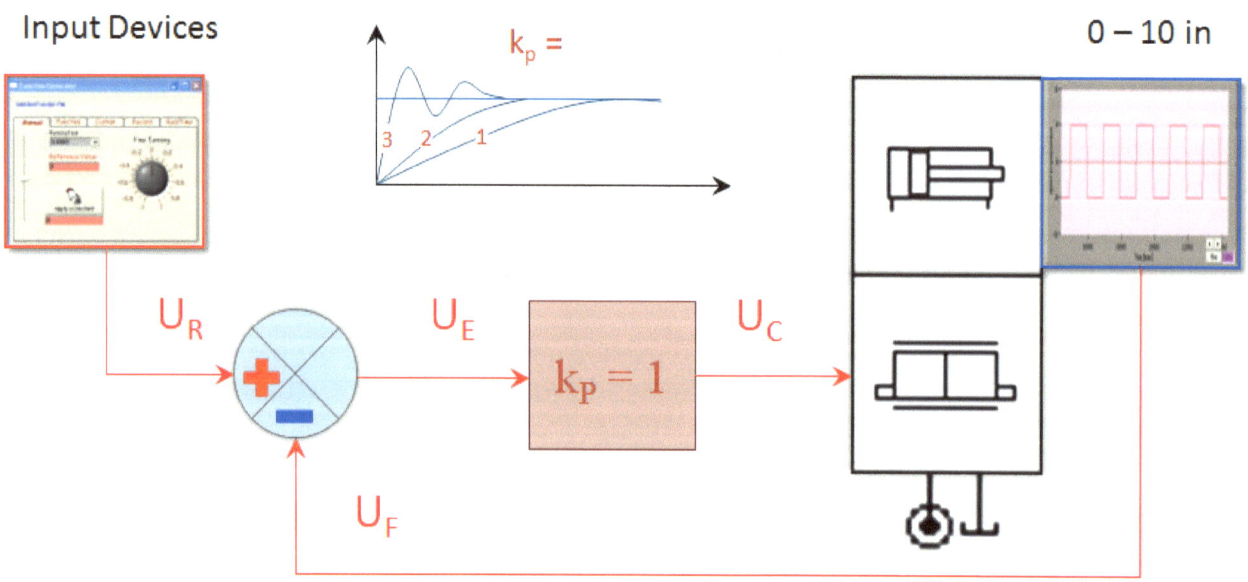

Fig. 9.24 - Effect of Proportional Gain on System Response

Figure 9.25 shows an example of electronic control units that offer the feature of adjusting the gain. The figure shows one potentiometer for adjusting the gain assuming the hydraulic actuator is either a synchronous cylinder or a hydraulic motor. Some control electronics may have two pots in order to adjust the gain for the actuator's two directions of motion separately. This is beneficial for differential cylinders and in case if loading conditions are different even for synchronous cylinders and hydraulic motors.

For better control resolution, the pots turn about 50 turns to produce value from 0-100% gain value.

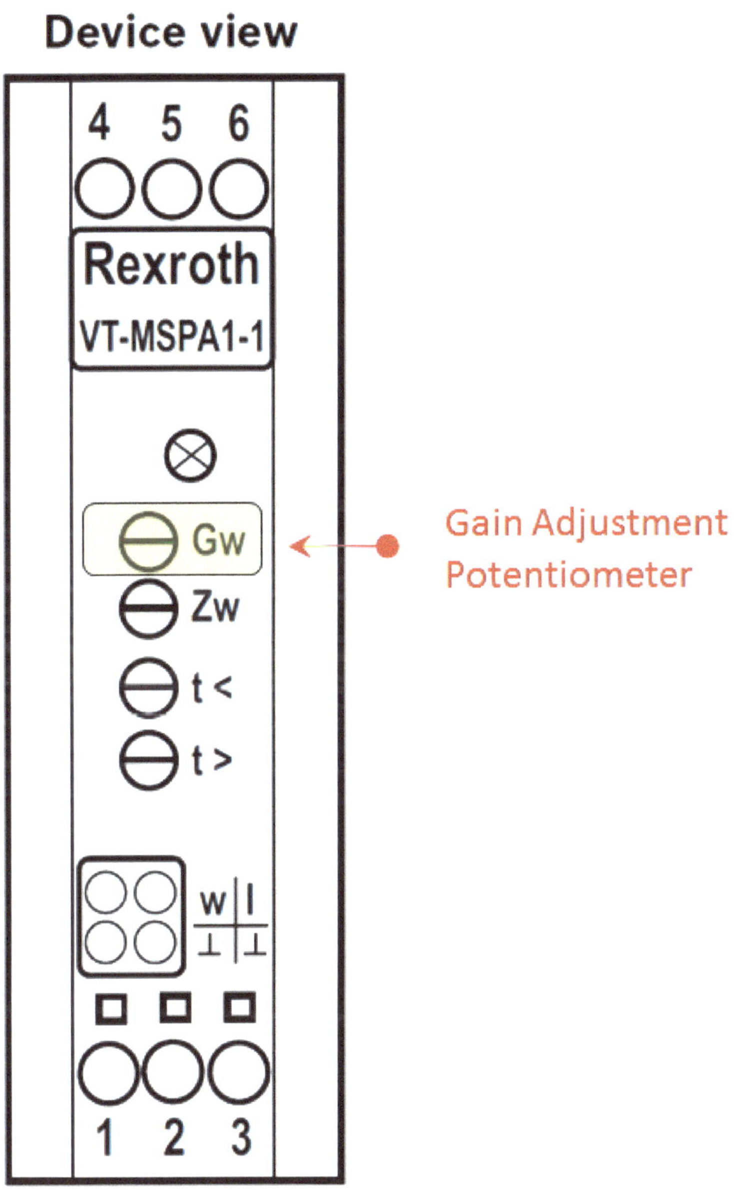

Fig. 9.25 - Example of Proportional Gain Adjustment (Courtesy from Bosch Rexroth)

Regulation Functions

9.10- Ramp Adjustor

As it has been discussed in the previous section, by adjusting the controller' gain, a control signal is produced and supplied to the valve. The valve responds to the change in the control signal as fast as its dynamics allow. Then the change in the actuator's speed will become sudden. For example, based on the assumptions shown in the previous example where $k_p = 1$, the valve will suddenly open and the cylinder will start harshly with maximum speed as if it was driven by a switching valve!

The question now becomes how to control the shifting time of the valve to ramp the actuator's speed nicely? The answer is to enable the *Ramp* function.

As shown in Fig. 9.26, if a valve receives a control signal to open a certain % of its full stroke, by adjusting the *Ramp Rate*, the shifting time is adjusted accordingly. The spool will eventually displace as required but along certain time that is controlled by the ramp rate. Obviously, a large load requires lower ramp rate to avoid jerky motion and to overcome the inertial effect of the load.

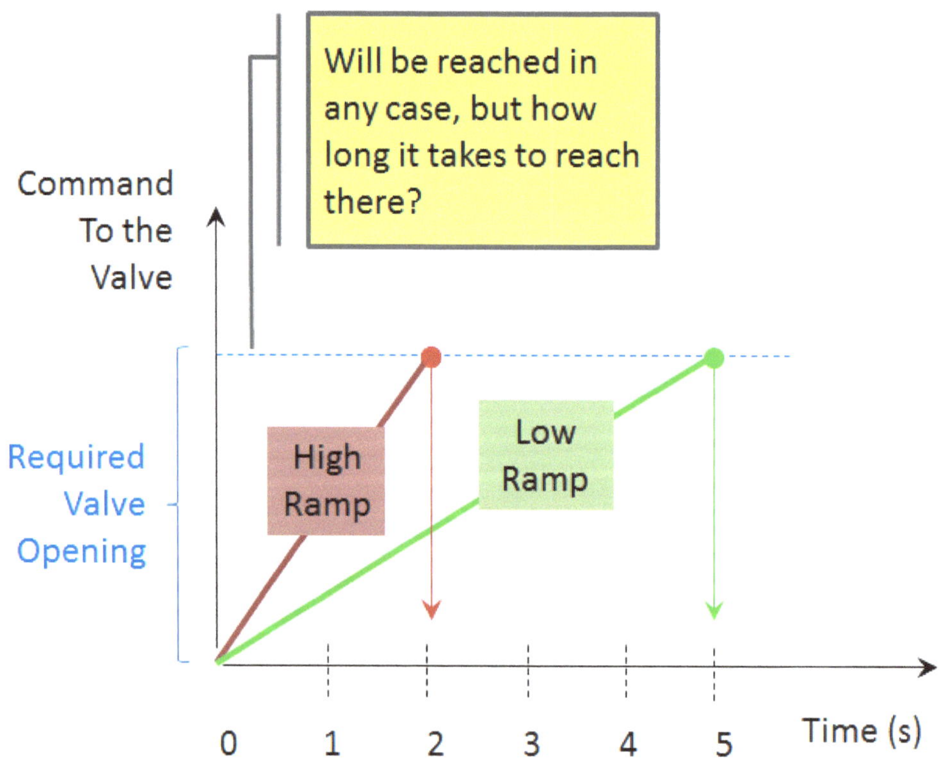

Fig. 9.26 - Concept of Ramp Adjustment

Figure 9.27 shows the symbol and the construction of an analog Ramp Generator. The figure shows that the ramp generator consists of some electrical standard elements as variable resistors and capacitance, etc. Figure 9.28 shows an example of electronic control units that offer the feature of adjusting the ramp. Ramp rates for accelerating and decelerating the actuator are adjusted differently.

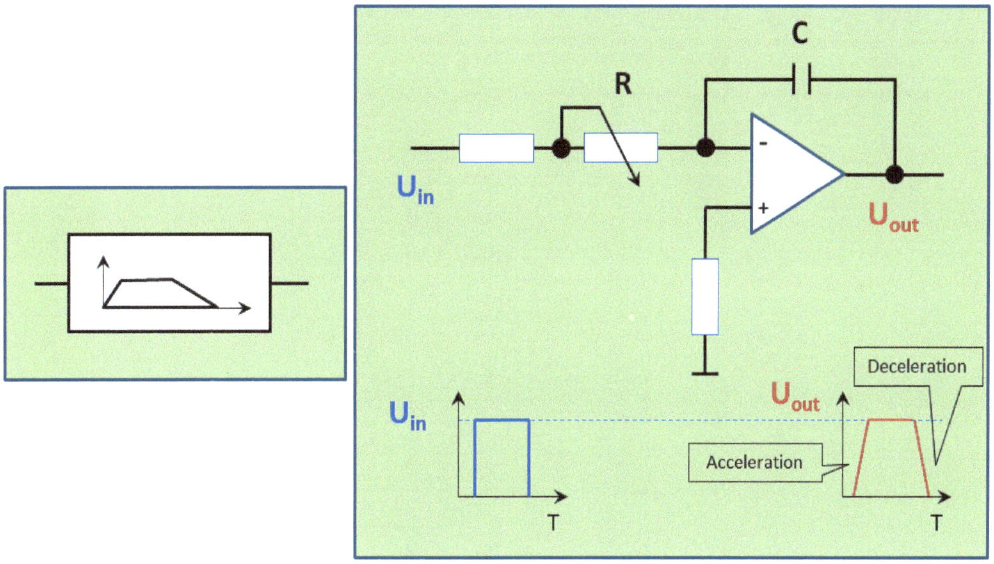

Fig. 9.27 - Analog Ramp Adjustor

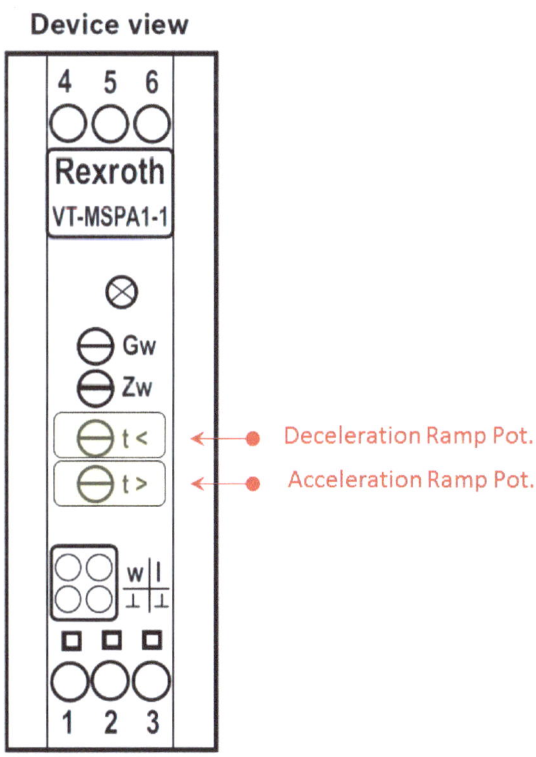

Fig. 9.28 - Example of Ramp Adjustment (Courtesy from Bosch Rexroth)

Figure 9.29 presents an example of building a digital ramp function using Matlab-Simulink.

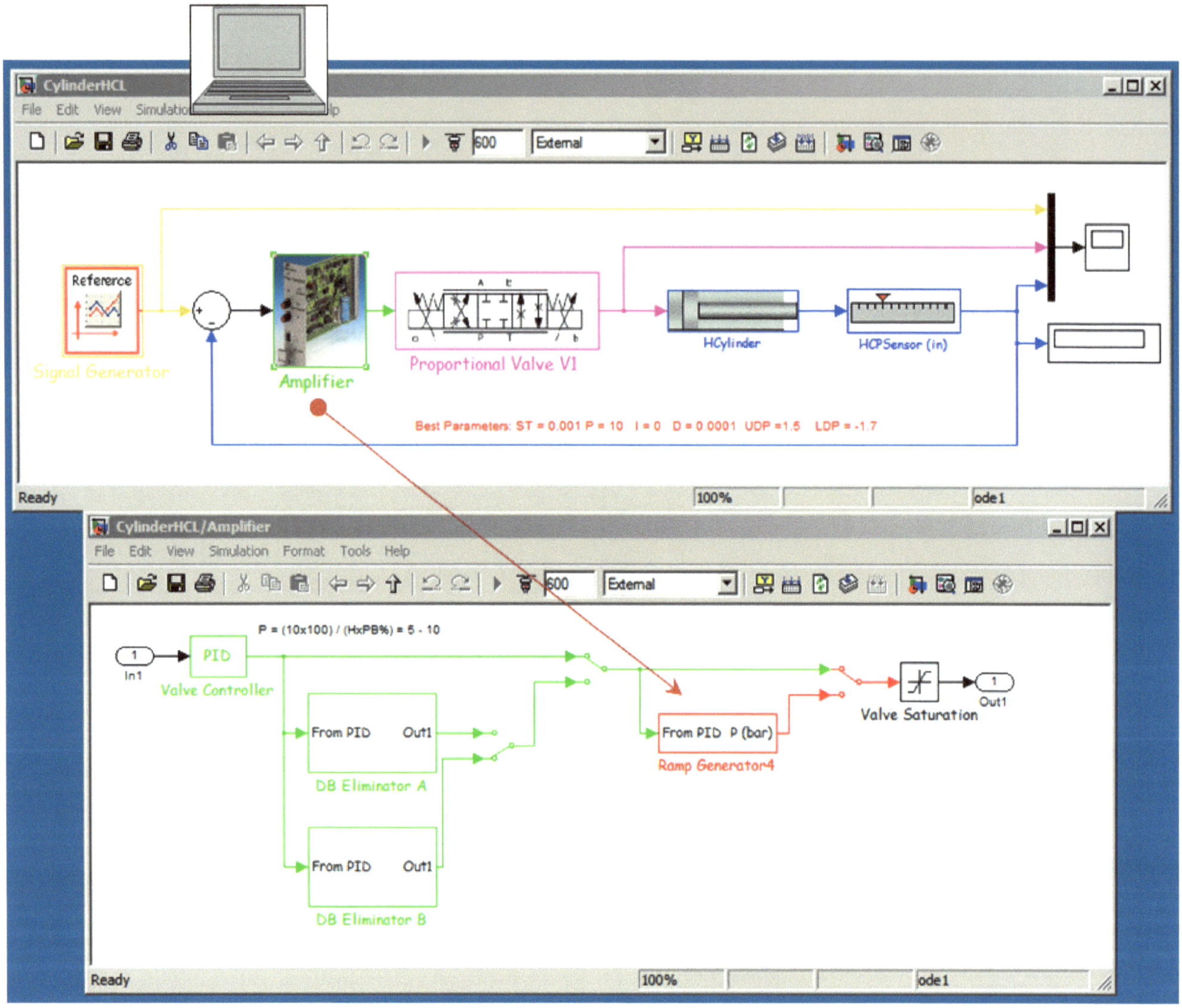

Fig. 9.29 - Example of using a Digital Ramp Function in a Cylinder Position Control System

9.11- Null Adjustor (I_{Bias})

If a valve spool deviates from perfect null position, particularly critical-center and open-center valves, load may creep when the valve is de-energized. Valves are typically being factory set for perfect *Null*. Despite that, there should be feature for in-filed tuning the valve null position if needed. Null adjustment is also referred to as *Offset Setting*.

Null adjustment can be done mechanically or electrically. The servo valve shown in fig. 9.30 has two ways to adjust the null position mechanically. First through the nozzles adjusting screws (7). This done only at the time of the valve assembly or maintenance at the factory or by certified personnel. Some manufacturers publish instruction not to touch these screws otherwise the user will void the warranty.

Second, null adjustment can be done through spool position adjusting screws (8). This is can be easily done in field depending on the need of application.

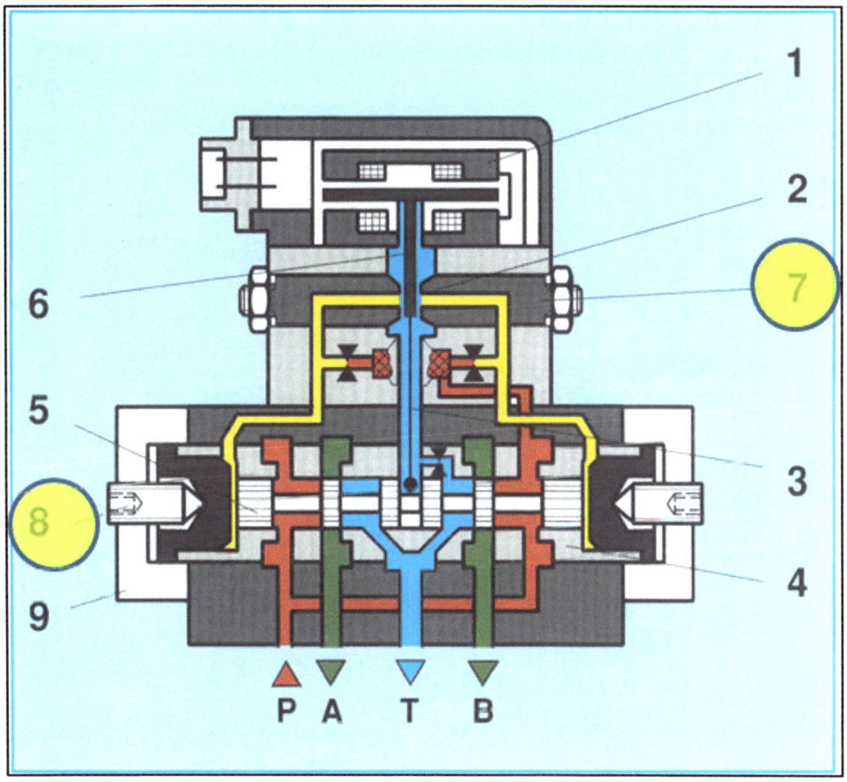

Fig. 9.30 - Mechanical Null Adjustment for a Servo Valve (Courtesy of Bosch Rexroth)

Electrical null adjustment is done by adding *Bias Current*. Bias current, as shown in Fig. 9.31, shifts the valve performance towards the null position. Bias current is typically 5% of the input signal and referred to as *Basic Current*. Figure 9.32 shows an example of electrical null adjustment screw in a servo valve. Figures 9.33 and 9.34 show two other examples for electrical null adjustment.

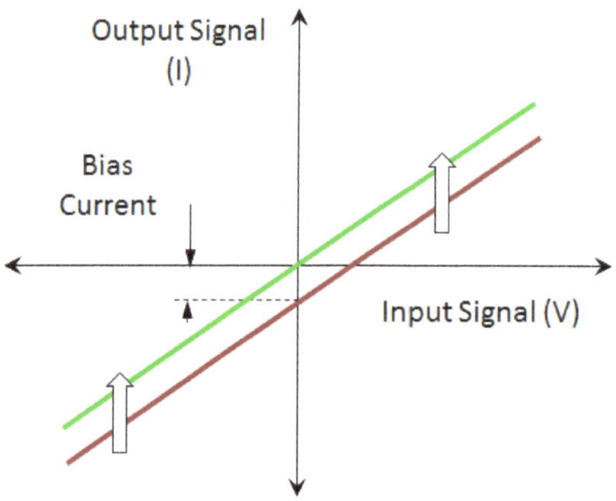

Fig. 9.31 - Electrical Null Adjustment by Bias Current

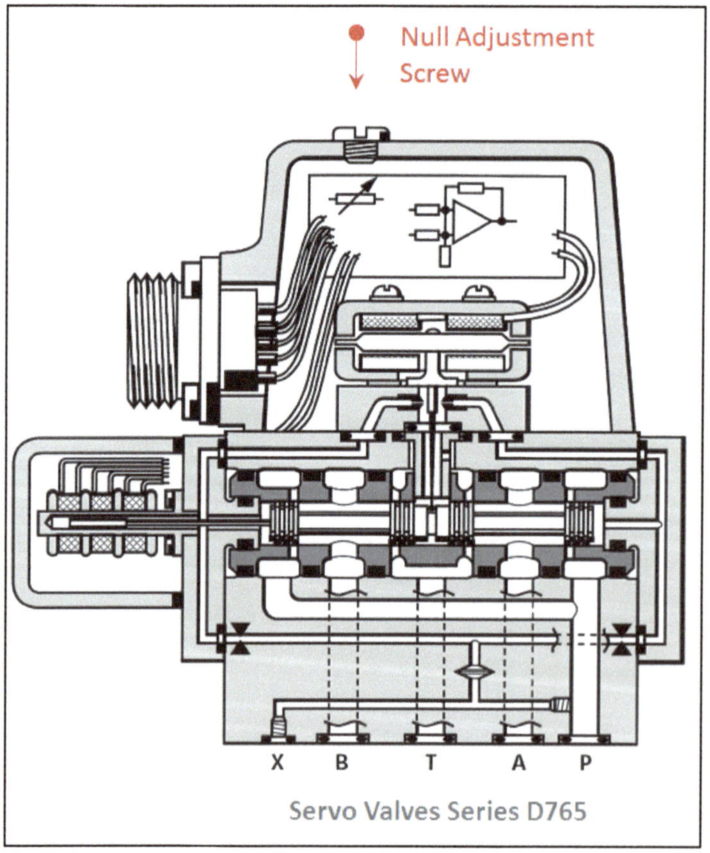

Fig. 9.32 - Electrical Null Adjustment, Example 1 (Courtesy of Moog)

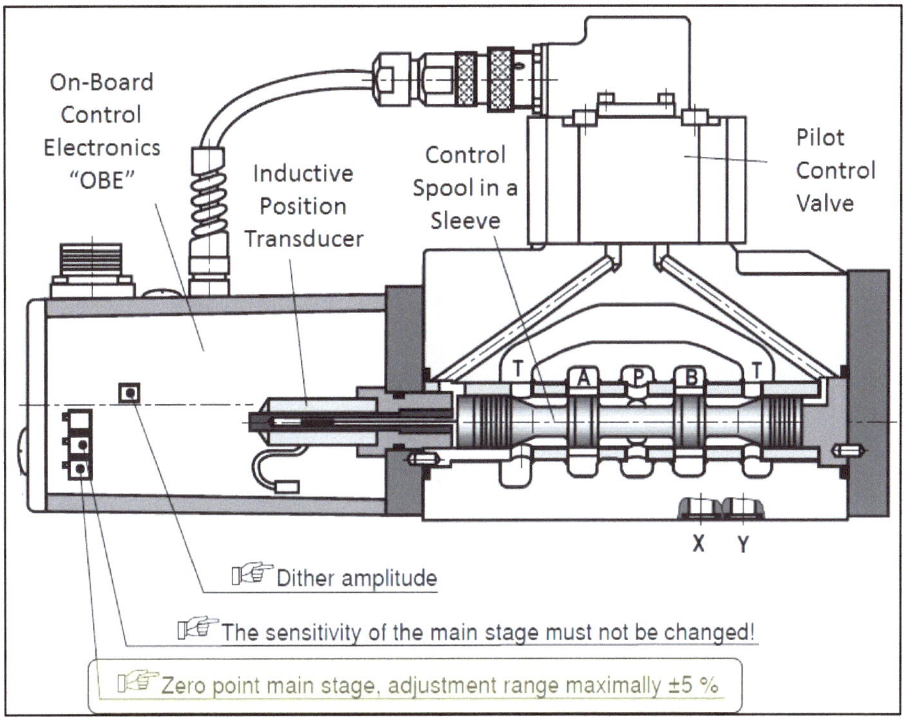

Fig. 9.33 - Electrical Null Adjustment, Example 2 (Courtesy of Bosch Rexroth)

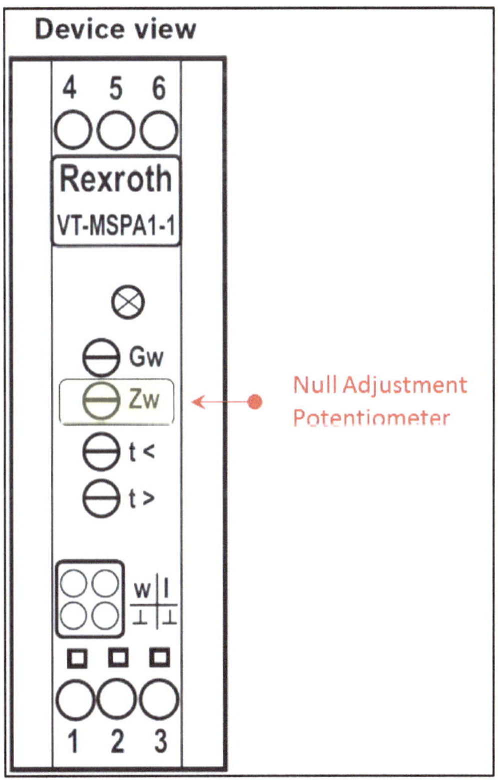

Fig. 9.34 - Electrical Null Adjustment, Example 3 (Courtesy of Bosch Rexroth)

9.12- Dead Band Eliminator (I_{min})

Overlapped (closed-center) valves result in steady state error. To explain this issue, Fig. 9.35 presents a block diagram for a closed-loop cylinder position control. As it has been discussed earlier, regardless of the valve null conditions, $U_C = k_p \times U_E$. If an overlapped valve with dead band (DB) is used, the cylinder will continue to move until $U_C \leq DB$. Once the valve enters into its dead band, it terminates the power and the cylinder stops. In mathematical form, the actuator stops when $U_E \leq (DB/k_p)$.

Assuming the dead band is 20% of the valve full stroke and $k_p = 1$, then the steady state error will be 20% of the controlled parameter.

Then, to reduce the SSE, one easy way is to increase the gain k_p. Mathematically, hypothetically, the gain should be increased to infinity to get rid of the SSE. Practically, the gain can be increased within the stability limits of the system. If that does not help, then the *Dead Band Eliminator* feature must be enabled.

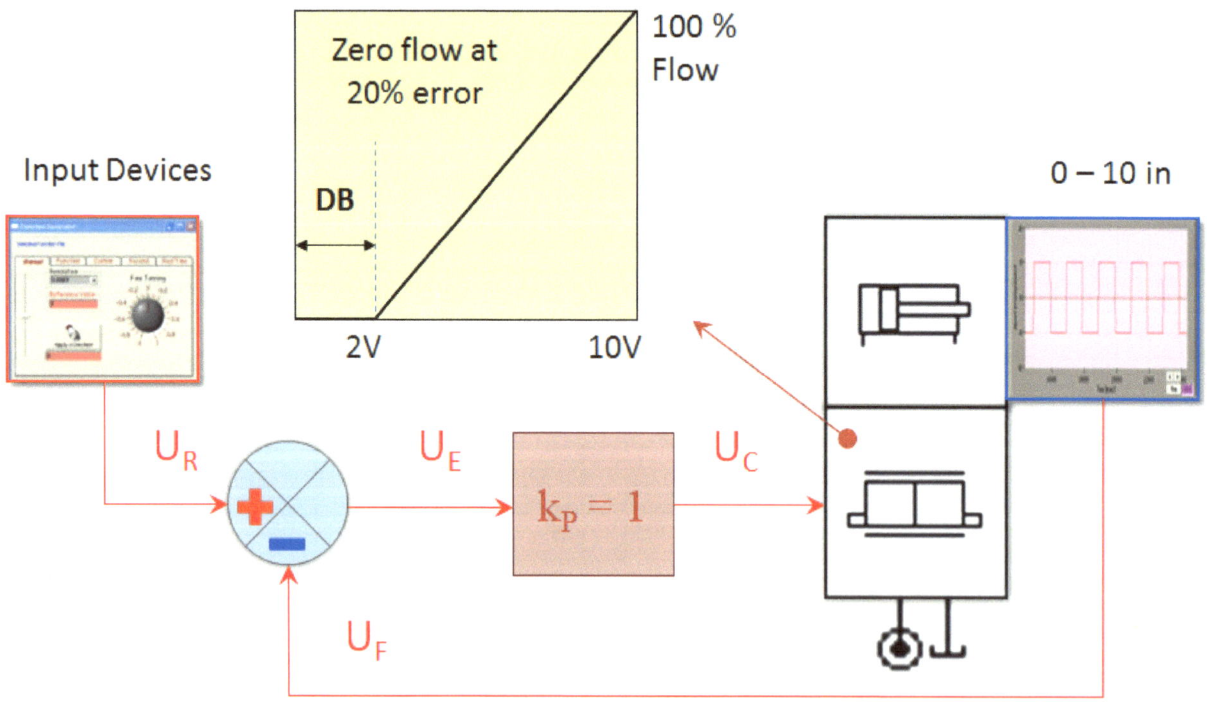

Fig. 9.35 - Effect of Dead Band on Steady State Error

By enabling the dead band elimination function, certain *Minimum Current* I_{min} is added to the input signal of the valve so that the valve will never fall in the DB zone. I_{min} is also referred to as *Jump Current*. Figure 9.36 shows that enabling the DB elimination function converts the valve to zero-lapped (critical-center) valve. Figure 9.37 shows that by applying the DB eliminating function the valve performance is shifted towards the null position. This will make the valve partially lose control resolution by making the valve saturated for input signal less than 100%. The figure also shows that the dead band is adjusted by analog pots, A and B, for both sides of the spool separately.

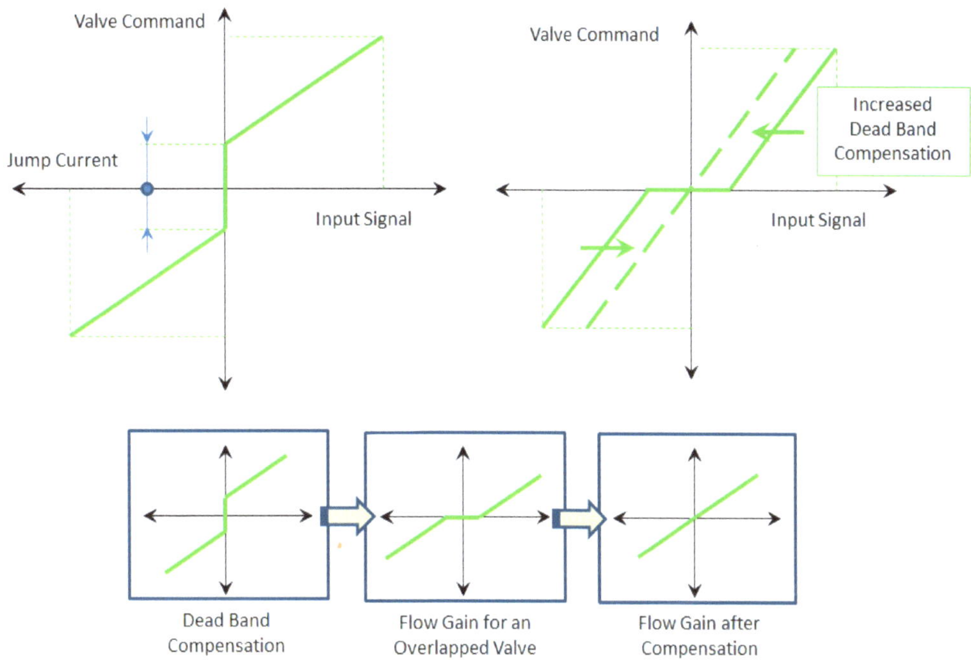

Fig. 9.36 - Concept of Applying Jump Current as a Dead Band Eliminator

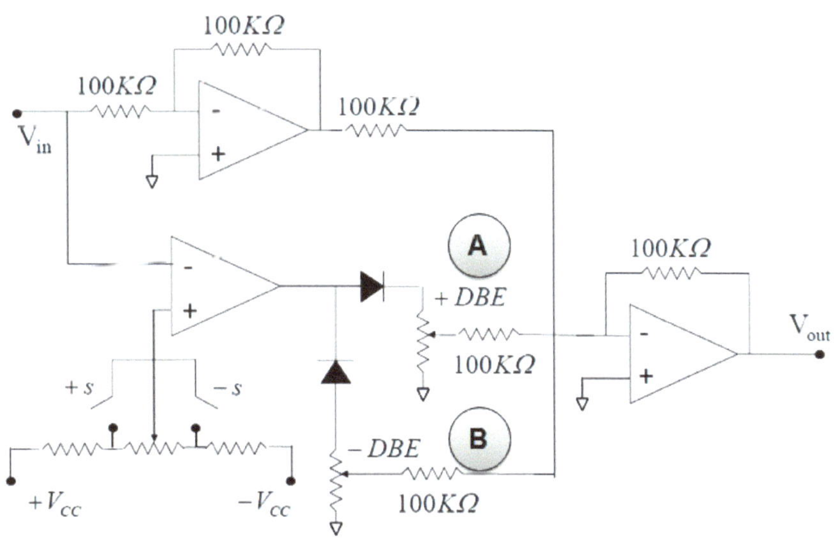

Fig. 9.37 - Example of Analog Dead Band Eliminator (Courtesy of Parker)

Figure 9.38 shows an example of building a digital dead band eliminator using Matlab-Simulink. The figure shows two different logic methods being used, one by adding jump current and the other by magnifying the error signal by a factor greater than one.

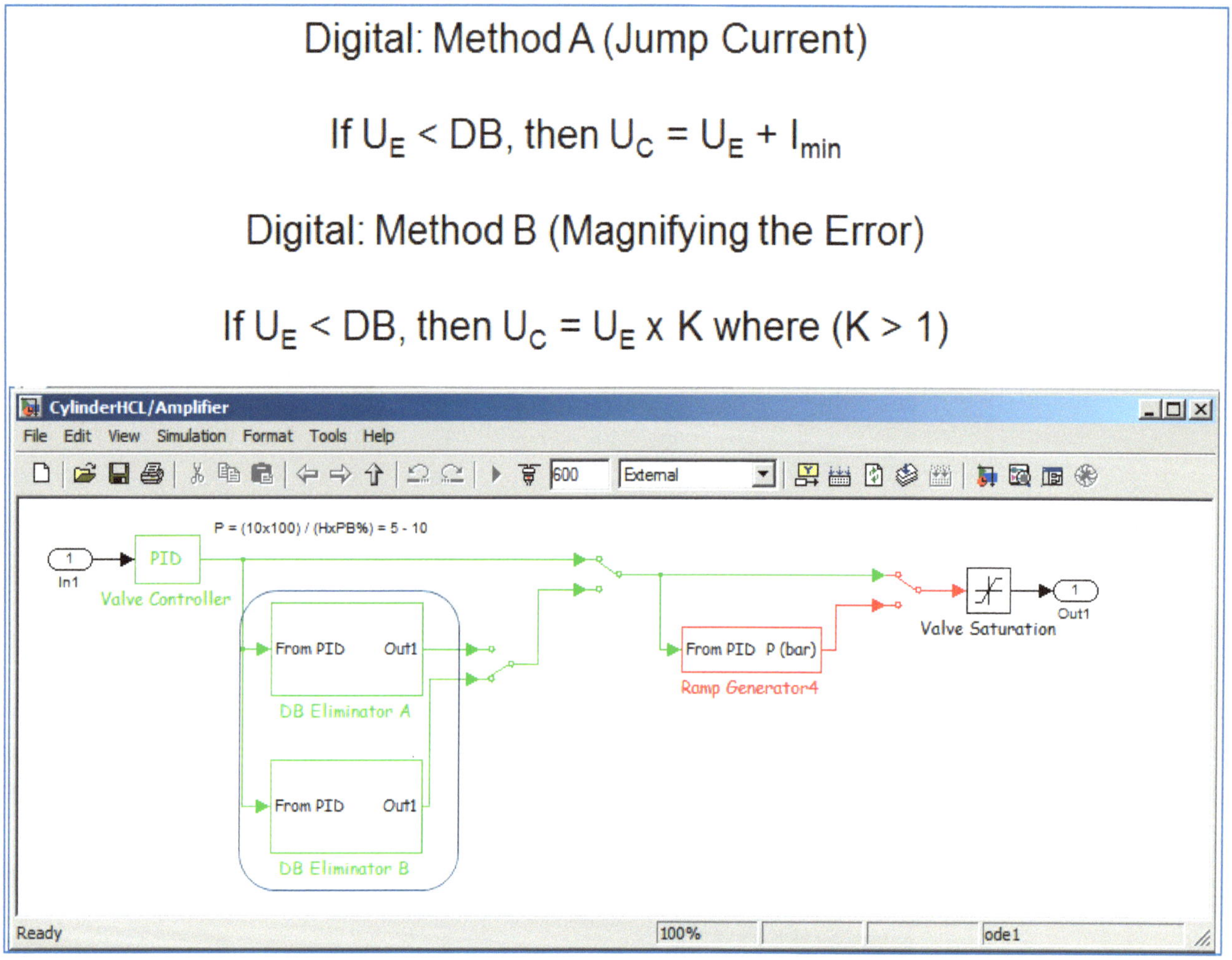

Fig. 9.38 - Example of Digital Dead Band Eliminator

9.13- Saturation Adjustor (I_{max})

Solenoids for proportional valves and torque motors for servo valves are designed for a certain maximum current I_{max}. If the control current exceeds the maximum value, it may burn out the electrical part of the valve. Therefore, the valve must be overload protected. The function that takes care of that has various names such as *Overload Protection*, *Saturation Adjustor*, and *Maximum Current Limiter*. Regardless of the name of the function, as shown in Fig. 9.39, it limits the control current to a maximum value. Figure 9.40 shows an example of analog overload protection. The figure shows that the function is factory set and an overload detection LED is used.

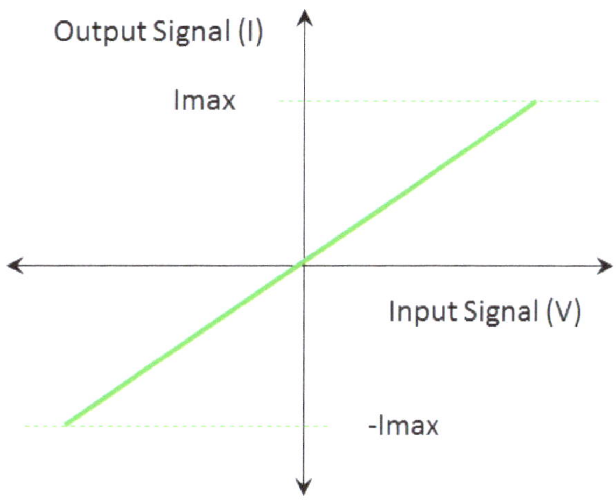

Fig. 9.39 - Maximum Current Limiter

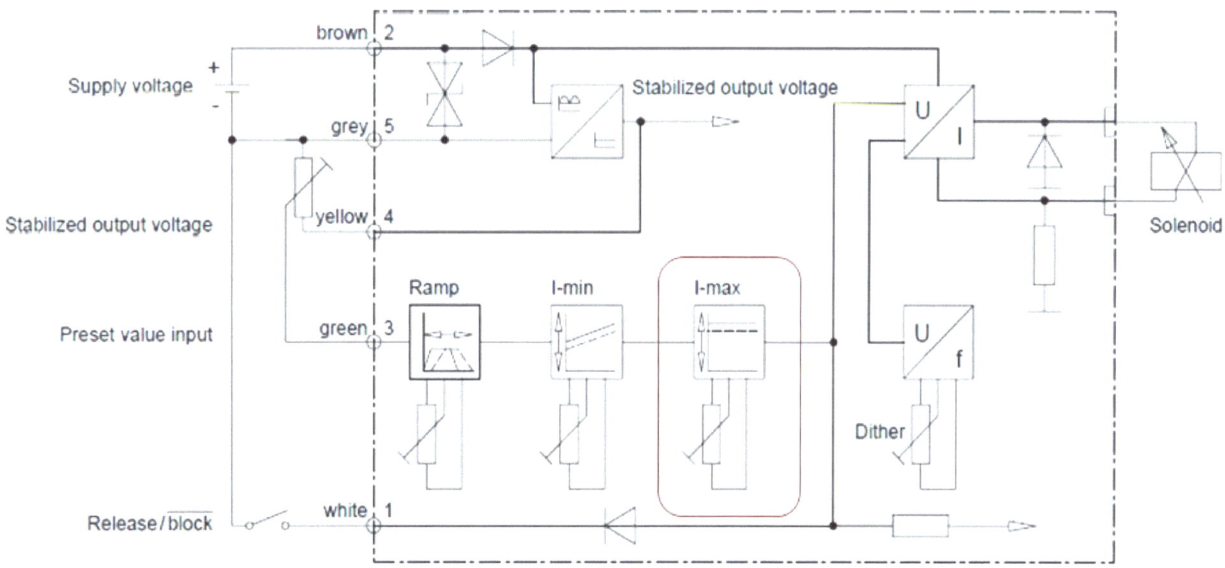

Fig. 9.40- Example of Analog Overload Protection (Courtesy from Wandfluh)

Figure 9.41 shows an example of building a maximum current limiter using Matlab-Simulink.

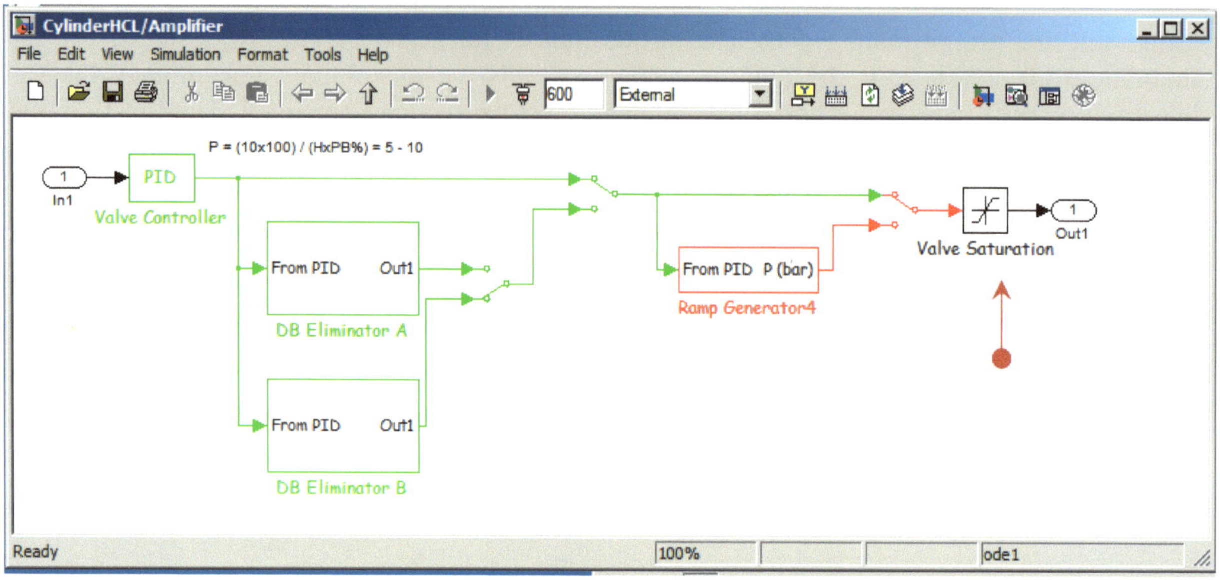

Fig. 9.41 - Example of Digital Overload Protection

Table 9.1 concludes the previously mentioned functions and the purpose of their use.

Function	Purpose of Enabling the Function
Proportional Gain Adjustor	Controls valve shifting displacement (actuator speed of response)
Ramp Generator	Valve shifting time (load acceleration and deceleration)
Dead Band Eliminator	Use I_{min} to remove SSE
Saturation Limiter	Consider I_{max} to protect solenoid and torque motors
Null Adjustor	Use I_{Bias} to adjust spool null position
Dither Signal	Remove spool sticking and improve hysteresis performance
Pulse Width Modulation	Reduce power consumption and heat generation

Table 9.1 - Summary of the Control Electronics Basic Functions for EH Valves

9.14- Control Electronics Basic Configurations

Electric control units that drive EH valves are configured differently. Figure 9.42 presents a brief classification of the basic configurations of the control electronics including legacy configurations and most recent ones.

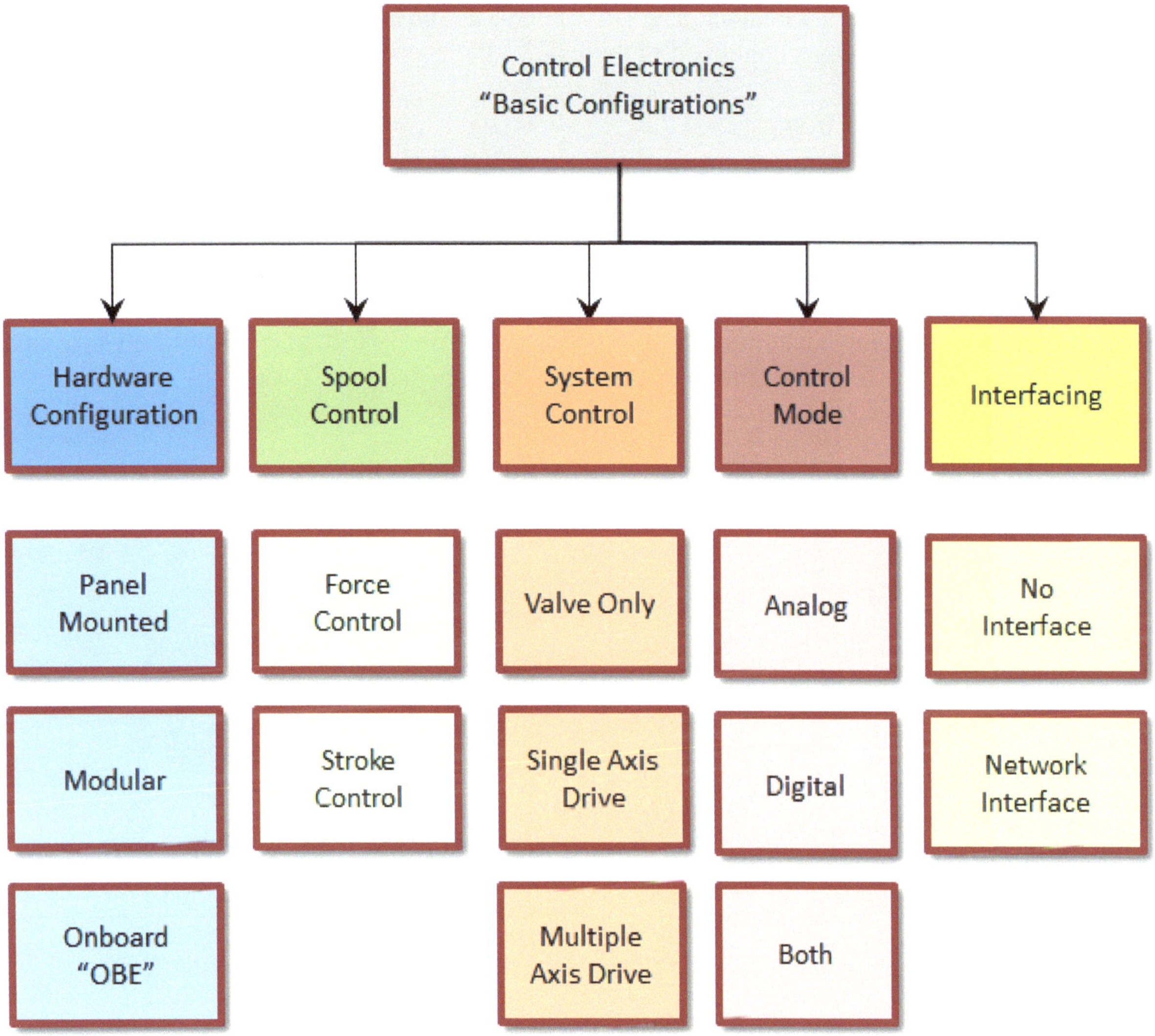

Fig. 9.42 - Control Electronics Basic Configurations

9.15- Hardware Configuration

As shown in Fig. 9.43, based on the hardware configuration, control electronics are classified as follows:

Panel Mounted Control electronics:
Panel Mounted ECUs are referred to as Control Cards. They are kept apart from the valve in a conditioned and guarded cabinet. The front panel, as shown in the figure, gives access to the end user to tune the functions contained in the ECU.

Modular Control Electronics:
Modular ECUs are contained inside a customized enclosure. It can be used for industrial/mobile applications. It can be used to control a single valve or a multi-axis system.

Integrated Control Electronics:
Integrated Control Electronics are referred to as *Onboard Electronics (OBE).* Using this configuration results in a compact system, reduced wiring effort, and ease of network interfacing. There are performance limitations at high temperature (good for T < 60 ^{O}C).

Panel Mounted Type | Modular Type | OBE

Fig. 9.43 - Control Electronics Hardware Configurations (Courtesy of Bosch Rexroth)

9.16- Typical Electronic Schematics for Continuous Valves

Figures 9.44 through 9.58, including the sub-figures, present examples of control electronics for real valves. With each case, a table was developed to summarize the main feature of the control electronics. With all the presented knowledge base in this chapter, it is due to the reader to go line by line to familiarize himself with the published data on each valve.

Example 1: ECU for controlling force-Controlled proportional pressure valves

Type VT-MSPA1

Hardware Configuration	
▪ Panel Mounted	
▪ Modular	✔
▪ OBE	
Spool Control	
▪ Force Controlled	✔
▪ Stroke Controlled	
System Control	
▪ Valve Only	✔
▪ Single Axis	
▪ Multiple Axis	
Control Mode	
▪ Analog	✔
▪ Digital	
Interfacing	
▪ No Interface	✔
▪ Network Interfacing	

Key Features:

- Inverse-polarity protection of the operating voltage.
- Differential input for command value voltage +10 V.
- Ramp generator with separate up and down adjustability.
- Zero point potentiometer.
- LED display: • Ready for operation (green).
- Measuring sockets for pressure command value and actual value.
- Dither generator with command value- and operating voltage-dependent frequency.

Fig. 9.44A - Typical Control Electronics - Example 1 (Courtesy of Bosch Rexroth

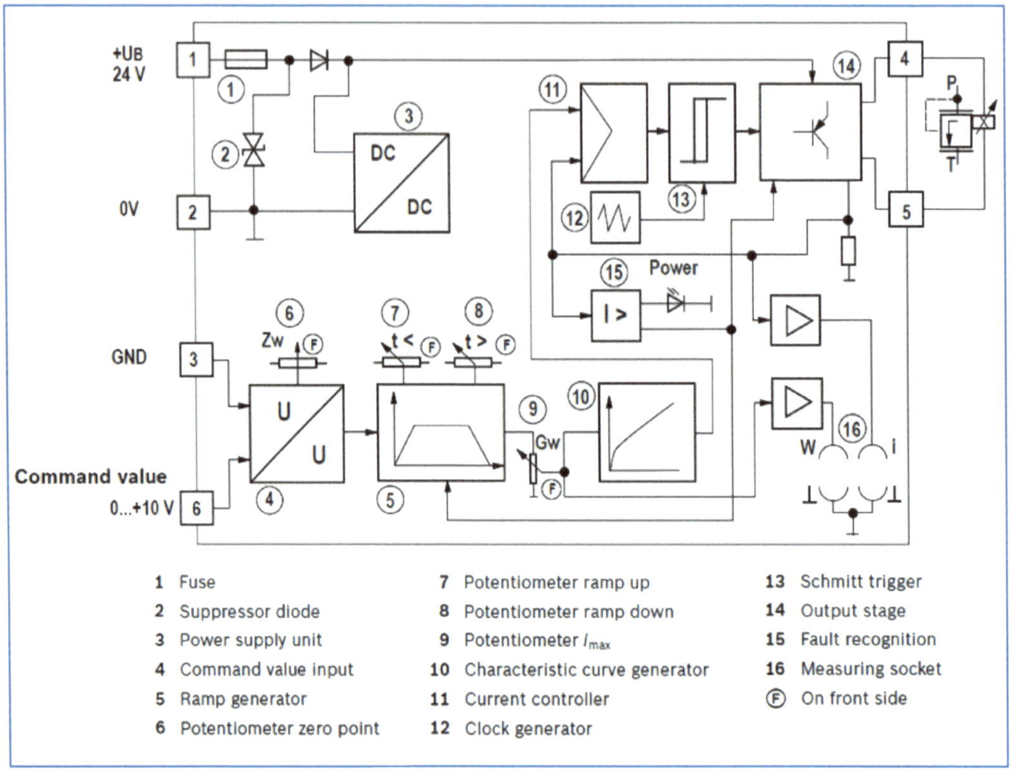

Fig. 9.44B - Block Diagram for Example 1 (Courtesy of Bosch Rexroth)

Device view

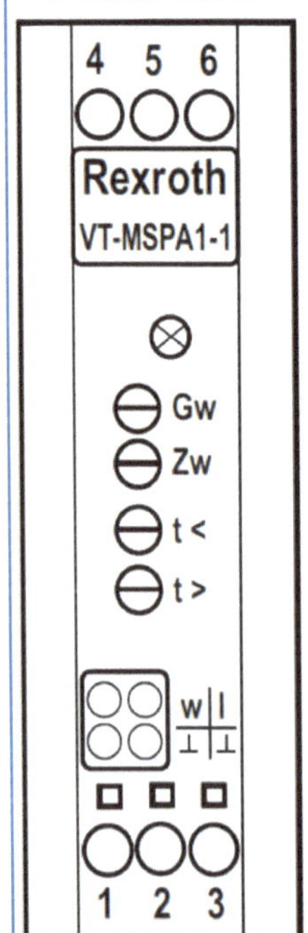

Potentiometer:

"Gw" Pressure command value
"Zw" Zero point
"t <" Ramp time up
"t >" Ramp time down

Sockets:

"w" Pressure command value
"I" Actual current value
"⊥" Measurement zero

Terminal assignment

Terminal	
1	$+U_B$
2	Ground
3	$-U_{command}$
4	Solenoid +
5	Solenoid −
6	$+U_{command}$

Fig. 9.44C - Terminal Assignment for Example 1 (Courtesy of Bosch Rexroth)

Example 2: OBE for controlling single-solenoid force-controlled proportional valves

Hardware Configuration	
▪ Panel Mounted	
▪ Modular	
▪ OBE	✔
Spool Control	
▪ Force Controlled	✔
▪ Stroke Controlled	
System Control	
▪ Valve Only	✔
▪ Single Axis	
▪ Multiple Axis	
Control Mode	
▪ Analog	✔
▪ Digital	
Interfacing	
▪ No Interface	✔
▪ Network Interfacing	

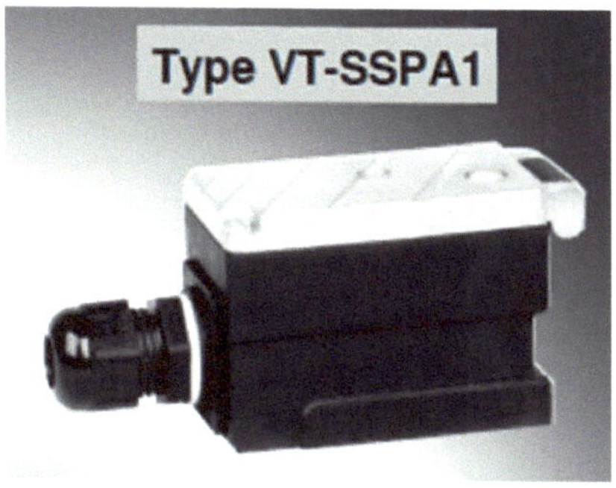

❑ **Key Features:**

- It is directly attached to the solenoid plug of the valve.
- The command value is specified as voltage 0...10 V or as current 4...20 mA.
- An adjustable ramp (60 ms – 5 sec).
- Adjustable Dither amplitude.
- Upon delivery, the dither amplitude has been set to a preset value.

Fig. 9.45A - Typical Control Electronics - Example 2 (Courtesy of Bosch Rexroth)

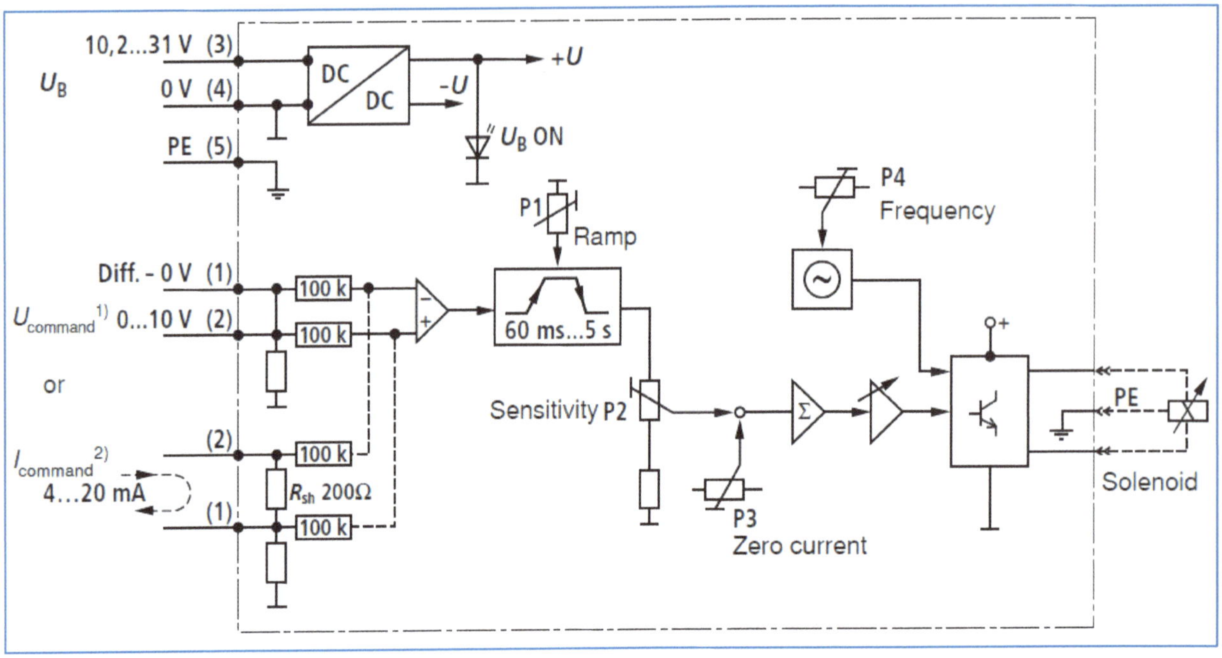

Fig. 9.45B - Block Diagram for Example 2 (Courtesy of Bosch Rexroth)

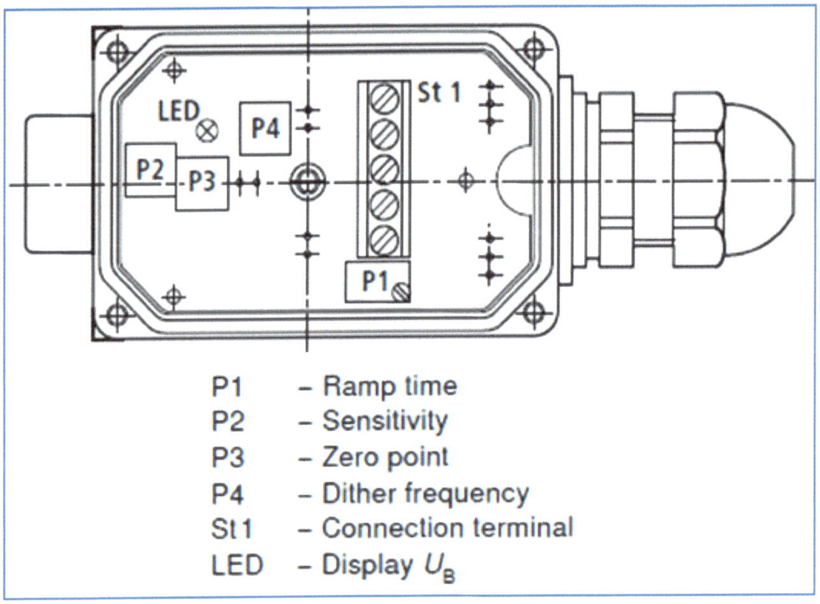

Fig. 9.45C - Terminal Assignment for Example 2 (Courtesy of Bosch Rexroth)

Example 3: OBE for controlling single-solenoid force-controlled proportional valves

Hardware Configuration	
▪ Panel Mounted	
▪ Modular	
▪ OBE	✓
Spool Control	
▪ Force Controlled	✓
▪ Stroke Controlled	
System Control	
▪ Valve Only	✓
▪ Single Axis	
▪ Multiple Axis	
Control Mode	
▪ Analog	✓
▪ Digital	✓
Interfacing	
▪ No Interface	✓
▪ Network Interfacing	

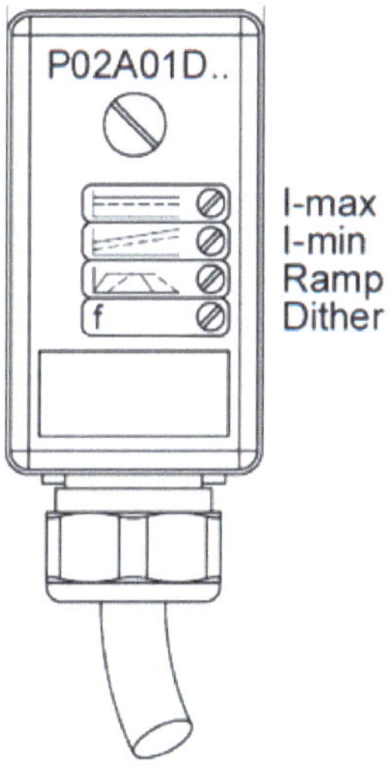

❑ **Key Features:**

- It is directly attached to the solenoid plug of the valve.
- The command value is specified as voltage 0...10 V or as current 4...20 mA.
- An adjustable ramp (60 ms – 5 sec).
- Adjustable Dither amplitude.
- Upon delivery, the dither amplitude has been set to a preset value.

Fig. 9.46A - Typical Control Electronics - Example 3 (Courtesy of Wandfluh)

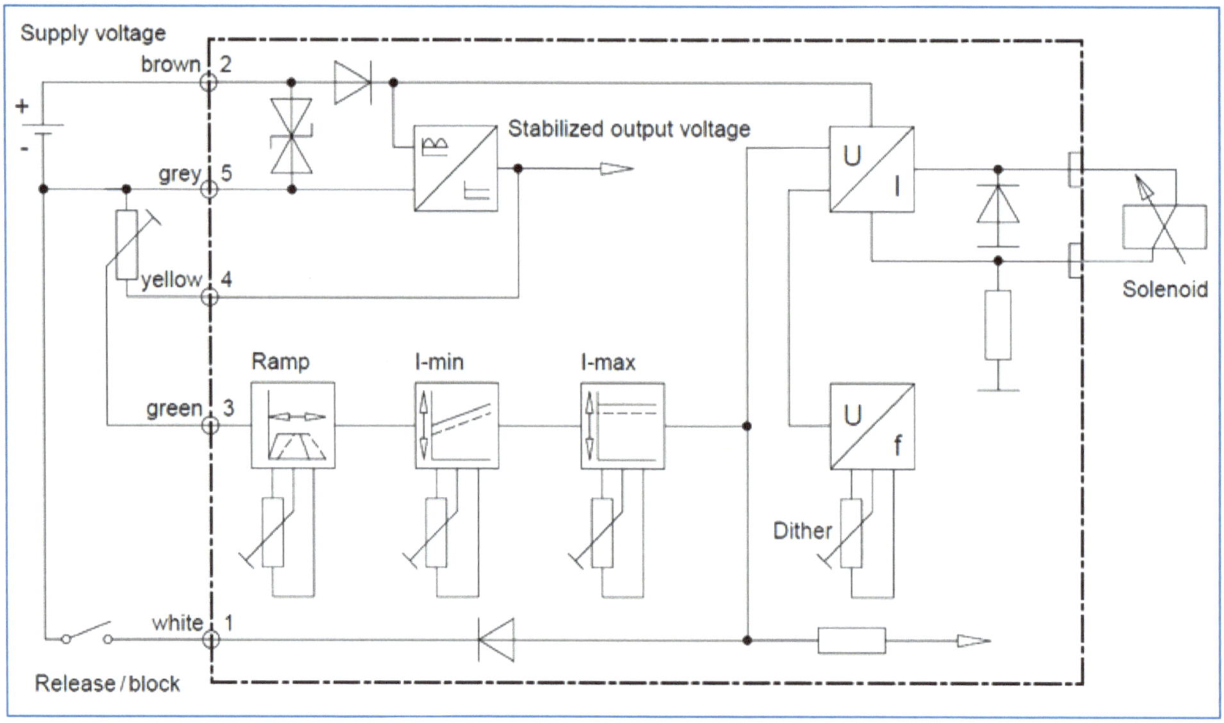

Fig. 9.46B - Block Diagram for Example 3 (Courtesy of Wandfluh)

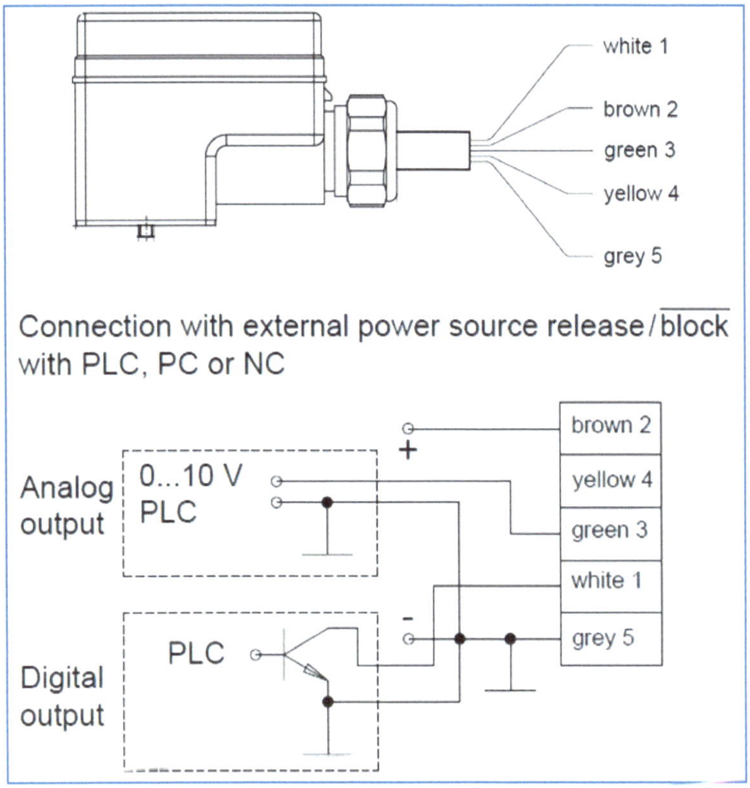

Fig. 9.46C - Terminal Assignment for Example 3 (Courtesy of Wandfluh)

Example 4: Valve amplifier for controlling double-solenoid force-controlled proportional valves

Type VT 11118

Hardware Configuration	
▪ Panel Mounted	
▪ Modular	✔
▪ OBE	
Spool Control	
▪ Force Controlled	✔
▪ Stroke Controlled	
System Control	
▪ Valve Only	✔
▪ Single Axis	
▪ Multiple Axis	
Control Mode	
▪ Analog	✔
▪ Digital	
Interfacing	
▪ No Interface	✔
▪ Network Interfacing	

❑ **Key Features:**

- Selection of the valve type by means of change-over switch at the front.
- Differential command value voltage ±10 V.
- Enable inputs.
- Adjustable ramp generator.
- 2 output stages with fixed-frequency clocking.
- DC/DC converter (L0 = M0).
- Reverse polarity protection for operating voltage.
- Short-circuit-proof outputs.

Fig. 9.47A - Typical Control Electronics - Example 4 (Courtesy of Bosch Rexroth)

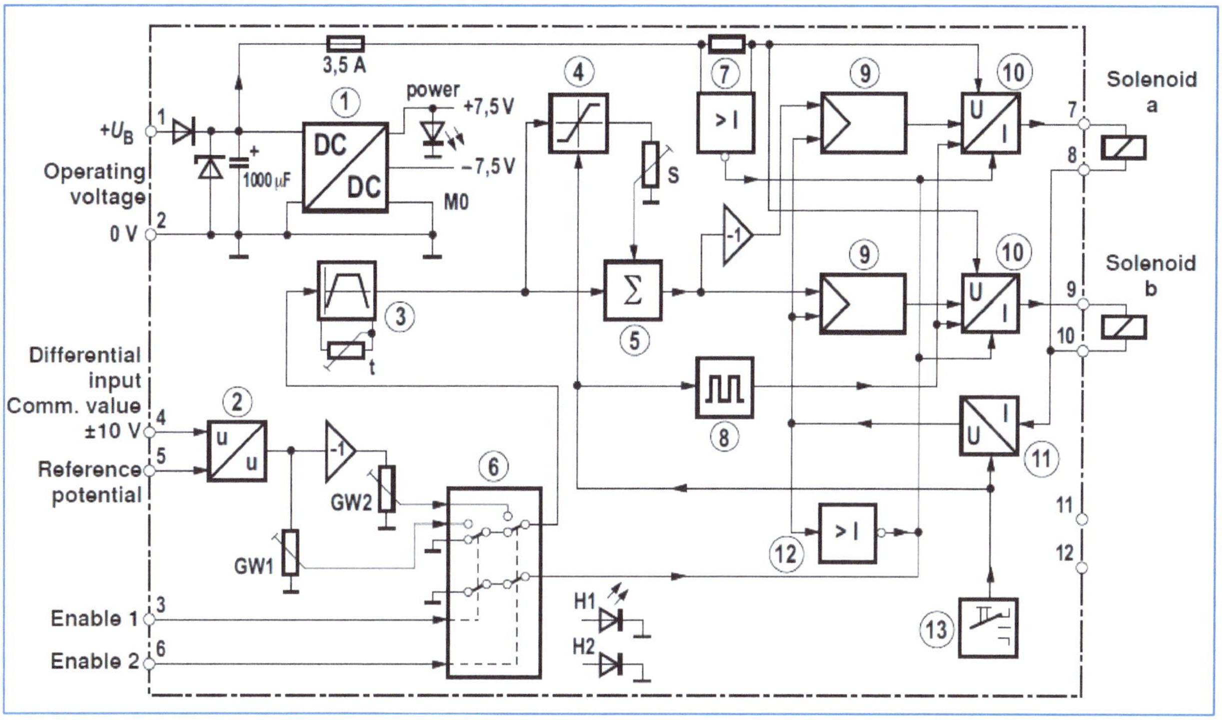

1 Power supply
2 Differential amplifier
3 Ramp generator
4 Step function generator
5 Summator
6 Command value changeover and output stage enable
7 Short-circuit detector
8 Clock-pulse generator
9 Current regulator
10 Output stage
11 Solenoid current measurement
12 Overcurrent detector
13 Valve type selector switch
GW1 Command value attenuator 1
GW2 Command value attenuator 2
t Ramp time setting
S Jump height at U_{Comm} = ±10 V
H1 Enable logic mode 1
H2 Enable logic mode 2

Fig. 9.47B - Block Diagram for Example 4 (Courtesy of Bosch Rexroth)

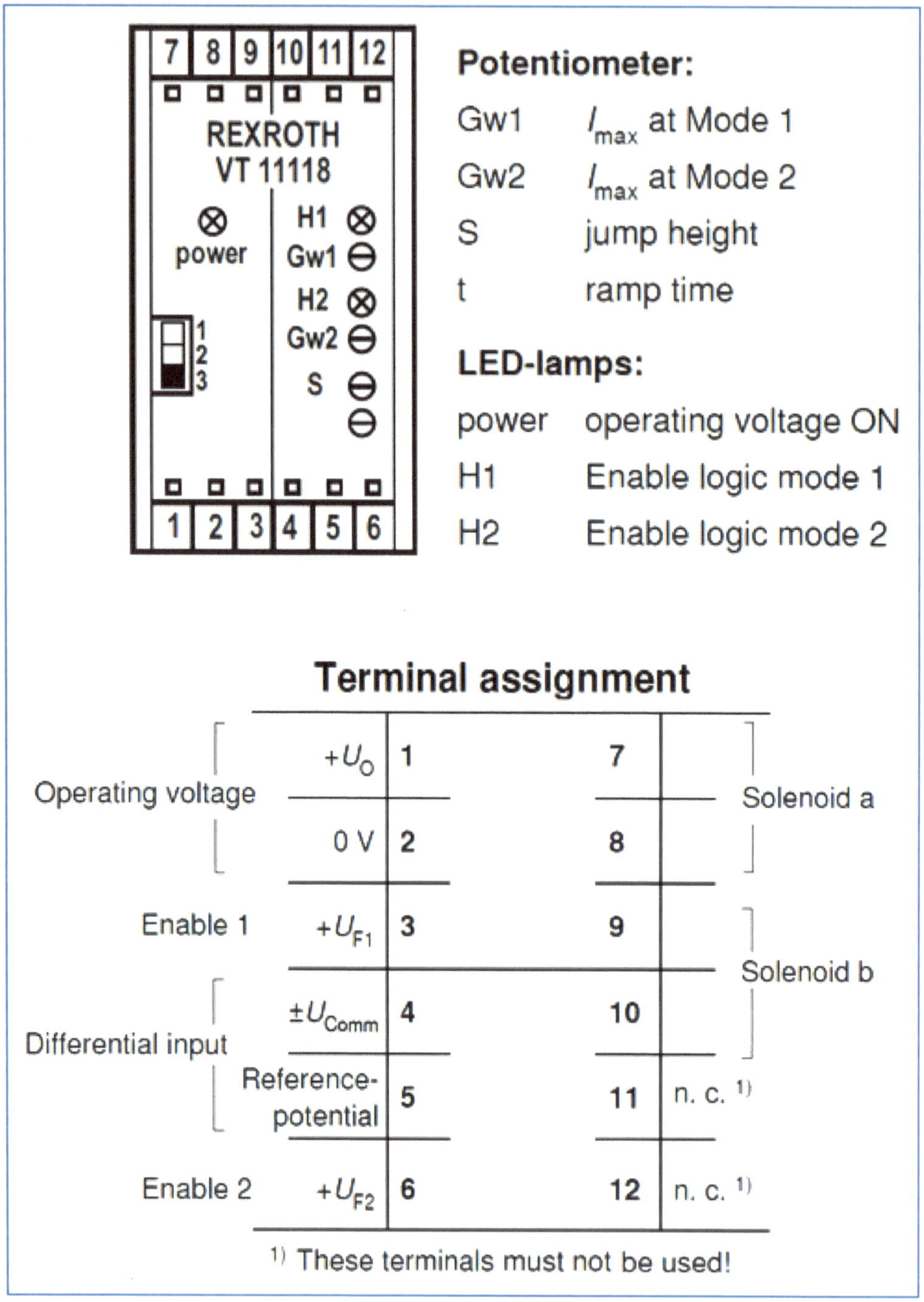

Fig. 9.47C - Terminal Assignment for Example 4 (Courtesy of Bosch Rexroth)

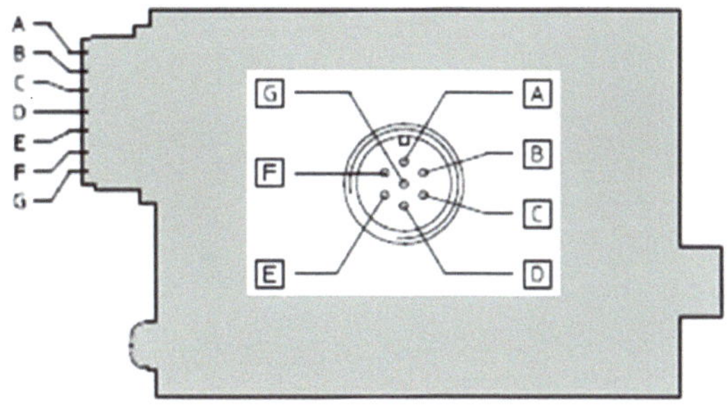

Pin A = Red, 24VDC valve power
Pin B = Black, Zero VDC valve power common
Pin C = Orange, LVDT feedback reference
Pin D = White, command +/-10 VDC
Pin E = Green, command reference 0 VDC
Pin F = Blue, LVDT feedback +/-10 VDC
Pin G = Silver, Shield

Fig. 9.48 - Terminal Assignment for Stroke-Controlled Valves

Example 5: Onboard Electronics for Proportional Valves

Hardware Configuration	
▪ Panel Mounted	
▪ Modular	
▪ OBE	✓
Spool Control	
▪ Force Controlled	
▪ Stroke Controlled	✓
System Control	
▪ Valve Only	✓
▪ Single Axis	
▪ Multiple Axis	
Control Mode	
▪ Analog	✓
▪ Digital	
Interfacing	
▪ No Interface	✓
▪ Network Interfacing	

Series D1FH
Proportional Directional Control Valves

❑ **Key Features:**

- On-Board Electronic Drive Amplifier
- High Frequency Response
- Four Position Spool Capability
- 315 Bar Pressure Capability
- Spool Position Feedback
- Contaminant Sensitivity
- Drive Enable Feature

Fig. 9.49A - Typical Control Electronics - Example 5 (Courtesy of Parker)

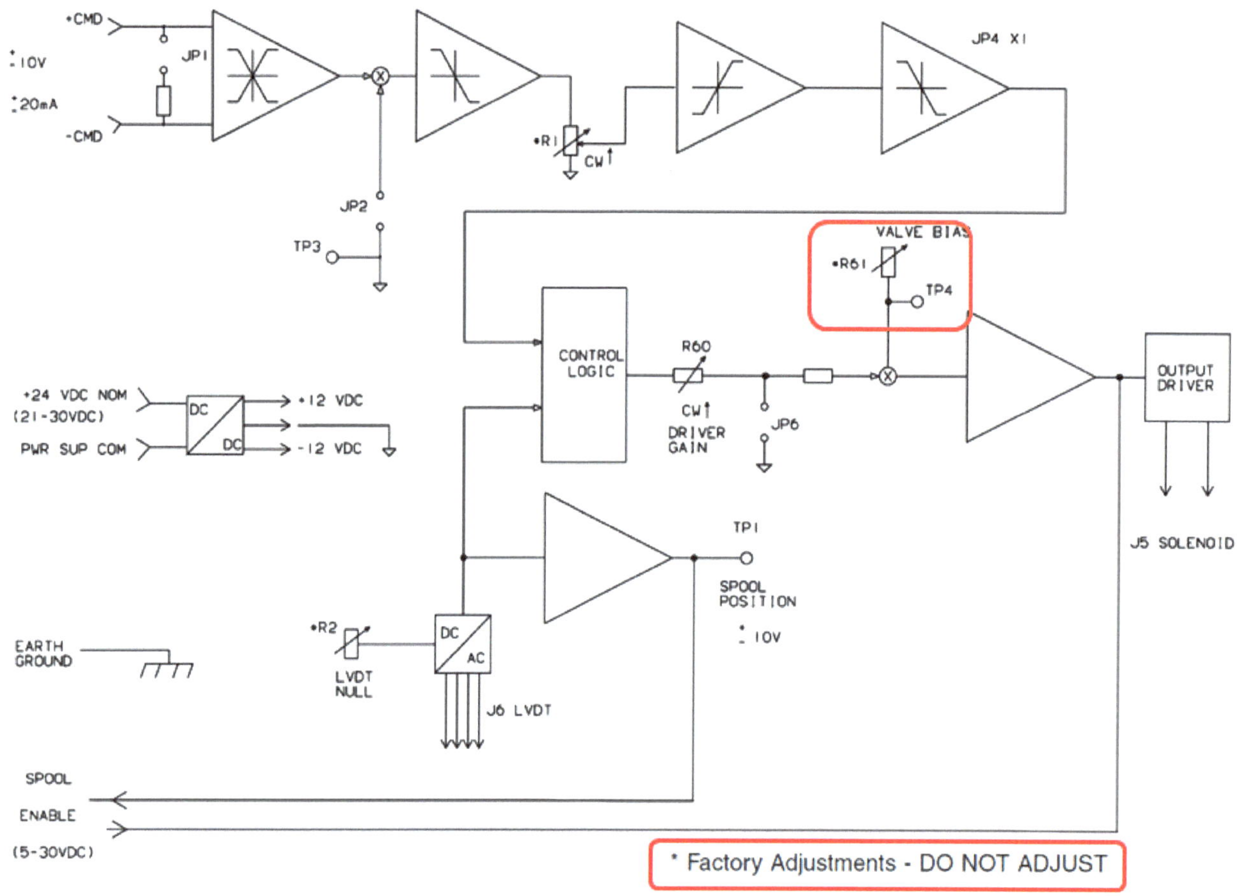

Fig. 9.49B - Block Diagram for Example 5 (Courtesy of Parker)

Jumpers

Command Input	Jumper Selection
±10 VDC	JP2 IN JP4 IN JP7 SPARE
±20 mA	JP1 IN JP2 IN JP4 IN

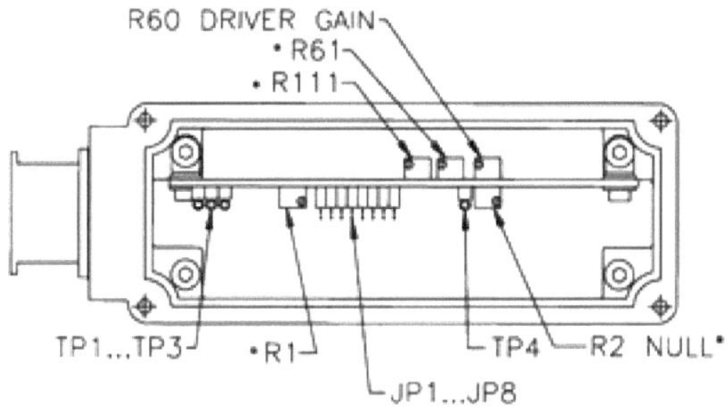

• Factory Adjustments - DO NOT ADJUST

Interface Wiring - 7 Pin I/O Connector

Power Supply	24 VDC Nominal 2.0 Amps 4.0 Amps Peak (<10 ms) + to Pin A - to Pin B
Enable	5 to 30 VDC at Pin C No signal disables the valve.
Command Input	±10 VDC between Pin D and Pin E If D is more positive than E, flow is from P→A. Note: If command source is not differential tie the unused input to the command source common.
Spool Position Output (optional)	Pin F ±10 VDC Positive voltage is P→A. Negative voltage is P→B.
Chassis Ground	Pin G, internally wired to the valve body.

EHC 8G Cable Wiring

PIN	FUNCTION	COLOR
A	+Pwr Sup	Red
B	Pwr Sup Com	Blk
C	Enable	Yel
D	+Cmd	Blue
E	-Cmd	Orn
F	Spool	Wht
G	Chassis Gnd	Grn

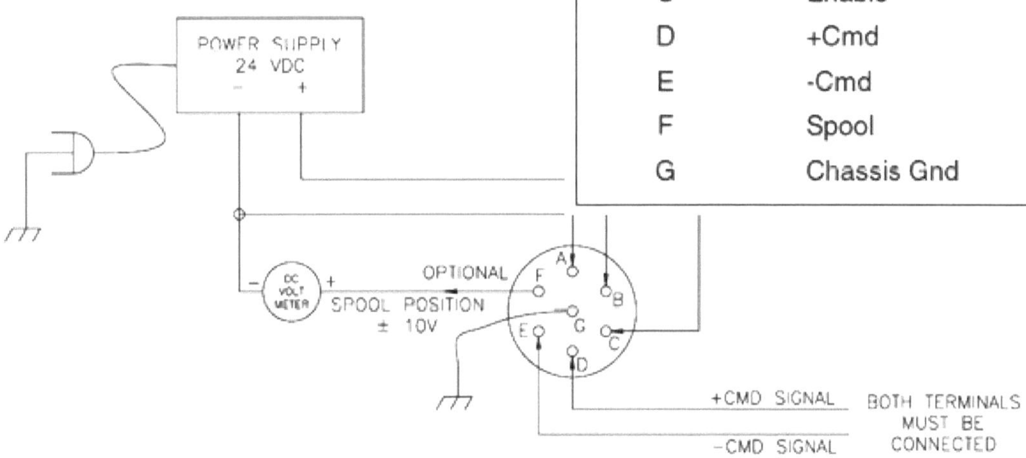

Fig. 9.49C - Terminal Assignment for Example 5 (Courtesy of Parker)

Example 6: Digital valve amplifier for Stroke-Controlled Proportional Valves

Hardware Configuration	
▪ Panel Mounted	✔
▪ Modular	
▪ OBE	
Spool Control	
▪ Force Controlled	
▪ Stroke Controlled	✔
System Control	
▪ Valve Only	✔
▪ Single Axis	
▪ Multiple Axis	
Control Mode	
▪ Analog	
▪ Digital	✔
Interfacing	
▪ No Interface	
▪ Network Interfacing	✔

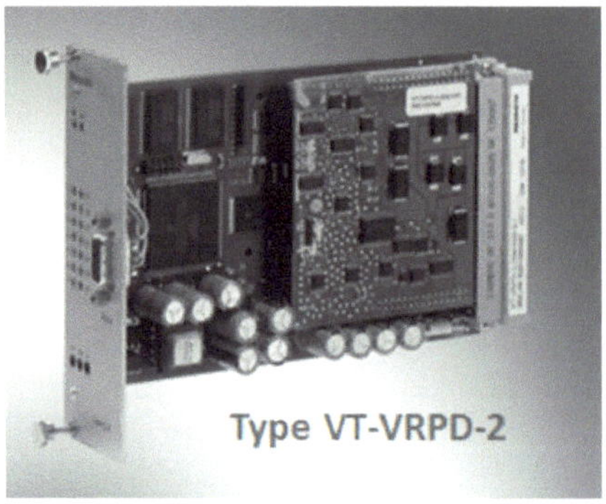

❑ **Key Features:**

- Use of a powerful microcontroller.
- Valve selection using operating software.
- Command value input as voltage or current.
- Ramp generator
- Digital inputs for calling pre-set command values.
- Enable input and fault output.
- Freely configurable measuring sockets
- Serial interface using PC software

Fig. 9.50A - Typical Control Electronics - Example 6 (Courtesy of Bosch Rexroth)

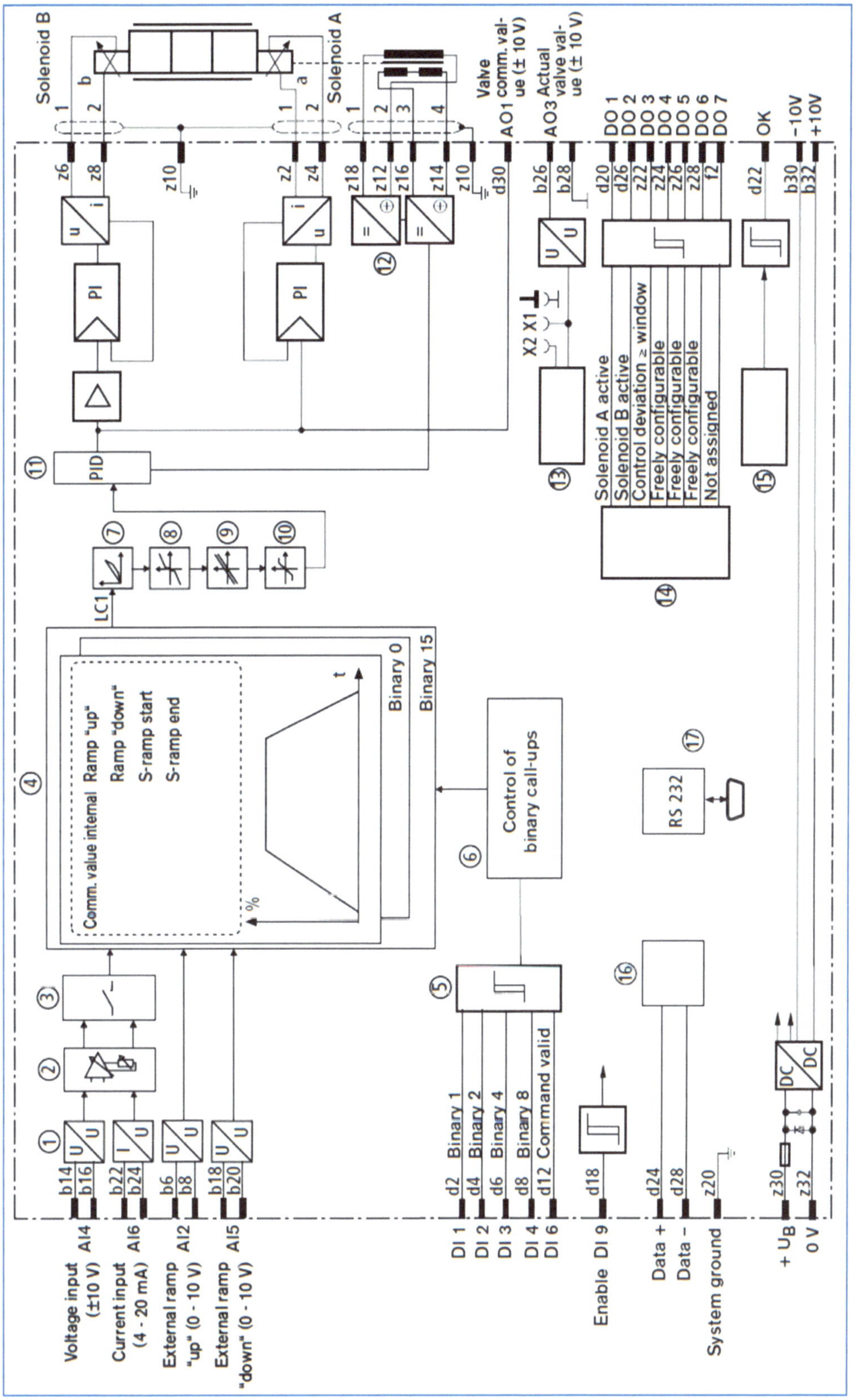

Fig. 9.50B - Block Diagram for Example 6 (Courtesy of Bosch Rexroth)

1	U/U or I/U converter	13	Measuring sockets
2	Analog input adjustment	14	Digital outputs
3	Switching matrix	15	Fault logic
4	Switched-mode power supply unit	16	Local bus
5	Binary command value call-ups	17	Serial interface
6	Control of binary call-ups		
7	Characteristic curve adaptation		**Measuring sockets:**
8	Step function generator	X1	Actual valve value
9	Offset	X2	Valve command value (default)
10	Limitation	⊥	Reference potential (0 V)
11	PID controller		
12	Oscillator / demodulator		

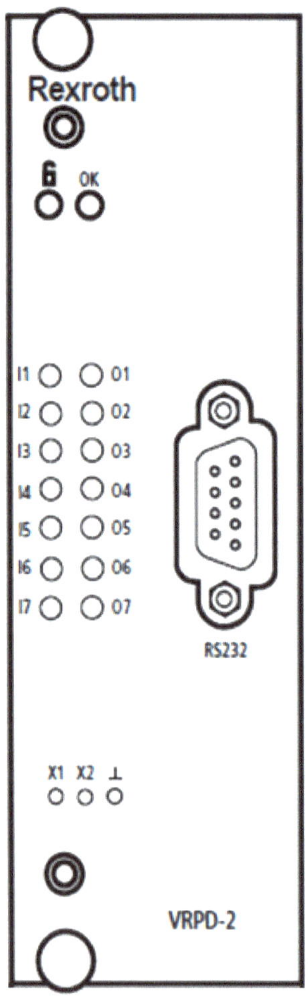

Fig. 9.50C - Terminal Assignment for Example 6 (Courtesy of Bosch Rexroth)

Example 7: Electric amplifier for displacement control of Piston Pumps.

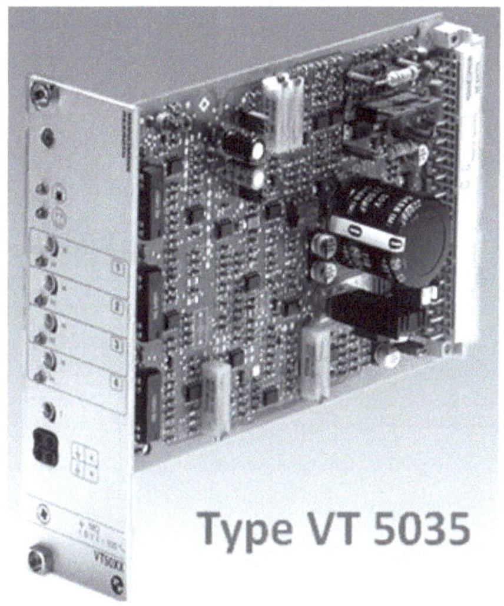

Hardware Configuration	
▪ Panel Mounted	✓
▪ Modular	
▪ OBE	
Spool Control	
▪ Force Controlled	✓
▪ Stroke Controlled	
System Control	
▪ Valve Only	
▪ Single Axis	✓
▪ Multiple Axis	
Control Mode	
▪ Analog	✓
▪ Digital	
Interfacing	
▪ No Interface	
▪ Network Interfacing	✓

❑ **Key Features:**

- Differential input.
- Enable input with LED display.
- "Ready for operation" message by LED display.
- Potentiometer adjustable ramp time.
- Potentiometer adjustable four values.
- Call-ups indicated by LEDs
- Controller for the pump swivel angle
- Oscillator and demodulator for inductive position measurement.
- Reverse polarity protection for the voltage supply

Fig. 9.51A - Typical Control Electronics - Example 7 (Courtesy of Bosch Rexroth)

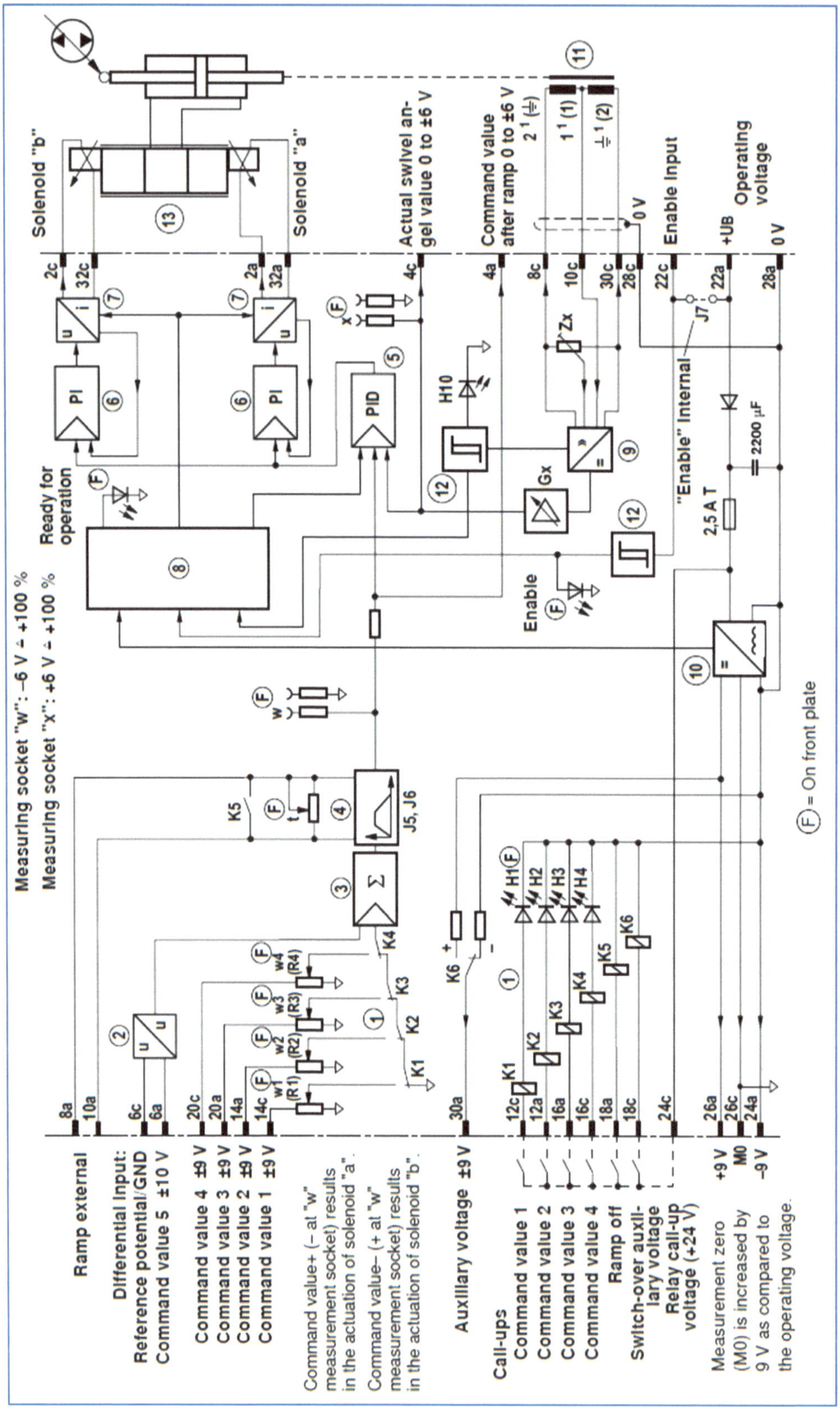

Fig. 9.51B - Block Diagram for Example 7 (Courtesy of Bosch Rexroth)

1 Command values
2 Differential amplifier
3 Summing device
4 Ramp generator
5 Swivel angle controller
6 Flow controller
7 Output stage
8 Monitoring
9 Oscillator/demodulator
10 Power supply unit
11 Position transducer
12 Actual value identification
13 Proportional valve

H1 to **H4** = LED displays for command value call-ups
K1 to **K6** = Call-up relay
R1 to **R4** = Command value potentiometer
t = Ramp time

Notice for connection of the position transducer:

1 Applies to pump with clockwise rotation
( ) Applies to pump with counterclockwise rotation

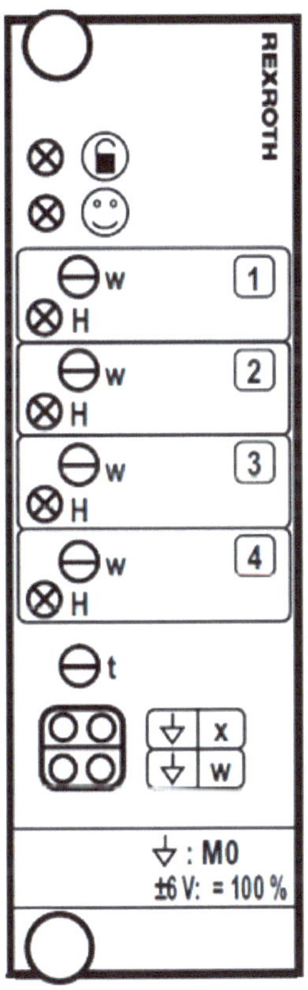

LED display "Enable" (yellow)
LED display "Ready for operation" (green)

w1/H1 Command value 1 with LED display
w2/H2 Command value 2 with LED display
w3/H3 Command value 3 with LED display
w4/H4 Command value 4 with LED display
t Ramp time (in the condition as supplied set to the minimum value)
x Measuring sockets for actual swivel angle value
w Measuring sockets for swivel angle command value

Fig. 9.51C - Terminal Assignment for Example 7 (Courtesy of Bosch Rexroth)

Example 8: Onboard electronics for High Performance Proportional Valve.

Hardware Configuration	
▪ Panel Mounted	
▪ Modular	
▪ OBE	✔
Spool Control	
▪ Force Controlled	
▪ Stroke Controlled	✔
System Control	
▪ Valve Only	
▪ Single Axis	✔
▪ Multiple Axis	
Control Mode	
▪ Analog	
▪ Digital	✔
Interfacing	
▪ No Interface	
▪ Network Interfacing	✔

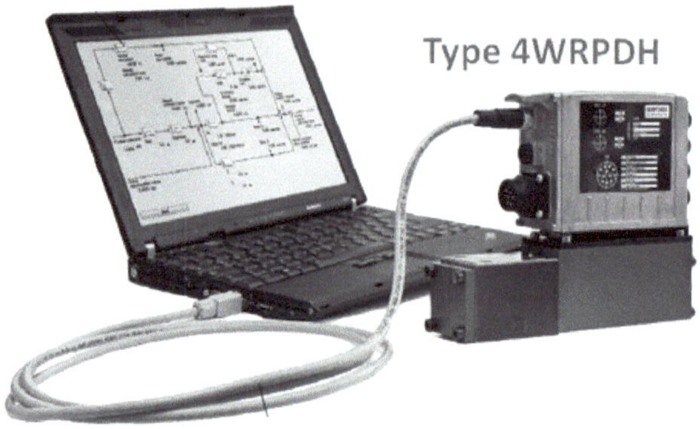

❑ **Key Features:**

- Integrated digital axis control functionality
- Multi Ethernet/Bus interface.
- 2 configurable analog sensor inputs.
- 1 input for linear position measurement system.
- Internal safety function.
- Best-in-class hydraulic controller.
- High response sensitivity and little hysteresis.

Fig. 9.52A - Typical Control Electronics - Example 8 (Courtesy of Bosch Rexroth)

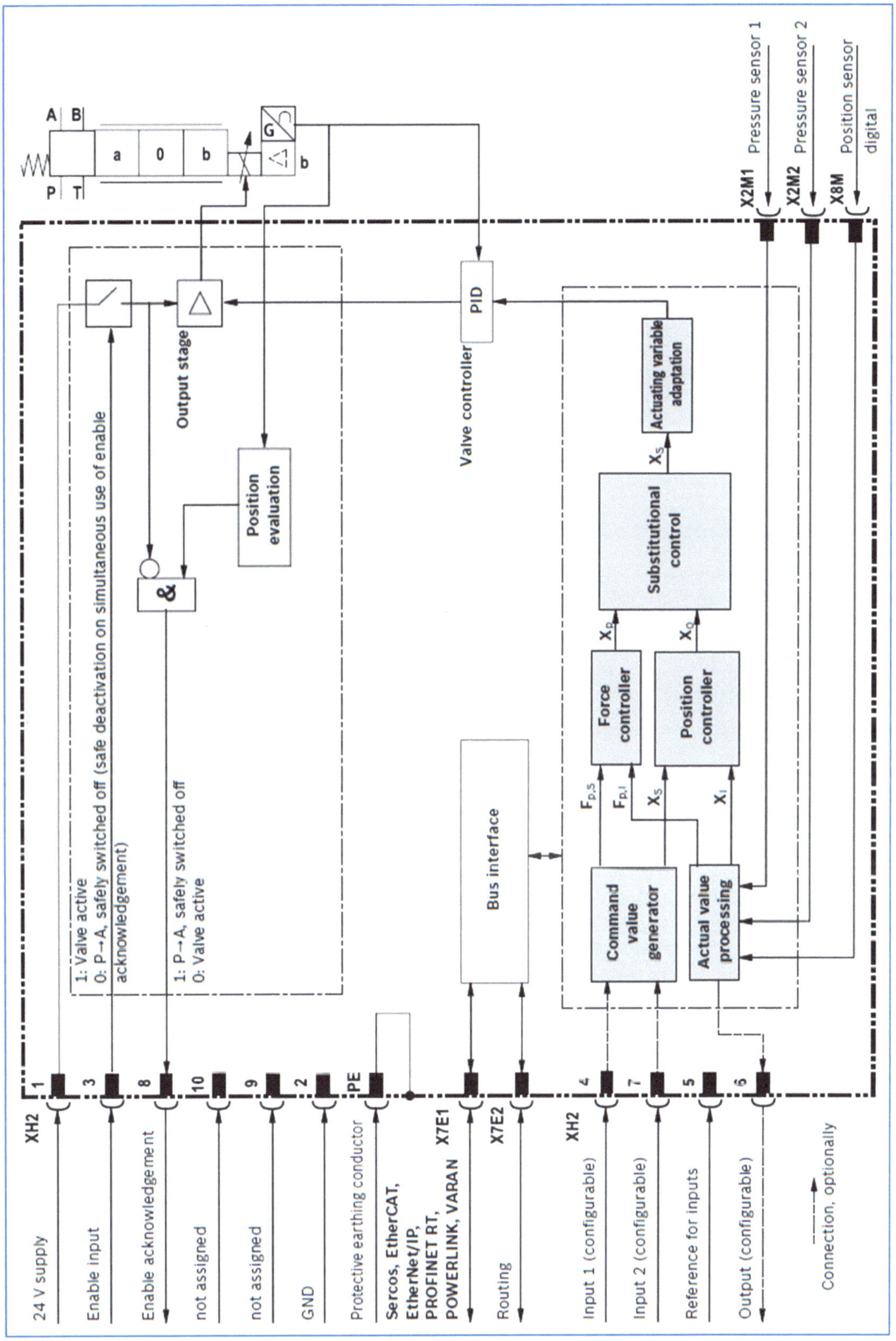

Fig. 9.52B - Block Diagram for Example 8 (Courtesy of Bosch Rexroth)

Connector pin assignment XH2, 11-pole + PE according to EN 175201-804			
Pin	**Core marking** Cable, one-part [1)]	Cable, split [2)]	**Interface D6 assignment**
1	1	1	24 V DC supply voltage
2	2	2	GND
3	3	White	Enable input 24 V DC (high ≥ 15 V; low < 2 V)
4	4	Yellow	Command values 1 (4 ... 20 mA/±10 V) [3)]
5	5	Green	Reference for command values
6	6	Violet	Actual value/4 ... 20 mA/±10 V) [3; 4)]
7	7	Pink	Command value 2 (4 ... 20 mA/±10 V) [3)]
8	8	Red	Enable acknowledgement 24 V DC (I_{max} 50 mA) [5)]
9	9	Brown	not assigned
10	10	Black	not assigned
11	11	Blue	Switching output 24 V, configurable (fault-free operation (24 V)/error (0 V) or power circuit signal), maximum 1.5 A [3; 5)]

Fig. 9.52C - Terminal Assignment for Example 8 (Courtesy of Bosch Rexroth)

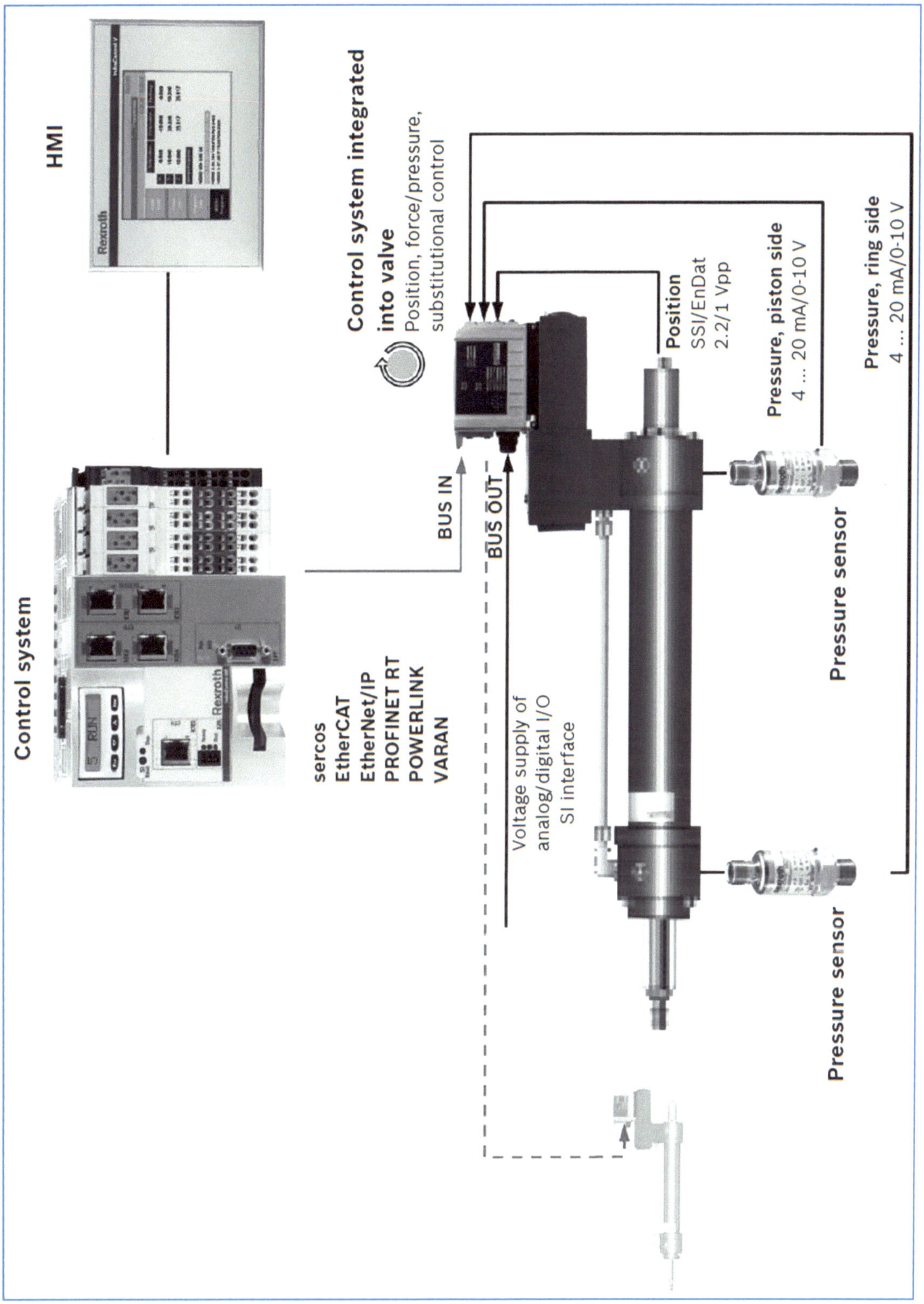

Fig. 9.52D - Typical Application for Example 8 (Courtesy of Bosch Rexroth)

Example 9: Onboard electronics for High Performance Proportional Valve.

Hardware Configuration	
▪ Panel Mounted	
▪ Modular	
▪ OBE	✔
Spool Control	
▪ Force Controlled	
▪ Stroke Controlled	✔
System Control	
▪ Valve Only	
▪ Single Axis	
▪ Multiple Axis	✔
Control Mode	
▪ Analog	✔
▪ Digital	✔
Interfacing	
▪ No Interface	
▪ Network Interfacing	✔

❑ Key Features:

- Zero overlap spools for accurate closed loop position controls.
- Pilot operated valves are equipped with 2 position transducers for double closed loop control

Fig. 9.53A - Typical Control Electronics - Example 9 (Courtesy of Atos)

servocylinder ① with position transducer ②, servoproportional valve ③ with integral TEZ controller ④, connections to position transducer ⑤, to electric power, electronic signals and fieldbus ⑥

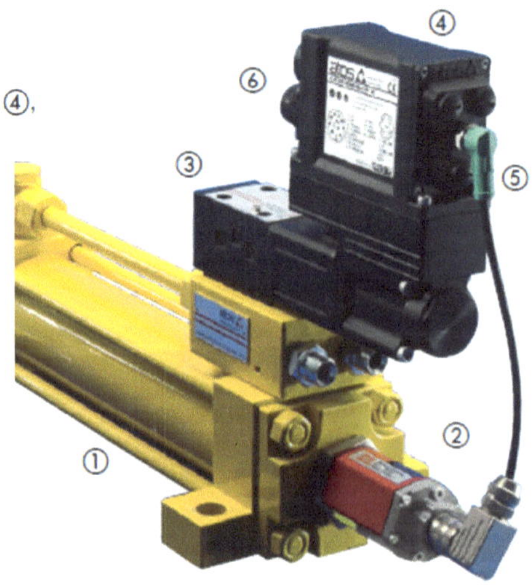

Fig. 9.53B - Typical Application for Example 9 (Courtesy of Atos)

Example 10: Digital Single-Axis Controller for driving proportional valves with two solenoids.

Hardware Configuration	
▪ Panel Mounted	
▪ Modular	✔
▪ OBE	
Spool Control	
▪ Force Controlled	✔
▪ Stroke Controlled	
System Control	
▪ Valve Only	
▪ Single Axis	✔
▪ Multiple Axis	
Control Mode	
▪ Analog	✔
▪ Digital	✔
Interfacing	
▪ No Interface	
▪ Network Interfacing	✔

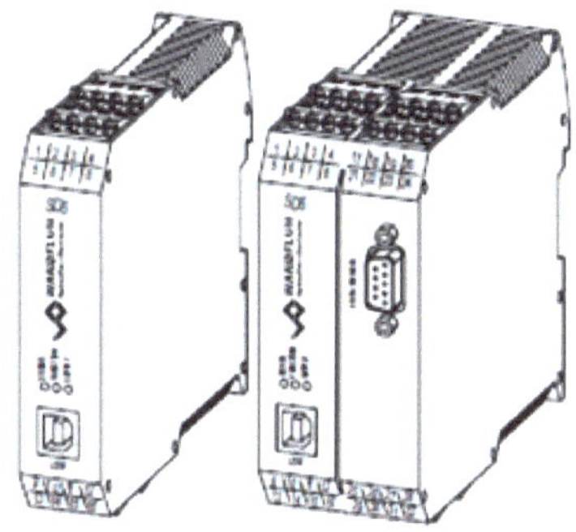

Type SD6

❑ **Key Features:**

- Controls pressure, flow or position.
- Menu-controlled parameterization and diagnostics.
- Software programming.
- USB-interface.
- Analog inputs.

Fig. 9.54A - Typical Control Electronics - Example 10 (Courtesy of Wandfluh)

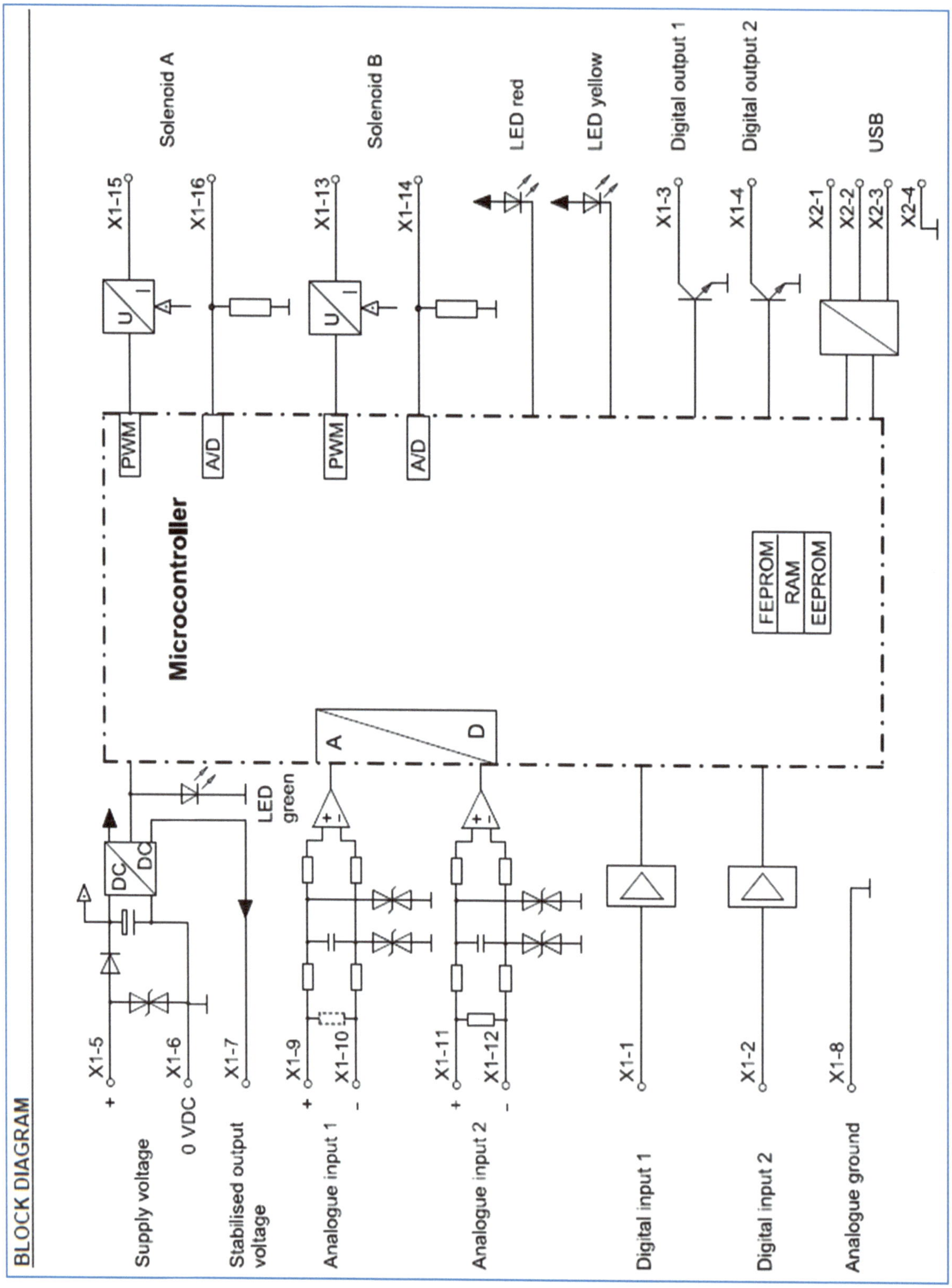

Fig. 9.54B - Block Diagram for Example 10 - (Courtesy of Wandfluh)

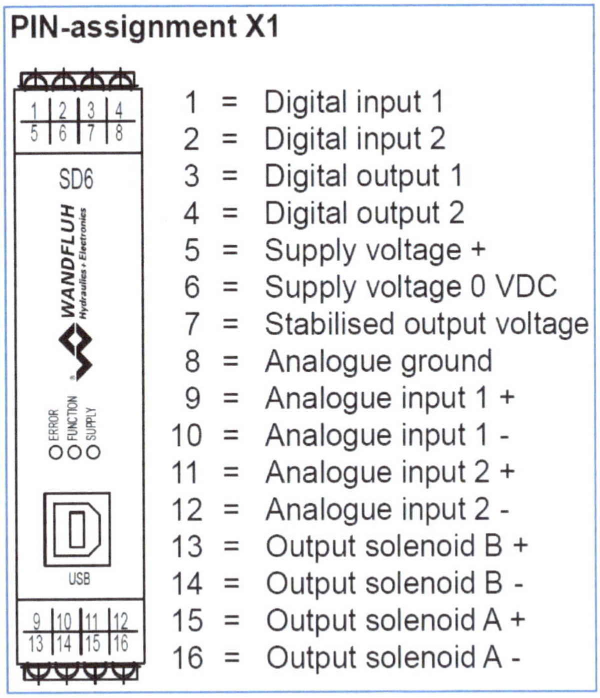

Fig. 9.54C - Terminal Assignment for Example 10 (Courtesy of Bosch Rexroth)

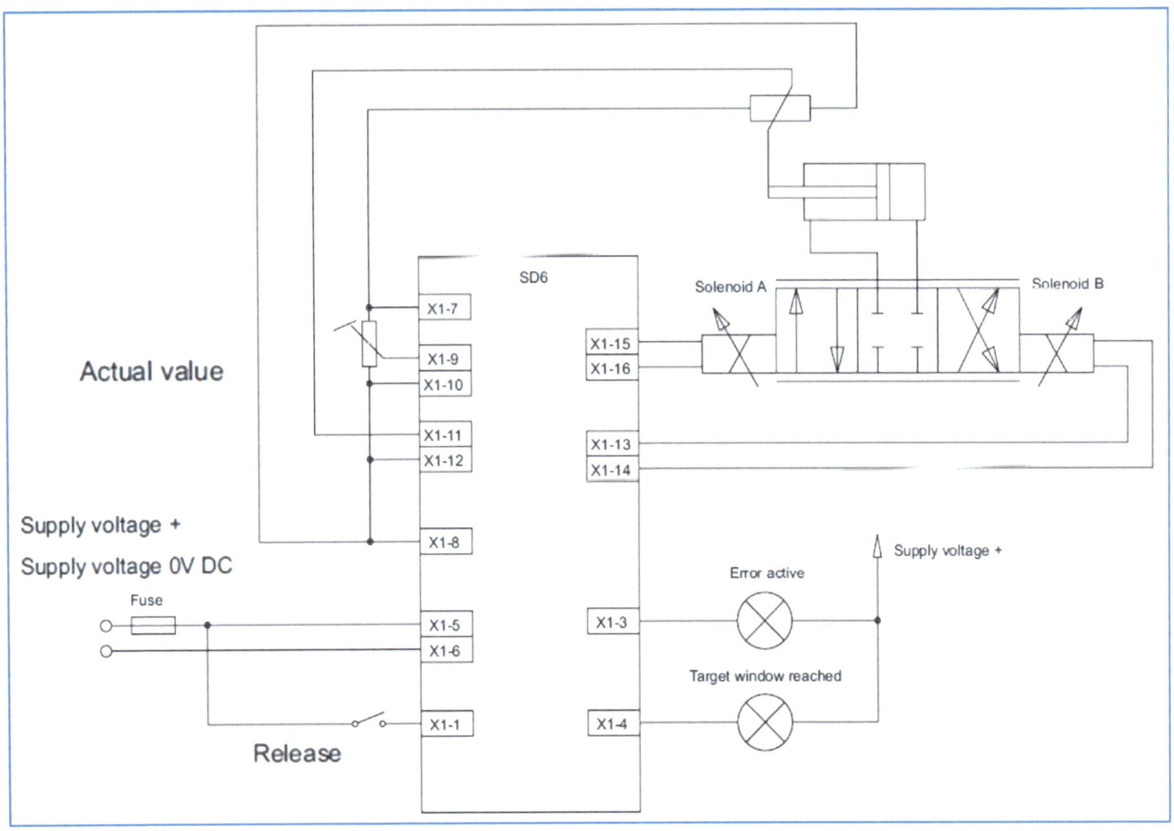

Fig. 9.54D - Typical Application for Example 10 (Courtesy of Wandfluh)

Example 11: Digital Multiple-Axis Controller.

Type VT-HNC100

Hardware Configuration	
▪ Panel Mounted	
▪ Modular	✔
▪ OBE	
Spool Control	
▪ Force Controlled	✔
▪ Stroke Controlled	✔
System Control	
▪ Valve Only	
▪ Single Axis	✔
▪ Multiple Axis	✔
Control Mode	
▪ Analog	✔
▪ Digital	✔
Interfacing	
▪ No Interface	
▪ Network Interfacing	✔

❑ **Key Features:**

- Analog/Digital Inputs/Outputs.
- Command value input as voltage or current.
- PC programmable controller for up to 4 axes.
- Complies with the requirements for closed-loop control of hydraulic drives.
- Multi Ethernet/Bus interface
- Protection: vibration and climate resistance.
- Password protection.
- Fields of application: Machine tools, injection molding, presses.

Fig. 9.55A - Typical Control Electronics - Example 11 (Courtesy of Bosch Rexroth)

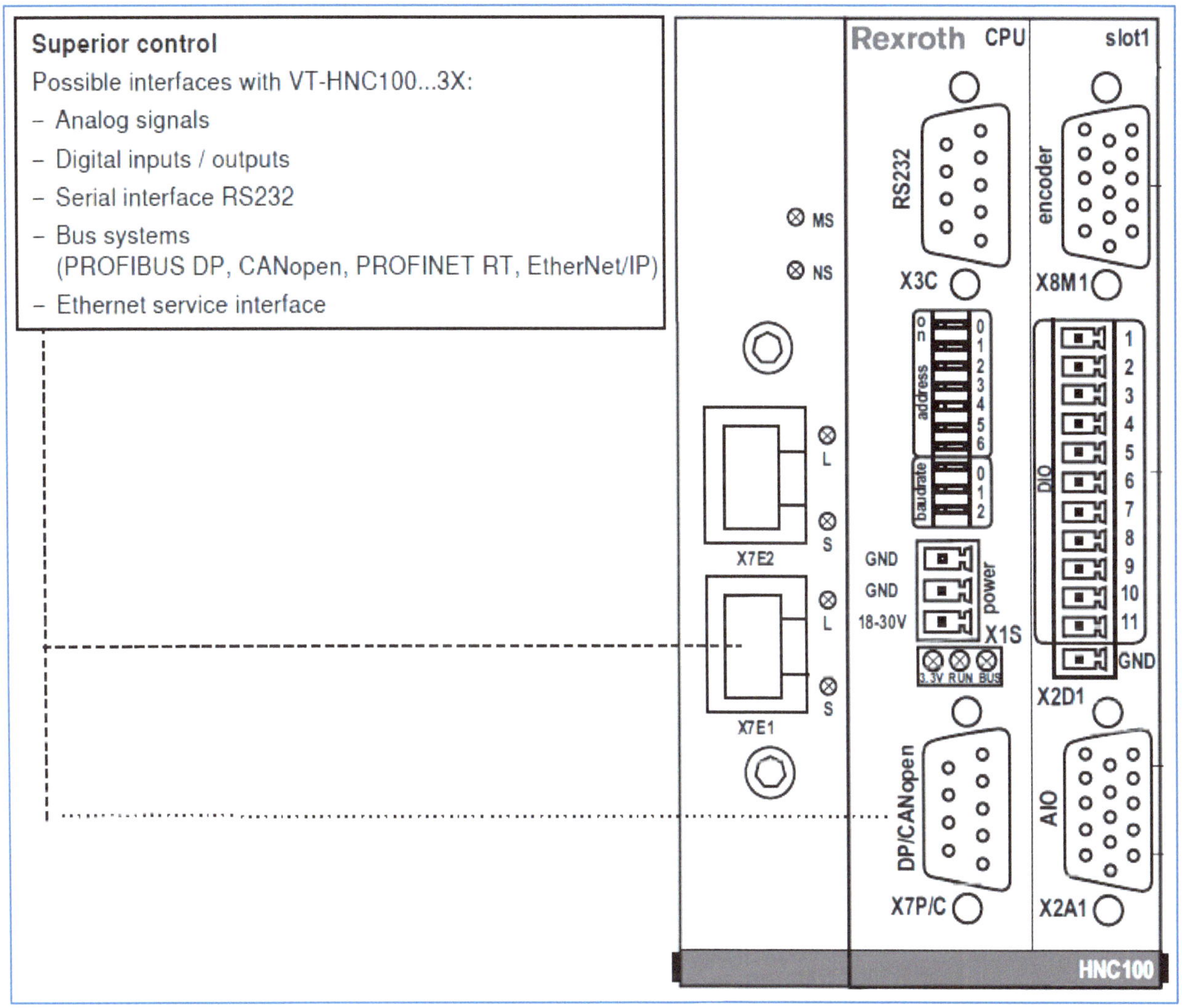

Fig. 9.55B - Terminal Assignment for Example 11 (Courtesy of Bosch Rexroth)

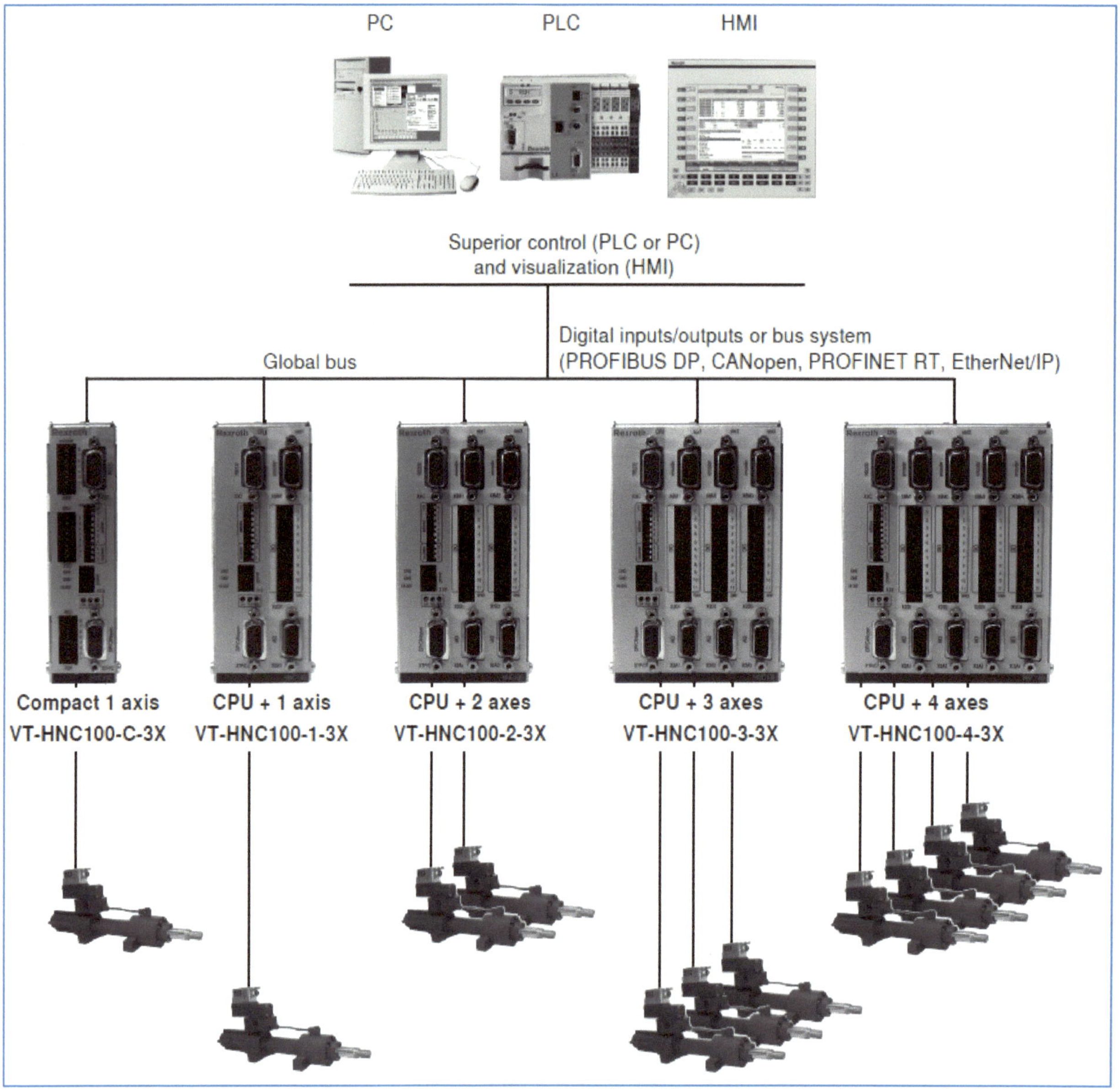

Fig. 9.55C - Typical Application for Example 11 (Courtesy of Bosch Rexroth)

Example 12: Digital Controller for mobile applications.

Hardware Configuration	
▪ Panel Mounted	
▪ Modular	✔
▪ OBE	
Spool Control	
▪ Force Controlled	
▪ Stroke Controlled	✔
System Control	
▪ Valve Only	
▪ Single Axis	
▪ Multiple Axis	✔
Control Mode	
▪ Analog	✔
▪ Digital	✔
Interfacing	
▪ No Interface	
▪ Network Interfacing	✔

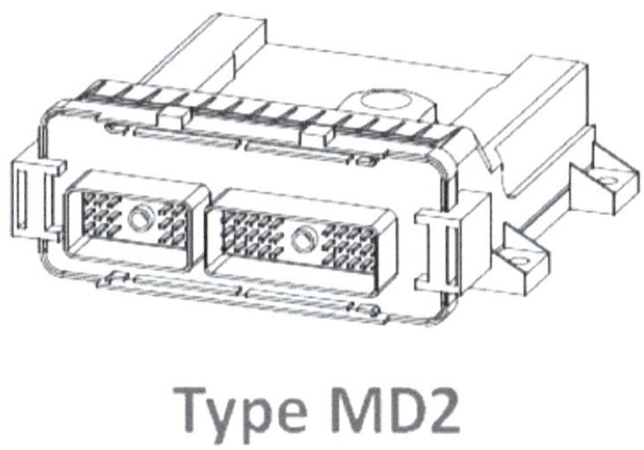

❑ **Key Features:**

- Digital controller with 4 or 8 solenoid outputs.
- Robust and compact construction.
- The extensive supply voltage range 12-24VDC.
- Menu-controlled parameterization and diagnostics.
- Optional CANinterface.

Fig. 9.56A - Typical Control Electronics - Example 12 (Courtesy of Wandfluh)

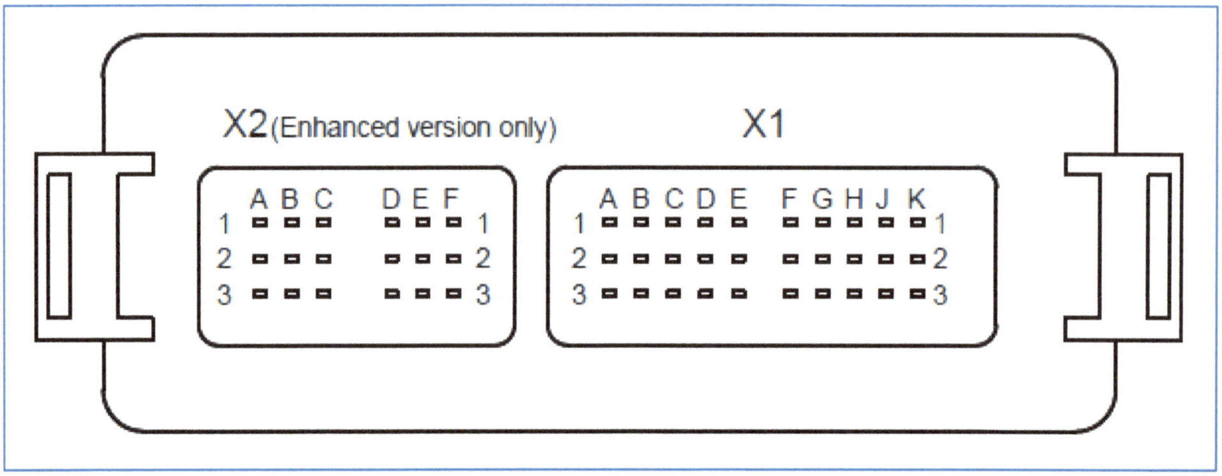

DEVICE PLUG (X1)

A1 = Stabilised output voltage
A2 = Supply voltage + (Solenoid outputs)
A3 = Supply voltage 0 VDC (Solenoid outputs)
B1 = Stabilised output voltage
B2 = Supply voltage + (Logic part)
B3 = Supply voltage 0 VDC (Logic part)
C1 = Analogue ground
C2 = Digital input 1
C3 = Digital input 2
D1 = Analogue ground
D2 = VBUS (USB)
D3 = GND (USB)
E1 = Analogue input 1
E2 = D- (USB)
E3 = D+ (USB)
F1 = Analogue input 2
F2 = Digital output 1
F3 = Digital output 2
G1 = Output solenoid A +
G2 = Output solenoid A -
G3 = Digital ground
H1 = Output solenoid B +
H2 = Output solenoid B -
H3 = Reserved
J1 = Output solenoid C +
J2 = Output solenoid C -
J3 = Reserved
K1 = Output solenoid D +
K2 = Output solenoid D -
K3 = Reserved

DEVICE PLUG (X2)

A1 = Output solenoid E +
A2 = Output solenoid E -
A3 = CAN High
B1 = Output solenoid F +
B2 = Output solenoid F -
B3 = CAN Low
C1 = Output solenoid G +
C2 = Output solenoid G -
C3 = CAN Gnd
D1 = Output solenoid H +
D2 = Output solenoid H -
D3 = CAN High
E1 = Digital input 3
E2 = Digital input 4
E3 = CAN Low
F1 = Analogue input 3
F2 = Analogue input 4
F3 = CAN Gnd

Fig. 9.56B - Terminal Assignment for Example 12 (Courtesy of Wandfluh)

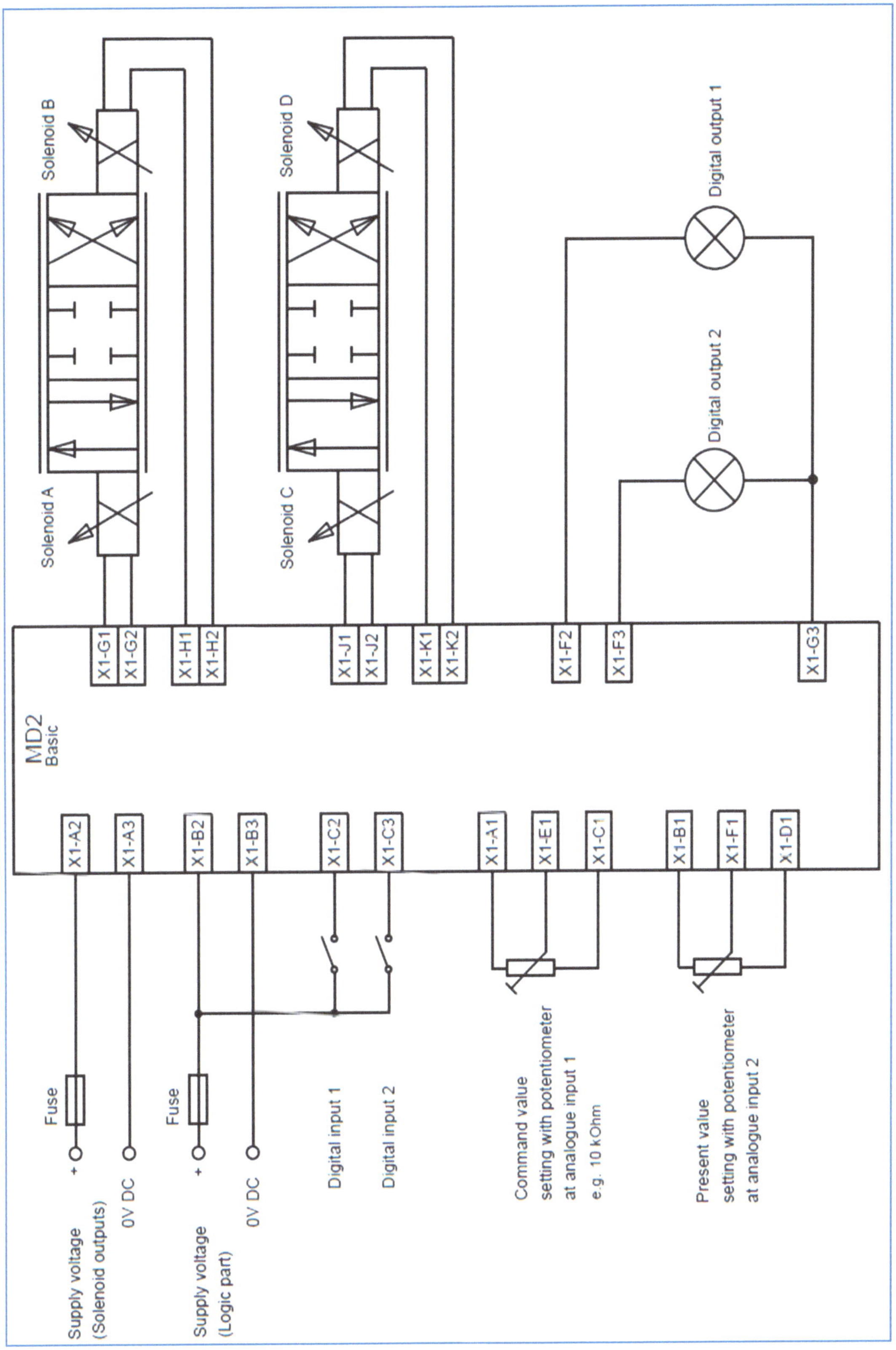

Fig. 9.56C - Typical Application for Example 12 (Courtesy of Wandfluh)

Example 13: Digital Controller for mobile applications.

CoreTek™ Benefits

- Controllers are optimized for Electro-Hydraulic System Integration
- Programmed with the powerful CoDeSys™ software tool, Backbone Configuration Tool and Impulse Service Tool.
- Fully-sealed, compact cast-aluminum housing is rated to IP67
- Advanced electronic design for reliability and accuracy.
- Reliable operation in real-world temperature conditions from -40° to +70°C (-40° to 158° F).
- Inputs and Outputs are protected against shorts.
- Outputs have diagnostic capability.
- No external cooling or heat dissipation required.

Fig. 9.57A - Typical Control Electronics - Example 13 (Courtesy of Hydraforce)

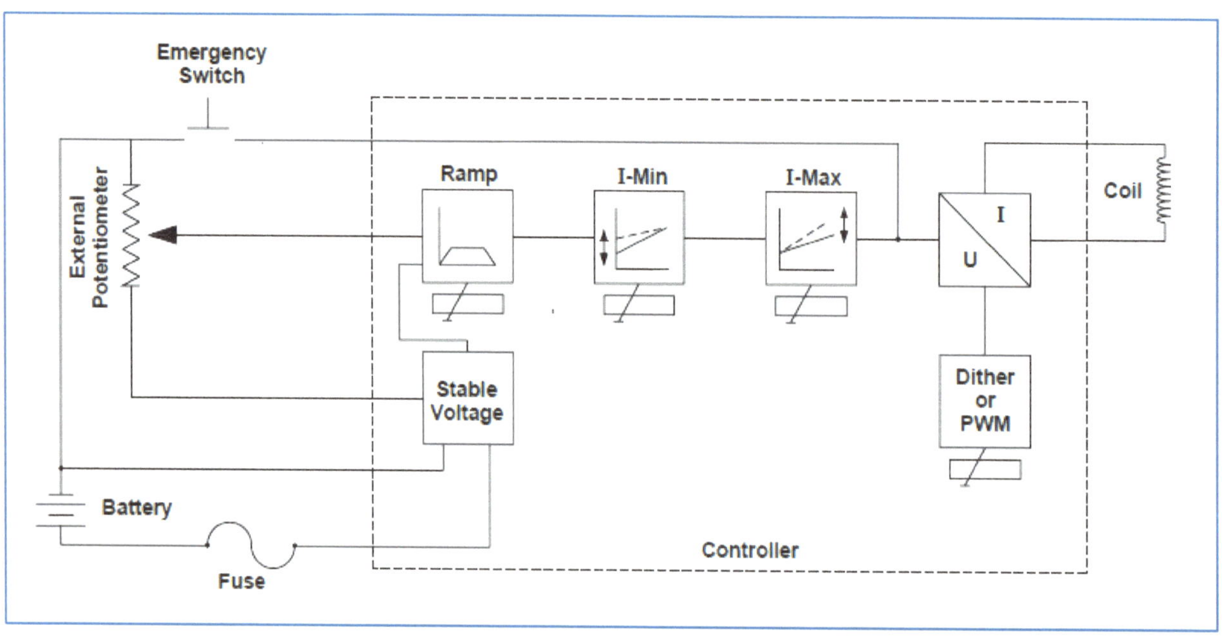

Fig. 9.57B - Block Diagram for Example 13 - (Courtesy of Hydraforce)

Example 14: Digital Display Proportional Valve Driver for mobile applications.

Features & Benefits

- Compatible with any proportional valves.
- Multi inputs & outputs; per model.
- Independent adjustments.
- Field adjustable without specialty tools.
- Clear readings in any light.
- 3 to 4 digit extra bright displays.
- Displays actual values.
- Suitable for extreme -40°C conditions.
- Dual and single valve drivers available.
- Wide ramp time range (0 to 99.5 Sec).
- Wide dither range (40 to 450Hz).
- Short circuit reverse polarity protection.
- Command input overcurrent protection.

Fig. 9.58 - Block Diagram for Example 14 - (Excerption from Lynch)

Chapter 10

Electro-Hydraulic Valves Commissioning and Maintenance

Objectives:

This chapter introduces guidelines for commissioning and maintenance of EH valves.

Brief Contents:

Chapter 10 - Electro-Hydraulic Valves Commissioning and Maintenance

10.1- Installation and Commissioning Instructions

This section presents some guidelines for installation and commissioning of an EH valve.

10.1.1- Cleanliness

Surroundings:
Wherever an EH valve is used, particularly proportional and servo ones, it is highly advisable to keep the surrounding area clean and organized. It's worth it to hire a person dedicated to cleaning the outside surfaces of the valve and the surrounding area on daily basis. That helps prevent dirt from getting into inside the valve or metallic chips from being attracted by the electric part of the valve.

Cleaning Towels:
Towels for cleaning purposes must be lint-free tissues or special material, e.g. 100% cotton.

Protective Plates and Plastic Caps:
If a valve is shipped between departments or stored somewhere, it should be covered by suitable protective plates or plastic caps. Such protective elements are removed only immediately prior to installation.

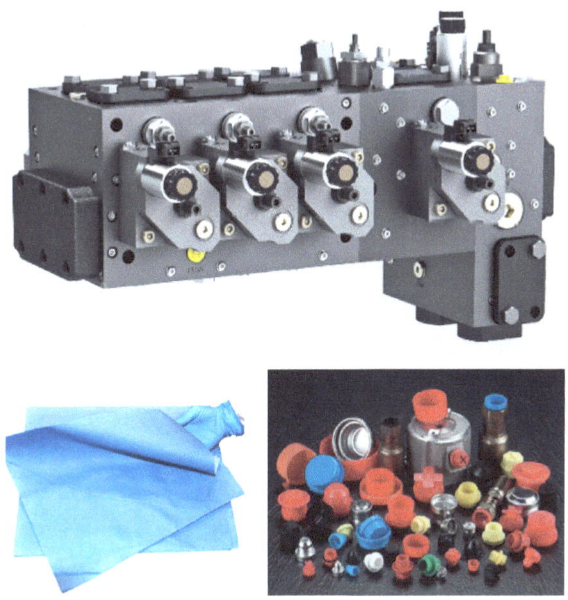

Fig.10.1- Instructions to Keep the Valves Clean

10.1.2- Installation Position

It is recommended to install the valve in a perfect horizontal position. Avoid installing a valve, particularly large one, in a vertical or even tilted position. That may cause drifting the spool from the central position due its weight.

If a valve is mounted on an actuator, particularly cylinders, the valve should be positioned so that the spool is positioned perpendicular to the acceleration direction of the actuator. That reduces the effect of induced vibration by the actuator on the valve performance.

10.1.3- Mounting Screws

Mounting screws must be tightened to the torque specified in the data sheets from the manufacturer. Exceeding the specified torque my distant the body of the valve and cause spool seizure.

10.1.4- Seals

Make sure that the valve is equipped with proper sealing elements based on the type of fluid and working temperature. It is highly advisable to review the manufacturer specifications.

10.1.5- Electrical Connections

The following bullets present some guidelines when connecting the valve to the control electronics:

- Associated electronics provide protection for the solenoids & torque motors against overload current. Make sure overload protection feature is enabled.

- Avoid running electrical cables within hydraulic plumbing or near sources of electrical noise such as electrical motors and variable frequency drivers (VFDs).

- Caution: a valve MUST NOT be tested with electrical signal only before hydraulic power is supplied including pilot pressure to the first stage. Otherwise, the wire spring in a flapper nozzle stage may be severely stretched. Same is applicable for proportional valves in order to prevent spool friction without lubrication.

10.1.6- Storage Locations

The storage locations for EH valves must be dry with low humidity and dirt-free. For storage longer than 3 months, fill the housing with preservative oil and seal the valve with protective covers. The stored valves must be checked from time to time to make sure that they are kept in the right position.

It is always advisable to refer to the manufacturer's specifications for longer term storage. Always store the valves with protective covers in place

10.2- Filtration Requirements

10.2.1-Reservoir Breather Filter

Air breather filter is recommended with feature of water vapor absorption ability. Air breather filter should have the same beta ratio as the oil filter.

10.2.2-Filtration

Non-Bypass Pressure Filters:
They are recommended to assure long service life for proportional and servo valve. For non-bypass pressure filters, filter element collapse rating must be greater than system operating pressure.

Last Chance Filters:
Last Chance Filter is built inside servo valves. It should be cleaned or replaced frequently as per the manufacturer instructions. The valve is constructed such that filters can be externally removed and replaced.

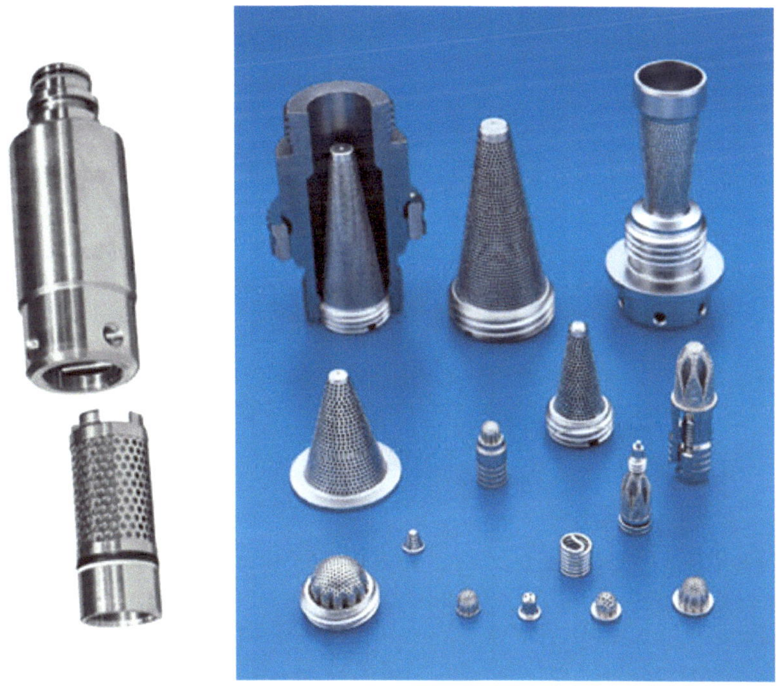

Fig.10.2- Examples of Non-bypass Last Chance Filters

Clogging Indicator:

Filter Clogging Indicators could be with visual indicator only. They can also be equipped with an electrical signal to alert the operator or shut the machine down.

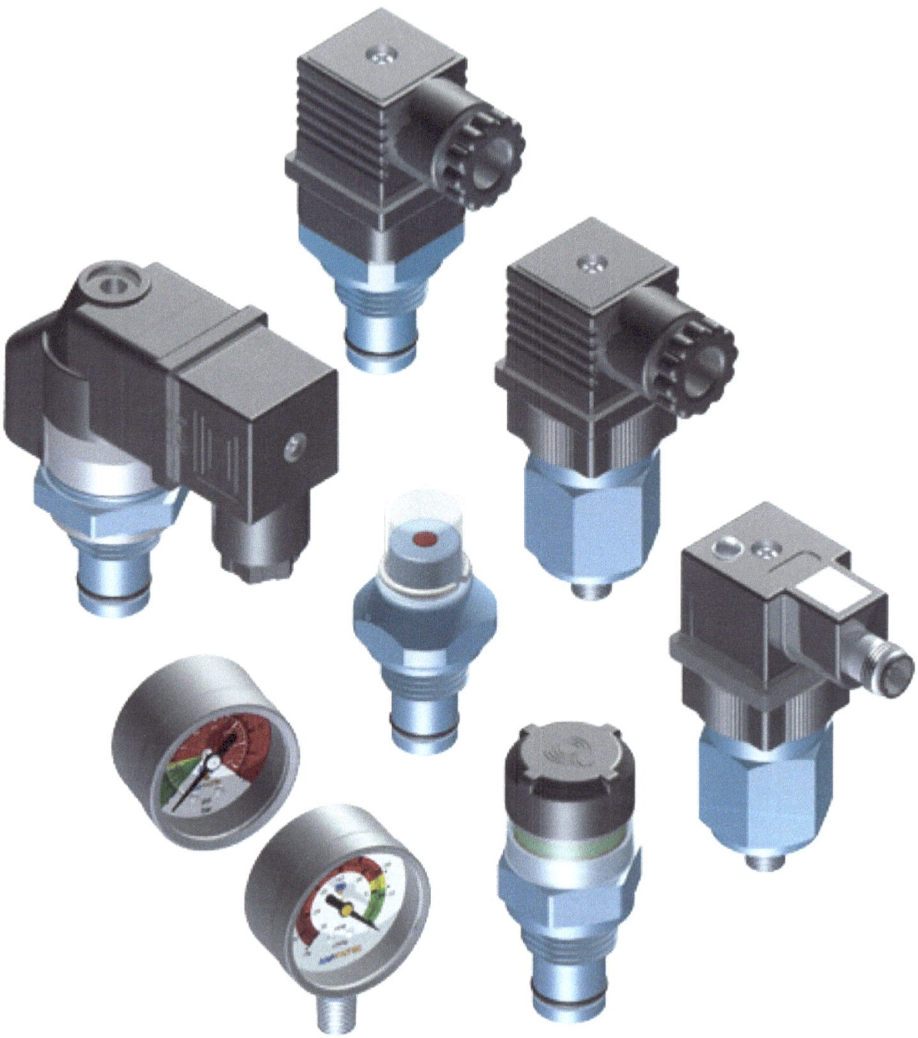

Fig.10.3- Examples of Filter Clogging Indicators (Excerption from MP Filtri)

Filter Rating:

It is always advisable to refer to manufacturers maximum recommended ISO 4406 contamination level. If no information is found, 5 microns absolute or better (β_5 = 75 i.e. Efficiency = 98.7%) is recommended as minimum. If the work environment is dirty, e.g. foundries, additional off-line filtration may be necessary.

Fluid Cleanliness:

If no specific recommendation is provided by the manufacturer, Fig. 10.4 shows ISO codes that are ok for proportional and servo valves. The code shows the allowable number of contaminants of size 4, 6, and 14 microns in milliliter of the fluid.

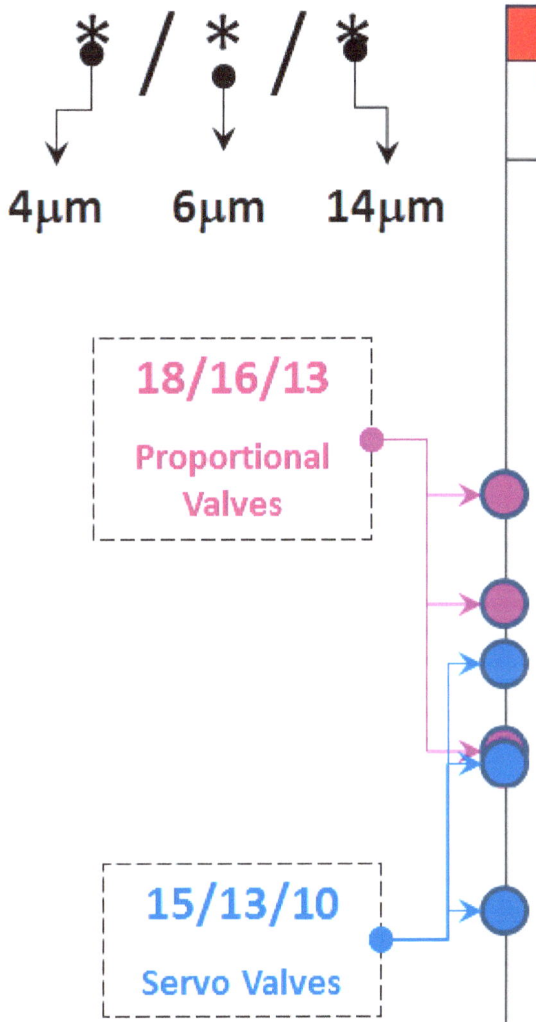

ISO 4406 Chart		
Range Code	Particles per milliliter	
	More than	Up to/including
24	80000	160000
23	40000	80000
22	20000	40000
21	10000	20000
20	5000	10000
19	2500	5000
18	1300	2500
17	640	1300
16	320	640
15	160	320
14	80	160
13	40	80
12	20	40
11	10	20
10	5	10
9	2.5	5
8	1.3	2.5
7	0.64	1.3
6	0.32	0.64

Fig. 10.4- ISO Contamination Code* 4406-1999

Figure 10.5 shows class 6 is recommended on standard NAS 1638. This standard shows the maximum number of dirt particles in 100 ml of fluid. This standard is being phased out in favor of ISO 4406. It has some limited use in aerospace industry but not used in current mobile and Industrial applications.

Contamination class	Particle size in μm				
	5 – 15	15 – 25	25 – 50	50 – 100	> 100
00	125	22	4	1	0
0	250	44	8	2	0
1	500	89	16	3	1
2	1000	178	32	6	1
3	2000	356	63	11	2
4	4000	712	126	22	4
5	8000	1425	253	45	8
6	16000	2850	506	90	16
7	32000	5700	1012	180	32
8	64000	11400	2025	360	64
9	128000	22800	4050	720	128
10	256000	45600	8100	1440	256
11	512000	91200	16200	2880	512
12	1024000	182400	32400	5760	1024

Fig. 10.5- NAS Contamination Code1638

10.2.3- System Flushing

Before installing one of these close clearance valves on a new or a refurbished system, it is essential to carry out an effective flushing procedure on the entire system to remove "built in" contamination particles (for example: weld scale, rust, dust, lint, metal shavings, blast-sand, paint chips, pipe dope, etc.). There are standard protocols that describe the flushing process. In brief, before flushing, valves and other critical components should be removed and replaced with flushing modules (sub-plates). The system should be flushed using high flow of less viscous fluid at high temperature. Turbulent flow with Reynolds number greater than 5000 is required for successful flushing

During flushing process, oil samples must be taken, and cleanliness class must to be checked using an electronic particle counter. The oil in the system should be flushed through the filter at least 150 to 300 times. Equation 10.1 shows how to calculate the time for flushing a system.

$$t = \frac{V}{Q} \times X \qquad 10.1$$

Where:

t = Time in hours
V = Tank volume in liters
Q = Pump delivery rate in L/min
X = [(150-300)/60] =2.5 to 5

This is just a guide and may not be sufficient depending on filter efficiency and contamination level prior to testing. Flushing must continue until the ISO 4406 level is below the manufacturer recommendations. Conductors must be flushed and dried prior to installation. Tanks must be sealed after cleaning. Figure 10.6 shows an example of filtration unit.

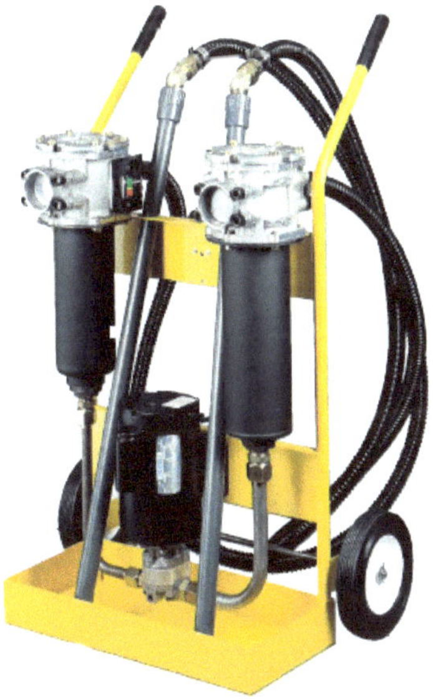

Fig. 10.6- Example of Hydraulic Filtration Unit

10.3- Electro-Hydraulic Valve Tests and Maintenance

There are number of standard tests that can be conducted to check the performance of the valve. These tests are conducted if required to diagnose some malfunctions of the valve. Or it can be conducted frequently as part of the preventive maintenance plans.

The following are the standard tests:

- Flow Gain
- Pressure Gain
- Hysteresis
- Null leakage
- Step response
- Frequency response

There are different models of valve test machines. Figure 10.7 shows a typical one that can be used to test static and dynamic characteristics of the valve.

Fig. 10.7- Example of Commercial Valve Testing Machine

APPENDIXES

APPENDIX A: LIST OF FIGURES

Chapter 1 - Hydraulic versus Electrical Systems

Chapter 2 - Hydro-Mechanical versus Electro-Hydraulic Solutions

Chapter 3 - Switching Valves Construction and Operation

Chapter 4 - Electrical Circuits for Switching Valves

Chapter 5 - Proportional Valves

Chapter 6 - Servo Valves

Chapter 7 - Electro-Hydraulic Valve Selection Criteria

Chapter 8 - Open-Loop versus Closed-Loop EH Applications

Chapter 9 - Chapter 9-Control Electronics for Electro-Hydraulic Valves

Chapter 10 - Electro-Hydraulic Valves Commissioning and Maintenance

APPENDIX B: LIST OF TABLES

APPENDIX C:

LIST OF ANIMATED CIRCUITS (AUTOMATION STUDIO)

Sign as a Member at (www.compudraulic.com)
Download the animated circuits for the textbook from the Member page.
Automation Studio Viewer is required to open the files (www.famictech.com)

APPENDIX D: LIST OF REFERENCES

Hydraulic Systems Volume 1- Introduction to Hydraulics for Industry Professionals
Author: Dr. Medhat Kamel Bahr Khalil, 2016.
Publisher: Compudraulic, USA.
ISBN: 978-0-692-62236-0

R01- Basic Electronics for Hydraulic Motion Control
Author: Jack L. Johnson, PE 1992.
Publisher: Penton Publishing Inc. 1100 Superior Avenue. Cleveland, OH 44114.
ISBN No. 0-932905-07-2.

R02- Closed Loop Electro-Hydraulics Systems Manual
Author: Vickers/Eaton.
Publisher: Vickers Inc. 1992.
Training Center, 2730 Research Drive, Rochester Hills, MI 48309-3570.
ISBN 0-9634162-1-9

R03- Bosch Automation Technology
Author: Werner Gotz, Steffen Haack, Ralph Mertlick.
Publisher: Bosch.
ISBN 3-933698-05-7.

R04- Electrohydraulic Proportional and Control Systems
Publisher: Bosch Automation 1999.
ISBN 0-7680-0538-8.

R05- Proportional and Servo Valve Technology – The Hydraulic Trainer Volume 2
Author: R. Edwards, J. Hunter, D. Kretz, F. Liedhegener, W. Schenkel, A. Schmitt.
Publisher: Mannesman Rexroth AG 1988. D-8770 Lohr a. Main.
ISBN 3-8023-0266-4.

R06- Proportional Hydraulics
Author: D. Scholz.
Publisher: Festo Didactic KG, Esslingen, Germany.

R07- Electricity, Fluid Power, and Mechanical Systems for Industrial Maintenance
Author: Thomas Kissell.
Publisher: Prentice Hall, Inc. 1999, Upper Saddle River, NJ 07458.
ISBN 0-13-896473-4.

R08- Fluid Power in Plant and Field – First Edition
Author: Charles S. Hedges, R.C. Womack.

Publisher: Womack Machine Supply Co. 1968.
Womack Educational Publication, 2010 Shea Road, Dallas, TX 75235.
ISBN 68-22573 (Library of Congress Card Catalog No.).

R09- Hydraulics, Fundamentals of Service
Author: Deere and Company.
Publisher: John Deere Publishing 1999.
Almon TIAC Bldg. Suite 104, 1300-19th Street, East Moline, IL 61244.
ISBN 0-86691-265-7.

R10- Industrial Hydraulics Troubleshooting
Author: James E. Anders, Sr.
Publisher: McGraw-Hill, Inc.
ISBN 0-07-001592-9.

R11- Power Hydraulics
Author: John Ashby.
Publisher: Prentice Hall 1989. Prentice Hall International, (UK) Ltd.
66 Wood Lane End, Hemel Hempstead, Hertfordshire, HP2 4RG.
ISBN 0-13-687443-6.

R12- Fluid Power with Application
Author: Anthony Esposito.
Publisher: Prentice Hall.
ISBN 0-13-060899-8.

R13- Hydraulic Component Design and Selection
Author: E.C. Fitch.
Publisher: BarDyne Inc. 5111 North Perkins Rd. Stillwater, OK 74075.
ISBN 0-9705922-3-X.

R14- Planning and Design of Hydraulic Power Systems – The Hydraulic Trainer, Volume 3
Author: Mannesmann Rexroth GmbH.
Publisher: Mannesman Rexroth AG 1988.
D-97813 Lhr a. Main, Jahnsrtrabe 3-5 D-97816 Lohr a. Main.
ISBN 3-8023-0266-4.

R15- Logic Element Technology: Hydraulic Trainer, Volume 4
Author: Mannesmann Rexroth GmbH.
Publisher: Mannesmann Rexroth GmbH 1989.
.Postfach 340, D 8770 Lohr am Main, Telefon (09352) 180.
ISBN 3-8023-0291-5.

R16- Hydrostatic Drives with Control of the Secondary Unit. The Hydraulic Trainer, Volume 6
Author: Dr. Alfred Feuser, Rolf Kordak, Gerold Liebler.
Publisher: Mannesmann Rexroth GmbH 1989.
Postfach 340, D 8770 Lohr am Main.

R17- Control Strategies for Dynamic Systems: Design and Implementation
Author: John H. Lumkes, Jr.
Publisher: Marcel Dekker, Inc. 2002.
Marcel Dekker, Inc. 270 Madison Avenue, New York, NY 10016.
ISBN 0-8247-0661-7.

R18- Feedback Control Of Dynamic Systems
Author: Gene F. Franklin, J. David Powell, Abbas Emami-Naeini.
Publisher: Prentice-Hall, Inc.
Upper Saddle River, New Jersey.
ISBN 0-13-032393-4.

R19- Modeling and Analysis of Dynamic Systems
Author: Charles M. Close, Dean. Frederick
Rensselaer Polytechnic Institute
Publisher: John Wiley & Sons, Inc.
ISBN 0-471-12517-2.

R20- Design of Electrohydraulic Systems for Industrial Motion Control
Author: Jack L. Johnson, PE.
Milwaukee School of Engineering.
Publisher: Parker.
Copyright © Jack L. Johnson, PE 1991.

R21- Basic Pneumatics
Author: Kjell Evensen & Jul Ruud.
Publisher: AB Mecmann Stockholm 1991.
S-125 81 Stockholm, Sweden.
ISBN 91-85800*21-X.

R22- Basic Pneumatics: The Pneumatic Trainer, Volume 1
Author: Ing. –Buro J.P. Hasebrink.
D7761 Moos.
Editor: Mannesmann Rexroth Pneumatik GmbH.
Bartweg 13, W 3000 Hannover 91.

R23- Electro-Pneumatics: The Pneumatic Trainer, Volume 2
Author: Rolf Balla.
Publisher: Mannesmann Rexroth 1990, Pneumatik GmbH.
Publication No: RE 00 262/01.92.

R24- Pneumatics Theory and Applications
Author: Bosch Automation.
Publisher: Robert Bosch GmbH 1998.
Automation Technology Division, Training (AT/VSZ)
ISBN 1-85226-135-8.

R25- Fluid Power Engineering
Author: M. Galal Rabie.
Publisher: McGraw-Hill.
ISBN 978-0-07-162246-2.

R26- Air Motors Ideas with Air
Author: GAST Mfg. Co.
Publisher: GAST Mfg. Co. 1978.
P.O. Box 97, Benton Harbor, MI 49022.
Book No: Booklet #100.

R27- Air Motor Handbook
Author: GAST Mfg. Co.
Publisher: GAST Mfg. Co. 1978.
P.O. Box 117, Benton Harbor, MI 49022.

R28- Troubleshooting Hydraulic Components: Using Leakage Path Analysis Methods
Author: Rory S. McLaren.
Publisher: Rory McLaren Fluid Power Training 1993.
562 East 7200 South, Salt Lake City, UT 84171.
ISBN No. 0-9639619-1-8.

R29- Hydraulics Theory and Application from Bosch
Author: Werner Gotz.
Publisher: Robert Bosch GmbH.
Hydraulics Division K6, Postfach 30 02 40, D-7000 Stuttgart 30.
Federal Republic of Germany, Technical Publications Department, K6/VKD2.

R30- A Complete Guide to ISO and ANSI Fluid Power Symbols
Author: Fluid Power Training Institute.
Publisher: Fluid Power Training Institute 200.
562 East Fort Union Boulevard, Midvale, Utah 84047.

R31- How to Work Safely with Hydraulics
Author: Fluid Power Training Institute.
Publisher: Fluid Power Training Institute 2004.
562 East7200 South, Midvale, Utah 84047.

R32- How to Interpret Fluid Power Symbols
Author: Rory S. McLaren.
Publisher: Fluid Power Training Institute.
Rory S. McLaren 1995.
ISBN 0-9639619-2-6.

R33- Safe Hydraulics
Editor: Gates Rubber Company.
Copyright 1995.
Denver, CO 80217.

R34- Electronically Controlled Proportional Valves. Selection and Application
Author: Michael J. Tonyan.
Publisher: Marcel Dekker, Inc. 1985.
Marcel Dekker, Inc., 270 Madison Avenue, New York, NY 10016.
ISBN 0-8247-7431-0.

R35- Introduction to Closed-Loop Oil Systems
Author: Rory S. McLaren.
Publisher: Rory McLaren Fluid Power Training Institute.
7050 Cherry Tree Lane, P.O. Box 711201, Salt Lake City, UT 84171.

R36- Industrial Hydraulic Technology, Second Edition
Author: Parker Hannifin Corporation.
Publisher: Parker Hannifin Corporation 1997.
6035 Parkland Blvd, Cleveland, OH 44124-4141.
Publication No: Bulletin 0231-B1.

R37- Basic Principle and Components of Fluid Technology – The Hydraulic Trainer, Volume 1
Author: Mannesman Rexroth.
Publisher: Mannesman Rexroth AG 1988.
D-97813 Lhr a. Main, Jahnsrtrabe 3-5 D-97816 Lohr a. Main.
ISBN 3-8023-0266-4.

R38- Safe-T-Bleed Corporation Catalog
Publisher: Safe-T-Bleed Corporation 2001.
Catalog No. STB-PC-1201-1

R39- Industrial Hydraulics Manual – EATON
Publisher: Eaton Fluid Power Training.
ISBN: 0-9788022-0-9.

R40- Vickers-Mobile Hydraulic Manual – Fourth Edition 1998
Author: Vickers.
Publisher: Vickers Inc. 1999.
Training Center, 2730 Research Drive, Rochester Hills, MI 48309-3570.
ISBN No. 0-9634162-5-1.

R41- Industrial Fluid Power Text, Volume 2
Author: Charles S. Hedges, R.C. Womack.
Publisher: Womack Machine Supply Company 1972.
Womack Educational Publications, 2010 Shea Road, Dallas, TX 75235.
ISBN 66-28254 (Library of Congress Card Catalog No.).

R42- Fluid Power Hydraulics and Pneumatics
Author: R. Daines.
Publisher: The Good-heart Willcox Company, Inc.

R43- Hydraulics in Industrial and Mobile Applications
Publisher: ASSOFLUID, Italian Association of Manufacturing and Trading Companies in Fluid Power Equipment and Components

R44- Fluid Power in Plant and Field – Second Edition
Author: Charles S. Hedges, R.C. Womack.
Publisher: Womack Machine Supply Co. 1968.
Womack Educational Publication, 2010 Shea Road, Dallas, TX 75235.
ISBN 68-22573.

R45- Mobile Hydraulics Manual
Author: Eaton.
Publisher: Eaton Corporation Training.
Eden Prairie, Minnesota.
ISBN 0-9634162-5-1.

R46- EH Control Systems
Author: F.D. Norvelle.

R47- Fluid Power Journal
Publisher: International Fluid Power Society.

R48- Fundamentals of Industrial Controls and Automation
Author: Lonnie L. Smith and Mike J. Rowlett.
Publisher: Womack Educational Publications.
Dallas, Texas.
ISBN: 0-943719-04-6.

R49- Lightning Reference Handbook
Publisher: Berendsen Fluid Power.

R50- Pneumatics Basic Level
Author: P. Croser, F. Ebel.
Publisher: Festo Didactic GmbH & Co.

R51- Electro-pneumatics Basic Level
Author: F. Ebel, G. Prede, D. Scholz.
Publisher: Festo Didactic GmbH & Co.

R52- Mechanical System Components
Author: James F. Thorpe.
Publisher: Allyn and Bacon.
Needham Heights, Massachusetts.
ISBN: 0-205-11713-9.

R53- Electrical Motor Controls for Integrated Systems, Third Edition
Author: Gary J. Rockis, Glen A. Mazur.
Publisher: American Technical Publishers, Inc.
ISBN: 0-8269-1207-9.

R54- Instrumentation, Fourth Edition
Author Franklyn W. Kirk, Thomas A. Weedon, Philip Kirk.
Publisher American Technical Publishers, Inc.
ISBN: 0-8269-3423-4.

R55- Introduction to Mechatronics and Measurement Systems, Second Edition
Author David G. Alciatore, Michael B. Histand.
Publisher McGraw-Hill, Inc.
ISBN: 0-07-240241-5.

R56- Study Guides for IFPS Certification

R57- Work Books from Coastal Training Technologies

R58- Industrial Hydraulic Manual – Fourth Edition 1999

Author: Vickers.
Publisher: Vickers Inc. 1999.
Training center, 2730 Research Drive, Rochester hills, Michigan 48309-3570.
ISBN 0-9634162-0-0.

R59- Industrial Automation and Process Control
Author: John Stenerson.
Publisher: Prentice Hall.
ISBN 0-13-033030-2.

R60- Industrial Automated Systems
Author: Terry Bartelt.
Publisher: Delmar Cengage Learning.
ISBN: 10-1-4354-888-1.

R61- Introduction to Fluid Power
Author: James L. Johnson.
Publisher: Delmar Cengage Learning.
ISBN: 10-0-7668-2365-2.

R62- Summary for Engineers
Author: Dr. Abdel Nasser Zayed.
Publisher: Dr. Abdel Nasser Zayed.
ISBN: 977-03-0647-9.

R63- Mechanics of Materials
Author: Ferdinand P.Beer, E. Russell Johnston Jr., John T DeWolf.
Publisher: McGraw Hill Publishing.
ISBN: 0-07-365935-5.

R64- Oil Hydraulic System, Principles and Maintenance
Author: S. R. Majumdar.
Publisher: McGraw Hill.
ISBN 10: -0-07-140669-7.

R65- Contamination Control in Hydraulic and Lubricating Systems
Publisher: Pall

R66- Diagnosing Hydraulic Pump Failure
Publisher: Caterpillar.

R67- Oil Service Products Catalog
Publisher: Schroder Industries.

R68- Industrial Fluid Power Volume 1
Author: Charles S. Hedges.
Publisher: Womack Educational Publication.
ISBN: 0-9605644-5-4.

R69- Industrial Fluid Power Volume 2
Author: Charles S. Hedges.
Publisher: Womack Educational Publication.
ISBN: 0-943719-01-1.

R70- Industrial Fluid Power Volume 3
Author: Charles S. Hedges.
Publisher: Womack Educational Publication.
ISBN: 0-943719-00-3.

R71- Electrical Control of Fluid Power
Author: Charles S. Hedges.
Publisher: Womack Educational Publication.
ISBN 0-9605644-9-7.

R72- Hydraulic Cartridge Valve Technology
Author: John J. Pippenger, P.E.
Publisher: Amalgam Publishing Company.
Post Office Box 617, Jenks, OK 74037 USA.
ISBN: 0-929276-01-9.

R73- Noise Control of Hydraulic Machinery
Author: Stan Skaistis.
Publisher: Marcel Dekker, 270 Madison Avenue, New York, NY 10016.
ISBN: 0-8247-7934-7.

R74-Solenoid Valves
Author: Hydraforce

R75-HF Proportional Valve Manual
Author: Hydraforce

R76-Automatic Control for Mechanical Engineers
Author: M. Galal Rabie, Professor of Mechanical Engineering
ISBN: 977-17-9869-3,2010.

R77-Fluid Power System Dynamics
Author: W. Durfee, Z. Sun

Index

D

E

F

G

H

I

J

L

M

N

O

P

R

S

T

U

V

W

Y

Z

www.ingramcontent.com/pod-product-compliance
Lightning Source LLC
Chambersburg PA
CBHW041432150926
51501CB00015B/258
9780997763423